T0302046

Quantum Mechanics for Scientists and Engineers

If you need a book that relates the core principles of quantum mechanics to modern applications in engineering, physics, and nanotechnology, then *Quantum Mechanics for Scientists and Engineers* is it.

The book's applied emphasis means that the key concepts are introduced in the context of their use in science and technology, with examples of nanostructured materials, optics, and semiconductor devices. Clear explanations of quantum mechanics's basic physics and mathematics, along with algorithms for the computational analysis of simple structures, make this an ideal introductory text. The many worked examples and 160 homework problems help students with problem solving and application of theory. No prior knowledge of high-level physics, difficult mathematics, or classical mechanics is assumed, and supporting math is included in appendices. Students of engineering, physics, nanotechnology, and other disciplines will be able to access more advanced texts on specific aspects of quantum mechanics after using this book.

The text introduces Schrödinger's equation, operators, and approximation methods. Systems, including the hydrogen atom and crystalline materials, are analyzed in detail. More advanced subjects, such as density matrices, quantum optics, and quantum information, are also covered. Online resources include instructor solutions, animations of figures to reinforce key concepts, illustrations, presentation materials, and worked examples. The text's pedagogy and accompanying materials provide a complete teaching package, which clearly develops and delivers one of physics's foundational topics.

Additional resources for this title are available from www.cambridge.org/9780521897839.

David A. B. Miller is the W. M. Keck Foundation Professor of Electrical Engineering at Stanford University, California, where he is also a professor of Applied Physics by courtesy, Director of the Solid State and Photonics Laboratory, and co-director of the Stanford Photonics Research Center. Before moving to Stanford, he was the head of advanced photonics research at AT&T Bell Laboratories. He is a Member of the US National Academy of Sciences, a Member of the US National Academy of Engineering, and a Fellow of the Royal Society, the Royal Society of Edinburgh, the Institute of Electrical and Electronics Engineers, the Optical Society of America, and the American Physical Society. He has received several awards for his work on the physics and applications of quantum-confined semiconductor structures.

Quantum Mechanics for Scientists and Engineers

David A. B. Miller
Stanford University

CAMBRIDGE
UNIVERSITY PRESS

CAMBRIDGE
UNIVERSITY PRESS

University Printing House, Cambridge CB2 8BS, United Kingdom

One Liberty Plaza, 20th Floor, New York, NY 10006, USA

477 Williamstown Road, Port Melbourne, VIC 3207, Australia

314-321, 3rd Floor, Plot 3, Splendor Forum, Jasola District Centre, New Delhi - 110025, India

103 Penang Road, #05-06/07, Visioncrest Commercial, Singapore 238467

Cambridge University Press is part of the University of Cambridge.

It furthers the University's mission by disseminating knowledge in the pursuit of education, learning and research at the highest international levels of excellence.

www.cambridge.org
Information on this title: www.cambridge.org/9780521897839

First published 2008
Reprinted with corrections 2009
14th printing 2020

A catalogue record for this publication is available from the British Library

Library of Congress Cataloging in Publication data
Miller, D. A. B.
Quantum mechanics for scientists and engineers / David A. B. Miller.
p. cm.
Includes bibliographical references and index.
ISBN 978-0-521-89783-9 (hardback)
1. Quantum theory. I. Title.
QC174.13.M556 2008
530.12–dc22 2008001249

ISBN 978-0-521-89783-9 Hardback

To Pat, Andrew, and Susan

Contents

Preface xiii

How to use this book xvi

Chapter 1 Introduction 1

1.1 Quantum mechanics and real life 1
1.2 Quantum mechanics as an intellectual achievement 4
1.3 Using quantum mechanics 6

Chapter 2 Waves and quantum mechanics – Schrödinger's equation 8

2.1 Rationalization of Schrödinger's equation 8
2.2 Probability densities 11
2.3 Diffraction by two slits 12
2.4 Linearity of quantum mechanics: multiplying by a constant 16
2.5 Normalization of the wavefunction 17
2.6 Particle in an infinitely deep potential well ("particle in a box") 18
2.7 Properties of sets of eigenfunctions 23
2.8 Particles and barriers of finite heights 26
2.9 Particle in a finite potential well 32
2.10 Harmonic oscillator 39
2.11 Particle in a linearly varying potential 42
2.12 Summary of concepts 50

Chapter 3 The time-dependent Schrödinger equation 54

3.1 Rationalization of the time-dependent Schrödinger equation 55
3.2 Relation to the time-independent Schrödinger equation 57
3.3 Solutions of the time-dependent Schrödinger equation 58
3.4 Linearity of quantum mechanics: linear superposition 59
3.5 Time dependence and expansion in the energy eigenstates 60
3.6 Time evolution of infinite potential well and harmonic oscillator 61
3.7 Time evolution of wavepackets 66
3.8 Quantum mechanical measurement and expectation values 73
3.9 The Hamiltonian 77
3.10 Operators and expectation values 77
3.11 Time evolution and the Hamiltonian operator 78
3.12 Momentum and position operators 81
3.13 Uncertainty principle 83
3.14 Particle current 85
3.15 Quantum mechanics and Schrödinger's equation 88
3.16 Summary of concepts 89

Chapter 4	Functions and operators	93
4.1	Functions as vectors	94
4.2	Vector space	100
4.3	Operators	103
4.4	Linear operators	104
4.5	Evaluating the elements of the matrix associated with an operator	107
4.6	Bilinear expansion of linear operators	108
4.7	Specific important types of linear operators	110
4.8	Identity operator	110
4.9	Inverse operator	113
4.10	Unitary operators	114
4.11	Hermitian operators	119
4.12	Matrix form of derivative operators	124
4.13	Matrix corresponding to multiplying by a function	125
4.14	Summary of concepts	125
Chapter 5	Operators and quantum mechanics	129
5.1	Commutation of operators	129
5.2	General form of the uncertainty principle	131
5.3	Transitioning from sums to integrals	135
5.4	Continuous eigenvalues and delta functions	136
5.5	Summary of concepts	150
Chapter 6	Approximation methods in quantum mechanics	154
6.1	Example problem – potential well with an electric field	155
6.2	Use of finite matrices	157
6.3	Time-independent nondegenerate perturbation theory	161
6.4	Degenerate perturbation theory	170
6.5	Tight binding model	172
6.6	Variational method	176
6.7	Summary of concepts	180
Chapter 7	Time-dependent perturbation theory	182
7.1	Time-dependent perturbations	182
7.2	Simple oscillating perturbations	185
7.3	Refractive index	192
7.4	Nonlinear optical coefficients	195
7.5	Summary of concepts	205
Chapter 8	Quantum mechanics in crystalline materials	207
8.1	Crystals	207
8.2	One electron approximation	209
8.3	Bloch theorem	209
8.4	Density of states in *k*-space	213
8.5	Band structure	214
8.6	Effective mass theory	216
8.7	Density of states in energy	220
8.8	Densities of states in quantum wells	221

8.9	k·p method	226
8.10	Use of Fermi's Golden Rule	231
8.11	Summary of concepts	239
Chapter 9	**Angular momentum**	**242**
9.1	Angular momentum operators	242
9.2	L squared operator	247
9.3	Visualization of spherical harmonic functions	250
9.4	Comments on notation	253
9.5	Visualization of angular momentum	254
9.6	Summary of concepts	255
Chapter 10	**The hydrogen atom**	**257**
10.1	Multiple-particle wavefunctions	258
10.2	Hamiltonian for the hydrogen atom problem	259
10.3	Coordinates for the hydrogen atom problem	260
10.4	Solving for the internal states of the hydrogen atom	264
10.5	Solutions of the hydrogen atom problem	270
10.6	Summary of concepts	275
Chapter 11	**Methods for one-dimensional problems**	**277**
11.1	Tunneling probabilities	277
11.2	Transfer matrix	280
11.3	Penetration factor for slowly varying barriers	288
11.4	Electron emission with a potential barrier	289
11.5	Summary of concepts	295
Chapter 12	**Spin**	**297**
12.1	Angular momentum and magnetic moments	298
12.2	State vectors for spin angular momentum	300
12.3	Operators for spin angular momentum	302
12.4	The Bloch sphere	303
12.5	Direct product spaces and wavefunctions with spin	305
12.6	Pauli equation	307
12.7	Where does spin come from?	307
12.8	Summary of concepts	308
Chapter 13	**Identical particles**	**311**
13.1	Scattering of identical particles	311
13.2	Pauli exclusion principle	315
13.3	States, single-particle states, and modes	316
13.4	Exchange energy	316
13.5	Extension to more than two identical particles	321
13.6	Multiple-particle basis functions	323
13.7	Thermal distribution functions	328
13.8	Important extreme examples of states of multiple identical particles	329
13.9	Quantum mechanical particles reconsidered	330
13.10	Distinguishable and indistinguishable particles	331

13.11 Summary of concepts 332

Chapter 14 The density matrix 335

14.1 Pure and mixed states 335
14.2 Density operator 338
14.3 Density matrix and ensemble average values 339
14.4 Time evolution of the density matrix 341
14.5 Interaction of light with a two-level "atomic" system 343
14.6 Density matrix and perturbation theory 350
14.7 Summary of concepts 351

Chapter 15 Harmonic oscillators and photons 354

15.1 Harmonic oscillator and raising and lowering operators 354
15.2 Hamilton's equations and generalized position and momentum 360
15.3 Quantization of electromagnetic fields 361
15.4 Nature of the quantum mechanical states of an electromagnetic mode 366
15.5 Field operators 367
15.6 Quantum mechanical states of an electromagnetic field mode 370
15.7 Generalization to sets of modes 373
15.8 Vibrational modes 378
15.9 Summary of concepts 379

Chapter 16 Fermion operators 383

16.1 Postulation of fermion annihilation and creation operators 384
16.2 Wavefunction operator 393
16.3 Fermion Hamiltonians 395
16.4 Summary of concepts 403

Chapter 17 Interaction of different kinds of particles 406

17.1 States and commutation relations for different kinds of particles 406
17.2 Operators for systems with different kinds of particles 407
17.3 Perturbation theory with annihilation and creation operators 409
17.4 Stimulated emission, spontaneous emission, and optical absorption 411
17.5 Summary of concepts 422

Chapter 18 Quantum information 424

18.1 Quantum mechanical measurements and wavefunction collapse 424
18.2 Quantum cryptography 425
18.3 Entanglement 431
18.4 Quantum computing 434
18.5 Quantum teleportation 437
18.6 Summary of concepts 440

Chapter 19 Interpretation of quantum mechanics 441

19.1 Hidden variables and Bell's inequalities 441
19.2 The measurement problem 448
19.3 Solutions to the measurement problem 449
19.4 Epilogue 454

19.5	Summary of concepts	455
Appendix A	**Background mathematics**	**457**
A.1	Geometrical vectors	457
A.2	Exponential and logarithm notation	460
A.3	Trigonometric notation	460
A.4	Complex numbers	461
A.5	Differential calculus	464
A.6	Differential equations	468
A.7	Summation notation	474
A.8	Integral calculus	475
A.9	Matrices	478
A.10	Product notation	489
A.11	Factorial	490
Appendix B	**Background physics**	**491**
B.1	Elementary classical mechanics	491
B.2	Electrostatics	494
B.3	Frequency units	495
B.4	Waves and diffraction	495
Appendix C	**Vector calculus**	**499**
C.1	Vector calculus operators	499
C.2	Spherical polar coordinates	504
C.3	Cylindrical coordinates	506
C.4	Vector calculus identities	507
Appendix D	**Maxwell's equations and electromagnetism**	**509**
D.1	Polarization of a material	509
D.2	Maxwell's equations	511
D.3	Maxwell's equations in free space	512
D.4	Electromagnetic wave equation in free space	512
D.5	Electromagnetic plane waves	513
D.6	Polarization of a wave	514
D.7	Energy density	514
D.8	Energy flow	514
D.9	Modes	516
Appendix E	**Perturbing Hamiltonian for optical absorption**	**519**
E.1	Justification of the classical Hamiltonian	519
E.2	Quantum mechanical Hamiltonian	520
E.3	Choice of gauge	521
E.4	Approximation to linear system	522

Appendix F Early history of quantum mechanics 523

Appendix G Some useful mathematical formulae 525
G.1 Elementary mathematical expressions 525
G.2 Formulae for sines, cosines, and exponentials 526
G.3 Special functions 529

Appendix H Greek alphabet 533

Appendix I Fundamental constants 534

Bibliography 535

Memorization list 539

Index 544

Preface

This book introduces quantum mechanics to scientists and engineers. It can be used as a text for junior undergraduates onward through to graduate students and professionals. The level and approach are aimed at anyone with a reasonable scientific or technical background looking for a solid but accessible introduction to the subject. The coverage and depth are substantial enough for a first quantum mechanics course for physicists. At the same time, the level of required background in physics and mathematics has been kept to a minimum to suit those from other science and engineering backgrounds.

Quantum mechanics has long been essential for all physicists and for other physical science subjects, such as chemistry. With the growing interest in nanotechnology, quantum mechanics has recently become increasingly important for an ever-widening range of engineering disciplines, such as electrical and mechanical engineering, and for subjects such as materials science that underlie many modern devices. Many physics students also find that they are increasingly motivated in the subject as the everyday applications become clear.

Nonphysicists have a particular problem finding a suitable introduction to the subject. The typical physics quantum mechanics course or text deals with many topics that, though fundamentally interesting, are useful primarily to physicists doing physics; that choice of topics also means omitting many others that are just as truly quantum mechanics but that have more practical applications. Too often, the result is that engineers or applied scientists cannot afford the time or sustain the motivation to follow such a physics-oriented sequence. As a result, they never have a proper grounding in the subject. Instead, they pick up bits and pieces in other courses or texts. Learning quantum mechanics in such a piecemeal approach is especially difficult; students then never properly confront the many fundamentally counterintuitive concepts of the subject. Those concepts need to be understood quite deeply if students are ever going to apply the subject with any reliability in any novel situation. Too often, also, even after working hard in a quantum mechanics class and even after passing the exams, students are still left with the depressing feeling that they do not understand the subject at all.

To address the needs of its broad intended readership, this book differs from most others in three ways. First, it presumes as little as possible in prior knowledge of physics. Specifically, it does not presume the advanced classical mechanics (including concepts such as Hamiltonians and Lagrangians) that is often a prerequisite in physics quantum mechanics texts and courses. Second, in two background appendices, it summarizes all of the key physics and mathematics beyond the high school level that the reader needs to start the subject. Third, it introduces the quantum mechanics that underlies many important areas of application, including semiconductor physics, optics, and optoelectronics. Such areas are usually omitted from quantum mechanics texts, but this book introduces many of the quantum mechanical principles and models that are exploited in those subjects.

It is also my belief and experience that using quantum mechanics in several different and practical areas of application removes many of the difficulties in understanding the subject. If quantum mechanics is illustrated only through examples that are found in the more esoteric branches of physics, the subject itself can seem irrelevant and obscure. There is nothing like designing a real device with quantum mechanics to make the subject tangible and meaningful.

Even with its deliberately limited prerequisites and its increased discussion of applications, this book offers a solid foundation in the subject. That foundation should well prepare the reader for the quantum mechanics in either advanced physics or deeper study of practical applications in other scientific and engineering fields. The emphasis is on understanding the ideas and techniques of quantum mechanics rather than attempting to cover all possible examples of their use. A key goal of this book is that the reader should subsequently be able to pick up texts in a broad range of areas – including, for example, advanced quantum mechanics for physicists, solid-state and semiconductor physics and devices, optoelectronics, quantum information, and quantum optics – and find they already have all the necessary basic tools and conceptual background in quantum mechanics to make rapid progress.

It is possible to teach quantum mechanics in many different ways, though most sequences will start with Schrödinger's wave equation and work forward from there. Even though the final emphasis in this book may be different from some other quantum mechanics courses, I have deliberately chosen to not take a radical approach here. This is for three reasons: first, most college and university teachers will be most comfortable with a relatively standard approach because that is the one they have most probably experienced themselves; second, taking a core approach that is relatively conventional will make it easier for readers (and teachers) to connect with the many other good physics quantum mechanics books; and third, this book should also be accessible and useful to professionals who have previously studied quantum mechanics to some degree but who need to update their knowledge or connect to the modern applications in engineering or applied sciences.

The background requirements for the reader are relatively modest and should represent little problem for students or professionals in engineering, applied sciences, physics, or other physical sciences. This material has been taught with apparent success to students in applied physics, electrical engineering, mechanical engineering, materials science, and other science and engineering disciplines, from third-year undergraduate level up to the graduate level. In mathematics, readers should have a basic knowledge of calculus, complex numbers, elementary matrix algebra, geometrical vectors, and simple and partial differential equations. In physics, readers should be familiar with ordinary Newtonian classical mechanics and elementary electricity and magnetism. The key requirements are summarized in two background appendices in case readers want to refresh some background knowledge or fill in gaps. A few other pieces of physics and mathematics are introduced as needed in the main body of the text. It is helpful if students have had some prior exposure to elementary modern physics, such as the ideas of electrons, photons, and the Bohr model of the atom, but no particular results are presumed here. The necessary parts of Hamiltonian classical mechanics are introduced briefly when required in later chapters.

This book goes deeper into certain subjects, such as the quantum mechanics of light, than most introductory physics texts. For the later chapters on the quantum mechanics of light, additional knowledge of vector calculus and electromagnetism to the level of Maxwell's equations is presumed, though again these are summarized in appendices.

One intent is for students to acquire a strong understanding of the concepts of quantum mechanics at the level beyond mere mathematical description. As a result, I have chosen to try to explain concepts with limited use of mathematics wherever possible. With the ready availability of computers and appropriate software for numerical calculations and simulations, it is progressively easier to teach principles of quantum mechanics without as heavy an emphasis on analytical techniques. Such numerical approaches are also closer to the methods that an engineer will likely use for calculations in real problems anyway, and access to some form of computer and high-level software package is assumed for some of the problems. This approach substantially increases the range of problems that can be examined both for tutorial examples and for applications.

Finally, I will make one personal statement on handling the conceptual difficulties of quantum mechanics in texts and courses. Some texts are guilty of stating quantum mechanical postulates, concepts, and assumptions as if they should be obvious, or at least obviously acceptable, when in fact they are far from obvious even to experienced practitioners or teachers. In many cases, these are subjects of continuing debate at the highest level. I try throughout to be honest about those concepts and assumptions whose obviousness or even correctness is genuinely unclear. I believe it is a particularly heinous sin to pretend that some concept should be clear to students when it is, in fact, not even clear to the professor (an overused technique that preserves professorial ego at the expense of the students'!).

It is a pleasure to acknowledge the many teaching assistants who have provided much useful feedback and correction of my errors in this material as I have taught it at Stanford, including Aparna Bhatnagar, Julien Boudet, Eleni Diamanti, Onur Fidaner, Martina Gerken, Noah Helman, Ekin Kocabas, Bianca Nelson, Vincent Revol, Jean-Christophe Richard, Tomas Sarmiento, and Scott Sharpe. I would like to thank Emel Tasyurek for a particularly careful reading of the manuscript, Ingrid Tarien for much help in preparing many parts of the course material, and Marjorie Ford for many helpful comments on writing.

I am also pleased to acknowledge my many professorial colleagues at Stanford, including, in particular, Steve Harris, Walt Harrison, Jelena Vuckovic, and Yoshi Yamamoto for many stimulating, informative, and provocative discussions about quantum mechanics. I would especially like to thank Jelena Vuckovic, who successfully taught the subject to many students despite having to use much of this material as a course reader, and who consequently corrected numerous errors and clarified many points. All remaining errors and shortcomings are, of course, my sole responsibility, and any further corrections and suggestions are most welcome.

David A. B. Miller

Stanford, California

September 2007

How to use this book

For teachers

The entire material in this book could be taught in a one-year course. More likely, depending on the interests and goals of the teacher and students and the length of time available, only some of the more advanced topics will be covered in detail. In a two-quarter course sequence for senior undergraduates and for engineering graduate students at Stanford, the majority of the material here will be covered, with a few topics omitted and some covered in lesser depth.

The core material (Chapters 1–5) on Schrödinger's equation and on the mathematics behind quantum mechanics should be taught in any course. Chapter 4 gives a more explicit introduction to the ideas of linear operators than is found in most texts. Chapter 4 also explains and introduces Dirac notation, which is used from that point onward in the book. This introduction of Dirac notation is earlier than in many older texts, but it saves considerable time thereafter in describing quantum mechanics. Experience teaching engineering students in particular, most of whom are quite familiar with linear algebra and matrices from other applications in engineering, shows that they have no difficulties with this concept.

Aside from that core, there are many possible choices about the sequence of material and about what material needs to be included in a course. The prerequisites for each chapter are clearly stated at the beginning of the chapter. There are also some sections in several of the chapters that are optional or that may only need to be read through when first encountered. These sections are clearly marked. The discussion of methods for one-dimensional problems in Chapter 11 can come at any point after the material on Schrödinger's equations (Chapters 2 and 3). The core transfer matrix part could even be taught directly after the time-independent equation (Chapter 2). The material is optional in that it is not central to later topics, but, in my experience, students usually find it stimulating and empowering to be able to do calculations with simple computer programs based on these methods. This can make students comfortable with the subject and begin to give them some intuitive feel for many quantum mechanical phenomena. (These methods are also used, in practice, for the design of real optoelectronic devices.)

For a broad range of applications, the approximation methods of quantum mechanics (Chapters 6 and 7) are probably the next most important after Chapters 1 through 5. The specific topic of the quantum mechanics of crystalline materials (Chapter 8) is particularly important for many applications and can be introduced at any point after Chapter 7; it is not, however, required for subsequent chapters (except for a few examples and some optional parts at the end of Chapter 11), so teachers can choose how far they want to progress through this chapter. For fundamentals, angular momentum (Chapter 9) and the hydrogen atom (Chapter 10) are the next most central topics, both of which can be taught directly after Chapter 5, if desired. After these, the next most important fundamental topics are spin (Chapter 12) and identical particles (Chapter 13), and these should probably be included in the second quarter or semester, if not before.

Chapter 14 introduces the important technique of the density matrix for connecting to statistical mechanics, and it can be introduced at any point after Chapter 5; preferably, students would also have covered Chapters 6 and 7 so they are familiar with perturbation theory, though that is not required. The density matrix material is not required for subsequent chapters, so Chapter 14 is optional.

The sequence of Chapters 15–17 introduces the quantum mechanics of electromagnetic fields and light as well as the important technique of second quantization in general, including fermion operators (a technique that is also used extensively in more advanced solid-state physics). The inclusion of this material on the quantum mechanics of light is the largest departure from typical introductory quantum mechanics texts. It does, however, redress a balance in material that is important from a practical point of view; we cannot describe even the simplest light emitter (including an ordinary light bulb) or light detector without it, for example. This material is also very substantial quantum mechanics at the next level of the subject. These chapters do require almost all of the preceding material, with the possible exceptions of Chapters 8, 11, and 14.

The final two chapters, Chapter 18 on a brief introduction to quantum information concepts and Chapter 19 on the interpretation of quantum mechanics, could conceivably be presented with only Chapters 1–5 as prerequisites. Preferably also Chapters 9, 10, 12, and 13 would have been covered, and it is probably a good idea that students have been working with quantum mechanics successfully for some time before attempting to grapple with the tricky conceptual and philosophical aspects in these final chapters. The material in these chapters is well suited to the end of a course, when it is often unreasonable to include any further new material in a final exam, but teachers want to keep the students' interest with stimulating ideas.

Problems are introduced directly after the earliest possible sections rather than being deferred to the end of the chapters, thus giving the greatest flexibility in assigning homework. Solutions to selected problems are openly available from www.cambridge.org/9780521897839 (these problems are marked with an asterisk [*] in the text); such problems could be used as additional practice material by students or as worked examples in class. Some of the problems can be used as substantial assignments, and all such problems are clearly marked. These are suitable for "take-home" problems or exams or as extended exercises coupled with tutorial "question-and-answer" sessions. These assignments may necessarily involve some more work, such as significant amounts of (relatively straightforward) algebra or calculations with a computer. I have found, though, that students gain a much greater confidence in the subject once they have used it for something beyond elementary exercises – exercises that are necessarily often artificial. At least, these assignments tend to approach the subject from the point of view of a problem to be solved rather than an exercise that just uses the last technique that was studied. Some of these larger assignments deal with quite realistic uses of quantum mechanics.

At the very end of the book, I also include a suggested list of simple formulae to be memorized in each chapter. These lists could also be used as the basis of simple quizzes or as required learning for "closed-book" exams.

For students

Necessary background

Students will come to this book with very different backgrounds. You may recently have studied a lot of physics and mathematics at the college level. If so, then you are ready to start. I suggest you have a quick look at Appendices A and B just to see the notations used in this book before starting Chapter 2.

For others, your mathematical or physics background may have been less complete, or it may be some time since you have seen or used some of the relevant parts of these subjects. Rest assured, first of all, that in writing this book I have presumed the least possible knowledge of mathematics and physics consistent with teaching quantum mechanics, and much less than the typical quantum mechanics text requires. Ideally, I expect you have had the physics and mathematics typical of first- or second-year college-level general engineering or physical science students. You do absolutely have to know elementary algebra, calculus, and physics to a good precollege level, however. I suggest you read the Background Mathematics Appendix A and the Background Physics Appendix B to see if you understand most of that. If not too much of that is new to you, then you should be able to proceed into the main body of this book. If you find some new topics in these Appendices, there is, in principle, enough material there to "patch over" those holes in knowledge so that you can use the mathematics and physics needed to start quantum mechanics; Appendices A and B are not, however, meant to be a substitute for learning these topics in greater depth.

Study aids in this book

Lists of concepts introduced

Because there are many concepts that students need to understand in quantum mechanics, I have summarized the most important ones at the end of the chapters in which they are introduced. These summaries should help in both following the "plot" of the book and revising the material.

Appendices

The book is as reasonably self-contained as I can make it. In addition to the background Appendices A and B covering the overall prerequisite mathematics and physics, background material needed later on is introduced in Appendices C and D (vector calculus and electromagnetism), and one specific detailed derivation is given in Appendix E. Appendix F summarizes the early history of quantum mechanics and Appendix G collects and summarizes most of the mathematical formulae that will be needed in the book, including the most useful ones from elementary algebra, trigonometric functions, and calculus. Appendix H gives the Greek alphabet (every single letter of it is used somewhere in quantum mechanics), and Appendix I lists all the relevant fundamental constants.

Problems

There are 160 problems and assignments, collected at the end of the earliest possible sections, rather than at the end of the chapters. Solutions to thirty-six of these are openly available from www.cambridge.org/9780521897839 (these problems are marked with an asterisk [*] in the text), giving additional worked examples for practice or study.

Memorization list

Quantum mechanics, like many aspects of physics, is not primarily about learning large numbers of formulae but rather understanding the key concepts clearly and deeply. It will, however, save a lot of time (including in exams!) to learn a few basic formulae by heart; certainly, if you also understand these well, you should have a good command of the subject.

At the very end of the book, there is a list of formulae worth memorizing in each chapter of the book. None of these formulae are particularly complicated – the most complicated ones are the two forms of the Schrödinger wave equation. Many of the formulae are simply short definitions of key mathematical concepts. If you learn these formulae chapter-by-chapter as you work through the book, there are not very many formulae to learn at any one time.

The list here is not of the formulae themselves but rather descriptions of them, so you can use this list as an exercise to test how successfully you have learned these key results.

Self-teaching

If you are teaching yourself quantum mechanics using this book, first, congratulations for having the courage to tackle what most people typically regard as a daunting subject. For someone with elementary college-level physics and mathematics, I believe it is quite an accessible subject, in fact. *But*, the most important point is that you must not start learning quantum mechanics "on the fly" by picking and choosing just the bits you need from this book or any other. Trying to learn quantum mechanics like that would be like trying to learn a language by reading a dictionary. You cannot treat quantum mechanics as just a set of formulae to be substituted into problems, just as you cannot translate a sentence from one language to another just by looking up the individual words in a dictionary and writing down their translation. There are just so many counterintuitive aspects about quantum mechanics that you will never understand it in that piecemeal way and, most likely, you would not use the formulae correctly anyway. Make yourself work on all of the first several chapters, through at least Chapter 5; that will get you to a first plateau of understanding. You can be somewhat more selective after that. For the next level of understanding, you need to study angular momentum, spin, and identical particles (Chapters 9, 12, and 13). Which other chapters you use will depend on your interests or needs. Of course, it is worthwhile studying all of them if you have the time!

Especially if you have no tutor of whom you can ask questions, then I also expect that you should be looking at other quantum mechanics books as well. Use this one as your core and, when I have just not managed to explain something clearly enough or to get it to "click" for you, look at some of the others, such as the ones listed in the Bibliography. My personal experience is that a difficult topic finally becomes clear to me once I have five books on it open on my desk. One hope I have for this book is that it enables readers to access the more specialized physics texts, if necessary. Their alternative presentations may well succeed where mine fail, and those other books can certainly cover a range of specific topics impossible to include here.

Chapter 1

Introduction

1.1 Quantum mechanics and real life

Quantum mechanics, we might think, is a strange subject, one that does not matter for daily life. Only a few people, therefore, should need to worry about its difficult details. These few, we might imagine, run about in the small dark corners of science, at the edge of human knowledge. In this unusual group, we would expect to find only physicists making ever larger machines to look at ever smaller objects, chemists examining the last details of tiny atoms and molecules, and perhaps a few philosophers absently looking out the window as they wonder about free will. Surely, quantum mechanics therefore should not matter for our everyday experience. It could not be important for designing and making real things that make real money and change real lives. Of course, we would be wrong.

Quantum mechanics is everywhere. We do not have to look far to find it. We only have to open our eyes. Look at some object, say a flowerpot or a tennis ball. Why is the flowerpot a soothing terra-cotta orange color and the tennis ball a glaring fluorescent yellow? We could say each object contains some appropriately colored pigment or dye, based on a material with an intrinsic color, but we are not much further forward in understanding. (Our color technology would also be stuck in medieval times, when artists had to find all their pigments in the colors in natural objects, sometimes at great cost.[1]) The particularly bright yellow of our modern tennis ball would also be quite impossible if we restricted our pigments to naturally occurring materials.

Why does each such pigment have its color? We have no answer from the "classical" physics and chemistry developed before 1900. But quantum mechanics answers such questions precisely and completely.[2] Indeed, the beginning of quantum mechanics comes from one

---

[1] They had to pay particularly dearly for their ultramarine blue, a pigment made by grinding up the gemstone *lapis lazuli*. The Spanish word for blue, *azul*, and the English word *azure* both derive from this root. The word *ultramarine* refers to the fact that the material had to be brought from "beyond (ultra) the sea (marine)" – i.e., imported, presumably also at some additional cost. Modern blue coloring is more typically based on copper phthalocyanine, a relatively cheap, manmade chemical.

[2] In quantum mechanics, photons, the quantum mechanical particles of light, have different colors depending on their tiny energies; materials have energy levels determined by the quantum mechanics of electrons, energy levels separated by similarly tiny amounts. We can change the electrons from one energy level to another by absorbing or emitting photons. The specific color of an object comes from the specific separations of the energy levels in the material. A few aspects of color can be explained without quantum mechanics. Color can sometimes result from scattering (e.g., the blue of the sky or the white of

particular aspect of color. Classical physics famously failed to explain the color of hot objects,[3] such as the warm yellow of the filament in a light bulb or the glowing red of hot metal in a blacksmith's shop. Max Planck realized in 1900 that if the energy in light existed only in discrete steps, or quanta, he could get the right answer for these colors. And so, quantum mechanics was born.

The impact of quantum mechanics in explaining our world does not end with color. We have to use quantum mechanics in explaining most properties of materials. Why are some materials hard and others soft? For example, why can a diamond scratch almost anything, but a pencil lead will slide smoothly, leaving a black line behind it?[4] Why do metals conduct electricity and heat easily but glass does not? Why is glass transparent? Why do metals reflect light? Why is one substance heavy and another light? Why is one material strong and another brittle? Why are some metals magnetic and others are not? We need, of course, a good deal of other science, such as chemistry, materials science, and other branches of physics, to answer such questions in any detail; but, in doing so, all of these sciences will rely on our quantum mechanical view of how materials are put together.

So, we might now believe, the consequences of quantum mechanics are essential for understanding the ordinary world around us. But is quantum mechanics useful? If we devote our precious time to learning it, will it let us make things we could not make before? One science in which the quantum mechanical view is obviously essential is chemistry, the science that enables most of our modern materials. No one could deny that chemistry is useful.

Suppose even that we set chemistry and materials themselves aside and ask a harder question: Do we need quantum mechanics when we design devices – objects intended to perform some worthwhile function? After all, the washing machines, eyeglasses, staplers, and automobiles of everyday life need only nineteenth century physics for their basic mechanical design, even if we employ the latest alloys, plastics, or paints to make them. Perhaps we can concede such macroscopic mechanisms to the classical world. But when, for example, we look at the technology to communicate and process information, we have simply been forced to move to quantum mechanics. Without quantum theory as a practical technique, we would not be able to design the devices that run our computers and our Internet connections.

The mathematical ideas of computing and information had begun to take their modern shape in the 1930s, 1940s and 1950s. By the 1950s, telephones and broadcast communication were well established and the first primitive electronic computers had been demonstrated. The transistor and integrated circuit were the next key breakthroughs. These devices made complex computers and information switching and processing practical. These devices relied heavily on the quantum mechanical physics of crystalline materials.

---

some paints), diffraction (e.g., by a finely ruled grating or a hologram), or interference (e.g., the varied colors of a thin layer of oil on the surface of water), all of which can be explained by classical wave effects. All such classical wave effects are also explained as limiting cases of quantum mechanics, of course.

[3] This problem was known as the "ultraviolet catastrophe" because classical thermal and statistical physics predicted that any warm object would emit ever-increasing amounts of light at ever shorter wavelengths. The colors associated with such wavelengths would necessarily extend past the blue into the ultraviolet – hence, the name.

[4] Even more surprising here is that diamonds and pencil lead are both made from exactly the same element, carbon.

A well-informed devil's advocate could still argue, though, that the design of transistors and integrated circuits themselves was initially still an activity using classical physics. Designers would still use the idea of resistance from nineteenth-century electricity, even if they added the ideas of charged electrons as particles carrying the current, and would add various electrical barriers (or "potentials") to persuade electrons to go one way or another. No modern transistor designer can ignore quantum mechanics, however. For example, when we make small transistors, we must also make very thin electrical insulators. Electrons can manage to penetrate through the insulators because of a purely quantum mechanical process known as *tunneling*. At the very least, we have to account for that tunneling current as an undesired, parasitic process in our design.

As we try to shrink transistors to ever smaller sizes, quantum mechanical effects become progressively more important. Naively extrapolating the historical trend in miniaturization would lead to devices the size of small molecules in the first few decades of the twenty-first century. Of course, the shrinkage of electronic devices as we know them cannot continue to that point. But, as we make ever tinier devices, quantum mechanical processes become ever more important. Eventually, we may need new device concepts beyond the semiclassical transistor; it is difficult to imagine how such devices would not involve yet more quantum mechanics.

We might argue, at least historically, about the importance of quantum mechanics in the design of transistors. We could have no comparable debate when we consider two other technologies crucial for handling information: optical communications and magnetic data storage.

Today, nearly all the information we send over long distances is carried on optical fibers – strands of glass about the thickness of a human hair. We very carefully put a very small light just at one end of that fiber. We send the "ones" and "zeros" of digital signals by rapidly turning that light on and off and looking for the pattern of flashes at the fiber's other end. To send and receive these flashes, we need optoelectronic devices – devices that will change electrical signals into optical pulses and vice versa. All of these optoelectronic devices are quantum mechanical on many different levels. First, they mostly are made of crystalline semiconductor materials, just like transistors, and hence rely on the same underlying quantum mechanics of such materials. Second, they send and receive photons, the particles of light Einstein proposed to expand upon Planck's original idea of quanta. Here, these devices are exploiting one of the first of many strange phenomena of quantum mechanics, the photoelectric effect. Third, most modern semiconductor optoelectronic devices used in telecommunications employ very thin layers of material, layers called *quantum wells*. The properties of these thin layers depend exquisitely on their thicknesses through a textbook piece of quantum mechanics known as the "particle-in-a-box" problem. That physics allows us to optimize some of the physical processes we already had in thicker layers of material and also to create some new mechanisms only seen in thin layers. For such devices, engineering using quantum mechanics is both essential and very useful.

When we try to pack more information onto the magnetic hard disk drives in our computers, we first have to understand exactly how the magnetism of materials works. That magnetism is almost entirely based on a quantum mechanical attribute called *spin* – a phenomenon with no real classical analog. The sensors that read the information off the drives are also often now based on sophisticated structures with multiple thin layers that are designed completely with quantum mechanics.

Quantum mechanics is, then, a subject increasingly necessary for engineering devices, especially as we make small devices or exploit quantum mechanical properties that only occur in small structures. The examples given here are only a few from a broad and growing field that can be called *nanotechnology*. Nanotechnology exploits our expanding abilities to make very small structures or patterns. The benefits of nanotechnology come from the new properties that appear at these very small scales. We get most of those new properties from quantum mechanical effects of one kind or another. Quantum mechanics is, therefore, essential for nanotechnology.

1.2 Quantum mechanics as an intellectual achievement

Any new scientific theory has to give the same answers as the old theories everywhere these previous models worked and yet successfully describe phenomena that previously we could not understand. The prior theories of mechanics, Newton's Laws, worked very well in a broad range of situations. Our models for light similarly were quite deep and had achieved a remarkable unification of electricity and magnetism (in Maxwell's equations). But, when we would try to make a model of the atom, for example, with electrons circling around some charged nucleus like satellites in orbit around the earth, we would meet major contradictions. Existing mechanics and electromagnetic theory would predict that any such orbiting electron would constantly be emitting light; but, atoms simply do not do that.

The challenge for quantum mechanics was not an easy one. To resolve these problems of light and the structure of matter, we actually had to tear down much of our view of the way the world works, to a degree never seen since the introduction of natural philosophy and the modern scientific method in the Renaissance. We were forced to construct a completely new set of principles for the physical world. These were, and still are in many cases, completely bizarre and certainly different from our intuition. Many of these principles simply have no analogs in our normal view of reality.

We mentioned previously one of the bizarre aspects of quantum mechanics: the process of tunneling allows particles to penetrate barriers that are classically too high for them to overcome. This process is, however, actually nothing like the act of digging a tunnel; we are confronting here the common difficulty in quantum mechanics of finding words or analogies from everyday experience to describe quantum mechanical ideas. We will often fail.

There are many other surprising aspects of quantum mechanics. The typical student starting quantum mechanics is confused when told, as he or she often will be, that some question simply does not have an answer. The student will, for example, think it perfectly reasonable to ask what are the position and momentum (or, more loosely, speed) of some particle, such as an electron. Quantum mechanics (or, in practice, its human oracle, the professor) will enigmatically reply that there is no answer to that question. We can know one or the other precisely but not both at once. This particular enigma is an example of Heisenberg's uncertainty principle.

Quantum mechanics does raise more than its share of deep questions, and it is arguable that we still do not understand quantum mechanics. In particular, there are still major questions about what a measurement really is in the quantum world. Erwin Schrödinger famously dramatized the difficulty with the paradox of his cat. According to quantum mechanics, an object may exist in a superposition state, in which it is, for example, neither definitely on the left nor on the right. Such superposition states are not at all unusual – in fact, they occur all the time for electrons in any atom or molecule. Though a particle might be in a superposition state, when we try to measure it, we always find that the object is at some specific position (e.g., definitely on the left or on the right). This mystical phenomenon is known as "collapse of the wavefunction." We might find that to be a bizarre idea, but one that – for something really tiny like an electron – we could perhaps accept.

But now Schrödinger proposes that we think not about an electron but instead about his cat. We are likely to care much more about the welfare of this "object" than we did about some electron. An electron is, after all, easily replaced with another just the same[5]; there are plenty of them – in fact, something like 10^{24} electrons in every cubic centimeter of any solid material. And Schrödinger constructs a dramatic scenario. His cat is sealed in a box with a lethal mechanism that may go off as a result of, for example, radioactive decay. Before we open the box to check on it, is the cat alive, dead, or, as quantum mechanics might seem to suggest, in some "superposition" of the two?

The superposition hypothesis now seems absurd. In truth, we cannot check it here; we do not know how to set up an experiment to test such quantum mechanical notions with macroscopic objects. In trying to repeat such an experiment, we cannot set up the same starting state exactly enough for something as complex as a cat. Physicists disagree about the resolution of this paradox. It is an example of a core problem of quantum mechanics: the process of measurement, with its mysterious "collapse of the wavefunction," cannot be explained by quantum mechanics.[6] The proposed solutions to this measurement problem can be extremely bizarre; in the "many worlds" hypothesis, for example, the world is supposed continually to split into multiple realities, one for each possible outcome of each possible measurement.

Another important discussion centers around whether quantum mechanics is complete. When we measure a quantum mechanical system, there is at least in practice some randomness in the result. If, for example, we tried to measure the position of an electron in an atom, we would keep getting different results. Or, if we measured how long it took a radioactive nucleus to decay, we would get different numbers each time. Quantum mechanics would correctly predict the average position we would measure for the electron and the average decay time of the nucleus, but it would not tell us the specific position or time yielded by any particular measurement.

We are, of course, quite used to randomness in our ordinary classical world. The outcome of many lotteries is decided by which numbered ball appears out of a chute in a machine. The various different balls are all bouncing around inside the machine, driven probably by some air blower. The process is sufficiently complicated that we cannot practically predict which ball will emerge, and all have equal chance. But we do tend to believe classically that if we knew the initial positions and velocities of all the air molecules and the balls in the machine, we could, in principle, predict which ball would emerge. Those variables are, in practice, hidden from us, but we do believe they exist. Behind the apparent randomness of quantum mechanics, then, are there just similarly some hidden variables? Could we actually predict outcomes precisely if we knew what those hidden variables were? Is the apparent randomness of quantum mechanics just because of our lack of understanding of some deeper theory and its starting conditions, some "complete" theory that would supersede quantum mechanics?

---

[5] Indeed, in quantum mechanics, electrons can be absolutely identical, much more identical than the so-called "identical" toys from an assembly line or "identical" twins in a baby carriage.

[6] If, at this point, the reader raises an objection that there is an inconsistency in saying that quantum mechanics will only answer questions about things we can measure but quantum mechanics cannot explain the process of measurement, the reader would be quite justified in doing so!

Einstein believed that, indeed, quantum mechanics was not complete, that there were some hidden variables that, once we understood them, would resolve and remove its apparent randomness. Relatively recent work, centered around a set of relations called Bell's inequalities, shows rather surprisingly that there are no such hidden variables (or, at least, not local ones that propagate with the particles) and that, despite its apparent absurdities, quantum mechanics may well be a complete theory in this sense.

It also appears that quantum mechanics is "nonlocal": two particles can be so "entangled" quantum mechanically that measuring one of them can apparently instantaneously change the state of the other one, no matter how far away it is (though it is not apparently possible to use such a phenomenon to communicate information faster than the velocity of light).[7]

Despite all its absurdities and contradictions of common sense, and despite the initial disbelief and astonishment of each new generation of students, quantum mechanics works. As far as we know, it is never wrong; we have made no experimental measurement that is known to contradict quantum mechanics, and there have been many exacting tests. Quantum mechanics is both stunningly radical and remarkably right. It is an astonishing intellectual achievement.

The story of quantum mechanics itself is far from over. We are still trying to understand exactly what are all the elementary particles and just what are the implications of such theories for the nature of the universe.[8] Many researchers are working on the possibility of using some of the strange possibilities of quantum mechanics for applications in handling information transmission. One example would send messages whose secrecy was protected by the laws of quantum physics, not just the practical difficulty of cracking classical codes. Another example is the field of quantum computing, in which quantum mechanics might allow us to solve problems that would be too hard ever to be solved by any conventional machine.

1.3 Using quantum mechanics

At this point, the poor student may be about to give up in despair. How can one ever understand such a bizarre theory? And, if one cannot understand it, how can one even think of using it? Here is the good news: whether we think we understand quantum mechanics, and whether there is yet more to discover about how it works, quantum mechanics is surprisingly easy to use.

The prescriptions for using quantum mechanics in a broad range of practical problems and engineering designs are relatively straightforward. They use the same mathematical techniques most engineering and science students will already have mastered to deal with the "classical" world.[9] Because of a particular elegance in its mathematics,[10] quantum mechanical calculations can actually be easier than those in many other fields.

---

[7] This nonlocality is often known through the original "EPR" thought experiment or paradox proposed by Einstein, Podolsky, and Rosen.

[8] Such theories require relativistic approaches that are, unfortunately, beyond the scope of this book.

[9] In the end, most calculations require performing integrals or manipulating matrices. Many of the underlying mathematical concepts are ones that are quite familiar to engineers used to Fourier analysis, e.g., or other linear transforms.

[10] Quantum mechanics is based entirely and exactly on linear algebra. Unlike most other uses of linear algebra, the fundamental linearity of quantum mechanics is apparently *not* an approximation.

The main difficulty the beginning student has with quantum mechanics lies in knowing which of our classical notions of the world have to be discarded, and what new notions we have to use to replace them.[11] The student should expect to spend some time in disbelief and conflict with what is being asserted in quantum mechanics – that is entirely normal! In fact, a good fight with these propositions is perhaps psychologically necessary, like the clarifying catharsis of an old-fashioned barroom brawl.

And, there is a key point that simplifies all the absurdities and apparent contradictions: provided we only ask questions about quantities that can be measured, there are no philosophical problems that need worry us or, at least, that would prevent us from calculating anything that we could measure.[12] As we use quantum mechanical principles in tangible applications, such as electronic or optical devices and systems, the apparently bizarre aspects become simply commonplace and routine. The student may soon stop worrying about quantum mechanical tunneling and Heisenberg's uncertainty principle. In the foreseeable future, such routine comprehension and acceptance may also extend to concepts such as nonlocality and entanglement as we press them increasingly into practical use.

Understanding quantum mechanics does certainly mark a qualitative change in one's view of how the world actually works.[13] That understanding gives students the opportunity to apply this knowledge in ways that others cannot begin to comprehend.[14] Whether the goal is basic understanding or practical exploitation, learning quantum mechanics is, in this author's opinion, certainly one of the most fascinating things one can do with one's brain.

---

[11] The associated teaching technique of breaking down the student's beliefs and replacing them with the professor's "correct" answers has a lot in common with brainwashing!

[12] This philosophical approach of dealing only with questions that can be answered by measurement (or that are purely logical questions within some formal system of logic) and regarding all other questions as meaningless is essentially what is known in the philosophical world as "logical positivism." It is the most common approach taken in dealing with quantum mechanics, at least at the elementary philosophical level, and, by allowing university professors to dismiss most student questions as meaningless, saves a lot of time in teaching the subject!

[13] It is undoubtedly true that if one does not understand quantum mechanics, one does not understand how the world actually works. It may also, however, be true that even if one does understand quantum mechanics, one still may not understand how the world works.

[14] Despite the inherent sense of superiority such an understanding may give the student, it is, however – as many physicists have already regrettably found – not particularly useful to point this out at parties.

Chapter 2

Waves and quantum mechanics – Schrödinger's equation

Prerequisites: Appendix A Background mathematics; Appendix B Background physics.

If the world of quantum mechanics is so different from everything we have been taught before, how can we even begin to understand it? Miniscule electrons seem so remote from what we see in the world around us that we do not know what concepts from our everyday experience we could use to get started. There is, however, one lever from our existing intellectual toolkit that we can use to pry open this apparently impenetrable subject, and that lever is the idea of waves. If we just allow ourselves to suppose that electrons might be describable as waves and follow the consequences of that radical idea, the subject can open up before us. Astonishingly, we will find we can then understand a large fraction of those aspects of our everyday experience that can only be explained by quantum mechanics, such as color and the properties of materials. We will also be able to engineer novel phenomena and devices for quite practical applications.

On the face of it, proposing that we describe particles as waves is a strange intellectual leap in the dark. There is apparently nothing in our everyday view of the world to suggest we should do so. Nevertheless, it was exactly such a proposal historically (i.e., de Broglie's hypothesis) that opened up much of quantum mechanics. That proposal was made before there was direct experimental evidence of wave behavior of electrons. Once that hypothesis was embodied in the precise mathematical form of Schrödinger's wave equation, quantum mechanics took off.

Schrödinger's equation remains to the present day one of the most useful relations in quantum mechanics. Its most basic application is to model simple particles that have mass, such as a single electron, though the extensions of it go much further than that. It is also a good example of quantum mechanics, exposing many of the more general concepts. We use these concepts as we go on to more complicated systems, such as atoms, or to other quite different kinds of particles and applications, such as photons and the quantum mechanics of light. Understanding Schrödinger's equation, therefore, is a very good way to start understanding quantum mechanics. In this chapter, we introduce the simplest version of Schrödinger's equation – the time-independent form – and explore some of the remarkable consequences of this wave view of matter.

2.1 Rationalization of Schrödinger's equation

Why do we have to propose wave behavior and Schrödinger's equation for particles such as electrons? After all, we are quite sure electrons are particles because we know that they have definite mass and charge. And we do not see directly any wave-like behavior of matter in our everyday experience. It is, however, now a simple and incontrovertible experimental fact that electrons can behave like waves or, at least, in some way are "guided" by waves. We know this for the same reasons we know that light is a wave – we can see the interference and diffraction that are so characteristic of waves. At least in the laboratory, we see this behavior routinely.

We can, for example, make a beam of electrons by applying a large electric field in a vacuum to a metal, pulling electrons out of the metal to create a monoenergetic electron beam (i.e., all with the same energy). We can then see the wave-like character of electrons by looking for the effects of diffraction and interference, especially the patterns that can result from waves interacting with particular kinds or shapes of objects.

One common situation in the laboratory is, for example, to shine such a beam of electrons at a crystal in a vacuum. Davisson and Germer did exactly this in their famous experiment[1] in 1927, diffracting electrons off a crystal of nickel. We can see the resulting diffraction if, for example, we let the scattered electrons land on a phosphor screen as in a television tube (cathode ray tube); we will see a pattern of dots on the screen. We would find that this diffraction pattern behaved rather similarly to the diffraction pattern we might get in some optical experiment; we could shine a monochromatic (i.e., single frequency) light beam at some periodic structure[2] whose periodicity was of a scale comparable to the wavelength of the waves (e.g., a diffraction grating). The fact that electrons behave both as particles (e.g., they have a specific mass and a specific charge) and as waves is known as a *wave-particle duality*.[3]

The electrons in such wave diffraction experiments behave as if they have a wavelength

$$\lambda = \frac{h}{p} \tag{2.1}$$

where p is the electron momentum and h is Planck's constant

$$h \cong 6.626 \times 10^{-34} \text{ Joule} \cdot \text{seconds}$$

(This relation, Eq. (2.1), is known as de Broglie's hypothesis). For example, the electron can behave as if it were a plane wave, with a wavefunction ψ, propagating in the z direction, and of the form[4]

$$\psi \propto \exp(2\pi i z / \lambda) \tag{2.2}$$

If it is a wave, or is behaving as such, we need a wave equation to describe the electron. We find empirically[5] that the electron behaves like a simple scalar wave (i.e., not like a vector wave, such as electric field, **E**, but like a simple acoustic [sound] wave with a scalar amplitude; in acoustics, the scalar amplitude could be the air pressure). We therefore propose that the

---

[1] C. Davisson and L. H. Germer, "Diffraction of Electrons by a Crystal of Nickel," Phys. Rev. **30**, 705 – 740 (1927)

[2] That is, a structure whose shape repeats itself in space, with some spatial "period" or length.

[3] This wave-particle duality is the first and one of the most profound of the apparently bizarre aspects of quantum mechanics that we will encounter.

[4] We have chosen a complex wave here, $\exp(2\pi i z / \lambda)$, rather than a simpler real wave, e.g., $\sin(2\pi z / \lambda)$ or $\cos(2\pi z / \lambda)$, because the mathematics of quantum mechanics is set up to require the use of complex numbers. The choice does also make the algebra easier.

[5] At least in the absence of magnetic fields or other magnetic effects.

electron wave obeys a scalar wave equation, and we choose the simplest one we know: the *Helmholtz wave equation* for a monochromatic wave. In one dimension, the Helmholtz equation is

$$\frac{d^2\psi}{dz^2} = -k^2\psi \tag{2.3}$$

This equation has solutions such as sin(*kz*), cos(*kz*), and exp(*ikz*) (and sin(-*kz*), cos(-*kz*), and exp(-*ikz*)) that all describe the spatial variation in a simple wave. In three dimensions, we can write this as

$$\nabla^2\psi = -k^2\psi \tag{2.4}$$

where the symbol ∇^2 (known variously as the *Laplacian operator*, *del squared*, and *nabla squared*, and sometimes written Δ) means

$$\nabla^2 \equiv \frac{\partial^2}{\partial x^2} + \frac{\partial^2}{\partial y^2} + \frac{\partial^2}{\partial z^2}, \tag{2.5}$$

where x, y, and z are the usual Cartesian coordinates, all at right angles to one another. This has solutions such as sin(**k.r**), cos(**k.r**), and exp(i**k.r**) (and sin(-**k.r**), cos(-**k.r**), and exp(-i**k.r**)), where **k** and **r** are vectors. The wavevector magnitude, k, is defined as

$$k = 2\pi / \lambda \tag{2.6}$$

or, equivalently, given the empirical wavelength exhibited by the electrons (i.e., de Broglie's hypothesis, Eq. (2.1))

$$k = p / \hbar \tag{2.7}$$

where

$$\hbar = h / 2\pi \cong 1.055 \times 10^{-34}\,\text{Joule} \cdot \text{seconds}$$

(a quantity referred to as *h bar*). With our expression for k (Eq. (2.7)), we can rewrite our simple wave equation (Eq. (2.4)) as

$$-\hbar^2\nabla^2\psi = p^2\psi \tag{2.8}$$

We can now choose to divide both sides by $2m_o$, where, for the case of the electron, m_o is the free electron rest mass

$$m_o \cong 9.11 \times 10^{-31}\,\text{kg}$$

to obtain

$$-\frac{\hbar^2}{2m_o}\nabla^2\psi = \frac{p^2}{2m_o}\psi \tag{2.9}$$

But we know for Newtonian classical mechanics, where $p = m_o v$ (with v as the velocity), that

$$\frac{p^2}{2m_o} \equiv \text{kinetic energy of an electron} \tag{2.10}$$

and, in general,

$$\text{Total energy } (E) = \text{Kinetic energy} + \text{Potential energy } (V(\mathbf{r})) \tag{2.11}$$

Note that this potential energy $V(\mathbf{r})$ is the energy that results from the physical position (i.e., the vector $\mathbf{r}$ in the usual coordinate space) of the particle.[6]

Hence, we can postulate that we can rewrite our wave equation (Eq. (2.9)) as

$$-\frac{\hbar^2}{2m_o}\nabla^2\psi = \left(E - V(\mathbf{r})\right)\psi \tag{2.12}$$

or, in a slightly more standard way of writing this, as

$$\left(-\frac{\hbar^2}{2m_o}\nabla^2 + V(\mathbf{r})\right)\psi = E\psi \tag{2.13}$$

which is the time-independent Schrödinger equation for a particle of mass m_o.

Note that we have not "derived" Schrödinger's equation. We have merely suggested it as an equation that agrees with at least one experiment. There is, in fact, no way to derive Schrödinger's equation from first principles; there are no "first principles" in the physics that precedes quantum mechanics that predict anything like such wave behavior for the electron. Schrödinger's equation has to be postulated, just like Newton's laws of motion were originally postulated. The only justification for making such a postulate is that it works.[7]

2.2 Probability densities

We find, in practice, that the probability $P(\mathbf{r})$ of finding the electron near any specific point $\mathbf{r}$ in space is proportional to the modulus squared, $|\psi(\mathbf{r})|^2$, of the wave $\psi(\mathbf{r})$. The fact that we work with the squared modulus for such a probability is not so surprising. First, it ensures that we always have a positive quantity (we would not know how to interpret a negative probability). Second, we are already aware of the usefulness of squared amplitudes with waves. The squared amplitude typically tells us the intensity (i.e., power per unit area) or energy density in a wave motion such as a sound wave or an electromagnetic wave. Given that we know that the intensity of electromagnetic waves also corresponds to the number of photons arriving per unit area per second, we would also find in the electromagnetic case that the probability of finding a photon at a specific point was proportional to the squared wave amplitude. If we chose to use complex notation to describe an electromagnetic wave, we would find that we would use the modulus squared of the wave amplitude to describe the wave intensity and, hence, also the probability of finding a photon at a given point in space.[8]

---

[6] Though the symbol V is used, it does not refer to a voltage, despite the fact that the potential energy can be (and often is) an electrostatic potential. It (and other energies in quantum mechanical problems) is often expressed in electron-Volts, this being a convenient unit of energy, but it is always an energy, not a voltage.

[7] The reader should get used to this statement. Again and again, we simply postulate things in quantum mechanics, with the only justification being that it works.

[8] Though the analogy between electromagnetic waves and quantum mechanical wave amplitudes may be helpful here, the reader is cautioned not to take this too far. The wave amplitude in a classical wave, e.g., an acoustic wave or the classical description of electromagnetic waves, is a measurable and meaningful quantity, e.g., air pressure or electric field (which, in turn, describes the actual force that would be experienced by a charge). The wave amplitude of a quantum mechanical wave does not describe any real

The fact that the probability is given by the modulus squared of some quantity – in this case, the wavefunction ψ – leads us also to call that quantity the *probability amplitude* or *quantum mechanical amplitude*. Note that this probability amplitude is quite distinct from the probability itself; to repeat, the probability is proportional to the modulus squared of the probability amplitude. The probability amplitude is one of those new concepts that is introduced in quantum mechanics that has little or no precedent in classical physics or, for that matter, classical statistics. For the moment, we think of that probability amplitude as being the amplitude of a wave of some kind; we will find later that the concept of probability amplitudes extends into quite different descriptions, well beyond the idea of quantum mechanical waves, while still retaining the concept of the modulus squared representing a probability.

This use of probability amplitudes in quantum mechanics is an absolutely central concept and a crucial but subtle one for the student to absorb. In quantum mechanics, we always first calculate this amplitude (here, a wave amplitude) by adding up all contributions to it (e.g., all the different scattered waves in a diffraction experiment) and then take the squared modulus of the result to come up with some measurable quantity. We do not add the measurable quantities directly. The effect of adding the amplitudes is what gives us interference, allowing cancellation between two or more amplitudes, for example. We will not see such a cancellation phenomenon if we add the measurable quantities or probabilities from two or more sources (e.g., electron densities) directly. A good example to understand this point is the diffraction of electrons by two slits.

2.3 Diffraction by two slits

With our postulation of Schrödinger's equation, Eq. (2.13), and our interpretation of $|\psi(\mathbf{r})|^2$ as proportional to the probability of finding the electron at position $\mathbf{r}$, we are now in a position to calculate a simple electron diffraction problem, that of an electron wave being diffracted by a pair of slits.[9] We need some algebra and wave mechanics to set up this problem, but it is well worth the effort. This behavior is not only one we can use relatively directly to see and verify the wave nature of electrons; it is also a conceptually important "thought experiment" in understanding some of the most bizarre aspects of quantum mechanics, and we will keep coming back to it.

---

quantity, and it is actually highly doubtful that it has any meaning at all other than as a way of calculating other quantities. A second very important difference is that, whereas a classical wave with higher intensity would be described by a larger amplitude of wave, in general, the states of quantum mechanical systems with many electrons (or with many photons) cannot be described by quantum mechanical waves simply with larger amplitudes. Instead, the description of multiparticle systems involves products of the wave amplitudes of the waves corresponding to the individual particles and sums of those products, in a much richer set of possibilities than a simple increase of the amplitude of a single wave. A third problem is that, in a proper quantum mechanical description of optics, there are many situations possible in which we have photons, but in which there is not anything very like the classical electromagnetic wave. Electromagnetic waves are not then actually analogous in a rigorous sense to the quantum mechanical wave that describes electron behavior.

[9] In optics, this apparatus is known as Young's slits, demonstrated by Thomas Young in about 1803. It is a remarkable experiment that enables us both to see the wave nature directly and to measure the wavelength without having to have some object at the same size scale as the wavelength itself. An instrument on a size scale of the wavelength of light (less than 1 micron) would have been unimaginable in 1803.

We consider two open slits, separated by a distance s, in an otherwise opaque screen (see Fig. 2.1). We are shining a monochromatic electron beam of wavevector k at the screen, in the direction normal to the screen. For simplicity, we presume the slits to be very narrow compared to both the wavelength $\lambda = 2\pi / k$ and the separation s. We also presume the screen is far away from the slits for simplicity – that is, $z_o >> s$, where z_o is the position of the screen relative to the slits.

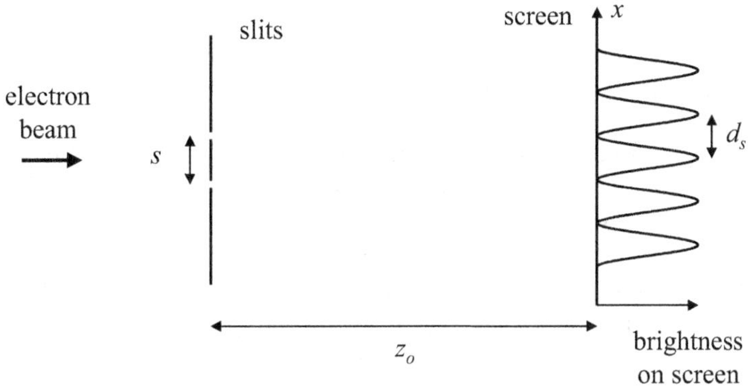

Fig. 2.1 A top view of diffraction from two slits, showing the form of the brightness of the interference pattern on a phosphorescent screen.

For simplicity of analysis, we regard the slits as essentially point sources[10] of expanding waves, in the spirit of Huygens's principle. We write these waves in the form $\exp(ikr)$, where r is the radius from the source point.[11] We have, therefore, one source (slit) at $x = s/2$ and another at $x = -s/2$. The net wave should be the sum of the waves from these two sources. Remembering that in the x-z plane the equation of a circle of radius r centered about a point $x = a$ and $z = 0$ is $r^2 = (x-a)^2 + z^2$, the net wave at the screen is

$$\psi_s(x) \propto \exp\left[ik\sqrt{(x-s/2)^2 + z_o^2}\right] + \exp\left[ik\sqrt{(x+s/2)^2 + z_o^2}\right] \tag{2.14}$$

where the first term corresponds to a wave expanding from the upper slit, and the second corresponds similarly with the wave from the lower slit. Note that we are adding the wave amplitudes here. If we presume we are only interested in the pattern on the screen for relatively small angles – that is, $x << z_o$ – then[12]

$$\begin{aligned}\sqrt{(x-s/2)^2 + z_o^2} &= z_o\sqrt{1+(x-s/2)^2/z_o^2} \cong z_o + (x-s/2)^2/2z_o \\ &= z_o + x^2/2z_o + s^2/8z_o - sx/2z_o\end{aligned} \tag{2.15}$$

[10] To be strictly correct, we should actually consider them as line sources because they extend in the direction out of the paper in Fig. 2.1. To consider such line sources rigorously would complicate our mathematics to no real benefit in this explanation, however.

[11] More rigorously, for spherically expanding waves, we ought to write them in the form $\exp(ikr)/r$ because they must get weaker with distance, but because we are considering essentially one large distance for the position of the screen, we can neglect the r in the denominator.

[12] Remember the power series expansion $\sqrt{1+a} \simeq 1 + a/2 + \ldots$ for small a, which can be proved by Taylor (or Maclaurin) expansion.

and similarly for the other exponent (though with opposite sign for the term in s). Hence, using $2\cos(\theta) = \exp(i\theta) + \exp(-i\theta)$, we obtain

$$\psi_s(x) \propto \exp(i\phi)\cos(ksx/2z_o) = \exp(i\phi)\cos(\pi sx/\lambda z_o) \tag{2.16}$$

where ϕ is a real number ($\phi = k(z_o + x^2/2z_o + s^2/8z_o)$), so $\exp(i\phi)$ is simply a phase factor. Hence, on the screen

$$|\psi_s(x)|^2 \propto \cos^2(\pi sx/\lambda z_o) = \frac{1}{2}[1+\cos(2\pi sx/\lambda z_o)] \tag{2.17}$$

So, if we shine a beam of monoenergetic electrons at the slits and put some phosphorescent screen (like our cathode ray tube screen) some distance behind the slits, we should expect to see a (co)sinusoidal interference pattern, or *fringes*, on the screen, with the fringes separated by a distance $d_s = \lambda z_o / s$.[13] This simple fringe pattern is somewhat idealized; with a more sophisticated diffraction model and with finite width for the slits, the intensity of the fringes falls off for larger x, but the basic interference fringes we predict here will be observed near the axis as long as the slit separation is much larger than the slit width.

The existence of these interference effects for the quantum mechanical amplitudes has some bizarre consequences that we simply cannot understand classically. For example, suppose that we block one of the slits so the electrons can only go through one slit; then we would not see the interference fringes. Near the axis, we would see a broad featureless band[14] that is readily understood from wave diffraction from a single slit. Such a broad band is already difficult to explain based on a classical model of a particle; in a classical model, with the electron particles all traveling from left to right in straight lines, we would expect to see a relatively sharp spot on the screen. If we were determined to explain this broad band classically, we might come up with some explanation, involving electrons bouncing off the edges of the slit, that would at least be qualitatively plausible (if ultimately incorrect). If we now uncover the second slit, however, we see something that cannot be explained by a classical particle picture – parts of the screen that were formerly bright now become dark (i.e., the minima of the (co)sinusoidal interference pattern described previously). How can we explain that opening a second source of particles actually reduces the number of particles arriving at some point in the screen?

We might try to argue that the particles from the second slit were somehow bouncing off the ones from the first slit and, hence, avoiding some particular part of the screen because of these collisions. If we repeat the experiment with extremely low electron currents so that there are never two electrons in the apparatus at a given time and take a time-exposure picture of the phosphorescent screen, we will, however, see exactly the same interference pattern emerge, and we cannot now invoke some explanation that involves particles colliding with one

---

[13] Note, incidentally, that because the distance z_o is much greater than the slit separation, we can effectively measure the wavelength in such an experiment without ever having a measuring device that is the size of the wavelength, which is one of the other beauties of this experiment.

[14] The actual intensity diffraction pattern from a single slit of width w is, in this simple Huygens diffraction model, of the form $\{[2z_o/kx][\sin(kxw/2z_o)]\}^2$, which has a central "bright" band of width $4\pi z_o/kw$ and progressively weaker bands each half this width. This band will be much larger in size than the interference fringe separation in the two-slit interference experiment as long as the slit width is much less than the slit separation. Note, incidentally, that the function $(\sin x)/x$ is equal to 1 for $x = 0$. If the reader is not familiar with this function, also often known as the "sinc" function, now would be a good time to graph it and understand how it behaves (it will come up again later).

another.[15] Hence, we are forced even qualitatively to describe the behavior of the electrons in terms of some process involving interference of amplitudes, and we also find that the wave description postulated herein does explain the behavior quantitatively.

Though a two-slit diffraction experiment of exactly the form described here might be quite difficult to perform, in practice, with electrons, diffraction phenomena such as this can be seen quite readily with electrons. Such diffraction is routinely used as a diagnostic and measurement tool. Electrons can be accelerated by electric fields and, if necessary, focused using magnetic and electric techniques. The wavelength associated with such accelerated electrons can be very small (e.g., an Ångstrom [1 Å], which is 0.1 nm). Diffractive effects are particularly strong when the wavelength is comparable to the size of an object (e.g., comparable to the slit spacing, s, shown previously). Electrons can diffract quite strongly off crystal surfaces, for example, where the spacings between the atoms are on the order of Ångstroms or fractions of a nanometer.

One diagnostic technique, reflection high-energy electron diffraction (RHEED), for example, monitors the form of a crystal surface during the growth of crystalline layers; an electron beam incident at a shallow angle relative to the surface (i.e., nearly parallel with the crystal surface) is reflected and diffracted onto a phosphorescent screen to give a diffraction pattern characteristic of the precise form of the surface. Electron diffraction is also intrinsic to the operation of some kinds of electron microscope. In general, the fact that the electron wavelength can be so small means that electron microscopes can be used to view very small objects. It is practically difficult to image objects much smaller than a wavelength with any optical or wave-based technique because of diffractive effects, but the small wavelength possible in electron beams means that small objects or features can be seen.

Problems

2.3.1 Suppose we have a screen that is opaque to electrons except for two thin slits separated by 5 nm. (We might imagine a plane of atoms with two missing rows of atoms, 5 nm apart, for example, as one way we might make such a structure.) We accelerate electrons, which are initially stationary, through 1 V of potential. These electrons arrive at the back of the screen perpendicular to the surface. A phosphorescent surface (e.g., like a cathode ray tube screen) is placed 10 cm away from the other side of the screen and parallel to it.

(i) What is the spatial period of the bright and dark stripes seen on the phosphorescent surface (i.e., the distance between the centers of the bright stripes)?

(ii) What is the period of the stripes if we use protons (hydrogen nuclei) instead of electrons? (The mass of a proton is $\sim 1.67 \times 10^{-27}$ kg.)

2.3.2 In an electron diffraction experiment, consider a screen with a single vertical slit of finite width, d, in the x direction, "illuminated" from behind by a plane monochromatic electron wave of wavelength λ, with the wave fronts parallel to the plane of the slit (i.e., the wave is propagating perpendicularly to the screen). Presume that we take the simple Huygens's principle model of diffraction and, as shown previously, model the slit as a source of a spherically expanding (complex) wave (and, hence, each vertical line in the slit is a source of a circularly expanding wave).

---

[15] Such experiments with single electrons and two slits raise other very interesting questions that probe the heart of the bizarre nature of quantum mechanics. Classically, we are tempted to ask, for any given electron passing through the slits to give one of the flashes on our screen that goes to make up the interference pattern over time on our photographic plate, which slit did the electron go through? In quantum mechanics, that is apparently a meaningless question. Any attempt to measure which slit the electron goes through apparently destroys the interference pattern, so there is no measurable situation that corresponds to the question we wish to ask; hence, the question can be thrown out!

(i) Find an approximate analytic expression for the form of the wave amplitude (by "form," we mean here that you may neglect any constant factor multiplying the wave amplitude) at a plane a distance z_o from the screen (assuming $z_o >> d$).

(ii) For a slit of width $d = 1$ μm, with an electron wavelength of $\lambda = 50$ nm, plot the magnitude of the light intensity we would see on a phosphorescent screen placed 10 cm in front of the slit, as a function of the lateral distance in the x direction. Continue the plot sufficiently far in x to illustrate all of the characteristic behavior of this intensity.

(iii) Consider now two such slits in the screen, positioned symmetrically a distance 5 μm apart in the x direction, but with all other parameters identical to part (ii). Plot the intensity pattern on the phosphorescent screen for this case.

(Notes: You may presume that, in the denominator, the distance r from a slit to a point on the screen is approximately constant at $r \simeq z_o$ [though you must not make this assumption for the numerator in the calculation of the phase]. You may also presume that for all x of interest on the screen, $x << z_o$, a so-called paraxial approximation. You will probably want to use a computer program or high-level mathematical computer package to plot the required functions. With this particular problem, you may find that you want to avoid asking the program to calculate the amplitude or brightness at exactly $x = 0$ because there may be a formal [though not actual] problem with evaluating the function there.)

2.4 Linearity of quantum mechanics: multiplying by a constant

Note that in Schrödinger's equation (2.15), we could multiply both sides by a constant a and the equation would still hold. In other words, if ψ is a solution of Schrödinger's equation, so also is $a\psi$. This may seem a trivial property to point out, but the reason why this is possible is because Schrödinger's equation is *linear*. The wavefunction only appears in first order (i.e., to the power one) in the equation; there are no second-order terms, such as ψ^2, or any other terms that are higher order in ψ. The linearity (in this particular sense) of equations in quantum mechanics is of profound importance and generality. As far as we understand it, all quantum mechanical equations are linear in this sense (i.e., linear in the quantum mechanical amplitude for which the equation is being solved).

With classical fields, we often use linear equations, such as the differential equations that allow us to solve for small oscillatory motion of, say, a pendulum. In such a classical case, the linear equation is an approximation; a pendulum with twice the amplitude of oscillation will not oscillate at exactly the same frequency, for example. Hence, we cannot take the solution derived at one amplitude of oscillation of the pendulum and merely scale it up for larger amplitudes of oscillation, except as a first approximation. We should emphasize right away, however, that, in quantum mechanics, this linearity of the equations with respect to the quantum mechanical amplitude is not an approximation of any kind; it is apparently an absolute property. Among other things, this linearity allows us to use the full power of linear algebra to handle the mathematics of quantum mechanics, a point to which we return later in some detail.

Problems

2.4.1 Which of the following differential equations is linear in the sense that, if some function $\psi(z)$ is a solution (and this may well be a different function for each equation), so also is the function $\phi(z) = a\psi(z)$, where a is an arbitrary constant? Justify your answers.

(i) $z\dfrac{d\psi(z)}{dz} + g(z)\psi(z) = 0$, where $g(z)$ is some specific function

(ii) $\psi(z)\dfrac{d\psi(z)}{dz} + \psi(z) = 0$

(iii) $\frac{d^2\psi(z)}{dz^2}+b\frac{d\psi(z)}{dz}=c\psi(z)$, where b and c are constants

(iv) $\frac{d^3\psi(z)}{dz^3}=1$

(v) $\frac{d^2\psi(z)}{dz^2}+\left(1+|\psi(z)|^2\right)\frac{d\psi(z)}{dz}=g\psi(z)$, where g is a constant

(Note: You do not need to solve these equations for the function $\psi(z)$. Merely show that if $\psi(z)$ is a solution, then $a\psi(z)$ is or is not also a solution. Hint: Is the equation for $\phi(z)$ identical to the equation for $\psi(z)$, or does a unavoidably appear in the equation for $\phi(z)$, meaning $\phi(z)$ is necessarily a solution of a different equation if we insist on arbitrary a?)

2.5 Normalization of the wavefunction

We have postulated previously that the probability $P(\mathbf{r})$ of finding a particle near a point $\mathbf{r}$ in space is $\propto|\psi(\mathbf{r})|^2$. So that we can use the concept of probability in its exact statistical sense, we should be more precise about this definition. Specifically, let us define $P(\mathbf{r})$ as the probability per unit volume of finding the particle near the point $\mathbf{r}$; $P(\mathbf{r})$ can then rigorously be viewed as a *probability density*. Then, for some very small (infinitesimal) volume $d^3\mathbf{r}$ around $\mathbf{r}$,[16] the probability of finding the particle in that volume is $P(\mathbf{r})d^3\mathbf{r} \propto |\psi(\mathbf{r})|^2\, d^3\mathbf{r}$. Presumably, we know that the particle is somewhere in the total volume of interest. Hence, the sum of such probabilities, considering all possible such infinitesimal volumes, should equal unity. That is

$$\int P(\mathbf{r})d^3\mathbf{r}=1 \tag{2.18}$$

where the integral is over the whole volume of interest. In general, unless we have been very cunning or very lucky, we will find that our first attempt at solving Schrödinger's equation will lead to a solution ψ for which $\int|\psi(\mathbf{r})|^2\, d^3\mathbf{r}\neq 1$. This integral will, however, be real (because it is an integral of a real quantity, $|\psi(\mathbf{r})|^2$), so, in general, we will have

$$\int|\psi(\mathbf{r})|^2\, d^3\mathbf{r}=\frac{1}{|a|^2} \tag{2.19}$$

where a is some number (possibly complex).[17] But we know from the previous discussion on linearity that if ψ is a solution, so also is $\psi_N = a\psi$, and we now have

$$\int|\psi_N(\mathbf{r})|^2\, d^3\mathbf{r}=1 \tag{2.20}$$

---

[16] Here $d^3\mathbf{r}$ is just shorthand for the volume element $dxdydz$ (in conventional Cartesian coordinates x, y, and z).

[17] Functions that can be normalized in this way, i.e., that give a finite answer for an integral of their squared modulus, are sometimes referred to as *square integrable* or, slightly more specifically in mathematical language, as being L^2 functions. Here the L refers to the name Lebesgue and the superscript 2 refers to the fact that we are talking about an integral of the square of the function. The Lebesgue, in turn, refers to a formal method of integration that is more tolerant to some kinds of badly behaved functions than the "Riemann" integration that we normally perform, though gives the same answers for all normal situations. These details do not concern us here and we mention these various terms just in case the reader comes across them in other contexts. See Appendix A.

This wavefunction solution ψ_N is referred to as a *normalized* wavefunction, and now there is a direct correspondence between probability density and the modulus squared of the wavefunction; that is, $P(\mathbf{r}) = |\psi_N(\mathbf{r})|^2$. The use of such normalized wavefunctions is quite convenient in the algebra of quantum mechanics.

Note, incidentally, that our plane wave, Eq. (2.2), cannot always be normalized in this way if we take the space of interest to be infinite. There is no way we can normalize a plane wave over an infinite space, for example, using this definition of normalization. Such problems can often be removed by considering that the particle is confined to some finite box, even if we take that box to be very large.[18]

2.6 Particle in an infinitely deep potential well ("particle in a box")

Now that we have introduced the (time-independent) Schrödinger equation and some basic concepts like probability density and wavefunction normalization, we can proceed to solve some simple problems, starting with the so-called particle in a box. In following sections, we look at solutions for waves incident on steps, the particle in a box of finite depth, the harmonic oscillator, and various problems related to linearly varying potentials. These problems, and the hydrogen atom problem to which we return in Chapter 10, are some of the main, exactly solvable problems in quantum mechanics. There are, unfortunately, relatively few other useful problems that can be solved exactly. Nearly all other practical problems have to be solved by approximation techniques, to which we return in Chapter 6.[19]

The exactly solvable problems do give us much insight into quantum mechanics in general, and we will use these problems to illustrate a number of basic concepts in quantum mechanics. These problems are also at the root of the solution of many other actual practical problems. The particle-in-a-box problem is used routinely to design the so-called quantum well optoelectronic structures that are at the core of a large fraction of modern semiconductor optoelectronic devices, for example. The harmonic oscillator problem allows us to understand vibrating systems of many kinds, including, for example, acoustic vibrations in solids and also electromagnetic waves, where it leads to the concept of photons (see Chapter 15). The linearly varying potential is important for understanding the quantum mechanics of accelerating particles in fields and has direct practical uses in semiconductor optical modulators and biased semiconductor devices generally.

We consider the simple problem of a particle, of mass m, with a spatially varying potential $V(z)$ in the z direction. We will not consider yet the fact that in a real structure, the particle may also be able to move in the x and y directions. In fact, for a simple problem like a particle in a cubic box (or, more generally, a cuboidal box; that is, one with rectangular faces), that motion can be considered separately and its consequences added in later.[20]

---

[18] We return to another way of normalizing functions like the plane wave in Section 5.4.

[19] The fact that there are only a few problems that are exactly solvable in quantum mechanics is more a statement about the limitations of mathematics than about the limitations of quantum mechanics. There are also relatively few problems that can be solved exactly in classical mechanics.

[20] Strictly, such a simple quantum mechanical problem is "separable" mathematically in the three directions.

The (time-independent) Schrödinger equation for the particle's motion in the z-direction in our "one-dimensional" box is, then, the simple differential equation

$$-\frac{\hbar^2}{2m}\frac{d^2\psi(z)}{dz^2}+V(z)\psi(z)=E\psi(z) \tag{2.21}$$

where E is the energy of the particle and $\psi(z)$ is the wavefunction.

We are particularly interested in the case where the potential is a simple "rectangular" potential well (sometimes known also as a "square" potential well) – that is, one in which the potential energy is constant inside the well and rises abruptly at the walls. We choose the thickness of the well to be L_z. We can choose the value of V in the well to be zero for simplicity (this is only a choice of energy origin and makes no difference to the physical meaning of the final results).

On either side of the well (i.e., for $z < 0$ or $z > L_z$), the potential, V, for this first problem is presumed infinitely high. (Such a structure is sometimes called an *infinite potential well.*) Because these potentials are infinitely high but the particle's energy E is presumably finite, we presume there can be no possibility of finding the particle in these regions outside the well. Hence, the wavefunction ψ must be zero inside the walls of the well, and, to avoid a discontinuity in the wavefunction, we therefore reasonably ask that the wavefunction must also go to zero inside the well at the walls.[21] Formally putting this infinite-well potential into Eq. (2.21), we have

$$-\frac{\hbar^2}{2m}\frac{d^2\psi(z)}{dz^2}=E\psi(z) \tag{2.22}$$

within the well, subject to the boundary conditions

$$\psi=0; \quad z=0,\ L_z \tag{2.23}$$

The reader may well recognize the form of the equation (2.22). The general solution to this equation can be written simply as

$$\psi(z)=A\sin(kz)+B\cos(kz) \tag{2.24}$$

where A and B are constants and $k=\sqrt{2mE/\hbar^2}$. The requirement that the wavefunction goes to zero at $z=0$ implies that $B=0$. Because we are now left only with the sine part of Eq. (2.24), the requirement that the wavefunction goes to zero also at $z=L_z$ then means that $k=n\pi/L_z$, where n is an integer. Hence, we find that the solutions to this equation are, for the wave

$$\psi_n(z)=A_n\sin\left(\frac{n\pi z}{L_z}\right) \tag{2.25}$$

where A_n is a constant that can be any real or complex number, with associated energies

[21] If the reader is bothered by the arguments here to justify these boundary conditions based on infinities (and is perhaps mathematically troubled by the discontinuities we are introducing in the wavefunction derivative), the reader can be assured that, if we take a well of finite depth and solve that problem with boundary conditions that are more mathematically reasonable, the present infinite-well results are recovered in the limit as the walls of the well are made arbitrarily high.

$$E_n = \frac{\hbar^2}{2m}\left(\frac{n\pi}{L_z}\right)^2 \tag{2.26}$$

We can restrict n to being a positive integer – that is,

$$n = 1, 2, \ldots \tag{2.27}$$

for the following reasons. Because $\sin(-a) = -\sin(a)$ for any real number a, the solutions with negative n are actually the same solutions as those with positive n; all we would have to do to turn one into the other is change the sign of the constant A_n, and the sign of that is arbitrary anyway. The solution with $n = 0$ is a trivial case; the wavefunction would be zero everywhere. If the wavefunction is zero everywhere, the particle is simply not anywhere, so the $n = 0$ solution can be discarded. The resulting energy levels and wavefunctions are sketched in Fig. 2.2.

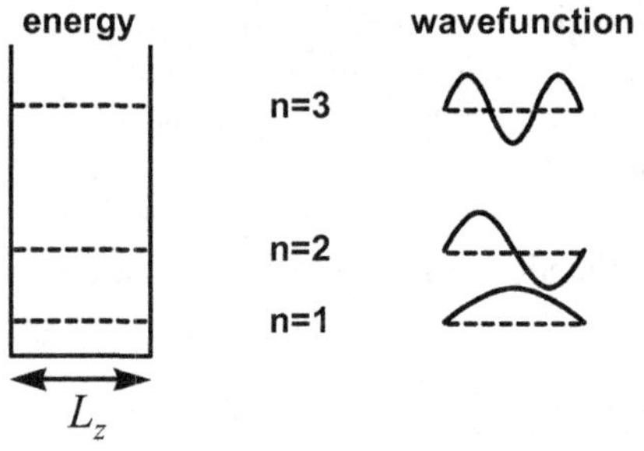

Fig. 2.2. Sketch of the energy levels in an infinitely deep potential well and the associated wavefunctions.

Solutions such as these, with a specific set of allowed values of a parameter (here, energy) and with a particular function solution associated with each such value, are called *eigen solutions*; the parameter value is called the *eigenvalue*, the equation that gives rise to such solutions (here, Eq. (2.21)) is called the *eigenequation* and the function is called the eigenfunction. It is possible to have more than one eigenfunction associated with a given eigenvalue, a phenomenon known as *degeneracy*. The number of such states with the same eigenvalue is sometimes called the degeneracy. Here, because the parameter is an energy, we can also call these eigenvalues the *eigenenergies* and can refer to the eigenfunctions as the energy eigenfunctions.

Incidentally, we can see that the eigenfunctions in Fig. 2.2 have a very definite symmetry with respect to the middle of the well. The lowest ($n = 1$) eigenfunction is the same on the right as on the left. Such a function is sometimes called an "even" function or, equivalently, is said to have "even parity." The second ($n = 2$) eigenfunction is an exact inverted image, with the value at any point to the right of the center being exactly minus the value of the mirror image point on the left of the center. Such a function is correspondingly called an "odd" function or has "odd parity." For this particular problem, the functions alternate between being even and odd, and all of the solutions are either even or odd – that is, all the solutions have a definite parity. It is quite possible for solutions of quantum mechanical problems not to have either odd or even behavior; such a situation could arise if the form of the potential was not itself symmetric. In situations where the potential is symmetric, however, such odd and even behavior is very common and can be quite useful because it can enable us to conclude that certain integrals and other quantum mechanical calculated properties will vanish exactly, for example.

For completeness in this solution, we can normalize the eigenfunctions. We have

$$\int_0^{L_z} |A_n|^2 \sin^2\left(\frac{n\pi z}{L_z}\right) dz = |A_n|^2 \frac{L_z}{2} \tag{2.28}$$

To have this integral equal one for a normalized wavefunction, we therefore should choose $|A_n| = (2/L_z)^{1/2}$. Note that A_n can, in general, be complex and the eigenfunctions are arbitrary within a constant complex factor; that is, even the normalized eigenfunction can be arbitrarily multiplied by any factor $\exp(i\theta)$ (where θ is real). By convention, here we choose these eigenfunctions to be real quantities for simplicity, so the normalized wavefunctions become

$$\psi_n(z) = \sqrt{\frac{2}{L_z}} \sin\left(\frac{n\pi z}{L_z}\right) \tag{2.29}$$

Now we have mathematically solved this problem. The question is, what does this solution mean physically? We started out here by considering the known fact that electrons behave in some ways like propagating waves, as shown by electron diffraction effects. We constructed a simple wave equation that could describe such effects for monochromatic (and, hence, monoenergetic) electrons. What we have now found is that if we continue with this equation that assumes the particle has a well-defined energy and put that particle in a box, there are only discrete values of that energy possible, with specific wavefunctions associated with each such value of energy. We are now going beyond the wave-particle duality we discussed before. This problem is showing us our first truly "quantum" behavior in the sense of the discreteness (or "quantization") of the solutions and the "quantum" steps in energy between the different allowed states.

There are three basic points about quantum confinement that emerge from this particle-in-a-box behavior that are qualitatively generally characteristic of such systems where we confine a particle in some region, and are very different from what we expect classically.

First, there is only a discrete set of possible values for the energy (Eq. (2.26)).

Second, there is a minimum possible energy for the particle, which is above the energy of the classical "bottom" of the box. In this problem, the lowest energy corresponds to $n = 1$, with the corresponding energy being $E_1 = (\hbar^2/2m)(\pi/L_z)^2$. This kind of minimum energy is sometimes called a *zero point energy*.

Third, the particle, as described by the modulus squared, $|\psi_n|^2$, of the appropriate eigenfunction is not uniformly distributed over the box, and its distribution is different for different energies. It is never found very near to the walls of the box. In general, the probability of finding the electron at a particular point in the box obeys a kind of standing wave pattern. In the lowest state ($n = 1$), it is most likely to be found near the center of the box. In higher states, there are points inside the box, away from the walls and corresponding to the other zeros of the sinusoidal eigenfunctions, where the particle will never be found.

All of these behaviors are very unlike the classical behavior of a classical particle (e.g., a billiard ball) inside a box.

We can also note that each successively higher energy state has one more "zero" in the eigenfunction (i.e., one more point where the function changes sign from positive to negative or vice versa). This is a very common behavior in quantum mechanics.

We can use this simple example to get some sense of orders of magnitude in quantum mechanics. Suppose we confine an electron in a box that is 5 Å (0.5 nm) thick, a characteristic size for an atom or a unit cell in a crystal. Then, the first allowed level for the electron is found at $E_1 = (\hbar^2 / 2m_o)(\pi / 5\times10^{-10})^2 \cong 2.4\times10^{-19}$ J.

In practice, in quantum mechanics, it is usually inconvenient to work with energies in Joules. A more useful practical unit is the electron-Volt (eV). An electron-Volt is the amount of energy acquired by an electron in moving through an electric potential change of 1 V. Because the magnitude of the electronic charge is

$$e \cong 1.602\times10^{-19} \text{ C}$$

and the energy associated with moving such a charge through an electrostatic potential change of U is $e\times U$, then 1 eV corresponds to an energy $\cong 1.602\times10^{-19}$ J. With this practical choice of energy units, the first allowed level in our 5 Å wide well is $2.4\times10^{-19}\text{J} \cong 1.5$ eV above the energy of the bottom of the well. The separation between the first and second allowed energies ($E_2 - E_1 = 3E_1$) is ~ 4.5 eV, which is of the same magnitude as major energy separations between levels in an atom. (Of course, this one-dimensional particle-in-a-box model is hardly a good one for an atom, but it does give a sense of energy and size scales.)

Problems

2.6.1* An electron is in a potential well of thickness 1 nm, with infinitely high potential barriers on either side. It is in the lowest possible energy state in this well. What would be the probability of finding the electron between 0.1 and 0.2 nm from one side of the well?

2.6.2 Which of the following functions have a definite parity relative to the point $x = 0$ (i.e., we are interested in their symmetry relative to $x = 0$)? For those that have a definite parity, state whether it is even or odd.

(i) $\sin(x)$

(ii) $\exp(ix)$

(iii) $(x-a)(x+a)$

(iv) $\exp(ix)+\exp(-ix)$

(v) $x(x^2-1)$

2.6.3 Consider the problem of an electron in a one-dimensional "infinite" potential well of width L_z in the z direction (i.e., the potential energy is infinite for $z < 0$ and for $z > L_z$, and, for simplicity, zero for other values of z). For each of the following functions, in exactly the form stated, is this function a solution of the time-independent Schrödinger equation?

(i) $\sin(7\pi z / L_z)$

(ii) $\cos(2\pi z / L_z)$

(iii) $0.5\sin(3\pi z / L_z)+0.2\sin(\pi z / L_z)$

(iv) $\exp(-0.4i)\sin(2\pi z / L_z)$

2.6.4 Consider an electron in a three-dimensional cubic box of side length L_z. The walls of the box are presumed to correspond to infinitely high potentials.

(i) Find an expression for the allowed energies of the electron in this box. Express the result in terms of the lowest allowed energy, E_1^∞, of a particle in a one-dimensional box.

(ii) State the energies and describe the form of the wavefunctions for the four lowest energy states.

(iii) Are any of these states degenerate? If so, say which, and also give the degeneracy associated with any of the eigenenergies you have found that are degenerate.

(Note: This problem can be formally separated into three uncoupled one-dimensional equations, one for each direction, with the resulting wavefunction being the product of the three solutions and the total energy being the sum of the three energies. This is easily verified by presuming this separation does work and finding that the product wavefunction is, indeed, a solution of the full three-dimensional equation.)

2.7 Properties of sets of eigenfunctions

Completeness of sets - Fourier series

As we can see from Eq. (2.29), the set of eigenfunctions for this problem is the set of sine waves that includes all the harmonics of a sine wave that has exactly one half period within the well (i.e., sine waves with two half periods [one full period], three half periods, and so on). This set of functions has a very important property, very common in the sets of functions that arise in quantum mechanics, called *completeness*. We discuss completeness in greater detail later, but we can illustrate it now through the relation of this particular set of functions to Fourier series.

The reader may well be aware that we could describe, for example, the movement of the loudspeaker in an audio system either in terms of the actual displacements of the loudspeaker cone at each successive instant in time or, equivalently, in terms of the amplitudes (and phases) of the various frequency components that constitute the music being played. These two descriptions are entirely equivalent, and both are "complete"; any conceivable motion can be described by the list of actual positions in time (so that approach is complete) or, equivalently, by the list of the amplitudes and phases of the frequency components.

The calculation of the frequency components required to describe the motion from the actual displacements in time is called *Fourier analysis*, and the resulting way of representing the motion in terms of these frequency components is called a *Fourier series*.

There are a few specific forms of Fourier series.[22] For a situation in which we are interested in the behavior from time zero to time t_o, an appropriate Fourier series to represent the loudspeaker displacement, $f(t)$, would be

$$f(t)=\sum_{n=1}^{\infty} a_n \sin\left(\frac{n\pi t}{t_o}\right) \tag{2.30}$$

where the a_n are the relevant amplitudes.[23]

We can see now that we could similarly represent any function $f(z)$ between the positions $z=0$ and $z=L_z$ as what we will now call, using a more general notation, an "expansion in the set of (eigen)functions," $\psi_n(z)$ from Eq. (2.29)

[22] Fourier series can be constructed with combinations of sine functions, combinations of cosine functions, combinations of sine and cosine functions, and combinations of complex exponential functions.

[23] Strictly, with this choice of sine Fourier series, we have to exclude the end points $t = 0$ and $t = t_o$ because there the function would have to be zero if we use this expansion. We can use this expansion to deal with functions that have finite values at any finite distance from these end points, however, so if we exclude the end points, this expansion is complete.

$$f(z)=\sum_{n=1}^{\infty}a_n \sin\left(\frac{n\pi z}{L_z}\right)=\sum_{n=1}^{\infty}b_n\psi_n(z) \tag{2.31}$$

where $b_n=\sqrt{L_z/2}\ a_n$ to account for our formal normalization of the ψ_n. The coefficients a_n are the so-called *expansion coefficients* in the expansion of the function $f(z)$ in the functions $\sin(n\pi z/L_z)$. Similarly the coefficients b_n are the expansion coefficients of $f(z)$ in the functions $\psi_n(z)$.

Thus, we have found that we can express any function between positions $z=0$ and $z=L_z$ as an expansion in terms of the eigenfunctions of this quantum mechanical problem. We justified this expansion through our understanding of Fourier analysis. There are many other sets of functions that are also complete, and we return to generalize these concepts later.

A set of functions such as the ψ_n that can be used to represent an arbitrary function $f(z)$ is referred to as a *basis set of functions* or, more simply, a *basis*. The set of coefficients (amplitudes), b_n, would then be referred to as the *representation* of $f(z)$ in the basis ψ_n. Because of the completeness of the set of basis functions ψ_n, this representation is just as good a one as the set of the amplitudes at every point z between zero and L_z required to specify or "represent" the function $f(z)$ in ordinary space.

The eigenfunctions of differential equations are very often complete sets of functions. We will find quite generally that the sets of eigenfunctions we encounter in solving quantum mechanical problems are complete sets, a fact that turns out to be mathematically very useful, as we will see in later chapters.

Orthogonality of eigenfunctions

The set of functions $\psi_n(z)$ have another important property, which is that they are *orthogonal*. In this context, two functions $g(z)$ and $h(z)$ are orthogonal[24] (formally, on the interval 0 to L_z) if[25]

$$\int_0^{L_z} g^*(z)h(z)dz=0 \tag{2.32}$$

It is easy to show for the specific ψ_n sine functions (Eq. (2.29)) that

$$\int_0^{L_z}\psi_n^*(z)\psi_m(z)dz=0 \text{ for } n\neq m \tag{2.33}$$

and, hence, that the different eigenfunctions are orthogonal to one another. Indeed, it is obvious from parity considerations, without performing the integral algebraically, that this integral will vanish if ψ_n and ψ_m have opposite parity. In such a case, the product function will have odd parity with respect to the center of the well, and the net integral of any odd function is zero. Hence, all the cases where n is an even number and m is an odd number, or where n is an odd number and m is an even number, lead to a net zero integral. The other cases are not

---

[24] We formally presume that neither of these functions is zero everywhere (which would have made this orthogonality integral trivial).

[25] $g^*(z)$ is the complex conjugate of $g(z)$. This explicit use of the complex conjugate may seem redundant given that the specific eigenfunctions we have considered so far have all been real, but this gives a more general statement of the orthogonality condition.

quite so obvious, but performing the actual integration shows zero net integral for all cases where $n \neq m$.[26] For $n = m$, the integral reduces to the normalization integral already performed (Eq. (2.28)). Introducing the notation known as the *Kronecker delta*

$$\begin{aligned} \delta_{nm} &= 0,\ n \neq m \\ \delta_{nn} &= 1 \end{aligned} \tag{2.34}$$

we can therefore write

$$\int_0^{L_z} \psi_n^*(z)\psi_m(z)\,dz = \delta_{nm} \tag{2.35}$$

The relation Eq. (2.35) expresses both the fact that all different eigenfunctions are orthogonal to one another, and that the eigenfunctions are also normalized. A set of functions obeying a relation like Eq. (2.35) is said to be *orthonormal* (i.e., both orthogonal and normalized), and Eq. (2.35) is sometimes described as the orthonormality condition. Orthonormal sets turn out to be particularly convenient mathematically, so most basis sets are chosen to be orthonormal.

The property of the orthogonality of different eigenfunctions is again a very common one in quantum mechanics and is not at all restricted to this specific simple problem where the eigenfunctions are sine waves.

Expansion coefficients

The orthogonality (and orthonormality) of a set of functions makes it very easy to evaluate the expansion coefficients. Suppose we want to write the function $f(x)$ in terms of a complete set of orthonormal functions $\psi_n(x)$; that is,

$$f(x) = \sum_n c_n \psi_n(x) \tag{2.36}$$

In general, incidentally, it is simple to evaluate the expansion coefficients c_n in Eq. (2.36). Explicitly, multiplying Eq. (2.36) on the left by $\psi_m^*(x)$ and integrating, we have

$$\begin{aligned} \int \psi_m^*(x) f(x)\,dx &= \int \psi_m^*(x)\left[\sum_n c_n \psi_n(x)\right]dx \\ &= \sum_n c_n \int \psi_m^*(x)\psi_n(x)\,dx = \sum_n c_n \delta_{mn} \\ &= c_m \end{aligned} \tag{2.37}$$

Problems

2.7.1 Which of the following pairs of functions are orthogonal on the interval −1 to +1?

(i) $x,\ x^2$

(ii) $x,\ x^3$

(iii) $x,\ \sin x$

---

[26] Note that $\sin(n\theta)\sin(m\theta) = (1/2)[\cos(n-m)\theta - \cos(n+m)\theta]$. With $\theta = \pi z / L_z$, the integration limits for θ become 0 to π. For a function cos $p\theta$, except for $p = 0$, the function is either "odd" round about the middle of the integration interval (i.e., round about $\pi/2$), so its integral is zero; or the integration is over a complete number of periods of the cosine functions, so its integral is again zero. $p = 0$ occurs only for $n = m$ in the first cosine term, and then the integration reduces to the normalization integral already performed.

(iv) x, $\exp(i\pi x/2)$
(v) $\exp(-2\pi ix)$, $\exp(2\pi ix)$

2.7.2 Suppose we wish to construct a set of orthonormal functions so that we can use them as a basis set. We wish to use them to represent any function of x on the interval between −1 and +1. We know that the functions $f_0(x)=1$, $f_1(x)=x$, $f_2(x)=x^2$, ..., $f_n(x)=x^n$, ... are all independent; that is, we cannot represent one as a combination of the others, and in this problem we will form combinations of them that can be used as this desired orthonormal basis.

(i) Show that not all of these functions are orthogonal on this interval. (You may prove this by finding a counter example.)

(ii) Construct a set of orthogonal functions by the following procedure:

(a) Choose $f_0(x)$ as the (unnormalized) first member of this set, and normalize it to obtain the resulting normalized first member, $g_0(x)$.

(b) Find an (unnormalized) linear combination of $g_0(x)$ and $f_1(x)$ of the form $f_1(x)+a_{10}g_0(x)$ that is orthogonal to $g_0(x)$ on this interval (this is actually trivial for this particular case), and normalize it to give the second member, $g_1(x)$, of this set.

(c) Find a linear combination of the form $f_2(x)+a_{20}g_0(x)+a_{21}g_1(x)$ that is orthogonal to $g_0(x)$, and $g_1(x)$ on this interval, and normalize it to obtain the third member, $g_2(x)$ of this set.

(d) Write a general formula for the coefficient a_{ij} in the $i+1$ th unnormalized member of this set.

(e) Find the normalized fourth member, $g_3(x)$, of this set, orthogonal to all the previous members.

(f) Is this the only set of orthogonal functions for this interval that can be constructed from the powers of x? Justify your answer.

(Note: This kind of procedure is known as Gram–Schmidt orthogonalization, and you should succeed in constructing a version of the Legendre polynomials by this procedure.)

2.8 Particles and barriers of finite heights

Boundary conditions

Thus far, we have only considered a potential V that is either zero or infinite, which led to very simple boundary conditions for the problem (i.e., the wavefunction was forced to be zero at the boundary with the infinite potential). We would like to consider problems with more realistic, finite potentials, though for simplicity of mathematical modeling, we would still like to be able to deal with abrupt changes in potential, such as a finite potential step.[27]

In particular, we would like to know what the boundary conditions should be on the wavefunction, ψ, and its derivative, $d\psi/dz$, at such a step. We know from the basic theory of second-order differential equations that if we know both of these quantities on the boundaries,

[27] Any abrupt change in potential should really be regarded as unphysical. It is, however, practically useful to set up problems with such abrupt steps, essentially as simple models of more realistic systems with relatively steep changes in potential. We then, however, have to find mathematical constructions that get us out of the mathematical problems we have created by this abruptness, and these boundary conditions are such a construction. The choice of boundary conditions is really one that appears not to create any physical problems (e.g., losing particles or particle current) and would be a limiting case as a potential was made progressively more abrupt. The boundary conditions given here are not quite as absolute as one might presume, however. E.g., if the mass of the particle varies in space (as does happen in some semiconductor problems), the boundary condition given here on the derivative of the wavefunction is not correct. The boundary condition of continuity of $(1/m)(d\psi/dz)$ is then often substituted instead of continuity of $d\psi/dz$.

we can solve the equation. We are interested in solutions of Schrödinger's equation, Eq. (2.21), for situations where V is presumably finite everywhere and where the eigenenergy E is also a finite number. If E and V are to be finite, then for $\psi(z)$ to be a solution to the equation, $d^2\psi / dz^2$ must also be finite everywhere. For $d^2\psi / dz^2$ to be finite

$$d\psi / dz \text{ must be continuous} \quad (2.38)$$

(if there were a jump in $d\psi / dz$, $d^2\psi / dz^2$ would be infinite at the position of the jump), and $d\psi / dz$ must be finite (otherwise, $d^2\psi / dz^2$ could also be infinite, being a limit of a difference involving an infinite quantity). For $d\psi / dz$ to be finite

$$\psi \text{ must be continuous} \quad (2.39)$$

These two conditions, Eqs. (2.38) and (2.39), are the boundary conditions we will use to solve problems with finite steps in the potential.

Reflection from barriers of finite height

Let us first remind ourselves of what a classical particle, such as a ball, does when it encounters a finite potential barrier. If the barrier is abrupt, like a wall, the ball is quite likely to reflect off the wall, even if the kinetic energy of the ball is more than the potential energy it would have at the top of the wall. (We loosely refer to the potential energy the ball or particle would have at the top of the barrier as the "height" of the barrier and, hence, we express this height in energy units [usually electron-Volts] rather than distance units.) If the barrier is a smoothly rising one, such as a gentle slope, the ball will probably continue over the barrier if its kinetic energy exceeds the (potential energy) height of the barrier. We certainly would not expect that the ball could get to the other side of the barrier if its kinetic energy was less than the barrier height. We also would never expect that the ball could be found inside the barrier region in that case. The behavior of a quantum mechanical particle at a potential barrier is quite different. As we will see, it can both be found within the barrier and get to the other side of the barrier, even if its energy is less than the height of the potential barrier.

We start by considering a barrier of finite height, V_o, but of infinite thickness, as shown in Fig. 2.3. For convenience, we choose the potential to be zero in the region to the left of the barrier (it would not matter if we chose it to be something different because only energy differences actually matter in these kinds of quantum mechanical calculations).

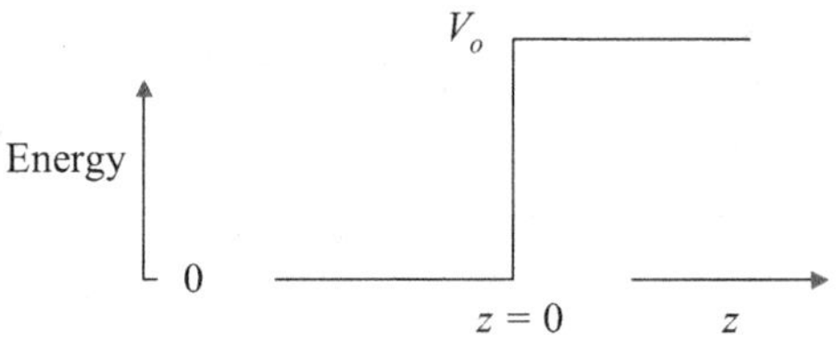

Fig. 2.3 Potential barrier of finite height but infinite thickness.

We presume that a quantum mechanical wave is incident from the left on the barrier, and we presume that the energy, E, associated with this wave, is positive (i.e., $E > 0$). We are not looking for eigenfunction solutions in this problem; we are merely considering what will happen to a monoenergetic particle wave as it interacts with the barrier, presuming that the energy E that we will consider is a valid one for the system overall.

We will also allow for possible reflection of the wave from the barrier into the region on the left. We can allow for both of these possibilities by allowing the wave on the left-hand side to be the general solution of the wave equation in this region. That equation is the same as the Eq.

(2.22) we used previously for the region inside the potential well because, in both cases, the potential is presumed to be zero in this region. The general solution could be written as in Eq. (2.24), but here we choose instead to write the solution, ψ_{left}, for $z < 0$ in terms of complex exponential waves

$$\psi_{left}(z) = C\exp(ikz) + D\exp(-ikz) \tag{2.40}$$

where we have, as before, $k = \sqrt{2mE/\hbar^2}$. Such a way of writing the solution can, of course, be exactly equivalent mathematically to that of Eq. (2.24).[28] The complex exponential form is conventionally used to represent running waves. In the convention we will use, $\exp(ikz)$ represents a wave traveling to the right (i.e., in the positive z direction), and $\exp(-ikz)$ represents a wave traveling to the left (i.e., in the negative z direction). The right-traveling wave, $C\exp(ikz)$, is the incident wave, and the left-traveling wave, $D\exp(-ikz)$, represents the wave reflected from the barrier.

Now let us presume that $E < V_o$; that is, we are presuming that the particle represented by the wave does not have enough energy classically to get over this barrier. Inside the barrier, the wave equation therefore becomes

$$\frac{-\hbar^2}{2m}\frac{d^2\psi}{dz^2} = -(V_o - E)\psi \tag{2.41}$$

The mathematical solution of this equation is straightforward, being for the wave, ψ_{right}, on the right (i.e., for $z > 0$) in the general form

$$\psi_{right}(z) = F\exp(\kappa z) + G\exp(-\kappa z) \tag{2.42}$$

where $\kappa = (2m(V_o - E)/\hbar^2)^{1/2}$.

We presume that $F = 0$. Otherwise, the wave amplitude would increase exponentially to the right forever, which does not appear to correspond to any classical or quantum mechanical behavior we see for particles incident from the left. (Also, the particle would never be found on the left because all of the probability amplitude would be arbitrarily far to the right because of the growing exponential.) Hence, we are left with

$$\psi_{right}(z) = G\exp(-\kappa z) \tag{2.43}$$

Even this solution is strange. It proposes that the wave inside the barrier is not identically zero but rather falls off exponentially as we move inside the barrier. Let us formally complete the mathematical solution here by using the boundary conditions (2.38) and (2.39). Continuity of the wavefunction, Eq. (2.39), gives us

$$C + D = G \tag{2.44}$$

and continuity of the derivative, Eq. (2.38), gives us

$$C - D = \frac{i\kappa}{k}G \tag{2.45}$$

Addition of Eqs. (2.44) and (2.45) gives us

---

[28] Formally, $B = C + D$, $A = -i(C - D)$.

$$G = \frac{2k}{k+i\kappa}C = \frac{2k(k-i\kappa)}{k^2+\kappa^2}C = 2\frac{E - i\sqrt{(V_o - E)E}}{V_o}C \tag{2.46}$$

Subtraction of Eqs. (2.44) and (2.45) gives us

$$D = \frac{k-i\kappa}{k+i\kappa}C = \frac{2E - V_o - 2i\sqrt{(V_o - E)E}}{V_o}C \tag{2.47}$$

Just as a check here, we find from Eq. (2.47) that $|D/C|^2 = 1$, so any incident particle is completely reflected. D/C is, however, complex, which means that there is a phase shift on reflection from the barrier, an effect with no classical precedent or meaning.

The most unusual aspect of this solution is the exponential decay of the wavefunction into the barrier. The fact that this exponential decay exists means that there must be some probability of finding the particle inside the barrier. This kind of behavior is sometimes called *tunneling* or *tunneling penetration*, by loose analogy with the classical idea that we could get inside or through a barrier that was too high simply by digging a tunnel. There is, however, little or no mathematical connection between the classical idea of a tunnel and this quantum mechanical process.[29]

The wavefunction has fallen off to $1/e$ of its initial amplitude[30] in a distance $1/\kappa$. That distance is short when $E \ll V_o$, becoming longer as E approaches V_o; the less is the energy deficit, $V_o - E$, the longer is the tunneling penetration into the barrier.

Let us look at some example numbers. Suppose that the barrier is $V_o = 2$ eV high and that we are considering incident electrons with 1 eV energy. Then

$$\kappa = \sqrt{2\times 9.1095\times 10^{-31}\times(2-1)\times 1.602\times 10^{-19}/\left(1.055\times 10^{-34}\right)^2} \cong 5\times 10^9 \text{ m}^{-1}$$

In other words, the attenuation length of the wave amplitude into the barrier (i.e., the length to fall to $1/e$ of its initial value) is $1/\kappa \cong 0.2$ nm $\equiv 2$ Å. Note that the probability density falls off twice as fast (i.e., $|\psi(z)|^2 \propto \exp(-2\kappa z)$), so the penetration depth of the electron into the

[29] Note, incidentally, that though tunneling penetration is not a concept that has any meaning or precedent in the classical mechanics of classical particles, the phenomenon of tunneling is quite common when dealing with waves in general, including classical waves. Perhaps the best-known example is found in total internal reflection in optics. A wave inside a high refractive index medium, e.g., water or glass, at a sufficiently steep angle of incidence at the interface to the outside air (past the so-called critical angle), is totally reflected, just as the wave here is totally reflected by the barrier for energies below the barrier height. This phenomenon is well known to any swimmer who opens his or her eyes under water, with the water surface looking quite "silvery" and reflective for directions off at some angle. Though the reflection is total, there is an exponentially decaying field amplitude inside the air. Just as there is a phase shift here in reflecting the quantum mechanical wave from the barrier, so also is there a phase shift in this optical reflection. In optics, that phase shift leads to an effective sideways shift in the beam on reflection from this surface, known as the *Goos–Hänchen shift*. This optical tunneling is a routine part of the design and operation of optical waveguides, e.g., optical fibers. Once the mathematical equations are separated formally, leaving the forward waveguide propagation in another equation, the remaining equation for the wave in transverse direction is of the same form as we are discussing in this section, with exactly analogous types of behavior. The propagating bound modes of an optical fiber are essentially the same concept as the bound states in a quantum potential well of finite depth, for example.

[30] e here is the base of the natural logarithms not, obviously, the electronic charge.

barrier is $\sim 1/2\kappa \cong 1$ Å. This calculation also gives a sense of why the units of electron-Volts and Ångstroms are commonly used in quantum mechanics for electrons.

Fig. 2.4 shows wavefunctions, $\psi(z)$, for an electron incident on such a barrier, and Fig. 2.5 shows the corresponding probability amplitudes, $|\psi(z)|^2$. (The real part of the wavefunction is shown in Fig. 2.4.[31]) Note in Figs. 2.4 and 2.5 that the reflection from the barrier (which is a total reflection, in this case) leads to a standing wave pattern in the electron wavefunction and probability density. The position of the standing wave pattern depends on the phase change on reflection from the barrier, and this changes as the electron energy changes. For example, for a barrier that is very high compared to the electron energy, the phase change on reflection is π (i.e., 180° or, equivalently, phase reversal), and when the electron energy approaches the barrier energy, the phase change becomes ~ 0.

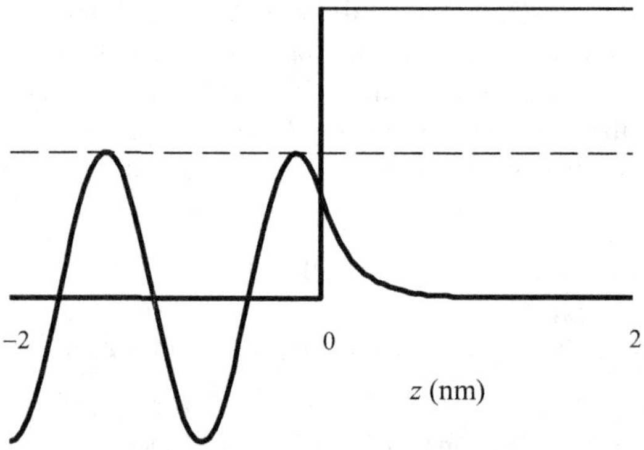

Fig. 2.4. Real part of the wavefunction $\psi(z)$ for electrons with 1 eV of energy when incident on a barrier that is 2 eV high.

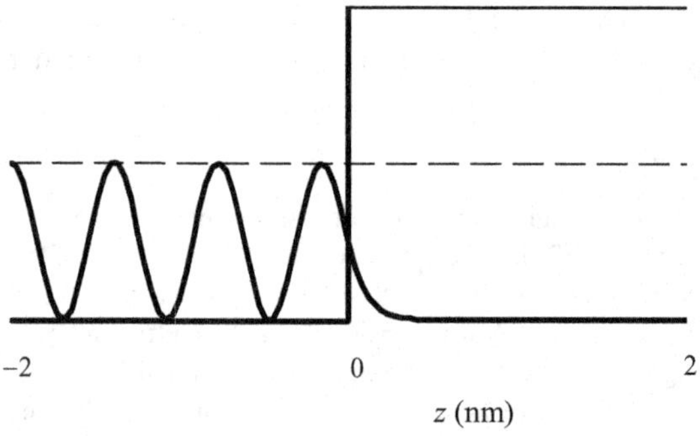

Fig. 2.5. Probability density $\propto |\psi(z)|^2$ for electrons with the 1 eV of energy when incident on a barrier that is 2 eV high.

[31] For a solution of the time-independent Schrödinger equation, the complex conjugate of any solution $\psi(z)$ is also a solution, as can be seen by taking the complex conjugate of both sides. Hence, the sum of a solution and its complex conjugate is also a solution, so the real part of any solution is also a solution. This works because the terms that multiply or operate on the wavefunction $\psi(z)$ are themselves real in the time-independent Schrödinger equation.

Problems

2.8.1* An electron of energy 1 eV is incident perpendicularly from the left on an infinitely high potential barrier. Sketch the form of the probability density for the electron, calculating a value for any characteristic distance you find in your result.

2.8.2 An electron wave of energy 0.5 eV is incident on an infinitely thick potential barrier of height 1 eV. Is the electron more likely to be found (a) within the first 1 Å of the barrier, or (b) somewhere farther into the barrier?

2.8.3* Graph the (relative) probability density as a function of distance for an electron wave of energy 1.5 eV incident from the left on a barrier of height 1 eV. Continue your graph far enough in distance on both sides of the barrier to show the characteristic behavior of this probability density.

2.8.4 Consider the one-dimensional problem, in the z direction, of an infinitely thick barrier of height V_o, at $z = 0$, beside an infinitely thick region with potential $V = 0$. We are interested in the behavior of an electron wave with electron energy E, where $E > V_o$.

(i) For the case where the barrier is to the right (i.e., the barrier is for $z > 0$), as shown as follows

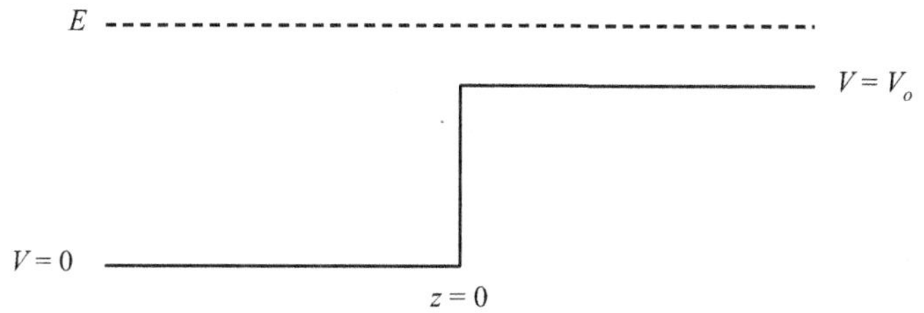

and the electron wave is incident from the left,

(a) solve for the wavefunction everywhere, within one arbitrary constant for the overall wavefunction amplitude, and

(b) sketch the resulting probability density, giving explicit expressions for any key distances in your sketch, and being explicit about the phase of any standing wave patterns you find.

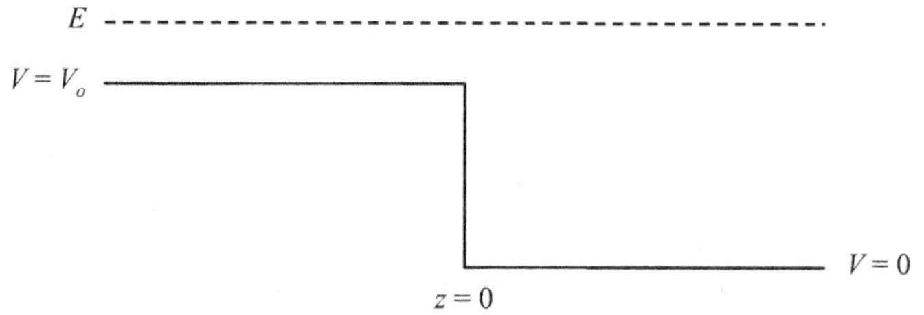

(ii) Repeat (i) but for the case where the barrier is on the left (i.e., for $z < 0$), the potential is V_o, and for $z > 0$ the potential is $V = 0$, as shown in the second figure. The electron is still "incident" from the left (i.e., from within the barrier region, in this case).

2.8.5 Electrons with energy E are incident, in the direction perpendicular to the barrier, on an infinitely thick potential barrier of height V_o, where $E > V_o$. Show that the fraction of electrons reflected from this barrier is

$$R = \left[(1-a)/(1+a)\right]^2$$

where $a = \sqrt{(E - V_o)/E}$.

2.8.6 An electron wave of unit amplitude is incident from the left on the potential structure shown in the following figure. In this structure, the potential barrier at $z = 0$ is infinitely high, and there is a potential step of height V_o and width b just to the left of the infinite potential barrier. The potential may be taken to be zero elsewhere on the left. For the purposes of this problem, we only consider electron energies $E > V_o$.

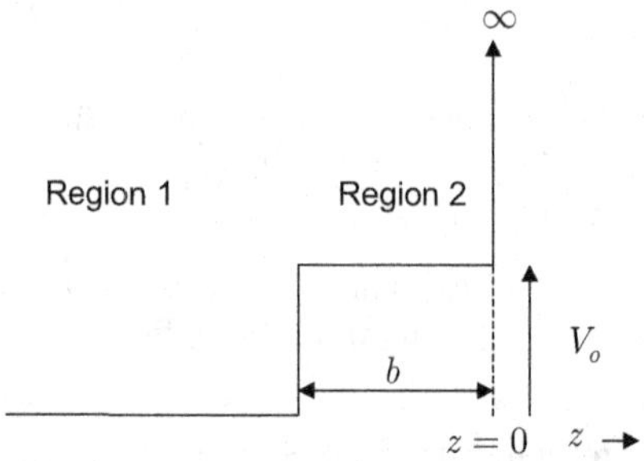

(i) Show that the wavefunction in Region 2 may be written in the form $\psi(z) = C\sin(fz)$, where C is a complex constant and f is a real constant.
(ii) What is the magnitude of the wave amplitude of the reflected wave (i.e., the wave propagating to the left)?
(iii) Find an expression for C in terms of E, V_o, and b.
(iv) Taking $V_o =$ 1 eV and $b =$ 10 Å, sketch $|C|^2$ as a function of energy from 1.1 to 3 eV.
(v) Sketch the (relative) probability density in the structure at $E =$ 1.356 eV.
(vi) Provide an explanation for the form of the curve in part (iv).

2.8.7* Consider an electron in the infinitely deep one-dimensional "stepped" potential well shown in the following figure.

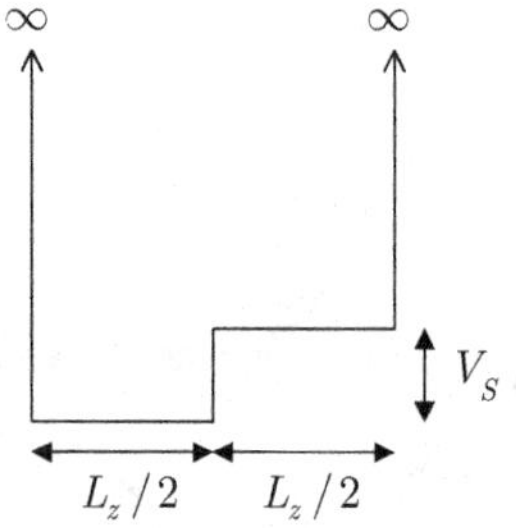

The potential step is of height V_S and is located in the middle of the well, which has total width L_z. V_S is substantially less than $(\hbar^2/2m_o)(\pi/L_z)^2$.

(i) Presuming that this problem can be solved and that it results in a solution for some specific eigenenergy E_S, state the functional form of the eigenfunction solutions in each half of the well, being explicit about the values of any propagation constants or decay constants in terms of the eigenenergy E_S, the step height V_S, and the well width L_z. (Note: Do not attempt a full solution of this problem – it does not have simple closed-form solutions for the eigenenergies. Merely state what the form of the solutions in each half would be if we had found an eigenenergy E_S.)
(ii) Sketch the form of the eigenfunctions (presuming we have chosen to make them real functions) for each of the first two states of this well. In your sketch, be explicit about whether any zeros in these functions are in the left half, the right half, or exactly in the middle. (For clarity, you may exaggerate differences between these wavefunctions and those of a simply infinitely deep well.)
(iii) State whether each of these first two eigenfunctions has definite parity with respect to the middle of the structure and, if so, whether that parity is even or odd.
(iv) Sketch the form of the probability density for each of the two states.
(v) State for each of these eigenfunctions whether the electron is more likely to be found in the left or the right half of the well.

2.9 Particle in a finite potential well

Now that we understand the interaction of a quantum mechanical wave with a finite barrier, we can consider a particle in a "square" potential well of finite depth. This is a more realistic problem than the "infinite" (i.e., infinitely deep or with infinitely high barriers) square potential well. We presume a potential structure as shown in Fig. 2.6.

Here, we have chosen the origin for the z position to be in the middle of the potential well (in contrast to the infinite well where we chose one edge of the well). Such a choice makes no difference to the final results, but is mathematically more convenient now.

Such a problem is relatively straightforward to solve. Indeed, it is one of the few nontrivial quantum mechanical problems that can be solved analytically with relatively simple algebra and elementary functions, so it is a useful example to go through completely. It also has a close correspondence with actual problems in the design of semiconductor quantum well structures.

We consider for the moment the case where $E < V_o$. Such solutions are known as *bound states*. For such energies, the particle is in some sense "bound" to the well. It certainly does not have enough energy classically to be found outside the well.

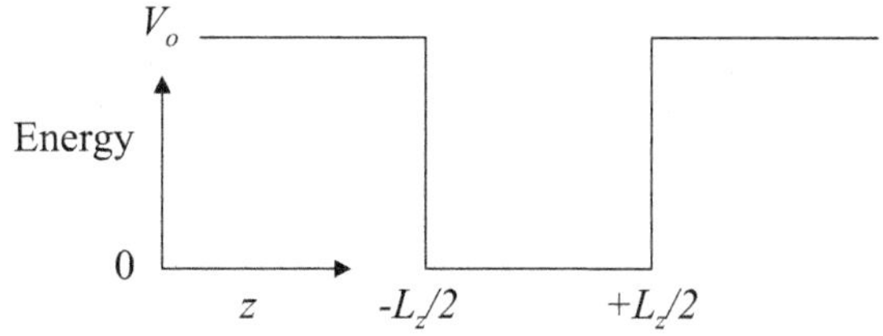

Fig. 2.6. A finite square potential well.

We know the nature of the solutions in the barriers (exponential decays away from the potential well) and in the well (sinusoidal), and we know the boundary conditions that link these solutions. We first need to find the values of the energy for which there are solutions to the Schrödinger equation, then deduce the corresponding wavefunctions.

The form of Schrödinger's equation in the potential well is the same as we had for the infinite well (i.e., Eq. (2.22)), and the solutions are of the same form (i.e., Eq. (2.24)), though the valid energies E and the corresponding values of k ($= \sqrt{2mE/\hbar^2}$) will be different from the infinite well case. The form of the solution in the barrier is an exponential one as discussed previously, except that the solution in the left barrier will be exponentially decaying to the left so that it does not grow as we move farther away from the well. Hence, formally, the solutions are of the form

$$\psi(z) = G\exp(\kappa z),\ z < -L_z/2$$

$$\psi(z) = A\sin kz + B\cos kz,\ -L_z/2 < z < L_z/2 \tag{2.48}$$

$$\psi(z) = F\exp(-\kappa z),\ z > L_z/2$$

where the amplitudes A, B, F, G, and the energy E (and, consequently, k and $\kappa = (2m(V_o - E)/\hbar^2)^{1/2}$) are constants to be determined. For simplicity of notation, we choose to write

$$X_L = \exp(-\kappa L_z/2),\ S_L = \sin(kL_z/2),\ C_L = \cos(kL_z/2)$$

so the boundary conditions give, from continuity of the wavefunction

$$GX_L = -AS_L + BC_L \tag{2.49}$$

$$FX_L = AS_L + BC_L \tag{2.50}$$

and from continuity of the derivative of the wavefunction

$$\frac{\kappa}{k} G X_L = A C_L + B S_L \tag{2.51}$$

$$-\frac{\kappa}{k} F X_L = A C_L - B S_L \tag{2.52}$$

Adding Eqs. (2.49) and (2.50) gives

$$2 B C_L = (F + G) X_L \tag{2.53}$$

Subtracting Eq. (2.52) from Eq. (2.51) gives

$$2 B S_L = \frac{\kappa}{k} (F + G) X_L \tag{2.54}$$

As long as $F \neq -G$, we can divide Eq. (2.54) by Eq. (2.53) to obtain

$$\tan(k L_z / 2) = \kappa / k \tag{2.55}$$

Alternatively, subtracting Eq. (2.49) from Eq. (2.50) gives

$$2 A S_L = (F - G) X_L \tag{2.56}$$

and adding Eqs. (2.51) and (2.52) gives

$$2 A C_L = -\frac{\kappa}{k} (F - G) X_L \tag{2.57}$$

Hence, as long as $F \neq G$, we can divide Eq. (2.57) by Eq. (2.56) to obtain

$$-\cot(k L_z / 2) = \kappa / k \tag{2.58}$$

For any situation other than $F = G$ (which leaves Eq. (2.55) applicable but Eq. (2.58) not) or $F = -G$ (which leaves Eq. (2.58) applicable but Eq. (2.55) not), the two relations Eqs. (2.55) and (2.58) would contradict each other, so the only possibilities are (i) $F = G$ with Eq. (2.55), and (ii) $F = -G$ with Eq. (2.58).

For $F = G$, we see from Eqs. (2.56) and (2.57) that $A = 0$,[32] so we are left with only the cosine wavefunction in the well, and the overall wavefunction is symmetrical from left to right (i.e., has even parity). Similarly, for $F = -G$, $B = 0$, we are left only with the sine wavefunction in the well, and the overall wavefunction is antisymmetric from left to right (i.e., has odd parity). Hence, we are left with two sets of solutions.

To write these solutions more conveniently, we change notation. We define a useful energy unit, the energy of the first level in the infinite potential well of the same width L_z

$$E_1^\infty = \frac{\hbar^2}{2m} \left(\frac{\pi}{L_z} \right)^2 \tag{2.59}$$

and define a dimensionless energy

---

[32] Note formally that C_L and S_L cannot both be zero at the same time, so the only way of satisfying both of these equations is for A to be zero.

$$\varepsilon \equiv E / E_1^\infty \tag{2.60}$$

and a dimensionless barrier height

$$v_o \equiv V_o / E_1^\infty \tag{2.61}$$

Consequently,

$$\frac{\kappa}{k} = \sqrt{\frac{V_o - E}{E}} = \sqrt{\frac{v_o - \varepsilon}{\varepsilon}} \tag{2.62}$$

$$\frac{kL_z}{2} = \frac{\pi}{2}\sqrt{\frac{E}{E_1^\infty}} = \frac{\pi}{2}\sqrt{\varepsilon} \tag{2.63}$$

$$\frac{\kappa L_z}{2} = \frac{\pi}{2}\sqrt{\frac{V_o - E}{E_1^\infty}} = \frac{\pi}{2}\sqrt{v_o - \varepsilon} \tag{2.64}$$

We can also conveniently define two quantities that will appear in the wavefunctions

$$c_L = \frac{C_L}{X_L} = \frac{\cos(kL_z/2)}{\exp(-\kappa L_z/2)} = \frac{\cos\left(\pi\sqrt{\varepsilon}/2\right)}{\exp\left(-\pi\sqrt{v_o - \varepsilon}/2\right)} \tag{2.65}$$

$$s_L = \frac{S_L}{X_L} = \frac{\sin(kL_z/2)}{\exp(-\kappa L_z/2)} = \frac{\sin\left(\pi\sqrt{\varepsilon}/2\right)}{\exp\left(-\pi\sqrt{v_o - \varepsilon}/2\right)} \tag{2.66}$$

and it will be convenient to define a dimensionless distance

$$\zeta = z / L_z \tag{2.67}$$

We can therefore write the two sets of solutions as follows.

Symmetric solution

The allowed energies satisfy

$$\sqrt{\varepsilon}\tan\left(\frac{\pi}{2}\sqrt{\varepsilon}\right) = \sqrt{v_o - \varepsilon} \tag{2.68}$$

The wavefunctions are

$$\begin{aligned}
\psi(\zeta) &= Bc_L \exp\left(\pi\sqrt{v_o - \varepsilon}\,\zeta\right),\ \zeta < -1/2 \\
\psi(\zeta) &= B\cos\left(\pi\sqrt{\varepsilon}\,\zeta\right),\ -1/2 < \zeta < 1/2 \\
\psi(\zeta) &= Bc_L \exp\left(-\pi\sqrt{v_o - \varepsilon}\,\zeta\right),\ \zeta > 1/2
\end{aligned} \tag{2.69}$$

Antisymmetric solution

The allowed energies satisfy

$$-\sqrt{\varepsilon}\cot\left(\frac{\pi}{2}\sqrt{\varepsilon}\right)=\sqrt{v_o-\varepsilon} \tag{2.70}$$

The wavefunctions are

$$\begin{aligned}
\psi(\zeta) &= -As_L \exp\left(\pi\sqrt{v_o-\varepsilon}\zeta\right), \zeta<-1/2 \\
\psi(\zeta) &= A\sin\left(\pi\sqrt{\varepsilon}\zeta\right),\ -1/2<\zeta<1/2 \\
\psi(\zeta) &= As_L \exp\left(-\pi\sqrt{v_o-\varepsilon}\zeta\right),\ \zeta>1/2
\end{aligned} \tag{2.71}$$

Here, A and B are normalization coefficients that will, in general, be different for each different solution.

The relations Eqs. (2.68) and (2.70) do not give simple formulae for the allowed energies; these relations have to be solved to deduce the allowed energies, though this is straightforward, in practice. A graphical illustration of the solutions of Eqs. (2.68) and (2.70) is shown in Fig. 2.7. Allowed energies ε correspond to the points where the appropriate solid curve (corresponding to the right-hand side of these relations) intersects with one of the broken curves (corresponding to the left-hand sides of these relations). Intersections with the dashed curves are solutions of Eq. (2.68) (corresponding to a symmetric solution), and intersections with the dot-dashed curves are solutions of Eq. (2.70) (corresponding to an antisymmetric solution).

We can see, for example, that for $v_o=8$, there are three possible solutions: (1) a symmetric solution at $\varepsilon=0.663$; (2) an antisymmetric solution at $\varepsilon=2.603$; and (3) a symmetric solution at $\varepsilon=5.609$. These three solutions are shown in Fig. 2.8.

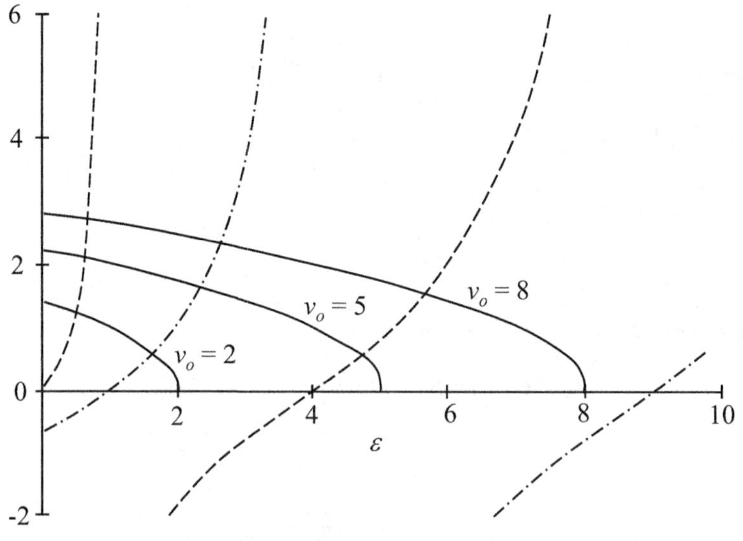

Fig. 2.7. Graphical solutions for Eqs. (2.68) and (2.70) for the allowed energies in a finite potential well. The solid curves correspond to different values of the height of the potential barrier. The simple dashed lines correspond to Eq. (2.68), the symmetrical solutions, and the dot-dashed lines correspond to Eq. (2.70), the antisymmetric solutions. Allowed energy solutions correspond to the intersections of the solid and the various dashed curves.

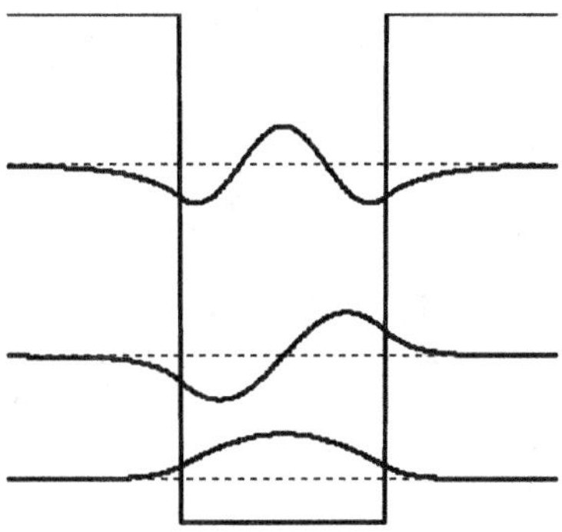

Fig. 2.8. First three solutions for a finite potential well of depth eight units (the energy unit is the first confinement energy of an infinitely deep potential well of the same width). The dotted lines indicate the energies corresponding to the three states. For convenience, these are used as the zeros for plotting the three eigenfunctions. Note that the first and third levels have symmetric wavefunctions, and the second has an antisymmetric wavefunction.

Note that these solutions for $E < V_o$ have two important characteristics. First, there are solutions of the time-independent Schrödinger equation only for specific discrete energies. We already saw such behavior for the infinite potential well, and here we have found this kind of behavior for a finite number of states in this finite well.[33] A second characteristic of these bound states is that the particle is indeed still largely found in the vicinity of the potential well, in correspondence with the classical expectation, though there is some probability of finding the particle in the barriers near the well. (Note the penetration of the wavefunction into the barrier rises for the higher energy states, as we would have expected from the behavior with the single barrier discussed in the previous section.)

We have considered solutions to this problem for energies only below the top of the potential well (i.e., for $E < V_o$). This problem can also be solved for energies above the top of the barrier, though we omit this here. In that case, there are solutions possible for all energies, a so-called continuum of energy eigenstates, just as there are solutions possible for all energies in the simple problem where V is a constant everywhere (i.e., the well-known plane waves we have been using to discuss diffraction and waves reflecting from single barriers).

Problems

2.9.1* Consider a one-dimensional problem with a potential barrier as shown here. A particle wave is incident from the left, but no wave is incident from the right. The energy, E, of the particle is less than the height, V_o, of the barrier.

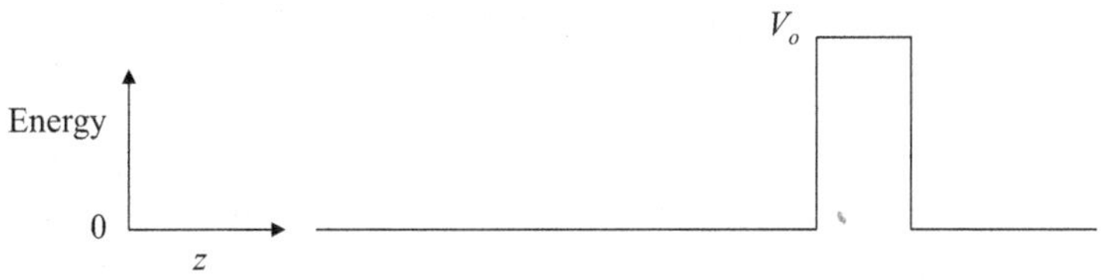

(i) Describe and sketch the form of the probability density in all three regions (i.e., on the left, in the barrier, and on the right). (Presume that the situation is one in which the transmission

---

[33] It is not generally true that there are only finite numbers of bound states in problems with bound states. The hydrogen atom, e.g., has an infinite number of bound states, though each of those has a specific eigenenergy, and there are separations between the different eigenenergies for these bound states.

probability of the particle through the barrier is sufficiently large that the consequences of this finite transmission are obvious in the sketched probability density.)

(ii) Show qualitatively how the probability density on the right of the barrier can be increased without changing the energy of the particle or the amplitude of the incident wave, solely by *increasing* the potential in some region to the left of the barrier. (This may require some creative thought!)

2.9.2* A one-dimensional potential well has a barrier of height 1.5 eV (relative to the energy of the bottom of the well) on the right-hand side and a barrier higher than this on the left-hand side. We happen to know that this potential well has an energy eigenstate for an electron at 1.3 eV (also relative to the energy at the bottom of the well).

State the general form of the wavefunction solution (i.e., within a normalizing constant that you need not attempt to determine) in each of the following two cases, giving actual values for any wavevector magnitude k and/or decay constant κ in these wavefunctions

(a) within the well

(b) in the barrier on the right-hand side

2.9.3 Consider a barrier, 10 Å thick and 1 eV high. An electron wave is incident on this barrier from the left (perpendicular to the barrier).

(i) Plot the probability of the transmission of an electron from one side of this barrier to the other as a function of energy from 0 to 3 eV.

(ii) Plot the probability density for the electron from 1 Å to the left of the barrier to 1 Å to the right of the barrier at an energy corresponding to the first maximum in the transmission for energies above the barrier.

(iii) Attempt to provide a physical explanation for the form of the transmission as a function of energy for energies above the top of the barrier.

Hints:

(1) The probability of transmission of the electron can be taken to be $|\psi_{RF}|^2/|\psi_{LF}|^2$, where $\psi_{LF}(z) \propto \exp(ikz)$ is the forward-going wave (i.e., the wave propagating to the right) on the left of the barrier, and $\psi_{RF}(z) \propto \exp(ikz)$ is the forward-going wave on the right of the barrier.

(2) Presume that there is a specific amplitude for the forward-going wave on the right and no backward-going wave on the right (there is no wave incident from the right). This enables you to work the problem mathematically "backward" from the right.

(3) You may wish to use a computer program or high-level mathematical programming package to deal with this problem, or, at least, a programmable calculator. This problem can be done by hand, though it is somewhat tedious to do that.

2.9.4 In semiconductors, it is possible to make actual potential wells, quite similar to the finite potential well discussed previously, by sandwiching a "well" layer of one semiconductor material (e.g., InGaAs) between two "barrier" layers of another semiconductor material (e.g., InP). In this structure, the electron has lower energy in the "well" material and sees some potential barrier height V_o at the interface to the "barrier" materials. This kind of structure is used extensively in, for example, the lasers for telecommunications with optical fibers. In semiconductors, such potential wells are called *quantum wells*. In these semiconductors, the electrons in the conduction band behave as if they had an effective mass, m^*, that is different from the free electron mass, m_o, and this mass is different in the two materials (e.g., m_w^* in the well and m_b^* in the barrier). Because the electron effective mass differs in the two materials, the boundary condition that is used at the interface between the two materials for the derivative of the wavefunction is not continuity of the derivative $d\psi/dz$; instead, a common choice is continuity of $(1/m)(d\psi/dz)$, where m is different for the materials in the well and in the barrier. (Without such a change in boundary conditions, there would not be conservation of electrons in the system as they moved in and out of the regions of different mass.) The wavefunction itself is still taken to be continuous across the boundary.

(i) Rederive the relations for the allowed energies of states in this potential well (treating it like the one-dimensional well analyzed previously) – i.e., relations like Eqs. (2.68) and (2.70) – using this different boundary condition.

(ii) InGaAs has a so-called bandgap energy of ~ 750 meV. The bandgap energy is approximately the photon energy of light that is emitted in a semiconductor laser. This energy corresponds to a wavelength that is too long for optimum use with optical fibers. (The relation between photon energy, E_{photon}, in electron-Volts and wavelength, λ, in meters is $E_{photon} = hc/e\lambda$, which becomes, for wavelengths in microns, $E_{photon} \cong 1.24/\lambda_{(microns)}$, a very useful relation to memorize.) For use with optical fibers, we would prefer light with wavelength ~ 1.55 microns. We wish to change the photon energy of emission from the InGaAs by making a quantum well structure with InGaAs between InP barriers. The confinement of electrons in this structure will raise the lowest possible energy for an electron in the conduction band by the "zero-point" energy of the electron (i.e., the energy of the first allowed state in the quantum well). Assuming for simplicity in this problem that the entire change in the bandgap is to come from this zero-point energy of the electron, what thickness should the InGaAs layer be made? (For InGaAs, the electron effective mass is $m^*_{InGaAs} \cong 0.041 m_o$, and for InP, it is $m^*_{InP} \cong 0.08 m_o$. The potential barrier seen by the electrons in the InGaAs at the interface with InP is $V_o \cong 260$ meV.)

2.10 Harmonic oscillator

The second, relatively simple quantum mechanical problem that we will solve exactly is the harmonic oscillator. This system is one of the most useful in quantum mechanics, being the first approximation to nearly all oscillating systems. One of its most useful applications is in describing photons, and we return to this point in Chapter 15. For the moment, we will consider a simple mechanical oscillator.

Classical harmonic oscillators are ones that give a simple, sinusoidal oscillation in time. Such behavior results, for example, from linear springs whose (restoring) force, F, is proportional to distance, z, with some spring constant, s (i.e., $F = -sz$). With a mass m, we obtain from Newton's second law ($F = ma$ where a is acceleration, d^2z/dt^2)

$$m\frac{d^2z}{dt^2} = -sz \tag{2.72}$$

The solutions to such a classical motion are sinusoidal with angular frequency

$$\omega = \sqrt{s/m} \tag{2.73}$$

(e.g., of the form $\sin \omega t$). To analyze such an oscillator quantum mechanically using Schrödinger's equation, we need to cast the problem in terms of potential energy. The potential energy, $V(z)$, is the integral of force exerted on the spring (i.e., $-F$) times distance; that is,

$$V(z) = \int_0^z -F\,dz = \frac{1}{2}sz^2 = \frac{1}{2}m\omega^2 z^2 \tag{2.74}$$

Hence, for a quantum mechanical oscillator, we have a Schrödinger equation

$$-\frac{\hbar^2}{2m}\frac{d^2\psi}{dz^2} + \frac{1}{2}m\omega^2 z^2\psi = E\psi \tag{2.75}$$

To make this more manageable mathematically, we define a dimensionless unit of distance

$$\xi = \sqrt{\frac{m\omega}{\hbar}}\,z \tag{2.76}$$

Changing to this variable and dividing by $-\hbar\omega$, we obtain

$$\frac{d^2\psi}{d\xi^2} - \xi^2\psi = -\frac{2E}{\hbar\omega}\psi \tag{2.77}$$

The reader might be astute enough to spot[34] that one specific solution to this equation is of the form $\psi \propto \exp(-\xi^2/2)$ (with a corresponding energy $E = \hbar\omega/2$). This suggests that we make a choice of form of function

$$\psi_n(\xi) = A_n \exp\left(-\xi^2/2\right) H_n(\xi) \tag{2.78}$$

where $H_n(\xi)$ is some set of functions still to be determined. Substituting this form in the Schrödinger equation Eq. (2.77), we obtain, after some algebra, the equation

$$\frac{d^2 H_n(\xi)}{d\xi^2} - 2\xi \frac{dH_n(\xi)}{d\xi} + \left(\frac{2E}{\hbar\omega} - 1\right) H_n(\xi) = 0 \tag{2.79}$$

The solutions to this equation are known. This equation turns out to be the defining differential equation for the Hermite polynomials. Solutions exist provided

$$\frac{2E}{\hbar\omega} - 1 = 2n \text{ , } n = 0, 1, 2, \ldots \tag{2.80}$$

that is,

$$E = \left(n + \frac{1}{2}\right)\hbar\omega \tag{2.81}$$

(Note that here n starts from 0, not 1.) Here, we see the first remarkable property of the harmonic oscillator – the allowed energy levels are equally spaced, separated by an amount $\hbar\omega$, where ω is the classical oscillation frequency. Like the potential well, there is also a "zero point energy" – the first allowed state is not at zero energy but instead here at $\hbar\omega/2$ compared to the classical minimum energy.

The first few Hermite polynomials are as follows:

$$H_0 = 1 \tag{2.82}$$

$$H_1(\xi) = 2\xi \tag{2.83}$$

$$H_2(\xi) = 4\xi^2 - 2 \tag{2.84}$$

$$H_3(\xi) = 8\xi^3 - 12\xi \tag{2.85}$$

$$H_4(\xi) = 16\xi^4 - 48\xi^2 + 12 \tag{2.86}$$

Note that the functions are either entirely odd or entirely even (i.e., they have a definite parity).

The polynomials have some other useful properties. In particular, they satisfy a recurrence relation

$$H_n(\xi) = 2\xi H_{n-1}(\xi) - 2(n-1) H_{n-2}(\xi), \tag{2.87}$$

---

[34] It is, in fact, quite unlikely that any normal reader could possibly be astute enough to spot this. The reason why the author is astute enough to spot this is, of course, because he knows the answer already.

which means that the successive Hermite polynomials can be calculated from the previous two.

The normalization coefficient, A_n, in the wavefunction Eq. (2.78) is

$$A_n = \sqrt{\frac{1}{\sqrt{\pi}\,2^n n!}} \tag{2.88}$$

and the wavefunction can be written explicitly in the original coordinate system as

$$\psi_n(z) = \sqrt{\frac{1}{2^n n!}}\sqrt{\frac{m\omega}{\pi\hbar}} \exp\left(-\frac{m\omega}{2\hbar}z^2\right) H_n\left(\sqrt{\frac{m\omega}{\hbar}}z\right) \tag{2.89}$$

The first several harmonic oscillator eigenfunctions are shown in Fig. 2.9, together with the parabolic potential $V\,(=\xi^2/2$ in the dimensionless units).

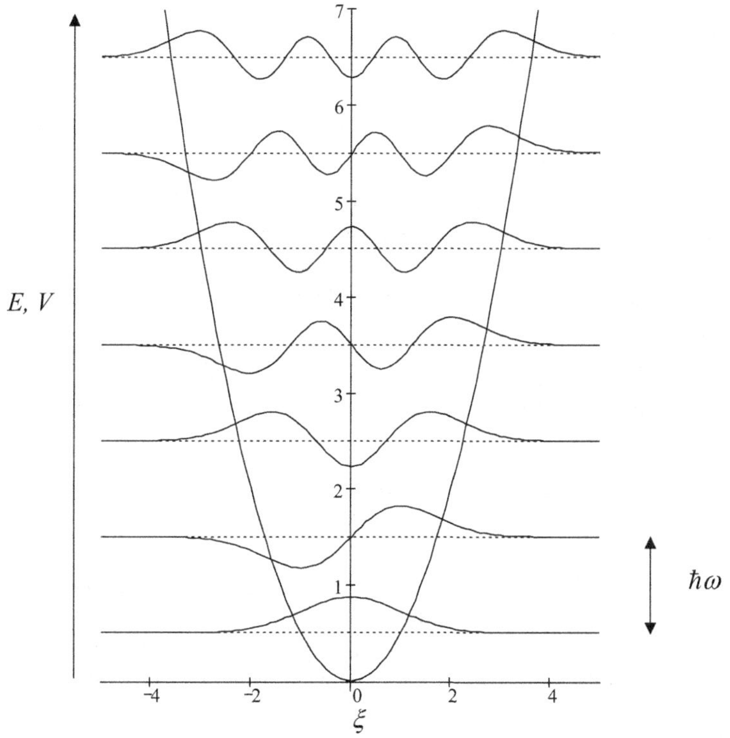

Fig. 2.9. Illustration of the eigenfunctions of a harmonic oscillator. Each eigenfunction is plotted relative to an origin that corresponds to the eigenenergy (i.e., $\hbar\omega/2$, $3\hbar\omega/2$, and so on). The parabolic harmonic oscillator potential is also shown.

The reader may be content that we have found the solution to Schrödinger's time-independent wave equation for the case of the harmonic oscillator, just as we did for the infinite and finite potential wells. But, on reflection, the reader may well now be perplexed. Surely, this is meant to be an oscillator; then why is it not oscillating? We have calculated stationary states for this oscillator, including stationary states in which the oscillator has energy much greater than zero. This would appear to be meaningless classically; an oscillator that has energy ought to oscillate. To understand how we recover oscillating behavior and, indeed, to understand the true meaning of the stationary eigenstates we have calculated, we need first to understand the time-dependent Schrödinger equation, which is the subject of the next chapter.

Problem

2.10.1 Suppose we have a "half harmonic oscillator" potential, for example, exactly half of a parabolic potential on the right of the "center" and an infinitely high potential barrier to the left of the center. Compared to the normal harmonic oscillator, what are the (normalized) energy eigenfunctions and eigenvalues? (Hint: There is very little you have to solve here; this problem mostly requires thought, not mathematics.)

2.11 Particle in a linearly varying potential

This topic can be omitted on a first reading, though it does give some very useful insights into wave mechanics and is useful for several practical problems.

Another situation that occurs frequently in quantum mechanics is that we have applied a uniform electric field, E, in some direction – say, the z direction. For a charged particle, this leads to a potential that varies linearly in distance. For example, an electron, which is negatively charged with a charge of magnitude e, will see a potential energy, relative to that at $z = 0$, of

$$V = e\mathsf{E}z \qquad (2.90)$$

In practice, we find this kind of potential in many semiconductor devices. We use it when we are calculating the quantum mechanical penetration (tunneling) through the gate oxide in Metal-Oxide-Semiconductor (MOS) transistors, for example. We see it in semiconductor optical modulators,[35] which use optical absorption changes that result from electric fields. It is of basic interest also if we want to understand how an electron is accelerated by a field, a point to which we return in the next chapter.

The technique for solving for the electron states is just the same as before; we merely have to put this potential into the Schrödinger equation and solve the equation. The Schrödinger equation then becomes

$$-\frac{\hbar^2}{2m}\frac{d^2\psi(z)}{dz^2} + e\mathsf{E}z\psi(z) = E\psi(z) \qquad (2.91)$$

The solutions to this equation are not obvious combinations of well-known functions. When one finds an equation such as this, the most productive practical technique for solving it is to look up a mathematical reference book[36] to see if someone has solved this kind of equation before. This particular kind of equation, with a linearly varying potential, has solutions that are so-called Airy functions. The standard form of differential equation that defines Airy functions is

$$\frac{d^2 f(\zeta)}{d\zeta^2} - \zeta f(\zeta) = 0 \qquad (2.92)$$

---

[35] There are two closely related electroabsorption mechanisms used in semiconductors: the Franz-Keldysh effect in bulk semiconductors and the quantum-confined Stark effect in quantum wells, both of which rely on this underlying physics.

[36] A very comprehensive reference is M. Abramowitz and I. A. Stegun, *Handbook of Mathematical Functions* (National Bureau of Standards, Washington, D.C.,1972).

The solutions of this equation are formally the Airy functions $Ai(\zeta)$ and $Bi(\zeta)$; that is, the general solution to this equation is

$$f(\zeta) = a\,Ai(\zeta) + b\,Bi(\zeta) \tag{2.93}$$

To get Eq. (2.91) into the form of Eq. (2.92), we make a change of variable to

$$\zeta = \left(\frac{2me\mathrm{E}}{\hbar^2}\right)^{1/3} \left(z - \frac{E}{e\mathrm{E}}\right) \tag{2.94}$$

The reader can verify that this change of variable does work by substituting the right-hand side of Eq. (2.94) each time for ζ in Eq. (2.92), which will give Eq. (2.91) after minor manipulations.

The functions *Ai* and *Bi* are plotted in Fig. 2.10. Though they are not very common functions, they are usually available in advanced mathematics programs as built-in functions.[37] Note that

(1) both functions are oscillatory for negative arguments, with a shorter and shorter period as the argument becomes more negative.

(2) The *Ai* function decays in an exponential-like fashion for positive arguments.

(3) The *Bi* function diverges for positive arguments.

Now let us examine solutions to some specific problems.

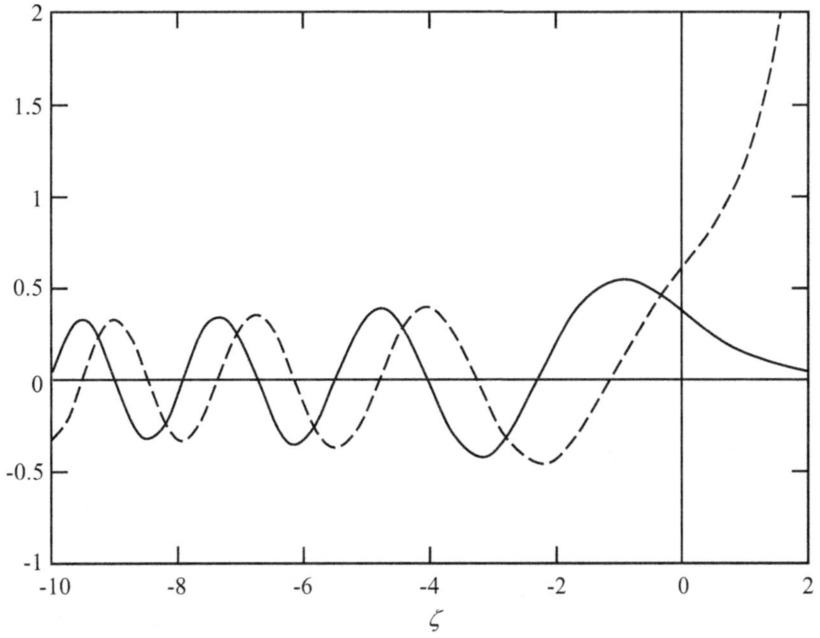

Fig. 2.10. Airy functions $Ai(\zeta)$ (solid line) and $Bi(\zeta)$ (dashed line).

[37] The Airy functions are also related to Bessel functions of one third fractional order.

Linear potential without boundaries

The simplest situation mathematically is just a potential that varies linearly without any additional boundaries or walls. This is somewhat unphysical because it presumes some source of potential, such as an electric field, that continues forever in some direction, but it is a simple idealization when we are far from any boundary. The mathematics allows for two possible solutions, one based on the *Ai* function and the other based on the *Bi* function. Physically, we discard the *Bi* solution here because it diverges for positive arguments, becoming larger and larger. Any attempt at normalizing this function would fail, and the particle would have increasing probability of being found at arbitrarily large positive *z*. This is the same argument we used to ignore the exponentially growing solution when considering penetration into an infinitely thick barrier. We are left only with the *Ai* function in this case. Substituting back from the change of variable, Eq. (2.94), the $Ai(\zeta)$ solution becomes explicitly

$$\psi_E(z) = Ai\left(\left(\frac{2me\mathrm{E}}{\hbar^2}\right)^{1/3}\left(z - \frac{E}{e\mathrm{E}}\right)\right) \tag{2.95}$$

This solution is sketched, for a specific eigenenergy E_o, together with the potential energy, in Fig. 2.11.

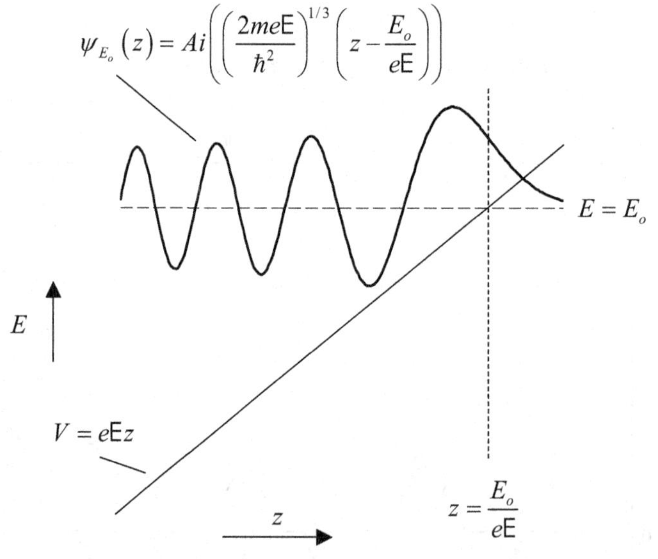

Fig. 2.11. Sketch of a linearly varying potential and the Airy function solution of the resulting Schrödinger equation.

There are several interesting aspects about this solution.

(1) Because we have introduced no additional boundary conditions, there are mathematical solutions for any possible value of the eigenenergy *E*. This behavior reminds us of the simple case of a uniform zero potential (i.e., $V = 0$ everywhere), which leads to plane wave solutions for any positive energy. In the present case also, the allowed values of the eigenenergies are continuous, not discrete. Like the case of a uniform potential, the eigenstates are not bound within some finite region (at least for negative *z*).

(2) The solution is oscillatory when the eigenenergy is greater than the potential energy, which occurs on the left of the point $z = E_o / e\mathrm{E}$, and it decays to the right of this point.

This point is known as the classical turning point, because it is the farthest to the right that a classical particle of energy E_o could go.

(3) The eigenfunction solutions for different energies are the same except they are shifted sideways (i.e., shifted in z).

(4) Unlike the uniform potential case, the solutions here are not running waves; rather, they are standing waves, which is more like the case of the particle in a box.

We again find, just like in the harmonic oscillator case, that we have been able to derive eigenstates (i.e., states that are stable in time). Just as in the harmonic oscillator case, where classically we would have expected to get an oscillation if we have finite energy, here we would have expected classically to get states that correspond to the electron being accelerated. Simply put, we have put an electron in an electric field, and the electron is not moving. Again, to resolve this paradox, we need to consider the time-dependent Schrödinger equation, which we cover in the next chapter.

How is it that we could even have a standing wave in this case? In the case of a particle in a box, we can easily rationalize that the particle is reflecting off the walls. In the present case, we could presumably readily accept that the particle should bounce off the increasing potential seen at or near the classical turning point, so it is relatively easy to see why there is a reflection at the right. There is also here reflection from the left; the reason for this reflection is that from the point of view, in general, of wave mechanics, any change in potential (or change of impedance in the case of acoustic or electromagnetic waves), even if it is smooth rather than abrupt, leads to reflections. Effectively, there is a distributed reflection on the left from the continuously changing potential there. The fact that there is such a distributed reflection explains why the wave amplitude decreases progressively as we go to the left. The fact that we have a standing wave is apparently because, integrated up, that reflection does eventually add up to 100 percent.

Why does the period of the oscillations in the wave decrease (i.e., the oscillations become faster) as we move to the left? Suppose in Schrödinger's equation we divide both sides by ψ; then we have

$$\frac{-\hbar^2}{2m}\frac{1}{\psi}\frac{d^2\psi}{dz^2}+V(z)=E \tag{2.96}$$

For any eigenstate of the Schrödinger equation, E is a constant (the eigenenergy). In such a state, if V decreases, then $-(1/\psi)(d^2\psi/dz^2)$ (which we can visualize as the degree of curvature of the wavefunction) must increase. If we imagine that we have an oscillating wave, which we presume is locally approximately sinusoidal, of the form $\sim\sin(kz+\theta)$ for some phase angle θ

$$\frac{-1}{\psi}\frac{d^2\psi}{dz^2}\simeq k^2 \tag{2.97}$$

Hence, if V decreases, the wavevector k must increase – that is, the period must decrease. Viewed from the perspective of the particle, we could imagine that the particle is going increasingly fast as it goes toward the left, being accelerated by the field, or equivalently, is going increasingly slowly as it goes toward the right, being decelerated by the field, either of which is consistent with smaller periods as we go to the left. This view of particle motion is not very rigorous, though there is a kernel of truth to it, but for a full understanding in terms of particle motion, we need the time-dependent Schrödinger equation of the next chapter.

Triangular potential well

If we put a hard barrier on the left, we again get a discrete set of eigenenergies. Formally, we can consider an electron, still in a uniform electric field E, with an infinitely high potential barrier at $z = 0$ (i.e., the potential is infinite for all negative z), as shown in Fig. 2.12, with the potential taken to be zero at $z = 0$ (or, at least, just to the right of $z = 0$).

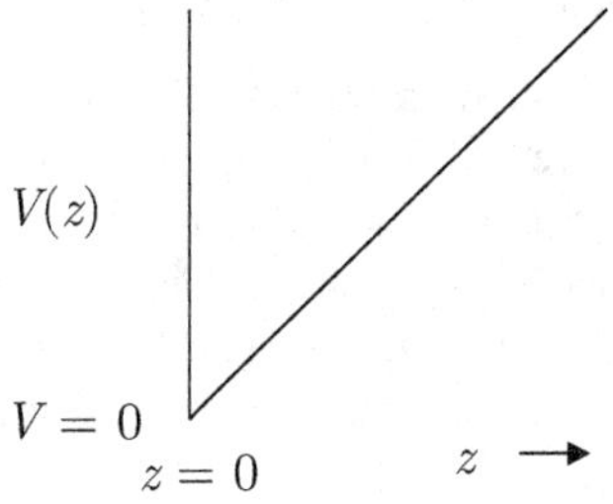

Fig. 2.12 Triangular potential well.

For all $z > 0$, we have the same potential as we considered previously. Again we can discard the Bi solution because it diverges, so we are left with the Ai solution. Now we have the additional boundary condition imposed by the infinitely high potential at $z = 0$, which means the wavefunction must go to zero there. This is easily achieved with the Ai function if we position it laterally so that one of its zeros is found at $z = 0$. The $Ai(\zeta)$ function will have zeros for a set of values ζ_i. These can be found in mathematical tables or are relatively easily calculated numerically in advanced mathematics programs.

The first few of these zeros are

$$\begin{aligned} \zeta_1 &\simeq -2.338 \\ \zeta_2 &\simeq -4.088 \\ \zeta_3 &\simeq -5.521 \\ \zeta_4 &\simeq -6.787 \\ \zeta_5 &\simeq -7.944 \end{aligned} \tag{2.98}$$

To get the solution Eq. (2.95) to be zero at $z = 0$ means, therefore, that

$$Ai\left(\left(\frac{2me\mathsf{E}}{\hbar^2}\right)^{1/3}\left(0-\frac{E}{e\mathsf{E}}\right)\right)=0 \tag{2.99}$$

that is, the argument of this function must be one of the zeros of the Ai function

$$\left(\frac{2me\mathsf{E}}{\hbar^2}\right)^{1/3}\left(-\frac{E}{e\mathsf{E}}\right)=\zeta_i \tag{2.100}$$

or, equivalently, the possible energy eigenvalues are

$$E_i = -\left(\frac{\hbar^2}{2m}\right)^{1/3}\left(e\mathsf{E}\right)^{2/3}\zeta_i \tag{2.101}$$

Fig. 2.13 shows the results of a specific calculation for the case of an electric field of 1 V/Å (10^{10} V/m). As can be seen, the wavefunctions for the different levels are simple shifted versions of one another, with the wavefunction being truncated at the infinitely high potential barrier at position 0, at which point each wavefunction is zero in amplitude. As in the simple rectangular potential wells, the lowest energy function has no zeros (other than at the left and right ends), and each successive, higher-energy solution has one more zero.

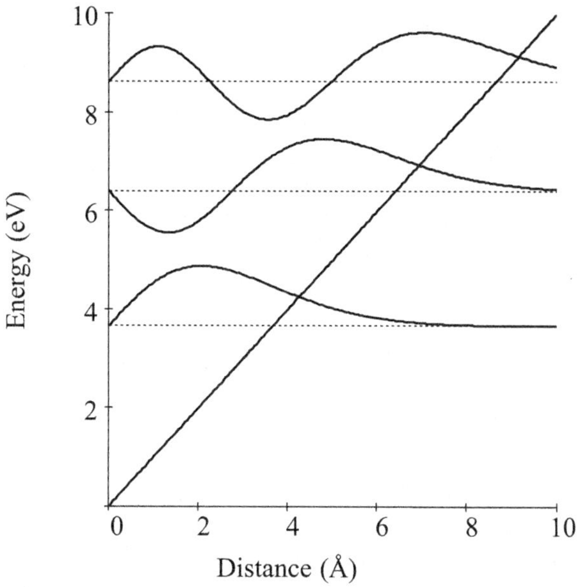

Fig. 2.13. Graphs of wavefunctions and energy levels for the first three levels in a triangular potential well, for a field of 1V/Å.

Infinite potential well with field

We can take the triangular well one step further by including also an infinitely high barrier on the right. This makes the potential structure into an infinite potential well with field (or a skewed infinite potential well).

The equations remain the same, except we have the additional boundary condition that the potential is infinite and, hence, that the wavefunction is zero, at $z = L_z$. Now we cannot discard the *Bi* solution; the potential forces the wavefunction to zero at the right wall, so there will be no wavefunction amplitude to the right of this wall, and so the divergence of the *Bi* function no longer matters for normalization (we would only be normalizing inside the box). Hence, we have to work with the general solution, Eq. (2.93), with both *Ai* and *Bi* functions.

This solution requires some more mathematical work but is ultimately straightforward. The two boundary conditions are that the wavefunction must be zero at $z = 0$ and at $z = L_z$, or, equivalently, at $\zeta = \zeta_0$ and $\zeta = \zeta_L$, where

$$\zeta_0 \equiv -\left(\frac{2m}{\hbar^2 e^2 \mathrm{E}^2}\right)^{1/3} E \tag{2.102}$$

$$\zeta_L \equiv \left(\frac{2me\mathrm{E}}{\hbar^2}\right)^{1/3} \left(L_z - \frac{E}{e\mathrm{E}}\right) \tag{2.103}$$

These boundary conditions will establish what the possible values of E are – i.e., the energy eigenvalues. The conditions result in two equations

$$a\,Ai(\zeta_0)+b\,Bi(\zeta_0)=0 \tag{2.104}$$

$$a\,Ai(\zeta_L)+b\,Bi(\zeta_L)=0 \tag{2.105}$$

or, in matrix form

$$\begin{bmatrix} Ai(\zeta_0) & Bi\zeta_0 \\ Ai(\zeta_L) & Bi(\zeta_L) \end{bmatrix}\begin{bmatrix} a \\ b \end{bmatrix}=0 \tag{2.106}$$

The usual condition for a solution of such equations is

$$\begin{vmatrix} Ai(\zeta_0) & Bi(\zeta_0) \\ Ai(\zeta_L) & Bi(\zeta_L) \end{vmatrix}=0 \tag{2.107}$$

or, equivalently,

$$Ai(\zeta_0)Bi(\zeta_L)-Ai(\zeta_L)Bi(\zeta_0)=0 \tag{2.108}$$

The next mathematical step is to find for what values of ζ_L Eq. (2.108) can be satisfied. This can be done numerically.

First, we change to appropriate dimensionless units. In this problem, there are two relevant energies. One is the natural unit for discussing potential well energies, which is the energy of the lowest state in an infinitely deep potential well, $(\hbar^2/2m)(\pi/L_z)^2$ (as in Eq. (2.26)), which here we call E_1^∞ to avoid confusion with the final energy eigenstates for this problem; we will use this as the energy unit. Hence, we use the dimensionless "energy"

$$\varepsilon \equiv E/E_1^\infty$$

The second energy in the problem is the potential drop from one side of the well to the other resulting from the electric field, which is

$$V_L = e\mathrm{E}L_z \tag{2.109}$$

or, in dimensionless form

$$v_L = V_L/E_1^\infty \tag{2.110}$$

With these definitions, we can rewrite Eqs. (2.102) and (2.103), respectively, as

$$\zeta_0 \equiv -\left(\frac{\pi}{v_L}\right)^{2/3}\varepsilon \tag{2.111}$$

$$\zeta_L = \left(\frac{\pi}{v_L}\right)^{2/3}(v_L-\varepsilon) \tag{2.112}$$

Now we choose a specific v_L, which corresponds to choosing the electric field for a given well width. Suppose, for example, that we consider a 6 Å wide well with a field of 1 V/Å. Then $E_1^\infty \simeq 1.0455$ eV and $v_L \simeq 5.739$ (i.e., the potential change from one side of the well to the other is $\simeq 5.739E_1^\infty$). Next, we numerically find the values of ε that make the determinant function from Eq. (2.108)

$$D(\varepsilon)=Ai(\zeta_0(\varepsilon))Bi(\zeta_L(\varepsilon))-Ai(\zeta_L(\varepsilon))Bi(\zeta_0(\varepsilon)) \tag{2.113}$$

equal to zero. One way to do this is to graph this function from $\varepsilon = 0$ upward to find the approximate position of the zero crossings and use a numerical root finder to find more accurate estimates. With these eigenvalues of ε, it is now straightforward also to evaluate the wavefunctions. From Eq. (2.104), we have for the coefficients a and b of the general solution, Eq. (2.93), for each eigenenergy ε_i

$$\frac{b_i}{a_i} = -\frac{Ai(\zeta_0(\varepsilon_i))}{Bi(\zeta_0(\varepsilon_i))} \tag{2.114}$$

Given that we know the ratio b_i/a_i, we can normalize the wavefunction by integrating from 0 to L_z to find the specific values of both a_i and b_i for each i if we wish. The resulting wavefunction is, therefore, for a given energy eigenstate, using the same notation as for Eqs. (2.111) and (2.112) with the dimensionless energies

$$\psi_i(z) = a_i Ai\left(\left(\frac{\pi}{\nu_L}\right)^{2/3}\left(\nu_L \frac{z}{L_z} - \varepsilon\right)\right) + b_i Bi\left(\left(\frac{\pi}{\nu_L}\right)^{2/3}\left(\nu_L \frac{z}{L_z} - \varepsilon\right)\right) \tag{2.115}$$

For the example numbers here, we have

	ε_i	E_i (eV)	b/a
First level (i = 1)	3.53	3.69	-0.04
Second level (i = 2)	6.95	7.27	-2.48
Third level (i = 3)	11.93	12.47	-0.12

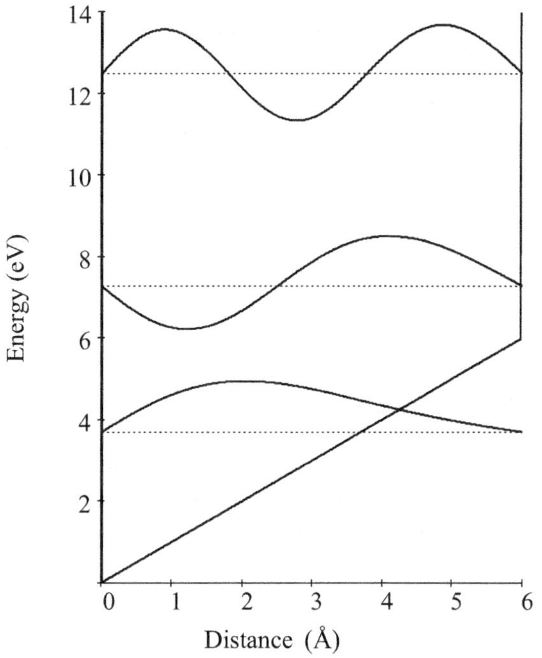

Fig. 2.14. First three eigenstates in a 6 Å potential well with infinitely high barriers at each side, for an electron in a field of 1 V/Å. The potential is also sketched.

The resulting solutions are plotted in Fig. 2.14. Note that

(i) All the wavefunctions go to zero at both sides of the well, as required by the infinitely high potential energies there.

(ii) The lowest solution is almost identical in energy and wavefunction to that of the lowest state in the triangular well. (The fraction of the *Bi* Airy function is very small, -0.04). The energy is actually slightly higher because the wavefunction is slightly more confined.

(iii) The second solution is now quite strongly influenced by the potential barrier at the right, with a significantly higher energy than in the triangular well.

(iv) The third solution is very close in form to that of the third level of a simple rectangular well. To the eye, it looks to be approximately sinusoidal, though the period is slightly shorter on the left-hand side, consistent with our previous discussion of the effect on the wavefunction oscillation period from changes in potential.

We can see in the lowest state that the electron has been pulled closer to the left-hand side, which is what we would expect classically from such an electric field. Note, though, that our classical intuition does not work for the higher levels. In fact, in the second level, a detailed calculation shows that the electron is actually substantially more likely to be in the right half (~ 64 percent) of the well than in the left half (~ 36 percent).[38]

Problems

2.11.1* Give actual energy levels in electron-Volts to three significant figures for the first three levels in the triangular well as in Fig. 2.13 (i.e., with a field of 1 V/Å).

2.11.2 Repeat the calculation of Problem 2.11.1 for electrons in the semiconductor GaAs, for specific electric fields of 1 V/μm and 10 V/μm. Instead of using the free electron mass, use the effective mass of an electron in GaAs, $m_{eff} \simeq 0.07 m_o$. Also, calculate the distance from the interface to the classical turning point in each case.

2.11.3 For the following two fields, calculate the first three energy levels (in electron-Volts) for an electron in a 100 Å GaAs potential well with infinitely high barriers, and plot the probability densities in units of $Å^{-1}$ for each of the three states. State the energies relative to the energy of the center of the well (*not* relative to the lower corner). Presume that the electron can be treated as having an effective mass of $m_{eff} \simeq 0.07 m_o$. (For this problem, mathematical software will be required. You need to be able to find roots numerically, evaluate the Airy functions, and perform numerical integration for normalization.)
(i) zero field
(ii) 20 V/μm

2.12 Summary of concepts

This chapter has seen the introduction, mostly by example, of various unusual concepts in quantum mechanics and of some key results and equations. We list these briefly here. (See also the memorization list at the end of the book for those formulae particularly worth learning by heart.)

---

[38] The reader is likely surprised at the large difference in probabilities between the right and left halves of the well; by eye, the wavefunction perhaps does not look so unbalanced between the two halves. Remember, though, that it is the modulus squared of the wavefunction that gives the probability. The left half contains the zero crossing, and there is little contribution to the square from the region near the zero crossing. The small difference in the magnitude of the peak of the wavefunction in the two halves is also magnified by taking the square.

Wave-particle duality

This is the idea that particles, such as electrons, also behave in some ways as if they were waves (e.g., showing diffraction).

Time-independent Schrödinger wave equation

Single particles with mass, such as a nonrelativistic electron in the absence of a magnetic field, often obey the (time-independent) Schrödinger wave equation

$$\left(-\frac{\hbar^2}{2m_o}\nabla^2 + V(\mathbf{r})\right)\psi = E\psi \tag{2.13}$$

Probability density and probability amplitude (or quantum mechanical amplitude)

For a particle with wave behavior $\psi(\mathbf{r})$, the probability of finding a particle near a given point $\mathbf{r}$ in space is proportional to $|\psi(\mathbf{r})|^2$, which can then be described as a probability density, and $\psi(\mathbf{r})$ can be called a probability amplitude or quantum mechanical amplitude. Such probability amplitudes or quantum mechanical amplitudes occur throughout quantum mechanics, not merely as wave amplitudes in Schrödinger wave equations. Quantum mechanical calculations proceed by first adding all possible contributions to the quantum mechanical amplitude and then taking the modulus squared to calculate probabilities. There is no analogous concept of probability amplitude in classical probability.

Normalization

A wavefunction $\psi(\mathbf{r})$ is normalized if $\int|\psi(\mathbf{r})|^2 d^3\mathbf{r} = 1$, meaning the sum of all the probabilities adds to 1.

Linearity of quantum mechanics and probability amplitude

Quantum mechanics is apparently exactly linear in the addition of probability amplitudes for a given particle, which allows linear algebra to be used as a rigorous mathematical foundation for quantum mechanics.

Eigenfunctions and eigenvalues

Solutions of some equations in quantum mechanics, including in particular the time-independent Schrödinger equation, lead to functions (eigenfunctions) each associated with a particular value of some parameter (the eigenvalue). For the case of the time-independent Schrödinger equation, the parameter (eigenenergy) is the energy corresponding to the eigenfunction.

Discrete energy states

Often, solutions of the time-independent Schrödinger equation are associated with discrete energy values or "states", with solutions possible only for those discrete values.

Parity and odd and even functions

Often in quantum mechanics, functions are encountered that are either even – that is, symmetrically the same on either side of some point in space – or odd – that is, exactly antisymmetric on either side of some point in space so that they have exactly the opposite value at such symmetric points. Functions with such even or odd character are described as having even or odd parity.

Zero point energy

The lowest energy solution in quantum mechanical problems often has an energy higher than the energy at the bottom of the potential. Such an energy is called a zero point energy, and has no classical analog.

Solutions for a particle in an infinitely deep potential well ("particle in a box")

$$E_n = \frac{\hbar^2}{2m}\left(\frac{n\pi}{L_z}\right)^2, \quad n = 1, 2, \ldots \tag{2.26}$$

$$\psi_n(z) = \sqrt{\frac{2}{L_z}} \sin\left(\frac{n\pi z}{L_z}\right) \tag{2.29}$$

Basis set

A basis set is a set of functions that can be used, by taking appropriate combinations, to represent other functions in the same space.

Representation

The coefficients of the basis functions required to represent some other specific function would be the representation of that specific function in this basis.

Orthogonality

This is a mathematically useful property of two functions $g(z)$ and $h(z)$, in which their "overlap" integral $\int g^*(z)h(z)dz$ is zero. It can also be applied as a condition for a set of functions, in which case any pair of different functions in the set is found to be orthogonal.

Completeness

This is the condition that a basis set of functions can be used to represent any function in the same space.

Orthonormality

The condition that a set of functions is orthogonal and the functions in the set are also normalized.

Expansion coefficients

In representing a function $f(z)$ in a complete orthonormal basis set of functions, $\psi_n(x)$, using the expression

$$f(x) = \sum_n c_n \psi_n(x) \tag{2.36}$$

the expansion coefficient c_m is given by

$$c_m = \int \psi_m^*(x) f(x)\,dx \tag{2.37}$$

Degeneracy

This is the condition in which two or more orthogonal states have the same eigenvalue (usually energy). The number of such states with the same eigenvalue is sometimes called the degeneracy.

Bound states

Bound states are states with energy less than the energy classically required to "escape" from the potential. They usually also have discrete allowed energies, with finite energy separations between any two (nondegenerate) bound states.

Unbound states

These are states with energy greater than the energy classically required to "escape" from the potential. They commonly have continuous ranges of allowed energies.

Chapter 3

The time-dependent Schrödinger equation

Prerequisite: Chapter 2.

Thus far with Schrödinger's equation, we have considered only situations where the spatial probability distribution was steady in time. In our rationalization of the time-independent Schrödinger equation, we imagined we had, for example, a steady electron beam, where the electrons had a definite energy; this beam was diffracting off some object, such as a crystal or through a pair of slits. The result was some steady diffraction pattern (at least, the probability distribution did not vary in time). We then went on to use this equation to examine some other specific problems, including potential wells, where this requirement of definite energy led to the unusual behavior that only specific, "quantized" energies were allowed.

In particular, we analyzed the problem of the harmonic oscillator and found stationary states of that oscillator. On the face of it, stationary states of an oscillator (other than the trivial one of the oscillator having zero energy) make little sense given our classical experience with oscillators – a classical oscillator with energy oscillates.

Clearly, we must expect quantum mechanics to model situations that are not stationary. The world about us changes, and if quantum mechanics is to be a complete theory, it must handle such changes. To understand such changes, at least for the kinds of systems where Schrödinger's equation might be expected to be valid, we need a time-dependent extension of Schrödinger's equation.

We start off this chapter by rationalizing such an equation. The equation we find is somewhat different from the ones we know from classical waves, though it is still straightforward. We then introduce a very important concept in quantum mechanics: *superposition states*. Superposition states allows us to handle the time evolution of quantum mechanical systems rather easily. Then we examine some specific examples of time evolution.

We will also be able to use this discussion of Schrödinger's time-dependent equation to illustrate many core concepts in quantum mechanics – concepts that we will be able to generalize and extend later – and to introduce intriguing topics such as the uncertainty principle and the issue of measurement in quantum mechanics. By the end of this chapter, we will have quite a complete view of quantum mechanics as illustrated by this wave equation approach for particles.

3.1 Rationalization of the time-dependent Schrödinger equation

The key to understanding time-dependence and Schrödinger's equation is to understand the relation between frequency and energy in quantum mechanics. One very well-known example of a relation between frequency and energy is the case of electromagnetic waves and photons. We can imagine doing a simple pair of experiments with a monochromatic electromagnetic wave. In one experiment, we would measure the frequency of the oscillation in the wave.[1] In a second experiment, we would measure the power in the wave and also count the number of photons per second. At optical frequencies, a simple photodiode can count the photons; it is relatively easy to make a photodetector that will generate one electron per absorbed photon, and we can count electrons (and, hence, photons) by measuring current. Hence, we can count how many photons per second correspond to a particular power at this frequency. We would find in such an experiment that the energy per photon was

$$E = h\nu = \hbar\omega \tag{3.1}$$

That is, the photon energy, E, is proportional to the frequency, ν, with proportionality constant h, or equivalently, it is proportional to the angular frequency, $\omega = 2\pi\nu$, with proportionality constant $\hbar = h/2\pi$.

Of course, this discussion is for photons, not the electrons or other particles with mass[2] for which the Schrödinger equation supposedly applies. Entities such as hydrogen atoms emit photons as they transition between allowed energy levels, however, and we might reasonably expect there is, therefore, some oscillation in the electrons at the corresponding frequency during the emission of the photon, which, in turn, might lead us to expect that there is a similar relation between energy and frequency associated with the separation of electron levels. Our question now is how to construct a wave equation that both has this kind of relation between energy and frequency (Eq. (3.1)) yet still allows, for example, a simple wave of the form $\exp[i(kz - \omega t)]$, to be a solution in a uniform potential (i.e., for constant $V(\mathbf{r})$). The answer that Schrödinger postulated to this question is the equation

$$-\frac{\hbar^2}{2m}\nabla^2\Psi(\mathbf{r},t) + V(\mathbf{r},t)\Psi(\mathbf{r},t) = i\hbar\frac{\partial\Psi(\mathbf{r},t)}{\partial t} \tag{3.2}$$

which is the time-dependent Schrödinger equation. It is easy to check, for example, that waves of the form

$$\exp\left[-i\left(\frac{Et}{\hbar} \pm kz\right)\right] \equiv \exp\left(-i\frac{Et}{\hbar}\right)\exp(\mp ikz)$$

with $E = \hbar\omega$ and $k = \sqrt{2mE/\hbar^2}$ would be solutions when $V = 0$ everywhere.

---

[1] At optical frequencies, this can be tricky but is essentially straightforward; alternatively, we could always measure the wavelength, which is relatively easy to do in an optical experiment, and deduce the frequency from the known relation between frequency and wavelength for an electromagnetic wave.

[2] Particles with mass, e.g., electrons, are sometimes referred to as massive particles even though their mass may be $\sim 10^{-30}$ kg! Even at that mass, they are much more massive than photons, which have zero mass.

Schrödinger also made a specific choice of sign for the imaginary part on the right-hand side, which means that a wave with a spatial part $\propto \exp(ikz)$ is quite definitely a wave propagating in the positive z direction for all positive energies E (i.e., the wave, including its time-dependence, would be of the form[3] $\exp[i(kz - Et/\hbar)]$).

The time-dependent Schrödinger equation is a different kind of wave equation from a more common wave equation encountered in classical physics, which is typically of the form

$$\nabla^2 f = \frac{k^2}{\omega^2}\frac{\partial^2 f}{\partial t^2} \tag{3.3}$$

and for which $f \propto \exp[i(kz - \omega t)]$ would also be solution. Eq. (3.3) has a *second* derivative with respect to time, as opposed to the *first* derivative in the time-dependent Schrödinger equation Eq. (3.2).

Note, incidentally, that the choice by Schrödinger to use complex notation here means that the wavefunction is required to be a complex entity. Unlike the use of complex notation with classical waves, it is not the case that the "actual" wave is taken at the end of the calculation to be the real part of the calculated "complex" wave.

Problems

3.1.1 Consider Schrödinger's time-dependent equation for an electron, with a potential that is uniform and constant at a value V_o, with a solution of the form $\exp[i(kz - \omega t)]$. Deduce the relationship giving k in terms of ω and V_o, and deduce under what conditions there is a solution for real k.

3.1.2* Presuming that the potential is constant in time and space and has a zero value, which of the following are possible solutions of the time-dependent Schrödinger equation for some positive (nonzero) real values of k and ω?

(i) $\sin(kz - \omega t)$

(ii) $\exp(ikz)$

(iii) $\exp\left[-i(\omega t + kz)\right]$

(iv) $\exp\left[i(\omega t - kz)\right]$

3.1.3 Consider the problem of an electron in a one-dimensional "infinite" potential well of width L_z in the z direction (i.e., the potential energy is infinite for $z < 0$ and for $z > L_z$ and, for simplicity, zero for other values of z). For each of the following functions, in exactly the form stated, state whether the function is a solution of the time-dependent Schrödinger equation (with time variable t).

(i) $\exp\left(-i\frac{\hbar\pi^2}{2m_o L_z^2}t\right)\sin\left(\frac{\pi z}{L_z}\right)$

(ii) $\exp\left(i\frac{4\hbar\pi^2}{2m_o L_z^2}t\right)\sin\left(\frac{2\pi z}{L_z}\right)$

---

[3] Unfortunately for engineers, and especially electrical engineers, the choice made by Schrödinger is the opposite of that commonly used by engineers. Schrödinger's choice means that a forward propagating wave is (and has to be) represented as $\exp[i(kz - \omega t)]$, whereas in engineering, it is much more common to have $\exp[i(\omega t - kz)]$ represent a forward wave (or to write $\exp[j(\omega t - kz)]$ with $j \equiv \sqrt{-1}$). In engineering, it does not matter which convention one uses because the quantities finally calculated are only the real part of such complex representations of waves; but, in quantum mechanics, we have to keep the full complex wave.

(iii) $\exp\left(-i\left(\frac{\hbar\pi^2}{2m_oL_z^2}t+\frac{\pi}{2}\right)\right)\cos\left(\frac{\pi z}{L_z}+\frac{\pi}{2}\right)$

(iv) $2\exp\left(-i\frac{\hbar\pi^2}{2m_oL_z^2}t\right)\sin\left(\frac{\pi z}{L_z}\right)-i\exp\left(-i\frac{9\hbar\pi^2}{2m_oL_z^2}t\right)\sin\left(\frac{3\pi z}{L_z}\right)$

3.2 Relation to the time-independent Schrödinger equation

Suppose that we had a solution where the spatial behavior of the wavefunction did not change its form with time (and in which, of necessity, the potential V did not change in time). In such a case, we could allow for some time-varying multiplying factor, $A(t)$, in front of the spatial part of the wavefunction. That is, we could write

$$\Psi(\mathbf{r},t)=A(t)\psi(\mathbf{r}) \tag{3.4}$$

where, explicitly, we are presuming that $\psi(\mathbf{r})$ is not changing in time. In our previous discussions on the time-independent Schrödinger equation, we had asserted that solutions whose spatial behavior was steady in time should satisfy the time-independent equation (Eq. (2.13))

$$-\frac{\hbar^2}{2m}\nabla^2\psi(\mathbf{r})+V(\mathbf{r})\psi(\mathbf{r})=E\psi(\mathbf{r}) \tag{3.5}$$

Merely adding the factor $A(t)$ in front of $\psi(\mathbf{r})$ makes no difference in Eq. (3.5); $\Psi(\mathbf{r},t)$ would also be a solution of Eq. (3.5), regardless of what function we chose for $A(t)$. That is, we would have

$$A(t)\left[-\frac{\hbar^2}{2m}\nabla^2\psi(\mathbf{r})+V(\mathbf{r})\psi(\mathbf{r})\right]=EA(t)\psi(\mathbf{r}) \tag{3.6}$$

Now let us see what happens if we also want to make the kind of solution in Eq. (3.4) work for the time-dependent Schrödinger equation. Substituting the form Eq. (3.4) into the time-dependent Schrödinger equation (3.2) (presuming the potential V is constant in time) then gives

$$A(t)\left[-\frac{\hbar^2}{2m}\nabla^2\psi(\mathbf{r})+V(\mathbf{r})\psi(\mathbf{r})\right]=i\hbar\psi(\mathbf{r})\frac{\partial A(t)}{\partial t} \tag{3.7}$$

which therefore means that if we want the same solution to work for both the time-independent and the time-dependent Schrödinger equation

$$EA(t)=i\hbar\frac{\partial A(t)}{\partial t} \tag{3.8}$$

that is,

$$A(t)=A_o\exp(-iEt/\hbar) \tag{3.9}$$

where A_o is a constant. Hence, for any situation in which the spatial part of the wavefunction is steady in time (and for which the potential V does not vary in time), the full time-dependent wavefunction can be written in the form

$$\Psi(\mathbf{r},t) = A_o \exp(-iEt/\hbar)\psi(\mathbf{r}) \tag{3.10}$$

In other words, if we have a solution $\psi(\mathbf{r})$ of the time-independent Schrödinger equation, with corresponding eigenenergy E, then multiplying by the factor $\exp(-iEt/\hbar)$ gives us a solution of the time-dependent Schrödinger equation.

The reader might be worried that we have a time-dependent part to the wavefunction when we are supposed to be considering a situation that is stable in time. If, however, we consider a meaningful quantity, such as the probability density, we will find that it is stable in time. Explicitly,

$$|\Psi(\mathbf{r},t)|^2 = \left[\exp(+iEt/\hbar)\psi^*(\mathbf{r})\right]\times\left[\exp(-iEt/\hbar)\psi(\mathbf{r})\right] = |\psi(\mathbf{r})|^2 \tag{3.11}$$

Again, we see that the wavefunction itself is not necessarily a meaningful quantity – introducing the factor $\exp(-iEt/\hbar)$ has given us an apparent time dependence in a time-independent problem; the wavefunction is merely a means to calculate other quantities that do have meaning, such as the probability density.

3.3 Solutions of the time-dependent Schrödinger equation

The time-dependent Schrödinger equation, Eq. (3.2), unlike the time-independent one, is not an eigenvalue equation; it is not an equation that only has solutions for a particular set of values of some parameter. In fact, it is quite possible to have *any* spatial function as a solution of the time-dependent Schrödinger equation at a given time (as long as it is a mathematically well-behaved function whose second derivative is defined and finite). That spatial function also determines exactly how the wavefunction will subsequently evolve in time (presuming we also know the potential V as a function of space and time). This ability to predict the future behavior of the wave from its current spatial form is a rather important property of the time-dependent Schrödinger equation.

In general, we can see that if we knew the wavefunction at every point in space at some time t_o – that is, if we knew $\Psi(\mathbf{r},t_o)$ for all $\mathbf{r}$ – we could evaluate the left-hand side of Eq. (3.2) at that time for all $\mathbf{r}$. We would know $\partial\Psi(\mathbf{r},t)/\partial t$ for all $\mathbf{r}$ and could, in principle, integrate the equation to deduce $\Psi(\mathbf{r},t)$ at all times. Explicitly, we would have for some small advance in time by an amount δt

$$\Psi(\mathbf{r},t_o+\delta t) \cong \Psi(\mathbf{r},t_o) + \left.\frac{\partial\Psi}{\partial t}\right|_{\mathbf{r},t_o}\delta t \tag{3.12}$$

Because Schrödinger's equation tells us $\partial\Psi/\partial t$ at time t_o if we know $\Psi(\mathbf{r},t_o)$, we have everything we need to know to calculate $\Psi(\mathbf{r},t_o+\delta t)$. In other words, the whole subsequent evolution of the wavefunction could be deduced from its spatial form at some given time.

Some physicists view this ability to deduce the wavefunction at all future times as being the reason why this equation has only a first derivative in time (as opposed to the second derivative in the more common wave equations in classical physics). In the ordinary classical wave equation, like Eq. (3.3), taking a snapshot of a wave on a string at some time allows us to know the second spatial derivative, from which we can deduce the second time derivative; the ordinary wave equation is simply a relation between the second spatial derivative and the second time derivative. But that is not enough to tell us what will happen next. Specifically, this snapshot does not tell us in which direction the wave is moving. By contrast, with the time-dependent Schrödinger equation, if we know the full complex form of the wavefunction in space at some point in time, we can know exactly what is going to happen next and at all subsequent times (at least, if we know the potential V everywhere for all subsequent times also). Hence, we can view the wavefunction $\psi(\mathbf{r})$ at some time as being a complete description of the particle being modeled for all situations for which the Schrödinger equation is appropriate. This is a core idea in quantum mechanics: the wavefunction or – as we will generalize it later, the quantum mechanical state – contains all the information required about the particle for the calculation of any observable quantity.

Note, of course, that if the spatial wavefunction is in an eigenstate, there is no subsequent variation in time of the wavefunction other than the oscillation $\exp(-iEt/\hbar)$ that, on its own, leads to no variation of the measurable properties of the system in time.

Problem

3.3.1 Consider the problem of an electron in a one-dimensional "infinite" potential well of width L_z in the z direction (i.e., the potential energy is infinite for $z < 0$ and for $z > L_z$ and, for simplicity, zero for other values of z). (In the functions below, t refers to time.)

(i) For each of the following functions, state whether it could be a solution to the time-*independent* Schrödinger equation for this problem.

(a) $\sin(3\pi z/L_z)$

(b) $\exp\left(-i\dfrac{(7.5)^2\hbar\pi^2}{2m_oL_z^2}t\right)\sin(7.5\pi z/L_z)$

(c) $A\sin(\pi z/L_z)+B\sin(4\pi z/L_z)$, where A and B are arbitrary complex constants.

(d) $A\exp\left(-i\dfrac{\hbar\pi^2}{2m_oL_z^2}t\right)\sin(\pi z/L_z)+B\exp\left(-i\dfrac{8\hbar\pi^2}{m_oL_z^2}t\right)\sin(4\pi z/L_z)$, where A and B are arbitrary complex constants.

(ii) For each of these functions, state whether it could be a solution of the time-*dependent* Schrödinger equation for this problem.

3.4 Linearity of quantum mechanics: linear superposition

The time-dependent Schrödinger equation is also linear in the wavefunction Ψ, just as the time-independent Schrödinger equation was. Again, no higher powers of Ψ appear anywhere in the equation. In the time-independent equation, this allowed us to say that if ψ was a solution, then so also was $A\psi$, where A is any constant, and the same kind of behavior holds here for the solutions Ψ of the time-dependent Schrödinger equation. Another consequence of linearity is the possibility of linear superposition of solutions, which we can state as follows:

$$\begin{gathered}\text{If } \Psi_a(\mathbf{r},t) \text{ and } \Psi_b(\mathbf{r},t) \text{ are solutions,}\\ \text{then so also is } \Psi_{a+b}(\mathbf{r},t)=\Psi_a(\mathbf{r},t)+\Psi_b(\mathbf{r},t).\end{gathered} \tag{3.13}$$

This is easily verified by substitution into the time-dependent Schrödinger equation.[4]

---

[4] This kind of superposition property also formally applies to the time-independent Schrödinger equation, but it is much less useful there, essentially only being helpful for the case when a state is degenerate (i.e., where there is more than one eigenfunction possible for a given eigenvalue). Then, superpositions of those different degenerate states are also solutions of the same equation – i.e., corresponding to the same

We can also multiply the individual solutions by arbitrary constants and still have a solution to the equation; that is,

$$\Psi_c(\mathbf{r},t) = c_a \Psi_a(\mathbf{r},t) + c_b \Psi_b(\mathbf{r},t) \tag{3.14}$$

where c_a and c_b are (complex) constants, is also a solution.

The concept of linear superposition solutions is, at first sight, a very strange one from a classical point of view. There is no classical precedent for saying that a particle or a system is in a linear superposition of two or more possible states. In classical mechanics, a particle simply has a "state" that is defined by its position and momentum, for example. Here, we are saying that a particle may exist in a superposition of states each of which may have different energies (or possibly positions or momenta). Actually, however, it will turn out that to recover the kind of behavior we expect from particles in the classical picture of the world, we need to use linear superpositions, as we illustrate in the following section.

3.5 Time dependence and expansion in the energy eigenstates

A particularly interesting and useful way to look at the time evolution of the wavefunction, especially for the case where the potential V is constant in time, is to expand it in the energy eigenfunction basis. If V is constant in time, each of the energy eigenstates is separately a solution of the time-dependent Schrödinger equation (with different values of energy E if the solutions are not degenerate). Explicitly, the nth energy eigenfunction can be written, following Eq. (3.10), as

$$\Psi_n(\mathbf{r},t) = \exp(-iE_n t/\hbar)\psi_n(\mathbf{r}) \tag{3.15}$$

where E_n is the nth energy eigenvalue, and now we presume that the ψ_n (and, consequently, the Ψ_n) are normalized. This function is a solution of the time-dependent Schrödinger equation. Because of the linear superposition defined previously, any sum of such solutions is also a solution.

Suppose that we had expanded the original spatial solution at time $t = 0$ in energy eigenfunctions; that is,

$$\psi(\mathbf{r}) = \sum_n a_n \psi_n(\mathbf{r}) \tag{3.16}$$

where the a_n are the expansion coefficients (the a_n are simply fixed complex numbers). We know that any spatial function $\psi(\mathbf{r})$ can be expanded this way because the set of eigenfunctions $\psi_n(\mathbf{r})$ is believed to be complete for describing any spatial solution. We can now write a corresponding time-dependent function

---

value of energy E. We cannot, however, superpose solutions corresponding to different values of E in the case of the time-independent Schrödinger equation; such solutions are solutions to what are actually different equations (different values of E on the right-hand side). In the case of the time-independent Schrödinger equation, if there is only one spatial form of solution, $\psi(\mathbf{r})$, for a given energy E, then the only different possible solutions correspond to multiples of that one solution. Hence, then linear superposition is the same as multiplying by a constant; i.e.,

$$\psi_{a+b}(\mathbf{r}) = \psi_a(\mathbf{r}) + \psi_b(\mathbf{r}) = A\psi(\mathbf{r}) + B\psi(\mathbf{r}) = (A+B)\psi(\mathbf{r})$$

$$\Psi(\mathbf{r},t) = \sum_n a_n \Psi_n(\mathbf{r},t) = \sum_n a_n \exp(-iE_n t/\hbar)\psi_n(\mathbf{r}) \tag{3.17}$$

We know that this function is a solution to the time-dependent Schrödinger equation because it is constructed from a linear combination of solutions to the equation. At $t = 0$, it correctly gives the known spatial form of the solution. Hence, Eq. (3.17) is the solution to the time-dependent Schrödinger equation with the initial condition

$$\Psi(\mathbf{r},0) = \psi(\mathbf{r}) = \sum_n a_n \psi_n(\mathbf{r}) \tag{3.18}$$

for the case where the potential V does not vary in time.

Hence, if we expand the spatial wavefunction in terms of the energy eigenstates at $t = 0$, as in Eq. (3.16), we have solved for the time evolution of the state thereafter; we have no further integration to do, merely a calculation of the sum (Eq. (3.17)) at each time of interest to us. This is a remarkable result and enables us immediately to start examining the time dependence of quantum mechanical solutions for potentials V that are fixed in time.

3.6 Time evolution of infinite potential well and harmonic oscillator

Now we look at the time evolution of a number of simple examples for cases where the potential is fixed in time (i.e., $V(\mathbf{r},t) \equiv V(\mathbf{r})$) and the system is in a superposition state.

Simple linear superposition in an infinite potential well

Let us consider a very simple physical case in which the mathematics is also very simple. We suppose that we have an infinite potential well (i.e., one with infinitely high barriers), and that the particle within that well is in a linear superposition state with equal parts of the first and second states of the well; that is,

$$\Psi(z,t) = \frac{1}{\sqrt{L_z}}\left[\exp\left(-i\frac{E_1}{\hbar}t\right)\sin\left(\frac{\pi z}{L_z}\right) + \exp\left(-i\frac{E_2}{\hbar}t\right)\sin\left(\frac{2\pi z}{L_z}\right)\right] \tag{3.19}$$

where the well is of thickness L_z, and E_1 and E_2 are the energies of the first and second states, respectively. (We have chosen this linear combination so that it is normalized.) Then, the probability density is given by

$$|\Psi(z,t)|^2 = \frac{1}{L_z}\left[\sin^2\left(\frac{\pi z}{L_z}\right) + \sin^2\left(\frac{2\pi z}{L_z}\right) + 2\cos\left(\frac{E_2 - E_1}{\hbar}t\right)\sin\left(\frac{\pi z}{L_z}\right)\sin\left(\frac{2\pi z}{L_z}\right)\right] \tag{3.20}$$

Here, we see that the probability density has a part that oscillates at an angular frequency $\omega_{21} = (E_2 - E_1)/\hbar = 3E_1/\hbar$. Notice, incidentally, that the absolute energy origin does not matter here because the oscillation frequency only depends on the separation of energy levels. We could have added an arbitrary amount to both of the two energies E_1 and E_2 without making any difference to the resulting oscillation. Changing the energy origin in this way would have changed the oscillatory factors in the wavefunctions, but it would have made no difference to anything measurable, another illustration that the wavefunction itself is not necessarily meaningful. The oscillatory behavior of this particular system is illustrated in Fig. 3.1.

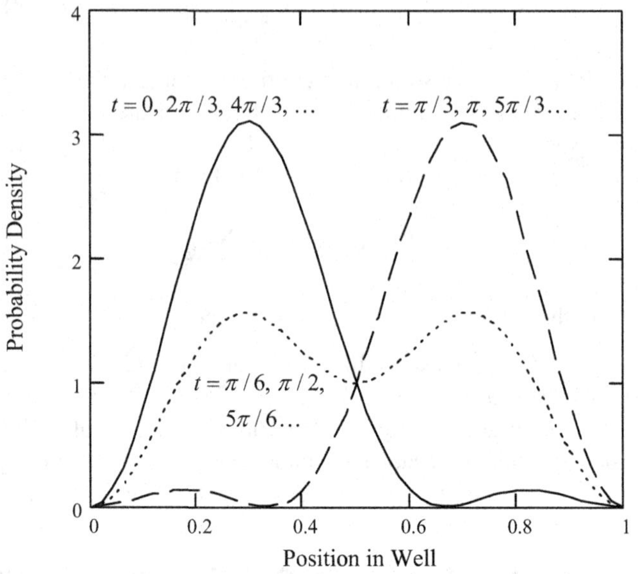

Fig. 3.1. Illustration of the oscillation resulting from the linear superposition of the first and second levels in a potential well with infinitely high barriers. For this illustration, the well is taken to have unit thickness, and the unit of time is taken to be $\hbar/E_1$. The oscillation angular frequency, ω_{21}, is three per unit time because the energy separation of the first and second levels is $3E_1$. The probability density, therefore, oscillates back and forward three times in 2π units of time.

Harmonic oscillator example

We can construct a linear superposition state for the harmonic oscillator to see what kind of time behavior we obtain. For example, we could construct a superposition with equal parts of the first and second states, just like we did for the potential well. Quite generally, if we make a linear superposition of two energy eigenstates with energies E_a and E_b, the resulting probability distribution will oscillate at the frequency $\omega_{ab} = |E_a - E_b|/\hbar$. That is, if we have a superposition wavefunction

$$\Psi_{ab}(\mathbf{r},t) = c_a \exp(-iE_a t/\hbar)\psi_a(\mathbf{r}) + c_b \exp(-iE_b t/\hbar)\psi_b(\mathbf{r}) \tag{3.21}$$

then the probability distribution is

$$\begin{aligned} |\Psi_{ab}(\mathbf{r},t)|^2 &= |c_a|^2 |\psi_a(\mathbf{r})|^2 + |c_b|^2 |\psi_b(\mathbf{r})|^2 \\ &\quad + 2\left|c_a^*\psi_a^*(\mathbf{r})c_b\psi_b(\mathbf{r})\right| \cos\left[\frac{(E_a - E_b)t}{\hbar} - \theta_{ab}\right] \end{aligned} \tag{3.22}$$

where $\theta_{ab} = \arg(c_a\psi_a(\mathbf{r})c_b^*\psi_b^*(\mathbf{r}))$.

Fig. 3.2 shows the resulting probability density for such an equal linear superposition of the first and second states of the harmonic oscillator. The probability density of this superposition oscillates at the (angular) frequency, ω, of the classical harmonic oscillator because the energy separation between the first and second states is $\hbar\omega$.

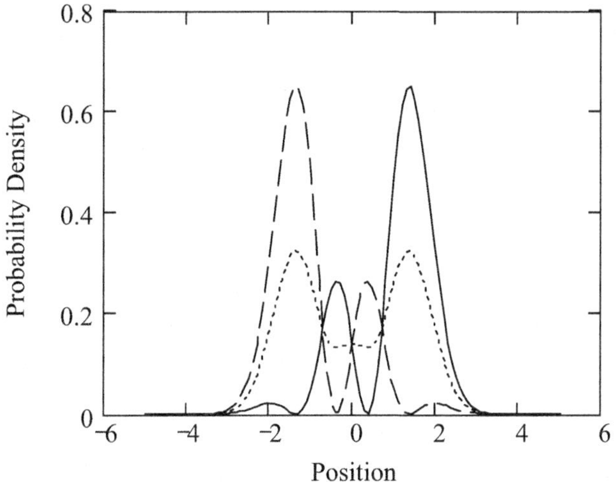

Fig. 3.2. Time evolution of an equal linear superposition of the first and second eigenstates of a harmonic oscillator. Here, the probability density is plotted against position in dimensionless units (i.e., distance units of $(\hbar / m\omega)^{1/2}$ where m is the particle's mass) at (i) the beginning of each period (solid line), (ii) one quarter and three quarters of the way through each period (dotted line), and (iii) halfway through each period (dashed line).

Now, we are beginning to get a correspondence with the classical case; at least, now we have the quantum mechanical oscillator actually oscillating and with the frequency we expect classically.

Coherent state

It turns out for the harmonic oscillator that the linear superpositions that correspond best to our classical understanding of harmonic oscillators are well known. They are known as *coherent states*. We do not look here in any mathematical detail at why these correspond to the classical behavior and from where this particular linear combination comes. For our purposes here, it is simply some specific superposition of all the harmonic oscillator eigenstates, one that happens – at least, for large total energies – to give a behavior that corresponds quite closely to what we expect a harmonic oscillator to do from our classical experience. We simply show some of the properties of this coherent state by example calculation; specifically, what happens to the probability density if we let this particular linear combination evolve in time.

The coherent state for a harmonic oscillator of frequency ω is, using the notation from Chapter 2

$$\Psi_N(\xi,t) = \sum_{n=0}^{\infty} c_{Nn} \exp\left[-i\left(n+\frac{1}{2}\right)\omega t\right]\psi_n(\xi) \tag{3.23}$$

where

$$c_{Nn} = \sqrt{\frac{N^n \exp(-N)}{n!}} \tag{3.24}$$

Incidentally, the reader may notice that

$$|c_{Nn}|^2 = \frac{N^n \exp(-N)}{n!} \tag{3.25}$$

is the well-known Poisson distribution from statistics, with mean N (and also standard deviation $\sqrt{N}$).[5] The parameter N is, in a way that will become clearer later, a measure of the amount of energy in the system overall – larger N corresponds to larger energy.

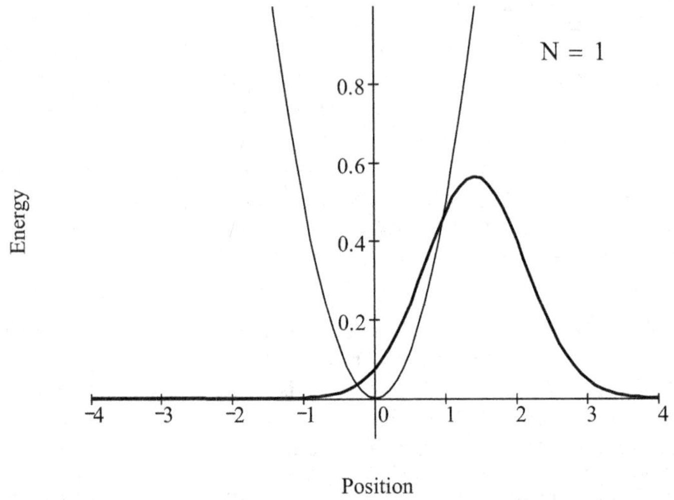

Fig. 3.3. Probability distribution (not to scale vertically) for a coherent state of a harmonic oscillator with $N = 1$ at time $t = 0$. Also shown is the parabolic potential energy in this case.

We can calculate the resulting probability density – for example, numerically – by simply including a finite but sufficient number of terms in the series in Eq. (3.23). Fig. 3.3 illustrates the probability density at time $t = 0$ for $N = 1$, and Fig. 3.4 and Fig. 3.5 illustrate this for $N = 10$ and $N = 100$, respectively. In each case, for subsequent times, the probability distribution essentially oscillates back and forth from one side of the potential to the other, with angular frequency ω, retaining essentially the same shape as it does so. For higher N, the spatial width of the probability distribution becomes a smaller fraction of the size of the overall oscillation. Hence, as we move to the classical scale of large oscillations, the probability distribution appears to be very localized (compared to the size of the oscillation), and so we can recover the classical idea of oscillation.

Now we have explicitly demonstrated that at least some particular quantum mechanical superposition state for a harmonic oscillator shows the kind of behavior we expected classically such an oscillator should have, with an object (here, an electron *wavepacket*)

---

[5] For those who are exasperated by their curiosity about where this coherent state superposition comes from and what it means, we give one real example. A laser under ideal conditions, operating in a single mode, will have the light field in a coherent state, and N here will be the average number of photons in the mode in question. The quantity that is oscillating in that case is the magnetic field amplitude rather than the position of some electron. Though an oscillating particle and an oscillating magnetic field seem very different at this point, the equations that govern both of these can turn out to be the same, a point to which we return in Chapter 15. It is also quite true, and easily checked experimentally, that the measured number of photons in such a light beam does have a Poissonian statistical distribution, in accord with the Poissonian distribution here.

oscillating back and forward, apparently sinusoidally, at the frequency we would expect classically. Quantum mechanics, though it works in a very different way, can reproduce the dynamics we have come to expect for objects in the classical world.

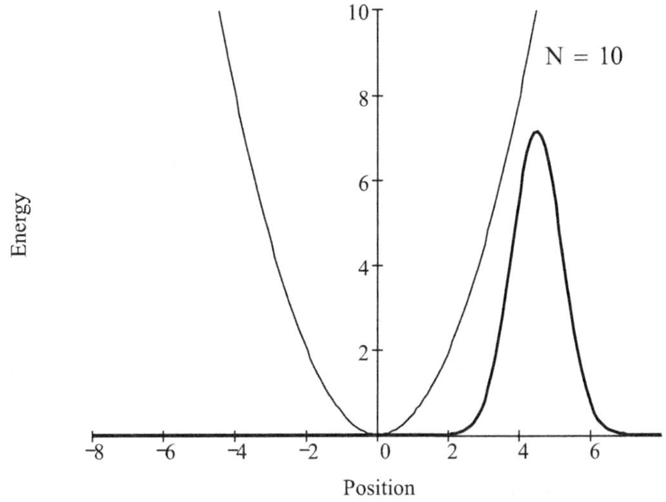

Fig. 3.4. Probability distribution (not to scale vertically) for a coherent state of a harmonic oscillator with $N = 10$ at time $t = 0$. Also shown is the parabolic potential energy in this case.

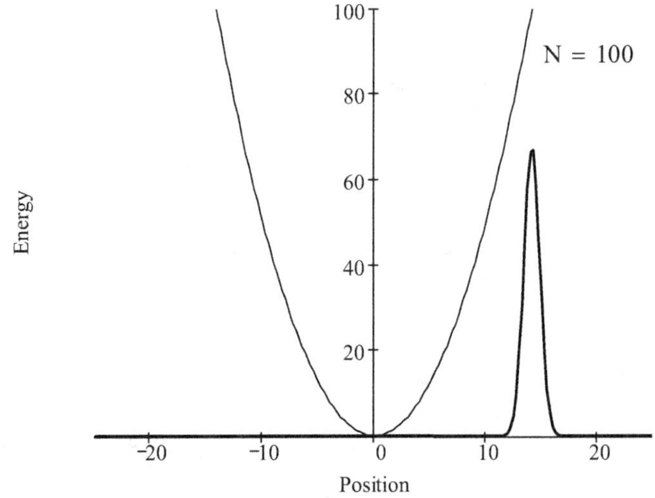

Fig. 3.5. Probability distribution (not to scale vertically) for a coherent state of a harmonic oscillator with $N = 100$ at time $t = 0$. Also shown is the parabolic potential energy in this case.

It is important to note, however, that, in general, a system in a linear superposition of multiple energy eigenstates does not execute a simple harmonic motion like this harmonic oscillator does. That harmonic motion is a special consequence of the fact that all the energy levels are equally spaced in the harmonic oscillator case. Fig. 3.6 shows the probability density at different times for the equal linear superposition of the first three bound levels of the finite well shown in Fig. 2.8. Because the energy separations between the levels are not in integer ratios, the resulting probability density does not repeat in time.

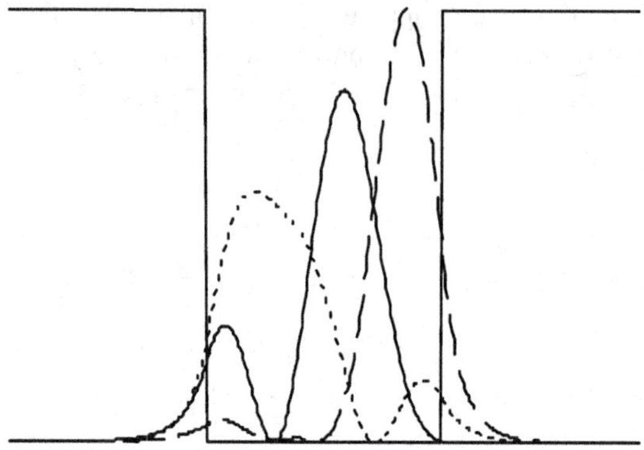

Fig. 3.6. Probability density at three different times for an equal linear superposition of the first three levels of a finite potential well as shown in Fig. 2.8. (i) $t = 0$ (solid line); (ii) $t = \pi/2$ (dotted line); (iii) $t = \pi$ (dashed line). The time units are $\hbar / E_1^\infty$, where E_1^∞ is the energy of the first level in a well of the same width but with infinitely high barriers.

Problems

3.6.1 An electron in an infinitely deep potential well of thickness 4 Å is placed in a linear superposition of the first and third states. What is the frequency of oscillation of the electron probability density?

3.6.2* A one-dimensional harmonic oscillator with a potential energy of the form $V(z) = az^2$ in its classical limit (i.e., in a coherent state with a large expectation value of the energy) would have a frequency of oscillation of f cycles per second.

(i) What is the energy separation between the first and second energy eigenstates in this harmonic oscillator?

(ii) If the potential energy was changed to $V(z) = 0.5az^2$, would the energy separation between these states increase or decrease?

3.6.3 Consider an electron in an infinitely deep one-dimensional potential well of width $L_z = 1$ nm. This electron is prepared in a state that is an equal linear superposition of the first three states. Presuming that oscillations in the squared modulus of the wavefunction amplitude lead to oscillations at the same frequency in the charge density, and that oscillations in the charge density at any point in the structure give rise to emitted electromagnetic radiation at the same frequency, list all of frequencies (in Hertz) of radiation that will be emitted by this system.

3.6.4 Consider an electron in a one-dimensional potential well of width L_z in the z direction, with infinitely high potential barriers on either side (i.e., at $z = 0$ and $z = L_z$). For simplicity, we assume the potential energy is zero inside the well. Suppose that at time $t = 0$, the electron is in an equal linear superposition of its lowest two energy eigenstates, with equal real amplitudes for those two components of the superposition.

(i) Write down the wavefunction at time $t = 0$ such that it is normalized.

(ii) Starting with the normalized wavefunction at time $t = 0$, write down an expression for the wavefunction valid for all times t.

(iii) Show explicitly whether this wavefunction is normalized for all such times t.

3.7 Time evolution of wavepackets

In looking at time evolution so far, we have explicitly considered the harmonic oscillator and found that, at least with a particular choice of linear superposition, we can recover the kind of behavior we associate with a classical harmonic oscillator. Another important example is the propagation of wavepackets to emulate the propagation of classical particles.

Imagine, for example, that the potential energy, V, is constant everywhere. For simplicity, we could take V to be zero. In such a situation, we know there is a solution of the time-independent Schrödinger equation possible for every energy E (greater than zero). In fact, there are two such solutions for every energy, a "right-propagating" one

$$\psi_{ER}(z) = \exp(ikz) \tag{3.26}$$

and a "left-propagating" one

$$\psi_{EL}(z) = \exp(-ikz) \tag{3.27}$$

where $k = \sqrt{2mE/\hbar^2}$ as usual.[6]

The corresponding solutions of the time-dependent Schrödinger equation are

$$\Psi_{ER}(z,t) = \exp\left[-i(\omega t - kz)\right] \tag{3.28}$$

and

$$\Psi_{EL}(z,t) = \exp\left[-i(\omega t + kz)\right] \tag{3.29}$$

where $\omega = E/\hbar$.

We want to understand the correspondence between the movement of such a "free" particle in the quantum mechanical description and in the classical one. We might at first be tempted to ask for the so-called *phase velocity* of the wave represented by either of Eqs. (3.28) or (3.29), which would be

$$v_p = \frac{\omega}{k} = \frac{E}{\hbar}\sqrt{\frac{\hbar^2}{2mE}} = \sqrt{\frac{E}{2m}} \tag{3.30}$$

That would lead to a relation between the particle's energy and this velocity v_p of $E = 2mv_p^2$, which does not correspond with the classical relation between kinetic energy and velocity of $E = (1/2)mv^2$. If we examine the $|\Psi_{EL}(z,t)|^2$ or $|\Psi_{ER}(z,t)|^2$ associated with either of these waves, we, however, find that they are uniform in space and time, and it is not meaningful to ask if there is any movement associated with them. To understand movement, we are going to have to construct a wavepacket – a linear superposition of waves that adds up to give a "packet" that is approximately localized in space at any given time. To understand what behavior we expect from such packets, we have to introduce the concept of group velocity.

Group velocity

Elementary wave theory, based on examining the behavior of linear superpositions of waves, says that the velocity of the center of a wavepacket or pulse is the *group velocity*

---

[6] We have deliberately not attempted to normalize these solutions. That is mathematically somewhat problematic for a uniform wave in an infinite space, though there is a solution to this problem (i.e., normalization to a delta function) to which we return in Chapter 5. This lack of normalization does not matter for the situations we examine here.

$$v_g = \frac{d\omega}{dk} \tag{3.31}$$

where ω is the frequency and k is the wavevector.

To understand this, consider a total wave consisting of a superposition of two waves, both propagating to the right, one at frequency $\omega + \delta\omega$ with a wavevector $k + \delta k$, and one at a frequency $\omega - \delta\omega$ with a wavevector $k - \delta k$. Then the total wave is

$$f(z,t) = \exp\left\{-i\left[(\omega+\delta\omega)t - (k+\delta k)z\right]\right\} + \exp\left\{-i\left[(\omega-\delta\omega)t - (k-\delta k)z\right]\right\} \tag{3.32}$$

We can rewrite this as

$$f(z,t) = 2\cos(\delta\omega t - \delta kz)\exp\left[-i(\omega t - kz)\right] \tag{3.33}$$

which can be viewed as an underlying wave $\exp[-i(\omega t - kz)]$ modulated by an envelope $\cos(\delta\omega t - \delta kz)$, as illustrated in Fig. 3.7.

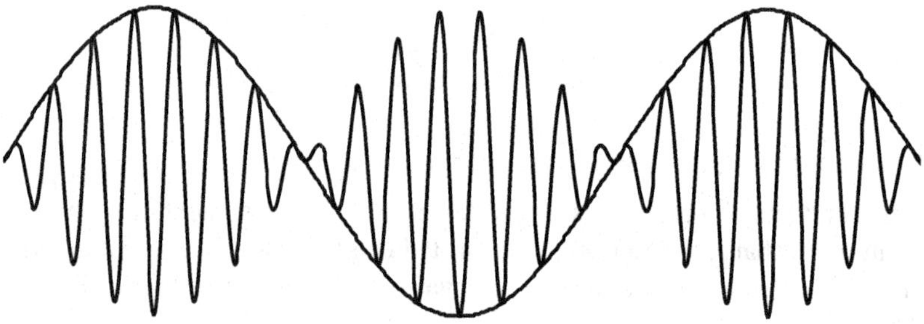

Fig. 3.7. Illustration of the "beating" between two waves in space (here, showing only the real part), leading to an envelope that propagates at the group velocity.

We know, in general, that some entity of the form $\cos(at - bz)$ is itself a wave that is moving at a velocity a/b. Hence, we see here that the envelope is moving at the group velocity

$$v_g = \frac{\delta\omega}{\delta k} \tag{3.34}$$

or, in the limit of very small $\delta\omega$ and δk, the expression Eq. (3.31).

We could extend this argument to some more complicated superposition of more than two waves, which would give some other form of the envelope. We would find similarly, as long as $d\omega/dk$ is approximately constant over the frequency or wavevector range we use to form this superposition, that this envelope would move at the group velocity of Eq. (3.31).

If the reader is new to the concept of group velocity, he or she might regard it as being some obscure phenomenon. After all, for waves such as light waves in free space or sound waves in ordinary air in a room, the velocity of the waves does not depend on the frequency – or, at least, not to any substantial degree – so $d\omega/dk = \omega/k$ and phase and group velocities are equal. There are many possible situations in which ω is not proportional to k, and we refer to this lack of proportionality as *dispersion*. For example, near to the center frequency of some optical absorption line, such as in an atomic vapor, the refractive index changes quite rapidly with frequency, the variation of refractive index with frequency (known as *material dispersion*) is not negligible, and the group and phase velocities are no longer the same. It is also true that in waveguides the different modes propagate with different velocities, so there is dispersion there also, this time from the geometry of the structure (an example of *structural dispersion*). In long optical fibers, for example, the effects of dispersion and of group velocity are far from negligible. Also, any structure whose physical properties, such as refractive index, change on a scale comparable to the wavelength will show structural dispersion.

In contrast to the normal optical or acoustic cases, for a particle such as an electron, the phase velocity and group velocity of quantum mechanical waves are almost never the same. For the simple free electron we have been considering, the frequency ω is not proportional to the wavevector magnitude k. For zero potential energy, the time-independent Schrödinger equation (e.g., in one dimension) tells us that, for any wave component, $\psi(z) \propto \exp(\pm ikz)$. In fact (for zero potential energy),

$$\frac{-\hbar^2}{2m_o}\frac{d^2\psi}{dz^2} = E\psi \tag{3.35}$$

that is,

$$E = \frac{\hbar^2 k^2}{2m_o} \tag{3.36}$$

So

$$\omega = \frac{E}{\hbar} = \frac{\hbar k^2}{2m_o}\text{, i.e., } \omega \propto k^2 \tag{3.37}$$

We see then that the propagation of the electron wave is always highly dispersive. Hence, for our present quantum mechanical case, we would have a velocity for a wavepacket consisting of a linear superposition of waves of energies near E

$$v_g = \frac{1}{dk/d\omega} = \frac{1}{\hbar dk/dE} = \sqrt{\frac{2E}{m}} \tag{3.38}$$

Fortunately, we now find that using this group velocity, we can write, using Eq. (3.37)

$$E = \frac{1}{2}mv_g^2 \tag{3.39}$$

Hence, the quantum mechanical description in terms of propagation as a superposition of waves does correspond to the same velocity as we would have expected from a classical particle of the same energy, as long as we correctly consider group velocity.[7]

Examples of motion of wavepackets

A real wavepacket that corresponds to a particle localized approximately in some region of space at a given time needs to be somewhat more complex than the simple sum of two

---

[7] Incidentally, situations can arise in quantum mechanics, e.g., with electrons in the valence band of semiconductors, in which the phase velocity and group velocity are even in opposite directions; this is not some bizarre and exceptional situation but is, in fact, a routine part of the operation of approximately half of all the transistors (the PMOS transistors that complement the NMOS transistors in complementary metal-oxide-semiconductor (CMOS) technology) in modern integrated circuits.

propagating waves. There are many forms that could give a wavepacket. A common one used as an example is a Gaussian wavepacket.

Freely propagating wavepacket

A Gaussian wavepacket propagating in the positive z direction in free space could be written as

$$\Psi_G(z,t) \propto \sum_k \exp\left[-\left(\frac{k-\bar{k}}{2\Delta k}\right)^2\right]\exp\left\{-i\left[\omega(k)t-kz\right]\right\} \tag{3.40}$$

where $\bar{k}$ is the value at the center of the distribution of k values, and the parameter Δk is a width parameter for the Gaussian function. The sum here runs over all possible values of k, which we presume to be evenly spaced.[8]

At this point, it is useful to introduce the idea of integration rather than summation when we are dealing with parameters that are continuous.[9] Instead of Eq. (3.40), we can choose to write

$$\Psi_G(z,t) \propto \int_k \exp\left[-\left(\frac{k-\bar{k}}{2\Delta k}\right)^2\right]\exp\left\{-i\left[\omega(k)t-kz\right]\right\}dk \tag{3.41}$$

Though this is now an integral rather than a sum, it is still just a linear combination of eigenfunctions of the time-dependent Schrödinger equation.

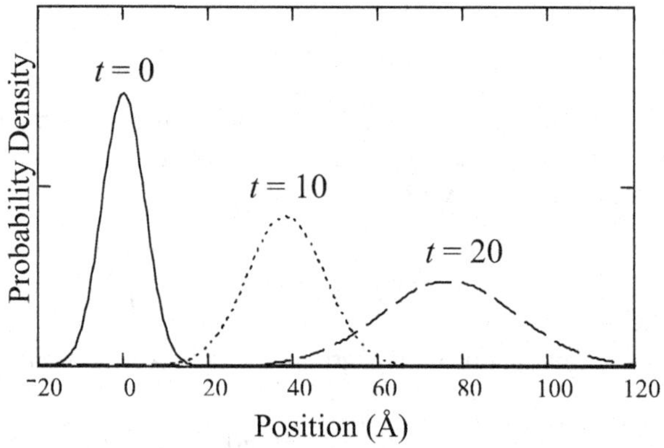

Fig. 3.8. Illustration of a wavepacket propagating in free space. The wavepacket is a Gaussian wavepacket in k-space, centered around a wavevector $\bar{k}$ = 0.5 Å^{-1}, which corresponds to an energy of ~ 0.953 eV, with a Gaussian width parameter Δk of 0.14 Å^{-1}. The units of time are $\hbar / e \simeq 0.66$ fs.

The motion of a Gaussian wavepacket for our free electron wave is illustrated in Fig. 3.8. We see first that the wavepacket does move to the right as we expect, with the center moving linearly in time (at the group velocity). We also see that the wavepacket gets broader in time. This increase in width is because the group velocity itself is not even the same for different wave components in the wavepacket, a dispersive phenomenon called *group-velocity*

[8] In practice for calculations, we work with a sum over an evenly spaced set of values of k, taking the spacings to be sufficiently close so that no substantial difference results in the calculations by choosing the spacing to be finer. In many real situations with finite space, the allowed k values are equally spaced (e.g., particles in large boxes or in crystals).

[9] See Section 5.3 for a more complete introduction to the transition from sums to integrals.

dispersion. Clearly, there will be group velocity dispersion if $d\omega / dk$ is not a constant over the region of wavevectors of interest (i.e., if $d^2\omega / dk^2 \neq 0$), which is certainly the case for our free electron, for which

$$\frac{dv_g}{dk} = \frac{d^2\omega}{dk^2} = \frac{\hbar}{m_o} \tag{3.42}$$

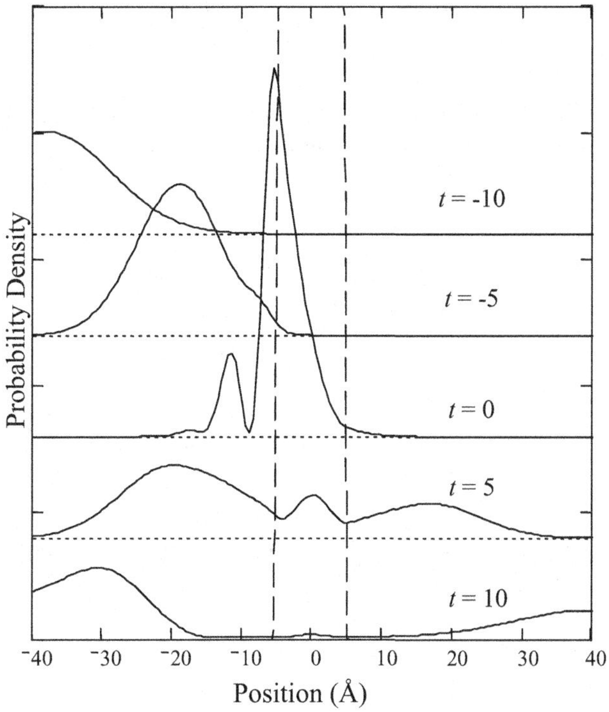

Fig. 3.9. Simulation of an electron wavepacket hitting a barrier. The barrier is 1 eV high and 10 Å thick and is centered around the zero position. The wavepacket is a Gaussian wavepacket in k-space, centered around a wavevector $\bar{k} = 0.5$ Å^{-1}, which corresponds to an energy of ~ 0.953 eV, with a Gaussian width parameter Δk of 0.14 Å^{-1}. The units of time are $\hbar / e \simeq 0.66$ fs.

Wavepacket hitting a barrier

A more complex example of wavepacket propagation is that of a wavepacket hitting a finite barrier. To analyze this problem, we can start by solving for the wavefunction of the time-independent Schrödinger equation in the presence of a finite barrier for the situation where there is no wave incident from the right.

We find that there are solutions for every energy. Each of these solutions contains a forward (right) propagating wave on the left of the barrier (as well as a reflected wave there), forward and backward waves within the barrier (which may be exponentially growing and decaying for energies below the top of the barrier), and a forward wave on the right. We can then form a linear superposition of these solutions with Gaussian weightings. The procedure is identical to that of Eq. (3.40) except that the waves are these more complicated solutions. The results are shown in Fig. 3.9.

Here, we see first the wavepacket approaching the barrier at times t = -10 and t = -5. Near $t = 0$, we see strong interference effects. These effects result from the incoming wave interfering with the wave reflected off the barrier and show "standing wave" phenomena. At times $t = 5$ and $t = 10$, we see a pulse propagating to the right on the right side of the barrier, corresponding to a pulse that has propagated through the barrier (in this case, mostly by tunneling),[10] as well as a reflected pulse propagating backward. It is important to emphasize here that all of these phenomena in the time-dependent behavior arise from the interference of the various energy eigenstates of the problem, with the time dependence itself arising from the change in phase in time between the various components as the $\exp(-iEt/\hbar)$ phase factors evolve in time. With the energy eigenstates already calculated for the problem, the time behavior arises simply from a linear sum of these different components with their time-dependent phase factors.

Simulating at a higher energy, such as the one corresponding to the first resonance above the barrier at an energy ~ 1.37 eV, shows similar kinds of behaviors but has a larger transmission and a smaller reflection.

Problems

3.7.1* Suppose that in some semiconductor material, the relation between the electron eigenenergies E and the effective wavevector k in the z direction is given by

$$E = -\frac{\hbar^2 k^2}{2b}$$

for some positive real constant b. If we consider a wavepacket made from waves with wavevectors in a small range around a given value of k, in what direction is the wavepacket moving

(i) for a positive value of k?

(ii) for a negative value of k?

3.7.2 In a crystalline solid material, the energy E of the electron energy eigenstates can be expressed as a function of an effective wavevector k_z for waves propagating in the z direction as shown in the following figure.

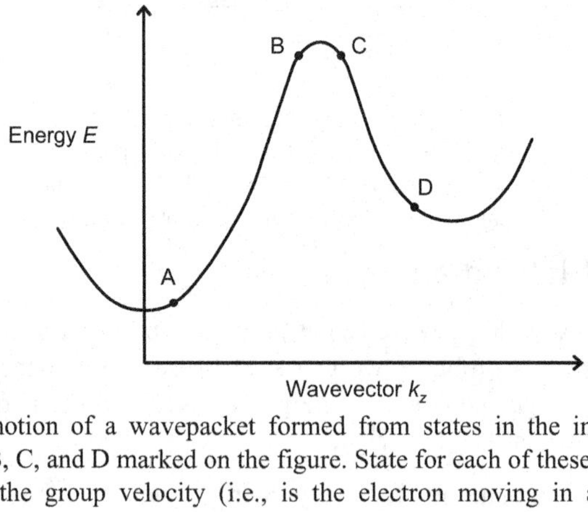

Consider now the motion of a wavepacket formed from states in the immediate vicinity of the particular points A, B, C, and D marked on the figure. State for each of these points

(i) the direction of the group velocity (i.e., is the electron moving in a positive or negative z direction?), and

(ii) the sign of the parameter, m_{eff}, known as the effective mass, where $\frac{1}{m_{eff}} = \frac{1}{\hbar^2}\frac{d^2E}{dk_z^2}$.

[10] The Gaussian distribution we have chosen does necessarily mean that there are some components of the wavepacket that have energies above the top of the barrier, and these likely do contribute to the small oscillatory component seen inside the barrier region in the simulation at $t = 5$ and $t = 10$. It is important to emphasize, however, that the penetration of the wavepacket through the barrier is substantially due to tunneling through the barrier in this case, not to propagation for energies above the barrier.

3.7.3 (This problem can be used as a substantial assignment.) (Notes: (i) See Section 2.11 before attempting this problem; (ii) some mathematical and/or numerical software will be required.) Consider the one-dimensional problem of an electron in a uniform electric field F, and in which there is an infinitely high potential barrier at $z = 0$ (i.e., the potential is infinite for all negative z) as sketched here, with the potential taken to be zero at $z = 0$ (or, at least, just to the right of $z = 0$).

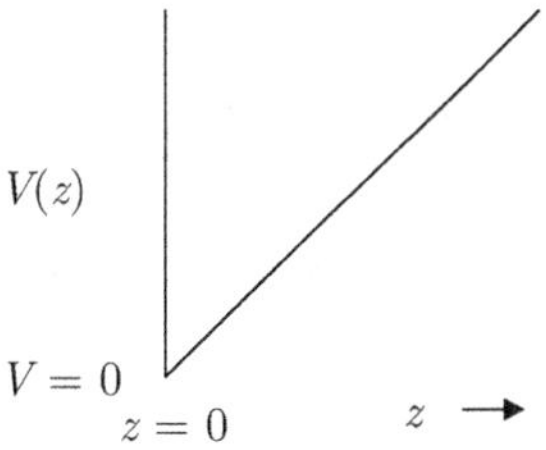

The field is in the +ve z direction and, consequently, the potential energy associated with the electron in that field is taken to be $V(z) = eFz$, for $z > 0$. (A positive field pushes the electron in the negative z direction, so the negative z direction is "downhill" for the electron; therefore, the electron has increasing potential energy in the positive z direction.)

(i) For an applied electric field of 10^{10} V/m, solve for the lowest twenty energy eigenvalues of the electron (in electron-Volts), and graph and state the explicit functional form of the first three energy eigenfunctions. (You need not normalize the eigenfunctions for this graph.)

(ii) Consider a wavepacket consisting of a linear superposition of such energy eigenfunctions. In particular, choose a wavepacket corresponding to an energy expectation value of about 17 eV, with a characteristic energy distribution width much less than 17 eV. (One convenient form is a wavepacket with a Gaussian distribution, in energy, of the weights of the [normalized] energy eigenfunctions.) Calculate and graph the time evolution of the resulting probability density, showing sufficiently many different times to show the characteristic behavior of this wavepacket. Also, describe in words the main features of the time evolution of this wavepacket. State explicitly whether the behavior of this system is exactly cyclic in time, and justify your answer.

(iii) Compare the quantum mechanical behavior to what one would expect classically, on the assumption that the electron would bounce perfectly off the barrier. In what ways is the behavior the same? In what ways is it different? (Note: Approximate numerical calculations and comparisons should be sufficient here.)

3.8 Quantum mechanical measurement and expectation values

When a normalized wavefunction is expanded in an orthonormal set, for example,

$$\Psi(\mathbf{r},t) = \sum_n c_n(t)\psi_n(\mathbf{r}) \tag{3.43}$$

then the normalization integral requires that

$$\int_{-\infty}^{\infty} |\Psi(\mathbf{r},t)|^2 \, d^3\mathbf{r} = \int_{-\infty}^{\infty} \left[\sum_n c_n^*(t)\psi_n^*(\mathbf{r})\right] \times \left[\sum_m c_m(t)\psi_m(\mathbf{r})\right] d^3\mathbf{r} = 1 \tag{3.44}$$

If we look at the integral over the sums, we see that because of the orthogonality of the basis functions, the only terms that survive after integration are for $n = m$, and because of the orthonormality of the basis functions, the result from any such term in the integration is simply $|c_n(t)|^2$. Hence, we have

$$\sum_n |c_n|^2 = 1 \tag{3.45}$$

When we make a measurement on a small system with a large measuring apparatus of some quantity such as energy, we find the following behavior, which is sometimes elevated to a postulate or hypothesis in quantum mechanics:[11]

> On measurement, the system collapses into an eigenstate of the quantity being measured, with probability
>
> $$P_n = |c_n|^2 \tag{3.46}$$
>
> where c_n is the expansion coefficient in the (orthonormal) eigenfunctions of the quantity being measured.

We see, though, that our conclusion Eq. (3.45) is certainly consistent with using the $|c_n|^2$ as probabilities, because they add up to 1.

Suppose now that we measure the energy of our system in such an experiment. We could repeat the experiment many times, and get a statistical distribution of results. Given the probabilities, we would find in the usual way that the average value of energy E that we would measure would be

$$\langle E \rangle = \sum_n E_n P_n = \sum_n E_n |c_n|^2 \tag{3.47}$$

where we are using the notation $\langle E \rangle$ to denote the average value of E, a quantity we call the "expectation value of E" in quantum mechanics. (In Eq. (3.47), the E_n are the energy eigenvalues.)

For example, for the coherent state discussed previously with parameter N, we have

$$\langle E \rangle = \sum_{n=0}^{\infty} E_n \frac{N^n \exp(-N)}{n!} = \hbar\omega \left[\sum_{n=0}^{\infty} n \frac{N^n \exp(-N)}{n!} \right] + \frac{1}{2}\hbar\omega = \left(N + \frac{1}{2} \right) \hbar\omega \tag{3.48}$$

We can show that having an energy $\approx N\hbar\omega$ for the large N implicit in a classical situation corresponds very well to our notions of energy, frequency, and oscillation amplitude in a classical oscillator. Note that N is not restricted to being an integer – it can take on any real value. Quite generally, the expectation value of the energy or any other quantum mechanically measurable quantity is not restricted to being one of a discrete set of values – it can take on any real value within the physically possible range of the quantity in question.

---

[11] There are many problems connected with this statement, especially if we try to consider it as anything other than an empirical observation for measurements by large systems on small ones. We postpone discussion of these difficulties until Chapter 19. The core difficulty is that it is not clear that we can explain the measurement process itself by quantum mechanics, at least in any way that is not apparently quite bizarre. Resolving these difficulties has, however, been a major activity in quantum mechanics up to the present day, and the modern pictures of these resolutions are much different from those originally envisaged in the early days of quantum mechanics. The branch of quantum mechanics that deals with these problems is known as *measurement theory*, and the core problem is known as the *measurement problem*.

Stern–Gerlach experiment

The measurement hypothesis is very strange, possibly even stranger than it already appears. It is important in quantum mechanics not to move past this point too lightly, and we should confront its true strangeness here. The Stern–Gerlach experiment is the first and the most classic experiment that shows just how strange it is.

An electron has another property, in addition to having a mass and a charge – it has a *spin*. We return to discuss spin in much greater depth in Chapter 12. For the moment, we can think of the electron spin as making the electron behave like a very small bar magnet. The strength of this magnet is exactly the same for all electrons.

If we pass a bar magnet through a region of space that has a uniform magnetic field, nothing will happen to the position of the bar because the north and south poles of the bar magnet are pulled with equal and opposite force.[12]

But, if the field is not uniform, then the pole that is in the stronger part of the field will experience more force, and the bar magnet will be deflected as it passes through the field. A magnet that would give such a nonuniform field is sketched in Fig. 3.10.[13] Because the shape of the north pole is "sharpened" so that one part of it is closer to the south pole than other parts, there will be greater magnetic field concentration near the "sharp" part, thereby giving us a nonuniform magnetic field.

Hence, we could imagine various situations with ordinary classical bar magnets fired, all at the same velocity, through this field horizontally. If the bar magnet started out oriented with the south pole facing up (Fig. 3.10(a)), there would be more force pulling the south pole up than is pulling the north pole down because the south pole is in a region of larger magnetic field. Hence, the magnet would be deflected upward, and would hit the screen at a point above the middle. If the bar magnet started out in the opposite orientation, it would instead be deflected downward (Fig. 3.10(b)). If the bar magnet started out oriented in the horizontal plane (in any direction), it would not be deflected at all (Fig. 3.10(c)). In any other orientation of the bar magnet, it would be deflected by some intermediate amount.

Now let us repeat this experiment again and again with bar magnets prepared in no particular orientation, each time marking where the bar magnet hits the screen. Then we would expect to see a fairly continuous line of points, as in Fig. 3.10(d). When we do this experiment with electrons,[14] however, we see that all the electrons land only at an upper position or at a lower position (Fig. 3.10(e)). This is very surprising. Remember that the electrons were not prepared in any way that always aligned their spins in the "up" or "down" directions. It also does not

---

[12] It might be that the bar magnet could be aligned by the field – i.e., twisted by the field to line up its south pole towards the north pole of the magnet producing the magnetic field – but it would not move up or down. In fact, we can presume in this experiment that there is negligible twisting, and it anyway could not lead to the actual result of the experiment. In the experiment, both possible extreme orientations appear to occur – the south pole of the bar magnet aligned to the north pole of the external magnet (which we could rationalize by the magnet twisting in the field), and also the north pole of the bar magnet aligned to the north pole of the external magnet (which could not be explained by the twisting in the field because the magnetic field will not twist north to align with north) – with apparently equal probability.

[13] The top end of the north pole is connected magnetically to the bottom end of the south pole, but we have omitted that connection for simplicity in the diagram.

[14] Stern and Gerlach actually did this experiment with silver atoms, which turn out to have the same spin properties as electrons, as far as this experiment is concerned.

matter if we change the direction of the magnets – the pattern of two dots just rotates as we rotate the external magnets.

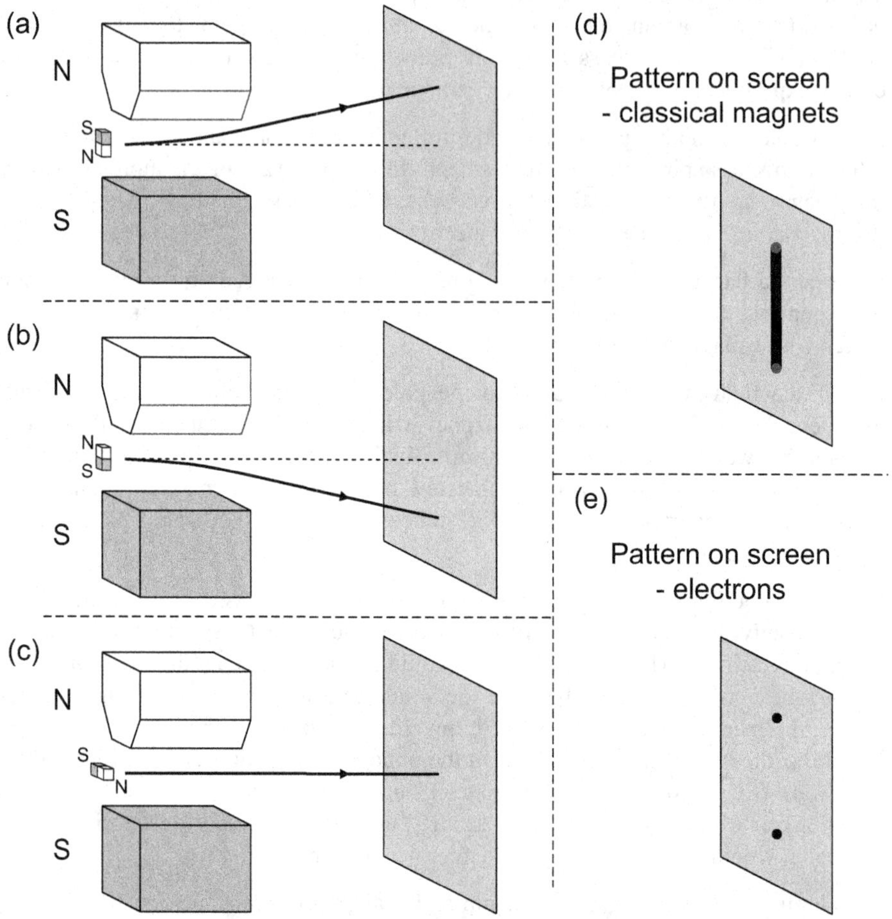

Fig. 3.10. Sketch of Stern–Gerlach experiment. A nonuniform magnetic field is imagined deflecting small bar magnets, entering horizontally from the left, in different orientations: (a) south pole up, (b) north pole up, and (c) both poles in a horizontal plane. For random initial orientations of the bar magnets, a solid line is expected for the accumulated arrival points on the screen, (d), for the arrival points of the magnets, but electrons in the same experiment, (e), show only two distinct arrival spots.

The quantum mechanical explanation is that this apparatus "measures" the vertical component of the electron spin. There are only two eigenfunctions of vertical electron spin – namely, "spin up" and "spin down." When we make a measurement, according to the measurement hypothesis, we "collapse" the state of the system into one of the eigen states of the quantity being measured (here, the vertical electron spin component) and, hence, we see only the two dots on the screen, corresponding respectively to spin-up and spin-down states. Again, if the reader thinks this is bizarre, then the reader is exactly correct – this measurement behavior is truly strange, and totally counter to our classical intuition.

Problem

3.8.1 As in Problem 3.6.4, consider an electron in a one-dimensional potential well of width L_z in the z direction, with infinitely high potential barriers on either side (i.e., at $z = 0$ and $z = L_z$). For simplicity, we assume the potential energy is zero inside the well. Suppose that, at time $t = 0$, the electron is in an equal linear superposition of its lowest two energy eigenstates, with equal real amplitudes for those two components of the superposition. What is the expectation value of the energy for an electron in this state? Does it depend on time t?

3.9 The Hamiltonian

Classical mechanics was put on a more sophisticated mathematical foundation in the late eighteenth and the nineteenth centuries. One particular important concept was the notion of the *Hamiltonian*, a function, usually of positions and momenta, essentially representing the total energy in the system. The Hamiltonian and some of the concepts surrounding it were important in the development of quantum mechanics and continue to be important in analyzing various topics. There are many formal links and correspondences between the Hamiltonian of classical mechanics and quantum mechanics. We avoid discussing these here except to make one definition. In quantum mechanics that can be analyzed by Schrödinger's equation, we can define the entity

$$\hat{H} = -\frac{\hbar^2}{2m}\nabla^2 + V(\mathbf{r},t) \tag{3.49}$$

so that we can write the time-dependent Schrödinger equation in the form

$$\hat{H}\Psi(\mathbf{r},t) = i\hbar\frac{\partial\Psi(\mathbf{r},t)}{\partial t} \tag{3.50}$$

or the time-independent Schrödinger equation as

$$\hat{H}\psi(\mathbf{r}) = E\psi(\mathbf{r}) \tag{3.51}$$

(where $\psi(\mathbf{r})$ is now restricted to being an eigenfunction with associated eigenenergy E).

We can regard this way of writing the equations as merely a shorthand notation, though this kind of approach turns out to be very useful in quantum mechanics. The entity $\hat{H}$ is not a number and it is not a function. It is instead an *operator*, just like the entity d/dz is a *spatial derivative operator*. We return to discuss operators in greater detail in Chapter 4. We use the notation with a "hat" above the letter here to distinguish operators from functions and numbers. The most general definition of an operator is an entity that turns one function into another.

The particular operator $\hat{H}$ is called the Hamiltonian operator because it is related to the total energy of the system. The idea of the Hamiltonian operator extends beyond the specific definition here that applies to single, nonmagnetic particles; in general, in nonrelativistic quantum mechanics, the Hamiltonian operator is the operator related to the total energy of the system.

3.10 Operators and expectation values

We can now show an important but simple relation involving the Hamiltonian operator, the wavefunction, and the expectation value of the energy. Consider the integral

$$I = \int \Psi^*(\mathbf{r},t)\hat{H}\Psi(\mathbf{r},t)d^3\mathbf{r} \tag{3.52}$$

where $\Psi(\mathbf{r},t)$ is the wavefunction of some system of interest. We can expand this wavefunction in energy eigenstates, as in Eq. (3.43). We know that with $\psi_n(\mathbf{r})$ as the energy eigenstates (of the time-independent Schrödinger equation)

$$\begin{aligned}\hat{H}\Psi(\mathbf{r},t) &= \left[-\frac{\hbar^2}{2m}\nabla^2 + V(\mathbf{r},t)\right]\Psi(\mathbf{r},t) = \left[-\frac{\hbar^2}{2m}\nabla^2 + V(\mathbf{r},t)\right]\sum_n c_n(t)\psi_n(\mathbf{r}) \\ &= \sum_n c_n(t)E_n\psi_n(\mathbf{r})\end{aligned} \tag{3.53}$$

and so

$$\int \Psi^*(\mathbf{r},t)\hat{H}\Psi(\mathbf{r},t)d^3\mathbf{r} = \int_{-\infty}^{\infty}\left[\sum_m c_m^*(t)\psi_m^*(\mathbf{r})\right]\times\left[\sum_n c_n(t)E_n\psi_n(\mathbf{r})\right]d^3\mathbf{r} \tag{3.54}$$

Given the orthonormality of the $\psi_n(\mathbf{r})$, we have

$$\int \Psi^*(\mathbf{r},t)\hat{H}\Psi(\mathbf{r},t)d^3\mathbf{r} = \sum_n E_n\left|c_n\right|^2 \tag{3.55}$$

But, comparing to the result (3.47), we therefore have

$$\langle E\rangle = \int \Psi^*(\mathbf{r},t)\hat{H}\Psi(\mathbf{r},t)d^3\mathbf{r} \tag{3.56}$$

This kind of relation involving the operator (here, $\hat{H}$), the quantum mechanical state (here, $\Psi(\mathbf{r},t)$), and the expected value of the quantity associated with the operator (here, E) is quite general in quantum mechanics.

The reader might ask, "But if we already knew how to calculate $\langle E\rangle$ from Eq. (3.47), what is the benefit of this new relation, Eq. (3.56)?" One major benefit is that we do not have to solve for the eigenfunctions of the operator (here, $\hat{H}$) to calculate the result. We used the decomposition into eigenfunctions to prove the result Eq. (3.56), but we do not have to do that decomposition to evaluate $\langle E\rangle$ from Eq. (3.56). All we need is the quantum mechanical state (here the wavefunction $\Psi(\mathbf{r},t)$), and the operator associated with the quantity $\langle E\rangle$ (here, $\hat{H}$).

Problem

3.10.1 Using a system of units in which the electron mass $m = 1$ and $\hbar = 1$, an electron in a potential $V(z) = z^2/2$ has a wavefunction at a given instant in time

$$\psi(z) = \frac{1}{\sqrt{2\sqrt{\pi}}}\left(1+\sqrt{2}\,z\right)\exp\left(-z^2/2\right)$$

What is the expectation value of the energy for the particle in this state? (You may use numerical integration to get a result, if you wish.)

3.11 Time evolution and the Hamiltonian operator

Let us look at Schrödinger's time-dependent equation in the form shown in Eq. (3.50) and rewrite it slightly as

$$\frac{\partial\Psi(\mathbf{r},t)}{\partial t} = -\frac{i\hat{H}}{\hbar}\Psi(\mathbf{r},t) \tag{3.57}$$

If we presume that $\hat{H}$ does not depend on time (i.e., the potential $V(\mathbf{r})$ is constant in time), it is tempting to wonder if it is "legal" and meaningful to integrate this equation directly to obtain

$$\Psi(\mathbf{r},t_1)=\exp\left(-\frac{i\hat{H}(t_1-t_0)}{\hbar}\right)\Psi(\mathbf{r},t_0) \tag{3.58}$$

Certainly, if $\hat{H}$ were replaced by a constant number (a rather trivial case of an operator), we could perform such an integration. If it were legal and meaningful to do this for an actual time-independent Hamiltonian, we would have an operator ($\exp[-i\hat{H}(t_1-t_0)/\hbar]$) that, in one operation, gave us the state of the system at time t_1 directly from its state at time t_0.

It is certainly not obvious that such an expression Eq. (3.58) is meaningful. What do we mean by the exponential of an operator? We can show, however, that provided we are careful to define what we mean here, this expression is meaningful. Understanding this expression is a useful exercise, introducing some of the concepts associated with operators that we will examine in more detail later. To think about this, first we note that because $\hat{H}$ is a linear operator, for any number a

$$\hat{H}\left[a\Psi(\mathbf{r},t)\right]=a\hat{H}\Psi(\mathbf{r},t) \tag{3.59}$$

We say that the operator $\hat{H}$ "commutes" with the scalar quantity (i.e., the number) a. Because this relation holds for any function $\Psi(\mathbf{r},t)$, we can write, as a shorthand

$$\hat{H}a=a\hat{H} \tag{3.60}$$

(Note that any time we have such an equation relating the operators themselves on either side, we are implicitly saying that this relation holds for these operators operating on any function in the space; that is, the relation

$$\hat{A}=\hat{B} \tag{3.61}$$

for any two operators $\hat{A}$ and $\hat{B}$ is really a shorthand for the statement

$$\hat{A}\Psi=\hat{B}\Psi \tag{3.62}$$

where Ψ is any arbitrary function in the space in question.)

Next, we have to define what we mean by an operator raised to a power. By $\hat{H}^2$, we mean

$$\hat{H}^2\Psi(\mathbf{r},t)=\hat{H}\left[\hat{H}\Psi(\mathbf{r},t)\right] \tag{3.63}$$

(i.e., $\hat{H}$ operating on the function that is itself the result of operating on $\Psi(\mathbf{r},t)$ with $\hat{H}$). Specifically, for example, for the energy eigenfunction $\psi_n(\mathbf{r})$

$$\hat{H}^2\psi_n(\mathbf{r})=\hat{H}\left[\hat{H}\psi_n(\mathbf{r})\right]=\hat{H}\left[E_n\psi_n(\mathbf{r})\right]=E_n\hat{H}\psi_n(\mathbf{r})=E_n^2\psi_n(\mathbf{r}) \tag{3.64}$$

We can proceed by an inductive process to define the meaning of all higher powers of an operator; that is,

$$\hat{H}^{m+1}\equiv\hat{H}\left[\hat{H}^m\right] \tag{3.65}$$

which will give for the case of an energy eigenfunction

$$\hat{H}^m\psi_n(\mathbf{r})=E_n^m\psi_n(\mathbf{r}) \tag{3.66}$$

Now let us look at the time evolution of some wavefunction $\Psi(\mathbf{r},t)$ between times t_0 and t_1. Suppose the wavefunction at time t_0 is $\psi(\mathbf{r})$, which we can expand in the energy eigenfunctions $\psi_n(\mathbf{r})$ as

$$\psi(\mathbf{r}) = \sum_n a_n \psi_n(\mathbf{r}) \tag{3.67}$$

Then we know (see, e.g., Eq. (3.17)) that

$$\Psi(\mathbf{r},t_1) = \sum_n a_n \exp\left[-\frac{iE_n(t_1-t_0)}{\hbar}\right]\psi_n(\mathbf{r}) \tag{3.68}$$

We can, if we wish, write the exponential factors as power series, noting that

$$\exp(x) = 1 + x + \frac{x^2}{2!} + \frac{x^3}{3!} + \cdots \tag{3.69}$$

so Eq. (3.68) can be written as

$$\Psi(\mathbf{r},t_1) = \sum_n a_n \left[1 + \left(-\frac{iE_n(t_1-t_0)}{\hbar}\right) + \frac{1}{2!}\left(-\frac{iE_n(t_1-t_0)}{\hbar}\right)^2 + \cdots\right]\psi_n(\mathbf{r}) \tag{3.70}$$

Because of Eq. (3.66), everywhere we have $E_n^m \psi_n(\mathbf{r})$, we can substitute $\hat{H}^m \psi_n(\mathbf{r})$, and so we have

$$\Psi(\mathbf{r},t_1) = \sum_n a_n \left[1 + \left(-\frac{i\hat{H}(t_1-t_0)}{\hbar}\right) + \frac{1}{2!}\left(-\frac{i\hat{H}(t_1-t_0)}{\hbar}\right)^2 + \cdots\right]\psi_n(\mathbf{r}) \tag{3.71}$$

Because the operator $\hat{H}$ and all its powers as defined here commute with scalar quantities (numbers), we can rewrite Eq. (3.71) as

$$\begin{aligned}\Psi(\mathbf{r},t_1) &= \left[1 + \left(-\frac{i\hat{H}(t_1-t_0)}{\hbar}\right) + \frac{1}{2!}\left(-\frac{i\hat{H}(t_1-t_0)}{\hbar}\right)^2 + \cdots\right]\sum_n a_n \psi_n(\mathbf{r}) \\ &= \left[1 + \left(-\frac{i\hat{H}(t_1-t_0)}{\hbar}\right) + \frac{1}{2!}\left(-\frac{i\hat{H}(t_1-t_0)}{\hbar}\right)^2 + \cdots\right]\Psi(\mathbf{r},t_0)\end{aligned} \tag{3.72}$$

So, provided we define the exponential of the operator in terms of a power series – that is,

$$\exp\left[-\frac{i\hat{H}(t_1-t_0)}{\hbar}\right] \equiv \left[1 + \left(-\frac{i\hat{H}(t_1-t_0)}{\hbar}\right) + \frac{1}{2!}\left(-\frac{i\hat{H}(t_1-t_0)}{\hbar}\right)^2 + \cdots\right] \tag{3.73}$$

with powers of operators as given by (3.63) and (3.65) – we can indeed write Eq. (3.58). Hence, we have established that there is a well-defined operator that, given the quantum mechanical wavefunction or state at time t_0, will tell us what the state is at a time t_1. The importance here is not so much that we have derived the form of the operator – this is not likely to be something that we use often for actual numerical calculations – but rather that we have deduced that there is such an operator, and we have understood how we can approach forming new operators that are "functions" of other operators.

The particular operator we have derived here is valid for situations where the Hamiltonian is not explicitly dependent on time (which usually means that the potential V does not depend on time). It is possible to derive operators that deal with more complex situations, though we do not consider those here.[15]

Problem

3.11.1 If the eigenenergies of the Hamiltonian $\hat{H}$ are E_n and the eigenfunctions are $\psi_n(\mathbf{r})$, what are the eigenvalues and eigenfunctions of the operator $\hat{H}^2 - \hat{H}$?

3.12 Momentum and position operators

Thus far, the only operator we have considered has been the Hamiltonian $\hat{H}$ associated with the energy E. In quantum mechanics, we can construct operators associated with many other measurable quantities. For the momentum operator, which we write as $\hat{p}$, we postulate the operator

$$\hat{p} \equiv -i\hbar\nabla \tag{3.74}$$

with

$$\nabla \equiv \mathbf{i}\frac{\partial}{\partial x} + \mathbf{j}\frac{\partial}{\partial y} + \mathbf{k}\frac{\partial}{\partial z} \tag{3.75}$$

where **i**, **j**, and **k** are unit vectors in the x, y, and z directions.

With this postulated form, Eq. (3.74), we find that

$$\frac{\hat{p}^2}{2m} \equiv -\frac{\hbar^2}{2m}\nabla^2 \tag{3.76}$$

and we have a correspondence between the classical notion of the energy E as

$$E = \frac{p^2}{2m} + V \tag{3.77}$$

and the corresponding Hamiltonian operator of the Schrödinger equation

$$\hat{H} = -\frac{\hbar^2}{2m}\nabla^2 + V = \frac{\hat{p}^2}{2m} + V \tag{3.78}$$

The plane waves $\exp(i\mathbf{k}\cdot\mathbf{r})$ are the eigenfunctions of the operator $\hat{p}$ with eigenvalues $\hbar\mathbf{k}$, because

$$\hat{p}\exp(i\mathbf{k}\cdot\mathbf{r}) = \hbar\mathbf{k}\exp(i\mathbf{k}\cdot\mathbf{r}) \tag{3.79}$$

(The eigenvalues, in this case, are vectors, which is quite acceptable mathematically.) We can therefore make the identification for these eigenstates that the momentum is

$$\mathbf{p} = \hbar\mathbf{k} \tag{3.80}$$

[15] See, e.g., J. J. Sakurai, *Modern Quantum Mechanics (Revised Edition)* (Addison Wesley, 1994), pp. 72–73, for a discussion of such operators for time-dependent Hamiltonians.

Note that the **p** in Eq. (3.80) is a vector, with three components with scalar values, not an operator.

For the position operator, the postulated operator is almost trivial when we are working with functions of position. It is simply the position vector, **r**, itself.[16] At least, when we are working in a representation that is in terms of position, we therefore typically do not write $\hat{\mathbf{r}}$, though rigorously perhaps we should. The operator for the z-component of position would, for example, also simply be z itself.

Problems

3.12.1* Suppose that a particle of mass m is in a one-dimensional potential well with infinitely high barriers and thickness L_z in the z direction. Suppose also that it is in a state that is an equal linear superposition of the first and second states of the well.

(Note that $\int_0^\pi \sin(\theta)\cos(2\theta)d\theta = -2/3$, $\int_0^\pi \sin(2\theta)\cos(\theta)d\theta = 4/3$.)

(i) At what frequency is this system oscillating in time?

(ii) Evaluate the expectation value of the z component of the momentum (i.e., $\langle p_z(t)\rangle$) as a function of time.

(iii) Suppose instead that the particle is in an equal linear superposition of the first and third states of the well. Deduce what now is $\langle p_z(t)\rangle$. (Hint: This should not need much additional algebra and may involve consideration of the consequences of odd functions in integrals.)

3.12.2 We perform an experiment in which we prepare a particle in a given quantum mechanical state and then measure the momentum of the particle. We repeat this experiment many times and obtain an average result for the momentum $\langle \mathbf{p}\rangle$ (the expectation value of the momentum). For each of the following quantum mechanical states, give the (vector) value of $\langle \mathbf{p}\rangle$ or, if appropriate, $\langle \mathbf{p}(t)\rangle$, where t is the time after the preparation of the state.

(i) $\psi(\mathbf{r}) \propto \exp(i\mathbf{k}\cdot\mathbf{r})$

(ii) a particle of mass m in an infinitely deep potential well of thickness L_z (here you need only give $\langle p_z\rangle$ or $\langle p_z(t)\rangle$, the z-component of the value, where z is the direction perpendicular to the walls of the well), in the lowest energy state.

(iii) Offer an explanation for the result of part (ii) based on the result from part (i).

3.12.3 Consider the equal linear superposition of the first two states of an infinitely deep potential well.

(i) Show by explicit substitution that this state is a solution of the time-dependent Schrödinger equation for a particle in such a well.

(ii) For this state, what is the expectation value of the position?

(Note: Take the expectation value of position as being given by the expression $\langle z\rangle = \int \Psi^* z \Psi dz$.)

3.12.4 In an experiment, an electron is prepared in the state described by the wavefunction $\Psi(\mathbf{r},t)$, where t is the time from the start of each run of the experiment. In this experiment, the momentum is measured at a specific time t_o after the start of the experiment. This experiment is then repeated multiple times. Give an expression, in terms of differential operators, fundamental constants, and this wavefunction, for the average value of momentum that would be measured in this set of experiments.

3.12.5 Suppose an electron is sitting in the lowest energy state of some potential, such as a one-dimensional potential well with finite potential depth (i.e., finite height of the potential barriers on either side). Suppose next we measure the momentum of the electron. What will have happened to the expectation value of the energy? That is, if we now measure the energy of the electron again, what will have happened to the average value of the result we now get? Has it increased, decreased, or stayed the same compared to what it was originally? Explain your answer.

---

[16] We return in Chapter 5 to consider the eigenfunctions of position. These are actually Dirac delta functions when expressed in terms of position.

3.13 Uncertainty principle

One of the most perplexing aspects of quantum mechanics from a classical viewpoint is the uncertainty principle. (There are actually several uncertainty principles with similar character.) The most commonly quoted form is to say that we cannot simultaneously know both the position and the momentum of a particle. This runs quite counter to our classical notion. Classical mechanics implicitly assumes that knowing both position and momentum is possible. In a practical sense for large objects, it is possible to know both at once; but more fundamentally according to quantum mechanics, it is not – a fact that has profound philosophical implications for any discussion of, for example, determinism.

We postpone a more formal discussion of uncertainty principles. Here, we simply illustrate the position-momentum uncertainty principle by example. We defined a Gaussian wavepacket in Eq. (3.41) as an integral over a set of waves with Gaussian weightings on their amplitudes about some central k value, $\bar{k}$. Indeed, we could rewrite Eq. (3.41) at time $t = 0$ as

$$\Psi(z,0) = \int_k \Psi_k(k)\exp(ikz)\,dk \tag{3.81}$$

where

$$\Psi_k(k) \propto \exp\left[-\left(\frac{k-\bar{k}}{2\Delta k}\right)^2\right] \tag{3.82}$$

We can regard $\Psi_k(k)$ as being the representation of the wavefunction in "k space." Specifically, we can regard $|\Psi_k(k)|^2$ as being the probability P_k (or, more strictly, the probability density) that if we measured the momentum of the particle (actually, in this case, the z component of the momentum), it was found to have value $\hbar k$. This probability would have a statistical distribution

$$P_k = |\Psi_k(k)|^2 \propto \exp\left[-\frac{(k-\bar{k})^2}{2(\Delta k)^2}\right] \tag{3.83}$$

The Gaussian in Eq. (3.83) corresponds to the standard statistical expression for a Gaussian probability distribution, with standard deviation Δk.

We also note that Eq. (3.81) is simply the Fourier transform of $\Psi_k(k)$. The result of this transform is well known. The Fourier transform of a Gaussian is a Gaussian.[17] Explicitly, if we were to formally perform the integral Eq. (3.81), we would find

$$\Psi(z,0) \propto \exp\left[-(\Delta k)^2 z^2\right] \tag{3.84}$$

Now considering the probability (or again, strictly, the probability density) of finding the particle at point z at time $t = 0$ as $|\Psi(z,0)|^2$, we have

$$|\Psi(z,0)|^2 \propto \exp\left[-2(\Delta k)^2 z^2\right] \equiv \exp\left[-\frac{z^2}{2(\Delta z)^2}\right] \tag{3.85}$$

[17] This is a very special (and very useful) property of Gaussian functions.

where we have chosen to define the quantity Δz so that it corresponds to the standard deviation of the probability distribution in real space. From Eq. (3.85), we find the relation

$$\Delta k \Delta z = \frac{1}{2} \tag{3.86}$$

or, with momentum (here, strictly the z component of momentum) $p = \hbar k$

$$\Delta p \Delta z = \frac{\hbar}{2} \tag{3.87}$$

where $\Delta p = \hbar \Delta k$.

We saw when we propagated the wavepacket in time that it got wider – that is, Δz became larger, though Δk had not changed; the same Gaussian distribution of magnitudes of amplitudes of k components remained, though their relative phases had now changed with time, leading to a broadening of the wavepacket, so we can also certainly have $\Delta p \Delta z > \hbar / 2$. It turns out that the Gaussian distribution and its Fourier transform have the minimum product $\Delta k \Delta z$ of any distribution (though we do not prove that here), and so we find the uncertainty principle

$$\Delta p \Delta z \geq \hbar / 2 \tag{3.88}$$

Though demonstrated here only for a specific example, this uncertainty principle turns out to be quite general. It expresses the nonclassical notion that if we know the position of a particle very accurately, we cannot know its momentum very accurately. For objects of everyday human scale and mass, this uncertainty is so small that it falls below our other measurement accuracies; but for very light objects such as electrons, this uncertainty is not negligible.

It is important to emphasize, too, that the modern understanding of quantum mechanics appears to say that it is not merely that we cannot simultaneously measure these two quantities or that quantum mechanics is only some incomplete statistical theory that does not tell us both momentum and position simultaneously even though they both exist to arbitrary accuracy. Quantum mechanics is apparently a complete theory, not merely a statistical "image" of some underlying deterministic theory; a particle simply does not have simultaneously both a well-defined position and a well-defined momentum, in this view.[18]

Uncertainty principles are well known to those who have done Fourier analysis of temporal functions. There one finds that one cannot simultaneously have both a well-defined frequency and a well-defined time for a signal. If a signal is a short pulse, it is necessarily composed of a range of frequencies. The shorter the pulse is, the larger the range of frequencies that must be used to construct it; that is,

$$\Delta \omega \Delta t \geq \frac{1}{2} \tag{3.89}$$

The mathematics of this well-known Fourier analysis result is identical to that for the uncertainty principle discussed previously.

Another common example of an uncertainty principle is found in the diffraction angle of a beam, propagating, for example in the x direction, emerging from a finite slit with some width in the z direction. Smaller slits correspond to more tightly defined position in the z direction

[18] This point is not absolutely settled, however. See the discussion in Chapter 19.

and give rise to larger diffraction angles. The diffraction angle corresponds to the uncertainty in the z component of the wavevector. If we think of light propagation as being due to momentum of photons, diffraction is understood as the uncertainty principle giving momentum uncertainty in the z direction for this example. The diffraction of an electron beam from a single slit shows exactly the same diffraction phenomenon; we can regard the fact that the beam gets wider as it propagates as being because it is constructed from a range of beams of different momenta, each going in somewhat different directions. The propagation of Gaussian light beams, commonly encountered with laser beams, corresponds exactly to this analysis if we define the beams with the correct parameters that correspond to the statistical definition of Gaussian distributions for the beam intensity.

We can see, therefore, that though the uncertainty principle seems at first a very strange notion, consequences of this kind of relation may actually be quite well known to us from classical situations with waves and time-varying functions. The unusual aspect is that it applies to properties of material particles also.

Problem

3.13.1 Suppose we have a 1g mass, whose position we know to a precision of 1 Å.
 (i) What would be the minimum uncertainty in its velocity in a given direction?
 (ii) What would be the corresponding uncertainty in velocity if the particle was an electron instead of a 1g mass?

3.14 Particle current

Additional prerequisite: Understanding of the divergence of a vector (see Appendix C).

Our classical intuition leads us to expect that particles with kinetic energy must be moving and, hence, there will be particle currents or current densities (i.e., particles crossing unit area per unit time). We have, however, apparently deduced that there are stationary states (energy eigenstates) in quantum mechanics where the particle may have energy that exceeds the potential energy, and we are now expecting that there may well be no current associated with such energy eigenstates. We need a meaningful way of calculating particle current in quantum mechanics so that we can check these notions.

In general, if we are to conserve particles, we expect that we have a relation of the form

$$\frac{\partial s}{\partial t} = -\nabla.\mathbf{j}_p \tag{3.90}$$

where s is the particle density and $\mathbf{j}_p$ is the particle current density (not the electrical current in this case, though if s was the charge density, this would be the form of the relation for conservation of charge with a charge current density). The reader may remember that this kind of vector calculus relation is justified by considering a small box and looking at the difference of particle currents in and out of the opposite faces of the box.

In our quantum mechanical case, the particle density for a particle in a state with wavefunction $\Psi(\mathbf{r},t)$ is $|\Psi(\mathbf{r},t)|^2$, so we are looking for a relation of the form of Eq. (3.90) but with $|\Psi(\mathbf{r},t)|^2$ instead of s. To do this requires a little algebra and a clever substitution.

We know that

$$\frac{\partial \Psi(\mathbf{r},t)}{\partial t} = \frac{1}{i\hbar}\hat{H}\Psi(\mathbf{r},t) \tag{3.91}$$

which is simply Schrödinger's equation. We can also take the complex conjugate of both sides; that is,

$$\frac{\partial \Psi^*(\mathbf{r},t)}{\partial t} = -\frac{1}{i\hbar}\hat{H}^*\Psi^*(\mathbf{r},t) \tag{3.92}$$

Hence, we can write

$$\frac{\partial}{\partial t}\left[\Psi^*\Psi\right] + \frac{i}{\hbar}\left(\Psi^*\hat{H}\Psi - \Psi\hat{H}^*\Psi^*\right) = 0 \tag{3.93}$$

If the potential is real (it is hard to imagine how it could not be) and does not depend on time, then we can rewrite Eq. (3.93) as

$$\frac{\partial}{\partial t}\left[\Psi^*\Psi\right] - \frac{i\hbar}{2m}\left(\Psi^*\nabla^2\Psi - \Psi\nabla^2\Psi^*\right) = 0 \tag{3.94}$$

Now, we use an algebraic "trick" to rearrange this; that is,

$$\begin{aligned}\Psi\nabla^2\Psi^* - \Psi^*\nabla^2\Psi &= \Psi\nabla^2\Psi^* + \nabla\Psi\nabla\Psi^* - \nabla\Psi\nabla\Psi^* - \Psi^*\nabla^2\Psi \\ &= \nabla\cdot\left(\Psi\nabla\Psi^* - \Psi^*\nabla\Psi\right)\end{aligned} \tag{3.95}$$

Hence, we have

$$\frac{\partial\left(\Psi^*\Psi\right)}{\partial t} = -\frac{i\hbar}{2m}\nabla\cdot\left(\Psi\nabla\Psi^* - \Psi^*\nabla\Psi\right) \tag{3.96}$$

which is an equation of the form of Eq. (3.90) if we identify

$$\mathbf{j}_p = \frac{i\hbar}{2m}\left(\Psi\nabla\Psi^* - \Psi^*\nabla\Psi\right) \tag{3.97}$$

as the particle current. Hence, we have found an expression for particle currents for situations where the potential does not depend on time.

Now we can use this to examine stationary states (i.e., energy eigenstates) to see what particle currents can be associated with them. The expression Eq. (3.97) for particle current applies regardless of whether the system is in an energy eigenstate. Explicitly presuming we are in the nth energy eigenstate, we have

$$\mathbf{j}_{pn}(\mathbf{r},t) = \frac{i\hbar}{2m}\left(\Psi_n(\mathbf{r},t)\nabla\Psi_n^*(\mathbf{r},t) - \Psi_n^*(\mathbf{r},t)\nabla\Psi_n(\mathbf{r},t)\right) \tag{3.98}$$

We can write $\Psi_n(\mathbf{r},t)$ explicitly as

$$\Psi_n(\mathbf{r},t) = \exp\left(-i\frac{E_n}{\hbar}t\right)\psi_n(\mathbf{r}) \tag{3.99}$$

The gradient operator ∇ has no effect on the exponential time factor, so the time factors in each term can be factored to the front of the expression and anyway multiply to unity because of the complex conjugation; that is,

$$\mathbf{j}_{pn}(\mathbf{r},t) = \frac{i\hbar}{2m}\exp\left(-i\frac{E_n}{\hbar}t\right)\exp\left(i\frac{E_n}{\hbar}t\right)\left(\psi_n(\mathbf{r})\nabla\psi_n^*(\mathbf{r}) - \psi_n^*(\mathbf{r})\nabla\psi_n(\mathbf{r})\right)$$
$$= \frac{i\hbar}{2m}\left(\psi_n(\mathbf{r})\nabla\psi_n^*(\mathbf{r}) - \psi_n^*(\mathbf{r})\nabla\psi_n(\mathbf{r})\right) \quad (3.100)$$

Hence, $\mathbf{j}_{pn}$ does not depend on time; that is, we can write for any energy eigenstate n

$$\mathbf{j}_{pn}(\mathbf{r},t) = \mathbf{j}_{pn}(\mathbf{r}) \quad (3.101)$$

Therefore, particle current is constant in any stationary state (i.e., energy eigenstate).

For a particle such as an electron, the electrical current density is simply $e\mathbf{j}_p$. A steady current does not radiate any electromagnetic radiation. This means that an electron in an energy eigenstate does not radiate electromagnetic radiation.

Hence, we have the quantum mechanical answer to the question, for example, about whether a hydrogen atom in an energy eigenstate, including any of the excited energy eigenstates, should be radiating. Classically, the electron orbiting around the nucleus would have a time-varying current; the electron in a classical orbit is continually being accelerated (and, hence, the associated current is being changed) because its direction is changing all the time to keep it in its circular or elliptical classical orbit, and so it would have to radiate electromagnetic energy. Regardless of the details of the energy eigenfunction solutions for a hydrogen atom, this quantum mechanical result says that the atom in such a state does not radiate electromagnetic energy because there is no changing current. The quantum mechanical picture agrees with reality for hydrogen atoms in states and the classical picture does not.

For the common case where the spatial part of the energy eigenstate (i.e., $\psi(\mathbf{r})$) is real or can be written as a real function multiplied by a complex constant, the right-hand side of Eq. (3.100) is zero, and there is zero particle current. Hence, for example, the energy eigenstates in a potential well or a harmonic oscillator have no particle current associated with them.

Problems

3.14.1 Suppose we have a particle in a wavepacket, where the spatial wavefunction at some time t is $\psi(\mathbf{r}) = A(\mathbf{r})\exp(i\mathbf{k}\cdot\mathbf{r})$. Here, $A(\mathbf{r})$ is a function that varies very slowly in space compared to the function $\exp(i\mathbf{k}\cdot\mathbf{r})$, describing the envelope of the wavepacket.

(i) Given that the particle current density is given by $\mathbf{j}_p = (i\hbar/2m)(\Psi\nabla\Psi^* - \Psi^*\nabla\Psi)$, show that $\mathbf{j}_p \cong |\psi(\mathbf{r})|^2 \mathbf{p}/m$, where $\mathbf{p}$ is the (vector) expectation value of the momentum.

(ii) With similar approximations, evaluate the expectation value of the energy on the assumption that the potential energy is constant in space.

(iii) Hence, show that the velocity of the probability density corresponds to the velocity we would expect classically.

3.14.2 There are situations in quantum mechanics in which the mass is not constant in space. This occurs specifically in analysis of semiconductor heterostructures where the effective mass is different in different materials. For the case where mass $m(z)$ varies with z, we can postulate the Hamiltonian

$$\hat{H} = -\frac{\hbar^2}{2}\frac{d}{dz}\left(\frac{1}{m(z)}\frac{d}{dz}\right) + V(z)$$

(For the sake of simplicity, we consider here only a one-dimensional case.)

(i) Show that this Hamiltonian leads to conservation of particle density if we postulate that the particle current (for the z direction) is given by

$$j_{pz} = \frac{i\hbar}{2m(z)}\left[\psi\frac{d\psi^*}{dz} - \psi^*\frac{d\psi}{dz}\right]$$

(actually, the same expression as in the situation in which mass did not depend on position). (Hint: Follow through the preceding argument for the particle current but with the new form of the Hamiltonian given here.)

(ii) Show that the boundary conditions that should be used at a potential step with this new Hamiltonian are continuity of $\frac{1}{m}\frac{d\psi}{dz}$ and continuity of ψ. (These are commonly used boundary conditions for analyzing such problems.) (Hint: Follow through the argument leading up to Eqs. (2.38) and (2.39) with the new Hamiltonian.)

3.15 Quantum mechanics and Schrödinger's equation

Thus far, all the quantum mechanics we have studied has been associated with Schrödinger's equation in its time-independent and time-dependent forms. This has introduced a very large number of the features of quantum mechanics to us. We have seen the emergence of "quantum" behavior, the idea of discrete states, with very specific energies associated with them. Though quantum mechanics operates in very different ways from classical mechanics, we have seen how quantum mechanics can describe moving particles in ways that can correspond to classical motion, such as particles moving at a constant velocity, oscillating particles, or accelerating particles. We have introduced a large number of the concepts of the mathematical approach to quantum mechanics, such as complete sets of eigenfunctions, states represented as linear superpositions of these states, and the general idea of linear algebra and quantum mechanics, including operators such as the Hamiltonian (for energy) and the momentum operator. We have also introduced quantum mechanical ideas like the uncertainty principle and wave-particle duality.

We introduced, too, the idea of quantum mechanical amplitudes, of which the wave amplitude in Schrödinger's equation is an example. We mentioned at various points that the wavefunction (and, indeed, any other quantum mechanical amplitude) is not necessarily a meaningful quantity on its own. In fact, it would actually cause us considerable problems if it were a measurable quantity; we would have solutions for measurable quantities that did not have the underlying symmetry of the problem, as in antisymmetric wavefunction solutions to symmetric problems (e.g., the second state in a square potential well), or had time dependence even when there was no real time dependence in the problem (as in any energy eigenstate). The wavefunction (or any other quantum mechanical amplitude we will encounter later) is arguably just a mathematical device that makes calculations more convenient. That might seem an odd concept, but another common example of a mathematical device of possibly no direct physical meaning is complex numbers themselves. The use of complex numbers makes many calculations in engineering and classical physics easier, but the imaginary numbers themselves arguably have no direct physical meaning.

The reader will also have noticed that we use complex numbers extensively in the quantum mechanics with Schrödinger's equation that we have discussed so far. This complex nature is retained as we go further in quantum mechanics. Quantum mechanical amplitudes, in general, are complex quantities.

The Schrödinger equation is a very powerful and important relation in quantum mechanics, but it is far from all of quantum mechanics. It is the equation that describes the behavior of a single particle with mass, under nonrelativistic conditions (i.e., everything moving much slower than the velocity of light) and in the absence of magnetic effects. Essentially, the Schrödinger equation describes the Hamiltonian of such a nonrelativistic particle in the absence of magnetic fields. We can extend it to cover some other situations, by adding further terms to it, and we will look at some of these situations later, but there is also important quantum mechanics, such as that describing light, in which we need to go beyond the ideas of Schrödinger's equation. The reader can be assured, however, that the underlying concepts we have already illustrated carry all the way through this additional quantum mechanics, especially the linear algebra aspects. Indeed, looking at quantum mechanics more generally in terms of linear algebra rather than only the mathematics of a differential equation such as Schrödinger's equation can be very liberating intellectually in understanding quantum mechanics and can also save us a lot of time. It is to that linear algebra description that we turn in the next chapter.

3.16 Summary of concepts

Relation between energy and frequency

In quantum mechanics, we find we can quite generally associate an energy E with a frequency ν or an angular frequency ω through the relation

$$E = h\nu = \hbar\omega \tag{3.1}$$

Time-dependent Schrödinger equation

The time-dependent Schrödinger equation is not an eigenequation. Knowing the solution in space at a given time (and any time dependence of the potential) is sufficient to calculate the solution at any subsequent (or, for that matter, previous) time.

$$-\frac{\hbar^2}{2m}\nabla^2\Psi(\mathbf{r},t)+V(\mathbf{r},t)\Psi(\mathbf{r},t)=i\hbar\frac{\partial\Psi(\mathbf{r},t)}{\partial t} \tag{3.2}$$

Superposition state

Because of the linearity of Schrödinger's time-dependent equation, if $\Psi_a(\mathbf{r},t)$ and $\Psi_b(\mathbf{r},t)$ are separately solutions of the equation, then so also is the sum $\Psi_a(\mathbf{r},t) + \Psi_b(\mathbf{r},t)$. This can be inductively extended to any linear superposition of an arbitrary number of solutions.

Group velocity and dispersion

Dispersion is the phenomenon in which the frequency ω and the magnitude of the wavevector k for some type of wave are not simply proportional to one another. The group velocity is the velocity at which some wavepacket, constructed from some superposition of waves, will move and it is defined as

$$v_g = \frac{d\omega}{dk} \tag{3.31}$$

Group velocity dispersion is the phenomenon in which the group velocity itself changes with ω or k. Such group velocity dispersion typically leads to a change in the width of the wavepacket as it propagates.

Quantum mechanical measurement and collapse of the wavefunction

When a small quantum mechanical system is measured by a large measuring apparatus, the system is always measured to be in one of the eigenstates of the quantity being measured, even if it was in a linear superposition of eigenstates before measurement. This is sometimes known as "collapse of the wavefunction." (If viewed as a fundamental postulate of quantum mechanics rather than an empirical behavior of small systems measured by large ones, it has various philosophical problems.)

The probability of finding the system in a particular eigenstate on measurement is proportional to the modulus squared of the expansion coefficient of that state in the original superposition; that is, if the expansion of the state in the (normalized) eigenstates of the quantity being measured is $\Psi(\mathbf{r},t)=\sum_n c_n(t)\psi_n(\mathbf{r})$, then the probability of finding the system in eigenstate $\psi_n(\mathbf{r})$ is

$$P_n = |c_n|^2 \tag{3.46}$$

Operators

An operator is an entity that changes one function into another. Operators are of central importance to the mathematical foundation of quantum mechanics. Operators are often indicated by having a "hat" over the corresponding letter – for example, $\hat{H}$ for the Hamiltonian operator (we adopt this notation consistently here).

Hamiltonian operator

The Hamiltonian operator is the operator that is associated with the energy of a quantum mechanical system. There is a very close link between Schrödinger's equation and the Hamiltonian operator (for those systems for which Schrödinger's equation is a good description). Quite generally, we write

$$\hat{H}\Psi(\mathbf{r},t) = i\hbar\frac{\partial\Psi(\mathbf{r},t)}{\partial t} \tag{3.50}$$

even when the system is more complex than that described by the simple Schrödinger equations discussed so far. We also write, quite generally, for Hamiltonians that do not depend on time

$$\hat{H}\psi(\mathbf{r}) = E\psi(\mathbf{r}) \tag{3.51}$$

Time evolution of a quantum mechanical state

The way in which a quantum mechanical state evolves in time is a key concept in quantum mechanics and is fundamentally unlike the time evolution of classical systems.

The evolution of a quantum mechanical system in time can be viewed as proceeding by the coherent addition of quantum mechanical amplitudes, each of which is evolving in time. This coherent addition is like the interference of different waves and, when the quantum mechanical amplitude we are considering is a wavefunction, there is an exact analogy here. Probability densities and other measurable quantities (e.g., expectation values of energy or momentum) can be deduced from the resulting sum of amplitudes at any particular time.

We have illustrated that the interference of quantum mechanical amplitudes can lead to the kind of behavior we see in the classical world, as in the harmonic oscillator with large energies and the propagation of wavepackets.

We can, of course, always simply integrate Schrödinger's time-dependent equation in time if we know the initial wavefunction and, consequently, we can deduce everything that happens subsequently.

There are also two specific methods for calculating the time evolution of quantum mechanical system, for the case in which the potential is constant in time, that give useful insight into quantum mechanical evolution of a system.

(i) If we express the initial spatial wavefunction as a linear superposition of the energy eigenstates

$$\Psi(\mathbf{r},0)=\psi(\mathbf{r})=\sum_n a_n\psi_n(\mathbf{r}) \tag{3.18}$$

then the evolution of the wavefunction in time is given by a simple linear superposition of these eigenstates with their oscillating prefactors ($\exp(-iE_n t/\hbar)$); that is,

$$\Psi(\mathbf{r},t)=\sum_n a_n\Psi_n(\mathbf{r},t)=\sum_n a_n\exp(-iE_n t/\hbar)\psi_n(\mathbf{r}) \tag{3.17}$$

(ii) Alternatively, and equivalently, we can define a time-evolution operator $\exp(-i\hat{H}t/\hbar)$ that enables us to deduce the state at time t_1, $\Psi(\mathbf{r},t_1)$, from that at time t_0, $\Psi(\mathbf{r},t_0)$, simply by applying this operator; that is,

$$\Psi(\mathbf{r},t_1)=\exp\left(-\frac{i\hat{H}(t_1-t_0)}{\hbar}\right)\Psi(\mathbf{r},t_0) \tag{3.58}$$

Momentum operator

It is possible also to define an operator associated with momentum

$$\hat{p}\equiv -i\hbar\nabla \tag{3.74}$$

which has associated eigenfunctions $\exp(i\mathbf{k}\cdot\mathbf{r})$ and eigenvalues $\hbar\mathbf{k}$.

Position operator

The position operator, at least when functions are expressed in terms of position, is simply the position vector itself, $\mathbf{r}$.

Expectation values

For quantum mechanical states that are not eigenstates corresponding to some measurable quantity (e.g., energy or momentum), it is still possible to define the average value of the quantity of interest. This is known as the expectation value. It is the average value that would be obtained after repeated measurements on the system if it were prepared in the same state each time.

The expectation value of a physical quantity, such as energy E, can be evaluated

(i) from the known expansion of the state (here, the wavefunction $\Psi(\mathbf{r},t)$) in the (normalized) eigenfunctions of the corresponding operator, for example, the Hamiltonian $\hat{H}$ for the energy E with eigenfunctions $\psi_n(\mathbf{r})$ and eigenvalues E_n leads to the expansion

$$\Psi(\mathbf{r},t)=\sum_n c_n(t)\psi_n(\mathbf{r}) \tag{3.43}$$

and the corresponding expectation value

$$\langle E \rangle = \sum_n E_n P_n = \sum_n E_n |c_n|^2 \tag{3.47}$$

or

(ii) directly from the known state of the system using the operator in the appropriate expression; for example, for the case of energy and the Hamiltonian operator, through an expression of the form

$$\langle E \rangle = \int \Psi^*(\mathbf{r},t)\hat{H}\Psi(\mathbf{r},\mathbf{t})d^3\mathbf{r} \tag{3.56}$$

Uncertainty principle

We find that quantum mechanics imposes a limit on the relative precision with which certain attributes of a particle can be defined. The best known such relation is the Heisenberg uncertainty principle between position and momentum for a particle, which is a relationship between the standard deviations of the probability distributions for position Δz in a given direction and momentum Δp in the same direction

$$\Delta p \Delta z \geq \hbar / 2 \tag{3.88}$$

This uncertainty principle is exactly analogous to the well-known relation

$$\Delta\omega\Delta t \geq \frac{1}{2} \tag{3.89}$$

between angular frequency and time in Fourier analysis of temporal functions and to the phenomenon of diffraction of waves.

Particle current

We find that for a potential that is constant in time, we can identify the quantity

$$\mathbf{j}_p = \frac{i\hbar}{2m}\left(\Psi\nabla\Psi^* - \Psi^*\nabla\Psi\right) \tag{3.97}$$

with particle current.

The particle current associated with an energy eigenstate is constant and that associated with an eigenstate with a real wavefunction (or a real function multiplied by a complex constant) is zero.

Meaninglessness of wavefunction

It is not clear that the wavefunction we have introduced has any meaning in itself; it is apparently not a measurable quantity. Like complex numbers themselves, however, the wavefunction is a very useful device for calculating other quite meaningful, mathematically real quantities. It is also a good example of a quantum mechanical amplitude, a concept that will recur many times in other aspects of quantum mechanics.

Chapter 4

Functions and operators

Prerequisites: Chapters 2 and 3.

So far, we have introduced quantum mechanics through the example of the Schrödinger equation and the spatial and temporal wavefunctions that are solutions to it. This has allowed us to solve some simple but important problems and to introduce many quantum mechanical concepts by example. Quantum mechanics does, however, go considerably beyond the Schrödinger equation. For example, photons are not described by the kind of Schrödinger equation we have considered so far, though they are undoubtedly very much quantum mechanical.[1]

To prepare for other aspects of quantum mechanics and to make the subject easier to deal with in more complex problems, we need to introduce a more general and extended mathematical formalism. This formalism is actually mostly linear algebra. Readers probably have encountered many of the basic concepts already in subjects such as matrix algebra, Fourier transforms, solutions of differential equations, possibly (though less likely) integral equations, or analysis of linear systems, in general. For this book, we assume that the reader is familiar with, at least, the matrix version of linear algebra – the other examples are not necessary prerequisites. The fact that the formalism is based on linear algebra is because of the basic observation that quantum mechanics is apparently absolutely linear in certain specific ways as we discussed previously.

Thus far, we have dealt with the state of the quantum mechanical system as the wavefunction $\Psi(\mathbf{r},t)$ of a single particle. More complex systems have more complex states to describe, but, in general, any quantum mechanical state can simply be written as a list (possibly infinitely long) of numbers. This list can be written as a vector, which is, after all, simply a list of numbers, and can be called the *state vector*. The operators of quantum mechanics can then be written as matrices (also possibly infinitely large), and the operation of the operator on the function corresponds to the multiplication of the state vector by the operator matrix. It is this generalized linear algebra approach that we discuss and develop in this chapter.

The linear algebra formalism is more generalized in quantum mechanics than in most of the other subjects that exploit linear algebra, and it is often presented in a rather abstract way. The shorthand notations that are introduced in these more abstract presentations are quite useful and worth learning, but the reader should be assured that the concepts are fundamentally the same as other manifestations of linear algebra. Here, we try to explain the concepts in a tangible and, hopefully, intelligible way. The mathematical approach here is deliberately

---

[1] In fact, arguably the first quantum mechanics was concerned with the photon, through Planck's solution of the problem of the spectral distribution of thermal radiation.

informal, with the emphasis being on grasping the core concepts and ways of visualizing the mathematical operations rather than on the more rigorous mathematical detail. The justification for this informality is that once the essence of the concepts is understood, the reader can come back to understand the detail in a more rigorous treatment, but the opposite approach is generally much less successful.

Our discussion so far of quantum mechanics has largely been one of breaking down classical beliefs and replacing them with a specific approach (i.e., Schrödinger's equation) that works for certain problems. A major goal of introducing this mathematical approach is to show a way of visualizing quantum mechanics, giving the reader an intuitive understanding of quantum mechanics that extends to a broad range of problems.

4.1 Functions as vectors

First, we look at functions as particular kinds of vectors. A function (e.g., $f(x)$) is essentially a mapping from one set of numbers (i.e., the "argument", x, of the function) to another (i.e., the "result" or "value", $f(x)$, of the function). The fundamentals of this concept are not changed for functions of multiple variables or for functions with complex number results or geometrical vector results (e.g., a position or momentum vector). We can imagine that the set of possible values of the argument is a list of numbers and the corresponding set of values of the function is another list.

One kind of list of arguments would be the list of all real numbers, which we could write in order as x_1, x_2, x_3 ... and so on. Of course, that is an infinitely long list and the adjacent values in the list are infinitesimally close together, but we regard these infinities as details and, for the moment, we ask the reader to assume that these details can be handled successfully from a more rigorous mathematical viewpoint.

If we presume that we know this list of possible arguments of the function, we can write out the function as the corresponding list of results or values and we choose to write this list as a column vector; that is,

$$\begin{bmatrix} f(x_1) \\ f(x_2) \\ f(x_3) \\ \vdots \end{bmatrix}$$

We can certainly imagine, for example, that as a reasonable approximation to an actual continuous function of a continuous variable, we could specify the function at a discrete set of points spaced by some small amount δx, with $x_2 = x_1 + \delta x$, $x_3 = x_2 + \delta x$ and so on; we would do this for sufficiently many values of x and over a sufficient range of x so that we would have useful representation of the function for the purposes of some calculation we wished to perform. For example, such an approximation to the function would be sufficient to calculate any integral of any reasonably well-behaved function to any desired degree of accuracy simply by taking δx sufficiently small. The integral of $|f(x)|^2$ could then be written as

$$\int |f(x)|^2 \, dx \cong \begin{bmatrix} f^*(x_1) & f^*(x_2) & f^*(x_3) & \cdots \end{bmatrix} \begin{bmatrix} f(x_1) \\ f(x_2) \\ f(x_3) \\ \vdots \end{bmatrix} \delta x \tag{4.1}$$

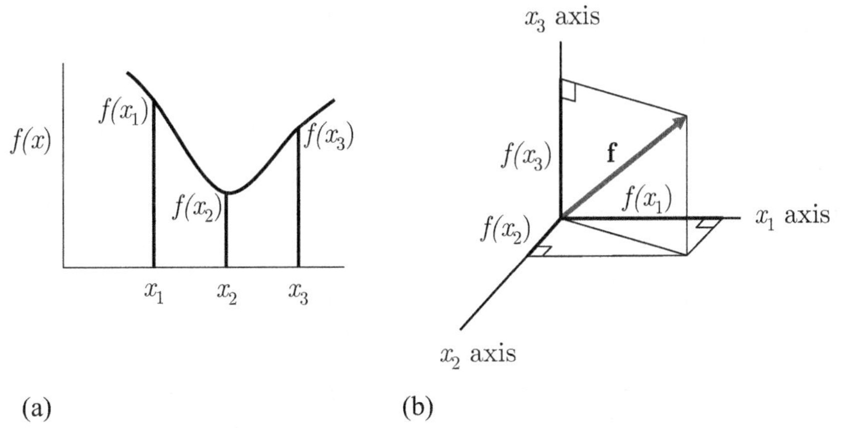

Fig. 4.1. (a) Representation of a function $f(x)$ approximately as a set of values at a discrete set of points, x_1, x_2, and x_3. (b) Illustration of how the function can be represented as a vector, at least for the case where there are only three points of interest, x_1, x_2, and x_3, and where the function is always real.

The fact that we can usefully write the function as a vector suggests that we might be able to visualize it as a "geometrical" vector. For example, in Fig. 4.1, the function $f(x)$ is approximated by its values at three points, x_1, x_2, and x_3, and is represented as a vector

$$\mathbf{f} \equiv \begin{bmatrix} f(x_1) \\ f(x_2) \\ f(x_3) \end{bmatrix}$$

in a three-dimensional space. We return later to discuss the mathematical space in which such vectors exist, but for the moment we merely note that we can visualize a function as a vector. Of course, because there are many elements in the vector (possibly an infinite number), we may need a space with a very large (possibly infinite) number of dimensions, but we still visualize the function (and, more generally, the quantum mechanical state) as a vector in a space.

Dirac bra-ket notation

Because we will be working with such vectors extensively, it will be useful to introduce a shorthand notation. In quantum mechanics, we use the *Dirac bra-ket notation*; this notation is a convenient way to represent linear algebra quite generally, though its main use is in quantum mechanics. In this notation, the expression $|f(x)\rangle$, called a *ket*, represents the column vector corresponding to the function $f(x)$. For the case of our function $f(x)$, we can define the ket as

$$|f(x)\rangle \equiv \begin{bmatrix} f(x_1)\sqrt{\delta x} \\ f(x_2)\sqrt{\delta x} \\ f(x_3)\sqrt{\delta x} \\ \vdots \end{bmatrix} \tag{4.2}$$

or, more strictly, the limit of this as $\delta x \to 0$. We have incorporated the factor $\sqrt{\delta x}$ into the vector so that we can handle normalization of the function, but this does not change the fundamental concept that we are writing the function as a vector list of numbers.

We can similarly define the *bra* $\langle f(x)|$ to refer to an appropriate form of our row vector, in this case

$$\langle f(x)| \equiv \begin{bmatrix} f^*(x_1)\sqrt{\delta x} & f^*(x_2)\sqrt{\delta x} & f^*(x_3)\sqrt{\delta x} & \cdots \end{bmatrix} \tag{4.3}$$

where again we more strictly mean the limit of this as $\delta x \to 0$.

Let us pause for a moment to consider the relation between bra and ket vectors. Note that, in our row vector, we have taken the complex conjugate of all the values. Quite generally, the vector

$$\begin{bmatrix} a_1^* & a_2^* & a_3^* & \cdots \end{bmatrix}$$

is what is called, variously, the Hermitian adjoint, the Hermitian transpose, the Hermitian conjugate, or, sometimes, simply the adjoint of the vector

$$\begin{bmatrix} a_1 \\ a_2 \\ a_3 \\ \vdots \end{bmatrix}$$

A common notation used to indicate the Hermitian adjoint is to use the "dagger" character "$\dagger$" as a superscript; that is,

$$\begin{bmatrix} a_1 \\ a_2 \\ a_3 \\ \vdots \end{bmatrix}^\dagger \equiv \begin{bmatrix} a_1^* & a_2^* & a_3^* & \cdots \end{bmatrix} \tag{4.4}$$

We can view the operation of forming the Hermitian adjoint as a reflection of the vector about a -45° line, coupled with taking the complex conjugate of all the elements, as shown here.

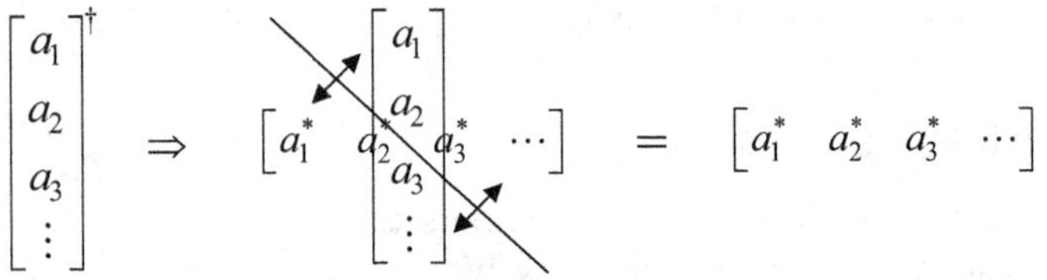

We see with this operation that the bra is the Hermitian adjoint of the ket and vice versa. Note that the Hermitian adjoint of a Hermitian adjoint takes us back to where we started; that is,

$$\left(\begin{bmatrix} a_1 \\ a_2 \\ a_3 \\ \vdots \end{bmatrix}^\dagger\right)^\dagger = \begin{bmatrix} a_1^* & a_2^* & a_3^* & \cdots \end{bmatrix}^\dagger = \begin{bmatrix} a_1 \\ a_2 \\ a_3 \\ \vdots \end{bmatrix} \tag{4.5}$$

We return later to consider the same kind of adjoint manipulation with matrices rather than vectors.

Returning to the discussion of $f(x)$ as a vector, with the definitions Eqs. (4.1), (4.2), and (4.3), we find

$$\int |f(x)|^2 dx \equiv \begin{bmatrix} f^*(x_1)\sqrt{\delta x} & f^*(x_2)\sqrt{\delta x} & f^*(x_3)\sqrt{\delta x} & \cdots \end{bmatrix} \begin{bmatrix} f(x_1)\sqrt{\delta x} \\ f(x_2)\sqrt{\delta x} \\ f(x_3)\sqrt{\delta x} \\ \vdots \end{bmatrix}$$

$$\equiv \sum_n f^*(x_n)\sqrt{\delta x} f(x_n)\sqrt{\delta x} \tag{4.6}$$

$$\equiv \langle f(x) | f(x) \rangle$$

where again the strict equality applies in the limit when $\delta x \to 0$. Writing this as a vector multiplication eliminates the need to write a summation or integral because it is implicit in the vector multiplication. The reader will note that we have written the vector product of the bra and ket in a specific shorthand way as

$$\langle g | \times | f \rangle \equiv \langle g | f \rangle \tag{4.7}$$

Of course, as suggested by Eq. (4.7), this notation is also useful when we are dealing with integrals of two different functions; that is,

$$\int g^*(x) f(x) dx \equiv \begin{bmatrix} g^*(x_1)\sqrt{\delta x} & g^*(x_2)\sqrt{\delta x} & g^*(x_3)\sqrt{\delta x} & \cdots \end{bmatrix} \begin{bmatrix} f(x_1)\sqrt{\delta x} \\ f(x_2)\sqrt{\delta x} \\ f(x_3)\sqrt{\delta x} \\ \vdots \end{bmatrix}$$

$$\equiv \sum_n g^*(x_n)\sqrt{\delta x} f(x_n)\sqrt{\delta x} \tag{4.8}$$

$$\equiv \langle g(x) | f(x) \rangle$$

In general this kind of "product" of two vectors is called an *inner product* in linear algebra. The geometric vector dot product is an inner product, the bra-ket product is an inner product, and the overlap integral on the left of Eq. (4.8) is an inner product. It is "inner" because it takes two vectors and turns them into a number, a "smaller" entity. The bra-ket notation can also be considered to give an inner "look" to this multiplication with the special parentheses at either end, giving a "closed" look to the expression.

We could also consider situations in which a function is not represented directly as a set of values for each point in ordinary geometrical space but is instead represented as an expansion in some complete orthonormal basis set, $\psi_n(x)$, in which case we would write

$$f(x) = \sum_n c_n \psi_n(x) \tag{4.9}$$

(An example of such an expansion would be a Fourier series.) In this case, we could also write the function as a vector or ket (which would also, in general, have an infinite number of elements)

$$|f(x)\rangle \equiv \begin{bmatrix} c_1 \\ c_2 \\ c_3 \\ \vdots \end{bmatrix} \tag{4.10}$$

This is just as valid a description of the function as is the list of values at successive points in space. In this case, the bra becomes

$$\langle f(x)| \equiv \begin{bmatrix} c_1^* & c_2^* & c_3^* & \cdots \end{bmatrix} \tag{4.11}$$

When we write the function in this different form, as a vector containing these expansion coefficients, we say we have changed its *representation*. The function $f(x)$ is still the same function as it was before and we visualize the vector $|f(x)\rangle$ as being the same vector in our space. We have merely changed the axes in that space that we use to represent the function; hence, the coordinates in our new representation of the vector have changed (now they are the numbers c_1, c_2, $c_3 \ldots$).[2] (This idea that the function is the same vector independent of how it is represented is an important one in this mathematics.)

Just as before, we could evaluate

$$\begin{aligned}\int |f(x)|^2 \, dx &\equiv \int f^*(x) f(x) dx \equiv \int \left[\sum_n c_n^* \psi_n^*(x) \right] \left[\sum_m c_m \psi_m(x) \right] dx \\ &\equiv \sum_{n,m} c_n^* c_m \int \psi_n^*(x) \psi_m(x) dx \equiv \sum_{n,m} c_n^* c_m \delta_{nm} \equiv \sum_n |c_n|^2 \\ &\equiv \begin{bmatrix} c_1^* & c_2^* & c_3^* & \cdots \end{bmatrix} \begin{bmatrix} c_1 \\ c_2 \\ c_3 \\ \vdots \end{bmatrix} \equiv \langle f(x)|f(x)\rangle \end{aligned} \tag{4.12}$$

Similarly, with

$$g(x) = \sum_n d_n \psi_n(x) \tag{4.13}$$

we have

$$\int g^*(x) f(x) dx \equiv \begin{bmatrix} d_1^* & d_2^* & d_3^* & \cdots \end{bmatrix} \begin{bmatrix} c_1 \\ c_2 \\ c_3 \\ \vdots \end{bmatrix} \equiv \langle g(x)|f(x)\rangle \tag{4.14}$$

with similar intermediate algebraic steps to those of Eq. (4.12).

Note that the result of a bra-ket expression like $\langle f(x)|f(x)\rangle$ or $\langle g(x)|f(x)\rangle$ is simply a number (in general, a complex one), which is easy to see if we think of this as a vector-vector multiplication. Note too that this number (i.e., inner product) is not changed as we change the representation, as we would expect by analogy with the dot product of two vectors, which is independent of the coordinate system. Again, this fact that the inner product does not depend on the representation is very important in this mathematics.

[2] The reader might ask, "What are the new coordinate axes?" The answer is that they are simply the functions $\psi_n(x)$, drawn as unit vectors in the same space, as we will discuss.

Expansion coefficients

Just as we did in Section 2.7, it is simple to evaluate the expansion coefficients c_n in Eq. (4.9) (or d_n in Eq. (4.13)) because we choose the set of functions $\psi_n(x)$ to be orthonormal. Because $\psi_n(x)$ is just another function, we can also write it as a ket. In the bra-ket notation, to evaluate the coefficient c_m, we premultiply by the bra $\langle\psi_m|$

$$\langle\psi_m(x)|f(x)\rangle = \sum_n c_n \langle\psi_m(x)|\psi_n(x)\rangle = c_m \tag{4.15}$$

(Here, remember that the different functions $|\psi_n(x)\rangle$ are all orthogonal to one another in such an orthonormal basis; hence, $\langle\psi_m(x)|\psi_n(x)\rangle \equiv \int \psi_m^*(x)\psi_n(x)dx = \delta_{nm}$.) Note that c_n is just a number, so it can be moved about in the product. Using the bra-ket notation, we can write Eq. (4.9) as

$$\begin{aligned} |f(x)\rangle &= \sum_n c_n |\psi_n(x)\rangle = \sum_n |\psi_n(x)\rangle c_n \\ &= \sum_n |\psi_n(x)\rangle\langle\psi_n(x)|f(x)\rangle \end{aligned} \tag{4.16}$$

Again, because c_n is just a number, it can be moved about in the product. (Formally, multiplication of a vector and a number is *commutative* – though, of course, multiplication of vectors or matrices generally is not.)

Often in using the bra-ket notation, we may drop arguments like x. Then, we can write Eq. (4.16) as

$$|f\rangle = \sum_n |\psi_n\rangle\langle\psi_n|f\rangle \tag{4.17}$$

Here, we see a key reason for introducing the Dirac bra-ket notation; it is a generalized shorthand way of writing the underlying linear algebra operations we need to perform and can be used whether we are thinking about representing functions as continuous functions in some space or as summations over basis sets. It will also continue to be useful as we consider quantum mechanical attributes that are not represented as functions in normal geometric space; an example (to which we return much later) is the spin of an electron, a magnetic property of the electron.

State vectors

In the quantum mechanical context where the function f represents the state of the quantum mechanical system (e.g., it might be the wavefunction), we think of the set of numbers represented by the bra ($\langle f|$) or ket ($|f\rangle$) vector as representing the state of the system; hence, we refer to the ket vector that represents f as the state vector of the system and the corresponding bra vector as the (Hermitian) adjoint of that state vector.

In quantum mechanics, the bra or ket always represents either the quantum mechanical state of the system (e.g., the spatial wavefunction $\psi(x)$) or some state that the system could be in (e.g., one of the basis states $\psi_n(x)$). The convention for what symbols we put inside the bra or ket is rather loose and usually one deduces from the context what exactly is being meant. For example, if it was quite obvious what basis we were working with, we might use the notation $|n\rangle$ to represent the nth basis function (or basis state) rather than the notation $|\psi_n(x)\rangle$ or $|\psi_n\rangle$. In general, which symbols we put inside the bra or ket should be enough to make it clear which state we are discussing in a given context. Beyond that, there are essentially no rules for the notation inside the bra or ket; the symbols inside the bra or ket are simply labels to remind us which quantum mechanical state is being represented. We could, if we wanted, write

$$\left| \begin{array}{l} \text{The state where the electron has the lowest} \\ \text{possible energy in a harmonic oscillator with} \\ \text{potential energy } 0.375x^2 \end{array} \right\rangle$$

but because we presumably already know we are discussing such a harmonic oscillator, it saves us time and space simply to write

$$|0\rangle$$

with the zero representing the quantum number of that state. Either would be correct mathematically.

Problems

4.1.1 Suppose we adopt a notation

$$|n\rangle \equiv \sqrt{\frac{2}{L_z}} \sin\left(\frac{n\pi z}{L_z}\right)$$

to label the states of a particle in a one-dimensional potential well of thickness L_z. Write the bra-ket notation form that is equivalent to each of the following integrals (do not evaluate the integrals – just change the notation).

(i) $\frac{2}{L_z}\int_0^{L_z} \sin\left(\frac{3\pi z}{L_z}\right) \sin\left(\frac{5\pi z}{L_z}\right) dz$

(ii) $\frac{2G}{L_z}\int_0^{L_z} \sin\left(\frac{3\pi z}{L_z}\right) \sin\left(\frac{3\pi z}{L_z}\right) dz$, where G is some constant

(iii) $\int_0^{L_z} \sin\left(\frac{5\pi z}{L_z}\right) \sin\left(\frac{5\pi z}{L_z}\right) dz$

4.1.2 Suppose that there are two quantum mechanically measurable quantities, c with associated operator $\hat{C}$ and d with associated operator $\hat{D}$. In particular, operator $\hat{C}$ has two eigenvectors $|\phi_1\rangle$ and $|\phi_2\rangle$, and similarly operator $\hat{D}$ has two eigenvectors $|\psi_1\rangle$ and $|\psi_2\rangle$. The relation between the eigenvectors is

$$|\phi_1\rangle = \frac{1}{5}\left(3|\psi_1\rangle + 4|\psi_2\rangle\right)$$

$$|\phi_2\rangle = \frac{1}{5}\left(4|\psi_1\rangle - 3|\psi_2\rangle\right)$$

Suppose a measurement is made of the quantity c, and the system is measured to be in state $|\phi_1\rangle$. Then a measurement is made of quantity d and, following that, the quantity c is again measured. What is the probability (expressed as a fraction) that the system will be found in state $|\phi_1\rangle$ on this second measurement of c? (Note: This is really a problem in quantum mechanical measurement discussed in the previous chapter, but it is a good exercise in the use of the Dirac notation.)

4.2 Vector space

Now we should try to understand more about the space in which the vectors representing functions exist. For a vector with three components

$$\begin{bmatrix} a_1 \\ a_2 \\ a_3 \end{bmatrix}$$

we can obviously imagine a conventional three-dimensional Cartesian space. The vector can be visualized as a line in that space, starting from the origin, with projected lengths a_1, a_2, and a_3 along the three Cartesian axes, respectively (with each of these axes being at right angles to each other axis), just as we did in Fig. 4.1.

In the case of a function expressed as its values at a set of points, instead of three axes labeled x_1, x_2, and x_3, we have more commonly an infinite number of different axes all orthogonal to one another. If we were only ever interested in functions expressed in terms of position x, we would create one axis for each possible position x and we would label the axes accordingly with those positions.

More generally, we represent a function as an expansion on a basis set, as in Eq. (4.16). In this generalized case, we have one axis for each element in the basis set of functions. We now label each axis with the name of the basis function with which it is associated (e.g., ψ_n). Just as we may formally label the axes in conventional space with unit vectors in the directions of the axes (e.g., one notation is $\hat{\mathbf{x}}$, $\hat{\mathbf{y}}$, and $\hat{\mathbf{z}}$ for the unit vectors), so also here we can label the axes with the kets associated with the basis functions $|\psi_n\rangle$; either notation is acceptable. Note that a basis function is itself a vector in this space and, if normalized, the basis function vectors are simply the unit vectors along the corresponding axis.

The geometrical space has a vector dot product that defines both the orthogonality of the axes; for example, with $\hat{\mathbf{x}}$, $\hat{\mathbf{y}}$, and $\hat{\mathbf{z}}$ as the unit vectors in the coordinate directions[3]

$$\hat{\mathbf{x}} \cdot \hat{\mathbf{y}} = 0 \tag{4.18}$$

and defines the components of a vector along those axes, for example, for

$$\mathbf{f} = f_x \hat{\mathbf{x}} + f_y \hat{\mathbf{y}} + f_z \hat{\mathbf{z}} \tag{4.19}$$

with

$$f_x = \mathbf{f} \cdot \hat{\mathbf{x}} \tag{4.20}$$

and similarly for the other components.

Our vector space has an inner product that defines both the orthogonality of the basis functions

$$\langle \psi_m | \psi_n \rangle = \delta_{nm} \tag{4.21}$$

as well as the components

$$c_m = \langle \psi_m | f \rangle \tag{4.22}$$

Note that the fact that the basis functions are mathematically orthogonal, as given by Eq. (4.21), corresponds with the fact that we can draw them as orthogonal axes in the space.

[3] Note that here we have used the symbol "ˆ" to indicate geometrical unit vectors, but we otherwise reserve it to indicate an operator.

The geometrical space and our vector space share a number of elementary mathematical properties that seem so obvious they are almost trivial. With respect to addition of vectors, both spaces are commutative

$$\mathbf{a}+\mathbf{b}=\mathbf{b}+\mathbf{a} \tag{4.23}$$

$$|f\rangle+|g\rangle=|g\rangle+|f\rangle \tag{4.24}$$

and associative

$$\mathbf{a}+(\mathbf{b}+\mathbf{c})=(\mathbf{a}+\mathbf{b})+\mathbf{c} \tag{4.25}$$

$$|f\rangle+(|g\rangle+|h\rangle)=(|f\rangle+|g\rangle)+|h\rangle \tag{4.26}$$

and linear with respect to multiplying by constants; for example,

$$c(\mathbf{a}+\mathbf{b})=c\mathbf{a}+c\mathbf{b} \tag{4.27}$$

$$c(|f\rangle+|g\rangle)=c|f\rangle+c|g\rangle \tag{4.28}$$

(The constants in our vector space case are certainly allowed to be complex.) The inner product is linear both in multiplying by constants

$$\mathbf{a}.(c\mathbf{b})=c(\mathbf{a}.\mathbf{b}) \tag{4.29}$$

$$\langle f|cg\rangle=c\langle f|g\rangle \tag{4.30}$$

and in superposition of vectors

$$\mathbf{a}.(\mathbf{b}+\mathbf{c})=\mathbf{a}.\mathbf{b}+\mathbf{a}.\mathbf{c} \tag{4.31}$$

$$\langle f|(|g\rangle+|h\rangle)=\langle f|g\rangle+\langle f|h\rangle \tag{4.32}$$

There is a well-defined "length" to a vector in both cases (formally, a norm), which we can write in a formal mathematical notation as

$$\|\mathbf{a}\|=\sqrt{\mathbf{a}.\mathbf{a}} \tag{4.33}$$

$$\|f\|=\sqrt{\langle f|f\rangle} \tag{4.34}$$

In both cases, any vector in the space can be represented to an arbitrary degree of accuracy as a linear combination of the basis vectors (this is essentially the completeness requirement on the basis set).

There is a slight difference between the geometrical space and our vector space in the inner product. Usually, in geometrical space, the lengths a_1, a_2, and a_3 of a vector are real, which would lead us to believe that the inner product (i.e., vector dot product) was commutative; that is,

$$\mathbf{a}.\mathbf{b}=\mathbf{b}.\mathbf{a} \tag{4.35}$$

In working with complex coefficients rather than real lengths, it is more useful to have an inner product (as we do) that has a complex conjugate relation

$$\langle f|g\rangle=(\langle g|f\rangle)^* \tag{4.36}$$

Such a relation ensures that $\langle f | f \rangle$ is real, as required for it to be a useful norm. (The existence of a norm is formally required to prove properties like completeness by showing that the norm of the difference of two vectors can be as small as desired.)

These kinds of requirements, plus a few others (the existence of a null or "zero" vector and the existence of an "antivector" that added to the vector gives the null vector) are sufficient to define these two spaces as *linear vector spaces* and specifically, with the properties of the inner product, what are called *Hilbert spaces*. The Hilbert space is the space in which the vector representation of the function exists, just as normal Cartesian geometrical space is the space in which a geometrical vector exists.[4]

The main differences between our vector space and the geometrical space are that (1) our components can be complex numbers rather than only real ones and (2) we can have more dimensions (possibly an infinite number). However, these differences are not so strong that we cannot use the idea of a geometrical space as a starting point for visualizing our vector space. In practice, we can carry over most of the intuition from our understanding of geometrical space and use it in visualizing the vector space in which we are representing functions.

Our vector space can also be called a *function space*. A vector in this space is a representation of a function. The set of basis vectors (i.e., basis functions) that can be used to represent vectors in this space is said in linear algebra to "span" the space.

Problem

4.2.1 We consider the function space that corresponds to all linear functions of a single variable; that is, functions of the form

$$f(x) = ax + b$$

defined over the range $-1 < x < 1$.

(i) Show that the functions $\psi_1(x) = 1/\sqrt{2}$ and $\psi_2(x) = \sqrt{\frac{3}{2}}x$ are orthonormal.

(ii) By showing that any arbitrary function $f(x) = ax + b$ can be represented as the linear combination $f(x) = c_1\psi_1(x) + c_2\psi_2(x)$, show that the functions $\psi_1(x)$ and $\psi_2(x)$ constitute a complete basis set for representing such functions.

(iii) Represent the function $2x+3$ as a vector in a two-dimensional function space by drawing that vector in a two-dimensional diagram with orthogonal axes corresponding to the functions $\psi_1(x)$ and $\psi_2(x)$, stating the values of appropriate coefficients or components.

4.3 Operators

A function turns one number (the argument) into another (the result). Most broadly stated,

an operator turns one function into another.

In the vector space representation of a function, an operator turns one vector into another. Operators are central to quantum mechanics and are encountered frequently in many forms in the mathematics that underlies much of science and engineering.

---

[4] Note that if we extended our notions of geometrical space to allow for complex lengths in the coordinate directions, and if we defined $(\mathbf{a} \cdot \mathbf{b})^* = \mathbf{b} \cdot \mathbf{a}$, then our geometrical space would also be, mathematically, a Hilbert space.

Suppose that we are constructing the new function $g(y)$ from the function $f(x)$ by acting on $f(x)$ with the operator $\hat{A}$. The variables x and y might actually be the same kind of variable, as in the case where the operator corresponds to differentiation of the function, for example,

$$g(x)=\left(\frac{d}{dx}\right)f(x) \tag{4.37}$$

or they might be quite different, as in the case of a Fourier transform operation where x might represent time and y might represent frequency; for example,

$$g(y)=\frac{1}{\sqrt{2\pi}}\int_{-\infty}^{\infty} f(x)\exp(-iyx)\,dx \tag{4.38}$$

A standard notation for writing such an operation on a function is

$$g(y)=\hat{A}f(x) \tag{4.39}$$

Note that this is not a multiplication of $f(x)$ by $\hat{A}$ in the normal algebraic sense but should rather be read as $\hat{A}$ operating on $f(x)$.

For $\hat{A}$ to be the most general operation we could imagine, it should be possible for the value of $g(y)$ – for example, at some particular value of $y=y_1$ – to depend on the values of $f(x)$ for all values of the argument x. This is the case, for example, in the Fourier transform operation of Eq. (4.38).

4.4 Linear operators

We are interested here solely in what are called *linear operators*. We only care about linear operators because they are essentially the only ones we use in quantum mechanics, again because of the fundamental linearity of quantum mechanics. A linear operator has the following characteristics:

$$\hat{A}\left[f(x)+h(x)\right]=\hat{A}f(x)+\hat{A}h(x) \tag{4.40}$$

$$\hat{A}\left[cf(x)\right]=c\hat{A}f(x) \tag{4.41}$$

for any complex number c and for any two functions $f(x)$ and $h(x)$.

To understand what this linearity implies about how we can represent $\hat{A}$, let us consider how, in the most general way, we could have $g(y_1)$ related to the values of $f(x)$ and still retain the linearity implied by Eqs. (4.40) and (4.41). Let us now go back to thinking of the function $f(x)$ as being represented by a list of values, $f(x_1)$, $f(x_2)$, $f(x_3)$, … , just as we did when considering $f(x)$ as a vector. Again, we can take the values of x to be as closely spaced as we want and we believe that this representation can give us as accurate a representation of $f(x)$ as we need for any calculation we need to perform. Then, we propose that for a linear operation represented by the operator $\hat{A}$, the value of $g(y_1)$ might be related to the values of $f(x)$ by a relation of the form

$$g(y_1)=a_{11}f(x_1)+a_{12}f(x_2)+a_{13}f(x_3)+\ldots \tag{4.42}$$

where the a_{ij} are complex constants. This form certainly has the linearity of the type required by Eqs. (4.40) and (4.41); that is, if we were to replace $f(x)$ by $f(x)+h(x)$, then we would have some other resulting function value

$$\begin{aligned}g(y_1) &= a_{11}\left[f(x_1)+h(x_1)\right]+a_{12}\left[f(x_2)+h(x_2)\right]+a_{13}\left[f(x_3)+h(x_3)\right]+\dots \\ &= a_{11}f(x_1)+a_{12}f(x_2)+a_{13}f(x_3)+\dots \\ &\quad +a_{11}h(x_1)+a_{12}h(x_2)+a_{13}h(x_3)+\dots\end{aligned} \tag{4.43}$$

which is just the sum as required by Eq. (4.40). Similarly, if we were to replace $f(x)$ by $cf(x)$, we would have for yet some other resulting function value

$$\begin{aligned}g(y_1) &= a_{11}cf(x_1)+a_{12}cf(x_2)+a_{13}cf(x_3)+\dots \\ &= c\left[a_{11}f(x_1)+a_{12}f(x_2)+a_{13}f(x_3)+\dots\right]\end{aligned} \tag{4.44}$$

as required by Eq. (4.41). Now let us consider whether the form Eq. (4.42) is the most general it could be. We can see this by trying to add other powers and "cross terms" of $f(x)$. Any more complicated function relating $g(y_1)$ to $f(x)$ could presumably be written as a power series in $f(x)$, possibly involving $f(x)$ for different values of x (i.e., cross terms). If we were to add higher powers of $f(x)$, such as $[f(x)]^2$, or cross terms such as $f(x_1)f(x_2)$ into the series in Eq. (4.42), it would, however, no longer have the required linear behavior of Eqs. (4.43) and (4.44). We also cannot add a constant term (i.e., one not dependent on $f(x)$) to the series in Eq. (4.42); that would violate the second linearity condition, Eq. (4.41), because the additive constant would not be multiplied by c. Hence, we conclude that the form Eq. (4.42) is the most general one possible for the relation between $g(y_1)$ and $f(x)$ if this relation is to correspond to a linear operator.

To construct the entire function $g(y)$, we should construct series like Eq. (4.42) for each other value of y (i.e., y_2, y_3, ...). It is now clear that if we write the functions $f(x)$ and $g(y)$ as vectors, then this general linear operation that relates the function $g(y)$ to the function $f(x)$ can be written as a matrix-vector multiplication

$$\begin{bmatrix} g(y_1) \\ g(y_2) \\ g(y_3) \\ \vdots \end{bmatrix} = \begin{bmatrix} a_{11} & a_{12} & a_{13} & \cdots \\ a_{21} & a_{22} & a_{23} & \cdots \\ a_{31} & a_{32} & a_{33} & \cdots \\ \vdots & \vdots & \vdots & \ddots \end{bmatrix} \begin{bmatrix} f(x_1) \\ f(x_2) \\ f(x_3) \\ \vdots \end{bmatrix} \tag{4.45}$$

with the operator

$$\hat{A} \equiv \begin{bmatrix} a_{11} & a_{12} & a_{13} & \cdots \\ a_{21} & a_{22} & a_{23} & \cdots \\ a_{31} & a_{32} & a_{33} & \cdots \\ \vdots & \vdots & \vdots & \ddots \end{bmatrix} \tag{4.46}$$

It is important to emphasize that any linear operator can be represented this way. At least, insofar as we presume functions can be represented as vectors, then linear operators can be represented by matrices. This now gives us a conceptual way of understanding what linear operators are and what they can do.

In bra-ket notation, we can write Eq. (4.39) as

$$|g\rangle = \hat{A}|f\rangle \tag{4.47}$$

Again, insofar as we regard the ket as a vector, we now regard the (linear) operator $\hat{A}$ as a matrix. In the language of vector (function) spaces, the operator takes one vector (function) and turns it into another. Now, we have a very general way of thinking about linear transformations of functions. All of the following linear mathematical operations can be described in this way: differentiation, rotation (and dilatation) of a vector, all linear transforms (Fourier, Laplace, Hankel, z-transform, …), Green's functions in integral equations, and linear integral equations generally.

In quantum mechanics, such linear operators are used as operators associated with measurable variables (e.g., the Hamiltonian operator for energy and the momentum operator for momentum), as operators corresponding to changing the representation of a function (i.e., changing the basis), and for a few other specific purposes, with the associated vectors representing quantum mechanical states.

A very important consequence is that the algebra for such operators is identical to that of matrices. In particular, operators do not, in general, commute; that is,

$$\hat{A}\hat{B}|f\rangle \text{ is not, in general, equal to } \hat{B}\hat{A}|f\rangle \tag{4.48}$$

for any arbitrary $|f\rangle$. If we understand that we are considering the operators to be operating on an arbitrary vector in the space, we can drop the vector itself and write relations between operators; for example, we can say instead of Eq. (4.48)

$$\hat{A}\hat{B} \text{ is not, in general, equal to } \hat{B}\hat{A} \tag{4.49}$$

which we would regard as an obvious statement if we are thinking of the operators as matrices.

There are specific operators that do commute just as there as specific matrices that do commute. Whether operators commute is also of central importance in quantum mechanics.

Of course, although we presented the previous argument for functions nominally of a variable x or y, it would have made no difference if we had been talking about expansion coefficients on basis sets. For example, we had expanded $f(x)$ on a basis set in Eq. (4.9), which gives a set of coefficients c_n that we can write as a vector. We similarly had expanded $g(x)$ on a basis set in Eq. (4.13) to get a vector of coefficients d_m. In that case, we had chosen the same basis for both expansions, but it would make no difference to the argument if $g(x)$ had been expanded on a different set. We could follow an argument identical to the previous one, requiring that each expansion coefficient d_i depend linearly on all the expansion coefficients c_n. The specific matrix we obtain for representing $\hat{A}$ depends on the choice of basis function sets for the expansions of $f(x)$ and $g(x)$, but we still obtain a matrix vector statement of the same form; that is,

$$\begin{bmatrix} d_1 \\ d_2 \\ d_3 \\ \vdots \end{bmatrix} = \begin{bmatrix} A_{11} & A_{12} & A_{13} & \cdots \\ A_{21} & A_{22} & A_{23} & \cdots \\ A_{31} & A_{32} & A_{33} & \cdots \\ \vdots & \vdots & \vdots & \ddots \end{bmatrix} \begin{bmatrix} c_1 \\ c_2 \\ c_3 \\ \vdots \end{bmatrix} \tag{4.50}$$

and the bra-ket statement of the relation between f, g, and $\hat{A}$, Eq. (4.47), remains unchanged. The statements in the forms either of the bra-ket relation, Eq. (4.47), or the matrix-vector notation, Eq. (4.50), could be regarded as being more general statements than the form Eq. (4.45), which applies only to representations in terms of specific variables x and y.

4.5 Evaluating the elements of the matrix associated with an operator

Now that we have established both the relation between linear operators operating on functions in linear spaces and the mathematics of matrix-vector multiplications, it will be useful to be able to evaluate the matrix associated with some specific operator for functions defined using specific basis sets.

Suppose we start with $f(x) = \psi_j(x)$ or, equivalently,

$$|f\rangle = |\psi_j\rangle \tag{4.51}$$

That is, we choose $f(x)$ to be the jth basis function. In the expansion Eq. (4.9), this means we are choosing $c_j = 1$ and setting all the other c's to be zero. Now we operate on this $|f\rangle$ with $\hat{A}$ as in Eq. (4.47) to get the resulting function $|g\rangle$. Suppose we want to know specifically what the resulting coefficient d_i is of the ith basis function in the expansion of this function $|g\rangle$ (i.e., as in Eq. (4.13)). It is obvious from the matrix form, Eq. (4.50), of the operation of $\hat{A}$ on this $|f\rangle$, with the choice $c_j = 1$ and all other c's zero, that

$$d_i = A_{ij} \tag{4.52}$$

For example, for the specific case of $j = 2$, we would have

$$\begin{bmatrix} d_1 \\ d_2 \\ d_3 \\ \vdots \end{bmatrix} = \begin{bmatrix} A_{12} \\ A_{22} \\ A_{32} \\ \vdots \end{bmatrix} = \begin{bmatrix} A_{11} & A_{12} & A_{13} & \cdots \\ A_{21} & A_{22} & A_{23} & \cdots \\ A_{31} & A_{32} & A_{33} & \cdots \\ \vdots & \vdots & \vdots & \ddots \end{bmatrix} \begin{bmatrix} 0 \\ 1 \\ 0 \\ \vdots \end{bmatrix} \tag{4.53}$$

and so, to take one example result, we know that

$$d_3 = A_{32} \tag{4.54}$$

But, from the expansions for $|f\rangle$ and $|g\rangle$, we have, for the specific case of $|f\rangle = |\psi_j\rangle$

$$|g\rangle = \sum_n d_n |\psi_n\rangle = \hat{A}|\psi_j\rangle \tag{4.55}$$

To extract d_i from this expression, we multiply by $\langle\psi_i|$ on both sides to obtain

$$d_i = \langle\psi_i|\hat{A}|\psi_j\rangle \tag{4.56}$$

and, hence, we conclude from Eq. (4.52)

$$A_{ij} = \langle\psi_i|\hat{A}|\psi_j\rangle \tag{4.57}$$

If we now think back to integrals considered as vector-vector multiplications, as in Eq. (4.14), then we can see that when the functions $\psi_i(x)$ are simple spatial functions, we have for the matrix elements corresponding to the operator $\hat{A}$

$$A_{ij} = \int \psi_i^*(x) \hat{A}\, \psi_j(x) dx \tag{4.58}$$

The formation of a matrix element considered in terms of functions represented as vectors in the Hilbert space is illustrated in Fig. 4.2. We can, if we wish, write out the matrix explicitly for the operator $\hat{A}$ obtaining with the notation of Eq. (4.57)

$$\hat{A} \equiv \begin{bmatrix} \langle\psi_1|\hat{A}|\psi_1\rangle & \langle\psi_1|\hat{A}|\psi_2\rangle & \langle\psi_1|\hat{A}|\psi_3\rangle & \cdots \\ \langle\psi_2|\hat{A}|\psi_1\rangle & \langle\psi_2|\hat{A}|\psi_2\rangle & \langle\psi_2|\hat{A}|\psi_3\rangle & \cdots \\ \langle\psi_3|\hat{A}|\psi_1\rangle & \langle\psi_3|\hat{A}|\psi_2\rangle & \langle\psi_3|\hat{A}|\psi_3\rangle & \cdots \\ \vdots & \vdots & \vdots & \ddots \end{bmatrix} \tag{4.59}$$

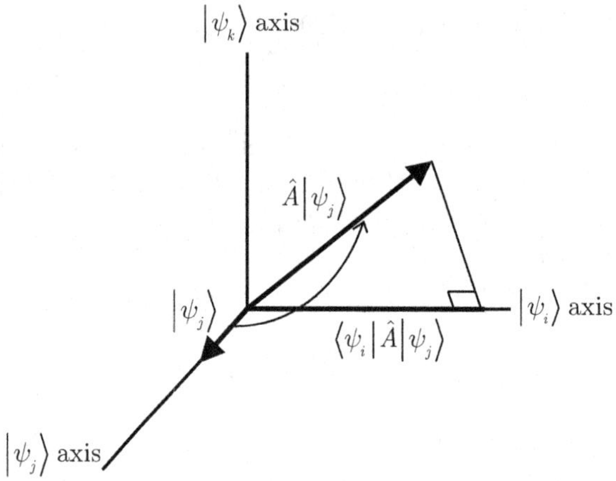

Fig. 4.2. Illustration of a matrix element of an operator $\hat{A}$ visualized in terms of vector components. Operator $\hat{A}$ acting on the unit vector $|\psi_j\rangle$ generates the vector $\hat{A}|\psi_j\rangle$, which, in general, has a different length and direction from the original vector $|\psi_j\rangle$. The matrix element $A_{ij} \equiv \langle\psi_i|\hat{A}|\psi_j\rangle$ is the projection of the vector $\hat{A}|\psi_j\rangle$ onto the $|\psi_i\rangle$ axis.

Now we have therefore deduced both how to set up the function as a vector in function space (establishing the components through Eq. (4.15)) and how to set up a linear operator as a matrix (through Eq. (4.58) or, equivalently, (4.57)) that operates on those vectors in the function space.

4.6 Bilinear expansion of linear operators

We know that we can expand functions in a basis set, as in Eqs. (4.9) and (4.13) or in bra-ket notation – for example, Eq. (4.16). What is the equivalent form of expansion for an operator? We can deduce this from the previous matrix representation. We do so by considering an arbitrary function f in the function space, written in ket form as $|f\rangle$, from which a function g (written as the ket $|g\rangle$) can be calculated by acting with a specific operator $\hat{A}$; that is,

$$|g\rangle = \hat{A}|f\rangle \tag{4.60}$$

The arbitrariness of the function $|f\rangle$ allows us to deduce a general expression for the operator $\hat{A}$. We presume that g and f are expanded on the basis set ψ_i; that is, in function space we have

$$|g\rangle = \sum_i d_i|\psi_i\rangle \tag{4.61}$$

$$|f\rangle = \sum_j c_j|\psi_j\rangle \tag{4.62}$$

From our matrix representation, Eq. (4.50), of the expression (4.60), we know that

$$d_i = \sum_j A_{ij} c_j \tag{4.63}$$

and, by definition of the expansion coefficient, we know that

$$c_j = \langle \psi_j | f \rangle \tag{4.64}$$

Hence, Eq. (4.63) becomes

$$d_i = \sum_j A_{ij} \langle \psi_j | f \rangle \tag{4.65}$$

and, substituting back into Eq. (4.61)

$$|g\rangle = \sum_{i,j} A_{ij} \langle \psi_j | f \rangle |\psi_i\rangle \tag{4.66}$$

Remember that $\langle \psi_j | f \rangle \equiv c_j$ is simply a number, so we can move it within the multiplicative expression. Hence, we have

$$|g\rangle = \sum_{i,j} A_{ij} |\psi_i\rangle \langle \psi_j | f \rangle \tag{4.67}$$

But $| f \rangle$ represents an arbitrary function in the space, so we therefore conclude that the operator $\hat{A}$ can be represented as

$$\hat{A} \equiv \sum_{i,j} A_{ij} |\psi_i\rangle \langle \psi_j | \tag{4.68}$$

This form, Eq. (4.68), is referred to as a *bilinear expansion* of the operator and is analogous to the linear expansion of a vector.

In integral notation for functions of a simple variable, we have, analogously, the relation

$$g(x) = \int \hat{A} f(x_1)\, dx_1 \tag{4.69}$$

which leads to the analogous form of the bilinear expansion

$$\hat{A} \equiv \sum_{i,j} A_{ij} \psi_i(x) \psi_j^*(x_1) \tag{4.70}$$

Note that these bilinear expansions can completely represent any linear operator that operates within the space; that is, for which the result of operating on a vector (function) with the operator is always a vector (function) in the same space.

Note, incidentally, that an expression of the form of Eq. (4.68) contains an *outer* product of two vectors. Whereas an inner product expression of the form $\langle g | f \rangle$ results in a single, complex number, an outer product expression of the form $|g\rangle\langle f|$ generates a matrix; that is, explicitly for the outer product

$$|g\rangle\langle f| = \begin{bmatrix} d_1 \\ d_2 \\ d_3 \\ \vdots \end{bmatrix} \begin{bmatrix} c_1^* & c_2^* & c_3^* & \cdots \end{bmatrix} = \begin{bmatrix} d_1c_1^* & d_1c_2^* & d_1c_3^* & \cdots \\ d_2c_1^* & d_2c_2^* & d_2c_3^* & \cdots \\ d_3c_1^* & d_3c_2^* & d_3c_3^* & \cdots \\ \vdots & \vdots & \vdots & \ddots \end{bmatrix} \tag{4.71}$$

The specific summation in Eq. (4.68) is actually, then, a sum of matrices, with the matrix $|\psi_i\rangle\langle\psi_j|$ having the element in the ith row and the jth column being one and all other elements being zero.[5] Such outer product expressions for operators are very common in quantum mechanics.

Problem

4.6.1 In the notation in which functions in a Hilbert space are expressed as vectors in that space and operators are expressed as matrices, for functions $|f\rangle$ and $|g\rangle$ and an operator $\hat{A}$, state where each of the following expressions corresponds to a column vector, a row vector, a matrix, or a complex number.

(i) $\langle f|g\rangle$

(ii) $\langle f|\hat{A}$

(iii) $|f\rangle\langle g|$

(iv) $\hat{A}|f\rangle\langle g|$

(v) $\hat{A}^\dagger|f\rangle(|f\rangle)^\dagger$

4.7 Specific important types of linear operators

In the use of Hilbert spaces, there are a few specific types of linear operators that are very important. Four of those are (1) the identity operator, (2) inverse operators, (3) unitary operators, and (4) Hermitian operators. The identity operator is relatively simple and obvious, but is important in the algebra of operators. Often, the mathematical solution to a physical problem involves finding an inverse operator and inverse operators are also important in operator algebra. Unitary operators are very useful for changing the basis for representing the vectors. They also occur, in a quite different usage, as the operators that describe the evolution of quantum mechanical systems. Hermitian operators are used to represent measurable quantities and they have some very powerful mathematical properties. In the following sections, we discuss these types of operators and their properties.

4.8 Identity operator

The identity operator $\hat{I}$ is that operator that, when it operates on a function, leaves the function unchanged. In matrix form, the identity operator is, obviously

$$\hat{I} = \begin{bmatrix} 1 & 0 & 0 & \cdots \\ 0 & 1 & 0 & \cdots \\ 0 & 0 & 1 & \cdots \\ \vdots & \vdots & \vdots & \ddots \end{bmatrix} \tag{4.72}$$

In bra-ket form, the identity operator can be written in the form

$$\hat{I} = \sum_i |\psi_i\rangle\langle\psi_i| \tag{4.73}$$

where the $|\psi_i\rangle$ form a complete basis for the function space of interest. Note, incidentally, that Eq. (4.73) holds for *any* complete basis, a fact that turns out to be very useful in the algebra of

[5] We presume here that we are working with the ψ_n as the basis functions for our vector space.

linear operators and to which we return later. Let us prove the statement Eq. (4.73). Consider the arbitrary function

$$|f\rangle = \sum_i c_i |\psi_i\rangle \tag{4.74}$$

By definition (or, explicitly, by multiplying on left and right by $\langle\psi_m|$), we know that

$$c_m = \langle\psi_m|f\rangle \tag{4.75}$$

so, explicitly

$$|f\rangle = \sum_i \langle\psi_i|f\rangle|\psi_i\rangle \tag{4.76}$$

Now consider $\hat{I}|f\rangle$, where we use the definition of $\hat{I}$ we had proposed in Eq. (4.73). We have

$$\hat{I}|f\rangle = \sum_i |\psi_i\rangle\langle\psi_i|f\rangle \tag{4.77}$$

But $\langle\psi_i|f\rangle$ is simply a number and so can be moved in the product. Hence

$$\hat{I}|f\rangle = \sum_i \langle\psi_i|f\rangle|\psi_i\rangle \tag{4.78}$$

and, hence, using Eq. (4.76), we have proved that for arbitrary $|f\rangle$

$$\hat{I}|f\rangle = |f\rangle \tag{4.79}$$

and so our proposed representation of the identity operator, Eq. (4.73), is correct.

The reader may be asking why we have gone to all the trouble of proving Eq. (4.73). The statement Eq. (4.72) of the identity operator seems sufficient. After all, the statement Eq. (4.73) is trivial if $|\psi_i\rangle$ is the basis being used to represent the space. Then

$$|\psi_1\rangle = \begin{bmatrix} 1 \\ 0 \\ 0 \\ \vdots \end{bmatrix},\ |\psi_2\rangle = \begin{bmatrix} 0 \\ 1 \\ 0 \\ \vdots \end{bmatrix},\ |\psi_3\rangle = \begin{bmatrix} 0 \\ 0 \\ 1 \\ \vdots \end{bmatrix},\ \ldots \tag{4.80}$$

so that

$$|\psi_1\rangle\langle\psi_1| = \begin{bmatrix} 1 & 0 & 0 & \cdots \\ 0 & 0 & 0 & \cdots \\ 0 & 0 & 0 & \cdots \\ \vdots & \vdots & \vdots & \ddots \end{bmatrix},\ |\psi_2\rangle\langle\psi_2| = \begin{bmatrix} 0 & 0 & 0 & \cdots \\ 0 & 1 & 0 & \cdots \\ 0 & 0 & 0 & \cdots \\ \vdots & \vdots & \vdots & \ddots \end{bmatrix},\ |\psi_3\rangle\langle\psi_3| = \begin{bmatrix} 0 & 0 & 0 & \cdots \\ 0 & 0 & 0 & \cdots \\ 0 & 0 & 1 & \cdots \\ \vdots & \vdots & \vdots & \ddots \end{bmatrix} \tag{4.81}$$

and obviously $\sum_i |\psi_i\rangle\langle\psi_i|$ gives the identity matrix of Eq. (4.72). Note, however, that the statement Eq. (4.73) is true *even if the basis being used to represent the space is not* $|\psi_i\rangle$. In that case, $|\psi_i\rangle$ is not a simple vector with the ith element equal to one and all other elements zero, and the matrix $|\psi_i\rangle\langle\psi_i|$, in general, has possibly all of its elements nonzero. Nonetheless, the sum of all of those matrices $|\psi_i\rangle\langle\psi_i|$ still leads to the identity matrix of Eq. (4.72). The important point for the algebra is that we can choose *any* convenient complete basis to write the identity operator in the form Eq. (4.73).

We can understand why the identity operator can be written this way for an arbitrary complete set of basis vectors (functions) $|\psi_i\rangle$. In an expression

$$|f\rangle = \sum_i |\psi_i\rangle\langle\psi_i|f\rangle \tag{4.82}$$

the bra $\langle\psi_i|$ projects out the component, c_i, of the vector (function) $|f\rangle$ of interest and multiplying by the ket $|\psi_i\rangle$ adds into the resulting vector (function) on the left an amount c_i of the vector (function) $|\psi_i\rangle$. Adding up all such components in the sum merely reconstructs the entire vector (function) $|f\rangle$.

An important point about thinking of functions as vectors is that, of course, the vector is the same vector regardless of which set of coordinate axes we choose to use to represent it. If we think about the identity operator in terms of vectors, then the identity operator is that operator that leaves any vector unchanged. Looked at that way, it is obvious that the identity operator is independent of which coordinate axes we use in the space. Our algebra here is merely showing that we have set up the rules for the vector space so that we get the behavior we wanted to have.

Trace of an operator

The identity matrix in the form we have defined in Eq. (4.73) can be very useful in formal proofs. The tricks are, first, that we can insert it, expressed on any convenient basis, within other expressions; and, second, we can often rearrange expressions to find identity operators buried within them that we can then eliminate to simplify the expressions. A good illustration of this is the proof that the sum of the diagonal elements of an operator $\hat{A}$ is independent of the basis on which we represent the operator; that sum of diagonal elements is called the "trace" of the operator and is written as $Tr(\hat{A})$. The trace itself can be quite useful in various situations related to operators and some of these will occur later.

Let us consider the sum, S, of the diagonal elements of an operator $\hat{A}$, on some complete orthonormal basis $|\psi_i\rangle$; that is,

$$S = \sum_i \langle\psi_i|\hat{A}|\psi_i\rangle \tag{4.83}$$

Now let us suppose we have some other complete orthonormal basis, $|\phi_m\rangle$. We can therefore write the identity operator as

$$\hat{I} = \sum_m |\phi_m\rangle\langle\phi_m| \tag{4.84}$$

We can insert an identity operator just before the operator $\hat{A}$ in Eq. (4.83), which makes no difference to the result because $\hat{I}\hat{A} = \hat{A}$, so we have

$$S = \sum_i \langle\psi_i|\hat{I}\hat{A}|\psi_i\rangle = \sum_i \langle\psi_i|\left(\sum_m |\phi_m\rangle\langle\phi_m|\right)\hat{A}|\psi_i\rangle \tag{4.85}$$

Rearranging gives

$$\begin{aligned} S &= \sum_m\sum_i \langle\psi_i|\phi_m\rangle\langle\phi_m|\hat{A}|\psi_i\rangle = \sum_m\sum_i \langle\phi_m|\hat{A}|\psi_i\rangle\langle\psi_i|\phi_m\rangle \\ &= \sum_m \langle\phi_m|\hat{A}\left(\sum_i |\psi_i\rangle\langle\psi_i|\right)|\phi_m\rangle \end{aligned} \tag{4.86}$$

where we have used the fact that $\langle \psi_i | \phi_m \rangle$ and $\langle \phi_m | \hat{A} | \psi_i \rangle$ are simply numbers and so can be swapped.

Now we see that we have another identity operator inside an expression in the bottom line; that is,

$$\hat{I} = \sum_i |\psi_i\rangle\langle\psi_i| \tag{4.87}$$

and so, because $\hat{A}\hat{I} = \hat{A}$, we can remove this operator from the expression, leaving

$$S = \sum_m \langle \phi_m | \hat{A} | \phi_m \rangle \tag{4.88}$$

Hence, from Eqs. (4.83) and (4.88), we have proved that the sum of the diagonal elements (i.e., the trace) of an operator is independent of the basis used to represent the operator, which is why the trace can be a useful property of an operator.

Problem

4.8.1 Prove that the sum of the modulus squared of the matrix elements of a linear operator $\hat{A}$ is independent of the complete orthonormal basis used to represent the operator.

4.9 Inverse operator

Now that we have defined the identity operator, we can formally consider an inverse operator. If we consider an operator $\hat{A}$ operating on an arbitrary function $|f\rangle$, then the inverse operator, if it exists, is that operator $\hat{A}^{-1}$ such that

$$|f\rangle = \hat{A}^{-1}\hat{A}|f\rangle \tag{4.89}$$

Because the function $|f\rangle$ is arbitrary, we can therefore identify

$$\hat{A}^{-1}\hat{A} = \hat{I} \tag{4.90}$$

that is, the identity operator. Viewed as a vector operation, the operator $\hat{A}$ takes an "input" vector and, in general, stretches it and reorients it. The inverse operator does exactly the opposite, restoring the original input vector.

Because the operator can be represented by a matrix, finding the inverse of the operator reduces to finding the inverse of a matrix, which is a standard linear algebra operation that is equivalent to solving a system of simultaneous linear equations.

Just as in matrix theory, not all operators have inverses. For example, the projection operator

$$\hat{P} = |f\rangle\langle f| \tag{4.91}$$

in general has no inverse because it projects all input vectors onto only one axis in the space, the one corresponding to the vector $|f\rangle$. This is a "many to one" mapping in vector space and there is now no way of knowing anything about the specific input vector other than its component along this axis. Hence, in general, we cannot go backward to the original input vector starting from this information alone.

4.10 Unitary operators

A *unitary operator*, $\hat{U}$, is one that obeys the following relation

$$\hat{U}^{-1} = \hat{U}^{\dagger} \tag{4.92}$$

That is, its inverse is its Hermitian transpose (or adjoint). The Hermitian transpose of a matrix is formed by reflecting the matrix about its diagonal and taking the complex conjugate, just as we discussed previously in the special case of a vector. Explicitly

$$\begin{bmatrix} u_{11} & u_{12} & u_{13} & \cdots \\ u_{21} & u_{22} & u_{23} & \cdots \\ u_{31} & u_{32} & u_{33} & \cdots \\ \vdots & \vdots & \vdots & \ddots \end{bmatrix}^{\dagger} = \begin{bmatrix} u_{11}^* & u_{21}^* & u_{31}^* & \cdots \\ u_{12}^* & u_{22}^* & u_{32}^* & \cdots \\ u_{13}^* & u_{23}^* & u_{33}^* & \cdots \\ \vdots & \vdots & \vdots & \ddots \end{bmatrix} \tag{4.93}$$

Conservation of length and inner product under unitary transformations

One important property of a unitary operator is that when it operates on a vector, it does not change the length of the vector. This is consistent with the "unit" part of the term unitary. In fact, more generally, when we operate on two vectors with the same unitary operator, we do not change their inner product (the conservation of length follows from this as a special case, as we will show).

Consider the unitary operator $\hat{U}$ and the two vectors $|f_{old}\rangle$ and $|g_{old}\rangle$. We form two new vectors by operating with $\hat{U}$

$$|f_{new}\rangle = \hat{U}|f_{old}\rangle \tag{4.94}$$

and

$$|g_{new}\rangle = \hat{U}|g_{old}\rangle \tag{4.95}$$

In conventional matrix (or matrix-vector) multiplication with real matrix elements, we know that

$$(AB)^T = B^T A^T \tag{4.96}$$

where the superscript "T" indicates the transpose (i.e., reflection about the diagonal). In matrix or operator multiplication with complex elements, we analogously obtain

$$\left(\hat{A}\hat{B}\right)^{\dagger} = \hat{B}^{\dagger}\hat{A}^{\dagger} \tag{4.97}$$

and, explicitly, for matrix-vector multiplication (because a vector is just a special case of a matrix)

$$\left(\hat{A}|h\rangle\right)^{\dagger} = \langle h|\hat{A}^{\dagger} \tag{4.98}$$

Hence,

$$\langle g_{new}|f_{new}\rangle = \langle g_{old}|\hat{U}^{\dagger}\hat{U}|f_{old}\rangle = \langle g_{old}|\hat{U}^{-1}\hat{U}|f_{old}\rangle = \langle g_{old}|\hat{I}|f_{old}\rangle = \langle g_{old}|f_{old}\rangle \tag{4.99}$$

so, as promised, the inner product is not changed if both vectors are transformed this way. In particular

$$\langle f_{new} | f_{new} \rangle = \langle f_{old} | f_{old} \rangle \tag{4.100}$$

That is, the length of a vector is not changed by a unitary operator.

Changing the representation of vectors

One major use of unitary operators is to change basis sets (or, equivalently, representations or coordinate axes). Suppose that we have a vector (function) $|f_{old}\rangle$ that is represented, when we express it as an expansion on the functions $|\psi_n\rangle$, as the mathematical column vector

$$|f_{old}\rangle = \begin{bmatrix} c_1 \\ c_2 \\ c_3 \\ \vdots \end{bmatrix} \tag{4.101}$$

These numbers c_1, c_2, c_3, … are the projections of $|f_{old}\rangle$ on the orthogonal coordinate axes in the vector space labeled with $|\psi_1\rangle$, $|\psi_2\rangle$, $|\psi_3\rangle$ … . Suppose now we wish to change to representing this vector on a new set of orthogonal coordinate axes, which we label $|\phi_1\rangle$, $|\phi_2\rangle$, $|\phi_3\rangle$, … . Changing the axes, which is equivalent to changing the basis set of functions, does not, of course, change the vector we are representing, but it does change the column of numbers used to represent the vector from those in Eq. (4.101) to some new set of numbers.

For example, suppose for simplicity that the original vector was actually the first basis vector in the old basis, $|\psi_1\rangle$ (which is simply the vector with 1 in the first element and zero in all the others). Then, in this new representation, the elements in the column of numbers would be the projections of this vector on the various new coordinate axes, each of which is simply $\langle \phi_m | \psi_1 \rangle$; that is, under this coordinate transformation (or change of basis)

$$\begin{bmatrix} 1 \\ 0 \\ 0 \\ \vdots \end{bmatrix} \Rightarrow \begin{bmatrix} \langle \phi_1 | \psi_1 \rangle \\ \langle \phi_2 | \psi_1 \rangle \\ \langle \phi_3 | \psi_1 \rangle \\ \vdots \end{bmatrix} \tag{4.102}$$

We could write out similar transformations for each of the other original basis vectors $|\psi_n\rangle$. We can see that we get the correct transformation if we define a matrix

$$\hat{U} = \begin{bmatrix} u_{11} & u_{12} & u_{13} & \cdots \\ u_{21} & u_{22} & u_{23} & \cdots \\ u_{31} & u_{32} & u_{33} & \cdots \\ \vdots & \vdots & \vdots & \ddots \end{bmatrix} \tag{4.103}$$

where

$$u_{ij} = \langle \phi_i | \psi_j \rangle \tag{4.104}$$

and define our new column of numbers $|f_{new}\rangle$ as

$$|f_{new}\rangle = \hat{U} |f_{old}\rangle \tag{4.105}$$

Note, incidentally, that, in this case, $|f_{old}\rangle$ and $|f_{new}\rangle$ refer to the same vector in the vector space; it is only the representation (the coordinate axes) and, consequently, the column of numbers that have changed, not the vector itself. This is rather like looking at, for example, a sculpture from different viewing positions. Suppose, for example, that we are in some modern art gallery in which an artist has made a sculpture in the form of an arrow sticking at an angle up out of the floor. As we change our position, our view of the arrow changes. We could write down a representation of the arrow's length and direction (e.g., the tip of the arrow is 2.5 m above the floor, leaning 50 cm to the left and 20 cm back toward us). If we move to another position, the representation we would write down would change, though the arrow remains the same.[6] (In fact, such a change in viewing position can be exactly described by a unitary operator.)

Now we can prove that $\hat{U}$ is unitary. Writing the matrix multiplication in its sum form, we have

$$\begin{aligned}\left(\hat{U}^\dagger\hat{U}\right)_{ij} &= \sum_m u_{mi}^* u_{mj} = \sum_m \langle\phi_m|\psi_i\rangle^* \langle\phi_m|\psi_j\rangle \\ &= \sum_m \langle\psi_i|\phi_m\rangle\langle\phi_m|\psi_j\rangle = \langle\psi_i|\left(\sum_m |\phi_m\rangle\langle\phi_m|\right)|\psi_j\rangle \\ &= \langle\psi_i|\hat{I}|\psi_j\rangle = \langle\psi_i|\psi_j\rangle \\ &= \delta_{ij}\end{aligned} \tag{4.106}$$

The statement $(\hat{U}^\dagger\hat{U})_{ij} = \delta_{ij}$ is equivalent to saying we have a matrix with one for every diagonal element and zeros for all the others (i.e., the identity matrix), so

$$\hat{U}^\dagger\hat{U} = \hat{I} \tag{4.107}$$

and, hence, $\hat{U}$ is unitary because its Hermitian transpose is therefore its inverse (Eq. (4.92). Hence, any change in basis can be implemented with a unitary operator. We can also say that any such change in representation to a new orthonormal basis is a unitary transform.

Note also, incidentally, that

$$\hat{U}\hat{U}^\dagger = \left(\hat{U}^\dagger\hat{U}\right)^\dagger = \hat{I}^\dagger = \hat{I} \tag{4.108}$$

Given that we concluded previously that a unitary transform did not change any inner product, we can now also conclude that a transformation to a new orthonormal basis does not change any inner product. Again, this is as we would have expected from thinking about the inner product being like a vector dot product of two geometrical vectors; of course, such an inner product does not depend on the coordinate axes, only on the directions and lengths of the vectors themselves.

Changing the representation of operators

We have discussed so far how to change the column of numbers that represents a vector (function) in vector space when we change the basis. What happens to the matrix of numbers that represents an operator when we change the basis?

---

[6] We can think of this arrow as being like a vector in Hilbert space. We just have to remember for our vectors in Hilbert space that the projections onto the various coordinate axes can be complex numbers and we can have as many coordinate axes as we need, not just the three of geometrical space.

We can understand what the new representation of $\hat{A}$ is by considering an expression such as

$$\begin{aligned}\langle g_{new}|\hat{A}_{new}|f_{new}\rangle &= \left(|g_{new}\rangle\right)^{\dagger}\hat{A}_{new}|f_{new}\rangle \\ &= \left(\hat{U}|g_{old}\rangle\right)^{\dagger}\hat{A}_{new}\left(\hat{U}|f_{old}\rangle\right) = \langle g_{old}|\hat{U}^{\dagger}\hat{A}_{new}\hat{U}|f_{old}\rangle\end{aligned} \tag{4.109}$$

where the vectors $|f\rangle$ and $|g\rangle$ are arbitrary. Note here also that the subscripts *new* and *old* refer to the representations, not the vectors (or operators). The actual vectors and operators are not changed by the change of representation; only the sets of numbers that represent them are changed. Hence, the result of such an expression should not be changed by changing the representation. So we believe that

$$\langle g_{new}|\hat{A}_{new}|f_{new}\rangle = \langle g_{old}|\hat{A}_{old}|f_{old}\rangle \tag{4.110}$$

Because this is to be true for arbitrary choices of vectors, we can write an equation for the operators themselves and from Eqs. (4.109) and (4.110), we can deduce that

$$\hat{A}_{old} = \hat{U}^{\dagger}\hat{A}_{new}\hat{U} \tag{4.111}$$

or, equivalently,

$$\hat{U}\hat{A}_{old}\hat{U}^{\dagger} = \left(\hat{U}\hat{U}^{\dagger}\right)\hat{A}_{new}\left(\hat{U}\hat{U}^{\dagger}\right) = \hat{A}_{new} \tag{4.112}$$

We can understand this expression directly. $\hat{U}$ is the operator that takes us from the old coordinate system to the new one. If we are in the new coordinate system, therefore, $\hat{U}^{-1} = \hat{U}^{\dagger}$ is the operator that takes us back to the old system. So, to operate with the operator $\hat{A}$ in the new coordinate system when we only know its representation, $\hat{A}_{old}$, in the old coordinate system, we first operate with $\hat{U}^{\dagger}$ to take us into the old system, then operate with $\hat{A}_{old}$, then operate with $\hat{U}$ to take us back to the new system.

Unitary operators that change the state vector

The unitary operator we discussed previously for changing basis sets is one important application of unitary operators in quantum mechanics. There is another one, which is different in character. In general, a linear operator in Hilbert space can change the "orientation" of a vector in the space and change its length by a factor. In the language of our modern art gallery analogy, a linear operator operating on the arrow itself would, in general, move the arrow sticking up out of the floor to a new angle and possibly lengthen it or shorten it. A unitary linear operator can rotate the arrow but leaves its length unchanged. Such operators are not changing the basis set – they are actually changing the state of the quantum mechanical system and changing the vector's orientation in vector space. Hence, in our modern art gallery, an operator that rotates the arrow to a new angle, while we stay standing exactly where we are, is also a unitary operator – this time actually changing the state of the arrow, not describing a change of our viewing position as previously.

Operators that change the quantum mechanical state of the system can quite often be unitary. One reason why such operators arise in quantum mechanics is simple to see. If we are working, for example, with a single particle, then presumably the sum of all the occupation probabilities of all the possible states is unity. That is, if the quantum mechanical state $|\psi\rangle$ is expanded on the complete orthonormal basis $|\psi_n\rangle$

$$|\psi\rangle = \sum_n a_n|\psi_n\rangle \tag{4.113}$$

then $\sum_n |a_n|^2 = 1$. If the particle is to be conserved, then this sum is retained as the quantum mechanical system evolves, for example in time. But this sum is just the square of the length of the vector $|\psi\rangle$. Hence, a unitary operator, which conserves length, is an appropriate operator for describing changes in that system that conserve the particle. For example, the time-evolution operator for a system in which the Hamiltonian does not change in time, $\exp(-i\hat{H}t/\hbar)$, can be shown to be unitary, the explicit proof of which is left as an exercise for the reader.

Problems

4.10.1* Consider the operator $\hat{M}_{old} = \begin{bmatrix} 0 & i \\ -i & 0 \end{bmatrix}$

(i) What are the eigenvalues and associated (normalized) eigenvectors of this operator?

(ii) What is the unitary transformation operator that will diagonalize this operator (i.e., the matrix that will change the representation from the old basis to a new basis in which the operator is now represented by a diagonal matrix)? Presume that the eigenvectors in the new basis are $\begin{bmatrix} 1 \\ 0 \end{bmatrix}$ and $\begin{bmatrix} 0 \\ 1 \end{bmatrix}$, respectively.

(iii) What is the operator $\hat{M}_{new}$ in this new basis?

4.10.2 Evaluate the unitary matrix for transforming a vector in two-dimensional space to a representation in a new set of axes rotated by an angle θ in an anticlockwise direction.

4.10.3 Consider the orthonormal basis functions $\psi_1(x) = 1/\sqrt{2}$ and $\psi_2(x) = \sqrt{3/2}\,x$ that are capable of representing any function of the form $f(x) = ax + b$ defined over the range $-1 < x < 1$.

(i) Consider now the new basis functions $\phi_1(x) = \dfrac{\sqrt{3}x}{2} + \dfrac{1}{2}$ and $\phi_2(x) = \dfrac{\sqrt{3}x}{2} - \dfrac{1}{2}$.

Represent the functions $\phi_1(x)$ and $\phi_2(x)$ in a two-dimensional diagram with orthogonal axes corresponding to the functions $\psi_1(x)$ and $\psi_2(x)$, respectively.

(ii) Construct the matrix that will transform a function in the old representation as a vector

$$\begin{bmatrix} c_1 \\ c_2 \end{bmatrix}$$

into a new representation in terms of these new basis functions as a vector

$$\begin{bmatrix} d_1 \\ d_2 \end{bmatrix}$$

where an arbitrary function $f(x) = ax + b$ is represented as the linear combination

$$f(x) = d_1\phi_1(x) + d_2\phi_2(x)$$

(iii) Show that the matrix from part (ii) is unitary.

(iv) Use the matrix of part (ii) to calculate the vector $\begin{bmatrix} d_1 \\ d_2 \end{bmatrix}$ for the specific example function $2x + 3$.

(v) Indicate the resulting vector on the same diagram as used for part (i).

4.10.4 Consider the so-called Pauli matrices

$$\hat{\sigma}_x = \begin{bmatrix} 0 & 1 \\ 1 & 0 \end{bmatrix}, \quad \hat{\sigma}_y = \begin{bmatrix} 0 & -i \\ i & 0 \end{bmatrix}, \quad \hat{\sigma}_z = \begin{bmatrix} 1 & 0 \\ 0 & -1 \end{bmatrix}$$

(which are used in quantum mechanics as the operators corresponding to the *x*, *y*, and *z* components of the spin of an electron, though for the purposes of this problem we can consider them simply as abstract operators represented by matrices). For this problem, find all the requested eigenvalues and eigenvectors by hand (i.e., not using a calculator or computer to find the eigenvalues and eigenvectors) and show your calculations.

(i) Find the eigenvalues and corresponding (normalized) eigenvectors $|\psi_{zi}\rangle$ of the operator $\hat{\sigma}_z$.

(ii) Find the eigenvalues and corresponding (normalized) eigenvectors $|\psi_{xi}\rangle$ of the operator $\hat{\sigma}_x$.

(iii) Show by explicit calculation that $\sum_i |\psi_{xi}\rangle\langle\psi_{xi}| = \hat{I}$, where $\hat{I}$ is the identity matrix in this two-dimensional space.

(iv) These operators have been represented in a basis that is the set of eigenvectors of $\hat{\sigma}_z$. Transform all three of the Pauli matrices into a representation that uses the set of eigenvectors of $\hat{\sigma}_x$ as the basis.

4.10.5 Consider an operator $\hat{W} = \sum_{i,j} a_{ij} |\phi_i\rangle\langle\psi_j|$ where $|\phi_i\rangle$ and $|\psi_j\rangle$ are two different complete sets of functions. Show that if the columns of the matrix representation of the operator are orthogonal (i.e., if $\sum_i a_{iq}^* a_{ij} = \delta_{qj}$), then the operator $\hat{W}$ is unitary. (Note: When multiplying operators represented by expansions, use different indices in the expansion summations for the different operators.)

4.10.6 Prove that the time-evolution operator $\hat{A} = \exp(-i\hat{H}t/\hbar)$ is unitary.

See also Problem 13.4.2.

4.11 Hermitian operators

A *Hermitian operator* is one that is its own Hermitian adjoint; that is,

$$\hat{M}^\dagger = \hat{M} \tag{4.114}$$

We can also equivalently say that a Hermitian operator is self-adjoint.

Expressed in matrix terms, we have, with

$$\hat{M} = \begin{bmatrix} M_{11} & M_{12} & M_{13} & \cdots \\ M_{21} & M_{22} & M_{23} & \cdots \\ M_{31} & M_{32} & M_{33} & \cdots \\ \vdots & \vdots & \vdots & \ddots \end{bmatrix} \tag{4.115}$$

that

$$\hat{M}^\dagger = \begin{bmatrix} M_{11}^* & M_{21}^* & M_{31}^* & \cdots \\ M_{12}^* & M_{22}^* & M_{31}^* & \cdots \\ M_{13}^* & M_{23}^* & M_{33}^* & \cdots \\ \vdots & \vdots & \vdots & \ddots \end{bmatrix} \tag{4.116}$$

so the Hermiticity condition, Eq. (4.114), implies

$$M_{ij} = M_{ji}^* \tag{4.117}$$

for all i and j. Incidentally, this condition, Eq. (4.117), implies that the diagonal elements of a Hermitian operator must be real.

Mathematically, statements about operators in vector spaces are only valid insofar as they apply to the operator operating on arbitrary functions. To understand what the Hermiticity statement (Eq. (4.114)) means for actions on functions in general, we can examine the result $\langle g|\hat{M}|f\rangle$. This result is from the multiplication of the vector $\langle g|$, the matrix $\hat{M}$, and the vector $|f\rangle$ and so we can consider the Hermitian adjoint of this result, $(\langle g|\hat{M}|f\rangle)^\dagger$, using the rules for the adjoints of the products of matrices (and vectors as special cases of matrices) – specifically, the relation Eq. (4.97) ($(\hat{A}\hat{B})^\dagger = \hat{B}^\dagger\hat{A}^\dagger$). Of course, in the specific case of the result $\langle g|\hat{M}|f\rangle$, the resulting matrix is a "one-by-one" matrix that can also be considered as simply a number and so

$$\left(\langle g|\hat{M}|f\rangle\right)^\dagger \equiv \left(\langle g|\hat{M}|f\rangle\right)^* \tag{4.118}$$

Hence, we have, using the rule for the adjoint of products of matrices, for any functions *f* and *g*,

$$\begin{aligned}\left(\langle g|\hat{M}|f\rangle\right)^* &\equiv \left(\langle g|\hat{M}|f\rangle\right)^\dagger = \left[\langle g|\left(\hat{M}|f\rangle\right)\right]^\dagger = \left(\hat{M}|f\rangle\right)^\dagger\left(\langle g|\right)^\dagger = \left(|f\rangle\right)^\dagger \hat{M}^\dagger\left(\langle g|\right)^\dagger \\ &= \langle f|\hat{M}^\dagger|g\rangle\end{aligned} \tag{4.119}$$

Now we use the Hermiticity of $\hat{M}$, Eq. (4.114), and obtain

$$\langle f|\hat{M}|g\rangle = \left(\langle g|\hat{M}|f\rangle\right)^* \tag{4.120}$$

which could be regarded as the most complete and general way of stating the Hermiticity of an operator $\hat{M}$. Note that this is true even if $|f\rangle$ and $|g\rangle$ are not orthogonal. The statement for the matrix elements, Eq. (4.117), is just a special case.

In integral form, for functions $f(x)$ and $g(x)$, the statement Eq. (4.120) of the Hermiticity of $\hat{M}$ can be written

$$\int g^*(x)\hat{M}f(x)dx = \left[\int f^*(x)\hat{M}g(x)dx\right]^* \tag{4.121}$$

We can rewrite the right-hand side using the property $(ab)^* = a^*b^*$ of complex conjugates to obtain

$$\int g^*(x)\hat{M}f(x)dx = \int f(x)\left\{\hat{M}g(x)\right\}^* dx \tag{4.122}$$

and a simple rearrangement leads to

$$\int g^*(x)\hat{M}f(x)dx = \int\left\{\hat{M}g(x)\right\}^* f(x)dx \tag{4.123}$$

Authors who prefer to introduce Hermitian operators in the integral form often use the form Eq. (4.123) to define the operator $\hat{M}$ as Hermitian. The forms Eqs. (4.114), (4.117), (4.120), and, for functions of a continuous variable, Eq. (4.123), can all be regarded as equivalent statements of the Hermiticity of the operator $\hat{M}$.

Note that the bra-ket notation is more elegant than the integral notation in one important way. In the bra-ket notation, the operator can also be considered to operate to the left; $\langle g|\hat{A}$ is just as meaningful a statement as the statement $\hat{A}|f\rangle$ and it does not matter how we group the multiplications in the bra-ket notation. That is,

$$\langle g|\hat{A}|f\rangle \equiv \left(\langle g|\hat{A}\right)|f\rangle \equiv \langle g|\left(\hat{A}|f\rangle\right) \tag{4.124}$$

because of the associativity of matrix multiplication. Conventional operators in the notation used in integration, such as a differential operator, *d/dx*, do not have any meaning when they operate "to the left"; hence, we end up with the somewhat clumsy form Eq. (4.123) for Hermiticity in this notation.

Properties of Hermitian operators

The eigenvalues and eigenvectors of Hermitian operators have some special properties, some of which are very easily proved.

Reality of eigenvalues

Suppose $|\psi_n\rangle$ is a normalized eigenvector of the Hermitian operator $\hat{M}$ with eigenvalue μ_n. Then, by definition

$$\hat{M}|\psi_n\rangle = \mu_n|\psi_n\rangle \tag{4.125}$$

Therefore

$$\langle\psi_n|\hat{M}|\psi_n\rangle = \mu_n\langle\psi_n|\psi_n\rangle = \mu_n \tag{4.126}$$

But, from the Hermiticity of $\hat{M}$, we know

$$\langle\psi_n|\hat{M}|\psi_n\rangle = \left(\langle\psi_n|\hat{M}|\psi_n\rangle\right)^* = \mu_n^* \tag{4.127}$$

and, hence, from the equality of Eqs. (4.126) and (4.127), μ_n must be real; that is, the eigenvalues of a Hermitian operator are real.[7] This suggests that such an operator may be useful for representing a quantity that is real, such as a measurable quantity.

Orthogonality of eigenfunctions for different eigenvalues

The eigenfunctions of a Hermitian operator corresponding to different eigenvalues are orthogonal, as can easily be proved in bra-ket notation. Trivially

$$0 = \langle\psi_m|\hat{M}|\psi_n\rangle - \langle\psi_m|\hat{M}|\psi_n\rangle \tag{4.128}$$

So, by associativity and the rule Eq. (4.97) ($(\hat{A}\hat{B})^\dagger = \hat{B}^\dagger\hat{A}^\dagger$)

$$\begin{aligned} 0 &= \left(\langle\psi_m|\hat{M}\right)|\psi_n\rangle - \langle\psi_m|\left(\hat{M}|\psi_n\rangle\right) \\ &= \left(\hat{M}^\dagger|\psi_m\rangle\right)^\dagger|\psi_n\rangle - \langle\psi_m|\left(\hat{M}|\psi_n\rangle\right) \end{aligned} \tag{4.129}$$

Now, using the Hermiticity of $\hat{M}$ ($\hat{M} = \hat{M}^\dagger$), the fact that the Hermitian adjoint of a complex number is simply its complex conjugate (the number is just a one-by-one matrix), and the fact that the eigenvalues of a Hermitian operator are real anyway, we have

$$0 = \mu_m\langle\psi_m|\psi_n\rangle - \mu_n\langle\psi_m|\psi_n\rangle = (\mu_m - \mu_n)\langle\psi_m|\psi_n\rangle \tag{4.130}$$

But, by assumption, μ_m and μ_n are different and, hence,

$$\langle\psi_m|\psi_n\rangle = 0 \tag{4.131}$$

---

[7] Note that the converse is not true; there are matrices with real eigenvalues that are not Hermitian matrices. E.g., the matrix $\begin{bmatrix} 1 & 2 \\ 3 & 2 \end{bmatrix}$ is not Hermitian, but has real eigenvalues, 4 and -1. It also worth noting that the resulting eigenvectors $\begin{bmatrix} 2 \\ 3 \end{bmatrix}$ and $\begin{bmatrix} 1 \\ -1 \end{bmatrix}$ are not orthogonal.

and we have proved that the eigenfunctions associated with different eigenvalues of a Hermitian operator are orthogonal.

Incidentally, it is quite possible (and actually common in problems that are highly symmetric in some way or another) to have more than one eigenfunction associated with a given eigenvalue. As discussed previously, this situation is known as *degeneracy*. It is provable, at least for a broad class of the operators that are used in quantum mechanics, that the number of such linearly independent degenerate solutions for a given finite, nonzero eigenvalue is itself finite, though we do not go into that proof here.[8]

Completeness of sets of eigenfunctions

A very important result for a broad class[9] of Hermitian operators is that, provided the operator is bounded (i.e., it gives a resulting vector of finite length when it operates on any finite input vector) the set of eigenfunctions is complete (i.e., it spans the space on which the operator operates). The proof of this result is understandable with effort but requires setting up a mathematical framework (e.g., for functional analysis) that is beyond what we can justify here.[10] This result means, in practice, that we can use the eigenfunctions of bounded Hermitian operators to expand functions. This greatly increases the available basis sets beyond the simple spatial or Fourier transform sets. For many problems, it means we can greatly simplify the description of them.

Hermitian operators and quantum mechanics

As we mentioned previously, a broad class of bounded Hermitian operators has the attractive properties of having real eigenvalues, orthogonal eigenfunctions, and complete sets of eigenfunctions. These properties make Hermitian operators powerful and quite easy to use for problems for which they are applicable. By a remarkable stroke of good fortune, it turns out that as far as we know, the physically measurable quantities in quantum mechanics can be represented by such Hermitian operators. In fact, some state this as an axiom of quantum mechanics. We have already seen momentum and energy (Hamiltonian) operators, both of which are apparently of this kind. We will encounter several other such operators corresponding to other physical quantities as we get further into quantum mechanics. All of these operators have the same algebra and properties as discussed here; hence, we have a very general, sound, and useful mathematical methodology for discussing quantum mechanics.

Problems

4.11.1 For each of the following matrices, state whether it is unitary and whether it is Hermitian.

$$\text{(i)}\begin{bmatrix}1 & 0\\ 0 & 1\end{bmatrix}\quad \text{(ii)}\begin{bmatrix}1 & i\\ -i & 1\end{bmatrix}\quad \text{(iii)}\begin{bmatrix}i & 0\\ 0 & i\end{bmatrix}\quad \text{(iv)}\begin{bmatrix}0 & 1\\ i & 0\end{bmatrix}$$

[8] It is certainly true for "compact" operators, which in practice are operators that can be approximated to any desired degree of accuracy by finite matrices. Because the sum of the modulus squared of the elements is one of the aspects that must converge as we move to a sufficiently large such finite matrix for a compact operator, then the sum of the eigenvalues must converge, which means that the degeneracy must be finite for any given degenerate eigenvalue.

[9] Again, the compact operators.

[10] See, e.g., David Porter and David S. G. Stirling, *Integral equations* (Cambridge, 1990), pp. 109–111 (proof of the spectral theorem) and pp. 112–113 (proof of completeness)

4.11.2 Prove that for two square matrices A and B, $(AB)^\dagger = B^\dagger A^\dagger$.

(Hint: Consider a general element (e.g., the ijth element) of the resulting matrix and write the result of the matrix multiplication for that element as a summation over appropriate terms.)

4.11.3 Consider the Hermiticity of the following operators:

(i) Prove that the momentum operator is Hermitian. For simplicity, you may perform this proof for a one-dimensional system (i.e., only consider functions of x and consider only the $\hat{p}_x$ operator).
(Hints: Consider $\int_{-\infty}^{\infty} \psi_i^*(x)\hat{p}_x\psi_j(x)dx$, where the $\psi_n(x)$ are a complete orthonormal set. You may want to consider an integration by parts. Note that the $\psi_n(x)$ must vanish at $\pm\infty$ because otherwise they could not be normalized.)

(ii) Is the operator $\dfrac{d}{dx}$ Hermitian? Prove your answer.

(iii) Is the operator $\dfrac{d^2}{dx^2}$ Hermitian? Prove your answer.

(Hints: You may want to consider another integration by parts and you may presume that the derivatives $\dfrac{d\psi_n(x)}{dx}$ also vanish at $\pm\infty$.)

(iv) Is the operator $\hat{H} = -\dfrac{\hbar^2}{2m}\dfrac{d^2}{dx^2} + V(x)$ Hermitian if $V(x)$ is real? Prove your answer.

4.11.4* Prove by operator algebra that a Hermitian operator transformed to a new coordinate system by a unitary transformation is still Hermitian.

4.11.5 A Hermitian operator $\hat{A}$ has a complete orthonormal set of eigenfunctions $|\psi_n\rangle$ with associated eigenvalues α_n. Show that we can always write

$$\hat{A} = \sum_i \alpha_i |\psi_i\rangle\langle\psi_i|$$

(This is known as the expansion of $\hat{A}$ in its eigenfunctions and is a very useful expansion.)

Now find a similar, simple expression for the inverse, $\hat{A}^{-1}$.

(This is also a very useful result. This result shows that if we can find the eigenfunctions of an operator, also known as *diagonalizing* the operator, we have effectively found the inverse and usually, in practice, we have solved the quantum mechanical problem of interest.)

4.11.6 Considering the expansion of $\hat{A}$ in its eigenfunctions (see Problem 4.11.5), show that the trace, $Tr(\hat{A})$, is always equal to the sum of the eigenvalues.

4.11.7* Prove the integral form of the definition of the Hermiticity of the operator $\hat{M}$

$$\int g^*(x)\hat{M}f(x)dx = \int\{\hat{M}g(x)\}^* f(x)dx$$

by expanding the functions f and g in a complete basis $|\psi_n\rangle$ and using the matrix element definition of Hermiticity

$$M_{ij} = M_{ji}^*$$

where

$$M_{ij} = \int \psi_i^*(x)\hat{M}\psi_j(x)dx$$

4.11.8 Prove for any Hermitian operator $\hat{M}$ and any arbitrary function or state $|f\rangle$ that the quantity $\langle f|\hat{M}|f\rangle$ is real. (Hence, the expectation value of any quantity represented by a Hermitian operator is always real, which is one good reason for using Hermitian operators to represent measurable quantities.)

4.12 Matrix form of derivative operators

So far, we have discussed operators as matrices and have dealt with these in the general case but have not related these matrices to the operators we have so far used in actual quantum mechanics, as in the Schrödinger equation or the momentum operator. The operators in those two cases happen to be differential operators, such as d^2/dx^2 or d/dx, and it may not be immediately obvious that those can be described as matrices. The reason for this discussion is not so that we can, in practice, use matrices to describe such operators; it is usually more convenient to handle such operators using the integral form of inner products and matrix elements. We merely wish to show how this can be done for conceptual completeness.

If we return to our original discussion of functions as vectors, we can postulate that an appropriate form for the differential operator d/dx would be

$$\frac{d}{dx} \equiv \begin{bmatrix} \ddots & & & & \\ \cdots -\frac{1}{2\delta x} & 0 & \frac{1}{2\delta x} & 0 \cdots & \\ \cdots 0 & -\frac{1}{2\delta x} & 0 & \frac{1}{2\delta x} \cdots & \\ & & & & \ddots \end{bmatrix} \tag{4.132}$$

where, as usual, we are presuming we can take the limit as $\delta x \to 0$. If we were to multiply the column vector whose elements are the values of the function $f(x)$ at a set of values spaced by an amount δx, then we would obtain

$$\begin{bmatrix} \ddots & & & & \\ \cdots -\frac{1}{2\delta x} & 0 & \frac{1}{2\delta x} & 0 \cdots & \\ \cdots 0 & -\frac{1}{2\delta x} & 0 & \frac{1}{2\delta x} \cdots & \\ & & & & \ddots \end{bmatrix} \begin{bmatrix} \vdots \\ f(x_i - \delta x) \\ f(x_i) \\ f(x_i + \delta x) \\ f(x_i + 2\delta x) \\ \vdots \end{bmatrix} = \begin{bmatrix} \vdots \\ \dfrac{f(x_i + \delta_x) - f(x_i - \delta x)}{2\delta x} \\ \dfrac{f(x_i + 2\delta_x) - f(x_i)}{2\delta x} \\ \vdots \end{bmatrix} = \begin{bmatrix} \vdots \\ \left.\dfrac{df}{dx}\right|_{x_i} \\ \left.\dfrac{df}{dx}\right|_{x_i + \delta x} \\ \vdots \end{bmatrix} \tag{4.133}$$

where again we understand that we are taking the limit as $\delta x \to 0$. Hence, we have a way of representing a derivative as a matrix.

Note that we have postulated a form that has a symmetry about the matrix diagonal. In this case, the matrix is antisymmetric in reflection about the diagonal. This matrix is not, however,

Hermitian, which reflects the fact that the operator d/dx is not a Hermitian operator, as can be verified from any of the previous definitions of Hermiticity. We can see from this matrix representation, by contrast, that the operator id/dx (or, for that matter, $-id/dx$) would be Hermitian (simply multiply all the matrix elements by i to see that we have a matrix that is Hermitian); hence, the momentum operator, such as its x component $p_x = -i\hbar d/dx$, would be Hermitian.

It is left as an exercise for the reader to show how the second derivative can be represented as a matrix and that the corresponding matrix is Hermitian.

Problem

4.12.1 Given that $d^2/dx^2 \equiv \lim_{\delta x \to 0}\left[\left(f(x-\delta x)-2f(x)+f(x+\delta x)\right)/(\delta x)^2\right]$, find an appropriate matrix that could represent such a derivative operator, in a form analogous to the first derivative operator matrix.

4.13 Matrix corresponding to multiplying by a function

One other situation we have already encountered is where the operator simply corresponds to multiplying each element in the input vector by a (different) number. For example, we can formally "operate" on the function $f(x)$ by multiplying it by the function $V(x)$ to generate another function $g(x) = V(x)f(x)$. Because the function $V(x)$ is performing the role of an operator (even though it is a particularly simple form of operator), we can, if we wish, represent it as a matrix, so that we can express it in the same form as all of our other operators. In that case, in the position representation, it is a simple diagonal matrix whose elements are the values of the function at each of the different points.

If the function is real, the corresponding matrix is Hermitian (though it is not if the function is complex). Hence, one can conclude that the Hamiltonian as used in Schrödinger's equation, being the sum of two Hermitian matrices (e.g., in the one-dimensional case, one corresponding to the Hermitian operator $(-\hbar^2/2m)\,\partial^2/\partial x^2$ and the other corresponding to the "operator" $V(x)$), is Hermitian, as long as $V(x)$ is real.

4.14 Summary of concepts

Functions as vectors

Functions can be regarded as vectors in a vector space, with the values of the function at the different coordinate points or the expansion coefficients of the function on some basis being the components of the vector along the corresponding coordinate axes in the space.

Hermitian adjoint

The Hermitian adjoint (also known as the Hermitian transpose, Hermitian conjugate, adjoint) of a vector or a matrix is the complex conjugate of the transpose of the vector or matrix. The Hermitian adjoint of the matrix $\hat{A}$ is written as $\hat{A}^\dagger$ (pronounced "A-dagger").

Dirac bra-ket notation

$|f\rangle$, called a ket, is the vector in function space that represents the function f. The Hermitian adjoint of that vector is the bra, $\langle f|$.

Inner product

The inner product in function space of two functions f and g is the vector product $\langle f|g\rangle$, which is, in general, a complex number, and we have

$$\langle f|g\rangle = \left(\langle g|f\rangle\right)^* \tag{4.36}$$

The inner product of a vector with itself gives the square of its length (also known as the norm of the vector) and always results in a real quantity. The inner product is linear in sums of functions and multiplying functions by constants.

Expansion coefficients as inner products

The expansion coefficients, c_n, of a function f on a basis ψ_n

$$f(x) = \sum_n c_n \psi_n(x) \tag{4.9}$$

are

$$c_m = \langle \psi_m | f\rangle \qquad \text{after Eq. (4.15)}$$

State vectors

Where a function f represents the quantum mechanical state of a system, the vector $|f\rangle$ is known as the state vector of the system.

Vector (or function) space

A vector or function space is a mathematical space in which the (usually multidimensional) vectors that represent functions exist.

Hilbert space

A Hilbert space is a vector space with an inner product. It is closely analogous to a conventional three-dimensional geometrical space, with two important differences: (1) the space may have any number of dimensions, including an infinite number; and (2) because coefficients can be complex in Hilbert space, the inner product is, in general, complex. It is a suitable vector space for representing vectors that are linear in both addition and in multiplication by a constant.

Operators

An operator is an entity that changes one function into another, with the value of the new function at any point possibly being dependent on the values of the original function at any or all values of its argument or arguments.

Linear operators are linear both in addition of functions and in multiplication by a constant. Linear operators can be represented by matrices that can operate on the vectors in function space and they obey the same algebra as matrices.

Elements of the matrix representation of an operator

For a matrix representing a linear operator in Hilbert space, the elements of the matrix are represented in the basis ψ_n as

$$A_{ij} = \langle \psi_i | \hat{A} | \psi_j \rangle \tag{4.57}$$

Bilinear expansion of linear operators

A linear operator in a Hilbert space can be written as

$$\hat{A} \equiv \sum_{i,j} A_{ij} |\psi_i\rangle\langle\psi_j| \quad (4.68)$$

where ψ_n is *any* complete basis in the space.

Identity operator

The identity operator acting on a function leaves the function unchanged and can be written as

$$\hat{I} = \sum_i |\psi_i\rangle\langle\psi_i| \quad (4.73)$$

where ψ_n is *any* complete basis in the space.

Trace of an operator

The trace of an operator $\hat{A}$, written as $Tr(\hat{A})$, is the sum of the diagonal elements of an operator. It is independent of the basis on which the operator is expressed.

Inverse operator

The inverse operator, $\hat{A}^{-1}$, if it exists, is that operator for which

$$\hat{A}^{-1}\hat{A} = \hat{I} \quad (4.90)$$

that is, the operator that exactly undoes the effect of the operator $\hat{A}$.

Unitary operator

A unitary operator $\hat{U}$ is an operator for which

$$\hat{U}^{-1} = \hat{U}^{\dagger} \quad (4.92)$$

or, equivalently,

$$\hat{U}^{\dagger}\hat{U} = \hat{I} \quad (4.107)$$

A unitary operator acting on a vector conserves the length of the vector. It can be used for coordinate transformations of vectors

$$|f_{new}\rangle = \hat{U}|f_{old}\rangle \quad (4.94)$$

and operators

$$\hat{A}_{new} = \hat{U}\hat{A}_{old}\hat{U}^{\dagger} \quad \text{after Eq. (4.112)}$$

Conservation of inner product under a unitary transformation of coordinate system

A unitary coordinate transformation conserves the inner product of any two vectors

$$\langle g_{new}|f_{new}\rangle = \langle g_{old}|f_{old}\rangle \quad (4.99)$$

and, hence, also conserves the length of any vector.

Identity for Hermitian adjoint of products of operators

A useful identity is that

$$\left(\hat{A}\hat{B}\right)^{\dagger} = \hat{B}^{\dagger}\hat{A}^{\dagger} \tag{4.97}$$

Unitary operators that change the state vector

Unitary operators representing physical processes (as opposed to the mathematical process of changing coordinate systems) can change the state vector of the quantum mechanical system and are useful in representing those quantum mechanical processes, such as the time evolution of the state of a particle, in which the particle is conserved (and, hence, in which the length of the state vector is constant).

Degeneracy

If there are multiple eigenfunctions corresponding to a particular eigenvalue, this condition is referred to as degeneracy, with the number of such multiple eigenfunctions being called the degeneracy.

Hermitian operators

A Hermitian operator is one for which

$$\hat{M}^{\dagger} = \hat{M} \tag{4.114}$$

or, equivalently, for its matrix elements on some complete basis set

$$M_{ij} = M_{ji}^{*} \tag{4.117}$$

or, equivalently, for functions of a spatial variable

$$\int g^{*}(x)\hat{M}f(x)dx = \int\left\{\hat{M}g(x)\right\}^{*} f(x)dx \tag{4.123}$$

Properties of Hermitian operators

For the Hermitian operators that we encounter in quantum mechanics

- the eigenvalues are real
- the eigenfunctions corresponding to different eigenfunctions are orthogonal
- the degeneracy of any given finite eigenvalue is finite
- the set of eigenfunctions of a bounded Hermitian operator is complete
- the diagonal elements are real
- for arbitrary functions $|f\rangle$ and $|g\rangle$, we have, for a Hermitian operator $\hat{M}$,

$$\langle f|\hat{M}|g\rangle = \left(\langle g|\hat{M}|f\rangle\right)^{*} \tag{4.120}$$

Hermitian operators and measurable quantities

Physically measurable quantities in quantum mechanics can be represented by Hermitian operators.

Chapter 5

Operators and quantum mechanics

Prerequisites: Chapters 2, 3, and 4.

In this chapter, we start to use and extend the mathematics of the previous chapter and relate it to quantum mechanics more directly.

Here, we first examine some of the important properties of operators that are associated with measurable quantities. Then, we discuss the uncertainty principle in greater mathematical detail. Finally, we introduce the δ-function, which is a very useful additional piece of mathematics, and consider some of its uses and consequences, especially in quantum mechanics.

5.1 Commutation of operators

It is considered a postulate of quantum mechanics that all measurable quantities can be associated with a Hermitian operator. We have seen the momentum and energy operators already and we will encounter others. It is not the case that all operators that are useful in quantum mechanics are Hermitian; for example, we will encounter later the non-Hermitian creation and annihilation operators that are used extensively in quantum optics.

A very important property of Hermitian operators representing physical variables is whether they commute; that is, whether

$$\hat{A}\hat{B} = \hat{B}\hat{A} \tag{5.1}$$

where $\hat{A}$ and $\hat{B}$ are two Hermitian operators. Remember that because these linear operators obey the same algebra as matrices, in general, operators do not commute. For quantum mechanics, we formally define an entity

$$\left[\hat{A},\hat{B}\right] = \hat{A}\hat{B} - \hat{B}\hat{A} \tag{5.2}$$

This entity is called the *commutator*. An equivalent statement to Eq. (5.1) is then[1]

$$\left[\hat{A},\hat{B}\right] = 0 \tag{5.3}$$

If the operators do not commute, then Eq. (5.3) does not hold and, in general, we can choose to write

[1] Technically, the zero on the right of Eq. (5.3) is the zero operator, which maps all functions to the function that is zero everywhere, not the number zero.

$$\left[\hat{A},\hat{B}\right]=i\hat{C} \quad (5.4)$$

where $\hat{C}$ is sometimes referred to as the *remainder of commutation* or the *commutation rest.* Eq. (5.4) is the *commutation relation* for operators $\hat{A}$ and $\hat{B}$. Let us now try to understand some of the consequences of operators commuting or not.

Commuting operators and sets of eigenfunctions

Operators that commute share the same set of eigenfunctions and operators that share the same set of eigenfunctions commute. We now prove both of these statements.

Suppose that operators $\hat{A}$ and $\hat{B}$ commute and suppose that the functions $|\psi_n\rangle$ are the eigenfunctions of $\hat{A}$ with eigenvalues A_i. Then

$$\hat{A}\hat{B}|\psi_i\rangle=\hat{B}\hat{A}|\psi_i\rangle=\hat{B}A_i|\psi_i\rangle=A_i\hat{B}|\psi_i\rangle \quad (5.5)$$

Hence, we have

$$\hat{A}\left[\hat{B}|\psi_i\rangle\right]=A_i\left[\hat{B}|\psi_i\rangle\right] \quad (5.6)$$

But this means that the vector $\hat{B}|\psi_i\rangle$ is also the eigenvector $|\psi_i\rangle$ or is proportional to it;[2] that is, for some number B_i

$$\hat{B}|\psi_i\rangle=B_i|\psi_i\rangle \quad (5.7)$$

This kind of relation holds for all the eigenfunctions $|\psi_i\rangle$ and so these eigenfunctions are also the eigenfunctions of the operator $\hat{B}$, with associated eigenvalues B_i. Hence, we have proved the first statement that operators that commute share the same set of eigenfunctions. Note that the eigenvalues A_i and B_i are not, in general, equal to one another.

Now we consider the second statement. Suppose that the Hermitian operators $\hat{A}$ and $\hat{B}$ share the same complete set of eigenfunctions $|\psi_n\rangle$ with associated sets of eigenvalues A_n and B_n, respectively. Then

$$\hat{A}\hat{B}|\psi_i\rangle=\hat{A}B_i|\psi_i\rangle=A_iB_i|\psi_i\rangle \quad (5.8)$$

and similarly

$$\hat{B}\hat{A}|\psi_i\rangle=\hat{B}A_i|\psi_i\rangle=B_iA_i|\psi_i\rangle \quad (5.9)$$

Hence, for any function $|f\rangle$, which can always be expanded in this complete set of functions

$$|f\rangle=\sum_i c_i|\psi_i\rangle \quad (5.10)$$

we have

$$\hat{A}\hat{B}|f\rangle=\sum_i c_iA_iB_i|\psi_i\rangle=\sum_i c_iB_iA_i|\psi_i\rangle=\hat{B}\hat{A}|f\rangle \quad (5.11)$$

Because we have proved this for an arbitrary function $|f\rangle$, we have proved that the operators commute, hence proving the second statement.

---

[2] For simplicity here, we neglect the case of degenerate eigenvalues, though this case can be handled relatively easily.

This equivalence of Hermitian operators commuting and having the same set of eigenfunctions has an important quantum mechanical consequence. Suppose that the operators represent different measurable quantities. An example of such a situation is the case of a free particle – that is, one for which the potential is constant everywhere. In this case, the energy operator (Hamiltonian) and the momentum operator have the same eigenfunctions (plane waves) and the operators for energy and momentum commute with one another. If the particle is in an energy eigenstate, then it is also in a momentum eigenstate and the particle, in this case, can simultaneously have both a well-defined energy and a well-defined momentum. We can measure both of these quantities and get perfectly well-defined values for both.

Of course, this raises the question of what happens when the operators do not commute and we deal with this next.

Problems

5.1.1 The Pauli spin matrices are quantum mechanical operators that operate in a two-dimensional Hilbert space and can be written as

$$\hat{\sigma}_x = \begin{bmatrix} 0 & 1 \\ 1 & 0 \end{bmatrix},\ \hat{\sigma}_y = \begin{bmatrix} 0 & -i \\ i & 0 \end{bmatrix},\ \hat{\sigma}_z = \begin{bmatrix} 1 & 0 \\ 0 & -1 \end{bmatrix}$$

Find the commutation relations between each pair of these operators, proving your answer by explicit matrix multiplication and simplifying the answers as much as possible.

5.1.2* Show, for Hermitian operators $\hat{A}$ and $\hat{B}$ that the product $\hat{A}\hat{B}$ is a Hermitian operator if and only if $\hat{A}$ and $\hat{B}$ commute.

5.1.3 Prove that the operator that is the commutator $\left[\hat{A},\hat{B}\right]$ of two Hermitian operators $\hat{A}$ and $\hat{B}$ is never Hermitian if it is nonzero.

5.2 General form of the uncertainty principle

Here, we give a general proof and definition of the uncertainty principle. The proof is somewhat mathematical, though in the end quite short and very powerful in defining one of the most nonclassical aspects of quantum mechanics.

First, we need to set up the concepts of the mean and variance of an expectation value. We discussed previously the mean value of a measurable quantity as being the expectation value of the operator in the quantum mechanical state. Using $\bar{A}$ to denote the mean value of such a quantity A, we have, in the bra-ket notation, for a measurable quantity associated with the Hermitian operator $\hat{A}$ when the state of the system is $|f\rangle$

$$\bar{A} = \langle A \rangle = \langle f | \hat{A} | f \rangle \tag{5.12}$$

Let us define a new operator $\Delta\hat{A}$ associated with the difference between the measured value of A and its average value; that is,

$$\Delta\hat{A} = \hat{A} - \bar{A} \tag{5.13}$$

$\bar{A}$ is just a real number,[3] and so this operator is also Hermitian.

[3] Technically, $\bar{A}$ in this expression is actually the identity operator multiplied by the number $\langle A \rangle$.

So that we can examine the variance of the quantity A, we examine the expectation value of the operator $(\Delta\hat{A})^2$. Expanding the arbitrary function $|f\rangle$ on the basis of the eigenfunctions, $|\psi_i\rangle$, of $\hat{A}$ (i.e., $|f\rangle = \sum_i c_i |\psi_i\rangle$), we can formally evaluate the expectation value of $(\Delta\hat{A})^2$. We have

$$\begin{aligned}\left\langle(\Delta\hat{A})^2\right\rangle &= \left(\sum_i c_i^* \langle\psi_i|\right)\left(\hat{A}-\bar{A}\right)^2\left(\sum_j c_j|\psi_j\rangle\right)\\ &= \left(\sum_i c_i^* \langle\psi_i|\right)\left(\hat{A}-\bar{A}\right)\left(\sum_j c_j\left(A_j-\bar{A}\right)|\psi_j\rangle\right)\\ &= \left(\sum_i c_i^* \langle\psi_i|\right)\left(\sum_j c_j\left(A_j-\bar{A}\right)^2|\psi_j\rangle\right)\\ &= \sum_i |c_i|^2\left(A_i-\bar{A}\right)^2\end{aligned} \tag{5.14}$$

Because the $|c_i|^2$ are interpreted in quantum mechanics as being the probabilities that the system is found on measurement to be in the state i (or, equivalently, $|\psi_i\rangle$) and the quantity $(A_i - \bar{A})^2$ simply represents the squared deviation of the value of the quantity A from its average value, then by definition

$$\overline{(\Delta A)^2} \equiv \left\langle(\Delta\hat{A})^2\right\rangle = \left\langle\left(\hat{A}-\bar{A}\right)^2\right\rangle = \langle f|\left(\hat{A}-\bar{A}\right)^2|f\rangle = \langle f|\hat{A}^2-\bar{A}^2|f\rangle \equiv \left\langle\hat{A}^2\right\rangle - \bar{A}^2 \tag{5.15}$$

is the mean squared deviation we will find for the quantity A on repeatedly measuring the system prepared in state $|f\rangle$. (Note: The algebraic step from $\langle f|(\hat{A}-\bar{A})^2|f\rangle$ to $\langle f|\hat{A}^2-\bar{A}^2|f\rangle$ is left as an exercise for the reader.) In statistical language, this quantity is called the *variance*, and the square root of the variance, which we can write as

$$\Delta A \equiv \sqrt{\overline{(\Delta A)^2}} \tag{5.16}$$

is the *standard deviation*. The standard deviation gives a well-defined measure of the width of a distribution.

We can also consider some other quantity B associated with the Hermitian operator $\hat{B}$

$$\bar{B} = \langle B\rangle = \langle f|\hat{B}|f\rangle \tag{5.17}$$

and, with similar definitions

$$\overline{(\Delta B)^2} \equiv \left\langle(\Delta\hat{B})^2\right\rangle = \left\langle\left(\hat{B}-\bar{B}\right)^2\right\rangle = \langle f|\left(\hat{B}-\bar{B}\right)^2|f\rangle \tag{5.18}$$

These two expressions, Eqs. (5.15) and (5.18), give us ways of calculating the uncertainty in the measurements of the quantities A and B when the system is in a state $|f\rangle$. Now we use these in our general proof of the uncertainty principle.

Suppose that the two operators $\hat{A}$ and $\hat{B}$ do not commute and have a commutation rest $\hat{C}$ as defined in Eq. (5.4). Consider,[4] for some arbitrary real number α, the number

[4] This treatment of the proof of the general uncertainty relation is similar to that of W. Greiner, *Quantum Mechanics (Third Edition)* (Springer-Verlag, Berlin, 1994), pp. 74–5.

$$G(\alpha) = \left\langle \left(\alpha\Delta\hat{A} - i\Delta\hat{B}\right)f \middle| \left(\alpha\Delta\hat{A} - i\Delta\hat{B}\right)f \right\rangle \geq 0 \tag{5.19}$$

(By $\left|(\alpha\Delta\hat{A} - i\Delta\hat{B})f\right\rangle$, we simply mean the vector $(\alpha\Delta\hat{A} - i\Delta\hat{B})|f\rangle$, but we wrote it in this form to emphasize that it is simply a vector and, as a result, has a positive inner product with itself, which must be greater than or equal to zero, as in Eq. (5.19)). Now we rearrange (5.19) to obtain

$$\begin{aligned} G(\alpha) &= \langle f|\left(\alpha\Delta\hat{A} - i\Delta\hat{B}\right)^\dagger \left(\alpha\Delta\hat{A} - i\Delta\hat{B}\right)|f\rangle \\ &= \langle f|\left(\alpha\Delta\hat{A}^\dagger + i\Delta\hat{B}^\dagger\right)\left(\alpha\Delta\hat{A} - i\Delta\hat{B}\right)|f\rangle \end{aligned} \tag{5.20}$$

By Hermiticity of the operators, we have then

$$\begin{aligned} G(\alpha) &= \langle f|\left(\alpha\Delta\hat{A} + i\Delta\hat{B}\right)\left(\alpha\Delta\hat{A} - i\Delta\hat{B}\right)|f\rangle \\ &= \langle f|\alpha^2\left(\Delta\hat{A}\right)^2 + \left(\Delta\hat{B}\right)^2 - i\alpha\left(\Delta\hat{A}\Delta\hat{B} - \Delta\hat{B}\Delta\hat{A}\right)|f\rangle \\ &= \langle f|\alpha^2\left(\Delta\hat{A}\right)^2 + \left(\Delta\hat{B}\right)^2 - i\alpha\left[\Delta\hat{A}, \Delta\hat{B}\right]|f\rangle \\ &= \langle f|\alpha^2\left(\Delta\hat{A}\right)^2 + \left(\Delta\hat{B}\right)^2 + \alpha\hat{C}|f\rangle \\ &= \alpha^2\overline{\left(\Delta A\right)^2} + \overline{\left(\Delta B\right)^2} + \alpha\overline{C} \\ &= \overline{\left(\Delta A\right)^2}\left[\alpha + \frac{\overline{C}}{2\overline{\left(\Delta A\right)^2}}\right]^2 + \overline{\left(\Delta B\right)^2} - \frac{\left(\overline{C}\right)^2}{4\overline{\left(\Delta A\right)^2}} \geq 0 \end{aligned} \tag{5.21}$$

The last step is a simple though not very obvious rearrangement. But this relation must be true for arbitrary α and so it is true for the specific value

$$\alpha = -\frac{\overline{C}}{2\overline{\left(\Delta A\right)^2}} \tag{5.22}$$

which sets the first term equal to zero in the last line of Eq. (5.21), and so we have

$$\overline{\left(\Delta A\right)^2}\,\overline{\left(\Delta B\right)^2} \geq \frac{\left(\overline{C}\right)^2}{4} \tag{5.23}$$

This is the general form of the uncertainty principle. It tells us the relative minimum size of the uncertainties in two quantities if we perform a measurement. Only if the operators associated with the two quantities commute (and, hence, give $\hat{C}$ and therefore $\overline{C} = 0$) is it possible for there to be no width to the distribution of results for both quantities for any arbitrary state. This is a very nonclassical result and is one of the core results of quantum mechanics that differs fundamentally from classical mechanics.

Position-momentum uncertainty principle

We can apply this result now to derive the most quoted uncertainty principle, which we introduced by example before: the position-momentum relation. Let us consider the commutator of $\hat{p}_x$ and x. (We treat the function x as the operator for position – this can be justified, as we discuss later.) To make the issue of differentiation clear, we explicitly consider this commutator operating on an arbitrary function $|f\rangle$. As we discussed previously, operator relations always implicitly assume that the operators are operating on an arbitrary function anyway. Hence, we have

$$\begin{aligned}[\hat{p}_x, x]|f\rangle &= -i\hbar\left(\frac{d}{dx}x - x\frac{d}{dx}\right)|f\rangle = -i\hbar\left\{\frac{d}{dx}(x|f\rangle) - x\frac{d}{dx}|f\rangle\right\} \\ &= -i\hbar\left\{|f\rangle + x\frac{d}{dx}|f\rangle - x\frac{d}{dx}|f\rangle\right\} \\ &= -i\hbar|f\rangle\end{aligned} \tag{5.24}$$

So, because $|f\rangle$ is arbitrary, we can write

$$[\hat{p}_x, x] = -i\hbar \tag{5.25}$$

and the commutation rest operator $\hat{C}$ is simply the number (strictly the identity matrix multiplied by the number)

$$\hat{C} = -\hbar \tag{5.26}$$

Hence

$$\overline{C} = -\hbar \tag{5.27}$$

and so from Eq. (5.23), we have

$$(\Delta p_x)^2 (\Delta x)^2 \geq \frac{\hbar^2}{4} \tag{5.28}$$

or, equivalently,

$$\Delta p_x \Delta x \geq \frac{\hbar}{2} \tag{5.29}$$

Energy-time uncertainty principle

We can proceed to calculate a similar relation between energy uncertainty and time uncertainty. The energy operator is the Hamiltonian, $\hat{H}$. From Schrödinger's time-dependent equation, we know that

$$\hat{H}|\psi\rangle = i\hbar\frac{\partial}{\partial t}|\psi\rangle \tag{5.30}$$

for an arbitrary state $|\psi\rangle$. If we take the time operator to be just the function t, then we have – using essentially identical algebra to that used for the momentum-position uncertainty principle

$$\left[\hat{H}, t\right] = i\hbar\left(\frac{\partial}{\partial t}t - t\frac{\partial}{\partial t}\right) = i\hbar \tag{5.31}$$

and so, similarly, we have

$$(\Delta E)^2 (\Delta t)^2 \geq \frac{\hbar^2}{4} \tag{5.32}$$

or

$$\Delta E \Delta t \geq \frac{\hbar}{2} \tag{5.33}$$

which is the energy-time uncertainty principle. We can relate this result mathematically to the frequency-time uncertainty principle that occurs in Fourier analysis. Noting that $E = \hbar\omega$ in quantum mechanics, we have

$$\Delta \omega \Delta t \geq \frac{1}{2} \tag{5.34}$$

Problems

5.2.1* Suppose that an operator $\hat{A}$ that does not depend on time (i.e., $\partial \hat{A} / \partial t = 0$ [where here we strictly mean the zero operator]) commutes with the Hamiltonian $\hat{H}$. Show that the expectation value of this operator, for an arbitrary state $|\psi\rangle$, does not depend on time (i.e., $\partial \langle A \rangle / \partial t = 0$). (Hint: Remember that $\hat{H} \equiv i\hbar \partial / \partial t$.)

5.2.2 Show that $\langle f | (\hat{A} - \bar{A})^2 | f \rangle = \langle f | \hat{A}^2 - \bar{A}^2 | f \rangle$ for any (normalized) function $|f\rangle$ and any Hermitian operator $\hat{A}$.

5.2.3 Consider the "angular momentum" operators

$$\hat{L}_x = y\hat{p}_z - z\hat{p}_y \,,\; \hat{L}_y = z\hat{p}_x - x\hat{p}_z \,, \text{ and } \hat{L}_z = x\hat{p}_y - y\hat{p}_x$$

where $\hat{p}_x$, $\hat{p}_y$, and $\hat{p}_z$ are the usual momentum operators associated with the x, y, and z directions. (Note that these momentum operators are all Hermitian.)

(i) Prove whether $\hat{L}_x$ is Hermitian.

(ii) Construct an uncertainty principle for $\hat{L}_x$ and $\hat{L}_y$.

5.3 Transitioning from sums to integrals

One additional piece of mathematics that we will need is how to change from sums to integrals. This change can be convenient because integrals are often easier to evaluate than sums. We can do this transition when the different states involved are closely spaced in some parameter (e.g., momentum or energy) and when all the terms in the sum vary smoothly with that parameter. This is relatively obvious from basic integral calculus, but we explicitly discuss this here for clarity and completeness and because we need to build on the formal results in discussing applications of delta functions.

We imagine for the moment that we have some states, indexed by an integer q, and that for each of those q, some quantity has the value f_q. Hence, summing all of those would give a result

$$S = \sum_q f_q \tag{5.35}$$

It could be that the quantity f_q can also equivalently be written as a function of some parameter u that itself takes on some value for each q; that is,

$$f_q \equiv f(u_q)$$

For example, the different q states could represent states of different momentum $\hbar k_q$, in which case u_q could be the momentum and f_q could be the energy associated with that momentum. Then, we could just as well write instead of Eq. (5.35)

$$S = \sum_q f(u_q) \tag{5.36}$$

Suppose now that the u_q and the f_q are very closely spaced as we change q and vary relatively smoothly with q. We suppose that this smooth change of u_q with q is such that we can represent u as some smooth and differentiable function of q. Hence

$$u_{q+1} - u_q \equiv \delta u \simeq \frac{\delta u}{\delta q}\delta q \simeq \frac{du}{dq}\delta q = \frac{du}{dq} \tag{5.37}$$

In Eq. (5.37), we have first defined δu as the difference between two adjacent values of u (this quantity may be different for different values of q). Then, we have multiplied top and bottom lines by δq. Next, we have approximated $\delta u / \delta q$ by du / dq. Finally, we have noted that δq, the separation in q between adjacent values of q, is just unity, because q is by choice an integer. So, if we were to consider some small range Δu, within which the separation δu between adjacent values of u was approximately constant, the number of different terms in the sum that would lie within that range is $\Delta u / \delta u \simeq \Delta u / (du / dq)$. Equivalently, defining a *density of states*

$$g(u) = \frac{1}{(du / dq)} \tag{5.38}$$

we could say, equivalently, that the number of terms in the sum that lie within Δu is $g(u)\Delta u$. Hence, instead of summing over q, we could instead consider a range of values of u, each separated by an amount Δu, and write the sum over all those values; that is,

$$S = \sum_q f_q \equiv \sum_q f(u_q) \simeq \sum_u f(u) g(u) \Delta u \tag{5.39}$$

Finally, we can formally let Δu become very small and approximate the sum by an integral, to obtain

$$S \simeq \int f(u) g(u)\, du \tag{5.40}$$

The rule, therefore, in going from a sum to an integral is to insert the density of states in the integration variable into the integrand; that is,

$$\sum_q \ldots \to \int \ldots g(u)\, du \tag{5.41}$$

Of course, the limits of the integral must correspond to the limits in the sum.

Incidentally, we will use this result explicitly in Chapter 8 when considering, for example, densities of states, both in momentum and in energy, in crystalline materials.

5.4 Continuous eigenvalues and delta functions

This section shows how we can resolve a number of problems, especially with position and momentum eigenfunctions, that we have so far carefully avoided. It also introduces a number of techniques that can be quite broadly useful. In the rest of this book, we make only occasional use of some of the results, though these results can come up often in quantum mechanics. Hence, this section need not be studied in depth at the moment and can be used as a reference when these techniques come up later, but we suggest at least reading through this section at this point.

We have so far dealt explicitly and completely only with discrete eigenvalues (i.e., ones that can take on only specific values, not continuous ranges of values) and the normalizable eigenfunctions associated with them. The astute reader may have noticed that this is not the only kind of situation we can encounter in quantum mechanics. For example, at the very beginning, we talked about plane waves as being solutions of Schrödinger's wave equation in empty space; such waves cannot be normalized in the way we have been discussing so far. Consider the simplest possible case, that of a plane wave in the z direction. Such a wave can be written in the form

$$\psi_k(z) = C_k \exp(ikz) \tag{5.42}$$

Obviously

$$|\psi_k(z)|^2 = |C_k|^2 \tag{5.43}$$

and so, if we integrate $|\psi(z)|^2$ over the infinite range of all possible z, we will get an infinite result for any finite value of C_k.[5] Hence, we cannot define a normalization coefficient C_k in the same way we did before.

Our same astute reader may also have noticed that these particular functions are the eigenfunctions of the momentum operator for the z direction, $\hat{p}_z \equiv -i\hbar\, \partial / \partial z$, with eigenvalues $\hbar k$, where the quantity k can take on any real value. It is a common situation in quantum mechanics that when eigenvalues can take on any value within a continuous range, the eigenfunctions cannot be normalized in the way we have discussed so far. Such situations are not unusual. They also occur for energy eigenvalues of unbounded systems, such as the states above the "top" of a finite potential well or states above the ionization energy of a hydrogen atom, for example.

The situation for energy eigenvalues can always be resolved mathematically by putting the whole system within a large but finite box, with infinitely high "walls," and letting the size of the box become arbitrarily large. That is not always mathematically convenient, however.[6] Furthermore, for the case of the momentum eigenfunctions, building a box with potential barriers makes no difference to the momentum eigenfunctions; the potential does not appear in the momentum eigenfunction equation and the solutions to that mathematical problem are still infinite plane waves no matter what potential box we build. So, the question is: How can we handle such situations in a mathematically tractable way? The key to this solution is to introduce the *Dirac delta function*.[7] The delta function has various other mathematical uses, in quantum mechanics and elsewhere and it is an important topic in its own right.

Dirac delta function

The Dirac delta function, $\delta(x)$, is essentially a very narrow peak of unit area, centered on $x = 0$. In fact, it is infinitesimally wide and infinitely high but still with unit area. It is not strictly a function because in the one place that it really matters ($x = 0$), its value is not strictly defined. The formal definition of the delta function is

---

[5] In mathematical parlance, these are not L^2 (pronounced "L two") functions (their squared modulus is not Lebesgue integrable).

[6] For a hydrogen atom, e.g., we would get a very inconvenient coordinate system; that problem is most conveniently solved using a coordinate system that only treats the relative position of the electron and proton, not the absolute position, as would be required if we put the atom in a box (see Chapter 10).

[7] Those familiar with, e.g., Fourier transforms or any of several other fields will be familiar also with Dirac's delta function, but it was apparently introduced by Dirac to solve this particular kind of problem in quantum mechanics.

$$\int_{-\infty}^{\infty} \delta(x)\,dx = 1\,,\ \delta(x) = 0 \text{ for } x \neq 0 \tag{5.44}$$

Its most important property is that for any continuous function $f(x)$

$$\int_{-\infty}^{\infty} f(x)\delta(x)\,dx = f(0) \tag{5.45}$$

This relation, Eq. (5.45), can be regarded as an operational definition of the delta function. From this relation, we can readily deduce

$$\int_{-\infty}^{\infty} f(x)\delta(x-a)\,dx = f(a) \tag{5.46}$$

Of course, $\delta(x-a)$ is essentially a very sharply peaked function around $x = a$. We can see, therefore, that a key property of the delta function is that it pulls out the value of the function at one specific value of the argument (i.e., at $x = a$ in Eq. (5.46)) as the result of this integral. This is exactly what we would expect a very sharply peaked function of unit area to do if we put it inside the integrand as in Eq. (5.46).

Representing the delta function

The delta function, in practice, can be defined as the limit of just about any symmetrical peaked function in the limit as the width of the peak goes to zero and the height goes to infinity, provided we make sure the function retains unit area as we take the limit. Several common examples are described here.

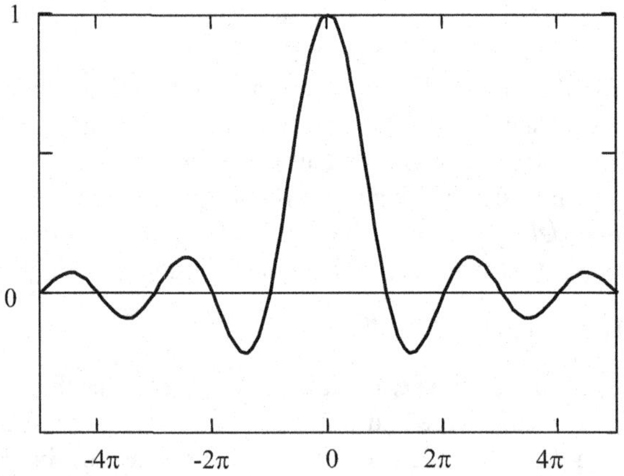

Fig. 5.1. Plot of the function $(\sin x)/x$, with x as the horizontal axis.

Sinc function representation

Based on the "sinc" function, graphed in Fig. 5.1

$$\frac{\sin x}{x} \equiv \operatorname{sinc} x \tag{5.47}$$

we can write

$$\delta(x) = \lim_{L\to\infty} \frac{\sin Lx}{\pi x} \tag{5.48}$$

where we have used the fact that

$$\int_{-\infty}^{\infty} \frac{\sin x}{x} dx = \pi \tag{5.49}$$

Exponential integral representation

A form that is very useful in formal evaluations of integrals is

$$\delta(x) = \frac{1}{2\pi} \int_{-\infty}^{\infty} \exp(ixt)\, dt \tag{5.50}$$

which can readily be proved using the result in Eq. (5.49).

Lorentzian representation

Based on the Lorentzian function, common as, for example, a line shape in atomic spectra, with a line width (half width at half maximum) of ε, we have

$$\delta(x) = \lim_{\varepsilon\to 0} \frac{1}{\pi\varepsilon} \frac{1}{1+(x/\varepsilon)^2} \tag{5.51}$$

where we have used the result

$$\int_{-\infty}^{\infty} \frac{1}{1+x^2} dx = \pi \tag{5.52}$$

Gaussian representation

Based on the Gaussian function of $1/e$ half width w, we have

$$\delta(x) = \lim_{w\to 0} \frac{1}{w\sqrt{\pi}} \exp\left(-\frac{x^2}{w^2}\right) \tag{5.53}$$

where we have used the result

$$\int_{-\infty}^{\infty} \exp(-x^2)\, dx = \sqrt{\pi} \tag{5.54}$$

Square pulse representation

One of the simplest representations is that of a "square pulse" function that we could define as

$$s(x) = \begin{cases} 0, & x < -\eta/2 \\ 1/\eta, & -\eta/2 \le x \le \eta/2 \\ 0 & x > \eta/2 \end{cases} \tag{5.55}$$

which is a function of width η and height $1/\eta$, centered at $x = 0$. With this square pulse function, we have

$$\delta(x) = \lim_{\eta\to 0} s(x) \tag{5.56}$$

Relation to Heaviside function

The square pulse function can be written in another equivalent way in terms of the Heaviside function. The Heaviside function is the "unit step" function

$$\Theta(x) = \begin{cases} 1, & x > 0 \\ 0, & x < 0 \end{cases} \tag{5.57}$$

in terms of which we have the square pulse from before

$$s(x) = \frac{\Theta(x+\eta/2) - \Theta(x-\eta/2)}{\eta} \tag{5.58}$$

In the limit as $\eta \to 0$, this is simply the definition of the derivative of Θ and so we have also

$$\begin{aligned} \delta(x) &= \lim_{\eta \to 0} \frac{\Theta(x+\eta/2) - \Theta(x-\eta/2)}{\eta} \\ &= \frac{d\Theta(x)}{dx} \end{aligned} \tag{5.59}$$

From this, we can immediately conclude that the Heaviside function is the integral of the delta function; that is,

$$\Theta(x) = \int_{-\infty}^{x} \delta(x_1)\, dx_1 \tag{5.60}$$

Basis function representation and closure

Thus far, we have been discussing representations of the delta function as limits of other mathematical functions. There is another kind of representation that is particularly general and useful, which is a representation in terms of any complete set.

Suppose we have a complete orthonormal set of functions, $\phi_i(x)$. Then, we can expand any function in this set; that is,

$$f(x) = \sum_n a_n \phi_n(x) \tag{5.61}$$

As usual, we can determine any specific expansion coefficients a_m by premultiplying by $\phi_m^*(x)$ and integrating over x; that is,

$$\int \phi_m^*(x) f(x)\, dx = \sum_n a_n \int \phi_m^*(x) \phi_n(x)\, dx = \sum_n a_n \delta_{nm} = a_m \tag{5.62}$$

Now we can use the far left of Eq. (5.62) to substitute for the expansion coefficients in Eq. (5.61); that is, writing

$$a_n = \int \phi_n^*(x') f(x')\, dx' \tag{5.63}$$

we have

$$f(x) = \sum_n \left(\int \phi_n^*(x') f(x')\, dx' \right) \phi_n(x) \tag{5.64}$$

Interchanging the order of the integral and the sum, we have

$$f(x) = \int f(x') \left(\sum_n \phi_n^*(x') \phi_n(x) \right) dx' \tag{5.65}$$

Comparing Eq. (5.65) to Eq. (5.46), we see that this sum is performing exactly as the delta function; that is,

$$\sum_n \phi_n^*(x') \phi_n(x) = \delta(x' - x) \ \left(= \delta(x - x')\right) \tag{5.66}$$

Hence, we have a general representation of the delta function in terms of any complete set. This can be formally useful.

This property, Eq. (5.66), of the set of functions is known as *closure* and is a consequence of the completeness of the set. We can also see that Eq. (5.66) is simply the expansion of the delta function in the set $\phi_i(x)$, with the expansion coefficients simply being the numbers $\phi_n^*(x')$. Hence, for example, the expansion of $\delta(x)$ would have expansion coefficients $\phi_n^*(0)$. We can understand intuitively that if a set of functions can manage to represent such an extreme function as a delta function, then it can represent any other reasonable function and so we can understand how this property of closure is related to completeness.

Delta function in 3 dimensions

It is straightforward to construct delta functions in higher dimensions. The result is merely the product of the various one-dimensional delta functions. For example, using the shorthand $\delta(\mathbf{r})$ to represent the delta function for three dimensions, we can write

$$\delta(\mathbf{r}) = \delta(x)\delta(y)\delta(z) \tag{5.67}$$

Normalizing to a delta function

Now that we have introduced the delta function, we can use it to perform a kind of normalization for those functions that are not normalizable in the previous sense. We can introduce this normalization through the example of the momentum eigenfunctions of Eq. (5.42).

Normalization of momentum eigenfunctions

Consider the "orthogonality" integral of two different momentum eigenfunctions, but where we deliberately restrict the range of integration to some large range $\pm L$; that is,

$$\begin{aligned}\int_{-L}^{L} \psi_{k1}^*(z)\psi_k(z)\,dz &= C_{k1}^* C_k \int_{-L}^{L} \exp(-ik_1 z)\exp(ikz)\,dz \\ &= C_{k1}^* C_k \int_{-L}^{L} \exp\left[i(k-k_1)z\right] dz \\ &= 2C_{k1}^* C_k \frac{\sin\left[(k-k_1)L\right]}{(k-k_1)}\end{aligned} \tag{5.68}$$

Hence, taking the limit as L becomes very large, we have

$$\int_{-\infty}^{\infty} \psi_{k1}^*(z)\psi_k(z)\,dz = 2\pi C_{k1}^* C_k \delta(k - k_1) \tag{5.69}$$

where we have used the sinc function representation, Eq. (5.48), of the delta function. So, if we choose[8]

$$C_k = \frac{1}{\sqrt{2\pi}} \tag{5.70}$$

that is, if we choose the momentum eigenfunctions to be defined as

$$\psi_k(z) = \frac{1}{\sqrt{2\pi}} \exp(ikz) \tag{5.71}$$

then, at least, we get a tidy form for the orthogonality integral. Specifically, instead of Eq. (5.69), we would have

$$\int_{-\infty}^{\infty} \psi_{k1}^*(z)\psi_k(z)\,dz = \delta(k - k_1) \tag{5.72}$$

This choice of normalization, leading to an orthogonality integral like Eq. (5.72) with only a delta function on the right, is called *normalization to a delta function*. Note that here we make the orthogonality relation Eq. (5.72) do the work of both the normalization and the orthogonality conditions – we do not write a separate normalization condition. It turns out we can construct a viable mathematics for handling such "unnormalizable" functions if we normalize in this way. We now develop this mathematics.

It is interesting to compare Eq. (5.72) with the orthonormality relation for conventional normalizable functions, Eq. (2.35). In that conventional case, the integral limits may be finite, but the equations are otherwise essentially identical, except that we now have a Dirac delta function, $\delta(k - k_1)$, instead of the Kronecker delta, δ_{nm}, that was the result for the orthogonality integral for two conventionally normalized basis functions, $\psi_n(z)$ and $\psi_m(z)$. We will find that this substitution of Dirac delta function for Kronecker delta is quite a general feature as we compare the results for the two classes of functions.

Using functions normalized to a delta function

A key point in using functions normalized to a delta function is that they can be handled provided we can work with integrals rather than sums and we must make careful use of the density of states in such sums. To see the basic mathematics of working with such functions, we first consider that we have an orthonormal basis set of functions $\psi_q(z)$ and we consider the expansion of some other function, $\phi(z)$, on this set. We have

$$\phi(z) = \sum_q f_q \psi_q(z) \tag{5.73}$$

where the f_q are the expansion coefficients. We note that the sum of the squares of the expansion coefficients gives

$$\int |\phi(z)|^2\,dz = \sum_{p,q} f_p^* f_q \int \psi_p^*(z)\psi_q(z)\,dz = \sum_q |f_q|^2 \tag{5.74}$$

so the normalization of the function of interest is the same as that of the expansion coefficients.[9]

---

[8] Note we could also have chosen to include any complex factor of unit magnitude in the definition Eq. (5.70) but, as usual, for simplicity we choose not to do so.

Remembering how we make the transition from sums to integrals, we presume that there is some quantity u_q (e.g., momentum) associated with the q that allows us to write, equivalently,

$$\phi(z)=\sum_q f(u_q)\psi(u_q,z) \tag{5.75}$$

where $f(u_q)\equiv f_q$ and we make the minor additional change in notation $\psi(u_q,z)\equiv\psi_q(z)$. From this we note, incidentally, that for any specific value of u_q, such as a value v, we can write

$$f(v)=\int\psi^*(v,z)\phi(z)dz \tag{5.76}$$

in the usual way of evaluating expansion coefficients for a function $\phi(z)$ on a basis set $\psi_q(z)\ (\equiv\psi(u_q,z))$.

Now, let us transform the sum, Eq. (5.75), into an integral, using the density of states, $g(u)=1/(du/dq)$ as in Eqs. (5.38) and (5.41). This gives

$$\phi(z)=\int f(u)\psi(u,z)g(u)du \tag{5.77}$$

Now we can substitute this form of $\phi(z)$ back into Eq. (5.76) to give, after exchanging the order of the integrals

$$f(v)=\int f(u)\left[\int\psi^*(v,z)\psi(u,z)g(u)dz\right]du \tag{5.78}$$

from which we see, by the definition of the delta function, Eq. (5.46), that the term in square brackets is performing as a delta function; that is,

$$\int\psi^*(v,z)\psi(u,z)g(u)dz=\delta(v-u) \tag{5.79}$$

The functions we are considering so far are all presumed to be normalized conventionally. Now, however, we have an interesting option in choosing other functions that will work with the delta function normalization to give useful and meaningful results. First, we have to make the restriction that the density of states is a constant; that is,

$$g(u)\equiv g \tag{5.80}$$

This restriction is appropriate, for example, for momentum eigenfunctions or plane waves in a large box. Now, let us define two new functions, in which we fold the square root of the density of states into each function; that is,

$$F(u)=\sqrt{g}f(u) \tag{5.81}$$

and

$$\Psi(u,z)=\sqrt{g}\psi(u,z) \tag{5.82}$$

Then we find, first, that the $\Psi(u,z)$ are basis functions normalized to a delta function; that is, Eq. (5.79) becomes

$$\int\Psi^*(v,z)\Psi(u,z)dz=\delta(v-u) \tag{5.83}$$

[9] Those used to Fourier series and transforms will recognize this as a form of Parseval's theorem.

Second, we find that we have a simple expression for the expansion in such functions normalized to a delta function; that is, Eq. (5.77) becomes

$$\phi(z) = \int F(u)\Psi(u,z)\,du \tag{5.84}$$

and we can also write for the expansion coefficient (or now expansion function), from Eq. (5.76)

$$F(v) = \int \Psi^*(v,z)\phi(z)\,dz \tag{5.85}$$

(where we have merely multiplied both sides of Eq. (5.76) by $\sqrt{g}$ and substituted from Eqs. (5.81) and (5.82)).

Third, we find that $F(u)$ has a simple normalization[10]

$$\int|\phi(z)|^2\,dz = \sum_q |f_q|^2 = \int|f(u)|^2\,g\,du = \int|F(u)|^2\,du \tag{5.86}$$

Thus, with functions normalized to delta functions, we recover a straightforward mathematics that allows us, for example, to write quite simple expressions for expansions in such functions. The only requirement is that the density of states be uniform.

This use of functions normalized to delta functions can be done any time the density of states is large and uniform. The fact that the final results do not depend on the density of states means that these expressions continue to be meaningful in the limit as the density of states becomes effectively infinite, as is the case for momentum eigenfunctions.

The incorporation of the square root of the density of states into each of the expansion coefficients and the basis functions essentially avoids two problems. As the density of states increases, (1) the expansion coefficients themselves would otherwise become very small and (2) so also would the amplitude of the basis functions. The incorporation of the square root of the density of states into both expansion coefficients and basis functions leaves them both quite finite and leaves us with a simple mathematics for handling the resulting functions, without infinities or other singularities.

In summary on the mathematics of working with functions normalized to a delta function as in Eq. (5.83), we can use them as a basis set, but we have Eqs. (5.84) and (5.85) as the expansion on the basis rather than Eqs. (5.76) and (5.73), and instead of expansion coefficients, we have an expansion "function" (here, $F(u)$) that obeys a modulus squared integral normalization condition Eq. (5.86) rather than the usual sum of squares of the expansion coefficients.

Example of normalization of plane waves

Let us return to the question of normalization of plane waves of the form $C_k\exp(ikz)$ and perform this in the two different approaches of normalization to unity in a box of finite length L and normalization to a delta function.

Box and delta function normalization

In a box of length L, normalizing such an exponential plane wave gives a normalization integral (taking symmetrical limits for simplicity)

[10] Again, readers used to Fourier transforms will recognize this as Parseval's theorem. The mathematics we are describing here is exactly the mathematics required to go from Fourier series to Fourier integrals.

$$\int_{-L/2}^{L/2} C_k^* \exp(-ikz) C_k \exp(ikz) dz = |C_k|^2 L = 1 \tag{5.87}$$

that is,

$$C_k = \frac{1}{\sqrt{L}} \tag{5.88}$$

so the box-normalized wavefunction is

$$\psi_k(z) \equiv \psi(k,z) = \frac{1}{\sqrt{L}} \exp(ikz) \tag{5.89}$$

To transform this to a wavefunction normalized to a delta function, our prescription Eq. (5.82) is to multiply this wavefunction by the square root of the density of states. Taking the density of states to be $g = L/2\pi$, corresponding to adjacent k values being spaced by $2\pi/L$ in such a box, we have

$$\Psi(k,z) = \sqrt{g}\psi(k,z) = \sqrt{\frac{L}{2\pi}} \frac{1}{\sqrt{L}} \exp(ikz) = \frac{1}{\sqrt{2\pi}} \exp(ikz) \tag{5.90}$$

which is exactly what we had proposed before in Eq. (5.71) when considering plane waves normalized to a delta function.

For a large box, therefore, we can use either a box-normalized approach or we can use functions normalized to a delta function. Either will give the same results in the end. As the size of the box becomes infinite, it is more common to work with the delta-function normalization because then the infinities (in densities of states and in box size) and zeros (in the amplitudes of the "normalized" functions) are avoided.

Relation to Fourier transforms

The classic example of the mathematics of basis functions normalized to delta functions is when our basis functions are the plane waves

$$\Psi(u,z) \equiv \frac{1}{\sqrt{2\pi}} \exp(-iuz) \tag{5.91}$$

in which case, the expansion of the function $F(u)$ in those functions is exactly equivalent to the mathematics of the Fourier transform; that is,

$$\phi(z) = \frac{1}{\sqrt{2\pi}} \int_{-\infty}^{\infty} F(u) \exp(-iuz) dz \tag{5.92}$$

where $\phi(z)$ is the Fourier transform of the function $F(u)$. Note that then Eq. (5.86) is simply a statement of Parseval's theorem, which, in turn, is saying that the Fourier transform is a transform that does not change the length of the vector in Hilbert space and it is, in fact, a unitary transform.[11]

---

[11] Viewed this way, we could say that in one sense, the Fourier transform does exactly nothing. I.e., it can be viewed as merely a change of basis in Hilbert space that makes no difference to the function or state being represented. In the language of Fourier transforms, the signal itself is not changed by Fourier transformation, just the representation of it in "frequency" rather than "time." The vector in Hilbert space

Periodic boundary conditions

Before leaving the discussion of plane waves, we should mention one other topic, which is the use of periodic boundary conditions. These boundary conditions are very commonly used in solid-state physics. We like to work with (complex) exponential waves rather than sines and cosines because the mathematics is easier to handle. Putting exponential waves in a box causes a minor formal problem. If we ask that the wavefunction reaches zero at the walls of the box, then we end up with sine waves as the allowed mathematical solutions (presuming we choose the origin at one end of the box), not exponentials. A mathematical trick is to pretend that the boundary conditions are periodic, with the length, L, of the box being the period; that is, to pretend that

$$\exp(ikz) = \exp[ik(z+L)] \tag{5.93}$$

This leads to the requirement that

$$\exp(ikL) = 1 \tag{5.94}$$

which, in turn, means that

$$k = \frac{2m\pi}{L} \tag{5.95}$$

where m is a positive or negative integer or zero. The allowed values of k are therefore spaced by $2\pi/L$ and the density of states in k (the number of states per unit k) is therefore

$$g = \frac{L}{2\pi} \tag{5.96}$$

We can then work with these functions just as we work with exponential plane waves normalized over a finite large box, using either box normalization or delta-function normalization as we wish.

This periodic boundary-condition trick is often stated as if it had some physical justification, but usually it does not. For example, it is used frequently in the physics of crystals (see Chapter 8). In one dimension in a crystal, we could say we are imagining the crystal is very long and that physically, we have bent it around into a circle, albeit one of very large radius; that might be an acceptable approach in such a case. In three dimensions, connecting it back around on itself in three directions at once is physically absurd and topologically impossible. The justification there is seldom stated explicitly, but it is simply that it makes the math easier. It is also true that if we were to solve the problem with hard-wall boundary conditions, with sine waves as the solutions, for example, we would, in fact, end up with the same number of states (except that the state with $k = 0$ is allowed in the exponential case but does not exist in the sine case), and essentially all measurable quantities would end up with the same results. The honest truth is that we use periodic boundary conditions because they are convenient, and experience tells us that we can get away with them, even though they are somewhat nonphysical, in fact.

---

is still the same length (and, hence, the power or energy in the signal is not changed by the Fourier transformation, which is Parseval's theorem), and is pointing in the same "direction"; only the coordinate axes have changed.

Position eigenfunctions

Thus far, the only quantum mechanical functions we have dealt with explicitly that are normalized to a delta function are plane waves, which are also the momentum eigenfunctions. The theory we derived (e.g., Eqs. (5.83) – (5.86)), however, was not restricted to any specific function that was normalized to a delta function. There is another very simple example: namely, the position eigenfunctions.

We stated before that the position operator – at least, in the representation where functions are described in terms of position – was simply the position, z, itself (in the one-dimensional case). We might ask, therefore, what are the functions that when operated on by the position operator give results that are simply an eigenvalue (which should be a "value" of position) times the function? We can see the answer to this by inspection. The eigenfunctions are delta functions. For example, consider the function

$$\psi_{z_o}(z) = \delta(z - z_o) \tag{5.97}$$

Then, we can see that

$$\hat{z}\psi_{z_o}(z) = z_o \psi_{z_o}(z) \tag{5.98}$$

where for emphasis and clarity, we have explicitly written the position operator as $\hat{z}$. The only value of z for which the eigenfunction is nonzero is the one $z = z_o$, so in any expression involving $\hat{z}\psi_{z_o}(z)$, we can simply replace it by $z_o\psi_{z_o}(z)$.

The delta function itself is normalized to a delta function. To see this, consider the integral

$$\int \delta(z_1 - z)\delta(z_2 - z)dz = \delta(z_1 - z_2) \tag{5.99}$$

To understand why this integral itself evaluates to a delta function, consider the first delta function as being one of its other representations, such as a Gaussian as in Eq. (5.53), before we have quite taken the limit. Then, by the definition of the delta function

$$\int \frac{1}{w\sqrt{\pi}} \exp\left(-\frac{(z_1 - z)^2}{w^2}\right)\delta(z_2 - z)\,dz = \frac{1}{w\sqrt{\pi}} \exp\left(-\frac{(z_1 - z_2)^2}{w^2}\right) \tag{5.100}$$

Then, take the limit of small w of the right-hand side, which is the delta function on the right of Eq. (5.99).

Hence, we have shown another example of a function that normalizes to a delta function and for which we can use the same general formalism.

Expansion of a function in position eigenfunctions

We expect that the position eigenfunctions form a complete set, and so we can expand other functions in them. Let us formally see what happens when we do that. Suppose that we have some set of expansion coefficients $F(z_o)$ used in an expansion of the form of Eq. (5.85), as appropriate for an expansion in functions that are normalized to a delta function. Then we have, using the position eigenfunctions as in Eq. (5.97)

$$\phi(z) = \int F(z_o)\delta(z - z_o)\,dz_o \tag{5.101}$$

The evaluation of this integral is trivial given the definition of the delta function; that is, we have

$$\phi(z) = F(z) \tag{5.102}$$

In other words, a function $\phi(z)$ of position is its own set of expansion coefficients in the expansion in position eigenfunctions. This point may seem trivial, but it shows that we can view all the wavefunctions we have been working with so far as being expansions on the position eigenfunctions. The wavefunction normalization integrals we have performed so far, for example, which have been of the form $\int |\phi(z)|^2 dz$, can now be seen as the normalization, Eq. (5.86), that we have deduced for expansions in functions normalized to a delta function. In other words, because the very beginning when we started working with wavefunctions, we have actually been using the concept of functions normalized to a delta function all along – we just did not know it.

Change of basis for basis sets normalized to a delta function

So far, we have only discussed change of basis sets for discrete sets normalized conventionally. We can, however, also change between basis sets normalized to delta functions. This is best illustrated by the example of changing between position and momentum basis sets, an example that is also, by far, the most common use of such a transformation anyway.

We can presume that we have some function $\phi_{old}(z)$ that is expressed in the position basis. The subscript *old* here refers to the old basis set, here the position basis. The new basis set, also normalized to a delta function, is the set of momentum eigenfunctions, $(1/2\pi)^{1/2}\exp(ikz)$, as in Eq. (5.90). Then, according to our expansion formula for functions normalized to a delta function, Eq. (5.85), we have

$$\phi_{new}(k) = \frac{1}{\sqrt{2\pi}}\int \phi_{old}(z)\exp(-ikz)\,dz \tag{5.103}$$

We can, if we wish, formally write this transformation in terms of an (integral) operator

$$\hat{U} \equiv \frac{1}{\sqrt{2\pi}}\int \exp(-ikz)\,dz \tag{5.104}$$

Note that $\hat{U}$ is an operator. One can only actually perform the integral once this operator operates on a function of z.

In this form, we can then write Eq. (5.103) in the form we have used before for basis transformations as

$$|\phi_{new}\rangle = \hat{U}|\phi_{old}\rangle \tag{5.105}$$

where in our notation we are anticipating that this operator $\hat{U}$ is unitary (a proof that is left to the reader).

Let us look at the specific case in which the function $\phi_{old}(z)$ is actually the position basis function $|\phi_{old}\rangle = \delta(z - z_o)$. Then we find that in what is now the momentum representation, that basis function is now expressed as

$$|\phi_{new}\rangle = \frac{1}{\sqrt{2\pi}}\int \delta(z - z_o)\exp(-ikz)\,dz = \frac{1}{\sqrt{2\pi}}\exp(-ikz_o) \tag{5.106}$$

In other words, a *position* eigenfunction in the *momentum* representation is $(1/2\pi)^{1/2}\exp(-ikz_o)$, where k takes on an unrestricted range of values, just as for a specific value of $k = k_o$ the *momentum* eigenfunction in the *position* representation is $(1/2\pi)^{1/2}\exp(ik_o z)$, where z takes on an unrestricted range of values.

The operator that will take us back to the position representation that we can guess by the symmetry of this particular problem is

$$\hat{U}^\dagger = \frac{1}{\sqrt{2\pi}}\int \exp(ikz)\,dk \tag{5.107}$$

Note that in constructing this adjoint, we have taken the complex conjugate and we have interchanged the roles of k and z, which is analogous to the formation of an adjoint in our conventionally normalizable basis representations, where we take the complex conjugate and interchange indices on the matrix elements or basis functions.

Using these definitions and as an illustration, we can now formally transform the position operator into the momentum basis, using the usual formula for such transformations; that is, formally operating on an arbitrary function $|f\rangle$

$$\begin{aligned}
\hat{z}_{new}|f\rangle &= \hat{U}\hat{z}_{old}\hat{U}^\dagger|f\rangle \\
&= \frac{1}{2\pi}\int \exp(-ikz)\int z\exp(ik'z)\,f(k')\,dk'dz \\
&= \frac{1}{2\pi}\int\int z\exp\left[-i(k-k')z\right]f(k')\,dk'dz \\
&= \frac{-1}{2\pi i}\frac{\partial}{\partial k}\int\int \exp\left[-i(k-k')z\right]dzf(k')\,dk' \\
&= i\frac{\partial}{\partial k}\int \delta(k'-k)\,f(k')\,dk' \\
&= i\frac{\partial}{\partial k}f(k) \\
&\equiv i\frac{\partial}{\partial k}|f\rangle
\end{aligned} \tag{5.108}$$

Note two points about this algebra. First, the index or variable of integration in the operation $\hat{z}_{old}\hat{U}^\dagger$, here chosen as k', is a different variable from the k in the final expression for $\hat{z}_{new}$ because we have to sum or integrate over the k' variable in performing the operation $\hat{z}_{old}\hat{U}^\dagger$. Second, we have used the algebraic trick that $-iz\exp[-i(k-k')z] \equiv (\partial/\partial k)\exp[-i(k-k')z]$.

Because $|f\rangle$ is arbitrary, we can then write the position operator in the momentum representation as

$$\hat{z}_{new} = i\frac{\partial}{\partial k} \tag{5.109}$$

Note the symmetry between this and the z momentum operator in the position representation, which is $\hat{p}_z = (-i\hbar)(\partial/\partial z)$.

With this exercise, we finally have a form of the position operator in which it is quite clearly an operator, in contrast to its form in the position representation where it was difficult to distinguish the operator $\hat{z}$ from the simple position z. Of course, now in the momentum representation, we have the similar difficulty of distinguishing the operator $\hat{p}$ from the simple momentum value p.

Problems

5.4.1 Prove that the operator

$$\hat{U} \equiv (1/\sqrt{2\pi})\int \exp(-ikz)\,dz$$

with Hermitian adjoint

$$\hat{U}^\dagger \equiv (1/\sqrt{2\pi})\int \exp(ik'z)\,dz$$

is unitary.

5.4.2* Demonstrate explicitly that the commutator $[\hat{z}, \hat{p}_z]$ is identical, regardless of whether it is evaluated in the position or the momentum representation.

5.4.3 Formally transform the momentum operator $\hat{p}_z$ into the momentum basis using algebra similar to that for the transformation of the position operator into the momentum basis.

5.5 Summary of concepts

Commutator

The commutator of two operators is defined as

$$\left[\hat{A}, \hat{B}\right] = \hat{A}\hat{B} - \hat{B}\hat{A} \tag{5.2}$$

Two operators are said to commute if

$$\left[\hat{A}, \hat{B}\right] = 0 \tag{5.3}$$

In general, we can write

$$\left[\hat{A}, \hat{B}\right] = i\hat{C} \tag{5.4}$$

where $\hat{C}$ is called the commutation rest or the remainder of commutation. Note, for example,

$$\left[\hat{p}_x, x\right] = -i\hbar \tag{5.25}$$

Commuting operators and eigenfunctions

Operators that commute share the same set of eigenfunctions and operators that share the same set of eigenfunctions commute.

General form of the uncertainty principle

$$\overline{(\Delta A)^2 (\Delta B)^2} \geq \frac{\left(\bar{C}\right)^2}{4} \tag{5.23}$$

Energy-time uncertainty principle

$$\Delta E \Delta t \geq \frac{\hbar}{2} \tag{5.33}$$

Transition from a sum to an integral

In transitioning from a sum to an integral, where the quantity f_q being summed is smoothly and slowly varying in the index q, we may make the approximation

$$S = \sum_q f_q \equiv \sum_q f(u_q) \simeq \sum_u f(u) g(u) \Delta u \simeq \int f(u) g(u) du \qquad \text{(5.39) and (5.40)}$$

where u is some variable in which the quantity being summed can also be expressed, where the limits on the integral must correspond to the limits on the sum, and where $g(u)$ is the density of states

$$g(u) = \frac{1}{(du/dq)} \tag{5.38}$$

Dirac delta function

A practical operational definition of the Dirac delta function is that it is an entity $\delta(x)$ such that for any continuous function $f(x)$

$$\int_{-\infty}^{\infty} f(x)\delta(x-a)dx = f(a) \tag{5.46}$$

$\delta(x)$ is essentially a sharply peaked function around $x = 0$, with unit area, in the limit as the width of that peak goes to zero.

Representing the delta function

Various peaked functions can be used to represent the delta function. Particularly common are the representation in terms of the sinc function

$$\frac{\sin x}{x} \equiv \operatorname{sinc} x \tag{5.47}$$

which gives

$$\delta(x) = \lim_{L\to\infty} \frac{\sin Lx}{\pi x} \tag{5.48}$$

and in terms of the square pulse representation

$$s(x) = \begin{cases} 0, & x < -\eta/2 \\ 1/\eta, & -\eta/2 \le x \le \eta/2 \\ 0 & x > \eta/2 \end{cases} \tag{5.55}$$

which gives

$$\delta(x) = \lim_{\eta\to 0} s(x) \tag{5.56}$$

A very useful formal representation is

$$\delta(x) = \frac{1}{2\pi} \int_{-\infty}^{\infty} \exp(ixt)\, dt \tag{5.50}$$

Basis function representation of the delta function, and closure

For any complete orthonormal set of functions $\phi_n(x)$

$$\sum_n \phi_n^*(x')\phi_n(x) = \delta(x'-x) \ \left(= \delta(x-x')\right) \tag{5.66}$$

This relation, for a set of functions $\phi_n(x)$ is called closure and is an important criterion for the completeness of the set.

Delta function in three dimensions

The delta function in three dimensions is simply the product of the delta functions in each dimension and can be written

$$\delta(\mathbf{r}) = \delta(x)\delta(y)\delta(z) \tag{5.67}$$

Normalization to a delta function

A function $\psi_\delta(u,z)$ of two continuous parameters u and z (e.g., momentum and position) is said to be normalized (in z) to a delta function if

$$\int \psi_\delta^*(v,z)\psi_\delta(u,z)dz = \delta(v-u) \tag{5.79}$$

Normalized momentum eigenfunctions

When normalized to a delta function, the momentum eigenfunctions become

$$\psi_k(z) = \frac{1}{\sqrt{2\pi}}\exp(ikz) \tag{5.71}$$

Expanding in functions normalized to a delta function

For a set of basis functions $\Psi(u,z)$ that are normalized to a delta function in z, we can expand an arbitrary function $\phi(z)$ in them using the expression

$$\phi(z) = \int F(u)\Psi(u,z)du \tag{5.84}$$

where the $F(u)$ serve as the expansion coefficients and

$$\int |\phi(z)|^2 dz = \int |F(u)|^2 du \qquad \text{after Eq. (5.86)}$$

Transition from conventional to delta function normalization

When functions $\psi_q(z)$ can be conventionally normalized, leading to a conventional expansion of some arbitrary function $\phi(z)$ in the form

$$\phi(z) = \sum_q f_q \psi_q(z) \tag{5.73}$$

presuming that we can make the integral approximation to this sum

$$\phi(z) = \int f(u)\psi(u,z)g(u)du \tag{5.77}$$

provided the density of states is uniform ($g(u) \equiv g$), we can choose to change to functions normalized to a delta function by choosing

$$F(u) = \sqrt{g}f(u) \tag{5.81}$$

and

$$\Psi(u,z) = \sqrt{g}\psi(u,z) \tag{5.82}$$

leading to the expansion as in Eq. (5.85).

Periodic boundary conditions

A set of boundary conditions often used in quantum mechanics for mathematical convenience is the periodic boundary conditions for a box of length L that the function should be the same at a point $z = L$ as they are at the point $z = 0$ (or possibly some shifted version of these same conditions). They are often used when the functions of interest are plane wave exponentials, in which case they imply

$$\exp(ikz) = \exp[ik(z+L)] \tag{5.93}$$

which leads to the condition

$$k = \frac{2m\pi}{L} \tag{5.95}$$

and a density of states

$$g = \frac{L}{2\pi} \tag{5.96}$$

These boundary conditions, though often not strictly correct physically, allow exponential functions to be used rather than sines or cosines, when considering boxes of finite size and can allow simpler mathematics are a result.

Position eigenfunctions

The position eigenfunctions in z are the delta functions $\delta(z - z_o)$ for each possible value of z_o. These eigenfunctions are themselves normalized to a delta function.

Expansion in position eigenfunctions

A function $\phi(z)$ of position is its own set of expansion coefficients in the expansion in position eigenfunctions.

Chapter 6

Approximation methods in quantum mechanics

Prerequisites: Chapters 2 – 5, including a first reading of Section 2.11.

We have seen how to solve some simple quantum mechanical problems exactly and, in principle, we know how to solve for any quantum mechanical problem that is the solution of Schrödinger's equation. Some extensions of Schrödinger's equation are important for many problems, especially those including the consequences of electron spin. Other equations also arise in quantum mechanics, beyond the simple Schrödinger equation description, such as appropriate equations to describe photons and relativistically correct approaches. We postpone discussion of any such more advanced equations.

For all such equations, however, there are relatively few problems that are simple enough to be solved exactly. This is not a problem peculiar to quantum mechanics; there are relatively few classical mechanics problems that can be solved exactly either. Problems that involve multiple bodies or that involve the interaction between systems are often quite difficult to solve.

One could regard such difficulties as being purely mathematical, say that we have done our job of setting up the necessary principles to understand quantum mechanics, and move on, consigning the remaining tasks to applied mathematicians or possibly to some brute-force computer technique. Indeed, the standard set of techniques that can be applied, for example, to the solution of differential equations can be (and are) applied to the solution of quantum mechanical differential equation problems. The problem with such an approach is that if we blindly apply the mathematical techniques, we may lose much insight as to how such more complicated systems work. Specifically, we would not understand just what were the important and often dominant aspects of such more complicated problems, aspects that often allow us to have a relatively simpler view of them. Hence, it is useful – both from the practical point of view (i.e., we can actually do the problems ourselves) and the conceptual one (i.e., we can know what we are doing) – to understand some of the key approximation methods of quantum mechanics.

There are several such techniques and it is quite common to invent new techniques or variants of old ones to tackle particular problems. In nearly all cases, the analysis in terms of expansions in complete sets of functions is central to the use or understanding of the approximation techniques and we rely heavily on the results from the Hilbert space view of functions and operators.

Among the most common techniques are (1) use of finite basis subsets or, equivalently, finite matrices; (2) perturbation theory (which comes in two flavors, time-independent and time-dependent); (3) the tight-binding approximation; and (4) the variational method. Each of these also offers some insight into the physical problem. In this chapter, we discuss all of these except the time-dependent perturbation theory, which we postpone until the next chapter. There are also some specific techniques that are very useful for one-dimensional problems and we devote Chapter 11 to those.

6.1 Example problem – potential well with an electric field

To illustrate the different methods, we analyze a particular problem, that of a one-dimensional, infinitely deep potential well for an electron with an applied electric field (i.e., a skewed infinite well). This problem is solvable exactly analytically and this was done in Section 2.11. These exact solutions are based on Airy functions that themselves have to be evaluated by numerical techniques (though these techniques are well understood and the error limits can be quantified). It is also straightforward to solve this problem by the various approximation methods without recourse to the Airy functions and, in practice, these methods can actually be easier than evaluating the "exact" solutions. Another reason for using this problem as an illustration is that because the solutions of the "unperturbed" problem (i.e., with no applied field) are mathematically simple, we can keep the mathematics simple in the illustrations of the various techniques.

This particular problem has a specific practical application, which is in the design of quantum well electroabsorption modulators. The shifts in the energy levels calculated here translate into shifts in the optical absorption edge in semiconductor quantum well structures with applied electric fields. This shift, in turn, is used to modulate the transmission of a light beam in high-speed modulators in optical communications systems. Aspects of this same problem occur also in analyzing the allowed states of carriers in silicon transistor structures and in tunneling in the presence of electric fields.

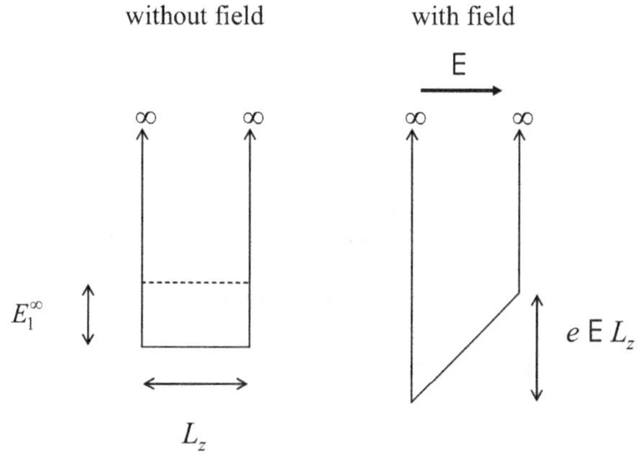

Fig. 6.1. Illustration of an infinitely deep potential well for an electron, without and with electric field E.

First, we set up this problem in an appropriate form. The potential structure with and without field is illustrated in Fig. 6.1. The energy of an electron in an electric field E simply increases linearly with distance. A positive electric field in the positive z direction pushes the electron in the negative z direction with a force of magnitude $e\mathsf{E}$ and so the potential energy of the electron increases in the positive z direction with the form $e\mathsf{E}z$. We are free to choose our potential energy origin wherever we please and, for convenience, we choose it to be zero in the middle of the well. Hence, within the well, the potential energy is

$$V(z) = e\mathsf{E}(z - L_z/2) \tag{6.1}$$

and the Hamiltonian becomes

$$\hat{H} = -\frac{\hbar^2}{2m}\frac{d^2}{dz^2} + e\mathsf{E}(z - L_z/2) \tag{6.2}$$

It is mathematically more convenient to define dimensionless units for this problem.[1] A convenient unit of energy, $E_1^\infty = (\hbar^2/2m)(\pi/L_z)^2$, is the confinement energy of the first state of the original infinitely deep well and in those units, the eigenenergy of the nth state is

$$\eta_n = \frac{E_n}{E_1^\infty} \tag{6.3}$$

A convenient unit of field is the field, E_o, that gives one unit of energy, E_1^∞, of potential change from one side of the well to the other; that is,

$$\mathsf{E}_o = \frac{E_1^\infty}{eL_z} \tag{6.4}$$

and in those units, the (dimensionless) field is

$$\mathsf{f} = \frac{\mathsf{E}}{\mathsf{E}_o} \tag{6.5}$$

A convenient unit of distance is the thickness of the well and so the dimensionless distance is

$$\xi = z/L_z \tag{6.6}$$

Dividing throughout by E_1^∞, the Hamiltonian within the well can now be written in these dimensionless units as

$$\hat{H} = -\frac{1}{\pi^2}\frac{d^2}{d\xi^2} + \mathsf{f}(\xi - 1/2) \tag{6.7}$$

with the corresponding time-independent Schrödinger equation

$$\hat{H}\phi(\xi) = \eta\phi(\xi) \tag{6.8}$$

For the original "unperturbed" problem without field, we write the "unperturbed" Hamiltonian within the well as

$$\hat{H}_o = -\frac{1}{\pi^2}\frac{d^2}{d\xi^2} \tag{6.9}$$

---

[1] The distance units that are most useful for illustrating the various approximation techniques are necessarily different from those that put this problem into the form required for the Airy function differential equation in Section 2.11, though we keep the energy units the same as in that section. One dimensionless unit of field here does correspond with one dimensionless unit of potential in Section 2.11; e.g., $\mathsf{f} = 3$ here is the same as $v_L = 3$ in Section 2.11.

The normalized solutions of the corresponding (unperturbed) Schrödinger equation

$$\hat{H}_o \psi_n = \varepsilon_n \psi_n \tag{6.10}$$

are then

$$\psi_n(\xi) = \sqrt{2}\sin(n\pi\xi) \tag{6.11}$$

This now completes the setup of this problem in dimensionless units so that we can conveniently illustrate the various approximation methods using it as an example.

6.2 Use of finite matrices

Most quantum mechanical problems, in principle, have an infinite number of functions in the basis set of functions that should be used to represent the problem.[2] In practice, we can very often make a useful and quite accurate approximate solution by considering only a few specific functions – that is, a finite subset of all of the possible functions, typically those with energies close to the state of the system in which we are most interested. Though the use of such finite basis subsets is quite common, especially given the common use of matrices in solving problems numerically on computers, it is not normally discussed explicitly in quantum mechanics texts. In practice, the use of such finite subsets of basis functions leads us to use matrices whose dimension corresponds with the number of basis functions used. With easy manipulation of matrices now routine with mathematical software, this may be the first numerical technique of choice. To give this technique a name, we can call it the "finite basis subset method" or the "finite matrix method," though the reader should be aware that both of these names are inventions of this author.[3]

As can be seen from the discussion of the Hilbert space view of functions and operators, quantum mechanical problems can often be conveniently reduced to linear algebra problems, with operators represented by matrices and functions by vectors. The practical solution of some problem, such as energy eigenvalues and eigenstates, then reduces to a problem of finding the eigenvectors of a matrix. Occasionally, such problems can be solved exactly, of course, but more often no exact analytic solution is known. Then, to solve the problem, we may have to solve numerically for eigenvalues and eigenvectors, which means we have to restrict the matrix to being a finite one, even if the problem, in principle, has a basis set with an infinitely large number of elements.

It is also quite common to consider analytically a finite matrix and solve that simpler problem exactly. Then one can have an approximate analytic solution. This approach is taken with great success, for example, in the so-called k.p ("k dot p") method of calculating band structures in semiconductors discussed in Section 8.9. The k.p method is the principal band structure

---

[2] Some problems do have quite finite numbers of required basis functions, however, e.g., the problem of a stationary electron in a magnetic field, which only needs two basis functions (i.e., one corresponding to spin-up and the other to spin-down).

[3] This approach is mathematically closely related to the problem of degenerate perturbation theory discussed in this chapter because it also involves similar manipulations with finite matrices. In some texts, this approach is even called degenerate perturbation theory when it is used in the way we discuss in this section. Because the approach in this section is definitely not perturbation theory, calling this approach degenerate perturbation theory in this context seems misleading to this author and possibly quite confusing to the reader, so we have to invent another name.

method used for calculating optical properties of semiconductors near the optical absorption edge. The results of this method are crucial in understanding the basic properties of modern semiconductor lasers, for example.

Why can we conclude that restricting to such a finite matrix is justifiable? The fundamental justification for why we can even consider such finite matrices is that the quantum mechanical operator we are dealing with is compact in the mathematical sense. If an operator is compact, then, as we add more basis functions into the set we are using in constructing the matrix for our operator, eventually our finite matrix will be as good an approximation as we want for any calculation to the infinite matrix we perhaps ought to have; this is essentially the mathematical definition of compactness of operators.

In practice, there is no substitute for intelligence in choosing the finite basis subset, however, and this is something of an art. If we choose the form of the basis subset badly or make a poor choice as to what elements to include in our finite subset, then we will end up with a poor approximation to the result or a matrix that is ill-conditioned. A very frequent choice of basis subset is to use the energy eigenfunctions of the *unperturbed*[4] problem (or possibly the energy eigenfunctions of a simpler, though related, problem) with energies closest to those of the states of interest.

Now we consider our specific example problem of an electron in a one-dimensional potential well in the z direction, with an electric field applied perpendicular to the well. We need to construct the matrix of the Hamiltonian. The matrix elements are

$$H_{ij} = -\frac{1}{\pi^2}\int_0^1 \psi_i^*(\xi)\frac{d^2}{d\xi^2}\psi_j(\xi)\,d\xi + \mathrm{f}\int_0^1 \psi_i^*(\xi)(\xi - 1/2)\psi_j(\xi)\,d\xi \tag{6.12}$$

(In this particular case, because the wavefunctions happen to be real, the complex conjugation makes no difference in the integrals.)

For our explicit example here, we consider a field of three dimensionless units (i.e., $\mathrm{f} = 3$) and we take as our finite basis only the first three energy eigenfunctions, from Eq. (6.11), of the unperturbed problem. Then, performing the integrals in Eq. (6.12) numerically, we obtain the approximate Hamiltonian matrix

$$\hat{H} = \begin{bmatrix} 1 & -0.54 & 0 \\ -0.54 & 4 & -0.584 \\ 0 & -0.584 & 9 \end{bmatrix} \tag{6.13}$$

Note that this matrix is Hermitian, as expected. (In this particular case, because of the reality of both the operators and the functions, the matrix elements are all real.) Now, we can numerically find the eigenvalues of this matrix, which are

$$\eta_1 \simeq 0.90437,\ \eta_2 \simeq 4.0279,\ \eta_3 \simeq 9.068 \tag{6.14}$$

These can be compared with the results from the exact Airy function solutions, which are

$$\varepsilon_1 \simeq 0.90419,\ \varepsilon_2 \simeq 4.0275,\ \varepsilon_3 \simeq 9.0173 \tag{6.15}$$

---

[4] The use of the term *unperturbed* does not mean that we are using perturbation theory here. It merely refers to the simpler problem before the *perturbation* (here, the applied electric field) was added.

We see first that by either the exact or approximate method, these are quite near to the unperturbed (zero field) values (which would be 1, 4, and 9, respectively). We see also that the lowest energy eigenvalue has reduced from its unperturbed value and the second and third eigenenergies are actually increased (which is counterintuitive but true). This finite basis subset approach has given quite an accurate answer for the lowest state and somewhat less accuracy for the others, especially the third level. This relative inaccuracy for the third level is not very surprising – we may well need to include at least one more basis function in the basis set to get an accurate answer here.[5]

The corresponding eigenvectors are solved numerically as

$$|\phi_1\rangle = \begin{bmatrix} 0.985 \\ 0.174 \\ 0.013 \end{bmatrix}, \; |\phi_2\rangle = \begin{bmatrix} -0.175 \\ 0.978 \\ 0.115 \end{bmatrix}, \; |\phi_2\rangle = \begin{bmatrix} 0.007 \\ -0.115 \\ 0.993 \end{bmatrix} \tag{6.16}$$

(These are normalized, with the sum of the squares of the elements of the vectors each adding to 1.) Explicitly, this means that, for example, the first eigenfunction is

$$\phi_1(\xi) = 0.985\sqrt{2}\sin(\pi\xi) + 0.174\sqrt{2}\sin(2\pi\xi) + 0.013\sqrt{2}\sin(3\pi\xi) \tag{6.17}$$

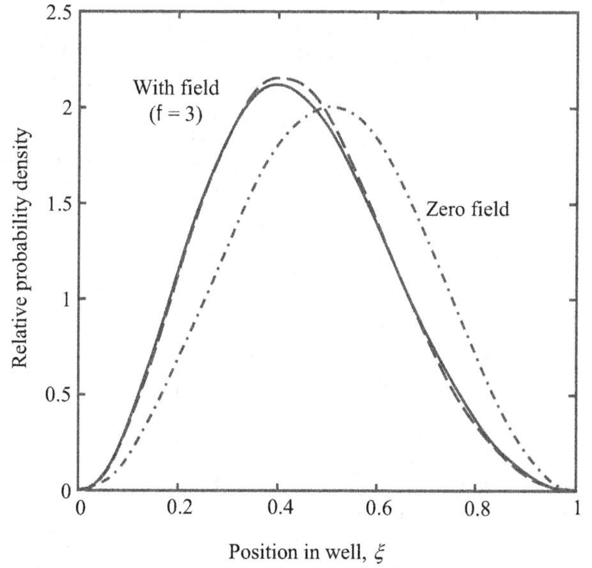

Fig. 6.2. Comparison of the unperturbed (zero field) relative probability density (dot-dashed line) and the probability density with three units of field for the first energy eigenstate in an infinitely deep potential well, calculated (1) using the finite basis subset method with a 3x3 matrix - solid line; and (2) using first-order perturbation theory – dashed line.

Fig. 6.2 compares the first approximate eigenfunction solution $\phi_1(\xi)$ with the "unperturbed" exact solution $\psi_1(\xi)$ (the comparison with the perturbation method is also shown in Fig. 6.2). We plot probability density (i.e., the squared modulus of the wavefunction) to make the differences between different calculation methods clearer. (Small differences in wavefunctions

[5] In fact, including the fourth basis function of the "unperturbed" problem gives 9.0175 for the energy of this third level, very close to the exact 9.0173

give larger relative differences in probability density.) The electron probability density with field has moved to the left somewhat, as would be expected for the electron being pushed by the field. On the scale of this graph, there is negligible difference between the probability density calculated here by this finite basis method and the exact Airy function result. Results for the first energy level for two, three, and four basis functions in the calculation are shown in Table 6.1. The energy eigenvalue, in this case, converges rapidly to the exact value, with the energy correct to within ~ 1 part in 10^5 with four basis functions.

To conclude on this approach, with the ease of handling matrices in modern mathematical computer programs, this technique can be quite convenient for simple problems and we see it can converge quite rapidly as we increase the number of basis functions, even with quite small numbers of functions (and, consequently, small matrices). The main art in this approach is choosing a good underlying set of basis functions and the right subset of them.

Method	*Result*
Exact (Airy functions)	0.90419
Finite basis – 2 functions (2x2 matrix)	0.90563
Finite basis – 3 functions (3x3 matrix)	0.90437
Finite basis – 4 functions (4x4 matrix)	0.90420
Perturbation theory – first order	1
Perturbation theory – second order	0.90253
Variational	0.90563

Table 6.1. Calculated energies (in dimensionless units) of an electron in the first level in a potential well with infinitely high barriers, for three units of electric field, calculated by various different methods.

Problems

6.2.1 Solve the problem of an electron in a potential well with three units of field using the first two energy eigenfunctions of the well without field as the finite basis subset. Give the energies (in the dimensionless units) and explicit formulae for the normalized eigenfunctions for the first two levels calculated by this method. Do the algebra of this problem by hand; that is, do not use mathematical software to evaluate matrix elements or to solve for the eigenvalues and eigenfunctions.
(Note: $\int_0^1 (\xi - 1/2)\sin(\pi\xi)\sin(2\pi\xi)d\xi = -(8/9\pi^2)$)

6.2.2 (This problem can be used as a substantial assignment.) Electrons in semiconductors can behave as if they had the charge of a normal electron but have a much different mass (the so-called effective mass). For GaAs, the electron effective mass is ~ 0.07 m_o. We are interested in a semiconductor device that could be a tunable detector for infrared wavelengths. We make this device using a 100 Å thick layer of GaAs, surrounded on either side by materials considered to behave as if the electron sees an infinitely high potential barrier on either side of the GaAs layer (this is an approximation to the actual behavior of AlGaAs barriers).

The concept in this device is that there will be an electron initially in the lowest state in this infinitely deep potential well and we are interested in the optical absorption of light polarized in the z direction (i.e., the optical electric field is in the z direction) that takes the electron from this lower state to one or other of the higher states in the well. We presume that the energy of the photons that can be absorbed corresponds to the energy separations of the states being considered. Once into these higher states, we have some other mechanism that we need not consider in detail that extracts any electrons from these higher states to give a photocurrent in the device (in practice, with actual finite barrier heights, this can be either a thermal emission or a tunneling through the finite barrier), and we also presume that we have another mechanism that puts another electron in the lower state again after any such photocurrent "emission."

The optical electric field is at all times very small, so we can presume its only effect is to cause transitions between levels, not otherwise to perturb the levels themselves. To tune the strengths and wavelengths of the optical transitions in this device, however, we apply an electric field F along the positive z direction, with a practical range from 0 to 10 V/μm (use this range for calculations.)

Consider the first, second, and third electron levels as a function of field F.

(i) Calculate the energy separations between the first and second electron levels and between the first and third energy levels over the stated range of fields and plot these on a graph.

(ii) Consider the energy eigenfunctions for each of the first three levels as a function of field. Specifically, calculate the approximate amplitudes of the first four infinite well basis functions in the expansion of each of these eigenfunctions and plot these amplitudes as a function of field, for each of the first three levels.

(iii) Relative to the strength (i.e., transition rate for a given optical field amplitude) of the optical absorption between the first and second levels at zero applied field, plot the strength of the optical absorption between the first and second levels and between the first and third levels as a function of field. (Note: The transition rate between states $|\psi_j(z)\rangle$ and $|\psi_i(z)\rangle$ for a given optical field amplitude can be taken to be proportional to $|z_{ij}|^2$, where $z_{ij} = \langle\psi_i(z)|z|\psi_j(z)\rangle$.)

(iv) Given that we presume this device is useful as a tunable detector only for optical absorption strengths that are at least 1/10 of that of the optical absorption between the first and second levels at zero applied field, what are the tuning ranges, in wavelength, for which this detector is useful given the stated range of fields F?

6.3 Time-independent nondegenerate perturbation theory

Many situations arise in quantum mechanics in which we want to know what happens when some system interacts with some other system or disturbance. An example might be a hydrogen atom in the presence of a weak electric field. Conceptually, we like to think that it is still meaningful to talk about the existence of the hydrogen atom but to imagine that the field somehow perturbs the hydrogen atom in some way. Strictly speaking, in the presence of a finite field, it is not clear that there is any such thing as a hydrogen atom in the ideal way we normally imagine it. The system has no bound states as we understood them before – it is possible for the electron to tunnel out of any of the formerly bound states.[6] In perturbation theory, however, when we are interested in small perturbations, we hold on to the concept of the unperturbed system as still being essentially valid and calculate corrections to that system caused by the small external perturbations. The effect of the small field is then viewed mathematically as a correction. We expect that energy levels might move as a result of the field, for example (an effect known in this particular case as the Stark effect), and we would like to be able to calculate such small perturbations. One key approach is time-independent perturbation theory (also known as stationary perturbation theory). This method is essentially one of successive approximations. Though it is often not the best method to calculate a numerical result to a given problem, especially with the easy availability of fast computers, it is conceptually quite a useful technique and is a way of looking at the interactions of physical systems.[7]

---

[6] The hydrogen atom in the presence of an electric field does still have energy eigenstates, though there is now a continuum of allowed energies, not the discrete spectrum of the former bound states.

[7] This idea of perturbations as the interactions of systems, especially the interactions of particles, is one that becomes particularly important as we go on to look at time-dependent perturbation theory because the perturbation approach lets us define processes in time, e.g., the absorption of a photon by an atom.

Nearly always in time-independent problems we are interested in energy levels and energy eigenstates.[8] We presume, therefore, that there is some unperturbed Hamiltonian, $\hat{H}_o$, that has known eigen solutions; that is,

$$\hat{H}_o |\psi_n\rangle = E_n |\psi_n\rangle \tag{6.18}$$

and we presume the eigenfunctions $|\psi_n\rangle$ are normalized. For example, this unperturbed problem could be that of an electron in an infinitely deep potential well, without an applied electric field.

In thinking about such perturbations, we can imagine that the perturbation we are considering could be progressively "turned on," at least in a mathematical sense. For example, we could imagine that we are progressively increasing the applied field, E, from zero. The core idea of perturbation theory is to look for the changes in the solutions (both the eigenfunctions and the eigenvalues) that are proportional first to E (so-called first-order corrections); then, if we want, those that are proportional to E^2 ("second-order corrections"); then, if we are still interested, those that are proportional to E^3; and so on. Presumably, the first-order corrections are the most important if this perturbation is a small one and we might just stop there. Sometimes the first-order correction is zero for some reason (e.g., because of some symmetry), in which case we may go on to the second-order correction. Usually in this perturbation theory method, we stop at the lowest order that gives us a nonzero result.

In general, in perturbation theory, we imagine that our perturbed system has some additional term in the Hamiltonian, the "perturbing Hamiltonian," $\hat{H}_p$. In our example case of an infinitely deep potential well with an applied field, that perturbing Hamiltonian would be $\hat{H}_p = e\mathsf{E}(z - L_z/2)$. We could construct the perturbation theory directly using the powers of E, but we may not always have a parameter like E in the problem that we can imagine we can increase smoothly. It will be more useful (though entirely equivalent) to generalize the way we write the theory by introducing a mathematical "housekeeping" parameter γ. In this way of writing the theory, we say that the perturbing Hamiltonian is $\gamma\hat{H}_p$, where $\hat{H}_p$ can be physically a fixed perturbation, and we imagine we can smoothly increase γ, looking instead for changes in the solutions that are proportional to γ (for first-order corrections), γ^2 (for second-order corrections), and so on. In the end, when we come to perform the actual calculation to a given order of approximation, having used the powers of γ to help separate out the different orders of corrections, we can set $\gamma = 1$ or, indeed, to any other value we like as long as $\gamma\hat{H}_p$ corresponds to the actual physical perturbation of the system.

If this concept is confusing at a first reading,[9] the reader can simply imagine that γ is essentially the strength of the electric field in our example problem. We could, for example, have a "γ" knob on the front of the voltage supply that provides the electric field for our system. We progressively turn this knob from 0 to 1 to ramp the voltage up to its full value. Conceptually, as we do so, we could be looking for responses of the system that are proportional to γ, to γ^2, to γ^3, and so on.

---

[8] In fact, the mere statement that we are interested in time-independent solutions actually guarantees that we have to deal ultimately with eigenstates of the Hamiltonian because they are the only states that do not change in time.

[9] Many students do find this concept of a "housekeeping" parameter confusing at first. It is not, in the end, a very difficult concept, but most students have not come across this kind of approach before.

With this way of thinking about the problem mathematically, we can write for our Hamiltonian (e.g., Schrödinger) equation

$$\left(\hat{H}_o + \gamma \hat{H}_p\right)|\phi\rangle = E|\phi\rangle \tag{6.19}$$

We now presume that we can express the resulting perturbed eigenfunction and eigenvalue as power series in this parameter; that is,

$$|\phi\rangle = \left|\phi^{(0)}\right\rangle + \gamma\left|\phi^{(1)}\right\rangle + \gamma^2\left|\phi^{(2)}\right\rangle + \gamma^3\left|\phi^{(3)}\right\rangle + \cdots \tag{6.20}$$

$$E = E^{(0)} + \gamma E^{(1)} + \gamma^2 E^{(2)} + \gamma^3 E^{(3)} + \cdots \tag{6.21}$$

Now we substitute these power series into the equation (6.19).

$$\begin{aligned}&\left(\hat{H}_0 + \gamma \hat{H}_p\right)\left(\left|\phi^{(0)}\right\rangle + \gamma\left|\phi^{(1)}\right\rangle + \gamma^2\left|\phi^{(2)}\right\rangle + \cdots\right)\\ &= \left(E^{(0)} + \gamma E^{(1)} + \gamma^2 E^{(2)} + \cdots\right)\left(\left|\phi^{(0)}\right\rangle + \gamma\left|\phi^{(1)}\right\rangle + \gamma^2\left|\phi^{(2)}\right\rangle + \cdots\right)\end{aligned} \tag{6.22}$$

A key point is that if this power series description is to hold for any γ (at least within some convergence range), then it must be possible to equate terms in given powers of γ on the two sides of the equation. Quite generally, if we had two power series that were equal, that is,

$$a_0 + a_1\gamma + a_2\gamma^2 + a_3\gamma^3 + \cdots = b_0 + b_1\gamma + b_2\gamma^2 + b_3\gamma^3 + \cdots = f(\gamma) \tag{6.23}$$

the only way this can be true for arbitrary γ is for the individual terms to be equal (i.e., $a_i = b_i$). This is the same as saying that the power series expansion of a function $f(\gamma)$ is unique. Of course, the equation (6.22) involves (column) vectors instead of the scalar coefficients a_i or b_i. But that simply means that Eq. (6.22) corresponds to a set of different equations like Eq. (6.23) – one for each element (or row) of the vector. Hence, in the vector case also, we must be able to equate powers of γ.

Hence, equating powers of γ, we can obtain from Eq. (6.22) a progressive set of equations. The first of these is equating terms in γ^0 (i.e., terms not involving γ), the "zeroth" order equation

$$\hat{H}_o\left|\phi^{(0)}\right\rangle = E^{(0)}\left|\phi^{(0)}\right\rangle \tag{6.24}$$

This is simply the original unperturbed Hamiltonian equation, with eigenfunctions $|\psi_n\rangle$ and eigenvalues E_n. We presume now that we are interested in a particular eigenstate $|\psi_m\rangle$ and how it is perturbed. We therefore write $|\psi_m\rangle$ instead of $\left|\phi^{(0)}\right\rangle$ and E_m instead of $E^{(0)}$. With this notation, our progressive set of equations, each equating a different power of γ, becomes

$$\hat{H}_o|\psi_m\rangle = E_m|\psi_m\rangle \tag{6.25}$$

$$\hat{H}_o\left|\phi^{(1)}\right\rangle + \hat{H}_p|\psi_m\rangle = E_m\left|\phi^{(1)}\right\rangle + E^{(1)}|\psi_m\rangle \tag{6.26}$$

$$\hat{H}_o\left|\phi^{(2)}\right\rangle + \hat{H}_p\left|\phi^{(1)}\right\rangle = E_m\left|\phi^{(2)}\right\rangle + E^{(1)}\left|\phi^{(1)}\right\rangle + E^{(2)}|\psi_m\rangle \tag{6.27}$$

and so on. We can choose to rewrite these equations, Eqs. (6.25) – (6.27), as

$$\left(\hat{H}_o - E_m\right)|\psi_m\rangle = 0 \tag{6.28}$$

$$\left(\hat{H}_o - E_m\right)\left|\phi^{(1)}\right\rangle = \left(E^{(1)} - \hat{H}_p\right)|\psi_m\rangle \tag{6.29}$$

$$\left(\hat{H}_o - E_m\right)\left|\phi^{(2)}\right\rangle = \left(E^{(1)} - \hat{H}_p\right)\left|\phi^{(1)}\right\rangle + E^{(2)}\left|\psi_m\right\rangle \tag{6.30}$$

and so on. Now we proceed to show how to calculate the various perturbation terms. It will become clear as we do this that perturbation theory is just a theory of successive approximations.

First-order perturbation theory

It is straightforward to calculate $E^{(1)}$ from Eq. (6.29). Premultiplying by $\langle\psi_m|$ gives

$$\begin{aligned}\left\langle\psi_m\left|\hat{H}_o - E_m\right|\phi^{(1)}\right\rangle &= \left(\left\langle\psi_m\left|\hat{H}_o - E_m\right.\right)\left|\phi^{(1)}\right\rangle = \left\langle\psi_m\right|\left(E_m - E_m\right)\left|\phi^{(1)}\right\rangle = 0 \\ &= \left\langle\psi_m\left|E^{(1)} - \hat{H}_p\right|\psi_m\right\rangle = E^{(1)} - \left\langle\psi_m\left|\hat{H}_p\right|\psi_m\right\rangle\end{aligned} \tag{6.31}$$

that is,

$$E^{(1)} = \left\langle\psi_m\left|\hat{H}_p\right|\psi_m\right\rangle \tag{6.32}$$

Hence, we have quite a simple formula for the first-order correction, $E^{(1)}$, to the energy of the mth state in the presence of our perturbation $\hat{H}_p$. Note that it depends only on the zeroth order (i.e., the unperturbed) eigenfunction.

To calculate the first-order correction, $\left|\phi^{(1)}\right\rangle$, to the wavefunction, we expand that correction in the basis set $\left|\psi_n\right\rangle$; that is,

$$\left|\phi^{(1)}\right\rangle = \sum_n a_n^{(1)}\left|\psi_n\right\rangle \tag{6.33}$$

Substituting this is in Eq. (6.29) gives

$$\begin{aligned}\left\langle\psi_i\left|\hat{H}_o - E_m\right|\phi^{(1)}\right\rangle &= \left(E_i - E_m\right)\left\langle\psi_i\left|\phi^{(1)}\right.\right\rangle = \left(E_i - E_m\right)a_i^{(1)} \\ &= \left\langle\psi_i\left|E^{(1)} - \hat{H}_p\right|\psi_m\right\rangle = E^{(1)}\left\langle\psi_i\left|\psi_m\right.\right\rangle - \left\langle\psi_i\left|\hat{H}_p\right|\psi_m\right\rangle\end{aligned} \tag{6.34}$$

We now make a restriction, which is that we presume that the energy eigenvalue, E_m, is not degenerate; that is, there is only one eigenfunction corresponding to this eigenvalue. In general, the whole approach we are discussing here requires this restriction and the perturbation theory we are discussing is therefore called *nondegenerate perturbation theory*. (Degeneracy needs to be handled somewhat differently and we consider that case later.) With this restriction, we still need to distinguish two cases in Eq. (6.34). First, for $i \neq m$, we have

$$a_i^{(1)} = \frac{\left\langle\psi_i\left|\hat{H}_p\right|\psi_m\right\rangle}{E_m - E_i} \tag{6.35}$$

For $i = m$, Eq. (6.34) gives us no additional information. Explicitly

$$\begin{aligned}\left(E_m - E_m\right)a_m^{(1)} &= 0a_m^{(1)} \\ &= E^{(1)} - \left\langle\psi_m\left|\hat{H}_p\right|\psi_m\right\rangle = E^{(1)} - E^{(1)} = 0\end{aligned} \tag{6.36}$$

This means we are free to choose $a_m^{(1)}$. The choice that makes the algebra simplest is to set $a_m^{(1)} = 0$, which is the same as saying that we choose to make $\left|\phi^{(1)}\right\rangle$ orthogonal to $\left|\psi_m\right\rangle$. An analogous situation occurs with all the higher order equations, such as Eq. (6.30). Adding an arbitrary amount of $\left|\psi_m\right\rangle$ into $\left|\phi^{(j)}\right\rangle$ makes no difference to the left-hand side of the equation. Hence, we make the convenient choice

$$\langle \psi_m | \phi^{(j)} \rangle = 0 \tag{6.37}$$

Hence, we obtain for the first-order correction to the wavefunction

$$|\phi^{(1)}\rangle = \sum_{n \neq m} \frac{\langle \psi_n | \hat{H}_p | \psi_m \rangle}{E_m - E_n} |\psi_n\rangle \tag{6.38}$$

Second-order perturbation theory

We can continue similarly to find the higher order terms. Premultiplying Eq. (6.30) on both sides by $\langle \psi_m |$ gives

$$\begin{aligned} \langle \psi_m | (\hat{H}_o - E_m) | \phi^{(2)} \rangle &= \langle \psi_m | (E_m - E_m) | \phi^{(2)} \rangle = 0 \\ &= \langle \psi_m | (E^{(1)} - \hat{H}_p) | \phi^{(1)} \rangle + \langle \psi_m | E^{(2)} | \psi_m \rangle \\ &= E^{(1)} \langle \psi_m | \phi^{(1)} \rangle - \langle \psi_m | \hat{H}_p | \phi^{(1)} \rangle + E^{(2)} \end{aligned} \tag{6.39}$$

Because we have chosen $|\psi_m\rangle$ orthogonal to $|\phi^{(j)}\rangle$ (Eq. (6.37)), we therefore have

$$E^{(2)} = \langle \psi_m | \hat{H}_p | \phi^{(1)} \rangle \tag{6.40}$$

or, explicitly, using Eq. (6.38)

$$E^{(2)} = \langle \psi_m | \hat{H}_p \left(\sum_{n \neq m} \frac{\langle \psi_n | \hat{H}_p | \psi_m \rangle}{E_m - E_n} |\psi_n\rangle \right) \tag{6.41}$$

that is,

$$E^{(2)} = \sum_{n \neq m} \frac{\left| \langle \psi_n | \hat{H}_p | \psi_m \rangle \right|^2}{E_m - E_n} \tag{6.42}$$

To find the second-order correction to the wavefunction, we can proceed similarly to before. We expand $|\phi^{(2)}\rangle$, noting now that by choice (Eq. (6.37)), $|\phi^{(2)}\rangle$ is orthogonal to $|\psi_m\rangle$, to obtain

$$|\phi^{(2)}\rangle = \sum_{n \neq m} a_n^{(2)} |\psi_n\rangle \tag{6.43}$$

We premultiply Eq. (6.30) by $\langle \psi_i |$ to obtain

$$\begin{aligned} \langle \psi_i | (\hat{H}_o - E_m) | \phi^{(2)} \rangle &= (E_i - E_m) a_i^{(2)} \\ &= \langle \psi_i | (E^{(1)} - \hat{H}_p) | \phi^{(1)} \rangle + \langle \psi_i | E^{(2)} | \psi_m \rangle \\ &= E^{(1)} a_i^{(1)} - \sum_{n \neq m} a_n^{(1)} \langle \psi_i | \hat{H}_p | \psi_n \rangle \end{aligned} \tag{6.44}$$

Note that we can write the summation in Eq. (6.44) excluding the term $n = m$ because we have chosen $|\phi^{(1)}\rangle$ to be orthogonal to $|\psi_m\rangle$ (i.e., we have chosen $a_m^{(1)} = 0$). Hence, for $i \neq m$, we have

$$a_i^{(2)} = \left(\sum_{n \neq m} \frac{a_n^{(1)} \langle \psi_i | \hat{H}_p | \psi_n \rangle}{E_m - E_i} \right) - \frac{E^{(1)} a_i^{(1)}}{E_m - E_i} \tag{6.45}$$

Note that the second-order wavefunction depends only on the first-order energy and wavefunction. We can write out Eq. (6.45) explicitly, using (6.32) to substitute for $E^{(1)}$ and (6.35) for $a_i^{(1)}$, to obtain

$$a_i^{(2)} = \left(\sum_{n \neq m} \frac{\langle \psi_i | \hat{H}_p | \psi_n \rangle \langle \psi_n | \hat{H}_p | \psi_m \rangle}{(E_m - E_i)(E_m - E_n)} \right) - \frac{\langle \psi_i | \hat{H}_p | \psi_m \rangle \langle \psi_m | \hat{H}_p | \psi_m \rangle}{(E_m - E_i)^2} \tag{6.46}$$

We can now gather these results together and write the perturbed energy and wavefunction up to second-order as

$$E \cong E_m + \langle \psi_m | \hat{H}_p | \psi_m \rangle + \sum_{n \neq m} \frac{\left| \langle \psi_n | \hat{H}_p | \psi_m \rangle \right|^2}{E_m - E_n} \tag{6.47}$$

and

$$|\phi\rangle \cong |\psi_m\rangle + \sum_{i \neq m} \frac{\langle \psi_i | \hat{H}_p | \psi_m \rangle}{E_m - E_n} |\psi_i\rangle + \sum_{i \neq m} \left[\left(\sum_{n \neq m} \frac{\langle \psi_i | \hat{H}_p | \psi_n \rangle \langle \psi_n | \hat{H}_p | \psi_m \rangle}{(E_m - E_i)(E_m - E_n)} \right) - \frac{\langle \psi_i | \hat{H}_p | \psi_m \rangle \langle \psi_m | \hat{H}_p | \psi_m \rangle}{(E_m - E_i)^2} \right] |\psi_i\rangle \tag{6.48}$$

that is,

$$|\phi\rangle \cong |\psi_m\rangle + \sum_{i \neq m} \left[\frac{\langle \psi_i | \hat{H}_p | \psi_m \rangle}{E_m - E_n} \left(1 - \frac{\langle \psi_m | \hat{H}_p | \psi_m \rangle}{E_m - E_i} \right) + \sum_{n \neq m} \frac{\langle \psi_i | \hat{H}_p | \psi_n \rangle \langle \psi_n | \hat{H}_p | \psi_m \rangle}{(E_m - E_i)(E_m - E_n)} \right] |\psi_i\rangle \tag{6.49}$$

Example of well with field

Now we consider the problem of the infinitely deep potential well with an applied field. We write the Hamiltonian as the sum of the unperturbed Hamiltonian, which is, in the well and in the dimensionless units we chose

$$\hat{H}_o = -\frac{1}{\pi^2} \frac{d^2}{d\xi^2} \tag{6.50}$$

and the perturbing Hamiltonian

$$\hat{H}_p = \mathrm{f}(\xi - 1/2) \tag{6.51}$$

where again we take $\mathrm{f} = 3$ for our explicit calculation.

Let us now calculate the various corrections. In first-order, the energy shift with applied field is

$$
\begin{aligned}
E^{(1)} &= \langle \psi_m | \hat{H}_p | \psi_m \rangle = \mathrm{f}\int_0^1 \sqrt{2}\sin(m\pi\xi)(\xi - 1/2)\sqrt{2}\sin(m\pi\xi)\,d\xi \\
&= 2\mathrm{f}\int_0^1 (\xi - 1/2)\sin^2(m\pi\xi)\,d\xi \\
&= 0
\end{aligned}
\tag{6.52}
$$

The integrals here are zero for all m because the sine-squared function is even with respect to the center of the well, whereas the $(\xi - 1/2)$ is odd. Hence, for this particular problem, there is no first-order energy correction (i.e., to first-order in perturbation theory, the energy is unchanged; hence, the result "1" in Table 6.1). Why is that? The answer is because of the symmetry of the problem. Suppose that there were an energy correction proportional to the applied field f. Then, if we changed the direction of f – that is, changed its sign – the energy correction would also have to change sign. But, by the symmetry of this problem, the resulting change in energy cannot depend on the direction of the field; the problem is symmetric in the + or - ξ directions, so there cannot be any change in energy linearly proportional to the field, f.

The general matrix elements that we need for further perturbation calculations are

$$
H_{puv} = \mathrm{f}\int_0^1 \sqrt{2}\sin(u\pi\xi)(\xi - 1/2)\sqrt{2}\sin(v\pi\xi)\,d\xi \tag{6.53}
$$

where u and v are integers. In general, we need u and v to have opposite parity (i.e., if one is odd, the other must be even) for these matrix elements to be nonzero because otherwise the overall integrand is odd about $\xi = 1/2$.

Now we can calculate the first-order correction to the wavefunction, which is for the first state

$$
\phi^{(1)}(\xi) = \sum_{u=2}^{q} \frac{H_{pu1}}{\varepsilon_{o1} - \varepsilon_{ou}} \psi_u(\xi) \tag{6.54}
$$

where here $\varepsilon_{ou} = u^2$ are the energies of the unperturbed states and q is a finite number that we must choose, in practice (we cannot in a general numerical calculation actually sum to infinity). For these calculations here, we chose $q = 6$, though a smaller number would probably be quite accurate (even $q = 2$ gives almost identical numerical answers, for reasons that will become apparent). Explicitly, for the expansion coefficients $a_u^{(1)} = H_{pu1}/(\varepsilon_{o1} - \varepsilon_{ou})$, we have numerically, for example

$$
a_2^{(1)} \cong 0.180\,,\ a_3^{(1)} = 0\,,\ a_4^{(1)} = 0.003 \tag{6.55}
$$

Here, the value of 0.180 for $a_2^{(1)}$ compares closely with the value of 0.174 for the second expansion coefficient in Eq. (6.16) obtained previously in the finite basis subset method. The wavefunction with the first-order correction is plotted in Fig. 6.2.

Because the first-order correction to the energy was zero, we have to go to second-order to get a perturbation correction to the energy. Explicitly, we have

$$
E^{(2)} \cong \sum_{u=2}^{q} \frac{|H_{pu1}|^2}{\varepsilon_{o1} - \varepsilon_{ou}} \tag{6.56}
$$

which numerically here gives

$$
E^{(2)} = -0.0975 \tag{6.57}
$$

or a final estimate of the total energy of

$$\eta_1 \cong \varepsilon_{o1} + E^{(1)} + E^{(2)} = 0.9025 \tag{6.58}$$

which compares with the exact result of 0.90419 (see Table 6.1).

Note that the second-order energy correction, $E^{(2)}$ (which is the first energy correction in this particular problem because $E^{(1)}$ is zero), is analytically proportional to the square of the field, f^2. Hence, perturbation theory gives us an approximate analytic result for the energy, which we can now use for any field without performing the perturbation theory calculation again. Explicitly, we can write

$$\eta_1 \cong \varepsilon_{o1} - 0.0108\,\mathrm{f}^2 \tag{6.59}$$

This is a typical kind of result from a perturbation calculation, allowing us to obtain an approximate analytic formula valid for small perturbations. We similarly find that the corrections to the wavefunction are approximately analytically proportional to the field and we have an approximate wavefunction of

$$\phi(\xi) \cong \sqrt{2}\sin(\pi\xi) + 0.06\mathrm{f}\sqrt{2}\sin(2\pi\xi) \tag{6.60}$$

We have dropped higher terms because the next nonzero term (i.e., the term in $\sin(4\pi\xi)$) is some sixty times smaller (see Eq. (6.55)). To a good degree of approximation, the perturbed wavefunction at low fields simply involves an admixture of the second basis function. Because it is the first-order wavefunction that is used to calculate the second-order energy, we can now see why even including only one term in the sums (i.e., setting $q = 2$ in the sums in Eqs. (6.54) and (6.56)) is quite accurate in this case.

Remarks on perturbation theory

Perturbation theory is particularly useful, as one might expect, for calculations involving small perturbations to the system. It can give simple analytic formulae and values of coefficients for various effects involving weak interactions. It is also conceptually useful in understanding interactions in general. Even if we are not performing an actual perturbation theory calculation, we can use perturbation theory to judge whether to include some level in, for example, a finite basis subset calculation. If a given level is far away in energy and/or has a small matrix element compared to that of some closer level, we can safely neglect that given level because of the energy separations that would appear in the denominators in the perturbation terms.

We have only shown here the first- and second-order perturbation formulae. Generally, perturbation calculations are most useful for the first nonzero order of correction. Specific effects sometimes require higher order calculations. For example, nonlinear optical effects of different types are associated with particular orders of perturbation theory calculations (though they are time-dependent perturbation calculations). Linear optics is based on first-order perturbation theory; linear electro-optic effects, second-harmonic generation, and optical parametric generation use second-order perturbation; nonlinear refraction and four-wave mixing (quite common effects in long-distance optical fiber systems) need third-order perturbation calculations.

One minor point about the wavefunction formulae we have mentioned here is that they are not quite normalized; we are merely adding the corrections to the original wavefunction in Eq. (6.20). This is not a substantial issue for small corrections. It is quite straightforward also to normalize the corrected wavefunctions if this is important.

Perturbation theory is a theory of successive approximations. As can be seen, we use the zeroth-order wavefunction to calculate the first-order energy correction and we use the first-order energy correction in calculating the second-order wavefunction correction. This process continues in time-dependent perturbation theory in the next chapter.

It is quite generally true of approximation methods that energies can be calculated reasonably accurately even with relatively poor wavefunctions. In perturbation theory, the nth approximation to the energy only requires the $(n-1)$th approximation to the wavefunction.

The particular kind of perturbation method we have discussed here (known as Rayleigh–Schrödinger perturbation theory) tends to lead to a series that does not converge very rapidly. Hence, trying to get a more accurate calculation by adding more terms to the series is often not very productive.[10] This is one reason why this kind of perturbation approach is most often used only up to the lowest nonzero terms in the perturbation expansion. Such an approach often exposes the dominant physics of the problem and gives physical insight, as well as returning a first reasonable estimate of the effect of interest. There are other numerical techniques, including other perturbation approaches (e.g., the Brillouin–Wigner theory) that give more accurate numerical answers. If one's goal is to understand the physics and obtain simple approximate results, the Rayleigh–Schrödinger perturbation approach is very useful. Once one understands the problem well physically, if one really wants accurate numerical answers, the problem becomes one of numerical analysis, and other perturbation techniques may be more useful.[11]

Problems

6.3.1* Consider a one-dimensional potential well of thickness L_z in the z direction, with infinitely high potential barriers on either side. Suppose we apply a fixed electric field in the z direction of magnitude F.

(i) Write down an expression, valid to the lowest order in F for which there is a nonzero answer, for the shift of the second electron energy level in this potential well as a function of field.

(ii) Suppose that the potential well is 10 nm thick and is made of the semiconductor GaAs, in which we can treat the electron as having an effective mass of 0.07 of the normal electron mass. What, approximately, is the shift of this second level relative to the potential in the center of the well, in electron-Volts (or milli-electron-Volts), for an applied electric field of 10^5 V/cm? (A numerical answer that one would reasonably expect to be accurate to better than 10 percent is sufficient.) Be explicit about the sign of the shift: Is this energy increasing or decreasing?

Note: You may need the expression

$$\int_0^{\pi}\left(\zeta-\frac{\pi}{2}\right)\sin(q\zeta)\sin(n\zeta)\,d\zeta = -\frac{4qn}{(n-q)^2(n+q)^2} \text{ for n + q odd}$$

$$= 0 \text{ for } n+q \text{ even}$$

6.3.2 Consider an electron in a one-dimensional potential well of width L_z, with infinitely high barriers on either side and in which the potential energy inside the potential well is parabolic, of the form

$$V(z) = u(z - L_z/2)^2$$

[10] E.g., in the problem of the infinitely deep potential well with a field of three dimensionless units analyzed in this Chapter, the second-order wavefunction correction actually leads to a wavefunction that is in poorer agreement with the exact result than is the first order correction.

[11] See, for example, H. Kroemer, *Quantum Mechanics for Engineering, Materials Science, and Applied Physics* (Prentice-Hall, Englewood Cliffs, 1994), Chapter 15.

where u is a real constant. This potential is presumed to be small compared to the energy E_1 of the first confined state of a simple rectangular potential well of the same width L_z. (Note for interest: This kind of situation can arise in semiconductor structures, where the parabolic curvature comes from the electrostatic potential of uniform background doping of the material.)

Find an approximate expression, valid in the limit of small u, for the transition energy between the first and second allowed states of this well in terms of u, L_z, and fundamental constants.

6.3.3 The polarization P can be considered as the average position of the charge density, $\rho(z)$; that is, for a particle of charge q, relative to some position z_o

$$\begin{aligned} P &= \int (z - z_o)\rho(z)dz \\ &= q\int (z - z_o)|\phi(z)|^2 dz \\ &= q\int \phi^*(z)(z - z_o)\phi(z)dz \\ &= q\langle\phi|(z - z_o)|\phi\rangle \end{aligned}$$

where $\phi(z)$ is the (normalized) wavefunction.

In the absence of applied electric field, the particle is presumed to be in the mth eigenstate, $|\psi_m\rangle$, of the unperturbed Hamiltonian. The symmetry of this unperturbed state is such that $\langle\psi_m|(z - z_o)|\psi_m\rangle = 0$ (e.g., it is symmetric about the point z_o). A field F is applied along the z direction so that the perturbing Hamiltonian is

$$\hat{H}_p = -qF(z - z_o)$$

(i) Evaluate P for the case $F = 0$.

(ii) Find an expression for P for the case of finite F. (Retain only the lowest order nonzero terms.)

(iii) Find an expression for the change in energy ΔE of this mth state of the system for finite F, again retaining only the lowest order nonzero terms.

(iv) Hence, show that, to lowest nonzero order

$$\Delta E \cong -\frac{1}{2}PF$$

6.4 Degenerate perturbation theory

Previously, we explicitly avoided above the degenerate case where there might be more than one eigenfunction associated with a given eigenvalue. Such degeneracy is not uncommon in quantum mechanics, especially in problems that are quite symmetric. For example, the three different P orbitals of a hydrogen atom, each corresponding to a different one of the directions x, y, and z, all have the same energy. It is quite important to understand such situations because often perturbations, such as an electric field, remove the degeneracy, causing some of the states to have different energies and defining the distinct eigenfunctions uniquely. We consider this case now, at least for first-order perturbation theory.

Suppose that there are r degenerate orthonormal eigenfunctions, $|\psi_{ms}\rangle$ (where $s = 1, 2, \ldots r$) associated with the eigenenergy E_m of the unperturbed problem. Then, in general, we can write a wavefunction corresponding to this eigenenergy as a linear combination of these; that is,

$$|\psi_{mtot}\rangle = \sum_{s=1}^{r} a_{ms}|\psi_{ms}\rangle \tag{6.61}$$

Now let us consider the first-order perturbation equation, Eq. (6.29), in a fashion similar to before but now with the unperturbed or zeroth-order wavefunction $|\psi_{mtot}\rangle$; that is,

$$(\hat{H}_o - E_m)|\phi^{(1)}\rangle = (E^{(1)} - \hat{H}_p)|\psi_{mtot}\rangle \tag{6.62}$$

Now let us premultiply by a specific one of the degenerate basis functions $|\psi_{mi}\rangle$ to obtain, analogously to Eq. (6.31)

$$\begin{aligned}\langle\psi_{mi}|\hat{H}_o - E_m|\phi^{(1)}\rangle &= \left(\langle\psi_{mi}|\hat{H}_o - E_m\right)|\phi^{(1)}\rangle = \langle\psi_{mi}|\left(E_m - E_m\right)|\phi^{(1)}\rangle = 0 \\ &= \langle\psi_{mi}|E^{(1)} - \hat{H}_p|\psi_{mtot}\rangle = E^{(1)}\langle\psi_{mi}|\psi_{mtot}\rangle - \langle\psi_{mi}|\hat{H}_p|\psi_{mtot}\rangle\end{aligned} \quad (6.63)$$

that is,

$$\langle\psi_{mi}|\hat{H}_p|\psi_{mtot}\rangle = E^{(1)}\langle\psi_{mi}|\psi_{mtot}\rangle \quad (6.64)$$

or, explicitly, in summation form

$$\sum_{s=1}^{r} H_{pmims}a_{ms} = E^{(1)}a_{mi} \quad (6.65)$$

where

$$H_{pmims} = \langle\psi_{mi}|\hat{H}_p|\psi_{ms}\rangle \quad (6.66)$$

We can repeat this for every $i = 1,2,\cdots r$ and so obtain a set of r equations of the form of Eq. (6.65). But, this set of equations is simply identical to the matrix-vector equation

$$\begin{bmatrix} H_{pm1m1} & H_{pm1m2} & \cdots & H_{pm1mr} \\ H_{pm2m1} & H_{pm2m2} & \cdots & H_{pm2mr} \\ \vdots & \vdots & \ddots & \vdots \\ H_{pmrm1} & H_{pmrm2} & \cdots & H_{pmrmr} \end{bmatrix} \begin{bmatrix} a_{m1} \\ a_{m2} \\ \vdots \\ a_{mr} \end{bmatrix} = E^{(1)} \begin{bmatrix} a_{m1} \\ a_{m2} \\ \vdots \\ a_{mr} \end{bmatrix} \quad (6.67)$$

This is just a matrix eigenequation. It generally has eigenvectors and eigenvalues. This first-order degenerate perturbation calculation has therefore reduced to a special case of the finite basis subset model (or finite matrix model) presented previously. In this case, the finite basis we choose is the set of r degenerate eigenfunctions corresponding to a particular unperturbed energy eigenvalue E_m.

The solution of Eq. (6.67) gives a set of r first-order corrections to the energy, which we could call $E_i^{(1)}$, each associated with a particular new eigenvector $|\phi_{mi}\rangle$ that is a linear combination of the degenerate basis functions $|\psi_{ms}\rangle$. All of these new eigenvectors $|\phi_{mi}\rangle$ are orthogonal to one another. To the extent that the energies $E_i^{(1)}$ are different from one another, the perturbation has "lifted the degeneracy."

Note that the eigenvectors $|\phi_{mi}\rangle$ are actually still zeroth-order wavefunctions, not first-order wavefunctions; each is an exact solution of the unperturbed problem with energy E_m. Indeed, any linear combination of the $|\psi_{mi}\rangle$ or the $|\phi_{mi}\rangle$ is a solution of the unperturbed problem with energy E_m. The perturbation theory has merely selected a particular set of linear combination of the unperturbed degenerate solutions. This is consistent with the result for the nondegenerate perturbation theory, in which the first-order energy correction depends only on the zeroth-order wavefunctions.

It is also possible that the perturbation does not lift the degeneracy of the problem to first-order. Degenerate perturbation theory can be extended to higher orders, though we will not do that here.

Problems

6.4.1 Consider an ideal cubical quantum box for confining an electron. The cube has length L on all three sides, with edges along the x, y, and z directions, and the walls of the box are presumed to correspond to infinitely high potential barriers. The resulting energy eigenfunctions in this box are simply the products of the particle-in-a-box wavefunctions in each of the three coordinate directions, as can be verified by substitution into the Schrödinger wave equation. (Note: We presume here for simplicity that $(x,y,z)=(0,0,0)$ is the point in the center of the box and it may be more convenient to write the wavefunctions centered on this point rather than on, say, a corner of the box.)

(i) Write down the normalized wavefunctions for the first three excited states for an electron in this box. (Note: In these states, the electron will be in the second state in one direction and the lowest state in the other two directions.)

(ii) Now presume that there is a perturbation $\hat{H}_p = eFz$ applied (e.g., from an electric field F in the z direction). What happens to the three states as a result of this perturbation, according to first-order degenerate perturbation theory?

(iii) Now presume that a perturbation $\hat{H}_p = \alpha z^2$ is applied instead. (Such a perturbation could result, e.g., from a uniform fixed background charge density in the box.) Using first-order degenerate perturbation theory, what are the new eigenstates and eigenenergies arising from the three originally degenerate states?

(Note: You may need the results

$$\int_{-\pi/2}^{\pi/2} \theta^2 \cos^2\theta \, d\theta = \frac{\pi^3}{24} - \frac{\pi}{4} \text{ and } \int_{-\pi/2}^{\pi/2} \theta^2 \sin^2 2\theta \, d\theta = \frac{\pi^3}{24} - \frac{\pi}{16} \text{)}$$

See also Problem 10.5.9, which can be attempted once the hydrogen atom wavefunctions are understood.

6.5 Tight binding model

An example of another problem in which the degeneracy is lifted is that of two identical potential wells with a finite barrier thickness between them (a "coupled potential well" problem). This has some similarities to a degenerate perturbation theory problem, so we consider it here, though it is awkward mathematically to force it into a form in which we are adding a simple perturbing potential. We can certainly think of it as a finite basis set approach using approximate starting basis functions. Solid-state physicists would refer to this particular calculation as a *tight-binding method.*[12] Regardless of what we call it, it is, however, quite straightforward to solve approximately and leads to simple approximate analytic results with interesting and important physical meaning.

We imagine two separate unperturbed potential wells, as shown in the second and third parts of Fig. 6.3. If we had the left potential well present on its own, with corresponding potential $V_{left}(z)$, we would have the wavefunction solution $\psi_{left}(z)$, with associated energy E_1 for the first state, a problem we already know how to solve exactly numerically. Similarly, if we considered the right potential well on its own, with potential $V_{right}(z)$, we would have the wavefunction solution $\psi_{right}(z)$ (which is the same as $\psi_{left}(z)$ except that it is shifted over to the right) and would have the same energy E_1. The actual potential for which we wish to

[12] The tight-binding here refers to the particle being tightly or deeply bound within one potential well or, in solid-state physics, within one unit cell of the crystal, not to a strong binding between states in adjacent wells or unit cells.

calculate the states is, however, the potential V at the top of Fig. 6.3, which we could call a coupled potential well.

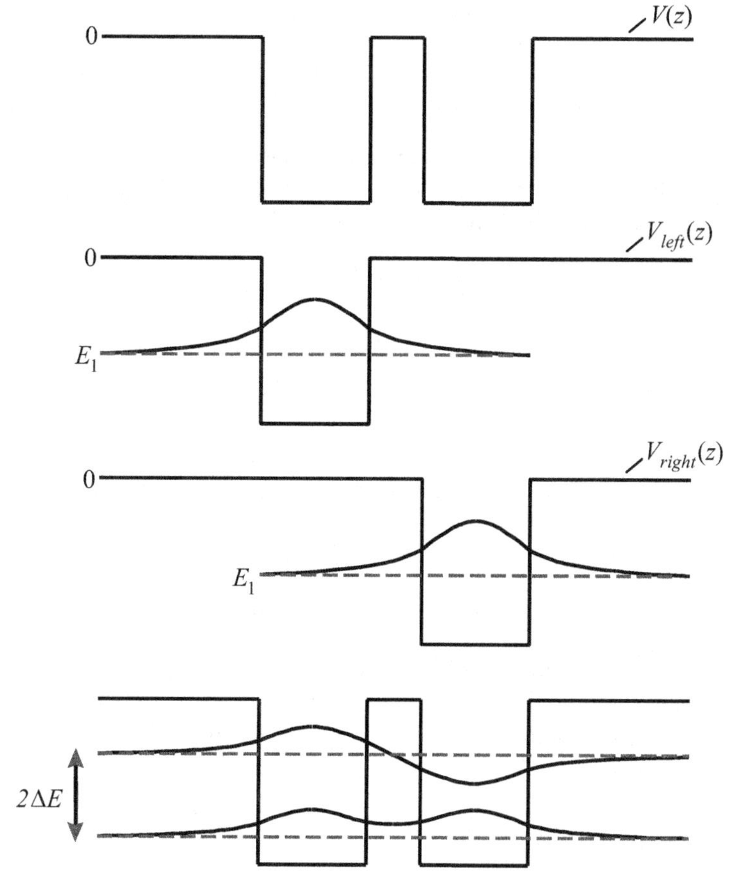

Fig. 6.3. Schematic illustration of a coupled potential well showing the two coupled states formed from the lowest states of the isolated wells. The lower state is symmetric and the upper state is antisymmetric.

Note that here we have chosen the origin for the potential at the top of the well. This particular choice means that we can say that $V(z) = V_{left}(z) + V_{right}(z)$ and our algebra is simplified somewhat. With our choice of energy origin, the Hamiltonian for this system is

$$\hat{H} = \frac{-\hbar^2}{2m}\frac{d^2}{dz^2} + V_{left}(z) + V_{right}(z) \tag{6.68}$$

We now solve using the finite basis subset method, choosing for our basis wavefunctions the wavefunctions in the isolated wells, ψ_{left} and ψ_{right}, respectively. These are two functions that are approximately orthogonal as long as the barrier is reasonably thick – hence, the term *tight-binding*. The basis wavefunctions are each assumed to be relatively tightly confined in one well, with little wavefunction "leakage" into the adjacent well. Hence, the wavefunction in this problem can be written approximately in the form

$$\psi = a\psi_{left} + b\psi_{right} \tag{6.69}$$

In matrix form, our finite basis subset approximate version of Schrödinger's equation is

$$\begin{bmatrix} H_{11} & H_{12} \\ H_{21} & H_{22} \end{bmatrix} \begin{bmatrix} a \\ b \end{bmatrix} = E \begin{bmatrix} a \\ b \end{bmatrix} \quad (6.70)$$

where we should have, for example

$$H_{11} = \int \psi_{left}^*(z) \left(\frac{-\hbar^2}{2m} \frac{d^2}{dz^2} + V_{left}(z) + V_{right}(z) \right) \psi_{left}(z) dz \quad (6.71)$$

Because we presume the barrier to be relatively thick, we are presuming that the amplitude of the left wavefunction is essentially zero inside the right well, so the integrand is essentially zero for all z inside the right-hand well; hence, the term

$$\int \psi_{left}^*(z) V_{right}(z) \psi_{left}(z) dz$$

can be neglected. Arguing similarly for H_{22}, neglecting $\int \psi_{right}^*(z) V_{left}(z) \psi_{right}(z) dz$

$$H_{11} = H_{22} \cong E_1 \quad (6.72)$$

For the same reason (that $\psi_{left}(z) \cong 0$ in the right-hand well) or the complementary one (that $\psi_{right}(z) \cong 0$ in the left-hand well), when integrating within either well we neglect all terms

$$\int \psi_{left}^*(z) V_{right}(z) \psi_{right}(z) dz \,, \quad \int \psi_{left}^*(z) V_{left}(z) \psi_{right}(z) dz \,,$$

$$\int \psi_{right}^*(z) V_{right}(z) \psi_{left}(z) dz \,, \text{ and } \int \psi_{right}^*(z) V_{left}(z) \psi_{left}(z) dz$$

(Remember that V_{left} and V_{right} are zero except within their respective wells.) We do, however, retain the interaction within the (middle) barrier where the wavefunctions, though small, are presumed not completely negligible; that is, we retain a result

$$\Delta E = \int_{barrier} \psi_{left}^*(z) \left(-\frac{\hbar^2}{2m} \frac{d^2}{dz^2} + V(z) (= 0 \text{ in the barrier}) \right) \psi_{right}(z) dz \quad (6.73)$$

Note, incidentally, that ΔE is a negative number if we have chosen $\psi_{left}(z)$ and $\psi_{right}(z)$ to be the positive real functions as in Fig. 6.3; in the barrier, each has have a positive second derivative and so the net result of the integral is negative.

By choosing to integrate only over the barrier thickness, we are neglecting any contributions to this integral that would have come from regions outside the barrier because, again, we presume one or other basis wavefunction to be essentially zero there. With these simplifications, we have

$$\begin{bmatrix} E_1 & \Delta E \\ \Delta E^* & E_1 \end{bmatrix} \begin{bmatrix} a \\ b \end{bmatrix} = E \begin{bmatrix} a \\ b \end{bmatrix} \quad (6.74)$$

(ΔE in this problem is real because the wavefunctions of this problem can [and will] be chosen to be real for simplicity, but the complex conjugate is shown here for completeness. ΔE is also a negative number here because the wavefunctions are each positively curved inside the barrier.) We find the energy eigenvalues of Eq. (6.74) in the usual way by setting

$$\det \begin{vmatrix} E_1 - E & \Delta E \\ \Delta E^* & E_1 - E \end{vmatrix} = 0 \quad (6.75)$$

that is,

$$\left(E_1 - E\right)^2 - \left|\Delta E\right|^2 = E^2 - 2EE_1 + E_1^2 - \left|\Delta E\right|^2 = 0 \tag{6.76}$$

obtaining eigenvalues

$$E = E_1 \pm \left|\Delta E\right| \tag{6.77}$$

Note that at least within the approximations here, the energy levels are split by the coupling between the wells, approximately symmetrically about the original "single-well" energy, E_1.

Substituting the eigenvalues back into Eq. (6.74) and taking the original wavefunctions and, hence, also ΔE to be real for simplicity[13] (and remembering that ΔE is negative) allows us to deduce the associated normalized wavefunctions

$$\psi_- = \frac{1}{\sqrt{2}}\left(\psi_{left} + \psi_{right}\right) \text{ and } \psi_+ = \frac{1}{\sqrt{2}}\left(\psi_{left} - \psi_{right}\right) \tag{6.78}$$

These wavefunctions are sketched in Fig. 6.3. The lower energy state is associated with a symmetric linear combination of the single-well eigenfunctions (i.e., the wavefunction has the same sign in both wells) and the upper energy state is associated with the antisymmetric combination (i.e., the wavefunction has the opposite sign in the two wells). Note now that we can no longer view the states as corresponding to an electron in the left well or an electron in the right well; in both states, the electron is equally in both wells. This general form of wavefunctions, one symmetric and the other antisymmetric, is characteristic of this kind of symmetric problem and is retained even as we perform more accurate calculations.

Hence, we can see from this simplified physical model that the physical perturbation of bringing two identical systems together leads to a splitting of the degenerate eigenvalues and a coupling of the states. This is a very general phenomenon in quantum mechanics. It occurs, for example, when we bring atoms together to form a crystalline solid and leads to the formation of energy bands of very closely spaced states rather than the discrete separated energy levels of the constituent atoms.

Note, incidentally, that this calculation has features that are also found in molecular bonding. Suppose that we have one electron to share between these two potential wells. As we bring these two potential wells together, two possible states emerge, one of which has lower energy than any of the states the system previously had. If we think of these potential wells as being analogous to atoms, we see that we can get lower energy in the system in this lowest state by bringing these "atoms" closer together. We can see, then, that if the system was in this lowest state, we would have to add energy to the electron if we were to try to pull the potential wells or "atoms" apart. Hence, this lowest state corresponds to a kind of molecular, chemically bonded state. The actual theory of molecular bonding is more complex than this because it has to account for multiple electrons in the system[14] and potentials that are not simply square wells. The symmetric and antisymmetric solutions we have found here for the coupled

[13] Remember that the solutions of the time-independent Schrödinger equation for a real potential can always be chosen to be real because if ψ is a solution, so also is ψ^*, as can be seen by taking the complex conjugate of both sides of the equation; hence, $\psi + \psi^*$, which is real, is also a solution.

[14] When there are multiple electrons, we have to account also for the so-called exchange energy, which is very important in determining actual chemical bonds in molecules.

quantum well are, however, sometimes referred to as "bonding" and "antibonding" states, respectively.

Problems

6.5.1 Consider three wells of equal thicknesses with equal barriers on either side of the middle well. Take the same tight-binding approach as used for two quantum wells, but now considering a 3 x 3 matrix. What are the approximate eigenenergies and wavefunctions of this coupled system? How many zeros are there in each of these wavefunctions (not counting the zeros at the extreme left and right of the system)?

6.5.2 Suppose we have a coupled potential well consisting of two weakly coupled identical potential wells with a barrier between them. We presume that we have solved this problem approximately using a tight-binding approach for the lowest two coupled states, giving approximate solutions

$$\psi_{-}(z)=\frac{1}{\sqrt{2}}\left(\psi_{left}(z)+\psi_{right}(z)\right) \text{ and } \psi_{+}(z)=\frac{1}{\sqrt{2}}\left(\psi_{left}(z)-\psi_{right}(z)\right)$$

with associated energies

$$E=E_1\pm|\Delta E|$$

where E_1 is the energy of the lowest solution in either of the potential wells considered separately, $\psi_{left}(z)$ is the corresponding wavefunction of the first state in the left well considered separately, $\psi_{right}(z)$ is the corresponding wavefunction of the first state in the right well considered separately, and ΔE is a number that has been calculated based on the coupling.

Suppose now that the coupled system is initially prepared, at time $t=0$, in the state such that the particle is in the left well, with initial wavefunction $\psi_{left}(z)$.

(i) Calculate expressions for the wavefunction and the probability density as a function of time after $t=0$.

(ii) Describe in words the time-dependence of this probability density.

(Note: This problem requires an understanding of Chapter 3)

6.6 Variational method

Consider for the moment an arbitrary quantum mechanical state, $|\phi\rangle$, of some system. We suppose that the Hamiltonian of the system is $\hat{H}$ and we want to evaluate the expectation value of the energy, $\langle E\rangle$. Because the Hamiltonian is presumably an appropriate Hermitian operator, it has a complete set of eigenfunctions, $|\psi_n\rangle$, with associated eigenenergies E_n; we may not know what they are – they may be mathematically difficult to calculate – but we do know that they exist. (For simplicity here, we assume the eigenvalues are not degenerate.) Consequently, we can certainly expand any arbitrary state in them and so we can write as usual, for some set of expansion coefficients a_i

$$|\phi\rangle=\sum_i a_i|\psi_i\rangle \tag{6.79}$$

We presume this representation of the state is normalized, so

$$\sum_i |a_i|^2=1 \tag{6.80}$$

Hence, the expectation value of the energy becomes, as usual

$$\langle E\rangle=\langle\phi|\hat{H}|\phi\rangle=\sum_i |a_i|^2 E_i \tag{6.81}$$

We also presume for convenience that we have ordered all of the eigenfunctions in order of the eigenvalues, starting with the smallest, E_1.

Now we ask the question, what is the smallest expectation value of the energy that we can have for any conceivable state $|\phi\rangle$? The answer is obvious from Eq. (6.81). The smallest energy expectation value we can have is E_1, with correspondingly $a_1 = 1$ and all the other expansion coefficients zero. If we made one of the other expansion coefficients a_j finite, then the energy expectation value would become

$$\begin{aligned}\langle E\rangle &= |a_1|^2 E_1 + |a_j|^2 E_j \\ &= \left(1-|a_j|^2\right)E_1 + |a_j|^2 E_j \\ &= E_1 + |a_j|^2\left(E_j - E_1\right) > E_1\end{aligned} \quad (6.82)$$

That is, because of the normalization sum, Eq. (6.80), the energy would have to increase. This simple property allows us to construct an approximate method of solution of quantum mechanical problems for the ground state (the lowest energy state) and especially for its energy. The key idea is that we choose some mathematical form of state, called the *trial wavefunction*, that is mathematically convenient for us (and which we believe reasonably fits, at least, the qualitative features we would expect for the ground state), and then vary some parameter or parameters in this mathematical form to minimize the resulting expectation value of the energy. As a result of this minimization with respect to variation, this approach is known as the variational method. If we use this method, we do not formally know how accurate our result is for the energy, but we do know that lower is better – we can never get an answer lower than the energy of the lowest actual state of the system as we proved previously – and we can, if we wish, keep refining our mathematical form so as to reduce the resulting calculated energy expectation value.

Why would we use such a method? One answer is that it allows us to calculate an approximation for the ground state energy without solving the exact eigenfunctions of any problem. A second reason is that with careful choice of the form of the function to be varied, so that the algebra of minimization gives simple analytic results, we may also be able to get approximate analytic results for the effect of some perturbation.

Given that the form of the wavefunction we are using is not the actual form of the exact solution, why does this method give even reasonable answers? The answer goes back to a point we discussed previously in relation to perturbation theory; we can often get quite good answers for energies even with approximate wavefunctions. Remember that the first-order energy correction uses the zeroth-order wavefunction, for example.

The variational approach can be progressively extended to higher levels of the system if we force the next trial wavefunction to be mathematically orthogonal to all the previous (i.e., lower energy) ones.[15] As far as numerical calculations are concerned, the variational method is nearly always used only for ground states. The discussion of the variational method does, however, point out a basic, exact property of eigenfunctions and eigenvalues that is actually obvious from Eq. (6.81). The eigenfunction corresponding to the lowest eigenvalue is that function that minimizes the expectation value. The eigenfunction corresponding to the second eigenvalue is that function that minimizes the expectation value, subject to the constraint that

---

[15] Such a mathematical approach is sometimes known as Gram–Schmidt orthogonalization.

the function is orthogonal to the first eigenfunction. This property extends to higher eigenfunctions, with each successive eigenfunction constrained to being orthogonal to all the previous ones. Indeed, this successive minimization property can be used mathematically to define the eigenfunctions and eigenvalues.[16]

As an example of the variational method, we can do a simple calculation on our example problem of an electron in an infinitely deep potential well with applied field. In this particular case, because it makes our mathematics tractable, we use as our trial function an unknown linear combination of the first two states of the infinitely deep quantum well – though, as mentioned previously, it is more common in variational calculations to choose some mathematical function unrelated to exact eigenfunctions of any problem. Hence, our trial function is

$$\phi_{trial}\left(\xi, a_{\mathrm{var}}\right) = \frac{\sqrt{2}}{\sqrt{1+a_{\mathrm{var}}^2}}\left(\sin \pi\xi + a_{\mathrm{var}} \sin 2\pi\xi\right) \tag{6.83}$$

where a_{var} is the parameter we vary to minimize the energy expectation value. Note that we have normalized this wavefunction by dividing by $\sqrt{1+a_{\mathrm{var}}^2}$. The expectation value of the energy then becomes, as a function of the parameter a_{var}

$$\begin{aligned}\left\langle E\left(a_{\mathrm{var}}\right)\right\rangle = \frac{1}{1+a_{\mathrm{var}}^2}&\left[\int_0^1 \left(\sqrt{2}\sin \pi\xi + a_{\mathrm{var}}\sqrt{2}\sin 2\pi\xi\right)\right] \\ &\times\left(-\frac{1}{\pi^2}\frac{\partial^2}{\partial \xi^2} + \mathrm{f}\left(\xi - 1/2\right)\right)\left(\sqrt{2}\sin \pi\xi + a_{\mathrm{var}}\sqrt{2}\sin 2\pi\xi\right)d\xi\end{aligned} \tag{6.84}$$

Using the result

$$\int_0^1 \sin \pi\xi \left(\xi - 1/2\right)\sin 2\pi\xi d\xi = -\frac{8}{9\pi^2} \tag{6.85}$$

the known eigenenergies of the unperturbed problem, and the orthogonality of the sine functions, Eq. (6.84) becomes

$$\left\langle E\left(a_{\mathrm{var}}\right)\right\rangle = \frac{1}{1+a_{\mathrm{var}}^2}\left[\varepsilon_1\left(1+4a_{\mathrm{var}}^2\right) - \frac{32 a_{\mathrm{var}}\mathrm{f}}{9\pi^2}\right] \tag{6.86}$$

Now, to find the minimum in this expectation value, we take the derivative with respect to a_{var}, to obtain

$$\frac{d\left\langle E\left(a_{\mathrm{var}}\right)\right\rangle}{da_{\mathrm{var}}} = \frac{2}{9\pi^2}\frac{16\mathrm{f}a_{\mathrm{var}}^2 + 27\pi^2 a_{\mathrm{var}} - 16\mathrm{f}}{\left(1+a_{\mathrm{var}}^2\right)^2} \tag{6.87}$$

This derivative is zero when the quadratic in the numerator is zero. The root that gives the lowest value of $\left\langle E\left(a_{\mathrm{var}}\right)\right\rangle$ is

[16] This rather general and important property of eigenfunctions and eigenvalues is, surprisingly, often not taught in courses on those subjects and is less well known than it ought to be.

$$a_{\text{var min}} = \frac{-27\pi^2 + \sqrt{\left(27\pi^2\right)^2 + 1024\text{f}^2}}{32\text{f}} \tag{6.88}$$

For $\text{f} = 3$ in our example, we find $a_{\text{var min}} \cong 0.175$. The normalized expansion coefficient of this second level, $a_{\text{var min}} / \sqrt{1 + a_{\text{var min}}^2} \cong 0.172$, which compares with 0.174 from the 3-basis-function finite basis subset method and 0.180 from the perturbation calculation. The corresponding energy expectation value, which is the approximation to the ground state energy in the presence of the field, is, substituting the value of $a_{\text{var min}}$ back into (6.86), $\langle E(0.175) \rangle \cong 0.90563$, which compares reasonably well with 0.90419 from the exact calculation.

Incidentally, it can be shown that a variational approach like this using the same basis functions as a finite basis subset calculation gives exactly the same results; if we had calculated the finite basis subset method using only the first two basis functions, we would get the same answer as our variational calculation here. We can see this explicitly from Table 6.1, where the two-basis-function finite basis subset method and the variational method with the same two basis functions give exactly the same answer for the energy (they also give identical wavefunctions). This is fundamentally because of the minimization property of eigenfunctions and eigenvalues discussed above. An explicit proof is given by Kroemer.[17]

Problems

6.6.1* Based on your understanding of the variational method and the principles behind it, prove that the finite basis subset method will always give an answer for the energy of the lowest energy eigenstate that is equal to or above the exact value.

6.6.2 (This problem may be used as a substantial assignment.) Solve this problem by any approximation method or methods you consider appropriate. Consider an electron in an infinitely deep potential well of thickness L_z = 30 Å, into which a potential barrier is introduced in the middle as shown in the figure. This barrier has thickness ΔL = 0.5Å and height V_o = 1 eV.

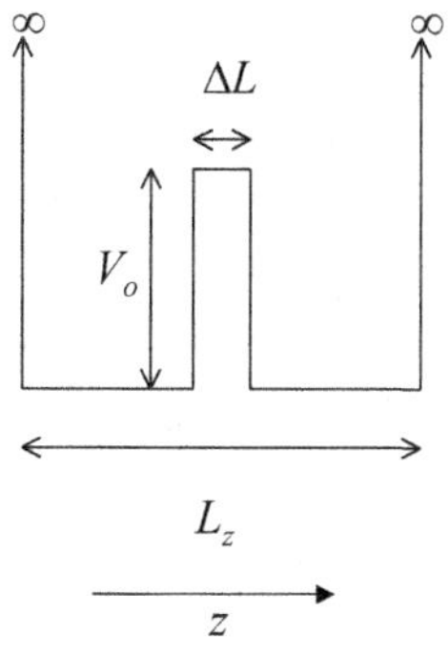

(i) Find the energies and plot the wavefunctions of the first two states (i.e., the ones with the lowest energies) in this potential. Though formal proof of accuracy is not required, you should have reasonable grounds for believing your energy calculation is accurate to ~5 percent or better for the first level.

(ii) Now we apply an electric field, F, to this structure along the z-direction. We consider the energy of the center of the well to be fixed as we apply field. We presume that for small fields, the energy E_1 of the first level changes quadratically with electric field; that is,

[17] See, for example, H. Kroemer, *Quantum Mechanics for Engineering, Materials Science, and Applied Physics* (Prentice-Hall, Englewood Cliffs, 1994), Chapter 14.

$$\Delta E_1 = -\frac{1}{2}\alpha F^2$$

(The quantity α is often called the *polarizability*.) Find a reasonable approximate numerical result for α, expressed in units appropriate for energies in eV and fields in V/Å.

6.7 Summary of concepts

Finite matrix/finite basis subsets method

We choose a finite number of basis functions to approximate the problem and solve by finding the eigenvalues and eigenfunctions of the corresponding finite matrix representing the operator of interest (usually, the Hamiltonian). The basis set usually chosen is the known solutions of the unperturbed problem and the finite set of functions chosen is most often those close in energy to the state of greatest interest.

Time-independent nondegenerate perturbation theory

Presuming that there is a perturbation $\gamma \hat{H}_p$ added to an unperturbed Hamiltonian $\hat{H}_o$ (whose eigen solutions E_n and $|\psi_n\rangle$ are presumed to be known), and presuming there are corresponding power series in γ for both the eigenenergies E

$$E = E^{(0)} + \gamma E^{(1)} + \gamma^2 E^{(2)} + \gamma^3 E^{(3)} + \cdots \tag{6.21}$$

and the eigenfunctions $|\phi\rangle$

$$|\phi\rangle = \left|\phi^{(0)}\right\rangle + \gamma\left|\phi^{(1)}\right\rangle + \gamma^2\left|\phi^{(2)}\right\rangle + \gamma^3\left|\phi^{(3)}\right\rangle + \cdots \tag{6.20}$$

a series of relations is found by equating powers of γ. Thereafter, γ is conventionally set to unity. The resulting first-order corrections to the mth level of the unperturbed system are

$$E^{(1)} = \langle \psi_m | \hat{H}_p | \psi_m \rangle \tag{6.32}$$

$$\left|\phi^{(1)}\right\rangle = \sum_{n \neq m} \frac{\langle \psi_n | \hat{H}_p | \psi_m \rangle}{E_m - E_n} |\psi_n\rangle \tag{6.38}$$

and the second-order correction to the energy is

$$E^{(2)} = \sum_{n \neq m} \frac{\left|\langle \psi_n | \hat{H}_p | \psi_m \rangle\right|^2}{E_m - E_n} \tag{6.42}$$

Higher order corrections to wavefunctions and energies are found progressively from the preceding orders of corrections. Usually, such perturbation corrections are used to the lowest nonzero order.

Degenerate perturbation theory

For the case where a given unperturbed energy level is degenerate (i.e., has more than one eigenfunction associated with it), the wavefunction is expanded on the basis of the r degenerate eigenfunctions

$$|\psi_{mtot}\rangle = \sum_{s=1}^{r} a_{ms} |\psi_{ms}\rangle \tag{6.61}$$

The resulting first-order correction to the eigenenergies is found as the eigenvalues of the matrix equation

$$\begin{bmatrix} H_{pm1m1} & H_{pm1m2} & \cdots & H_{pm1mr} \\ H_{pm2m1} & H_{pm2m2} & \cdots & H_{pm2mr} \\ \vdots & \vdots & \ddots & \vdots \\ H_{pmrm1} & H_{pmrm2} & \cdots & H_{pmrmr} \end{bmatrix} \begin{bmatrix} a_{m1} \\ a_{m2} \\ \vdots \\ a_{mr} \end{bmatrix} = E^{(1)} \begin{bmatrix} a_{m1} \\ a_{m2} \\ \vdots \\ a_{mr} \end{bmatrix} \tag{6.67}$$

where

$$H_{pmims} = \langle \psi_{mi} | \hat{H}_p | \psi_{ms} \rangle \tag{6.66}$$

and the corresponding eigenvectors give the lowest order eigenfunctions corresponding to the respective eigenvalues. The eigenfunctions found this way are zeroth-order eigenfunctions, but specific combinations with different energies may now have been forced by the perturbation, in which case the perturbation is said to have lifted the degeneracy.

Tight-binding model

A tight-binding model is one in which the electrons in adjacent systems are presumed to be quite tightly bound within those systems but with a weak overlap between adjacent systems. This weak interaction allows simplified analytic models to be constructed that can expose the nature of the interactions between the systems.

Variational method

The variational method relies on the provable notion that the lowest possible expectation value of the energy that can be attained for any proposed wavefunction of the system is exactly that of the lowest energy eigenstate. This allows mathematically convenient forms to be used for the wavefunction in approximate calculations, especially of the lowest state of the system. Usually, a parameter of the proposed approximate wavefunction is varied to minimize the resulting energy expectation value.

Chapter 7

Time-dependent perturbation theory

Prerequisites: Chapters 2–6.

Time-dependent perturbation theory is one of the most useful techniques for understanding how quantum mechanical systems respond in time to changes in their environment. It is especially useful for understanding the consequences of periodic changes; a classic example is understanding how a quantum mechanical system responds to light, which can often be usefully approximated as a periodically oscillating electromagnetic field. We develop time-dependent perturbation theory in this chapter, including some specific applications to interactions with light.

7.1 Time-dependent perturbations

For time-dependent problems, we are often interested in the situation where we have some time-dependent perturbation, $\hat{H}_p(t)$, to an unperturbed Hamiltonian $\hat{H}_o$ that is itself not dependent on time. The total Hamiltonian is then

$$\hat{H} = \hat{H}_o + \hat{H}_p(t) \tag{7.1}$$

To deal with such a situation, we return to the time-dependent Schrödinger equation

$$i\hbar\frac{\partial}{\partial t}|\Psi\rangle = \hat{H}|\Psi\rangle \tag{7.2}$$

where now the ket $|\Psi\rangle$ is time-varying in general. As before, a convenient way to represent a solution of the time-dependent Schrödinger equation is to expand it in the energy eigenfunctions of the unperturbed problem. With $|\psi_n\rangle$ and E_n as the energy eigenfunctions and eigenvalues of the time-independent equation

$$\hat{H}_o|\psi_n\rangle = E_n|\psi_n\rangle \tag{7.3}$$

we can expand $|\Psi\rangle$ as

$$|\Psi\rangle = \sum_n a_n(t)\exp(-iE_n t/\hbar)|\psi_n\rangle \tag{7.4}$$

Note that in Eq. (7.4), we chose to include the time-dependent factor $\exp(-iE_n t/\hbar)$ explicitly in the expansion. We could have left that out and merely included it in $a_n(t)$. As is often the case, however, it is better to take out the major underlying dependence if one knows it, which here leaves the time dependence of $a_n(t)$ to deal only with the changes that are in addition to the underlying unperturbed behavior.[1]

Now we can substitute the expansion Eq. (7.4) into the time-dependent Schrödinger equation Eq. (7.2), obtaining

$$\sum_n \left(i\hbar\dot{a}_n + a_n E_n\right)\exp\left(-iE_n t/\hbar\right)|\psi_n\rangle = \sum_n a_n\left(\hat{H}_o + \hat{H}_p(t)\right)\exp\left(-iE_n t/\hbar\right)|\psi_n\rangle \tag{7.5}$$

where

$$\dot{a}_n \equiv \frac{\partial a_n}{\partial t} \tag{7.6}$$

Using the time-independent Schrödinger equation (Eq. (7.3)) to replace $\hat{H}_o|\psi_n\rangle$ with $E_n|\psi_n\rangle$ leads to the cancellation of terms in $E_n|\psi_n\rangle$ from the two sides of the equation. Now premultiplying by $\langle\psi_q|$ on both sides of Eq. (7.5) leads to

$$i\hbar\dot{a}_q(t)\exp\left(-iE_q t/\hbar\right) = \sum_n a_n(t)\exp\left(-iE_n t/\hbar\right)\langle\psi_q|\hat{H}_p(t)|\psi_n\rangle \tag{7.7}$$

Note that we have made no approximations in going from Eq. (7.2) to Eq. (7.7); these are entirely equivalent equations.

Now, we consider a perturbation series in a manner closely analogous to the series we defined for the time-independent problem. We introduce the expansion parameter γ for the purposes of mathematical housekeeping, just as before, now writing our perturbation as $\gamma\hat{H}_p$. Just as in the time-independent perturbation theory, we can set this parameter to a value of 1 at the end. We presume that we can express the expansion coefficients a_n as a power series

$$a_n = a_n^{(0)} + \gamma a_n^{(1)} + \gamma^2 a_n^{(2)} + \cdots \tag{7.8}$$

and we substitute this expansion into Eq. (7.7). Because now in Eq. (7.7) we have replaced $\hat{H}_p$ by $\gamma\hat{H}_p$, the lowest power of γ we can possibly find on the right-hand side of Eq. (7.7) is γ^1 (i.e., γ); there is no term in γ^0. Hence, equating powers of γ on both sides of the equation, we obtain for this zeroth-order term

$$\dot{a}_q^{(0)}(t) = 0 \tag{7.9}$$

Not surprisingly, the zeroth-order solution simply corresponds to the unperturbed solution; hence, there is no change in the expansion coefficients in time. For the first-order term (obtained by equating terms in γ^1 on both sides of the equation), we have

$$\dot{a}_q^{(1)}(t) = \frac{1}{i\hbar}\sum_n a_n^{(0)}\exp\left(i\omega_{qn}t\right)\langle\psi_q|\hat{H}_p(t)|\psi_n\rangle \tag{7.10}$$

where we have introduced the notation

$$\omega_{qn} = \left(E_q - E_n\right)/\hbar \tag{7.11}$$

Note here that the $a_n^{(0)}$ are all constants; we deduced in Eq. (7.9) that they do not change in time. They represent the "starting" state of the system at time $t = 0$. We note now, therefore, that if we know the starting state, the perturbing potential, and the unperturbed eigenvalues and

---

[1] This approach is sometimes known as the *interaction picture*.

eigenfunctions, then everything on the right-hand side of Eq. (7.10) is known. Hence, we can integrate Eq. (7.10) to obtain the first-order time-dependent correction, $a_q^{(1)}(t)$, to the expansion coefficients. Because we now know the new approximate expansion coefficients

$$a_q \simeq a_q^{(0)} + a_q^{(1)}(t) \tag{7.12}$$

then we know the new wavefunction and can calculate the behavior of the system from this new wavefunction. Hence, we have our first approximation to the effect of this time-dependent perturbation to the system.

We can proceed to higher order in this time-dependent perturbation theory. In general, equating powers of progressively higher order, we obtain

$$\dot{a}_q^{(p+1)}(t) = \frac{1}{i\hbar}\sum_n a_n^{(p)} \exp(i\omega_{qn}t)\langle\psi_q|\hat{H}_p(t)|\psi_n\rangle \tag{7.13}$$

We see that this perturbation theory is also a method of successive approximations, just like the time-independent perturbation theory. We calculate each higher order correction from the preceding correction.

Just as for the time-independent perturbation theory, the time-dependent theory is often most useful for calculating some process to the lowest nonzero order. Higher order time-dependent perturbation theory is very useful, for example, for understanding nonlinear optical processes. First-order time-dependent perturbation theory gives the ordinary, linear optical properties of materials, as we will see in Section 7.3. Higher order time-dependent perturbation theory is used to calculate processes such as second harmonic generation and two-photon absorption in nonlinear optics; for example, processes that are seen routinely with the high optical intensities of modern lasers.

Problems

7.1.1* Consider a one-dimensional semiconductor potential well of width L_z with potential barriers on either side approximated as being infinitely high. An electron in this potential well is presumed to behave with an effective mass m_{eff}. Initially, there is an electron in the lowest state of this potential well.

We want to use this semiconductor structure as a detector for a very short electric field pulse. If the electron is found in the second energy state after the pulse, the electron is presumed to be collected as photocurrent by some mechanism and the pulse is therefore detected.

To model this device, we presume that the electric field pulse $F(t)$ can be approximated as a half-cycle pulse of length Δt; that is, a pulse of the form

$$F(t) = F_o \sin\left(\frac{\pi t}{\Delta t}\right)$$

for times t from 0 to Δt and zero for all other times.

(i) Find an approximate expression, valid for sufficiently small field amplitude F_o, for the probability of finding the electron in its second state after the pulse.

(ii) For a pulse of length Δt = 100 fs and a GaAs semiconductor structure with $m_{eff} = 0.07m_o$ and width L_z = 10 nm, for what minimum electric field magnitude F_o does this detector have at least a 1 percent chance of detecting the pulse?

(iii) For a full-cycle pulse (i.e., one of the form $F(t) = F_o \sin(2\pi t/\Delta t)$ for times t from 0 to Δt and zero for all other times), what is the probability of detecting the pulse with this detector? Justify your answer.

7.1.2 An electron is initially in the lowest state of an infinitely deep one-dimensional potential well of thickness L_z. An electric field pulse, with its field direction polarized perpendicular to the well, and of the form

$$F(t) = 0,\ t < 0;\quad F(t) = F_o \exp(-t/\tau),\ t \geq 0$$

is applied to the well. This pulse will create some probability for times $t >> \tau$ of finding the electron in the second state and we presume that electrons excited into the second state are subsequently swept out to give a photocurrent on some timescale long compared to τ.

(i) Find an expression, valid for sufficiently small values of F_o, for the probability of generating an electron of photocurrent from such a pulse.

(ii) Suppose now that we consider a pulse of a given fixed energy E_{pulse}, which we may take to be of the form $E_{pulse} = AF_o^2\tau$, where A is some constant. For what value of characteristic pulse length τ does this detector have the maximum sensitivity (i.e., maximum chance of generating an electron of photocurrent)?

(iii) Treating an electron in a GaAs quantum well as being modeled by such an infinitely deep well, with an electron effective mass of $m_{eff} = 0.07m_o$, for a well of thickness 10 nm, calculate the probability of generating an electron of photocurrent for a field value of $F_o = 10^3$ V/cm and a characteristic time $\tau = 100$ fs.

(iv) If we were now to make 10^{11} such single electron systems (or, equivalently, to put 10^{11} electrons in one actual GaAs quantum well, a number quite feasible for $\sim 1\ \text{cm}^2$ area), what now would be the average number of electrons of photocurrent generated per pulse?

(v) For the same pulse energy, what would be the optimum pulse length to maximize the photocurrent for the GaAs case of parts (iii) and (iv), and how much larger would the photocurrent be?

7.2 Simple oscillating perturbations

One of the most useful applications of time-dependent perturbation theory is the case of oscillating perturbations. We consider this problem here in first-order time-dependent perturbation theory. For example, the interaction of a monochromatic electromagnetic wave with a material is a situation in which the perturbation, the electromagnetic field, is varying sinusoidally in time. Such a sinusoidal perturbation is also called a harmonic perturbation – the same use of the term harmonic as in the harmonic oscillator.

One common form would be to have an electric field in, say, the z direction[2]

$$\mathsf{E}(t) = \mathsf{E}_o\left[\exp(-i\omega t) + \exp(i\omega t)\right] = 2\mathsf{E}_o \cos(\omega t) \tag{7.14}$$

where ω is a positive (angular) frequency. For an electron, as before, the resulting electrostatic energy in this field, relative to position $z = 0$, would lead to a perturbing Hamiltonian

$$\hat{H}_p(t) = e\mathsf{E}(t)z = \hat{H}_{po}\left[\exp(-i\omega t) + \exp(i\omega t)\right] \tag{7.15}$$

where, in this case

$$\hat{H}_{po} = e\mathsf{E}_o z \tag{7.16}$$

Note that this operator, $\hat{H}_{po}$, does not depend on time. This particular form of the perturbing Hamiltonian is called the *electric dipole approximation*. In this particular case, this operator is

[2] The use of complex exponentials is often preferred for the mathematics of handling such perturbations, though in the end it does not matter. Oscillating fields are often defined using this particular convention, with the factor of 2 in the cosine version of the field, though there is not unanimity on this convention.

just a scalar function of z, though in other formulations of this problem, it often has stronger operator character.[3] Our approach here is valid for any perturbing Hamiltonian that is of the general form of the far right of Eq. (7.15).[4]

Fermi's Golden Rule

For the purposes of our calculation, we presume that this perturbing Hamiltonian is only "on" for some finite time. For simplicity, we presume that the perturbation starts at time $t = 0$ and ends at time $t = t_o$, so formally we have

$$\begin{aligned}\hat{H}_p(t) &= 0, t < 0 \\ &= \hat{H}_{po}\left[\exp(-i\omega t) + \exp(i\omega t)\right], 0 < t < t_o \\ &= 0, t > t_o\end{aligned} \tag{7.17}$$

To be specific, we are interested in a situation where, for times before $t = 0$, the system is in some specific energy eigenstate, $|\psi_m\rangle$. We expect that the time-dependent perturbation theory will tell us with what probability the system will make transitions into other states of the system as a result of the perturbation. With this choice, all of the $a_n^{(0)}$ (i.e., the initial expansion coefficients) are zero except $a_m^{(0)}$, which has the value 1. With this simplification of the initial state, the first-order perturbation solution, Eq. (7.10), becomes

$$\dot{a}_q^{(1)}(t) = \frac{1}{i\hbar}\exp(i\omega_{qm}t)\langle\psi_q|\hat{H}_p(t)|\psi_m\rangle \tag{7.18}$$

Then we have, substituting the perturbing Hamiltonian, Eq. (7.17)

$$\begin{aligned}a_q^{(1)}(t > t_o) &= \frac{1}{i\hbar}\int_0^{t_0}\langle\psi_q|\hat{H}_p(t_1)|\psi_m\rangle\exp(i\omega_{qm}t_1)\,dt_1 \\ &= \frac{1}{i\hbar}\langle\psi_q|\hat{H}_{po}|\psi_m\rangle\int_0^{t_o}\left\{\exp\left[i(\omega_{qm}-\omega)t_1\right] + \exp\left[i(\omega_{qm}+\omega)t_1\right]\right\}dt_1 \\ &= -\frac{1}{\hbar}\langle\psi_q|\hat{H}_{po}|\psi_m\rangle\left\{\frac{\exp(i(\omega_{qm}-\omega)t_o)-1}{\omega_{qm}-\omega} + \frac{\exp(i(\omega_{qm}+\omega)t_o)-1}{\omega_{qm}+\omega}\right\} \\ &= \frac{t_o}{i\hbar}\langle\psi_q|\hat{H}_{po}|\psi_m\rangle\left\{\begin{aligned}&\exp\left[i(\omega_{qm}-\omega)t_o/2\right]\frac{\sin\left[(\omega_{qm}-\omega)t_o/2\right]}{(\omega_{qm}-\omega)t_o/2} \\ &+\exp\left[i(\omega_{qm}+\omega)t_o/2\right]\frac{\sin\left[(\omega_{qm}+\omega)t_o/2\right]}{(\omega_{qm}+\omega)t_o/2}\end{aligned}\right\}\end{aligned} \tag{7.19}$$

---

[3] Another common form is $\hat{H}_p \cong -(e/m_o)\mathbf{A}\cdot\hat{\mathbf{p}}$, where $\mathbf{A}$ is the magnetic vector potential and $\hat{\mathbf{p}}$ is the momentum operator. This form is formally more complete than the electric dipole approximation (e.g., it includes magnetic effects) and is favored by solid-state physicists, though it makes little or no difference for nearly all common linear optical problems. We use it in Chapter 8 and derive it in Appendix E.

[4] Another common occurrence of a harmonic perturbation is in the interaction of vibrating atoms with electrons, as in the interaction of phonons in solids with electrons (i.e., electron-phonon interactions), which is responsible for many of the limitations on the speed of electrons in semiconductors, for example.

The reader may remember that the function $\mathrm{sinc}(x) \equiv (\sin x)/x$ peaks at 1 for $x = 0$ and falls off in an oscillatory fashion on either side. This function is shown in Fig. 5.1. It is essentially only appreciably large for $x \cong 0$, which tells us we have a strongly resonant behavior, with relatively strong perturbations when the frequency ω is close to $\pm\omega_{qm}$. We will discuss the meaning of this in more detail later in this section.

What we have now calculated is the new quantum mechanical state for times $t > t_o$, which is, to first order,

$$|\Psi\rangle \simeq \exp(-iE_m t/\hbar)|\psi_m\rangle + \sum_q a_q^{(1)}(t > t_o)\exp(-iE_q t/\hbar)|\psi_q\rangle \tag{7.20}$$

with the $a_q^{(1)}(t > t_o)$ given by Eq. (7.19). Now that we have established our approximation to the new state, we can start calculating the time dependence of measurable quantities.

In our example here, we chose the system initially to be in the energy eigenstate $|\psi_m\rangle$. The application of the perturbation has changed that state and we would like to know, if we were to make a measurement of the energy after the perturbation is over (i.e., for $t > t_o$), what is the probability that the system will be found in some other state, $|\psi_j\rangle$. Another way of stating this is that we want to know the transition probability from state $|\psi_m\rangle$ to $|\psi_j\rangle$. Provided we are dealing with small perturbations (so that the correction to wavefunction normalization can be ignored), the probability, $P(j)$, of finding the system in state $|\psi_j\rangle$ is

$$P(j) = \left|a_j^{(1)}\right|^2 \tag{7.21}$$

that is,

$$P(j) \simeq \frac{t_o^2}{\hbar^2}\left|\langle\psi_j|\hat{H}_{po}|\psi_m\rangle\right|^2 \left\{ \begin{array}{l} \left[\dfrac{\sin\left[(\omega_{jm}-\omega)t_o/2\right]}{(\omega_{jm}-\omega)t_o/2}\right]^2 + \left[\dfrac{\sin\left[(\omega_{jm}+\omega)t_o/2\right]}{(\omega_{jm}+\omega)t_o/2}\right]^2 \\ +2\cos(\omega t_o)\dfrac{\sin\left[(\omega_{jm}-\omega)t_o/2\right]}{(\omega_{jm}-\omega)t_o/2}\dfrac{\sin\left[(\omega_{jm}+\omega)t_o/2\right]}{(\omega_{jm}+\omega)t_o/2} \end{array} \right\} \tag{7.22}$$

As we see from Fig. 5.1, the sinc function and, consequently, its square fall off rapidly for arguments >> 1. Hence, for sufficiently long t_o, either one or the other of the two sinc functions in the last term in Eq. (7.22) will be small. Essentially, as the time t_o is increased, these two sinc line functions get sharper and sharper and they will eventually not overlap for any value of ω. Presuming we take t_o sufficiently large, therefore, we are left with

$$P(j) \simeq \frac{t_o^2}{\hbar^2}\left|\langle\psi_j|\hat{H}_{po}|\psi_m\rangle\right|^2 \left\{ \left[\frac{\sin\left[(\omega_{jm}-\omega)t_o/2\right]}{(\omega_{jm}-\omega)t_o/2}\right]^2 + \left[\frac{\sin\left[(\omega_{jm}+\omega)t_o/2\right]}{(\omega_{jm}+\omega)t_o/2}\right]^2 \right\} \tag{7.23}$$

Clearly, we now have some finite probability that the system has changed state from its initial state, $|\psi_m\rangle$, to another "final" state, $|\psi_j\rangle$. We see that this probability depends on the strength of the perturbation squared and, specifically, on the modulus squared of the matrix element of the perturbation between the initial and final states.

In the case where the perturbation is the oscillating electric field acting on an electron, we see therefore that this probability is proportional to the square of the electric field amplitude, E_o^2, which, in turn, is proportional to the intensity I (power per unit area) of an electromagnetic field. Hence, in the case of the oscillating electric field perturbation, we see that the probability of making a transition is proportional to the intensity, I. This is the kind of behavior we expect for linear optical absorption.

What is the meaning of the two different terms in Eq. (7.23)? The first term is significant if $\omega_{jm} \approx \omega$; that is, if

$$\hbar\omega \approx E_j - E_m \tag{7.24}$$

Because we chose ω to be a positive quantity, this term is significant if we are absorbing energy into the system, raising the system from a lower energy state, $|\psi_m\rangle$, to a higher energy state, $|\psi_j\rangle$. We note that the amount of energy we are absorbing is $\approx \hbar\omega$. This term behaves as we would require for absorption of a photon.

By contrast, the second term is significant if $\omega_{jm} \approx -\omega$; that is, if

$$\hbar\omega \approx E_m - E_j \tag{7.25}$$

This can only be the case if the system is moving from a higher energy state, $|\psi_m\rangle$, to a lower energy state, $|\psi_j\rangle$. This term behaves as we would require for emission of a photon. In fact, the process associated with this term is stimulated emission, the process used in lasers.[5]

Now let us consider only the case associated with absorption, presuming we are starting in a lower energy state and transitioning to a higher energy one. (The treatment of the stimulated emission case is essentially identical,[6] with the energies of the states reversed.) Then, we have

$$P(j) \simeq \frac{t_o^2}{\hbar^2}\left|\langle\psi_j|\hat{H}_{po}|\psi_m\rangle\right|^2 \left[\frac{\sin\left[\left(\omega_{jm}-\omega\right)t_o/2\right]}{\left(\omega_{jm}-\omega\right)t_o/2}\right]^2 \tag{7.26}$$

Analyzing the case of a transition between one state and exactly one other state using this approach has some formal difficulties; as we let the time t_o become arbitrarily large, the form of the sinc-squared term becomes arbitrarily sharp in ω and unless we get the frequency exactly correct, we get no absorption. For any specific ω not mathematically exactly equal to ω_{jm}, the probability $P(j)$ of a transition into the state $|\psi_j\rangle$ always becomes essentially zero if

---

[5] Incidentally, this particular so-called semiclassical analysis does not correctly model spontaneous emission, the dominant kind of light emission in nearly all kinds of everyday situations (including, e.g., light bulbs). To model that correctly requires that the electromagnetic field is treated quantum mechanically, not approximated by a classical field. Spontaneous emission can be modeled using this semiclassical model if one presumes that for some reason, there is $\hbar\omega/2$ of energy in each electromagnetic mode, with an associated electromagnetic field, sometimes called the *vacuum field*, available to stimulate emission, though the only real justification for this approach is to solve the problem correctly by treating the electromagnetic field quantum mechanically. We return to this point in Chapter 16, where we correctly include spontaneous emission. Somewhat surprisingly, it is actually harder in quantum mechanics to explain a light bulb than it is to explain a laser.

[6] Note, incidentally, that we can see from this quantum mechanical approach that the two processes of optical absorption and stimulated emission fundamentally are equally strong processes – both have the same prefactors (i.e., matrix elements). We do not see stimulated emission as much as we see absorption just because quantum mechanical systems are not normally found starting out in upper, excited states, tending for thermodynamic reasons to be found in their lower, unexcited states. This equivalence of the microscopic strength of absorption and stimulated emission is exactly what is required by a statistical mechanics analysis of optical absorption and emission, as deduced by Einstein in his famous "A and B coefficient" argument. See, e.g., H. Haken, *Light*, Vol. 1 (North-Holland, Amsterdam, 1981), pp. 58–62.

we leave the perturbation on for long enough. This problem can be resolved for calculating, for example, transitions between states in atoms, though it requires a more sophisticated analysis using the density matrix that we postpone to Chapter 14. That approach gives a "width" to an optical absorption line. Essentially, we end up replacing the sinc-squared function with a Lorentzian line whose width in angular frequency is $\sim 1/T_2$, where T_2 is the time between scattering events (e.g., collisions with other atoms) that disrupt, at least, the phase of the quantum mechanical oscillation of the wavefunction.[7] We can rationalize such a change based on an energy-time uncertainty relation; if the system only exists in its original form for some time T_2, then we should expect that the energy of the transition is only defined in energy to $\sim \pm\hbar/2T_2$ or in angular frequency to $\sim \pm 1/2T_2$.

Fortunately, however, there is a major class of optical-absorption problems that can be analyzed using this approach. Suppose that we have not one possible transition with energy difference $\hbar\omega_{jm}$ but rather a whole dense set of such possible transitions in the vicinity of the photon energy $\hbar\omega$, all with essentially identical matrix elements. This kind of situation occurs routinely in solids, for example, which have very dense sets of possible transitions, all of which have rather similar properties in any given small energy range. We presume that this set is very dense, with a density[8] $g_J(\hbar\omega)$ per unit energy near the photon energy $\hbar\omega$. ($g_J(\hbar\omega)$ is sometimes known as a *joint density of states* because it refers to transitions between states, not the density of states of only the starting or ending states.) Then, adding up all the probabilities for absorbing transitions, we obtain a total probability of absorption by this set of transitions of

$$P_{tot} \simeq \frac{t_o^2}{\hbar^2}\left|\langle\psi_j|\hat{H}_{po}|\psi_m\rangle\right|^2 \int \left[\frac{\sin\left[(\omega_{jm}-\omega)t_o/2\right]}{(\omega_{jm}-\omega)t_o/2}\right]^2 g_J(\hbar\omega_{jm})\, d\hbar\omega_{jm} \tag{7.27}$$

g_J is essentially constant over any small energy range and the sinc-squared term is essentially quite narrow in ω_{jm}. Hence, we can take $g_J(\hbar\omega_{jm})$ out of the integral as, approximately, $g_J(\hbar\omega)$. Formally changing the variable in the integral to $x = (\omega_{jm}-\omega)t_o/2$, therefore, gives

$$P_{tot} \simeq \frac{t_o^2}{\hbar^2}\left|\langle\psi_j|\hat{H}_{po}|\psi_m\rangle\right|^2 \frac{2\hbar}{t_o} g_J(\hbar\omega)\int\left[\frac{\sin x}{x}\right]^2 dx \tag{7.28}$$

Using the mathematical result

$$\int_{-\infty}^{\infty}\left(\frac{\sin x}{x}\right)^2 dx = \pi \tag{7.29}$$

we obtain

$$P_{tot} \simeq \frac{2\pi t_o}{\hbar}\left|\langle\psi_j|\hat{H}_{po}|\psi_m\rangle\right|^2 g_J(\hbar\omega) \tag{7.30}$$

---

[7] The reader may be wondering if we introduced a time T_2, is there a time T_1? There is. In the density matrix approach (Chapter 14), T_1 is the total lifetime of the state and it is certainly true that any processes that cause a transition out of the state (e.g., from an excited state back to a ground state) also give a line width to the state. T_2 is introduced because there are collision processes other than such major transitions that also disrupt the relative phase of the quantum mechanical response of the system and give a width to the transition.

[8] See Section 5.3 for a discussion of the use of such densities and the transition from sums to integrals.

Now we see that we have a total probability of making some transition that is proportional to the time, t_o, that the perturbation is turned on. This allows us now to deduce a transition rate or rate of absorption of photons

$$W = \frac{2\pi}{\hbar}\left|\langle\psi_j|\hat{H}_{po}|\psi_m\rangle\right|^2 g_J(\hbar\omega) \tag{7.31}$$

This result is sometimes known as *Fermi's Golden Rule* or, more completely, *Fermi's Golden Rule No. 2*. It is one of the most useful results of time-dependent perturbation theory and forms the basis for calculation of, for example, the optical absorption spectra of solids or many scattering processes. Though we have discussed it here in the context of optical absorption, it applies to any simple harmonic perturbation.

This rule is sometimes also stated using the Dirac δ-function (see Section 5.4) in the form

$$w_{jm} = \frac{2\pi}{\hbar}\left|\langle\psi_j|\hat{H}_{po}|\psi_m\rangle\right|^2 \delta\left(E_{jm} - \hbar\omega\right) \tag{7.32}$$

where w_{jm} is the transition rate between the specific states $|\psi_m\rangle$ and $|\psi_j\rangle$. From Eq. (7.32), one calculates the total transition rate involving all the possible similar transitions in the neighborhood as

$$W = \int w_{jm} g_J\left(\hbar\omega_{jm}\right) d\hbar\omega_{jm} \tag{7.33}$$

which gives the expression (7.31). We give an explicit example of the use of Fermi's Golden Rule in Section 8.10 when we evaluate optical absorption in semiconductors.

Note that Fermi's Golden Rule is built on the presumption of a periodic perturbation. The perturbation has to be "on" for many cycles, otherwise the function in Eq. (7.27), $\{\sin[(\omega_{jm} - \omega)t_o / 2]\}/[(\omega_{jm} - \omega)/(t_o / 2)]$, will not be "sharp" enough to allow the step from Eq. (7.27) to Eq. (7.28). For perturbations that are short, we can still use time-dependent perturbation theory directly, as we discussed in Section 7.1, but just not Fermi's Golden Rule. Similarly, we must not leave the periodic perturbation on for too long so that the population of the final state starts to become significant; Fermi's Golden Rule is a first-order perturbation result that implicitly presumes only small changes in the expansion coefficients. In practice, relaxation processes not included in the perturbation theory analysis may well continually depopulate the final state and then Fermi's Golden Rule can still, in fact, be used to calculate, for example, absorption in the steady state with a periodically oscillating perturbation. A more complete theory would include such relaxation properly, such as the density matrix approach in Chapter 14.

Problems

7.2.1* An electron is in the *second* state of a one-dimensional, infinitely deep potential well, with potential $V(z) = 0$ for $0 < z < L_z$ and infinite otherwise. An oscillating electric field of the form

$$F(t) = F_o\left[\exp(-i\omega t) + \exp(i\omega t)\right] = 2F_o\cos(\omega t)$$

is applied along the z direction for a large but finite time, leading to a perturbing Hamiltonian during that time of the form

$$\hat{H}_p(t) = eF(t)z = \hat{H}_{po}\left[\exp(-i\omega t) + \exp(i\omega t)\right]$$

(i) Consider the first four states of this well and presume that we are able to tune the frequency ω arbitrarily accurately to any frequency we wish. For each conceivable transition to another one of those states, say whether that transition is possible or essentially impossible, given appropriate choice of frequency.

(ii) What qualitative difference, if any, would it make if the well was of finite depth (though still considering only the first four states, all of which we presume to be bound in this finite well)?

7.2.2 I wish to make a quantum mechanical device that will sense static electric field F based on the shift of an optical transition energy. I want to make an estimate of the sensitivity of this device by considering optical transitions in an idealized infinitely deep one-dimensional potential well of thickness L_z, with an electron initially in the first state of that well, and considering the transition between the first and second states of the well. The electric field to be sensed is presumed to be in the direction perpendicular to the walls of the potential well, which we call the z direction.

(i) In what direction should the optical electric field be polarized for this device? Justify your answer.

(ii) Give an approximate expression for the shift of the transition energy with field, numerically evaluating any summations. (You may make reasonable simplifying calculational assumptions, such as retaining only the most important terms in a sum if they are clearly dominant.)

(iii) So that this estimate would correspond approximately to the situation we might encounter in atoms, we make the potential well 3 Å thick. Presuming that the minimum change of optical transition energy that can be measured is 0.1 percent, calculate the corresponding minimum static electric field that can be sensed.

(iv) Still using the electron transition energy in a well of this thickness and the same measurable change in optical transition energy, suggest a method that would increase the sensitivity of this device for measuring changes in electric field.

7.2.3 (Note: This problem can be used as a substantial assignment. It is an exercise in both time-independent calculation techniques and in consequences of Fermi's Golden Rule.) We wish to make a detector for radiation in the terahertz frequency regime. The concept for this detector is first that we make a coupled potential well structure using GaAs wells each of 5 nm thickness, surrounded by $Al_{0.3}Ga_{0.7}As$ layers on either side and with a barrier of the same $Al_{0.3}Ga_{0.7}As$ material between the wells, of a thickness w to be determined.

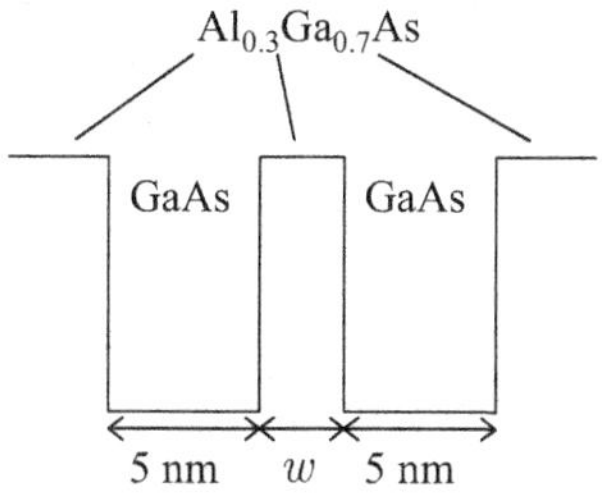

Initially, there will be electrons in the lowest level of this structure and, if we can raise the electrons by optical absorption from the first level into the second level, we presume that the carriers in the second level are then swept out as photocurrent by a mechanism not shown. The terahertz electric field is presumed to be polarized in the horizontal direction in the diagram. We presume that this detector is held at very low temperature so that, to a sufficient degree of approximation, all the electrons are initially in this lowest state. (The electrons are free to move in the other two directions, but this does not substantially change the result of this problem. The nature of optical absorption within the conduction band of semiconductors is such that, even if the electron does have momentum [and, hence, kinetic energy] in the other two directions, that momentum [and, hence, kinetic energy] is conserved in an optical transition, so the transition energy is unaffected by that initial momentum in this particular so-called *intersubband* transition.)

Assume that the electron can be treated like an ordinary electron, but with an effective mass, a mass that is different in different material layers (see Problem 2.9.4 for appropriate boundary conditions and a solution of the finite well problem in this case). Note the following parameters: separation between $Al_{0.3}Ga_{0.7}As$ and GaAs so-called "zone center" conduction band edges (i.e., the potential barrier height in this problem) = 0.235 eV; electron effective masses: 0.067 m_o in GaAs, 0.092 m_o in $Al_{0.3}Ga_{0.7}As$.

(i) Deduce what thickness w of barrier should be chosen if this detector is to work for approximately 0.5 THz (500 GHz) frequency. (Hint: You may assume that the coupling is relatively weak and that a tight binding approach would be a good choice. Note: You may have to discard a solution here that goes beyond the validity of the approximate solution method you use.)

(ii) We wish to tune this detector by applying a static electric field in the horizontal direction.

(a) Graph how the detected frequency changes with electric field up to a detected frequency of approximately 1 THz.

(b) Graph how the sensitivity of the detector (i.e., the relative size of the photocurrent) changes as a function of static electric field.

(c) Over what frequency range does this detector have a sensitivity that changes by less than 3 dB (i.e., a factor of two) as it is tuned?

7.3 Refractive index

This section can be omitted at a first reading, though it is a good exercise in the use of perturbation theory. Prerequisite: Appendix D, Section D.1, for an introductory discussion of the polarization **P**.

First-order time-dependent perturbation theory is sufficient to model all linear optical processes quantum mechanically. Fermi's Golden Rule, stated previously, shows how absorption (and stimulated emission) can be modeled. Here, we illustrate how another linear optical process, refractive index, can be calculated quantum mechanically using first-order time-dependent perturbation theory. This is a different kind of example of time-dependent perturbation theory because it does not involve a transition rate, as calculated using Fermi's Golden Rule. It also prepares us for the following calculation of nonlinear optical processes.

The key quantity we need to calculate is the polarization, P. In classical electromagnetism, the relation between electric field and polarization (here, for a simple isotropic medium so that the polarization and the electric field are in the same direction) for the linear case is

$$\mathsf{P} = \varepsilon_o \chi \mathsf{E} \tag{7.34}$$

where χ is the susceptibility and ε_o is the permittivity of free space. The refractive index, n_r, can be deduced through the relation

$$n_r = \sqrt{1+\chi} \tag{7.35}$$

(at least, if the material is transparent [nonabsorbing] at the frequencies of interest). Hence, if we can calculate the proportionality between P and E, we can deduce the refractive index.

We consider for simplicity here a system with a single electron or in which our interactions are only with a single electron. We also know classically that the dipole moment, μ_{dip}, associated with moving a single electron through a distance z, is, by definition

$$\mu_{dip} = -ez \tag{7.36}$$

(The minus sign arises because the electron charge is negative.) The polarization P is the dipole moment per unit volume and so we expect that for the quantum mechanical expectation value of the polarization, we have

$$\langle \mathsf{P} \rangle = \frac{-e\langle z \rangle}{V} \tag{7.37}$$

where V is the volume of the system. Our quantum mechanical task of calculating refractive index therefore reduces essentially to calculating $\langle \mathsf{P} \rangle$.

Because we are working in first-order perturbation theory, we can write the total state of the system as, approximately

$$|\Psi\rangle = |\Phi^{(0)}\rangle + |\Phi^{(1)}\rangle \tag{7.38}$$

where we note now that we are dealing with the full time-dependent state vectors (kets). Here, $|\Phi^{(0)}\rangle$ is the unperturbed (time-dependent) state vector and $|\Phi^{(1)}\rangle$ is the first-order (time-dependent) correction that we can write as

$$|\Phi^{(1)}\rangle = \sum_n a_n^{(1)}(t)\exp(-i\omega_n t)|\psi_n\rangle \tag{7.39}$$

where

$$\omega_n = E_n / \hbar \tag{7.40}$$

and $|\psi_n\rangle$ are the time-independent energy eigenfunctions of the unperturbed system. With such a state vector, Eq. (7.38), the expectation value of the polarization would be

$$\begin{aligned}\langle \mathsf{P} \rangle &= -\frac{e}{V}\langle \Psi | z | \Psi \rangle \\ &= -\frac{e}{V}\left[\langle \Phi^{(0)} | z | \Phi^{(0)} \rangle + \langle \Phi^{(1)} | z | \Phi^{(0)} \rangle + \langle \Phi^{(0)} | z | \Phi^{(1)} \rangle + \langle \Phi^{(1)} | z | \Phi^{(1)} \rangle\right]\end{aligned} \tag{7.41}$$

The first term $-e\langle \Phi^{(0)} | z | \Phi^{(0)} \rangle$ is just the static dipole moment of the material in its unperturbed state and is not of interest to us here,[9] so we do not consider it further. The fourth term, $-e\langle \Phi^{(1)} | z | \Phi^{(1)} \rangle$, is second-order in the perturbation (it would, e.g., correspond to a term proportional to the square of the electric field) and, hence, in this first-order calculation, we drop it also. So, noting that $\langle \Phi^{(1)} | z | \Phi^{(0)} \rangle = \langle \Phi^{(0)} | z | \Phi^{(1)} \rangle^*$ (which follows from the Hermiticity of z as an operator corresponding to a physical observable [the position]), we have

$$\langle \mathsf{P} \rangle = -\frac{2e}{V}\mathrm{Re}\left[\langle \Phi^{(0)} | z | \Phi^{(1)} \rangle\right] \tag{7.42}$$

For the sake of definiteness, we now presume that the system is initially in the eigenstate m; that is,

$$|\Phi^{(0)}\rangle = \exp(-i\omega_m t)|\psi_m\rangle \tag{7.43}$$

Hence, using the expansion Eq. (7.39) for $|\Phi^{(1)}\rangle$, we have from Eq. (7.42)

$$\langle \mathsf{P} \rangle = -\frac{2e}{V}\mathrm{Re}\left[\sum_n a_n^{(1)}(t)\exp(i\omega_{mn} t)\langle \psi_m | z | \psi_n \rangle\right] \tag{7.44}$$

[9] Most materials do not have a static polarization in them, but such a phenomenon is not unknown, being present in, e.g., ferroelectric materials.

We are interested here in the steady-state situation with a continuous oscillating field and we take the perturbing Hamiltonian Eq. (7.15) as valid for all times.[10] We can rewrite Eq. (7.18) as

$$\dot{a}_q^{(1)}(t) = \frac{e\mathrm{E}_o}{i\hbar}\langle\psi_q|z|\psi_m\rangle\exp(i\omega_{qm}t)\left[\exp(-i\omega t)+\exp(i\omega t)\right] \tag{7.45}$$

to obtain

$$a_q^{(1)}(t) = -\frac{e\mathrm{E}_o}{\hbar}\langle\psi_q|z|\psi_m\rangle\left[\frac{\exp\left[i(\omega_{qm}-\omega)t\right]}{(\omega_{qm}-\omega)}+\frac{\exp\left[i(\omega_{qm}+\omega)t\right]}{(\omega_{qm}+\omega)}\right] \tag{7.46}$$

Substituting into Eq. (7.44) gives

$$\begin{aligned}\langle \mathrm{P}\rangle &= \frac{2e^2\mathrm{E}_o}{\hbar V}\mathrm{Re}\sum_n\left|\langle\psi_m|z|\psi_n\rangle\right|^2\exp(i\omega_{mn}t)\\ &\quad\times\left[\frac{\exp\left[i(\omega_{nm}-\omega)t\right]}{(\omega_{nm}-\omega)}+\frac{\exp\left[i(\omega_{nm}+\omega)t\right]}{(\omega_{nm}+\omega)}\right]\\ &= \frac{2e^2\mathrm{E}_o}{\hbar V}\sum_n\left|\langle\psi_m|z|\psi_n\rangle\right|^2\left[\frac{\cos(-\omega t)}{(\omega_{nm}-\omega)}+\frac{\cos(\omega t)}{(\omega_{nm}+\omega)}\right]\\ &= \frac{2e^2\mathrm{E}_o\cos(\omega t)}{\hbar V}\sum_n\left|\langle\psi_m|z|\psi_n\rangle\right|^2\left[\frac{1}{(\omega_{nm}-\omega)}+\frac{1}{(\omega_{nm}+\omega)}\right]\end{aligned} \tag{7.47}$$

(where we have used the fact that $\omega_{mn} = -\omega_{nm}$) and so we have, from Eq. (7.34)

$$\chi = \frac{e^2}{\varepsilon_o\hbar V}\sum_n\left|\langle\psi_m|z|\psi_n\rangle\right|^2\left[\frac{1}{(\omega_{nm}-\omega)}+\frac{1}{(\omega_{nm}+\omega)}\right] \tag{7.48}$$

from which we can deduce the refractive index, n_r, if we wish, from Eq. (7.35).[11] This therefore completes our calculation from first-order perturbation theory of refractive index.

---

[10] Here, we do not have to use the device we used for the Fermi's Golden Rule derivation of having the Hamiltonian only be "on" for a finite window because we do not have to take a limit as we did there. Depending on how the oscillating perturbation is "started," one does get a different constant of integration in evaluating $a^{(1)}(t)$ from $\dot{a}^{(1)}(t)$, which we have simply ignored here. E.g., we would get a different answer for this constant of integration if we turned on a $\sin(\omega t)$ field at $t = 0$ than if we turned on the $\cos(\omega t)$ field we examined explicitly for the Fermi's Golden Rule derivation. This constant of integration is a term that does not vary with time and, therefore, does not concern us for this calculation of how the polarization changes in time in response to the electric field. It is a physically real phenomenon, however, reflecting how the starting transient can give rise to additional occupation of states in the system.

[11] The dependence of χ on the volume V might confuse the reader, who expects that the refractive index does not depend on the volume of the material – glass has an index of ~ 1.5 regardless of the size of the piece of glass. In practice, the number of states in the material – and, hence, the number of elements in the sum – does usually grow with the volume V, so these effects cancel. The expression does also correctly deal with the situation of a single atom, e.g., for which the dipole moment per unit volume will, indeed, shrink with the total volume of the system being considered.

Note a major difference between the absorption (i.e., Fermi's Golden Rule) and the refractive index. For absorption, the frequency ω must match the transition frequency ω_{nm} very closely for that particular transition to give rise to absorption of photons. For the refractive index, the contribution of a particular possible transition $|\psi_m\rangle \rightarrow |\psi_n\rangle$ to the susceptibility (and, hence, the refractive index) is finite even when the frequencies do not match exactly or even closely; that contribution to the susceptibility rises steadily as ω rises toward ω_{nm}.

We can also see that there is a very close relation between absorption and refractive index. If we have an absorbing transition at some frequency ω_{nm}, it contributes to the refractive index at all frequencies. In fact, in this view, the refractive index (in a region where the material is transparent) arises entirely because of the absorption at other frequencies. It is also clear that if there is a refractive index different from unity, then there must be absorption at some other frequency or frequencies.

The fundamental relation between refractive index and absorption is already well known from classical physics and is expressed through the so-called Kramers–Kronig relations. The derivation of those relations, though relatively brief, is entirely mathematical, shedding no light on the physical mechanism whereby absorption and refractive index are related. With the quantum mechanical expressions we have here for these two processes, we can attempt to understand any particular aspect that interests us in the physical relation between the two.

Unlike the case of absorption, calculated using Fermi's Golden Rule, the expansion coefficients do not grow steadily in time; there is no net transition rate. In this quantum mechanical picture of refraction, we find that even though we are in the transparent region of the material, there are, however, finite expansion coefficients, in general oscillating in time, and there are also, consequently, finite occupation probabilities for all of the states of the system. It is only because we have such probabilities that the material has a polarization. The polarization arises because the charges in the material change their physical wavefunctions in response to the field, mixing in some of the other states of the system in response to this perturbation. If we examined the expectation value of the energy of the material, we would also find quite real energy stored in the material as a result. By these kinds of quantum mechanical analyses, we can understand any specific measurable aspect of the process of refractive index and its relation to absorption without recourse to the somewhat arcane Kramers–Kronig relations.

7.4 Nonlinear optical coefficients

This section can be omitted on a first reading, though it is a good example of how quantum mechanics, which is based entirely on linear algebra, can calculate a nonlinear process. Prerequisite: Section 7.3.

The formalism developed in the preceding section for calculation of refractive index provides the basis for calculation of nonlinear optical effects, at least for the case where we are working in spectral regions where the material is transparent. Nonlinear optical effects are quite common phenomena now that we have relatively high powers and optical intensities routinely available from lasers.

Nonlinear optical effects are quite important in the engineering of long-distance fiber-optic communication systems, for example. Nonlinear refraction and related effects such as four-wave mixing are important mechanisms for degrading the transmission of optical pulses and causing interaction between data streams at different wavelengths. Nonlinear refraction can also cause so-called *soliton propagation* in which a short pulse can propagate long distances without degradation because, in that case, the nonlinear refractive effects can counteract the linear dispersion that otherwise would progressively take the pulse apart. *Raman amplification*, another nonlinear optical effect, is a potentially useful method of overcoming loss in optical fibers. Many other nonlinear optical phenomena exist and are exploited, including electric-field dependence of refractive index used in some optical modulators and a broad variety of effects that generate new optical frequencies by combining existing ones, such as second and third harmonic generations, difference frequency mixing, and optical parametric oscillators.

Nonlinear optical effects also provide an excellent example of higher order time-dependent perturbation theory in action. In particular, they show how the perturbation approach helps generate and classify different processes. Different effects are associated with different orders of perturbation. Second-order time-dependent perturbation theory leads, for example, to second harmonic generation and the linear electro-optic effect (i.e., a linear variation of refractive index with applied static electric field, known also as the Pockels effect) and to three-wave mixing phenomena where a new frequency emerges as the sum or difference of the other two frequencies. Third-order theory leads to intensity-dependent refractive index and refractive index changes proportional to the square of the static electric field (both known as *Kerr effect*" nonlinearities), as well as third harmonic generation and four-wave mixing. There are many other variants and combinations of such effects.

Nearly all nonlinear optical processes that are used for practical purposes are described by second- and third-order perturbation theory. Higher order processes are known and can be calculated by higher order perturbation theory, though they are usually quite weak by comparison. The strongest effects are generally the second-order ones, though to see those, as will become apparent later, the material needs to be asymmetric in a particular way. Isotropic materials or those with a *center of symmetry*, such as glass and nonpolar materials such as silicon, therefore do not show second-order phenomena,[12] and their lowest order nonlinear effects are therefore third-order phenomena.

We do not discuss here all the details of nonlinear optics, which would be beyond the scope of this work, deferring to other excellent texts.[13] For example, we do not deal here with the various macroscopic electromagnetic propagation effects that arise formally once we substitute the calculated polarization into Maxwell's equations (or the wave equation that results from those), though such effects (e.g., phase matching) are very important for calculating the final electromagnetic waves that result from microscopic nonlinear optical processes.

Formalism for nonlinear optical coefficients

In most cases, nonlinear optical phenomena are weak effects and it can therefore be useful to look upon them in terms of a power-series expansion. We are interested in the response of the material, characterized through the polarization $\mathsf{P}(t)$, which we expand as a power series in the electric field $\mathsf{E}(t)$; that is,

$$\frac{\mathsf{P}(t)}{\varepsilon_o} = \chi^{(1)}\mathsf{E}(t) + \chi^{(2)}\mathsf{E}^2(t) + \chi^{(3)}\mathsf{E}^3(t) + \ldots \tag{7.49}$$

---

[12] Such materials can, however, show second-order effects at their surfaces because the symmetry of the bulk material is broken there.

[13] See, e.g., R. W. Boyd, *Nonlinear Optics* (second ed.) (Academic Press, New York, 1992); and Y. R. Shen, *The Principles of Nonlinear Optics* (Wiley, New York, 1984).

In general, both the electric field E and the polarization P are vectors. Also, in general, because the polarization and the electric field may well not be in the same direction,[14] the susceptibility coefficients $\chi^{(1)}$, $\chi^{(2)}$, $\chi^{(3)}$, and so forth, are tensors. We neglect such anisotropic effects for simplicity here and treat the electric field and polarization as always being in the same direction, in which case we can treat them as scalars. In Eq.(7.49), $\chi^{(1)}$ is simply the linear susceptibility calculated as in Eq. (7.48). $\chi^{(2)}$ and $\chi^{(3)}$ are, respectively, the second- and third-order nonlinear susceptibilities. In handling higher order perturbations, it is particularly important to be very systematic in notations because the full expressions can become quite complicated. Many of the nonlinear optical effects involve multiple different frequencies in the fields. For example, with two frequency components, at (angular) frequencies ω_1 and ω_2, the total field would be

$$\begin{aligned} \mathsf{E}(t) &= 2\mathsf{E}_{o1}\cos(\omega_1 t + \delta_1) + 2\mathsf{E}_{o2}\cos(\omega_2 t + \delta_2) \\ &= \mathsf{E}_{o1}\left\{\exp\left[-i(\omega_1 t + \delta_1)\right] + \exp\left[i(\omega_1 t + \delta_1)\right]\right\} \\ &\quad + \mathsf{E}_{o2}\left\{\exp\left[-i(\omega_2 t + \delta_2)\right] + \exp\left[i(\omega_2 t + \delta_2)\right]\right\} \end{aligned} \tag{7.50}$$

where now we have formally allowed the two fields also to have different phase angles δ_1 and δ_2. Another way of writing Eq. (7.50) is

$$\mathsf{E}(t) = \sum_s \mathsf{E}(\omega_s)\exp(-i\omega_s t) \tag{7.51}$$

where

$$\mathsf{E}(\omega_s) = \mathsf{E}_{os}\exp(-i\delta_s) \tag{7.52}$$

and the sum now is understood to be not just over the frequencies ω_1 and ω_2 but also to include the "negative" frequencies, $-\omega_1$ and $-\omega_2$.[15] Hence, there are four terms in the sum Eq. (7.51) for this two-frequency case, corresponding to the four terms in the second line of Eq. (7.50). Note also that

$$\mathsf{E}(-\omega_s) = \mathsf{E}^*(\omega_s) \tag{7.53}$$

as can be deduced from Eq. (7.52). This property is required for the actual electric field to be real. This sum over positive and negative frequencies simplifies the algebra that follows. We can, of course, keep the form Eq. (7.51) as we extend to more frequency components in the electric field.

Formal calculation of perturbative corrections

We consider here nonlinearities up to third-order in the electric field (i.e., up to $\chi^{(3)}$) and, as a result, consider up to third-order time-dependent perturbation corrections. We proceed as

---

[14] We could imagine, e.g., that we had an electron on a spring along the x direction that was not free to move in any other direction. If we applied an electric field in the x direction, then the electron would be displaced in the x direction, giving a polarization in the x direction; but even if the field were applied in a somewhat different direction, the electron would still only move in the x direction. Hence, the polarization and the electric field, in this case, would not be in the same direction and the susceptibility would formally have to be described as a tensor.

[15] Note that the choice to use $\exp(-i\omega_n t)$ rather than $\exp(i\omega_n t)$ in Eq. (7.51) is entirely arbitrary, but this particular choice is a more typical notation in nonlinear optics texts.

before for time-dependent perturbation theory, now using the expression Eq. (7.51) for the electric field; hence, we have a perturbing Hamiltonian

$$\hat{H}_p(t) = e\mathrm{E}(t)z = ez\sum_s \mathrm{E}(\omega_s)\exp(-i\omega_s t) \tag{7.54}$$

Presuming as before that the system starts in some specific state m, the time derivative of the first-order perturbation correction to the wavefunction then becomes, as in Eq. (7.18) (or Eq. (7.45))

$$\dot{a}_q^{(1)}(t) = \frac{-\mu_{qm}}{i\hbar}\sum_s \mathrm{E}(\omega_s)\exp\left[i(\omega_{qm} - \omega_s)t\right] \tag{7.55}$$

where we have also introduced the more common notation of the electric dipole moment between states

$$\mu_{qm} = -e\langle\psi_q|z|\psi_m\rangle \tag{7.56}$$

Integrating over time, assuming as before that we can simply neglect any constant of integration because we are only considering oscillating effects, we have

$$a_q^{(1)}(t) = \frac{1}{\hbar}\sum_s \frac{\mu_{qm}\mathrm{E}(\omega_s)}{(\omega_{qm} - \omega_s)}\exp\left[i(\omega_{qm} - \omega_s)t\right] \tag{7.57}$$

We may then use the relation Eq. (7.13) that allows us to calculate subsequent levels of perturbative correction from the preceding one to calculate the second-order correction. Using Eq. (7.13), we have

$$\begin{aligned}\dot{a}_j^{(2)}(t) &= \frac{-1}{i\hbar}\sum_q a_q^{(1)}\mu_{jq}\sum_u \mathrm{E}(\omega_u)\exp\left[i(\omega_{jq} - \omega_u)t\right] \\ &= \frac{-1}{i\hbar^2}\sum_q\sum_{s,u}\frac{\mu_{jq}\mathrm{E}(\omega_u)\mu_{qm}\mathrm{E}(\omega_s)}{(\omega_{qm} - \omega_s)}\exp\left[i(\omega_{jm} - \omega_s - \omega_u)t\right]\end{aligned} \tag{7.58}$$

where we have noted that

$$\omega_{jq} + \omega_{qm} = \omega_{jm} \tag{7.59}$$

(This is simply a statement that the energy separation between levels j and m is equal to the energy separation between level q and level m plus the energy separation between level q and level j.) Hence

$$a_j^{(2)}(t) = \frac{1}{\hbar^2}\sum_q\sum_{s,u}\frac{\mu_{jq}\mathrm{E}(\omega_u)\mu_{qm}\mathrm{E}(\omega_s)}{(\omega_{jm} - \omega_s - \omega_u)(\omega_{qm} - \omega_s)}\exp\left[i(\omega_{jm} - \omega_s - \omega_u)t\right] \tag{7.60}$$

Similarly

$$\begin{aligned}\dot{a}_k^{(3)} &= \frac{-1}{i\hbar}\sum_j a_j^{(2)}\mu_{kj}\sum_v \mathrm{E}(\omega_v)\exp\left[i(\omega_{kj} - \omega_v)t\right] \\ &= \frac{-1}{i\hbar^3}\sum_{j,q}\sum_{s,u,v}\frac{\mu_{kj}\mathrm{E}(\omega_v)\mu_{jq}\mathrm{E}(\omega_u)\mu_{qm}\mathrm{E}(\omega_s)}{(\omega_{jm} - \omega_s - \omega_u)(\omega_{qm} - \omega_s)}\exp\left[i(\omega_{km} - \omega_s - \omega_u - \omega_v)t\right]\end{aligned} \tag{7.61}$$

and so

$$a_k^{(3)}(t) = \frac{1}{\hbar^3}\sum_{j,q}\sum_{s,u,v}\frac{\mu_{kj}\mathsf{E}(\omega_v)\mu_{jq}\mathsf{E}(\omega_u)\mu_{qm}\mathsf{E}(\omega_s)}{(\omega_{km}-\omega_s-\omega_u-\omega_v)(\omega_{jm}-\omega_s-\omega_u)(\omega_{qm}-\omega_s)} \times \exp\left[i(\omega_{km}-\omega_s-\omega_u-\omega_v)t\right] \quad (7.62)$$

Note in these sums, j and q are indices going over all possible states of the system and s, u, and v are indices going over all the frequencies of electric fields, including both their positive and negative versions.

Formal calculation of linear and nonlinear susceptibilities

In general, including all possible terms in the polarization up to third-order in the perturbation, we now formally write the expectation value of the polarization as being the observable quantity, using $\mu = -ez$ as the dipole moment (operator)

$$\begin{aligned}\langle \mathsf{P}(t)\rangle &= \frac{1}{V}\langle\Psi|\mu|\Psi\rangle \cong \frac{1}{V}\langle\Phi^{(0)}+\Phi^{(1)}+\Phi^{(2)}+\Phi^{(3)}|\mu|\Phi^{(0)}+\Phi^{(1)}+\Phi^{(2)}+\Phi^{(3)}\rangle \\ &\cong \langle\mathsf{P}^{(0)}(t)\rangle+\langle\mathsf{P}^{(1)}(t)\rangle+\langle\mathsf{P}^{(2)}(t)\rangle+\langle\mathsf{P}^{(3)}(t)\rangle\end{aligned} \quad (7.63)$$

where

$$\langle\mathsf{P}^{(0)}\rangle = \frac{1}{V}\langle\Phi^{(0)}|\mu|\Phi^{(0)}\rangle \quad (7.64)$$

is the static polarization of the material

$$\langle\mathsf{P}^{(1)}(t)\rangle = \frac{1}{V}\left(\langle\Phi^{(0)}|\mu|\Phi^{(1)}\rangle+\langle\Phi^{(1)}|\mu|\Phi^{(0)}\rangle\right) \quad (7.65)$$

is the linear polarization (first-order correction to the polarization), giving the linear refractive index

$$\langle\mathsf{P}^{(2)}(t)\rangle = \frac{1}{V}\left(\langle\Phi^{(0)}|\mu|\Phi^{(2)}\rangle+\langle\Phi^{(2)}|\mu|\Phi^{(0)}\rangle+\langle\Phi^{(1)}|\mu|\Phi^{(1)}\rangle\right) \quad (7.66)$$

is the second-order polarization, giving rise to phenomena such as second harmonic generation, and sum and difference frequency mixing, and

$$\langle\mathsf{P}^{(3)}(t)\rangle = \frac{1}{V}\left(\langle\Phi^{(0)}|\mu|\Phi^{(3)}\rangle+\langle\Phi^{(3)}|\mu|\Phi^{(0)}\rangle+\langle\Phi^{(1)}|\mu|\Phi^{(2)}\rangle+\langle\Phi^{(2)}|\mu|\Phi^{(1)}\rangle\right) \quad (7.67)$$

is the third-order polarization, giving rise to phenomena such as third harmonic generation, nonlinear refractive index, and four-wave mixing.

Now, we formally evaluate the different linear and nonlinear susceptibilities associated with these various phenomena.

Linear susceptibility

We have already calculated this, but we briefly repeat the result in the present notation as used in nonlinear optics. Because by choice $|\Phi^{(0)}\rangle = \exp(-i\omega_m t)|\psi_m\rangle$, and using the standard expansion notation Eq. (7.39) for $|\Phi^{(1)}\rangle$, we have, from the definition Eq. (7.65) above

$$\langle \mathsf{P}^{(1)}(t)\rangle = \frac{1}{V}\frac{1}{\hbar}\sum_q\sum_s\left\{\begin{array}{l}\dfrac{\mu_{mq}\mu_{qm}}{\omega_{qm}-\omega_s}\mathsf{E}(\omega_s)\exp(i\omega_{mq}t)\exp\left[i(\omega_{qm}-\omega_s)t\right]\\ +\dfrac{\mu_{qm}\mu_{mq}}{\omega_{qm}-\omega_s}\mathsf{E}^*(\omega_s)\exp(i\omega_{qm}t)\exp\left[-i(\omega_{qm}-\omega_s)t\right]\end{array}\right\}$$

$$= \frac{1}{V}\frac{1}{\hbar}\sum_q\sum_s \mu_{mq}\mu_{qm}\left\{\frac{\mathsf{E}(\omega_s)}{\omega_{qm}-\omega_s}\exp(-i\omega_s t)+\frac{\mathsf{E}(-\omega_s)}{\omega_{qm}-\omega_s}\exp(i\omega_s t)\right\} \tag{7.68}$$

Now we make a formal change, trivial for this case, but more useful in the higher order susceptibilities. We note that because we are summing over positive and negative values of ω_s we can change ω_s to $-\omega_s$ in any terms we wish without changing the final result for the sum. Hence, we can write

$$\langle \mathsf{P}^{(1)}(t)\rangle = \frac{1}{V}\frac{1}{\hbar}\sum_q\sum_s \mu_{mq}\mu_{qm}\left\{\frac{1}{\omega_{qm}-\omega_s}+\frac{1}{\omega_{qm}+\omega_s}\right\}\mathsf{E}(\omega_s)\exp(-i\omega_s t) \tag{7.69}$$

If we wish, we can now write

$$\frac{\langle \mathsf{P}^{(1)}(t)\rangle}{\varepsilon_o} = \sum_s \chi^{(1)}(\omega_s;\omega_s)\mathsf{E}(\omega_s)\exp(-i\omega_s t) \tag{7.70}$$

where by $\chi^{(1)}(\omega_s;\omega_s)$, we mean the (linear) susceptibility that gives rise to a polarization at frequency ω_s in response to a field at frequency ω_s. Of course, this notation is trivial for the case of linear polarization because there is no question that the frequency of the polarization will be the same as that of the incident field, but this will not necessarily be the case for higher order polarizations. From Eqs. (7.69) and (7.70), we must have the definition

$$\chi^{(1)}(\omega_s;\omega_s) = \frac{1}{\varepsilon_o\hbar V}\sum_q \mu_{mq}\mu_{qm}\left[\frac{1}{\omega_{mq}-\omega_s}+\frac{1}{\omega_{mq}+\omega_s}\right] \tag{7.71}$$

We can see directly from this, incidentally, that

$$\chi^{(1)}(\omega_s;\omega_s) = \chi^{(1)}(-\omega_s;-\omega_s) \tag{7.72}$$

so the latter is redundant here (an example of one of the many symmetries inherent in this approach to susceptibilities). This expression allows us to calculate the linear refractive index.

Second-order susceptibility

In the second-order case, we use Eq. (7.66). For the first pair of terms, we have

$$\frac{1}{V}\left(\langle\Phi^{(0)}|\mu|\Phi^{(2)}\rangle+\langle\Phi^{(2)}|\mu|\Phi^{(0)}\rangle\right) = \frac{1}{V}\frac{1}{\hbar^2}\sum_{j,q}\sum_{s,u}\mu_{mj}\mu_{jq}\mu_{qm}$$
$$\times\left[\frac{\mathsf{E}(\omega_u)\mathsf{E}(\omega_s)\exp\left[-i(\omega_u+\omega_s)t\right]}{(\omega_{jm}-\omega_u-\omega_s)(\omega_{qm}-\omega_s)}+\frac{\mathsf{E}^*(\omega_u)\mathsf{E}^*(\omega_s)\exp\left[i(\omega_u+\omega_s)t\right]}{(\omega_{jm}-\omega_u-\omega_s)(\omega_{qm}-\omega_s)}\right] \tag{7.73}$$

Making the formal substitution of $-\omega_s$ for ω_s and $-\omega_u$ for ω_u, just as we did for the linear case, makes no difference to the result of the sum because it is over positive and negative frequencies anyway and so we obtain

$$\frac{1}{V}\left(\langle\Phi^{(0)}|\mu|\Phi^{(2)}\rangle+\langle\Phi^{(2)}|\mu|\Phi^{(0)}\rangle\right)=\frac{1}{V}\frac{1}{\hbar^2}\sum_{j,q}\sum_{s,u}\mu_{mj}\mu_{jq}\mu_{qm}\mathrm{E}(\omega_u)\mathrm{E}(\omega_s)$$
$$\times\left[\frac{1}{(\omega_{jm}-\omega_u-\omega_s)(\omega_{qm}-\omega_s)}+\frac{1}{(\omega_{jm}+\omega_u+\omega_s)(\omega_{qm}+\omega_s)}\right]\exp\left[-i(\omega_u+\omega_s)t\right] \tag{7.74}$$

Now, examining the third term in Eq. (7.66) above, we similarly have

$$\frac{1}{V}\langle\Phi^{(1)}|\mu|\Phi^{(1)}\rangle$$
$$=\frac{1}{V}\frac{1}{\hbar^2}\sum_{j,q}\sum_{s,u}\mu_{mj}\mu_{jq}\mu_{qm}\frac{\mathrm{E}^*(\omega_u)\mathrm{E}(\omega_s)}{(\omega_{jm}-\omega_u)(\omega_{qm}-\omega_s)}\exp\left[i(\omega_u-\omega_s)t\right] \tag{7.75}$$
$$=\frac{1}{V}\frac{1}{\hbar^2}\sum_{j,q}\sum_{s,u}\mu_{mj}\mu_{jq}\mu_{qm}\frac{\mathrm{E}(\omega_u)\mathrm{E}(\omega_s)}{(\omega_{jm}+\omega_u)(\omega_{qm}-\omega_s)}\exp\left[-i(\omega_u+\omega_s)t\right]$$

where we made the formal substitution of $-\omega_u$ for ω_u with the same justification as before.

Hence, now having all terms arranged with the same formal time dependence of $\exp\left[-i(\omega_u+\omega_s)t\right]$, we can write

$$\frac{\langle\mathrm{P}^{(2)}(t)\rangle}{\varepsilon_o}=\sum_{s,u}\chi^{(2)}(\omega_u+\omega_s;\omega_u,\omega_s)\mathrm{E}(\omega_u)\mathrm{E}(\omega_s)\exp\left[-i(\omega_u+\omega_s)t\right] \tag{7.76}$$

where

$$\chi^{(2)}(\omega_u+\omega_s;\omega_u,\omega_s)=\frac{1}{\varepsilon_o V}\frac{1}{\hbar^2}\sum_{j,q}\mu_{mj}\mu_{jq}\mu_{qm}$$
$$\times\left\{\frac{1}{(\omega_{jm}-\omega_u-\omega_s)(\omega_{qm}-\omega_s)}+\frac{1}{(\omega_{jm}+\omega_u)(\omega_{qm}-\omega_s)}+\frac{1}{(\omega_{jm}+\omega_u+\omega_s)(\omega_{qm}+\omega_s)}\right\} \tag{7.77}$$

Second-order nonlinear optical phenomena

With this first nontrivial case, we see the usefulness of the notation. For example, if we consider $\omega_u=\omega_s$, we see that this $\chi^{(2)}(2\omega_s;\omega_s,\omega_s)$ gives the strength of the second harmonic generation process with input frequency ω_s. We can see, incidentally, that this effect would be relatively quite strong if we had an energy level j such that ω_{jm} was close to $2\omega_s$ and especially if there was another energy level q such that ω_{qm} was close to ω_s because then we would have two strong resonant denominators.

If we consider that our original electric field has two frequency components, ω_u and ω_s, we can easily see that $\chi^{(2)}(\omega_u+\omega_s;\omega_u,\omega_s)$ directly gives the strength of the sum frequency generation process. With such fields, we must also remember that the negative of the actual frequency should be considered as well because it is included in the sums over frequencies, and so we would also have a process whose strength was given by $\chi^{(2)}(\omega_u-\omega_s;\omega_u,-\omega_s)$, which is one of the difference frequency generation terms. We can proceed with any combination of input frequencies to calculate the strengths of the processes giving rise to all of the new generated frequencies given by this second-order perturbation correction, and the analysis of such effects is a core part of nonlinear optics.

Note, incidentally, that if all the states in a system have definite parity (in the single direction we are considering), there will be exactly no second-order nonlinear optical effects. We can see this by looking at the three matrix elements. If the states have definite parity, then for μ_{qm} to be finite, states q and m must have opposite parity, and for μ_{jq} to be finite, states j and q must have opposite parity, which then means that states j and m must have the same parity, and, hence, μ_{mj} must be zero. Hence, the product of these three matrix elements is always zero if all the states have definite parity. Hence, a certain asymmetry is required in the material if the second-order effects are to be finite.

Third-order susceptibility

The approach for third-order susceptibility is similar and we do not repeat the algebra here. We merely quote the result, which is left as an exercise for the reader. We can write

$$\frac{\langle \mathsf{P}^{(3)}(t)\rangle}{\varepsilon_o} = \sum_{s,u,v} \chi^{(3)}(\omega_v+\omega_u+\omega_s;\omega_v,\omega_u,\omega_s)\mathsf{E}(\omega_v)\mathsf{E}(\omega_u)\mathsf{E}(\omega_s) \times \exp\left[-i(\omega_v+\omega_u+\omega_s)t\right] \tag{7.78}$$

in which case we would have

$$\chi^{(3)}(\omega_v+\omega_u+\omega_s;\omega_v,\omega_u,\omega_s) = \frac{1}{\varepsilon_o V}\frac{1}{\hbar^3}\sum_{k,j,q}\mu_{mk}\mu_{kj}\mu_{jq}\mu_{qm} \times \left[\begin{aligned} &\frac{1}{(\omega_{km}-\omega_v-\omega_u-\omega_s)(\omega_{jm}-\omega_u-\omega_s)(\omega_{qm}-\omega_s)} \\ &+\frac{1}{(\omega_{km}+\omega_v)(\omega_{jm}-\omega_u-\omega_s)(\omega_{qm}-\omega_s)} \\ &+\frac{1}{(\omega_{km}+\omega_v)(\omega_{jm}+\omega_v+\omega_u)(\omega_{qm}-\omega_s)} \\ &+\frac{1}{(\omega_{km}+\omega_v)(\omega_{jm}+\omega_v+\omega_u)(\omega_{qm}+\omega_v+\omega_u+\omega_s)} \end{aligned}\right] \tag{7.79}$$

We can see here, as in the second-order case, how we can calculate the strength of various third-order processes. For example, setting $\omega_v = \omega_u = \omega_s$, as would be particularly relevant if there was only one input frequency, would give the strength of the process for third harmonic generation.

Problems

7.4.1 Consider a quantum mechanical system that has effectively only two levels of interest, levels 1 and 2, separated by some energy E_{21}. We presume that each of the levels has a spatial wavefunction with a definite parity. The system is subject to an oscillating electric field of the form

$$\mathsf{E}(t) = \mathsf{E}_o\left[\exp(-i\omega t)+\exp(i\omega t)\right] = 2\mathsf{E}_o\cos(\omega t)$$

leading to a perturbing Hamiltonian, in the electric dipole approximation, of

$$\hat{H}_p(t) = e\mathsf{E}(t)z = e\mathsf{E}_o z\left[\exp(-i\omega t)+\exp(i\omega t)\right]$$

We presume that $\hbar\omega \neq E_{21}$ (so we avoid steady absorption from the radiation field and may consider the steady-state case as in the consideration of linear susceptibility or refractive index), and we take the system to be completely in the lower state in the absence of any perturbation.

(i) Show that the second-order perturbation of the upper state (state 2) is zero (or, at least, constant in time and, hence, not of interest here) (i.e., $a_2^{(2)} = 0$ or, at least, is constant)

(ii) What can you say about the parities of the two states if there is to be any response at all to the perturbing electric field?

7.4.2 (Note: This problem can be used as a substantial assignment.) Consider a quantum mechanical system in which a single electron has only three levels of interest, levels 1, 2, and 3, with energies E_1, E_2, and E_3 and spatial wavefunctions $|\psi_1\rangle$, $|\psi_2\rangle$, and $|\psi_3\rangle$, respectively. The system is initially in its lowest level (level 1). (We could imagine this system is, for example, a molecule of some kind.). The system is illuminated by a light beam of angular frequency ω, polarized in the z direction, which we can write as

$$\mathsf{E}(t) = \mathsf{E}_o\left[\exp(-i\omega t) + \exp(i\omega t)\right] = 2\mathsf{E}_o \cos(\omega t)$$

and which perturbs the molecule through the electric dipole interaction.

(i) Derive an expression for the contributions to the expectation value of the dipole moment, μ_{dip}, that are second-order in this perturbation. At least for this derivation, you may presume that all the matrix elements between the spatial parts of the wavefunctions

$$z_{ij} = \langle \psi_i | z | \psi_j \rangle$$

are finite. (Note: A term of the form $\langle \phi_i^{(1)} | z | \phi_j^{(1)} \rangle$ would give a second-order contribution, as would a term $\langle \phi_i^{(0)} | z | \phi_j^{(2)} \rangle$, whereas a term $\langle \phi_i^{(1)} | z | \phi_j^{(2)} \rangle$ would be a third-order contribution.) (This derivation may take quite a lot of algebra, though it is straightforward and does lead to a relatively simple expression in the end.)

(ii) You should find in your result to part (i) a term or set of terms corresponding to second harmonic generation; that is, a term or terms $\propto \cos(2\omega t)$. You should also have another term or set of terms that behaves differently in time. This second effect is sometimes known as *optical rectification*. What is the physical meaning of this second term; that is, if we shine such a light beam at this "molecule," what is the physical consequence that results from this term?

(iii) What will be the consequence for these second-order effects if the states all have definite parity?

(iv) Calculate the amplitudes of the second-harmonic and optical-rectification electric fields generated under the following set of conditions: Take $z_{ij} = 1$ Å for all of the z_{ij}, except we choose $z_{11} = 0$ (i.e., no static dipole in the lowest state). $E_2 - E_1 = 1$ eV, $E_3 - E_1 = 1.9$ eV, $\hbar\omega = 0.8$ eV, presume that there are 10^{19} cm^{-3} of these "molecules" per unit volume and consider an optical intensity I of the field $\mathsf{E}(t)$ of 10^{10} W/m^2. (Such an intensity corresponds to that of a 1 pJ, 1 ps long light pulse focused to a 10 x 10 micron spot, a situation easily achieved in the laboratory.) (Note that the relation between optical intensity (i.e., power per unit area) I and the amplitude E_o is

$$I = \frac{2\mathsf{E}_o^2}{Z_o}$$

where $Z_o \cong 377\Omega$. Note also that the polarization P is the same as the dipole moment per unit volume and the magnitude of a polarization can always be viewed in terms of an equivalent pair of equal and opposite surface charge densities $\pm\sigma$, unit distance apart, of magnitude $\sigma = \mathsf{P}$. The electric field from such charge densities, assuming no background dielectric constant [i.e., $\varepsilon_r = 1$] is of magnitude $\mathsf{E}_{dip} = \sigma / \varepsilon_o$, and so the electric field from a dipole moment per unit volume of magnitude P is $\mathsf{E}_{dip} = \mathsf{P} / \varepsilon_o$. This field, incidentally, is negative in direction if the dipole moment is positive in direction [positive dipole moment corresponds to positive charge on the right, negative on the left, which corresponds to a field pointing away from the positive charge].)

(v) Repeat the calculations of part (iv) but for $\hbar\omega = 0.98$ eV. (Note: You should only now need to calculate a few terms because these will dominate over all of the others.)

7.4.3 (Note: This problem can be used as a substantial assignment.) Consider a quantum mechanical system in which a single electron has only four states of interest, states 1, 2, 3, and 4, with energies E_1, E_2, E_3, and E_4 (all distinct) and spatial wavefunctions $|\psi_1\rangle$, $|\psi_2\rangle$, $|\psi_3\rangle$, and $|\psi_4\rangle$, respectively, all with definite parities. The system is initially in its lowest level (level 1). (We could imagine this system is, e.g., a molecule of some kind.) The system is illuminated by a light beam of angular frequency ω, polarized in the z direction, which we can write as

$$\mathsf{E}(t) = \mathsf{E}_o\left[\exp(-i\omega t) + \exp(i\omega t)\right] = 2\mathsf{E}_o\cos(\omega t)$$

and which perturbs the molecule through the electric dipole interaction, with perturbing Hamiltonian

$$\hat{H}_p = \hat{H}_{po}\left[\exp(-i\omega t) + \exp(i\omega t)\right]$$

where

$$\hat{H}_{po} \equiv e\mathsf{E}_o z$$

We presume that $\hbar\omega$ and its multiples (e.g., $2\hbar\omega$, $3\hbar\omega$) do not coincide with any of the energy differences between the states 1, 2, 3, and 4. Throughout this problem, use a notation where $\omega_n \equiv E_n/\hbar$ and $(E_p - E_q)/\hbar \equiv \omega_{pq}$ and $H_{pq} \equiv \langle\psi_p|\hat{H}_{po}|\psi_q\rangle$.

We are interested here in calculating the lowest order nonlinear refractive index contribution from such systems, a contribution that corresponds to a dipole μ_{dip} (strictly, its expectation value) that is third-order in the field E_o (i.e., third-order overall in the perturbation) and is at frequency ω.

As we do this, we will be considering terms up to the third order in the time-dependent perturbation expansion of the wavefunction

$$|\Psi\rangle \cong |\Psi^{(0)}\rangle + |\Psi^{(1)}\rangle + |\Psi^{(2)}\rangle + |\Psi^{(3)}\rangle$$

where $|\Psi^{(n)}\rangle \equiv \sum_{s=1}^{4} a_s^{(n)}(t)\exp(-i\omega_s t)|\psi_s\rangle$.

For simplicity in handling this problem, we choose $E_1 = 0$ (which we can do arbitrarily because this is simply an energy origin for the problem). Given this choice and our choice of state 1 as the starting state, we have $|\Psi^{(0)}\rangle = |\psi_1\rangle$; that is, $a_1^{(0)} = 1$, $a_2^{(0)} = a_3^{(0)} = a_4^{(0)} = 0$.

(i) Show that the expression for the first-order expansion coefficients $a_j^{(1)}(t)$ is

$$a_j^{(1)}(t) = -\frac{H_{j1}}{\hbar}\left\{\frac{\exp\left[i(\omega_{j1}-\omega)t\right]}{\omega_{j1}-\omega} + \frac{\exp\left[i(\omega_{j1}+\omega)t\right]}{\omega_{j1}+\omega}\right\}.$$

(Note: In integrating over time to get the required result, you may neglect the formal constant of integration because we are only interested in the time varying parts here.)

(ii) Now show that the expression for the second-order expansion coefficients $a_m^{(2)}(t)$ is

$$a_m^{(2)}(t) = \frac{1}{\hbar^2}\sum_{j=1}^{4} H_{mj}H_{j1}$$
$$\times\left\{\frac{\exp\left[i(\omega_{m1}-2\omega)t\right]}{(\omega_{j1}-\omega)(\omega_{m1}-2\omega)} + \frac{\exp(i\omega_{m1}t)}{(\omega_{j1}-\omega)\omega_{m1}} + \frac{\exp\left[i\omega_{m1}t\right]}{(\omega_{j1}+\omega)\omega_{m1}} + \frac{\exp\left[i(\omega_{m1}+2\omega)t\right]}{(\omega_{j1}+\omega)(\omega_{m1}+2\omega)}\right\}$$

noting that successive orders in the perturbation calculation of the state can be calculated from the previous one. (Note that $\omega_{mj} + \omega_{j1} = \omega_{m1}$, which may keep the algebra slightly simpler.) (For simplicity here and later, you may ignore the formal problem that $\omega_{11} = 0$.)

(iii) Now, similarly, show that the expression for the third-order expansion coefficients $a_q^{(3)}(t)$ is

$$a_q^{(3)} = \frac{-1}{\hbar^3}\sum_{m=1}^{4}\sum_{j=1}^{4} H_{qm}H_{mj}H_{j1}$$
$$\times\left\{\frac{\exp\left[i(\omega_{q1}-3\omega)t\right]}{(\omega_{j1}-\omega)(\omega_{m1}-2\omega)(\omega_{q1}-3\omega)} + \frac{\exp\left[i(\omega_{q1}-\omega)t\right]}{(\omega_{j1}-\omega)(\omega_{m1}-2\omega)(\omega_{q1}-\omega)}\right.$$
$$+\frac{\exp\left[i(\omega_{q1}-\omega)t\right]}{(\omega_{j1}-\omega)\omega_{m1}(\omega_{q1}-\omega)} + \frac{\exp\left[i(\omega_{q1}+\omega)t\right]}{(\omega_{j1}-\omega)\omega_{m1}(\omega_{q1}+\omega)} + \frac{\exp\left[i(\omega_{q1}-\omega)t\right]}{(\omega_{j1}+\omega)\omega_{m1}(\omega_{q1}-\omega)} + \frac{\exp\left[i(\omega_{q1}+\omega)t\right]}{(\omega_{j1}+\omega)\omega_{m1}(\omega_{q1}+\omega)}$$
$$\left.+\frac{\exp\left[i(\omega_{q1}+\omega)t\right]}{(\omega_{j1}+\omega)(\omega_{m1}+2\omega)(\omega_{q1}+\omega)} + \frac{\exp\left[i(\omega_{q1}+3\omega)t\right]}{(\omega_{j1}+\omega)(\omega_{m1}+2\omega)(\omega_{q1}+3\omega)}\right\}$$

(iv) Now derive an expression for the contributions to the expectation value of the dipole moment, $\mu_{dip} \equiv e\langle\Psi|z|\Psi\rangle$, that are third-order in this perturbation and that are oscillating at frequencies ω or $-\omega$. (Note: A term of the form $\langle\Psi^{(1)}|z|\Psi^{(1)}\rangle$ would give a second-order contribution, as would a term $\langle\Psi^{(0)}|z|\Psi^{(2)}\rangle$, whereas a term $\langle\Psi^{(1)}|z|\Psi^{(2)}\rangle$ or a term $\langle\Psi^{(0)}|z|\Psi^{(3)}\rangle$ would be a third-order contribution. Note also that $\langle\Psi^{(n)}|z|\Psi^{(r)}\rangle = \langle\Psi^{(r)}|z|\Psi^{(n)}\rangle^*$ and that $e\langle\psi_p|z|\psi_q\rangle = H_{pq}/\mathsf{E}_o$. Note too that because by choice the states have definite parities, then $H_{pp} = 0$ for any p, which should help eliminate some terms in the sums here.)

(v) Now we restrict the problem to one where the parities of states 1 and 4 are the same and the parities of states 2 and 3 are the same as each other but opposite to those of states 1 and 4. For simplicity, we also choose all the nonzero perturbation matrix elements to be equal. Hence we have

$$H_{11} = H_{22} = H_{33} = H_{44} = H_{14} = H_{23} = 0\,;\; H_{12} = H_{13} = H_{24} = H_{34} = H_D \equiv e\mathsf{E}_o z_o$$

where H_D and z_o are constants characteristic of the specific system. We also presume that ω_{41} is very close (but not equal to) 2ω, though we also assume that ω_{21} and ω_{31} are not very close to ω (or $-\omega$). As a result, we may retain only the terms for which there is a term $(\omega_{41} - 2\omega)$ or its equivalent in the denominator. (Ignore again the formal problem that $\omega_{11} = 0$. In a full analysis, these terms cancel out, avoiding the apparent singularity here.)

(a) Write out the expression for μ_{dip} with these assumptions.

(b) Choose $\hbar\omega_{21} \equiv$ 3 eV, $\hbar\omega_{31} \equiv$ 3.5 eV, $\hbar\omega_{41} \equiv$ 5.05 eV, $\hbar\omega \equiv$ 2.5 eV, and $z_o =$ 1 Å. Presume that there are 10^{22} cm^{-3} of such "molecules" per unit volume. Calculate the nonlinear refraction coefficient n_2, which is the change of refractive index per unit optical intensity. (Note that (1) the polarization P is the same as the dipole moment per unit volume, (2) $\mathsf{P} = \varepsilon_o \chi \mathsf{E}$, where $\varepsilon_o \simeq$ 8.85 x 10^{-12} F/m, and we assume for simplicity here that the only contribution to P is the nonlinear contribution we are calculating [there is also a linear contribution which we could also calculate, though we neglect it for simplicity here], (3) the relative dielectric constant $\varepsilon_r = 1 + \chi$, (4) the refractive index $n = \sqrt{\varepsilon_r}$, (5) the relation between optical intensity [i.e., power per unit area] I and the amplitude E_o is [strictly, in free space, but we use this relation here for simplicity] $I = 2\mathsf{E}_o^2 / Z_o$ where $Z_o \cong 377\Omega$.)

(c) Suppose we imagine that this model approximately describes nonlinear refractive index in a hypothetical optical fiber intended for nonlinear optical switching applications. In such a fiber, the cross-sectional size of the optical mode is ~ 10 x 10 μm^2 and launched power into the fiber could be ~ 10 mW. Nonlinear refractive effects can have quite substantial consequences if they lead to changes in optical path length in the total length l of the fiber is of the order of half a wavelength. The change in index is ~ $n_2 I$, and the resulting change in optical path length is ~ $n_2 I l$. Considering a photon energy of 2.5 eV (as previously) with 10 mW power, what length l of fiber would lead to half a wavelength of optical path length change from this nonlinear effect?

(d) Repeat the calculation of part (b), changing $\hbar\omega_{31}$ to 2.0 eV. Explain your result.

7.5 Summary of concepts

Time-dependent perturbation theory

In time-dependent perturbation theory, the Hamiltonian is presumed to be the sum of a time-independent Hamiltonian, $\hat{H}_o$, whose eigenenergies E_n and eigenfunctions $|\psi_n\rangle$ are presumed known, and a time-dependent perturbation, $\hat{H}_p(t)$. The perturbed wavefunction is presumed to be written as

$$|\Psi\rangle = \sum_n a_n(t)\exp(-iE_n t/\hbar)|\psi_n\rangle \tag{7.4}$$

The time derivatives of the first-order corrections to the expansion coefficients are shown to be

$$\dot{a}_q^{(1)}(t) = \frac{1}{i\hbar}\sum_n a_n^{(0)} \exp(i\omega_{qn}t)\langle\psi_q|\hat{H}_p(t)|\psi_n\rangle \tag{7.10}$$

where $\omega_{qn} = (E_q - E_n)/\hbar$.

Higher order perturbation corrections to the state are derived from the immediately preceding correction using

$$\dot{a}_q^{(p+1)}(t) = \frac{1}{i\hbar}\sum_n a_n^{(p)} \exp(i\omega_{qn}t)\langle\psi_q|\hat{H}_p(t)|\psi_n\rangle \tag{7.13}$$

Fermi's Golden Rule

For an oscillating perturbation of the form

$$\hat{H}_p(t) = \hat{H}_{po}\left[\exp(-i\omega t) + \exp(i\omega t)\right] \qquad \text{after Eq. (7.15)}$$

the transition rate from a state $|\psi_m\rangle$ to a state $|\psi_j\rangle$ can be written as

$$w_{jm} = \frac{2\pi}{\hbar}\left|\langle\psi_j|\hat{H}_{po}|\psi_m\rangle\right|^2 \delta(E_{jm} - \hbar\omega) \tag{7.32}$$

or, equivalently, when there is a density of similar transitions per unit transition energy of $g_J(\hbar\omega)$, the total transition rate per unit transition energy is given as

$$W = \frac{2\pi}{\hbar}\left|\langle\psi_j|\hat{H}_{po}|\psi_m\rangle\right|^2 g_J(\hbar\omega) \tag{7.31}$$

Chapter 8

Quantum mechanics in crystalline materials

Prerequisites: Chapters 2–7, including the discussion of periodic boundary conditions in Section 5.4.

One of the most important practical applications of quantum mechanics is the understanding and engineering of crystalline materials. Of course, the full understanding of crystalline materials is a major part of solid-state physics and merits a much longer discussion than we give here. We will, however, try to introduce some of the most basic quantum mechanical principles and simplifications in crystalline materials. This will also allow us to perform many quantum mechanical calculations of important processes in semiconductors.

8.1 Crystals

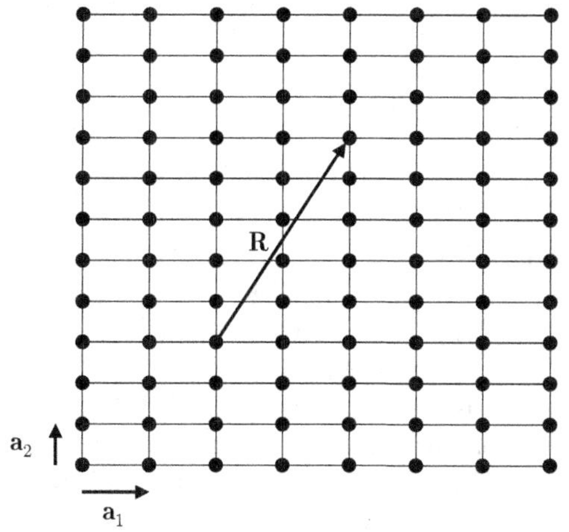

Fig. 8.1. Illustration of a simple rectangular lattice in two dimensions.

A crystal is a material whose measurable properties are periodic in space. We can think about it using the idea of a *unit cell*. If we think of the unit cell as a "block" or "brick," then a definition of a crystal structure is one that can fill all space by the regular stacking of these unit cells. If we imagine that we marked a black spot on the same position of the surface of each block, these spots or points would form a crystal lattice. We can, if we wish, define a set of vectors, $\mathbf{R}_L$, that we call *lattice vectors*. The set of lattice vectors consists of all of the vectors that link points on this lattice; that is,

$$\mathbf{R}_L = n_1\mathbf{a}_1 + n_2\mathbf{a}_2 + n_3\mathbf{a}_3 \tag{8.1}$$

Here, $\mathbf{a}_1$, $\mathbf{a}_2$, and $\mathbf{a}_3$ are the three linearly independent vectors[1] that take us from a given point in one unit cell to the equivalent point in the adjacent unit cell. In a simple cubic lattice, these three vectors lie along the x, y, and z directions. The numbers n_1, n_2, and n_3 range through all (positive and negative) integer values. Fig. 8.1 illustrates a simple rectangular lattice in two dimensions. In three dimensions, there are only fourteen distinct kinds of mathematical lattices of points (i.e., Bravais lattices) that can be made that fill all space by the stacking of identical blocks.[2]

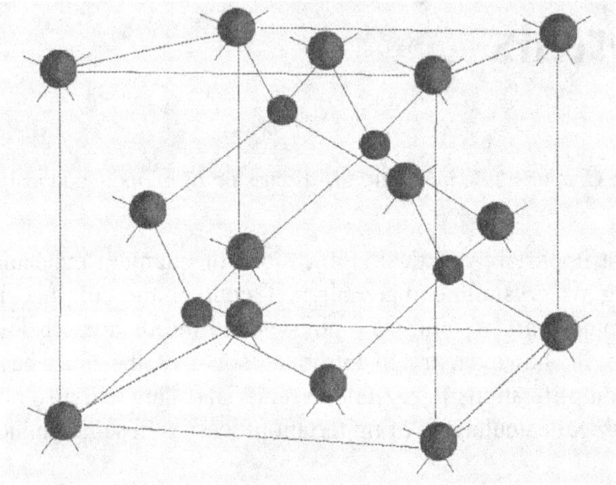

Fig. 8.2. Illustration of a zinc-blende lattice – i.e., two interlocking face-centered cubic lattices. The larger atoms are clearly on a face-centered cubic lattice. The smaller atoms are also on a face-centered cubic lattice that is displaced with respect to the other lattice, though this second lattice structure may be less obvious from the illustration. The diamond lattice has the same structure, but then both sets of atoms are of the same type. Each atom of one type is connected by bond lines to four other atoms of the other type and those four atoms lie on the corners of a regular tetrahedron (i.e., a four-sided structure with equal [triangular] sides). The bond lines are shown, as well as the edges of a cube (these edges are not bond lines).

We do not concern ourselves here with the details of the various crystal lattices and their properties and, for simplicity, we imagine that we are dealing with a cubic kind of lattice. A large fraction of the semiconductor materials of practical interest, such as silicon, germanium, and most of the III-V (e.g., GaAs) and II-VI (e.g., ZnSe) materials, have a specific form of cubic lattice. This lattice is based on two interlocking face-centered cubic lattices. A *face-centered cubic lattice* is one with an atom on each corner of a cube plus one in the middle of each face. In the case of the III-V and II-VI materials of this type, which are known as *zinc-blende structure materials*, the Group III (or II) atoms lie on one such face-centered cubic lattice and the Group V (or VI) lie on the interlocking face-centered cubic lattice. In the case of the Group IV materials (e.g., silicon, germanium), both interlocking lattices, of course, have the same atoms on them and this structure is called the *diamond lattice*. The basic quantum

---

[1] Linear independence means that none of these vectors can be expressed as any kind of linear combination of the other vectors. In practice, this means that we do not have all three vectors in the same plane.

[2] A group of atoms can be associated with each mathematical lattice point. That group can have its own symmetry properties, so there are more possible crystals than there are Bravais lattices.

mechanical properties we discuss here can, however, be generalized to all the Bravais lattices. The zinc-blende lattice is illustrated in Fig. 8.2.

8.2 One electron approximation

A crystal may well have a very large number of atoms or unit cells in it (e.g., 10^{23}). How can we start to deal with such a complex system? One key approximation is to presume that any given electron sees a potential $V_P(\mathbf{r})$ from all the other nuclei and electrons that is periodic with the same periodicity as the crystal lattice. Because of that presumed periodicity, we have

$$V_P(\mathbf{r}+\mathbf{R}_L) = V_P(\mathbf{r}) \tag{8.2}$$

In this potential, the charged nuclei are presumed to be fixed and the charge distribution from all the other electrons is also presumed to be effectively fixed, giving this net periodic potential.

It is important to recognize that this is an approximation. In truth, any given electron does not quite see a periodic potential from all the other nuclei and electrons because it interacts with them, pulling them slightly out of this hypothetical periodic structure. We treat those interactions, when we have to, as perturbations, however, so we start by pretending we can use this perfectly periodic potential to model the potential any one electron sees in this crystalline material.

If we do have to examine the interactions between the electron states and the nuclei that perturb this perfectly crystalline world, we describe nuclear motions through the modes of vibration of periodic collections of masses connected by the "springs" of chemical bonds. These modes are called *phonon modes*. Just as photons are the quanta associated with electromagnetic modes, so phonons are the quanta associated with these mechanical modes. The interaction of the electrons with the nuclei is, therefore, described through electron–phonon interactions, for example. Any given electron state, in reality, also interacts with other electrons and we can describe that through various electron-electron interactions such as electron–electron scattering. There are also many other interactions between particles that we can consider in solid-state physics. These interactions are very often handled as perturbations, starting with the one-electron model results as the unperturbed solutions.

In this one-electron approximation, we presume, then, that we can write an effective, approximate Schrödinger equation for the one electron in which we are interested; for example, neglecting magnetic effects for simplicity, we have

$$-\frac{\hbar^2}{2m_e}\nabla^2\psi(\mathbf{r}) + V_P(\mathbf{r})\psi(\mathbf{r}) = E\psi(\mathbf{r}) \tag{8.3}$$

8.3 Bloch theorem

The Bloch theorem is a very important simplification for crystalline structures. It essentially enables us to separate the problem into two parts: one that is the same in every unit cell and one that describes global behavior. For simplicity, we prove this theorem in one direction and then generalize to three dimensions. We know that the crystal is periodic, having the same potential at $x+sa$ as it has at x (where s is an integer). Any observable quantity must also have the same periodicity because the crystal must look the same in every unit cell. For example, charge density $\rho \propto |\psi|^2$ must be periodic in the same way. Hence

$$|\psi(x)|^2 = |\psi(x+a)|^2 \tag{8.4}$$

which means

$$\psi(x+a) = C\psi(x) \tag{8.5}$$

where C is a complex number of unit amplitude. Note that there is no requirement that the wavefunction itself be periodic with the crystal periodicity because it is not apparently an observable or measurable quantity.

To proceed further, we need to introduce boundary conditions on the wavefunction. As is often the case, the boundary conditions lead to the quantization of the problem. The introduction of boundary conditions for the crystal is a tricky problem. We have mathematically idealized the problem by presuming that the crystal is infinite, so we could get a simple statement of periodicity, such as Eq. (8.2) with the simple definition of the lattice vectors, Eq. (8.1). But, in practice, we know all crystals are actually finite. How can we introduce the concept of the finiteness of the crystal and corresponding finite countings of states without having to abandon our simple description in terms of infinite periodicity?

In one dimension, we could argue as follows. Suppose that we had a very long chain of N equally spaced atoms and that we joined the two ends of the chain together. Interpreting distance x as the distance along this loop, we could have the simple kind of definition of periodicity we have used previously – for example, $V_P(x+ma) = V_P(x)$, where m is any integer (even much larger than N). We could argue that if this chain is very long, we do not expect that its internal properties will be substantially different from an infinitely long chain and so this finite system is a good model. Such a loop introduces a boundary condition, however. We do expect that the wavefunction is a single-valued function (otherwise, how could we differentiate it, evaluate its squared modulus, and so forth), so when we go around the loop, we must get back to where we started; that is, explicitly

$$\psi(x) = \psi(x+Na) \tag{8.6}$$

This is known as a *periodic boundary condition*[3] (also known as a *Born–von Karman boundary condition*). Combining this with our condition Eq. (8.5), we have

$$\psi(x) = \psi(x+Na) = C^N\psi(x) \tag{8.7}$$

so

$$C^N = 1 \tag{8.8}$$

and so C is one of the N roots of unity; that is,

$$C = \exp(2\pi is/N);\ s = 0, 1, 2, \ldots N-1 \tag{8.9}$$

(We could also choose

$$C = \exp\left(2\pi i\left(\frac{s}{N}+m\right)\right);\ s = 0, 1, 2, \ldots N-1,\ m \text{ any integer} \tag{8.10}$$

so there is some arbitrariness here, though this will not matter in the end.)

---

[3] See the discussion of periodic boundary conditions in Section 5.4.

Substituting C from Eq. (8.9) in Eq. (8.7)

$$\psi(x+a)=\exp(ika)\psi(x) \tag{8.11}$$

where we could choose

$$k=\frac{2\pi s}{Na};\ s=0,1,2,\ldots N-1 \tag{8.12}$$

Note, we could also choose

$$k=\frac{2\pi s}{Na}+\frac{2m\pi}{a};\ s=0,1,2,\ldots N-1 \tag{8.13}$$

Conventionally, we choose

$$k=\frac{2\pi n}{Na}\quad \ldots n=0,\pm 1,\pm 2,\ldots \pm N/2 \tag{8.14}$$

which still gives essentially N states,[4] but now symmetrically disposed about $k=0$. Note that the allowed k values are evenly spaced by $2\pi/L$, where $L=Na$ is the length of the crystal in this dimension,[5] regardless of the detailed form of the periodic potential. Hence, one form of the Bloch theorem in one dimension is the statement that the wavefunction in a crystal can be written in the form Eq. (8.11), subject to the condition Eq. (8.14) on the allowed k values.

There is another more common and often more useful way to state the Bloch theorem. We multiply Eq. (8.11) by $\exp[-ik(x+a)]$ to obtain

$$\psi(x+a)\exp(-ik(x+a))=\psi(x)\exp(-ikx) \tag{8.15}$$

Hence, if we define a function

$$u(x)=\psi(x)\exp(-ikx) \tag{8.16}$$

we can restate Eq. (8.15) as

$$u(x+a)=u(x) \tag{8.17}$$

and, hence, $u(x)$ is periodic with the lattice periodicity. (By obvious extension of Eq. (8.17), we conclude $u(x+2a)=u(x+a)=u(x)$, and so on for every unit cell in our crystal.) Hence, we can rewrite the Bloch theorem Eq. (8.11) in the alternative form

$$\psi(x)=u(x)\exp(ikx) \tag{8.18}$$

where $u(x)$ is periodic with the lattice periodicity. (Note that the two forms Eq. (8.11) and Eq. (8.18) are entirely equivalent – we have just proved that Eq. (8.11) implies Eq. (8.18) and it is trivial to show by mere substitution that Eq. (8.18) implies Eq. (8.11).)

---

[4] Strictly, of course, Eq. (8.14) gives us $N+1$ states and we should actually take off one or the other of the end points (i.e., $+N/2$ or $-N/2$) to get a rigorous result; but, despite that, the result is commonly written as in Eq. (8.14). If that counting really matters, as it might in some short chain, then we should be rigorous here, removing one of these two from the counting.

[5] Note that we now take the length of the crystal to be the length of the loop presumed in the periodic boundary conditions.

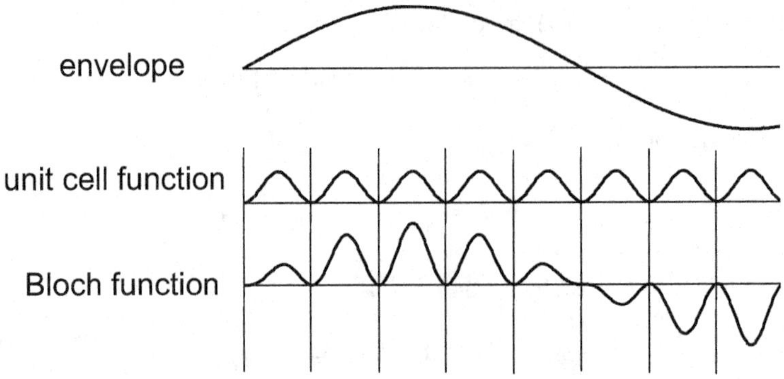

Fig. 8.3. Illustration of the Bloch function as a product of an envelope function, which is a sinusoidal wave, and a unit cell function, which is the same in every unit cell. (Here, we show the real part of both functions.)

Fig. 8.3 illustrates the concept of the Bloch functions in the form of Eq. (8.18). We can think of the $\exp(ikx)$ as being an example of an "envelope" function that multiplies the unit cell function $u(x)$.

In three dimensions, we can follow similar arguments. Periodic boundary conditions are then a strange concept if we treat them too literally – we would then need to imagine a crystal where each face is joined to the opposite one in a long loop, something we cannot do in three dimensions. Despite this absurdity, we do use periodic boundary conditions in three dimensions as being a way that allows our simple definition of periodicity and yet correctly counts the available states.[6]

The Bloch theorem in three dimensions is otherwise a straightforward extension of the one-dimensional version. We have

$$\psi(\mathbf{r}+\mathbf{a}) = \exp(i\mathbf{k}.\mathbf{a})\psi(\mathbf{r}) \tag{8.19}$$

or, equivalently

$$\psi(\mathbf{r}) = u(\mathbf{r})\exp(i\mathbf{k}.\mathbf{r}) \tag{8.20}$$

where $\mathbf{a}$ is any crystal lattice vector and $u(\mathbf{r}) = u(\mathbf{r}+\mathbf{a})$. Considering the three crystal basis vector directions, 1, 2, and 3, with lattice constants (repeat distances) a_1, a_2, and a_3, and numbers of atoms N_1, N_2, and N_3

$$k_1 = \frac{2\pi n_1}{N_1 a_1} \quad \ldots n_1 = 0, \pm 1, \pm 2, \ldots \pm N_1/2 \tag{8.21}$$

and similarly for the other two components of $\mathbf{k}$ in the other two crystal basis vector directions. Note that the number of possible values of $\mathbf{k}$ is the same as the number of atoms in the crystal.[7] Eqs. (8.20) and (8.21), therefore, constitute the Bloch theorem result that the electron wavefunction can be written in this form in a crystalline material.

[6] Again, see the discussion of periodic boundary conditions in Section 5.4.

[7] Again, strictly, with the definition in Eq. (8.21) we would actually appear to have $N+1$ allowed values of k in a given direction for a crystal N atoms thick in that direction. This is actually not correct and must be dealt with properly for some short chain, but it does not matter for large crystals.

Problem

8.3.1 Consider a ring of six identical "atoms" or "unit cells" (e.g., like a benzene ring) with a repeat length (center to center distance between the atoms) of 0.5 nm. Explicitly, write out each of the different allowed Bloch forms (i.e., $\psi(v) = u(v)\exp(ikv)$) for the effective one-dimensional electron wavefunctions $\psi(v)$, where v is the distance coordinate as we move around the ring and $u(v)$ is a function that is periodic with period 0.5 nm. Be explicit about the numerical values and units of the allowed values of k.

8.4 Density of states in *k*-space

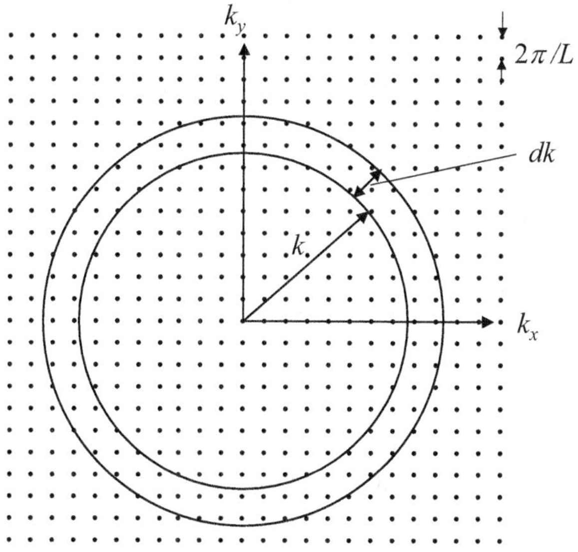

Fig. 8.4. Illustration of the allowed states in k-space, shown here in a two-dimensional section, with the allowed values illustrated by dots. L is the linear size of the "box" in real space (assumed the same in all three directions). Also shown is a thin annulus (or spherical shell) of radius k and thickness dk, as used in the calculation of the density of states in energy.

We see that the allowed values of k_1, k_2, and k_3 are each equally spaced, with separations

$$\delta k_1 = \frac{2\pi}{N_1 a_1} = \frac{2\pi}{L_1}, \ \delta k_2 = \frac{2\pi}{N_2 a_2} = \frac{2\pi}{L_2}, \text{ and } \delta k_3 = \frac{2\pi}{N_3 a_3} = \frac{2\pi}{L_3} \tag{8.22}$$

respectively, along the three axes.[8] Note that the lengths of the crystal along the three axes are, respectively, $L_1 = N_1 a_1$, $L_2 = N_2 a_2$, and $L_3 = N_3 a_3$. We could draw a three-dimensional diagram, with axes k_1, k_2, and k_3, and mark the allowed values of $\mathbf{k}$. We would then have a set of dots in space that themselves constitute a mathematical lattice. Fig. 8.4 shows this concept in a two-dimensional cross-section. This lattice is an example of a *reciprocal lattice*, which is a lattice in so-called k-space. We can then imagine that each point has a volume surrounding it,

[8] The axis directions in k-space are not, in general, the same as the axis directions in the real-space lattice, though they are the same for cubic lattices. The lattice vectors in k-space (i.e., the vectors that separate adjacent points in k-space) are, more formally, $\mathbf{b}_1 = 2\pi(\mathbf{a}_2 \times \mathbf{a}_3)/(\mathbf{a}_1 \cdot (\mathbf{a}_2 \times \mathbf{a}_3))$ and the cyclic permutations of this for the other directions.

with these volumes touching one another to completely fill all the space. For our cubic lattices, these volumes in k-space are of size $\delta V_k = \delta k_1 \delta k_2 \delta k_3$[9]; that is,

$$\delta V_k = \frac{(2\pi)^3}{V} \tag{8.23}$$

where $V = L_1 L_2 L_3$ is the volume of our crystal. Hence, the density of states in k-space is $1/\delta V_k$. Note that this density grows as we make the crystal larger. It is often more useful to work with the density of states per unit volume of the crystal. Hence we have the density of states in k-space per unit volume of the crystal

$$g(\mathbf{k}) = \frac{1}{(2\pi)^3} \tag{8.24}$$

The density of states is a very useful quantity for actual quantum mechanical calculations in crystalline materials.

Problem

8.4.1 A two-dimensional crystal has a rectangular unit cell, with spacings between the centers of the atoms of 0.5 nm in one direction (e.g., the x direction) and 0.4 nm in the other direction (e.g., the y direction). Presuming that the crystal has 1,000 unit cells in each direction, sketch a representative portion of the reciprocal lattice of all the k-states in k-space on a scale drawing, showing the dimensions, units, and directions (i.e., k_x and k_y).

8.5 Band structure

If we knew the potential $V_P(\mathbf{r})$ and could solve the one-electron Schrödinger equation (8.3), we could calculate the energies E of all of the various possible states. There are several ways of approaching such calculations from first principles and we do not go into those here. The results of such calculations give what is known as a *band structure*.

There are multiple bands in a band structure (in fact, an infinite number), but usually only a few are important in determining particular properties of a material. Fig. 8.5 illustrates a simple band structure. Each band has a total number of allowed k-states equal to the number of unit cells[10] in the crystal. These states are evenly spaced in k-space, as discussed previously. Each band loosely corresponds to a different atomic state in the constituent atoms – the bands can be viewed as being formed from the atomic states as the atoms are pushed together into the crystal.

Fig. 8.5 illustrates a simplified band structure, similar to structures encountered with some semiconductors. In each band, we only have to plot k-values from $-\pi/a$ to π/a. (This range is sometimes known as the [first] Brillouin zone.) The bands are usually plotted as if they were continuous but, in fact, k can only take on discrete (though evenly spaced) values; hence, the bands are really a set of very closely spaced points on this diagram. The lower band is like the

---

[9] These volumes are the unit cells of this reciprocal lattice. Note that in solid state physics, when we use the term "the reciprocal lattice" we usually mean the lattice that has complete Brillouin zones as the unit cells, not just these volumes round a k-state. Both of these are still reciprocal lattices, however.

[10] Strictly, we need to say the number of "primitive" unit cells, where the primitive unit cells are the smallest (in volume) unit cells we can construct.

highest valence band in a semiconductor and, in the unmodified or unexcited semiconductor, it is typically full of electrons. The upper band is like the lowest conduction band in some semiconductors and, again in the unmodified or unexcited semiconductor, it is typically empty of electrons. E_G is the band gap energy that separates the lowest point in the conduction band from the highest point in the valence band. The particular band structure in Fig. 8.5 corresponds to what is called a *direct gap semiconductor*; the lowest point in the conduction band is directly above the highest point in the valence band. Many III-V and II-VI semiconductors are of this type. It is also very common for there to be minima or maxima in the bands at $k = 0$. Also shown in the conduction band in Fig. 8.5 are subsidiary minima away from $k = 0$. It is possible that these minima, rather than any minimum at $k = 0$, are the lowest points in a semiconductor conduction band structure, in which case we have an *indirect gap semiconductor*. Silicon and germanium are both indirect gap semiconductors.

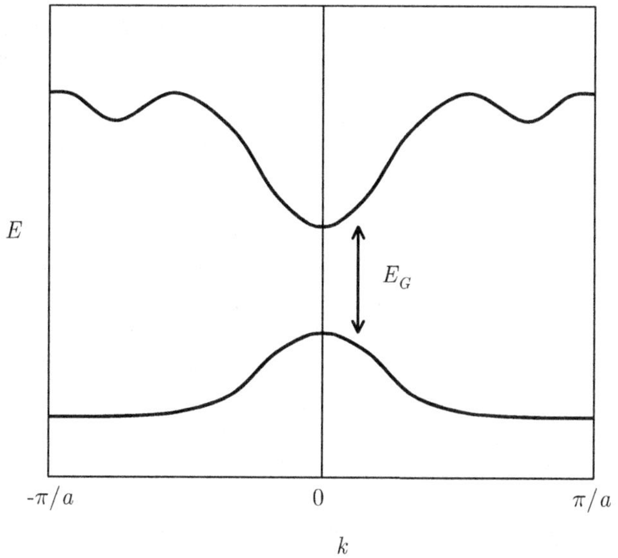

Fig. 8.5. Figurative illustration of a semiconductor band structure, plotted along one crystal direction. The upper "band" (line) is essentially empty of electrons and is called the *conduction band*; the lower band is essentially full of electrons and is called the *valence band*.

The band structure in Fig. 8.5 is also drawn to be symmetric about $k = 0$. Band structures are often symmetric in this way. In our simple one-electron model, neglecting magnetic effects, the existence of symmetries like this is easily proved. Suppose that $\psi(\mathbf{k},\mathbf{r}) = u_{\mathbf{k}}(\mathbf{r})\exp(i\mathbf{k}.\mathbf{r})$ is the Bloch function that satisfies the Schrödinger equation for the specific allowed value $\mathbf{k}$. We have now introduced $\mathbf{k}$ as an explicit notation in our Bloch function parts. (Note, incidentally, that the unit-cell part of the wavefunction, $u_{\mathbf{k}}(\mathbf{r})$, is, in general, different for every different $\mathbf{k}$.) Hence, we have

$$\hat{H}\psi(\mathbf{k},\mathbf{r}) = E_{\mathbf{k}}\psi(\mathbf{k},\mathbf{r}) \tag{8.25}$$

where $E_{\mathbf{k}}$ is the eigenenergy associated with this specific $\mathbf{k}$ and $\hat{H} = -(\hbar^2/2m_e)\nabla^2 + V_P(\mathbf{r})$. Now, take the complex conjugate of both sides of Eq. (8.25). We note that $\hat{H}$ expressed in this way does not change when we take the complex conjugate and we also know that $E_{\mathbf{k}}$ is real because it is the eigenvalue associated with a Hermitian operator. Hence, we have

$$\hat{H}\psi^*(\mathbf{k},\mathbf{r}) = E_{\mathbf{k}}\psi^*(\mathbf{k},\mathbf{r}) \tag{8.26}$$

But, $\psi^*(\mathbf{k},\mathbf{r}) = u_\mathbf{k}^*(\mathbf{r})\exp(-i\mathbf{k}.\mathbf{r})$, which is also a wavefunction in Bloch form, but for wavevector $-\mathbf{k}$. Hence, we are saying that for every Bloch function solution with wavevector **k** and energy $E_\mathbf{k}$, there is one with wavevector $-\mathbf{k}$ with the same energy. Hence the band structure is symmetric about $k = 0$.[11] We can, if we wish, choose to write

$$\psi^*(\mathbf{k},\mathbf{r}) = u_\mathbf{k}^*(\mathbf{r})\exp(-i\mathbf{k}.\mathbf{r}) \equiv u_{-\mathbf{k}}(\mathbf{r})\exp(-i\mathbf{k}.\mathbf{r}) = \psi(-\mathbf{k},\mathbf{r}) \tag{8.27}$$

This equivalence of the energies for **k** and $-\mathbf{k}$ is known as *Kramers degeneracy*.

Problem

8.5.1 Conventionally, we express Bloch functions within the first Brillouin zone, which for a simple one-dimensional crystal of repeat length a is the range $-\pi/a \le k \le \pi/a$. We could instead consider k values lying outside this range. Show, however, that any such Bloch function (i.e., a function of the form $\psi(x) = u(x)\exp(ikx)$, where $u(x)$ is a function periodic with repeat length a) for any k_{new} outside this first Brillouin zone can also be expressed as a Bloch function with a k value inside the first Brillouin zone. (Hint: Any k_{new} lying outside the first Brillouin zone can be written as $k_{new} = k + 2n\pi/a$ for some positive or negative integer n.)

8.6 Effective mass theory

As mentioned previously, it is very common to have minima or maxima in the bands at $\mathbf{k} = 0$. It is also quite common to have other minima or maxima in the band structure, as indicated in Fig. 8.5. The minima in the conduction band and the maxima in the valence band are very important in the operation of both electronic and optoelectronic semiconductor devices. Any extra electrons in the conduction band tend to fall into the lowest minimum. Any absences of electrons in the valence band tend to "bubble up" to the highest maximum in the valence band. Such absences of electrons are often described as if they were positively charged "holes." As a result, the properties of most electronic devices and many optoelectronic devices (especially light emitting devices, which involve recombination of electrons in the conduction band with holes in the valence band) are dominated by what happens in these minima and maxima. It is also the case in optoelectronics that many other devices, such as some optical modulators, work for photon energies very near to the band gap energy, E_G, and their properties are also determined by the behavior of electrons and holes in these minima and maxima.

Because of the importance of these minima and maxima in the band structure, it is very useful to have approximate models that give simplified descriptions of what happens in these regions. One of these models is the *effective mass approximation*.

We would expect, near a minimum or maximum, that to lowest order, the energy $E_\mathbf{k}$ would vary quadratically as **k** is varied along some direction in k-space. For simplicity here, we

---

[11] We have not yet included electron spin. The consequent more sophisticated version of the Kramers degeneracy proof involves time-reversing the Schrödinger equation (including spin) rather than merely taking the complex conjugate. In this case, the Kramers degeneracy remains, though we then find that the state with spin up at **k** is degenerate with the state with spin down at $-\mathbf{k}$ and vice versa. See O. Madelung, *Introduction to Solid State Theory* (Springer-Verlag, Berlin, 1978), p 91.

Often, the fact that the band structure is symmetric about $k = 0$ means that we have maxima or minima in the bands there, as in Fig. 8.5. The required symmetry can also, however, be provided by having "mirror image" bands cross at $k = 0$. It is the band structure overall that is symmetric, not necessarily a given band.

presume that this variation is isotropic and, also for simplicity, we presume that the minimum or maximum of interest is located at $k = 0$. Neither of these simplifications is necessary for this effective mass approach, though it keeps our algebra simpler. This isotropic $k = 0$ minimum or maximum is a good approximation for the lowest conduction band and a reasonable first approximation for the highest valence bands, in the direct gap semiconductors that are important in optoelectronics (e.g., GaAs, InGaAs). For the lowest conduction bands in silicon or germanium or other indirect gap semiconductors (e.g., AlAs), the minima are not at $k = 0$ and, though approximately parabolic in any given direction, they are not isotropic; the theory is, however, easily extended to cover those cases.

If the energy at the minimum or maximum itself is some amount V, then by assumption, we have $E_{\mathbf{k}} - V \propto k^2$. For reasons that will become obvious, we choose to write this as

$$E_{\mathbf{k}} = \frac{\hbar^2 k^2}{2m_{eff}} + V \tag{8.28}$$

where the quantity m_{eff} is, for the moment, merely a parameter that sets the appropriate proportionality. (Of course, as the reader has probably guessed, m_{eff} will turn out to be the quantity we call the effective mass.) A relation such as Eq. (8.28) between energy and k-value is sometimes called a *dispersion relation*. This particular approximation for the behavior of the energies in a band is called, for obvious reasons, an *isotropic parabolic band.*

The reader will note, incidentally, that the effective mass is actually a negative number for the case of electrons near the top of the valence band in Fig. 8.5, because the band is curved downward. At least in a semiconductor, however, this valence band is always almost entirely full of electrons and we are more interested in what happens to the few states in which the electrons are missing. We can, in fact, treat those states of missing electrons (i.e., holes) as if they were positively charged and as if they had a positive mass. The precise reasons why we can make this conceptual change are actually quite subtle. They can be deduced by a careful consideration of group velocity and the change of group velocity with applied field, though we do not repeat the arguments here.[12]

The next step in deriving this approximation is to consider a wavepacket – a linear superposition of different Bloch states. Because we are going to consider the time evolution, we also include the time-varying factor $\exp(-iE_{\mathbf{k}}t/\hbar)$ for each component in the superposition. Hence, we consider a wavefunction

$$\Psi(\mathbf{r},t) = \sum_{\mathbf{k}} c_{\mathbf{k}} u_{\mathbf{k}}(\mathbf{r}) \exp(i\mathbf{k}.\mathbf{r}) \exp(-iE_{\mathbf{k}}t/\hbar) \tag{8.29}$$

where $c_{\mathbf{k}}$ are the coefficients of the different Bloch states in this superposition. We have restricted this superposition to states within only one band. We make the further assumption that this superposition is only from a small range of k-states (near $k = 0$). This is what can be called a *slowly varying envelope approximation* because it means that the resulting wavepacket does not vary rapidly in space.

Because of this slowly varying envelope approximation, we can presume that for all the $\mathbf{k}$ of interest to us, all of the unit-cell functions $u_{\mathbf{k}}(\mathbf{r})$ are approximately the same. Though the $u_{\mathbf{k}}(\mathbf{r})$ are all, in principle, different and they do indeed vary substantially with important

---

[12] For a detailed discussion of this point, see W. A. Harrison, *Applied Quantum Mechanics* (World Scientific, Singapore, 2000), pp. 189–190

consequences for large changes in **k**, for a sufficiently small range of **k** we can take them all to be the same. Hence, we presume $u_{\mathbf{k}}(\mathbf{r}) \cong u_0(\mathbf{r})$ for the range of interest to us, which enables us to factor out this unit-cell part, writing

$$\Psi(\mathbf{r},t) = u_0(\mathbf{r})\Psi_{env}(\mathbf{r},t) \tag{8.30}$$

where the envelope function $\Psi_{env}(\mathbf{r},t)$ can be written

$$\Psi_{env}(\mathbf{r},t) = \sum_{\mathbf{k}} c_{\mathbf{k}} \exp(i\mathbf{k.r})\exp(-iE_{\mathbf{k}}t/\hbar) \tag{8.31}$$

Now, we are going to try to construct a Schrödinger-like equation for this envelope function. Differentiating with respect to time and then substituting $E_{\mathbf{k}}$ from (8.28) gives

$$\begin{aligned} i\hbar\frac{\partial\Psi_{env}}{\partial t} &= \sum_{\mathbf{k}} c_{\mathbf{k}} E_{\mathbf{k}} \exp(i\mathbf{k.r})\exp(-iE_{\mathbf{k}}t/\hbar) \\ &= \frac{\hbar^2}{2m_{eff}}\sum_{\mathbf{k}} c_{\mathbf{k}} k^2 \exp(i\mathbf{k.r})\exp(-iE_{\mathbf{k}}t/\hbar) + V\sum_{\mathbf{k}} c_{\mathbf{k}} \exp(i\mathbf{k.r})\exp(-iE_{\mathbf{k}}t/\hbar) \\ &= \frac{\hbar^2}{2m_{eff}}\sum_{\mathbf{k}}\left[-c_{\mathbf{k}}\nabla^2 \exp(i\mathbf{k.r})\right]\exp(-iE_{\mathbf{k}}t/\hbar) + V\Psi_{env} \end{aligned} \tag{8.32}$$

because $\nabla^2 \exp(i\mathbf{k.r}) = -k^2 \exp(i\mathbf{k.r})$. Hence, finally we have

$$-\frac{\hbar^2}{2m_{eff}}\nabla^2\Psi_{env}(\mathbf{r},t) + V(\mathbf{r})\Psi_{env}(\mathbf{r},t) = i\hbar\frac{\partial}{\partial t}\Psi_{env}(\mathbf{r},t) \tag{8.33}$$

We can see, therefore, that we have managed to construct a Schrödinger equation for this envelope function. All of the details of the periodic potential and the unit-cell wavefunction have been suppressed in this equation and their consequences are all contained in the single parameter, the effective mass m_{eff}. This effective mass model is a very powerful simplification and is at the root of a large number of models of processes in semiconductors.

Note, incidentally, that we have allowed the potential $V(\mathbf{r})$ (i.e., the energy of the band at $k = 0$) to vary with position $\mathbf{r}$ in Eq. (8.33). This is justifiable if the changes in that potential are very small compared to $\hbar^2k^2/2m_{eff}$ over the scale of a unit cell and over the wavelength $2\pi/k$. Technically, if that potential changes with position, then we no longer have a truly periodic structure and we might presume that we cannot use our crystalline theory to model it; in practice, however, we can presume that the material is to a good enough approximation still locally crystalline as long as that potential is slowly varying. In fact, practical calculations and comparisons with experiment show that this kind of approach remains valid even for some very rapid changes in potential; it does not apparently take many periods of the crystal structure to define the basic properties of the crystalline behavior. We can also handle abrupt changes in $V(\mathbf{r})$, in practice, through the use of appropriate boundary conditions. Changes in $V(\mathbf{r})$ with position can result, for example, from applying electric fields or from changes in material composition.

Effective mass approximation in semiconductor heterostructures

Structures involving more than one kind of material are called *heterostructures*. An example of a change in material composition would be changing the relative proportions of Ga and Al in the alloy semiconductor $Al_xGa_{1-x}As$. Such changes are made routinely in modern semiconductor structures, especially abrupt changes in material composition (e.g., the interface between GaAs and $Al_{0.3}Ga_{0.7}As$) in optoelectronic devices such as laser diodes and, in particular, in quantum well structures involving very thin (e.g., 10 nm) layers of semiconductor materials.

In analyzing semiconductor heterostructures, one also has to consider that the effective mass is, in general, different in different materials. It is then better to write Eq. (8.33) as

$$-\frac{\hbar^2}{2}\nabla\cdot\left[\frac{1}{m_{eff}}\nabla\Psi_{env}(\mathbf{r},t)\right]+V(\mathbf{r})\Psi_{env}(\mathbf{r},t)=i\hbar\frac{\partial}{\partial t}\Psi_{env}(\mathbf{r},t) \tag{8.34}$$

and to use boundary conditions such as

$$\Psi_{env} \text{ continuous} \tag{8.35}$$

and

$$\frac{1}{m_{eff}}\nabla\Psi_{env} \text{ continuous} \tag{8.36}$$

to handle abrupt changes in material and/or potential. The choice of Eq. (8.34) and of the boundary conditions Eqs. (8.35) and (8.36) is to some extent arbitrary. For constant mass, Eq. (8.34) is no different from Eq. (8.33). However, these new choices do conserve probability density if the mass changes with position.[13]

There is, however, apparently no way of deriving from first principles the boundary conditions we should use for the envelope functions and more than one set is possible. One reason why we cannot derive the boundary conditions is that by allowing abrupt changes in material composition, we have severely violated the assumption, implicit in the effective mass theory, that the material is locally crystalline. In practice, we find that Eq. (8.34) and boundary conditions Eqs. (8.35) and (8.36) work well in modeling many experimental situations; we should simply consider them to be a viable model for handling the quantum mechanics of semiconductor heterostructures, a model that is, at least, self-consistent (i.e., it does not lose particles).

Problems

8.6.1* Suppose we have a crystalline material that has an isotropic parabolic band with an energy minimum at some point $\mathbf{k}_{new}=\mathbf{k}-\mathbf{k}_o$ in the Brillouin zone; i.e., in general, *not* at $\mathbf{k}=0$. Show that we can also construct an effective mass Schrödinger equation of the form

$$-\frac{\hbar^2}{2m_{eff}}\nabla^2\Psi_{envnew}(\mathbf{r},t)+V(\mathbf{r})\Psi_{envnew}(\mathbf{r},t)=i\hbar\frac{\partial}{\partial t}\Psi_{envnew}(\mathbf{r},t)$$

where the full wavefunction for the electron can be written as

$$\Psi(\mathbf{r},t)=u_0(\mathbf{r})\exp(i\mathbf{k}_o\cdot\mathbf{r})\Psi_{envnew}(\mathbf{r},t)$$

You may presume that the unit-cell function in the range of $\mathbf{k}$ of interest around $\mathbf{k}_o$ is approximately the same for all such $\mathbf{k}$, with a form $u_o(\mathbf{r})$. (Hint: Follow through the derivation of the effective mass envelope function equation, but centering the energy "parabola" around $\mathbf{k}_o$.)

[13] Eq. (8.33) does not conserve probability density if the mass changes with position and neither does the boundary condition "$\nabla\Psi_{env}$ continuous." The conservation of probability density for Eq. (8.34) and the boundary conditions Eqs. (8.35) and (8.36) were proved in Problem 3.14.2.

8.6.2 Suppose that we have a material that has a different effective mass in each of the x, y, and z directions and, hence, has a dispersion relation (for a band minimum at $k = 0$)

$$E_{\mathbf{k}} = \frac{\hbar^2}{2}\left[\frac{k_x^2}{m_x} + \frac{k_y^2}{m_y} + \frac{k_z^2}{m_z}\right] + V$$

where m_x, m_y, and m_z are possibly different effective masses in the three different Cartesian directions and k_x, k_y, and k_z are the corresponding components of the $\mathbf{k}$ vector. Show that we can construct an effective mass Schrödinger equation for the envelope function, of the form

$$-\frac{\hbar^2}{2}\left[\frac{1}{m_x}\frac{\partial^2}{\partial x^2} + \frac{1}{m_y}\frac{\partial^2}{\partial y^2} + \frac{1}{m_z}\frac{\partial^2}{\partial z^2}\right]\Psi_{env}(\mathbf{r},t) + V(\mathbf{r})\Psi_{env}(\mathbf{r},t) = i\hbar\frac{\partial}{\partial t}\Psi_{env}(\mathbf{r},t)$$

8.7 Density of states in energy

We deduced the form of the density of states in k-space, which is constant (Eq. (8.24)) and quite generally independent of the form of the band structure. For many calculations, however, what we need is the density of states in energy. For example, we might want to know the thermal distribution of carriers (i.e., electrons or holes) in a band. The thermal occupation probability of a given state is a function of energy; hence, to know the number of electrons within some energy range, we need to know the number of states within that energy range.

The density of states in energy does, however, depend on the band structure. To deduce the density of states in energy per unit (real) volume, $g(E)$, we need to know the relation between the electron energy, E, and $\mathbf{k}$. Here, we work out that density of states for the case of the isotropic parabolic band.

Because the band is isotropic by assumption, states of a given energy E all have the same magnitude of k, so they are all found on a spherical surface in k-space. The number of states between energies E and $E + dE$ (i.e., $g(E)dE$) is then the number of states in k-space in a spherical shell between k and $k + dk$, where

$$dk = \left(\frac{dk}{dE}\right)dE \tag{8.37}$$

Using the parabolic band dispersion relation Eq. (8.28), we have

$$\frac{dk}{dE} = \frac{1}{2}\sqrt{\frac{2m_{eff}}{\hbar^2}}\frac{1}{\sqrt{E-V}} \tag{8.38}$$

Before proceeding further, we now introduce the fact that electrons can have two possible spin states. For each k state, we have two possible electron basis states, one corresponding to each of the two possible spins. Hence, we must multiply our density of states by a factor of 2 if we are to include both electron spins, giving the factor of 2 in front of $g(\mathbf{k})$ in Eq. (8.39).[14] Putting all of this together gives

$$g(E)dE = 2g(\mathbf{k})d^3\mathbf{k} = \frac{2}{(2\pi)^3}4\pi k^2 dk = \frac{2}{(2\pi)^3}4\pi k^2 \frac{1}{2}\sqrt{\frac{2m_{eff}}{\hbar^2}}\frac{1}{\sqrt{E-V}}dE \tag{8.39}$$

[14] We are making the simple assumption that introducing spin means that we double the number of transitions. This is essentially correct, though we need a more detailed analysis to correctly calculate the spin selection rules in the transitions.

that is,

$$g(E) = \frac{1}{2\pi^2}\left(\frac{2m_{eff}}{\hbar^2}\right)^{3/2}(E-V)^{1/2}. \tag{8.40}$$

This gives an "$E^{1/2}$" density of states. As the energy E rises above the energy of the bottom of the "parabola," the density of states rises as the square root of the extra energy, as shown in Fig. 8.6.

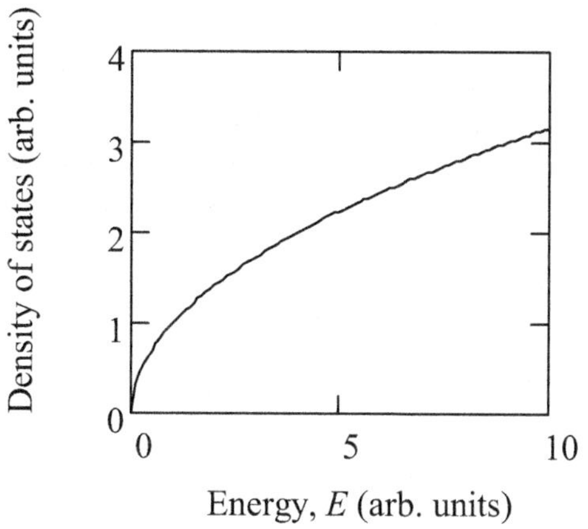

Fig. 8.6. Density of states as a function of energy above the bottom of the band for the case of a parabolic band ($V = 0$ for simplicity).

8.8 Densities of states in quantum wells

Semiconductor heterostructures made in the form of a thin layer of a narrow band gap material, such as GaAs, between two wider band gap (e.g., AlGaAs) layers, are a good example of a quantum-confined structure, in this case known as a quantum well. Such quantum confinement changes the form of the density of states from that of simple bulk materials. Those changes are particularly useful for engineering improved optoelectronic devices, such as lasers and optical modulators.

We previously discussed how to solve for the states of electrons (or holes) for the motion perpendicular to the layers (conventionally, the z direction).[15] Such solutions are classic examples of "particle-in-a-box" quantum mechanical behavior. Now we look at the formal quantum mechanical problem and the full density of states that follow from that behavior once we consider the motion of the electron or hole in the other two directions (x and y).

Formal separation of the quantum well problem

The simple picture of the states in a quantum well is that the eigenstates of a particle (electron or hole) correspond to the particle being in one of the states of the one-dimensional potential, with quantum number n and envelope wavefunction, $\psi_n(z)$, in the z direction and

[15] Now that we have formally developed the effective mass theory, we can justify the assumptions previously made that we can treat the electron or hole behavior through a Schrödinger equation for the envelope function of a particle with an effective mass.

unconstrained "free" plane-wave motion in the other two directions (which are in the plane of the quantum well layer), with wavevector $\mathbf{k}_{xy}$. (Note, incidentally, that the quantum well wavefunctions labeled with quantum number n are not necessarily the simple sinusoidal wavefunctions of the infinite well but instead are, in general, the solutions for the real quantum well potential of interest.)

Formally, for a particle of effective mass m_{eff}, the full Schrödinger equation for the envelope function for the quantum well electron is

$$-\frac{\hbar^2}{2m_{eff}}\nabla^2\psi(\mathbf{r})+V(z)\psi(\mathbf{r})=E\psi(\mathbf{r}) \tag{8.41}$$

where the potential $V(z)$ is, in the quantum well, only a function of z and corresponds to the effective potential energy change of an electron or hole as we move between the layers of different material.[16] For quantum-confined structures such as quantum wires (which have a confining potential in two dimensions) or quantum boxes or "dots" (with confinement in all three directions), we formally would have a potential that was a function of two or three directions, respectively.

To solve (8.41), we formally separate the variables in the equation. We have

$$-\frac{\hbar^2}{2m_{eff}}\nabla_{xy}^2\psi(\mathbf{r})-\frac{\hbar^2}{2m_{eff}}\frac{\partial^2}{\partial z^2}\psi(\mathbf{r})+V(z)\psi(\mathbf{r})=E\psi(\mathbf{r}) \tag{8.42}$$

where

$$\nabla_{xy}^2\equiv\frac{\partial^2}{\partial x^2}+\frac{\partial^2}{\partial y^2} \tag{8.43}$$

We postulate a separation

$$\psi(\mathbf{r})=\psi_n(z)\psi_{xy}(\mathbf{r}_{xy}) \tag{8.44}$$

where $\mathbf{r}_{xy}\equiv x\mathbf{i}+y\mathbf{j}$ is the position of the electron in the plane of the quantum wells. As usual in the technique of separation of variables, substituting this form and dividing by it throughout leads to

$$-\frac{\hbar^2}{2m_{eff}}\frac{1}{\psi_{xy}(\mathbf{r})}\nabla_{xy}^2\psi_{xy}(\mathbf{r})-\frac{\hbar^2}{2m_{eff}}\frac{1}{\psi_n(z)}\frac{\partial^2}{\partial z^2}\psi_n(z)+V(z)=E \tag{8.45}$$

We can formally separate the equation as

$$-\frac{\hbar^2}{2m_{eff}}\frac{1}{\psi_{xy}(\mathbf{r}_{xy})}\nabla_{xy}^2\psi_{xy}(\mathbf{r}_{xy})=E+\frac{\hbar^2}{2m_{eff}}\frac{1}{\psi_n(z)}\frac{\partial^2}{\partial z^2}\psi_n(z)-V(z) \tag{8.46}$$

---

[16] Such an equation is quite a good approximation for electrons in semiconductors such as GaAs and is a reasonable starting approximation for holes, though the valence band structure is somewhat more complicated. The effective mass of holes is anisotropic and formally to model the quantum well states properly, we also need to consider at least two different bands, the so-called heavy and light hole bands, complicating this envelope function approach.

where now the left-hand side depends on $\mathbf{r}_{xy}$ and the right-hand side depends only on z. Therefore, as usual in this technique, these must also equal a separation constant, here chosen to be E_{xy}, giving us

$$-\frac{\hbar^2}{2m_{eff}}\nabla_{xy}^2\psi_{xy}\left(\mathbf{r}_{xy}\right)=E_{xy}\psi_{xy}\left(\mathbf{r}_{xy}\right) \tag{8.47}$$

With the formal choice

$$E=E_{xy}+E_n \tag{8.48}$$

we have also

$$-\frac{\hbar^2}{2m_{eff}}\frac{d^2}{dz^2}\psi_n\left(z\right)+V\left(z\right)\psi_n\left(z\right)=E_n\psi_n\left(z\right) \tag{8.49}$$

Eq. (8.47) is easily solved to give

$$\psi_{xy}\left(\mathbf{r}_{xy}\right)\propto\exp\left(i\mathbf{k}_{xy}.\mathbf{r}\right) \tag{8.50}$$

with

$$E_{xy}=\frac{\hbar^2k_{xy}^2}{2m_{eff}} \tag{8.51}$$

and Eq. (8.49) is the simple one-dimensional quantum well equation we have already solved for simple "square" potentials for the particle-in-a-box energy levels and wavefunctions in the z direction. This separation is very simple for this quantum well case. (We can follow a similar procedure to analyze quantum wires and dots, separating in appropriate ways.)

Now that we are considering the motion possible in the plane of the layers, the total allowed energies are therefore no longer simply the energies E_n for the quantum well energies associated with state n, but now have the additional energy $\hbar^2k_{xy}^2/2m_{eff}$ associated with the in-plane motion. As a result, instead of discrete energy levels, we have so-called subbands, as sketched in Fig. 8.7. Note that the bottom of each subband has the energy E_n of the one-dimensional quantum well problem.

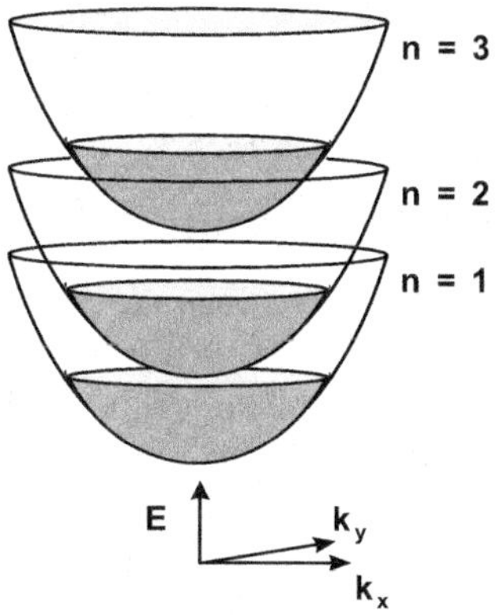

Fig. 8.7. Sketch of the first three subbands for one of the particles in a quantum well. (The solid regions merely emphasize that the subbands are paraboloids of revolution and have no physical significance.)

Quantum well density of states

Now we treat the density of states of a subband. Just as for the bulk case, we formally impose periodic boundary conditions in the x and y directions. This gives us allowed values of the wavevector in the x direction, k_x, spaced by $2\pi / L_x$, where L_x is the length of the crystal in the x direction; similarly, the allowed values of k_y are spaced by $2\pi / L_y$, with analogous definitions. Each $\mathbf{k}_{xy}$ state occupies an "area" of $\mathbf{k}_{xy}$-space of $(2\pi)^2 / A_{qw}$, where $A_{xy} = L_x L_y$, and there is altogether one allowed value of $\mathbf{k}_{xy}$ for each unit cell in the x-y plane of the quantum well. The number of states in a small area $d^2\mathbf{k}_{xy}$ of $\mathbf{k}_{xy}$-space is therefore $[A_{qw}/(2\pi)^2]d^2\mathbf{k}$. Hence, we can usefully define a ($\mathbf{k}_{xy}$-space) density of states per unit (real) area, $g_{2D}(\mathbf{k}_{xy})$, given by

$$g_{2D}\left(\mathbf{k}_{xy}\right) = \frac{1}{(2\pi)^2} \tag{8.52}$$

The number of $\mathbf{k}$ states between energies E_{xy} and $E_{xy} + dE_{xy}$ (i.e., $g_{2D}(E_{xy})dE_{xy}$) is then the number of states in $\mathbf{k}_{xy}$-space in an annular ring of area $2\pi k_{xy} dk_{xy}$, between k_{xy} and $k_{xy} + dk_{xy}$, where

$$dk_{xy} = \left(\frac{dk_{xy}}{dE_{xy}}\right) dE_{xy} \tag{8.53}$$

provided that E_{xy} is positive. Using the assumed parabolic relation between E_{xy} and k_{xy}, we have, therefore, now including the factor of 2 for spin

$$\begin{aligned} g_{2D}\left(E_{xy}\right)dE_{xy} &= 2g_{2D}\left(\mathbf{k}_{xy}\right)2\pi k_{xy}\frac{dk_{xy}}{dE_{xy}}dE_{xy} \\ &= \frac{2}{(2\pi)^2}2\pi\sqrt{\frac{2m_{eff}}{\hbar^2}}\sqrt{E_{xy}}\,\frac{1}{2}\sqrt{\frac{2m_{eff}}{\hbar^2}}\frac{1}{\sqrt{E_{xy}}}dE_{xy} \end{aligned} \tag{8.54}$$

that is,

$$g_{2D}\left(E_{xy}\right) = \frac{m_{eff}}{\pi\hbar^2} \tag{8.55}$$

This density of states therefore has the very simple form that it is constant for all $E_{xy} > 0$. (It is zero for $E_{xy} < 0$). It is therefore a "step" density of states, starting at $E_{xy} = 0$; that is, starting at $E = E_n$. Hence, the total density of states as a function of the energy E rises as a series of steps, with a new step starting as we reach each E_n.

Special case of "infinite" quantum well density of states

There is a particularly simple relation between the density of states in an "infinite" quantum well (i.e., a layer with infinitely high potential barriers on either side) and the density of states in a conventional bulk ("3D") semiconductor. This relation helps in visualizing the quantum well density of states. The "3D" density of states is

$$g(E) = \frac{1}{2\pi^2}\left(\frac{2m_{eff}}{\hbar^2}\right)^{3/2} E^{1/2} \tag{8.56}$$

Let us evaluate the density of states in a bulk semiconductor at the energy E_1 that corresponds to the first confined state of a quantum well of thickness L_z. We obtain

$$g(E_1) = \frac{m_{eff}}{\pi\hbar^2}\frac{1}{L_z} \tag{8.57}$$

which is the same as the density of states per unit volume (rather than per unit area) of an infinite quantum well; that is, dividing g_{2D} by L_z

$$g(E_1) = \frac{g_{2D}}{L_z} \tag{8.58}$$

In other words, if we plot the infinite quantum well density of states (per unit volume), it "touches" the bulk density of states (per unit volume) at the edge of the first step. Furthermore, because the steps are spaced quadratically in energy and the bulk density of states is a "parabola on its side," the quantum well (volume) density of states touches the bulk (volume) density of states at the corner of *each* step, as shown in Fig. 8.8.

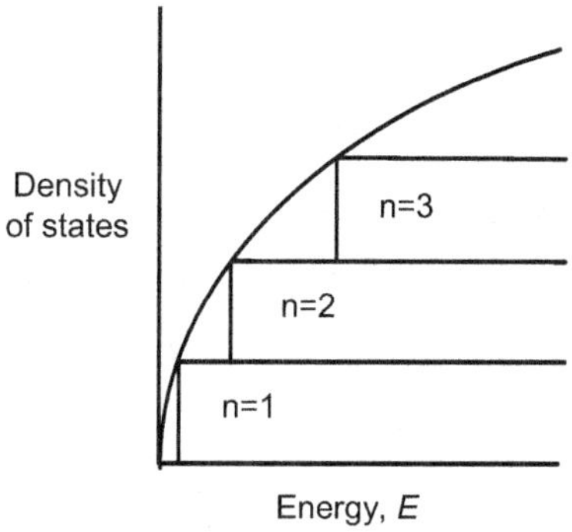

Fig. 8.8. Comparison of the bulk semiconductor density of states (i.e., the curve) with the density of states for an infinite quantum (per unit volume) (i.e., the steps).

Note, incidentally, that this relation between the quantum well and bulk density of states allows for a simple correspondence; if we started to increase the thickness of the quantum well, the steps would get closer and closer together, but their corners would still touch the bulk density of states so that as the quantum well became very thick, we would eventually not be able to distinguish its density of states from that of the bulk material.

Problems

8.8.1 Derive an expression for the form of the density of states of a subband in a quantum "wire" with a rectangular cross-section – a cuboidal structure that is small in two directions but large (approximately infinite) in the third. Take the "walls" of the quantum wire to be infinitely "high" (i.e., this is an infinite quantum wire in the same sense as an infinite quantum well). (Note that the quantum wire problem, in this case, is "separable" into separate Schrödinger equations for each of the three directions.)

8.8.2* In the simplest form of a quantum well structure, a "quantum well" layer of one material (e.g., GaAs) is sandwiched between two "barrier" layers of a different material (e.g., AlGaAs alloy), resulting in a "rectangular" potential well (e.g., for electrons). The electrons are free to move in two (x and y) directions, but the wavefunction in the third direction (z) behaves like that of a particle in a box. Suppose now that instead of a simple uniform quantum well layer, the material is actually smoothly graded in the z direction so that the potential well seen by the electrons in this direction is parabolic, with the potential increasing quadratically as we move in the z direction from the center of the well. (Such structures can be made, for example, by progressively grading the aluminum concentration in an AlGaAs alloy.) For simplicity, we presume that the electron mass does not change as we change the material composition. Sketch the form of the electron density of states in this quantum well, including the first several subbands. (Note: You should not have to perform any actual calculations for this.)

8.9 k·p method

This section is optional in that it is not required for subsequent sections, other than some problems that are clearly marked, but it is an interesting and practically useful exercise in quantum mechanics of crystalline materials and the finite basis subset method (Section 6.2).

We are particularly interested in the behavior of semiconductors near to maxima and minima. Though the calculation of the complete band structure is a difficult task, it is possible to construct simple, semi-empirical models useful near band minima and maxima. The *k·p method* is one useful approach to such models. It allows us to calculate how properties change near to those maxima and minima and allows us to relate various different phenomena, such as band gap energies, effective masses, and optical absorption strengths. Once we have used the $\mathbf{k}\cdot\mathbf{p}$ method to derive an appropriate model, only a few measurable parameters are required to define the most useful properties of the band structure.

We start by substituting the Bloch form, Eq. (8.20), into the Schrödinger equation, Eq. (8.3). Now, we explicitly label the band n of a given unit-cell function. Noting that

$$\nabla\left[u_{n\mathbf{k}}(\mathbf{r})\exp(i\mathbf{k}\cdot\mathbf{r})\right]=\left\{\left[\nabla u_{n\mathbf{k}}(\mathbf{r})\right]+i\mathbf{k}u_{n\mathbf{k}}(\mathbf{r})\right\}\exp(i\mathbf{k}\cdot\mathbf{r}) \tag{8.59}$$

and

$$\nabla^2\left[u_{n\mathbf{k}}(\mathbf{r})\exp(i\mathbf{k}\cdot\mathbf{r})\right]=\left\{\nabla^2 u_{n\mathbf{k}}(\mathbf{r})+2i\mathbf{k}\cdot\nabla u_{n\mathbf{k}}(\mathbf{r})-k^2u_{n\mathbf{k}}(\mathbf{r})\right\}\exp(i\mathbf{k}\cdot\mathbf{r}) \tag{8.60}$$

we have

$$\left[\hat{H}_o+\frac{\hbar}{m_o}\mathbf{k}\cdot\hat{\mathbf{p}}\right]u_{n\mathbf{k}}(\mathbf{r})=\left[E_n(\mathbf{k})-\frac{\hbar^2k^2}{2m_o}\right]u_{n\mathbf{k}}(\mathbf{r}) \tag{8.61}$$

where we have used $\hat{\mathbf{p}}=-i\hbar\nabla$, $E_n(\mathbf{k})$ is the energy eigenvalue for the state $\mathbf{k}$ in band n

$$\hat{H}_o=\frac{-\hbar^2}{2m_o}\nabla^2+V_L(\mathbf{r}) \tag{8.62}$$

and

$$\hat{H}_o u_{n0}(\mathbf{r})=E_n(0)u_{n0}(\mathbf{r}) \tag{8.63}$$

(Note that we are now using the notation m_o for the electron mass to clarify that this is the rest mass of the electron, not any electron effective mass.) These $u_{n0}(\mathbf{r})$ are, therefore, the solutions for the unit-cell functions at $k=0$ (to see this, set $k=0$ in Eq. (8.61)).

Now comes a key step in this approach to band structure. Because the $u_{n0}(\mathbf{r})$ are the solutions to an eigenequation, Eq. (8.63), they form a complete set for describing unit-cell functions and so we can, if we wish, expand the $u_{n\mathbf{k}}(\mathbf{r})$ in them; that is,

$$u_{n\mathbf{k}}(\mathbf{r})=\sum_{n'}a_{nn'k}u_{n'0}(\mathbf{r}) \tag{8.64}$$

In other words, we are expanding the unit-cell function in band n and wavevector $\mathbf{k}$ in the set of unit-cell functions of all the bands for $k = 0$. The expansion Eq. (8.64) is sometimes known as the *Luttinger–Kohn representation*. This expansion is over the bands n'. When we have to add in some finite amount of the unit-cell function from some particular band, $u_{n'0}(\mathbf{r})$, in the expansion Eq. (8.64), we say that we are mixing in some of that "band," though strictly we are adding in some of the zone-center unit-cell function from that band.

Thus far, we have made no approximations (beyond those used to get the Schrödinger equation (8.3) to start with). We have merely rewritten the Schrödinger equation for the electron given the known Bloch form of its wavefunction in a crystalline material. We can, however, see some interesting possibilities in this rearrangement. A common approach is simply to presume that we know the wavefunctions $u_{n0}(\mathbf{r})$ and energies $E_n(0)$ at $\mathbf{k} = 0$. These energies can often be determined experimentally – from measured bandgap energies, for example. Though we do not really know the wavefunctions, in the end, we only need some specific matrix elements using these wavefunctions and it will turn out that these matrix elements can often also be deduced from other experimental measurements. With this presumption, we can see that one approach to solving Eq. (8.61) – to understand what happens just away from $\mathbf{k} = 0$ – would be to treat the $\mathbf{k}.\mathbf{p}$ term as a perturbation and use perturbation theory to deduce effective masses and other properties of interest. A useful basis for the perturbation theory would be the complete set of unit-cell functions, $u_{n0}(\mathbf{r})$, at $\mathbf{k} = 0$. (This approach is readily amended to work at any other specific point in the Brillouin zone, so it is not restricted to the neighborhood of $\mathbf{k} = 0$ only.)

We do not, for the moment, take such a perturbative approach, though it can be done. Instead, more generally useful results can be obtained by "pretending" that we only need to consider a small number of basis functions, $u_{n0}(\mathbf{r})$, to analyze the problem; that is, we presume, as a starting point that we only need to include a small number of bands in the expansion, Eq. (8.64). This is what we referred to in Chapter 6 as the finite basis subset method. This approach enables us to get exact results for a somewhat artificial problem; if we have chosen our limited set of basis functions wisely, we would, however, have a good first approximation to the actual problem and then we could add in other terms as perturbations. In other words, we consider the effects of some bands exactly and we presume we could treat the remaining bands perturbatively, if necessary.[17]

There are at least three well-known and useful approaches based on this method: the Kane band model;[18] the Luttinger–Kohn model;[19] and the Pikus–Bir model.[20] The Kane model treats the lowest conduction band and the top three valence bands exactly and adds in the next most

---

[17] The inclusion of the effects of other bands as a perturbation is routinely done in actual band structure calculations. The technique used, Löwdin perturbation theory (see P. Löwdin, "A Note on the Quantum-Mechanical Perturbation Theory," *J. Chem. Phys.* **19**, 1396–1401 [1951]), is slightly more complex than the simple perturbation theory we used, because one is trying to calculate a perturbation to a matrix element in a finite basis set method rather than calculating a perturbation of the final solution.

[18] E. O. Kane, "Band Structure of Indium Antimonide," *J. Phys. Chem. Solids* **1**, 249–261 (1957). E. O. Kane, "The **k.p** Method," in *Semiconductors and Semimetals*, *Vol. 1*, *Physics of III-V Compounds*, Eds. R. K. Willardson and A. C. Beer (Academic, New York, 1966).

[19] J. M. Luttinger and W. Kohn, "Motion of Electrons and Holes in Perturbed Periodic Fields," *Phys. Rev.* **97**, 869–883 (1955).

[20] G. L. Bir and G. E. Pikus, *Symmetry and Strain-Induced Effects in Semiconductors* (Wiley, New York, 1974).

important effects as perturbations. The Luttinger–Kohn model focuses on the valence bands. The Pikus–Bir model includes the effects of strain. These three models form the basis for most calculations of valence (and conduction) band structure for zinc-blende materials near the minima and maxima in the band structure, both in bulk materials and, with some additional work, in quantum wells. We do not go into the detail of these models, which are best left to a more comprehensive discussion than space permits here. We do, however, look at the simplest model to show how such methods work and the kinds of results they can give. This simple model is for an idealized "two-band" semiconductor.

k·p model for a two-band semiconductor

The two-band $\mathbf{k}\cdot\mathbf{p}$ model is not one we can use to treat any particular real semiconductor,[21] but it illustrates the $\mathbf{k}\cdot\mathbf{p}$ approach. In general, if we substitute the expansion, Eq. (8.64) into the rewritten Schrödinger equation, Eq. (8.61), we have

$$\left[\hat{H}_o + \frac{\hbar}{m_o}\mathbf{k}\cdot\hat{\mathbf{p}}\right]\sum_{n'} a_{n'}u_{n'0}(\mathbf{r}) = \left[E_n(\mathbf{k}) - \frac{\hbar^2 k^2}{2m_o}\right]\sum_{n'} a_{n'}u_{n'0}(\mathbf{r}) \tag{8.65}$$

Multiplying on the left by $u_{q0}^*(\mathbf{r})$ and integrating over a unit cell, we have (at least for nondegenerate bands)

$$\sum_{n'}\left\{\left[E_n(0) + \frac{\hbar^2 k^2}{2m_o}\right]\delta_{nn'} + \frac{\hbar}{m_o}\mathbf{k}\cdot\mathbf{p}_{nn'}\right\}a_{n'} = E_n(\mathbf{k})a_n \tag{8.66}$$

where we have used the orthonormality of the $u_{q0}(\mathbf{r})$, and where

$$\mathbf{p}_{nn'} \equiv \int_{\substack{unit\\cell}} u_{n0}^*(\mathbf{r})\hat{\mathbf{p}}u_{n'0}(\mathbf{r})\,d^3\mathbf{r} \tag{8.67}$$

Up to this point, we have made no approximations.[22] Eq. (8.66) is a complete statement of the Schrödinger wave equation for a periodic potential and periodic boundary conditions. Now, we presume that only two bands are important. We presume (as is usually the case) that the $u_{n0}(\mathbf{r})$ have definite parity, so $\mathbf{p}_{nn} = 0$ (the gradient operator implicit in $\hat{\mathbf{p}}$ flips the parity of the wavefunction on which it operates; e.g., the derivative of an even function is an odd function, leading to a zero value for the integral (8.67) for $n = n'$). Hence, now writing Eq. (8.66) in matrix form for the two bands of interest, 1 and 2, we have

$$\begin{bmatrix} E_1(0) + \dfrac{\hbar^2 k^2}{2m_o} & \dfrac{\hbar}{m_o}\mathbf{k}\cdot\mathbf{p}_{12} \\ \dfrac{\hbar}{m_o}\mathbf{k}\cdot\mathbf{p}_{21} & E_2(0) + \dfrac{\hbar^2 k^2}{2m_o} \end{bmatrix}\begin{bmatrix} a_1 \\ a_2 \end{bmatrix} = E(\mathbf{k})\begin{bmatrix} a_1 \\ a_2 \end{bmatrix} \tag{8.68}$$

---

[21] It is actually not a bad model for the conduction band and the light hole band in zinc-blende materials near $k = 0$, though it certainly does not include the heavy hole band or split-off hole band in these materials and does not include the effects of spin and spin-orbit coupling that are needed in a more complete model. The procedure with those more complicated models is, however, essentially the same as that illustrated here for the simple two-band case.

[22] We have not included spin effects in the Hamiltonian so, in that sense, the original Schrödinger equation is itself not very complete, but we have made no approximations after postulating that equation.

To solve this equation for the eigenenergies of the two bands, we set the appropriate determinant to zero; that is,

$$\begin{vmatrix} E_1(0)+\frac{\hbar^2 k^2}{2m_o}-E(\mathbf{k}) & \frac{\hbar}{m_o}\mathbf{k}\cdot\mathbf{p}_{12} \\ \frac{\hbar}{m_o}\mathbf{k}\cdot\mathbf{p}_{21} & E_2(0)+\frac{\hbar^2 k^2}{2m_o}-E(\mathbf{k}) \end{vmatrix} = 0 \tag{8.69}$$

The operator $\hat{\mathbf{p}}$ is Hermitian and so $\mathbf{p}_{12} = \mathbf{p}_{21}^*$. Let us also presume for simplicity here that (1) that we know for some reason that $\mathbf{p}_{12}$ is, at least, approximately isotropic, and (2) $\hbar^2 k^2 / 2m_o$ is negligible compared to $E(\mathbf{k})$ (as will be the case if the resulting bands turn out to have very light effective masses). Hence, we have

$$\begin{vmatrix} E_1(0)-E(\mathbf{k}) & \frac{\hbar}{m_o}kp_{12} \\ \frac{\hbar}{m_o}kp_{12}^* & E_2(0)-E(\mathbf{k}) \end{vmatrix} \cong 0 \tag{8.70}$$

We choose the energy origin as $E_1(0) = 0$ (this choice makes no difference to the results) and we write $E_2(0) - E_1(0) = E_g$. We also define the parameter E_p as

$$E_p = \frac{2}{m_0}|p_{12}|^2, \tag{8.71}$$

Hence, we have, from Eq. (8.70)

$$-E(k)\left(E_g - E(k)\right) - E_p\frac{\hbar^2 k^2}{2m_o} = 0 \tag{8.72}$$

which is a quadratic with two solutions. For small k, we have

$$E(k) = E_g + \frac{E_p}{E_g}\frac{\hbar^2 k^2}{2m_o} \tag{8.73}$$

and

$$E(k) = -\frac{E_p}{E_g}\frac{\hbar^2 k^2}{2m_o} \tag{8.74}$$

Hence, the band structure calculated from this simple set of assumptions consists of two bands: a conduction-like band starting at energy E_g, with a positive effective mass

$$m_{eff\,2} = m_o\frac{E_g}{E_p} \tag{8.75}$$

and a valence-like band starting at energy 0, with a negative (i.e., hole-like) effective mass

$$m_{eff\,1} = -m_o\frac{E_g}{E_p} \tag{8.76}$$

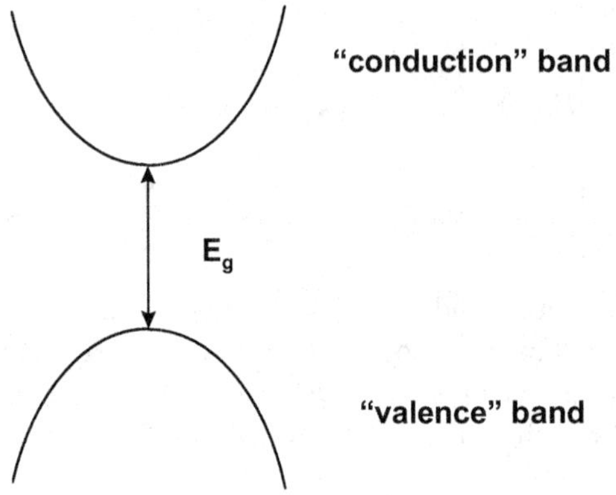

Fig. 8.9. Band structure calculated in the $\mathbf{k}\cdot\mathbf{p}$ model assuming two bands only are important. Conduction and valence bands result, with equal (and opposite) effective masses.

This band structure is sketched in Fig. 8.9. Note, for example, that the effective masses depend proportionately on the energy gap, E_g.

Note that the effective mass (as opposed to the bare electron mass) is "caused" by the "interaction between the bands". If the $\mathbf{p}_{12}$ and $\mathbf{p}_{21}$ matrix elements were zero, then the matrix in Eq. (8.68) would simply be diagonal and both bands would simply have the free electron mass (they would also both be curved in the same direction).

Having worked out the eigenenergies, it is now a simple matter to work out the eigenvectors or eigenfunctions. With the simplifying approximations and definitions we made, for example, the matrix equation, Eq. (8.68), is

$$\begin{bmatrix} 0 & \dfrac{\hbar}{m_o} k p_{12} \\ \dfrac{\hbar}{m_o} k p_{12}^* & E_g \end{bmatrix} \begin{bmatrix} a_1 \\ a_2 \end{bmatrix} = E(\mathbf{k}) \begin{bmatrix} a_1 \\ a_2 \end{bmatrix} \tag{8.77}$$

Hence, for the upper band (band 2), we have, using the eigenenergy solution, Eq. (8.73)

$$\begin{bmatrix} -E_g - \dfrac{E_p}{E_g}\dfrac{\hbar^2 k^2}{2m_o} & \dfrac{\hbar}{m_o} k p_{12} \\ \dfrac{\hbar}{m_o} k p_{12}^* & -\dfrac{E_p}{E_g}\dfrac{\hbar^2 k^2}{2m_o} \end{bmatrix} \begin{bmatrix} a_{21k} \\ a_{22k} \end{bmatrix} = 0 \tag{8.78}$$

where by a_{21k} we mean the coefficient of $u_{10}(\mathbf{r})$ in the expansion of the form Eq. (8.64) of the unit-cell function $u_{2k}(\mathbf{r})$; that is, explicitly, the expansion of Eq. (8.64) is, in this simple two-band case

$$u_{2\mathbf{k}}(\mathbf{r}) = a_{21k} u_{10}(\mathbf{r}) + a_{22k} u_{20}(\mathbf{r}) \tag{8.79}$$

To deduce the unit-cell wavefunctions as a function of k in this band 2, we solve Eq. (8.78) for the coefficients a_{21k} and a_{22k}. For the simple case of $k = 0$, the only nonzero term is the upper-left term, which is then equal to $-E_g$. As a result, we have $E_g a_{21k} = 0$, which means that $a_{21k} = 0$. Hence, at $k = 0$, the unit-cell wavefunction is $u_{20}(\mathbf{r})$ for the upper band, as we would have expected. Away from $k = 0$, however, the coefficients a_{21k} and a_{22k} deduced from Eq. (8.78) are both, in general, nonzero. We would find a growing admixture of $u_{10}(\mathbf{r})$ as we move away from $k = 0$. This is a simple example of *band mixing*. As we move away from $k = 0$, the unit-cell wavefunctions are no longer exactly the ideal functions, with simple exact parities and symmetry properties, found at $k = 0$. For "allowed" processes (e.g., one-photon optical absorption between valence and conduction bands), this mixing is not very important for small k, though we should understand that "forbidden" processes (i.e., ones disallowed at $k = 0$ by symmetry) can become progressively stronger as we move away from $k = 0$. Examples of forbidden processes are those that violate polarization selection rules, or two-photon absorption, which is parity forbidden at $k = 0$. This kind of approach to band structure enables us to calculate such forbidden processes if we wish.

8.10 Use of Fermi's Golden Rule

Now that we have a sufficient model for crystalline semiconductors, we can use the most important result of time-dependent perturbation theory we derived before, Fermi's Golden Rule (No. 2), to calculate the optical absorption in direct gap semiconductors. From Eq. (7.32), we know that the transition rate for absorption between an initial electron state $|\psi_i\rangle$ in the valence band, with energy E_i, and a final state $|\psi_f\rangle$ in the conduction band, with energy E_f, in the presence of an oscillating perturbation of angular frequency ω, is

$$w_{abs} = \frac{2\pi}{\hbar}\left|\langle\psi_f|\hat{H}_{po}|\psi_i\rangle\right|^2 \delta(E_f - E_i - \hbar\omega) \tag{8.80}$$

Here, $\hat{H}_{po}$ is of the form defined by Eq. (7.15); that is, in the present context where we are interested also in the spatial variation of the perturbation, $\hat{H}_{po}$ varies here through the spatial dependence of the amplitude of an electromagnetic wave

$$\hat{H}_p(\mathbf{r},t) = \hat{H}_{po}(\mathbf{r})\left[\exp(-i\omega t) + \exp(i\omega t)\right] \tag{8.81}$$

while the oscillatory field is "turned on." It represents the amplitude of the oscillatory perturbation. $\langle\psi_f|\hat{H}_{po}|\psi_i\rangle$ can now be written explicitly as

$$\langle\psi_f|\hat{H}_{po}|\psi_i\rangle = \int \psi_f^*(\mathbf{r})\hat{H}_{po}(\mathbf{r})\psi_i(\mathbf{r})\,d^3\mathbf{r} \tag{8.82}$$

where $\psi_i(\mathbf{r})$ and $\psi_f(\mathbf{r})$ are, respectively, the wavefunctions of the initial and final states.[23]

[23] Note that we are considering only the spatial parts of the initial and final states. In reality, they also have spin character, which strictly ought to be included and which does make a significant difference (e.g., a factor of 1/2 or 1/3 in the final result for optical absorption strength). Proper consideration of the spin is also somewhat complicated by the fact that the spin character of some of the valence band states in, e.g., zinc blende materials, is mixed and requires a more detailed understanding of the valence-band structure. The spin character also is important in determining the polarization dependence of optical absorption in the presence of strain or in layered structures, e.g., quantum wells, in both of which cases there is a special axis (e.g., the strain or growth direction).

Form of the perturbing Hamiltonian for an electromagnetic field

For the case of an electron in an electromagnetic field, the usual form[24] presumed for $\hat{H}_p(\mathbf{r},t)$ in semiconductors is (see Appendix E)

$$\hat{H}_p(\mathbf{r},t) \cong \frac{e}{m_o}\mathbf{A}\cdot\hat{\mathbf{p}} \tag{8.83}$$

where m_o is the free electron mass, $\hat{\mathbf{p}}$ is the momentum operator $-i\hbar\nabla$, and $\mathbf{A}$ is the electromagnetic vector potential corresponding to a wave of (angular) frequency ω

$$\mathbf{A} = \mathbf{e}\left\{\frac{A_0}{2}\exp\left[i\left(\mathbf{k}_{op}\cdot\mathbf{r}-\omega t\right)\right]+\frac{A_0}{2}\exp\left[-i\left(\mathbf{k}_{op}\cdot\mathbf{r}-\omega t\right)\right]\right\} \tag{8.84}$$

Here, $\mathbf{k}_{op}$ is the wavevector of the optical field inside the material and we have taken the field to be linearly polarized with its electric vector in the direction of the unit vector $\mathbf{e}$. Here, we presume that we are only dealing with absorption processes so, as discussed in Chapter 7, we can drop the term in $\exp(+i\omega t)$ in our perturbing Hamiltonian because it corresponds to emission. Hence, from this point on, we use

$$\hat{H}_{po}(\mathbf{r}) = -\frac{eA_o\exp\left(i\mathbf{k}_{op}\cdot\mathbf{r}\right)}{2m_o}\mathbf{e}\cdot\hat{\mathbf{p}} \tag{8.85}$$

so here we are formally retaining only the $\exp(-i\omega t)$ part of Eq. (8.81); that is, we are taking $\hat{H}_p(\mathbf{r},t) = \hat{H}_{po}\exp(-i\omega t)$ with $\hat{H}_{po}(\mathbf{r})$ as in Eq. (8.85).[25]

Direct valence to conduction-band absorption

To proceed further, we need to know $\psi_i(\mathbf{r})$ and $\psi_f(\mathbf{r})$. We presume that we can use the single-electron Bloch states already deduced. We are most interested in the transitions between an initial state in the valence band and a final state in the conduction band. We want to use normalized wavefunctions in the following calculation, so we define

$$\psi_i(\mathbf{r}) = B_i u_v(\mathbf{r})\exp(i\mathbf{k}_v\cdot\mathbf{r}) \tag{8.86}$$

and

$$\psi_f(\mathbf{r}) = B_f u_c(\mathbf{r})\exp(i\mathbf{k}_c\cdot\mathbf{r}) \tag{8.87}$$

---

[24] It is quite common to treat the interaction between the electromagnetic field and the electron in terms of the electric field and the electronic charge. This usually leads to a first approximation called the *electric dipole approximation*. Solid-state physicists in particular, however, tend to prefer a formally more complete treatment in terms of the magnetic vector potential. One main reason solid-state physicists use it is because the momentum matrix element that emerges from this form of calculation is the same momentum matrix element that emerges in the **k.p** calculations of the band structure. We give this theory in terms of the vector potential, though the reader should note that there is no substantial difference here in the result compared to an electric dipole calculation.

[25] In this semiclassical approach (i.e., with the electromagnetic field treated classically), we end up making this *ad hoc* change to the perturbing Hamiltonian to restrict to absorption processes only. In the full approach with quantized electromagnetic fields (see Chapter 17), the resulting algebra takes care of such restrictions automatically.

where B_i and B_f are normalization constants (from here on, we omit the subscript $\mathbf{k}$ on $u_c(\mathbf{r})$ and $u_v(\mathbf{r})$ for simplicity). Note that we are now explicitly allowing the conduction ($u_c(\mathbf{r})$) and valence ($u_v(\mathbf{r})$) unit-cell functions to be different. This is quite important – we would, in the end, get no optical absorption here if they were the same. We do, however, presume that they do not depend on $\mathbf{k}$, which is, in practice, a good approximation for an allowed process (i.e., one in which the matrix element is finite in the lowest order approximation).

In our normalization calculation, first we choose $u_v(\mathbf{r})$ and $u_c(\mathbf{r})$ to be normalized over a unit cell; that is,

$$\int_{unit\ cell} u_v^*(\mathbf{r})u_v(\mathbf{r})d^3\mathbf{r} = 1 \tag{8.88}$$

and, similarly, for $u_c(\mathbf{r})$. Hence, normalizing $\psi_i(\mathbf{r})$ and $\psi_f(\mathbf{r})$, we have, for ψ_i

$$\begin{aligned}\int_V \psi_i^*(\mathbf{r})\psi_i(\mathbf{r})d^3\mathbf{r} &= 1 \\ &= B_i^2 \int_V u_v^*(\mathbf{r})\exp(-i\mathbf{k}_v\cdot\mathbf{r})u_v(\mathbf{r})\exp(i\mathbf{k}_v\cdot\mathbf{r})d^3\mathbf{r} \\ &= B_i^2 \int_V u_v^*(\mathbf{r})u_v(\mathbf{r})d^3\mathbf{r} = B_i^2 N \int_{unit\ cell} u_v^*(\mathbf{r})u_v(\mathbf{r})d^3\mathbf{r} \\ &= B_i^2 N\end{aligned} \tag{8.89}$$

(where N is the number of unit cells in the crystal and V is the volume of the crystal) and similarly for $\psi_f(\mathbf{r})$. Hence, we have

$$B_i = B_f = \frac{1}{\sqrt{N}} \tag{8.90}$$

With the choice of a valence-band initial state and a conduction-band final state, the matrix element of interest is now, from Eq. (8.82)

$$\begin{aligned}&\left\langle \psi_f \left| \hat{H}_{po}(\mathbf{r}) \right| \psi_i \right\rangle = \\ &-\frac{eA_0}{2m_0 N}\int_V \left[u_c^*(\mathbf{r})\exp(-i\mathbf{k}_c\cdot\mathbf{r})\right]\exp(i\mathbf{k}_{op}\cdot\mathbf{r})\mathbf{e}\cdot\hat{\mathbf{p}}\left[u_v(\mathbf{r})\exp(i\mathbf{k}_v\cdot\mathbf{r})\right]d^3\mathbf{r}\end{aligned} \tag{8.91}$$

where the integral is over the volume V of the crystal. We are interested in transitions involving states near the center of the Brillouin zone, so $|\mathbf{k}_v|$ and $|\mathbf{k}_c|$ are both $<< \pi/a$ (because π/a corresponds to the edge of the zone). Strictly, $\hat{\mathbf{p}}$ operates on the product $u_v(\mathbf{r})\exp(i\mathbf{k}_v\cdot\mathbf{r})$. Because we assume small $|\mathbf{k}_v|$, $\exp(i\mathbf{k}_v\cdot\mathbf{r})$ changes very slowly compared to the rate of change of $u_v(\mathbf{r})$, however; that is, $|\nabla\exp(i\mathbf{k}_v\cdot\mathbf{r})| \sim k_v << |\nabla u_v(\mathbf{r})| \sim \pi/a$. Hence, we can neglect $\hat{\mathbf{p}}$ operating on $\exp(i\mathbf{k}_v\cdot\mathbf{r})$, at least as a first approximation.

For definiteness, we choose the direction $\mathbf{e}$ of polarization of the optical electric field to be in the x-direction (x is one of the directions perpendicular to the propagation of the electromagnetic wave). Rewriting, we have (neglecting the effect of $\hat{\mathbf{p}}$ on $\exp(i\mathbf{k}_v\cdot\mathbf{r})$)

$$\left\langle \psi_f \left| \hat{H}_{po}(\mathbf{r}) \right| \psi_i \right\rangle = -\frac{eA_0}{2m_0 N}\int_V \exp\left[i(\mathbf{k}_v - \mathbf{k}_c + \mathbf{k}_{op})\cdot\mathbf{r}\right]\left[u_c^*(\mathbf{r})\hat{p}_x u_v(\mathbf{r})\right]d^3\mathbf{r} \tag{8.92}$$

The optical wavevector $\mathbf{k}_{op}$ corresponds to a wavelength (even inside the material) of 100s of nm or more, which means that $|\mathbf{k}_{op}| \ll \pi/a$. In fact, compared to the size scale of the Brillouin zone, $|\mathbf{k}_{op}|$ is almost totally negligible; even with a 100 nm wavelength inside the material,

$|\mathbf{k}_{op}| \sim 2\pi/100\,\text{nm}^{-1}$, whereas the edge of the Brillouin zone is $\pi/a \sim \pi/0.5\,\text{nm}^{-1}$, corresponding to $|\mathbf{k}_{op}|/(\pi/a) \sim 1\%$. Hence, the entire factor $\exp[i(\mathbf{k}_v - \mathbf{k}_c + \mathbf{k}_{op})\cdot\mathbf{r}]$ varies very slowly over the length scale, a, of a unit cell. As a result, we can approximately separate the integral in Eq. (8.92) into a sum of integrals over a unit cell, treating the value of $\exp[i(\mathbf{k}_v - \mathbf{k}_c + \mathbf{k}_{op})\cdot\mathbf{r}]$ as being approximately constant within a given unit cell; that is,

$$\langle \psi_f | \hat{H}_{po}(\mathbf{r}) | \psi_i \rangle = -\frac{eA_0}{2m_0 N} \langle c | \hat{p}_x | v \rangle \sum_{\substack{m(i.e.,\\ \text{unit cells})}} \exp\left[i\left(\mathbf{k}_v - \mathbf{k}_c + \mathbf{k}_{op}\right)\cdot\mathbf{R}_m\right] \tag{8.93}$$

where $\mathbf{R}_m$ is the position of (the center of) the mth unit cell in the crystal and the summation is over all N unit cells. Here

$$\langle c | \hat{p}_x | v \rangle \equiv \int_{\substack{\text{unit}\\ \text{cell}}} u_c^*(\mathbf{r})\, \hat{p}_x u_v(\mathbf{r})\, d^3\mathbf{r} \equiv p_{cv} \tag{8.94}$$

(which is the same for any unit cell in the crystal).

In general, the separation of the slowly varying envelope function (i.e., in this case, the $\exp(i\mathbf{k}\cdot\mathbf{r})$ term) from the unit-cell part, treating the variation of the envelope function as being negligible within a unit cell, is known as the *slowly-varying envelope approximation*.

The summation in Eq. (8.93) averages approximately to zero unless

$$\mathbf{k}_v - \mathbf{k}_c + \mathbf{k}_{op} = 0 \tag{8.95}$$

because otherwise, the function $\exp[i(\mathbf{k}_v - \mathbf{k}_c + \mathbf{k}_{op})\cdot\mathbf{r}]$ is oscillatory. (Note, incidentally, that the condition Eq. (8.95) can be seen to correspond to conservation of "momentum" ($\hbar\mathbf{k}$)). In this case, the sum becomes

$$\sum_m \exp\left[i\left(\mathbf{k}_v - \mathbf{k}_c + \mathbf{k}_{op}\right)\cdot\mathbf{R}_m\right] = \sum_m \exp(0) = N \tag{8.96}$$

Hence, we have

$$\langle \psi_f | \hat{H}_{po}(\mathbf{r}) | \psi_i \rangle = -\frac{eA_0}{2m_0} p_{cv} \tag{8.97}$$

and the transition rate becomes

$$w_{abs} = \frac{2\pi}{\hbar} \frac{e^2 A_0^2}{4m_0^2} |p_{cv}|^2 \,\delta\left(E_f - E_i - \hbar\omega\right) \tag{8.98}$$

Eq. (8.98) is, therefore, the restatement of Eq. (8.80) after we have made use of the facts that (1) the electron wavefunctions in the crystalline semiconductor are in the Bloch form; and (2) the optical wavelength is much larger than the unit-cell size. Note that the concept of conservation of "momentum" emerged automatically and is consistent with $\hbar\mathbf{k}$ being the effective electron "momentum" in a Bloch state. Note also that the transition rate is proportional to a (squared) matrix element, $|p_{cv}|^2$, that only depends on the unit-cell functions. Finally, we note that the transition rate is proportional to the optical intensity (because it is proportional to the square of the electromagnetic field amplitude, as expressed by A_o^2), which, because it represents power per unit area, corresponds also to the average arrival rate of photons in the semiconductor (per unit area).

Now, Eq. (8.98) represents the transition rate from one particular initial valence-band state with **k** vector $\mathbf{k}_v$ to a particular conduction-band state – in this case, the state with wavevector $\mathbf{k}_c = \mathbf{k}_v + \mathbf{k}_{op}$. The total transition rate, W_{TOT}, that we need to calculate the optical absorption coefficient, α, is the sum of the transition rates between all possible initial states and all possible final states; that is,

$$W_{TOT} = \frac{2\pi}{\hbar} \sum_i \sum_f \left| \langle \psi_f | \hat{H}_{po}(\mathbf{r}) | \psi_i \rangle \right|^2 \delta(E_f - E_i - \hbar\omega) \tag{8.99}$$

We have shown here that for a given initial state with wavevector $\mathbf{k}_v$, the only final state possible is the conduction band state with $\mathbf{k}_c = \mathbf{k}_v + \mathbf{k}_{op}$. Because $|\mathbf{k}_{op}|$ is generally a very small fraction of the size of the Brillouin zone, we henceforth neglect it, taking $\mathbf{k}_c = \mathbf{k}_v$; this negligible size of the optical wavevector means that the direct optical transitions are essentially "vertical" on the energy-momentum diagram. Hence, for a given initial state $\mathbf{k}_v$, only one term remains in the sum over the final states – namely that with $\mathbf{k}_c = \mathbf{k}_v$. Hence, we have, substituting from Eq. (8.98) and dropping the suffix "v" (i.e., writing $\mathbf{k}$ instead of $\mathbf{k}_v$)

$$W_{TOT} = \frac{2\pi}{\hbar} \frac{e^2 A_0^2}{4m_0^2} |p_{cv}|^2 \sum_{\mathbf{k}} \delta\left[E_c(\mathbf{k}) - E_v(\mathbf{k}) - \hbar\omega\right] \tag{8.100}$$

In Eq. (8.100), we have additionally assumed $|p_{cv}|^2$ is independent of **k**, which turns out to be a reasonable approximation.

Now, we formally rewrite (considering unit volume)

$$\sum_{\mathbf{k}} \approx \int_{\mathbf{k}} g(\mathbf{k}) d^3\mathbf{k} \tag{8.101}$$

where $g(\mathbf{k})$ is the density of states in k-space. We next change variables in the integral to the energy $E_c(\mathbf{k}) - E_v(\mathbf{k})$. Assuming parabolic bands, we can define E_J, the energy separation between the valence and conduction bands at a particular **k**, as

$$E_J(\mathbf{k}) = E_c(\mathbf{k}) - E_v(\mathbf{k}) = \frac{\hbar^2 k^2}{2}\left(\frac{1}{m_{effe}} + \frac{1}{m_{effh}}\right) + E_g = \frac{\hbar^2 k^2}{2\mu_{eff}} + E_g \tag{8.102}$$

where m_{effe} is the effective mass of the electrons in the conduction band and m_{effh} is the effective mass of the holes in the valence band. Note that, typically, the electron effective mass at the top of the valence band is actually negative because we are at a maximum in the band. Holes have energy that increases as we go down into the valence band, just as the potential energy of a bubble in water increases as we push it farther under the surface. Hence, in terms of holes, the valence-band effective mass is positive and it is this positive mass that we mean by m_{effh}. The reduced effective mass we need for this problem, μ_{eff}, is given by

$$\frac{1}{\mu_{eff}} = \frac{1}{m_{effe}} + \frac{1}{m_{effh}} \tag{8.103}$$

Hence, we can follow the same argument as we followed previously in deriving the energy density of states. We can define a *joint density of states*, $g_J(E_J)$, and write, including the factor of 2 for the two different spin states

$$g_J(E_J) dE_J = 2g(\mathbf{k}) d^3\mathbf{k} \tag{8.104}$$

Hence, for $E_J \geq E_g$

$$g_J(E_J) = \frac{1}{2\pi^2}\left(\frac{2\mu_{eff}}{\hbar^2}\right)^{3/2}\left(E_J - E_g\right)^{1/2} \tag{8.105}$$

So, using Eqs. (8.100), (8.101), (8.104), and (8.105), we obtain (per unit volume)

$$W_{TOT} = \frac{2\pi}{\hbar}\frac{e^2 A_0^2}{4m_0^2}|p_{cv}|^2 \int_{E_J \geq E_g} \frac{1}{2\pi^2}\left(\frac{2\mu_{eff}}{\hbar^2}\right)^{3/2}\left(E_J - E_g\right)^{1/2}\delta\left(E_J - \hbar\omega\right)dE_J \tag{8.106}$$

that is,

$$W_{TOT}(\hbar\omega) = \frac{2\pi}{\hbar}\frac{e^2 A_0^2}{4m_0^2}|p_{cv}|^2 \frac{1}{2\pi^2}\left(\frac{2\mu_{eff}}{\hbar^2}\right)^{3/2}\left(\hbar\omega - E_g\right)^{1/2} \tag{8.107}$$

(for $\hbar\omega \geq E_g$).

The final step is to relate the absorption coefficient, α, to the total transition rate W_{TOT}. α is the probability of absorption of a photon per unit length (in the direction of propagation). The number of photons incident per unit area per second is

$$n_p = \frac{I}{\hbar\omega} \tag{8.108}$$

where I is the optical intensity (power per unit area), so

$$\alpha = \frac{W_{TOT}}{n_p} = \frac{\hbar\omega W_{TOT}}{I} \tag{8.109}$$

The intensity I can be written (see Appendix D, Section D.8 and Appendix F, Section F.3)

$$I = \frac{n_r c\varepsilon_0 \omega^2 A_0^2}{2} \tag{8.110}$$

where n_r is the refractive index, c is the velocity of light, and ε_o is the permittivity of free space. Hence

$$\begin{aligned}\alpha(\hbar\omega) &= \frac{2\pi}{\hbar}\frac{e^2 A_0^2}{4m_0^2}\frac{1}{2\pi^2}\left(\frac{2\mu_{eff}}{\hbar^2}\right)^{3/2}\hbar\omega\frac{2}{n_r c\varepsilon_0\omega^2 A_0^2}|p_{cv}|^2\left(\hbar\omega - E_g\right)^{1/2} \\ &= \frac{e^2}{2\pi m_0^2 c\varepsilon_o}\frac{1}{n_r\omega}|p_{cv}|^2\left(\frac{2\mu_{eff}}{\hbar^2}\right)^{3/2}\left(\hbar\omega - E_g\right)^{1/2}\end{aligned} \tag{8.111}$$

In actual calculations, the parameter

$$E_p = \frac{2}{m_0}|p_{cv}|^2 \tag{8.112}$$

which typically has a value of ~ 20 eV in many semiconductors, is often used instead of $|p_{cv}|^2$, in which case we can rewrite Eq. (8.111) as

$$\alpha(\hbar\omega) = \frac{\hbar e^2}{4\pi m_0 c\varepsilon_0}\frac{1}{n_r}\frac{E_p}{\hbar\omega}\left(\frac{2\mu_{eff}}{\hbar^2}\right)^{3/2}\left(\hbar\omega - E_g\right)^{1/2} \tag{8.113}$$

or

$$\alpha(\hbar\omega) = \frac{\pi\hbar e^2}{2m_0 c\varepsilon_0} \frac{1}{n_r} \frac{E_p}{\hbar\omega} g_J(\hbar\omega) \tag{8.114}$$

We can see that the absorption spectrum, $\alpha(\hbar\omega)$, follows the joint density of states, rising (in this model) as $(\hbar\omega - E_g)^{1/2}$ (and being zero for $\hbar\omega < E_g$).

This model is very often used as a starting point for optical calculations in semiconductors. The very simple model here does ignore some important spin effects, which can be qualitatively important for polarization dependence. It also completely neglects excitonic effects, which are quantitatively and qualitatively quite important in optical absorption in direct gap semiconductors, though we do not treat them here.

Problems

8.10.1 (Notes: This is a substantial problem that could be used as an assignment. Section 8.9 should also have been studied before attempting this problem.) In this problem, we progressively consider direct optical transitions between the two bands in the simple two-band $\mathbf{k}\cdot\mathbf{p}$ model constructed previously, assuming that the lower band (band 1) is initially full of electrons and the upper band (band 2) is initially empty of electrons. For this problem, do not assume the small k approximation of Eq. (8.72). Work with the full roots of Eq. (8.72) for finding the eigenenergies and unit-cell functions.

(i) Show that the wavefunction in the upper band may be written as

$$u_{2k}(\mathbf{r}) = \frac{1}{d_+} u_{10}(\mathbf{r}) + \frac{1+\sqrt{1+|b|^2}}{bd_+} u_{20}(\mathbf{r})$$

where

$$d_\pm = \sqrt{1 + \frac{1}{|b|^2}\left(1 \pm \sqrt{1+|b|^2}\right)^2}$$

where $b = 2\hbar k p_{12} / E_g m_o$ (so that $|b|^2 = 2E_p \hbar^2 k^2 / E_g^2 m_o$).

Hints:

(a) You may find it useful in the intermediate algebra to (1) work only in terms of p_{21} rather than E_P $(= 2|p_{21}|^2 / m_o)$, and (2) introduce the parameter b early in your manipulations.

(b) Note that if we find, for example, a_2 in terms of a_1, then we actually can deduce both coefficients through the normalization of the overall unit-cell wavefunction (i.e., $|a_1|^2 + |a_2|^2 = 1$).

(ii) Find a similar expression for $u_{1k}(\mathbf{r})$.

(iii) Show for the interband momentum matrix element p_{21k} at Bloch wavevector k

$$p_{21k} = \int_{unit\ cell} u_{2k}^*(\mathbf{r})\hat{\mathbf{p}}u_{1k}(\mathbf{r})d^3\mathbf{r}$$

that

$$|p_{21k}|^2 = \frac{m_o}{2} \frac{E_p}{1 + \dfrac{2E_p}{E_g^2}\dfrac{\hbar^2 k^2}{m_o}}$$

Note that $p_{nn} = 0$ because we assume the unit-cell functions at $k = 0$ to have definite parity. (This part may take a fair amount of algebra, though it is straightforward algebra.)

(iv) Construct a mathematical expression for the absorption coefficient, $\alpha(\hbar\omega)$ for interband direct optical absorption in this model, expressed in terms of E_P, the band gap energy, E_g, the refractive index, n_r, the photon energy, $\hbar\omega$, and fundamental constants. (Note: Do *not* assume a simple parabolic density of states.)

(v) Assuming the refractive index is constant at $n_r \cong 3.5$, $E_g = 1$ eV, and $E_p = 20$ eV, graph the absorption coefficient, expressed in units of cm^{-1}, calculated using this model and compare it with the results for the simple parabolic band model with constant $|p_{21}|^2$. Graph your results up to a photon energy of 2 eV.

8.10.2 Here, we consider the equivalence of the electric dipole and " $\mathbf{A}\cdot\mathbf{p}$ " versions of the Hamiltonian for certain optical absorption processes.

(i) Prove for a single-electron Hamiltonian of the form

$$\hat{H}_{el} = \frac{-\hbar^2}{2m_o}\nabla^2 + V(\mathbf{r})$$

that

$$\left[\hat{H}_{el}, \mathbf{r}\right] = -i\frac{\hbar}{m_o}\hat{\mathbf{p}}$$

(ii) When $|\psi_1\rangle$ and $|\psi_2\rangle$ are eigenfunctions of $\hat{H}_{el}$, with corresponding eigenenergies E_1 and E_2, find an expression relating $\mathbf{p}_{21} = \langle\psi_2|\hat{\mathbf{p}}|\psi_1\rangle$ to $\mathbf{r}_{21} = \langle\psi_2|\mathbf{r}|\psi_1\rangle$.

(iii) Consider the two possible time-dependent perturbing Hamiltonians

$$\hat{H}_{pA} = -\frac{e}{m_o}\mathbf{A}\cdot\hat{\mathbf{p}}\text{, with } \mathbf{A} = \mathbf{A}_o\cos\omega t$$

and

$$\hat{H}_{pE} = -e\mathbf{E}\cdot\mathbf{r}\text{, with } \mathbf{E} = \mathbf{E}_o\cos\omega t$$

Show that, in Fermi's Golden Rule, these two Hamiltonians lead to the same transition rate for the same strength of electromagnetic field for transitions between eigenstates of $\hat{H}_{el}$. (You may use the relation $\mathbf{E} = -\partial\mathbf{A}/\partial t$.)

8.10.3 (Note: This problem can be used as a substantial assignment and should be attempted only after completing Problem 8.10.1.) Consider two-photon absorption in the simple two-band $\mathbf{k}\cdot\mathbf{p}$ theory semiconductor considered previously. Assume the lower band is initially full of electrons and the upper band is initially completely empty.

For the purposes of this exercise, you need only consider the process corresponding to two-photon absorption from the lower band (band 1) to the upper band (band 2) in which states in the upper band are available as the intermediate states. You also need only consider the first contribution to the process; that is, in the previous notation, you may presume

$$\left|a_j^{(2)}(t_o)\right|^2 =$$

$$\frac{t_o}{\hbar^4}\left[\sum_n\sum_q \frac{\langle\psi_m|\hat{H}_{po}|\psi_q\rangle\langle\psi_q|\hat{H}_{po}|\psi_j\rangle\langle\psi_j|\hat{H}_{po}|\psi_n\rangle\langle\psi_n|\hat{H}_{po}|\psi_m\rangle}{(\omega_{nm}-\omega)(\omega_{qm}-\omega)} t_o\left[\frac{\sin\left(\omega_{jm}-2\omega\right)t_o/2}{\left(\omega_{jm}-2\omega\right)t_o/2}\right]^2\right]$$

(Note that at least as part of a two-photon process, it is possible to consider the optical transition from a state within band 2 to an adjacent state in the same band. Though the optical wavevector is usually negligible numerically, its finite magnitude ensures that such a transition is between two different states, albeit ones of nearly identical wavevector **k**.)

Specifically,

(i) Given these presumptions about the initial, intermediate, and final states; evaluate each of the matrix elements $H_{poij} = \langle\psi_i|\hat{H}_{po}|\psi_j\rangle$ in this expression in terms of the magnetic vector potential amplitude A_o of the oscillatory electromagnetic field, momentum matrix elements of the form $p_{nn'\mathbf{k}} = \int_{unit\ cell} u_{n\mathbf{k}}^*(\mathbf{r})\, p u_{n'\mathbf{k}}(\mathbf{r}) d^3\mathbf{r}$, and fundamental constants; and deduce any restrictions these impose on the terms surviving in the summations. (Consider the optical wavevector $\mathbf{k}_{op}$ to be negligibly small.)

(ii) For a particular initial state with wavevector $\mathbf{k}_m$ in band 1, derive a mathematical expression for the two-photon transition rate $w_{\mathbf{k}m} = \left|a_j^{(2)}(t_o)\right|^2/t_o$ in terms of the optical intensity, I; the photon energy $\hbar\omega$; the energy separation, $E_J(\mathbf{k}_m)$, between the states in bands 1 and 2 at wavevector $\mathbf{k}_m$; the quantities $|p_{21\mathbf{k}_m}|^2$ and $|p_{22\mathbf{k}_m}|^2$; a function of the form $t_o[(\sin x)/x]^2$; the refractive index n_r; and fundamental constants.

(iii) Find an approximate expression for the matrix element $|p_{22\mathbf{k}m}|^2$, valid to lowest order in k_m, and expressed in terms of $|p_{21}|^2$, the band gap energy E_g, and fundamental constants. (This will be an expression proportional to k_m^2.)

(iv) Find an expression for the total two-photon transition rate from all possible initial states to all possible final states, valid to lowest order in k_m, and in terms of optical intensity, I, the photon energy, $\hbar\omega$, the quantity $E_p = 2|p_{21}|^2 / m_o$, the refractive index n_r, and fundamental constants. (Note: Because we are only interested in the lowest order behavior with k_m, you may use the parabolic dispersion relations for the bands.)

(v) Consider a semiconductor with $n_r \cong 3.5$, $E_g = 1$ eV, and $E_p = 20$ eV. A short pulse laser has an average power of 1 mW and a photon energy of 0.6 eV. Its light is concentrated in pulses that are 100 fs long with a repetition rate of 100 MHz. For simplicity of calculation, you may assume that the light intensity is constant throughout the pulse. This light shines on a 10 x 10 micron square area of the surface of the semiconductor, approximately uniformly (you may assume the surface is antireflection coated so there is no reflection from the surface). The semiconductor is 1 micron thick. The semiconductor is structured like a photodiode so that every electron-hole pair generated in the semiconductor gives rise to one electron of current. Approximately what will be the average current generated in this diode using the two-photon absorption model constructed here?

8.11 Summary of concepts

Crystals

Crystals are structures whose measurable properties are periodic in space. Lattice vectors of the form

$$\mathbf{R}_L = n_1\mathbf{a}_1 + n_2\mathbf{a}_2 + n_3\mathbf{a}_3 \tag{8.1}$$

where the numbers n_1, n_2, and n_3 are integers, map between equivalent points in the structure. There are only fourteen distinct mathematical lattices of points (i.e., Bravais lattices) that form the corners of objects that can repetitively fill all space. Two very common forms of lattices are the diamond lattice, of which silicon is an example, and the zinc-blende lattice (interlocking face-centered cubic lattice), a lattice form very closely related to the diamond lattice and one that is characteristic of most III-V compounds (e.g., GaAs, InP).

One electron approximation

The starting point for most band theory of crystalline solids is to presume that an electron moves in an effective periodic potential given by all of the other nuclei and electrons in the structure.

Bloch theorem

The Bloch theorem states that the wavefunction solutions in a periodic structure must be of the form

$$\psi(\mathbf{r}+\mathbf{a}) = \exp(i\mathbf{k}.\mathbf{a})\psi(\mathbf{r}) \tag{8.19}$$

where $\mathbf{a}$ is any lattice vector or, equivalently (and more commonly)

$$\psi(\mathbf{r}) = u(\mathbf{r})\exp(i\mathbf{k}.\mathbf{r}) \tag{8.20}$$

where $u(\mathbf{r})$ is a function that is periodic with the lattice periodicity and $\mathbf{k}$ can only take on one of a discrete set of vector values; considering the three crystal basis vector directions, 1, 2, and 3, with lattice constants (repeat distances) a_1, a_2, and a_3, and numbers of atoms N_1, N_2, and N_3

$$k_1 = \frac{2\pi n_1}{N_1 a_1} \quad ...n_1 = 0, \pm 1, \pm 2, ... \pm N_1/2 \tag{8.21}$$

and similarly for the other two components of **k** in the other two crystal basis vector directions.

Density of states in *k*-space

k-states are equally spaced in *k* in each direction, regardless of the precise form of the crystal's potential, and have a density of states in *k*-space, per unit volume of the crystal, of

$$g(\mathbf{k}) = \frac{1}{(2\pi)^3} \tag{8.24}$$

Band structure

Solving for the possible states of an electron in a crystalline material leads to a band structure, which is plotted as electron energy as a function of **k**, in principle, in all three directions, though, in practice, in some projections along specific symmetry directions. There is essentially one line or band on such a diagram for every atomic state of the original constituent atoms or molecules and essentially one actual state on the line for every atom or molecule that makes up the piece of material. The highest filled band in a semiconductor is called the valence band and the lowest unfilled band is the conduction band.

Kramers degeneracy

Neglecting the effects of electron spin, for every state with a corresponding wavevector **k** in a band, there is one with equal energy of wavevector −**k**.

Effective mass theory

Specific points in the band structure are maxima or minima of energy in the bands. Because extra electrons and the absences of electrons (holes) tend to collect near these minima or maxima, many important properties, such as transport and optical emission, can be understood from the properties near such maxima and minima.

By presuming that the periodic parts of the wavefunctions in a given band are rather similar near such a maximum or minimum and treating the minima and maxima as approximately parabolic, we can postulate that the wavefunction can be written as

$$\Psi(\mathbf{r},t) = u_0(\mathbf{r})\Psi_{env}(\mathbf{r},t) \tag{8.30}$$

where the envelope function $\Psi_{env}(\mathbf{r},t)$ can be written

$$\Psi_{env}(\mathbf{r},t) = \sum_{\mathbf{k}} c_{\mathbf{k}} \exp(i\mathbf{k}.\mathbf{r})\exp(-iE_{\mathbf{k}}t/\hbar) \tag{8.31}$$

which leads to a Schrödinger equation for the envelope

$$-\frac{\hbar^2}{2m_{eff}}\nabla^2\Psi_{env}(\mathbf{r},t) + V(\mathbf{r})\Psi_{env}(\mathbf{r},t) = i\hbar\frac{\partial}{\partial t}\Psi_{env}(\mathbf{r},t) \tag{8.33}$$

where the energy of states near these minima or maxima can be written in this parabolic approximation as

$$E_{\mathbf{k}} = \frac{\hbar^2 k^2}{2m_{eff}} + V \tag{8.28}$$

thus defining the effective mass m_{eff} for each band minimum or maximum. Electrons and holes (absences of electrons) in crystalline solids, therefore, can behave as particles with effective masses different from the free electron mass.

Effective mass approximation in semiconductor heterostructures

Use of boundary conditions such as

$$\Psi_{env} \text{ continuous} \tag{8.35}$$

and

$$\frac{1}{m_{eff}} \nabla \Psi_{env} \text{ continuous} \tag{8.36}$$

allows problems, such as quantum wells, to be handled in semiconductor heterostructures using the effective masses of the electrons and holes in the constituent materials and considering the simpler envelope function Schrödinger equation.

Density of states in energy

Near a parabolic band maximum or minimum, the density of states in energy is

$$g(E) = \frac{1}{2\pi^2}\left(\frac{2m_{eff}}{\hbar^2}\right)^{3/2}(E-V)^{1/2} \tag{8.40}$$

One-electron Schrödinger equation for Bloch functions

In a crystal, we can rewrite the Schrödinger equation without further approximations as

$$\left[\hat{H}_o + \frac{\hbar}{m_o}\mathbf{k}\cdot\hat{\mathbf{p}}\right]u_{n\mathbf{k}}(\mathbf{r}) = \left[E_n(\mathbf{k}) - \frac{\hbar^2 k^2}{2m_o}\right]u_{n\mathbf{k}}(\mathbf{r}) \tag{8.61}$$

Luttinger–Kohn representation

The unit-cell functions at $k = 0$ in all the different bands form a complete set for representing unit-cell functions, so we can expand any unit-cell function in terms of them, as

$$u_{n\mathbf{k}}(\mathbf{r}) = \sum_{n'} a_{nn'k} u_{n'0}(\mathbf{r}) \tag{8.64}$$

Chapter 9

Angular momentum

Prerequisites: Chapters 2–5.

Thus far, we have dealt primarily with energy, position, and linear momentum and have proposed operators for each of these. One other quantity that is important in classical mechanics, *angular momentum*, is particularly important also in quantum mechanics. Here, we introduce angular momentum, its operators, eigenvalues, and eigenfunctions. If this discussion seems somewhat abstract, the reader can be assured that the concepts of angular momentum will become very concrete in the discussion of the hydrogen atom.

One aspect of angular momentum that is different from the quantities and operators discussed previously is that its operators *always* have discrete eigenvalues. Whereas linear momentum is associated with eigenfunctions that are functions of position along a specific spatial direction, angular momentum is associated with eigenfunctions that are functions of angle or angles about a specific axis. The fact that the eigenvalues are discrete is associated with the fact that for a single-valued spatial function, once we have gone an angle 2π about a particular axis, we are back to where we started. The wavefunction is presumably continuous and single-valued[1] and, hence, we must therefore have integral numbers of periods of oscillation with angle within this angular range; this requirement of integer numbers of periods leads to the discrete quantization of angular momentum.

Another surprising aspect of angular momentum operators is that the operators corresponding to angular momentum about different orthogonal axes (e.g., $\hat{L}_x$, $\hat{L}_y$, and $\hat{L}_z$) do *not* commute with one another (in contrast, e.g., to the linear momentum operators for the different orthogonal coordinate directions). We do, however, find that there is another useful angular momentum operator, $\hat{L}^2$, which does commute with each of $\hat{L}_x$, $\hat{L}_y$, and $\hat{L}_z$ separately.

Here, we propose the form of the angular momentum operators and solve for the eigenfunctions and eigenvalues. The eigenfunctions for $\hat{L}_x$, $\hat{L}_y$, and $\hat{L}_z$ are quite straightforward. Those for $\hat{L}^2$, the spherical harmonics, may at first appear to be mathematically rather obscure and complicated functions, though, in fact, they can be understood relatively simply.

9.1 Angular momentum operators

The reader may remember from elementary classical mechanics that the angular momentum of a small object of (vector) linear momentum **p** centered at a point given by the vector displacement **r** relative to some origin is

---

[1] That is, at any given angle there is only one answer for the amplitude of the wavefunction.

$$\mathbf{L} = \mathbf{r} \times \mathbf{p} \tag{9.1}$$

relative to that origin, where the vector cross product (see Appendix A, Section A.1) is defined as usual by

$$\begin{aligned} \mathbf{C} = \mathbf{A} \times \mathbf{B} \equiv \mathbf{c} AB \sin\theta \equiv \begin{vmatrix} \mathbf{i} & \mathbf{j} & \mathbf{k} \\ A_x & A_y & A_z \\ B_x & B_y & B_z \end{vmatrix} \\ \equiv \mathbf{i}(A_y B_z - A_z B_y) - \mathbf{j}(A_x B_z - A_z B_x) + \mathbf{k}(A_x B_y - A_y B_x) \end{aligned} \tag{9.2}$$

where here **i**, **j**, and **k** are unit vectors in the usual x, y, and z coordinate directions and A_x is the component of **A** in the x direction (and similarly for the other components).[2] **C** is perpendicular to the plane of **A** and **B** just as the z axis is perpendicular to the plane containing the x and y axes for normal right-handed axes.[3] θ is the angle between the vectors **A** and **B** and **c** is a unit vector in the direction of the vector **C**. Fig. 9.1 illustrates vectors **r** and **p** for the angular momentum cross product of Eq. (9.1).

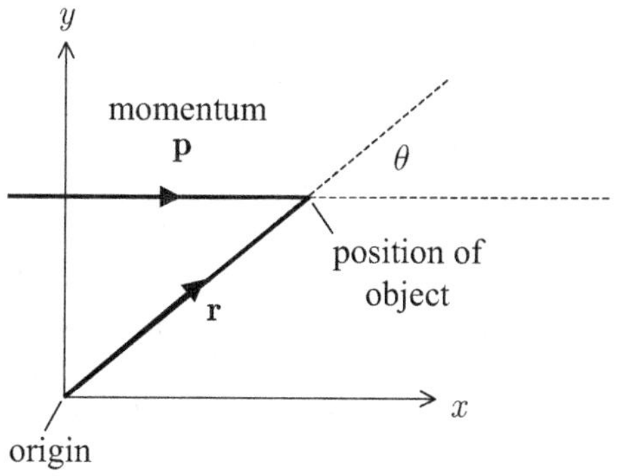

Fig. 9.1. Illustration of the momentum and position vectors for the case where the momentum is parallel to the x axis. In this case, the resulting angular momentum vector would point into the page away from the reader.

Because the angular momentum is a vector quantity, we can, if we wish, explicitly write out the various components, obtaining

$$L_x = yp_z - zp_y\,,\quad L_y = zp_x - xp_z\,,\quad L_z = xp_y - yp_x \tag{9.3}$$

In quantum mechanics, we can propose an angular momentum operator $\hat{\mathbf{L}}$ by analogy with the classical angular momentum[4]

---

[2] Note, incidentally, that the ordering of the multiplications in the second line of Eq. (9.2) is carefully chosen to work also when we use operators instead of numbers for one vector or the other; the sequence of multiplications in each term is always in the sequence of the rows from top to bottom.

[3] Turning a right-handed screw head from the x direction to the y direction drives it in the z direction in right-handed axes.

[4] It might seem pedantic here to use the position operator $\hat{\mathbf{r}}$ rather than the simple position **r**, and then merely confusing to use **r** later. It is, however, only in the position representation that the position operator $\hat{\mathbf{r}}$ is simply the position **r**. If we used some other representation – for example, a momentum representation (i.e., a spatial Fourier transform of the position representation) as discussed in Section 5.4 – the position operator is not simply the same as the position vector. The statement $\hat{\mathbf{L}} = \hat{\mathbf{r}} \times \hat{\mathbf{p}}$ is correct in

$$\hat{\mathbf{L}} = \hat{\mathbf{r}} \times \hat{\mathbf{p}} = -i\hbar(\mathbf{r} \times \nabla) \tag{9.4}$$

with corresponding components

$$\hat{L}_x = \hat{y}\hat{p}_z - \hat{z}\hat{p}_y = -i\hbar\left(y\frac{\partial}{\partial z} - z\frac{\partial}{\partial y}\right) \tag{9.5}$$

$$\hat{L}_y = \hat{z}\hat{p}_x - \hat{x}\hat{p}_z = -i\hbar\left(z\frac{\partial}{\partial x} - x\frac{\partial}{\partial z}\right) \tag{9.6}$$

$$\hat{L}_z = \hat{x}\hat{p}_y - \hat{y}\hat{p}_x = -i\hbar\left(x\frac{\partial}{\partial y} - y\frac{\partial}{\partial x}\right) \tag{9.7}$$

These individual components of the angular momentum operator $\hat{\mathbf{L}}$ are each Hermitian,[5] and so, correspondingly, is the operator $\hat{\mathbf{L}}$ itself. The operators corresponding to individual coordinate directions obey commutation relations

$$\hat{L}_x\hat{L}_y - \hat{L}_y\hat{L}_x = \left[\hat{L}_x, \hat{L}_y\right] = i\hbar\hat{L}_z \tag{9.8}$$

$$\hat{L}_y\hat{L}_z - \hat{L}_z\hat{L}_y = \left[\hat{L}_y, \hat{L}_z\right] = i\hbar\hat{L}_x \tag{9.9}$$

$$\hat{L}_z\hat{L}_x - \hat{L}_x\hat{L}_z = \left[\hat{L}_z, \hat{L}_x\right] = i\hbar\hat{L}_y \tag{9.10}$$

Note that these relations are cyclic permutations of one another (i.e., the first two subscript indices in Eq. (9.8) are x and y, the last subscript index on the right is z; the corresponding order of these same indices in Eq. (9.9) is y, z, x, and in Eq. (9.10) the order is z, x, y), which may make them easier to remember. These individual commutation relations can also be written in a more compact form as a shorthand

$$\hat{\mathbf{L}} \times \hat{\mathbf{L}} = i\hbar\hat{\mathbf{L}} \tag{9.11}$$

Eq. (9.11) may seem a strange way to write a set of commutation relations – it does not resemble the usual form of commutation relations very closely – but it is only a shorthand for writing or remembering Eqs. (9.8) through (9.10).[6]

---

any representation; the statement $\hat{\mathbf{L}} = -i\hbar(\mathbf{r} \times \nabla)$ is only correct in the position representation. We are using the position representation for the moment, however.

[5] The proof of the Hermiticity of the individual components of the angular momentum operator is left as an exercise for the reader – it follows from the known Hermiticity of the individual operator components and the commutation properties of the position and momentum operators corresponding to the different coordinate directions.

[6] Another shorthand way of writing the same set of commutation relations, which is closer to the form of a simple commutation relation, is $[\hat{L}_m, \hat{L}_n] = i\hbar\varepsilon_{mnl}\hat{L}_l$, where the indices m, n, and l can each be any of x, y, or z, and ε_{mnl} is an entity known as the totally antisymmetric tensor or, equivalently, the Levi–Civita symbol. This tensor has the following properties. First, $\varepsilon_{xyz} = 1$. For an ε_{mnl} formed by an even number of permutations of the indices (an even permutation), e.g., ε_{yzx} (which is formed by swapping x and y, followed by swapping x and z), the value is still +1. For an ε_{mnl} formed by an odd number of permutations of the indices (an odd permutation), e.g., ε_{xzy} (which is formed by swapping y and z), the value is –1. If two of the indices are the same, as in ε_{xxy}, the value is zero (as required, e.g., because an operator commutes with itself). The "multiplication" $\varepsilon_{mnl}\hat{L}_l$ is technically a tensor multiplication, which means that there is technically a summation over the repeated index l. For any given (different) m and n,

We started here by postulating an operator $\hat{\mathbf{L}}$ for angular momentum and deduced the corresponding operators for angular momentum components on each of the Cartesian directions. Note, however, that unlike our previously postulated vector operators, $\hat{\mathbf{r}}$ for position and $\hat{\mathbf{p}}$ for linear momentum, the different components of this angular momentum operator do not commute with one another. It was quite possible, for example, for a particle to have simultaneously a well-defined position in both the x and y directions, for example, or alternatively to have simultaneously a well-defined momentum in both the x and y directions. The commutation relations Eqs. (9.8) through (9.10) tell us that a particle cannot, in general, simultaneously have a well-defined angular momentum component in more than one direction.[7] This conclusion may appear bizarre, but it is nonetheless apparently correct.

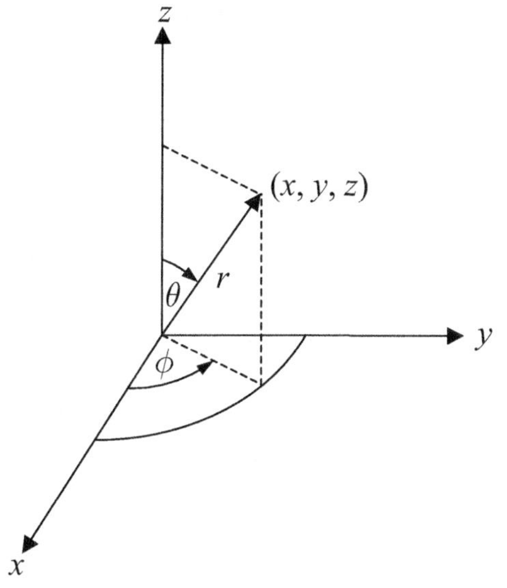

Fig. 9.2. Illustration of spherical polar coordinates, r, θ, and ϕ.

The eigenfunctions and eigenvalues of angular momentum operators are most usefully written using spherical polar coordinates. The standard definition of these coordinates is illustrated in Fig. 9.2 (see also Appendix C, Section C.2). The relation between spherical polar and Cartesian coordinates is

$$x = r\sin\theta\cos\phi \ , \ y = r\sin\theta\sin\phi \ , \ z = r\cos\theta \tag{9.12}$$

and, in inverse form

there is, however, only one value of l for which ε_{mnl} is not zero, so this tensor multiplication reduces in practice to a simple multiplication.

[7] These angular momentum commutation relations also differ from the previous commutation relations we have examined in that the quantity on the right is an operator, not simply a number. The practical result of this is that the uncertainty relation between the standard deviations of the expected values of, say, the x and y components of the angular momentum depends on the state of the quantum mechanical system. E.g., if the system is in an eigenstate of $\hat{L}_z$, the operator $\hat{L}_z$ is effectively replaced by its eigenvalue, m_z, in the $\hat{L}_x$ and $\hat{L}_y$ commutation relation and in the corresponding uncertainty principle for the x and y components of the angular momentum.

$$r=\sqrt{x^2+y^2+z^2}\ ,\ \theta=\sin^{-1}\left(\frac{\sqrt{x^2+y^2}}{\sqrt{x^2+y^2+z^2}}\right),\ \phi=\tan^{-1}\left(\frac{y}{x}\right) \tag{9.13}$$

θ is known as the polar angle and ϕ is the azimuthal angle.

With these definitions of spherical polar coordinates and with standard partial derivative relations of the form (see Appendix A, Section A.5)

$$\frac{\partial}{\partial x}\equiv\frac{\partial r}{\partial x}\frac{\partial}{\partial r}+\frac{\partial\theta}{\partial x}\frac{\partial}{\partial\theta}+\frac{\partial\phi}{\partial x}\frac{\partial}{\partial\phi} \tag{9.14}$$

for each of the Cartesian coordinate directions, it is straightforward (if a little tedious)[8] to show from Eqs. (9.5) through (9.7) that

$$\hat{L}_x=i\hbar\left(\sin\phi\frac{\partial}{\partial\theta}+\cot\theta\cos\phi\frac{\partial}{\partial\phi}\right) \tag{9.15}$$

$$\hat{L}_y=i\hbar\left(-\cos\phi\frac{\partial}{\partial\theta}+\cot\theta\sin\phi\frac{\partial}{\partial\phi}\right) \tag{9.16}$$

$$\hat{L}_z=-i\hbar\frac{\partial}{\partial\phi} \tag{9.17}$$

We can solve Eq. (9.17) for the eigenfunctions and eigenvalues of $\hat{L}_z$. The eigenequation is

$$\hat{L}_z\Phi(\phi)=m\hbar\Phi(\phi) \tag{9.18}$$

where $m\hbar$ is the eigenvalue to be determined. The solution of this equation is

$$\Phi(\phi)=\exp(im\phi) \tag{9.19}$$

The requirements that the wavefunction and its derivative are continuous when we return to where we started (i.e., for $\phi=2\pi$) mean that m must be an integer (positive or negative or zero). Hence we find that the angular momentum around the z axis is quantized, with units of angular momentum of $\hbar$.

From this discussion, we can understand that indeed the eigenfunctions of the angular momentum about the different axes are not the same, as required by the fact that the different angular momentum operators do not commute. Our choice of the z axis was quite arbitrary – we could equally well have chosen the x or y axes as the polar axes for our coordinate system, in which case we would see quite clearly that the eigenfunctions of the $\hat{L}_x$ or $\hat{L}_y$ operators are of similar form to Eq. (9.19) but in terms of the angles of rotation about the x or y axes,

[8] Remember that, e.g., the partial derivative $\partial r/\partial x$ is implicitly taken at constant y and z and is *not* the same as $1/\left(\frac{\partial x}{\partial r}\right)$, where the derivative is implicitly taken at constant θ and ϕ. Hence, the expression for r in terms of x, y, and z must be used to evaluate this derivative. Similar considerations apply to the other partial derivatives. The relations $\frac{d\sin^{-1}x}{dx}=\frac{1}{\sqrt{1-x^2}}$ and $\frac{d\tan^{-1}x}{dx}=\frac{1}{1+x^2}$ are useful here.

respectively. Hence, the eigenfunctions of the angular momentum operators $\hat{L}_x$, $\hat{L}_y$, and $\hat{L}_z$ are not the same.

Problems

9.1.1 Starting from the definition of the operators for the Cartesian components of the angular momentum, show that $\hat{\mathbf{L}} \times \hat{\mathbf{L}} = i\hbar\hat{\mathbf{L}}$. (Note that you should also explicitly derive any commutation relations that you need between the operators of the angular momentum components, starting from the known commutation properties of position and [linear] momentum operators.)

9.1.2 Starting from the definition in Cartesian coordinates and noting that

$$\frac{\partial}{\partial x} \equiv \frac{\partial r}{\partial x}\frac{\partial}{\partial r} + \frac{\partial \theta}{\partial x}\frac{\partial}{\partial \theta} + \frac{\partial \phi}{\partial x}\frac{\partial}{\partial \phi}$$

show that

$$L_x = i\hbar\left(\sin\phi\frac{\partial}{\partial \theta} + \cot\theta\cos\phi\frac{\partial}{\partial \phi}\right)$$

(Remember that $\dfrac{d\sin^{-1}x}{dx} = \dfrac{1}{\sqrt{1-x^2}}$ and $\dfrac{d\tan^{-1}x}{dx} = \dfrac{1}{1+x^2}$)

9.2 L squared operator

In quantum mechanics, it is also useful to consider another operator associated with angular momentum, the operator $\hat{L}^2$. This should be thought of as the "dot" product of $\hat{\mathbf{L}}$ with itself and is defined as

$$\hat{L}^2 = \hat{L}_x^2 + \hat{L}_y^2 + \hat{L}_z^2 \tag{9.20}$$

It is similarly straightforward (and, again, somewhat tedious) to show then that

$$\hat{L}^2 = -\hbar^2\nabla^2_{\theta,\phi} \tag{9.21}$$

where the operator $\nabla^2_{\theta,\phi}$ is given by

$$\nabla^2_{\theta,\phi} = \left[\frac{1}{\sin\theta}\frac{\partial}{\partial\theta}\left(\sin\theta\frac{\partial}{\partial\theta}\right) + \frac{1}{\sin^2\theta}\frac{\partial^2}{\partial\phi^2}\right] \tag{9.22}$$

This $\nabla^2_{\theta,\phi}$ operator is actually the θ and ϕ part of the Laplacian (∇^2) operator in spherical polar coordinates; hence, the notation. (We return to the full Laplacian in spherical polar coordinates when we examine the hydrogen atom later.) This relation between $\hat{L}^2$ and the Laplacian gives a mathematical hint as to why we have chosen to introduce this operator.

$\hat{L}^2$ has another property that is very important physically: it commutes with $\hat{L}_x$, $\hat{L}_y$, and $\hat{L}_z$. It is easy to see from Eq. (9.22) and the form of $\hat{L}_z$ (Eq. (9.17)) that, at least, $\hat{L}^2$ and $\hat{L}_z$ commute; the operation $\partial/\partial\phi$ has no effect on functions or operators depending on θ alone, there are no functions of ϕ in the $\hat{L}^2$ operator, and $\partial/\partial\phi$ commutes with $\partial^2/\partial\phi^2$. Of course, the choice of the z direction is quite arbitrary; we could equally well have developed this problem considering the polar axis along the x or y directions, in which cases it would similarly be obvious that $\hat{L}^2$ commutes with $\hat{L}_x$ or $\hat{L}_y$.[9]

---

[9] At this point, the reader should be confused. Did we not decide that $\hat{L}_x$, $\hat{L}_y$, and $\hat{L}_z$ do not commute with each other? How then can each of them commute with the same operator $\hat{L}^2$? The answer is that it is true

Now let us examine the eigenfunctions of this operator $\hat{L}^2$. We are looking for eigenfunctions of $\hat{L}^2$ (or, equivalently, $\nabla^2_{\theta,\phi}$) and so the equation we hope to solve is of the form

$$\nabla^2_{\theta,\phi} Y_{lm}(\theta,\phi) = -l(l+1) Y_{lm}(\theta,\phi) \tag{9.23}$$

We have anticipated the answer by writing the expected eigenvalue in the form $-l(l+1)$ but, for the moment, we can consider this as an arbitrary number to be determined. The notation $Y_{lm}(\theta,\phi)$ also anticipates the final answer, though again, for the moment, it is an arbitrary function to be determined.

Formally, we attempt separation of variables to divide the problem into smaller solvable units. The reader is probably familiar with this from previous problems in more than one variable. Indeed, the main reason for the choice of particular coordinate systems (e.g., here spherical polars) is to make such separations possible. We presume that the final eigenfunctions can be separated[10] in the form

$$Y_{lm}(\theta,\phi) = \Theta(\theta)\Phi(\phi) \tag{9.24}$$

where $\Theta(\theta)$ only depends on θ and $\Phi(\phi)$ only depends on ϕ. Substituting this form in Eq. (9.23) gives

$$\frac{\Phi(\phi)}{\sin\theta}\frac{\partial}{\partial\theta}\left(\sin\theta\frac{\partial}{\partial\theta}\right)\Theta(\theta) + \frac{\Theta(\theta)}{\sin^2\theta}\frac{\partial^2\Phi(\phi)}{\partial\phi^2} = -l(l+1)\Theta(\theta)\Phi(\phi) \tag{9.25}$$

Multiplying both sides by $\sin^2\theta/\Theta(\theta)\Phi(\phi)$ and rearranging terms, we have

$$\frac{1}{\Phi(\phi)}\frac{\partial^2\Phi(\phi)}{\partial\phi^2} = -l(l+1)\sin^2\theta - \frac{\sin\theta}{\Theta(\theta)}\frac{\partial}{\partial\theta}\left(\sin\theta\frac{\partial}{\partial\theta}\right)\Theta(\theta) \tag{9.26}$$

In the usual manner of separation of variables, we note that the left-hand side depends only on ϕ, whereas the right-hand side depends only on θ, and so these must both equal a constant (the separation constant). Again anticipating the answer with our notation, we choose a separation constant of $-m^2$, where m is still to be determined. Note that now that we have performed this separation, it is no longer necessary to use the partial derivative notation; on each side of the equation, the functions only depend on one variable, so henceforth we can use the normal total derivative notation for the resulting separated equations.

Taking the left-hand side first, we now have an equation

---

that the eigenfunctions of $\hat{L}^2$ can indeed be chosen so that they are also the eigenfunctions of $\hat{L}_x$, $\hat{L}_y$, or $\hat{L}_z$. But the version of the $\hat{L}^2$ eigenfunctions that is the same as the $\hat{L}_z$ eigenfunctions is different from the versions that are eigenfunctions of $\hat{L}_y$ or $\hat{L}_z$ (they each have different polar axes). Why is there this ambiguity? The answer lies in the spherical symmetry of the operator $\hat{L}^2$; it does not care what we choose as the polar axis and will happily have its eigenfunctions defined with respect to any polar axis.

In real physical problems, there is very often something about the problem that defines one axis as being special in some way; by convention, we typically call that axis the z axis. E.g., there might be an electric or magnetic field along that axis. In such a case, that axis defines one direction over the others and, in the limit as the perturbation along that axis tends towards zero, we typically recover the kinds of solutions we have here with the z axis having a special role.

[10] The justification for any such proposed separation is that, *a posteriori*, we find that it satisfies the equation and has enough arbitrary constants to be a general solution.

$$\frac{d^2\Phi(\phi)}{d\phi^2} = -m^2\Phi(\phi) \tag{9.27}$$

The solutions to an equation like this are of the form $\sin m\phi$, $\cos m\phi$, or $\exp im\phi$. In quantum mechanics, we choose the exponential form so that this is also a solution of the $\hat{L}_z$ eigenequation, Eq. (9.18). Again, as we mentioned previously, because this is to be a quantum mechanical spatial wavefunction, we expect that it and its derivative are continuous and single valued. As a result, this wavefunction must be cyclic every 2π of angle ϕ and m must be an integer.

Taking the right-hand side of Eq. (9.26), we can now write an equation

$$\frac{1}{\sin\theta}\frac{d}{d\theta}\left(\sin\theta\frac{d}{d\theta}\right)\Theta(\theta) - \frac{m^2}{\sin^2\theta}\Theta(\theta) + l(l+1)\Theta(\theta) = 0 \tag{9.28}$$

Fortunately, the solutions to this equation are known. This is the associated Legendre equation and the solutions to it are the associated Legendre functions, $\Theta(\theta) = P_l^m(\cos\theta)$. The solutions require that

$$l = 0, 1, 2, 3, \ldots \tag{9.29}$$

$$-l \le m \le l \ \ (m \text{ integer}) \tag{9.30}$$

The associated Legendre functions can conveniently be defined using Rodrigues's formula

$$P_l^m(x) = \frac{1}{2^l l!}\left(1-x^2\right)^{m/2}\frac{d^{l+m}}{dx^{l+m}}\left(x^2-1\right)^l \tag{9.31}$$

The first few of these functions are, explicitly

$$\begin{aligned}
P_0^0(x) &= 1 \\
P_1^0(x) &= x \\
P_1^1(x) &= \left(1-x^2\right)^{1/2} \\
P_1^{-1}(x) &= -\frac{1}{2}\left(1-x^2\right)^{1/2} \\
P_2^0(x) &= \frac{1}{2}\left(3x^2-1\right) \\
P_2^1(x) &= 3x\left(1-x^2\right)^{1/2} \\
P_2^{-1}(x) &= -\frac{1}{2}x\left(1-x^2\right)^{1/2} \\
P_2^2(x) &= 3\left(1-x^2\right) \\
P_2^{-2}(x) &= \frac{1}{8}\left(1-x^2\right)
\end{aligned} \tag{9.32}$$

We can see by inspection that these functions have the following properties: (1) the highest power of the argument x is always x^l; and (2) the functions for a given l for $+m$ and $-m$ are identical (other than for differences in numerical prefactors). Less obviously, between -1 and $+1$, and not including the values at those end points, the functions have $l-|m|$ zeros.

Now we can retrace our steps and construct the full eigenfunctions of $\hat{\mathbf{L}}^2$. From Eqs. (9.21) and (9.23), the eigenequation is

$$\hat{L}^2 Y_{lm}(\theta,\phi) = \hbar^2 l(l+1) Y_{lm}(\theta,\phi) \tag{9.33}$$

where $Y_{lm}(\theta,\phi)$ are referred to as the spherical harmonics, which, after normalization,[11] can be written

$$Y_{lm}(\theta,\phi) = (-1)^m \sqrt{\frac{2l+1}{4\pi}\frac{(l-m)!}{(l+m)!}}\, P_l^m(\cos\theta)\exp(im\phi) \tag{9.34}$$

The allowed eigenvalues of $\hat{L}^2$ are $\hbar^2 l(l+1)$. As is easily verified, these spherical harmonics[12] are also eigenfunctions of the $\hat{L}_z$ operator. Explicitly, we have the eigenequation

$$\hat{L}_z Y_{lm}(\theta,\phi) = m\hbar Y_{lm}(\theta,\phi) \tag{9.35}$$

with the eigenvalues of $\hat{L}_z$ being $m\hbar$.

Problems

9.2.1* Show explicitly in Cartesian (x, y, z) coordinates that the ∇^2 and $\hat{L}_z$ operators commute; that is,

$$\left[\nabla^2, \hat{L}_z\right] = 0$$

9.2.2 Write out the associated Legendre functions for $l = 0$, 1, and 2 with cos θ instead of x as the argument (possibly simplifying the results using $\cos^2\theta + \sin^2\theta = 1$) and, hence, state all the angles θ for which each of these associated Legendre functions has zeros.

9.3 Visualization of spherical harmonic functions

See also Appendix D, Section D.9, for a brief background discussion of modes.

These spherical harmonics may appear to be relatively obscure functions but, in fact, it is quite possible to visualize what they are like and there is quite a simple rule that allows one to describe an arbitrary spherical harmonic qualitatively.

It is easiest to visualize spherical harmonics if we step away from the quantum mechanical problem for the moment and think of another problem for which the spherical harmonics are the solutions – namely, the modes of vibration of a thin spherical shell.[13] The only substantial

---

[11] Note that the normalization factor is just what is required to make the θ dependence of these spherical harmonics identical for positive and negative values of m; it compensates for the different numerical prefactors in the P_l^m and P_l^{-m} functions.

[12] The spherical harmonics are the eigenfunctions of the $\hat{L}_z$ operator at least when defined using $\exp(im\phi)$ rather than an alternative definition using $\cos m\phi$ and $\sin m\phi$.

[13] One can readily construct this spherical shell problem, just as one constructs the usual wave equation for, say, a wave on a string. If an element of the surface is displaced radially by an amount ψ, it experiences an elastic restoring force proportional to the local second spatial derivative of that displacement in the plane (i.e., $\propto \nabla^2_{\theta\phi}\psi$), which leads to a proportionate restoring acceleration $\propto \partial^2\psi/\partial t^2$. The presumption of sinusoidal oscillations in time leads to the usual Helmholtz wave equation that here we can write in the form $\nabla^2_{\theta\phi}\psi = -a\psi$ for some positive eigenvalues a to be determined, which is the same form as our quantum mechanical equation.

mathematical difference between these two problems is that the shell's vibrations must have real amplitudes; and so, instead of using positive and negative m with complex functions $\exp(im\phi)$, we use only positive m but use both $\sin m\phi$ and $\cos m\phi$ solutions. (Remember that the θ dependence of the spherical harmonics is the same for positive and negative m.) Because the solutions corresponding to $\sin m\phi$ and $\cos m\phi$ are merely rotated by one quarter of a period (an angle of $\pi/2m$) around the polar axis with respect to one another, we only need to understand one of these to understand the nature of the spherical harmonics. It is simple to construct the $\exp(im\phi)$ solution from these.

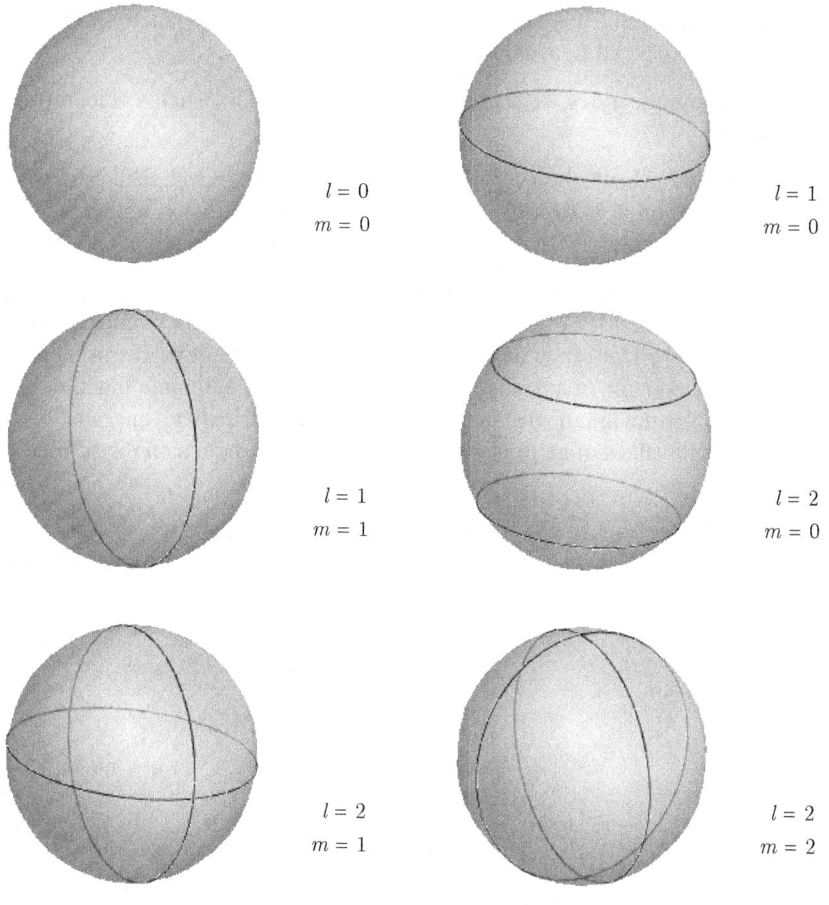

Fig. 9.3. Illustration of the nodal circles for the spherical harmonics corresponding to the vibration modes of a spherical shell, for various cases ($l = 0$, $m = 0$, $l = 1$, $m = 0$, $l = 1$, $m = 1$, $l = 2$, $m = 0$, $l = 2$, $m = 1$, $l = 2$, $m = 2$.)

Fig. 9.3 illustrates the nature of the solutions for the vibrating spherical shell problem, here showing the $\cos m\phi$ solutions (i.e., the real part of the complex spherical harmonics). The lowest solution ($l = 0, m = 0$) simply corresponds to the "breathing" mode of the spherical shell – the shell simply expands and contracts periodically while retaining its spherical shape. For all other solutions, there are one or more nodal circles on the sphere. A nodal circle is one that is exactly unchanged in that particular oscillating mode. To understand the oscillations of the sphere in a given spherical harmonic, we can use the following rules:

(i) The surfaces on opposite sides of a nodal circle oscillate in opposite directions – exactly as one would expect for a node in an oscillating surface.

(ii) The total number of nodal circles is equal to l.

(iii) The number of nodal circles passing through the poles is m and they divide the sphere equally in the azimuthal angle ϕ. (The precise positions of these circles are determined by the zeros of $\cos m\phi$.)

(iv) The remaining nodal circles are either equatorial or parallel to the equator and are symmetrically distributed between the top and bottom halves of the sphere. (Only the precise position of these circles is not immediately obvious from symmetry and is determined by the zeros of the associated Legendre functions.)

With these simple rules, it is quite straightforward to visualize the form of the spherical harmonics.

Of course, this visualization of vibrating modes of a spherical shell is not itself the spherical harmonic function. It is the *amplitude* of the oscillation of the vibration at any given angle that is the spherical harmonic function; spherical harmonics are functions of angle only, not radius.

Note that the amplitude of the spherical harmonic may be positive for one range of angles and negative for another; for example, for all the $l = 1$ spherical harmonics, the function is positive on one hemispherical range of angle and negative on the other hemispherical range. Fig. 9.4 illustrates an example of a parametric plot of a spherical harmonic function. Such plots, though often plotted in quantum mechanics texts, can be difficult to interpret because it is not so obvious which values of angles correspond to particular points on the surface.

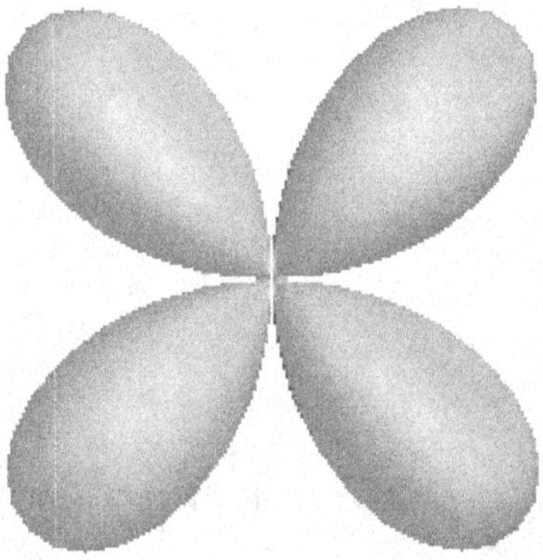

Fig. 9.4. A plot of the spherical harmonic for $l = 2$, $m = 1$, showing two "dumbbells." These dumbbells can be seen to be the consequence of a polar circle of value zero, which pinches the function to zero in the vertical direction, and an equatorial circle of value zero, which pinches the function to zero in the horizontal direction. This kind of plot is a polar parametric plot as a function of θ and ϕ of the surface given by the set of points (x, y, z) with $x = Y_{21}(\theta,\phi)\sin\theta\cos\phi$, $y = Y_{21}(\theta,\phi)\sin\theta\sin\phi$, and $z = Y_{21}(\theta,\phi)\cos\theta$. The view is along the y axis.

9.4 Comments on notation

Dirac notation

We also commonly use the Dirac notation in writing equations associated with angular momentum, in which case it is conventional to write

$$\hat{L}^2 |l,m\rangle = \hbar^2 l(l+1)|l,m\rangle \tag{9.36}$$

instead of Eq. (9.33) and

$$\hat{L}_z |l,m\rangle = m\hbar |l,m\rangle \tag{9.37}$$

instead of Eq. (9.35).

"s, p, d, f" notation

l	0	1	2	3	4	5
notation	s	p	d	f	g	h

In the context of quantum mechanics, the spherical harmonics come up in the solution of the hydrogen-atom problem. Different values of l give rise to different sets of spectral lines that spectroscopists in the nineteenth century had identified empirically in their work on hydrogen. In particular, they had identified what they called, respectively, *spectral* (s), *principal* (p), *diffuse* (d), and *fundamental* (f) groups of lines. Each of these is now identified with the specific values of l as follows, where we also indicate the alphabetic extension to higher l values.

It is also convenient to note that the s wavefunctions are all spherically symmetric, even though the s of the notation originally had nothing to do with spherical symmetry.

S, X, Y, Z functions

Many semiconductor crystals have crystalline structure that is based on a form of cubic crystal lattice (see Chapter 8). In particular, most of the semiconductors of technological importance, such as Si, Ge, GaAs, and InP, have such crystalline structures. For understanding many properties of these crystals, such as optical absorption, for example, we need to understand some of the attributes of the electron wavefunctions within the unit cell (the unit that when periodically repeated constructs the entire crystal lattice).

The detailed calculation of the wavefunctions within the unit cell is difficult, but often the qualitative aspects of the behavior, such as the polarization dependence of optical absorption, can be understood if we understand the symmetry of the unit-cell wavefunctions in the different bands. The wavefunctions within the unit cell can have symmetries that are the same as or similar to the symmetries of atomic states; the angular symmetries of atomic states are essentially those of the spherical harmonics.

For example, the unit-cell wavefunctions in the conduction band of materials such as GaAs and InP have approximately spherical symmetry (even though they are embedded in cubic unit cells), just like the $l = 0$ spherical harmonic function. In the analysis of such materials and their properties, this conduction-band unit-cell function is often written as $S(\mathbf{r})$, where here $\mathbf{r}$

refers to the position within the unit cell. We do not know the actual detailed form of $S(\mathbf{r})$, but we do know that it is approximately spherically symmetric within the unit cell.

If we look at the spherical harmonic for $l = 1$, $m = 0$, we see that it is antisymmetric about $z = 0$. The value of the function at a given point above the equator is minus that of the value of the function at the corresponding point below the equator. This is, somewhat trivially, the same symmetry as the function z itself. A function with this symmetry can, therefore, be written as $Z(\mathbf{r})$. For such a function, we may not know its detailed form, but we do know that it is antisymmetric about $z = 0$.

There are two other spherical harmonics for $l = 1$. If we consider the real form of the spherical harmonics (with $\cos m\phi$ and $\sin m\phi$ angular dependences) with $m = 1$, one of these (the $\cos m\phi$ one illustrated in Fig. 9.3) corresponds to a function that is antisymmetric about $x = 0$. A function with the same symmetry as this spherical harmonic can be written as $X(\mathbf{r})$. Similarly, the $l = 1$, $m = 1$ spherical harmonic with the $\sin m\phi$ dependence corresponds to a function that is antisymmetric about $y = 0$, and a function with this symmetry can be written as $Y(\mathbf{r})$.

Functions with these $X(\mathbf{r})$, $Y(\mathbf{r})$, and $Z(\mathbf{r})$ symmetries are the p atomic orbitals (often shown as three orthogonal dumbbell orbitals in chemistry texts). The valence band states of most importance in materials such as GaAs and InP can be characterized by unit-cell functions of these forms also, even though the detailed form of these functions is not known with any accuracy. This notation is extensively used in the analysis of the band structure properties of such materials to allow calculation of optical and other properties.

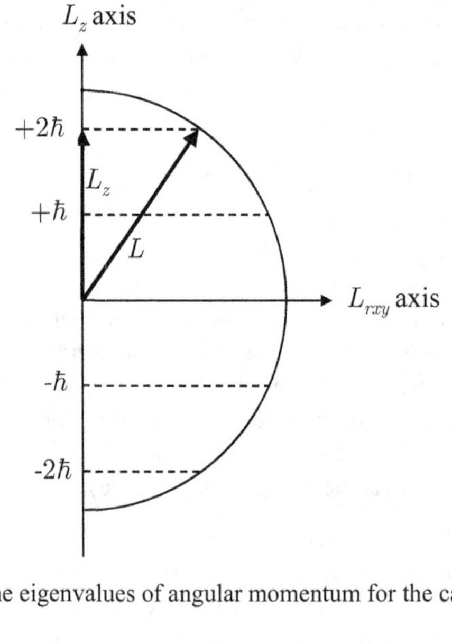

Fig. 9.5. Illustration of the eigenvalues of angular momentum for the case of $l = 2$.

9.5 Visualization of angular momentum

We can visualize the quantization of angular momentum as shown in Fig. 9.5. To make this visualization formal, we need to consider one more operator

$$\hat{L}_{rxy}^2 = \hat{L}^2 - \hat{L}_z^2 \tag{9.38}$$

where by the subscript rxy we are referring to the radius $r_{xy} = \sqrt{x^2 + y^2}$ in the x-y plane. Then we have

$$\hat{L}^2_{rxy}\left|l,m\right\rangle = \hat{L}^2\left|l,m\right\rangle - \hat{L}^2_z\left|l,m\right\rangle = \hbar^2\left[l(l+1) - m^2\right]\left|l,m\right\rangle = L^2_{rxy}\left|l,m\right\rangle \tag{9.39}$$

In other words, the state $\left|l,m\right\rangle$, in addition to being an eigenstate of $\hat{L}^2$ and $\hat{L}_z$, is also an eigenstate of this new operator $\hat{L}^2_{rxy}$, with eigenvalue $L^2_{rxy} = \hbar^2[l(l+1) - m^2]$. Because the two operators $\hat{L}_z$ and $\hat{L}^2_{rxy}$ have the same eigenstates, they both simultaneously can have exactly defined expectation values (i.e., eigenvalues) and we can plot the eigenstates and eigenvalues as on a circle in Fig. 9.5, which is drawn for a particular case of $l = 2$. The possible eigenstates correspond to the intersections of the horizontal straight lines, corresponding to the values of L_z and spaced by $\hbar$, and the semicircle of radius $\hbar\sqrt{l(l+1)}$ (i.e., $\sqrt{6}\hbar$ for this example).

9.6 Summary of concepts

Angular momentum operators

By extension from linear momentum, we can define angular momentum operators associated with each of the coordinate directions, $\hat{L}_x$, $\hat{L}_y$, and $\hat{L}_z$.

$$\hat{L}_x = \hat{y}\hat{p}_z - \hat{z}\hat{p}_y = -i\hbar\left(y\frac{\partial}{\partial z} - z\frac{\partial}{dy}\right) = i\hbar\left(\sin\phi\frac{\partial}{\partial\theta} + \cot\theta\cos\phi\frac{\partial}{\partial\phi}\right) \quad \text{(9.5) and (9.15)}$$

$$\hat{L}_y = \hat{z}\hat{p}_x - \hat{x}\hat{p}_z = -i\hbar\left(z\frac{\partial}{\partial x} - x\frac{\partial}{dz}\right) = i\hbar\left(-\cos\phi\frac{\partial}{\partial\theta} + \cot\theta\sin\phi\frac{\partial}{\partial\phi}\right) \quad \text{(9.6) and (9.16)}$$

$$\hat{L}_z = \hat{x}\hat{p}_y - \hat{y}\hat{p}_x = -i\hbar\left(x\frac{\partial}{\partial y} - y\frac{\partial}{dx}\right) = -i\hbar\frac{\partial}{\partial\phi} \quad \text{(9.7) and (9.17)}$$

or, in shorthand form

$$\hat{\mathbf{L}} = \hat{\mathbf{r}} \times \hat{\mathbf{p}} = -i\hbar(\mathbf{r} \times \nabla) \tag{9.4}$$

Each of these has possible eigenvalues $m\hbar$ where m is an integer, but these operators do not commute; it is not possible to have well-defined angular momentum about two different axes at once. Unlike linear momentum, angular momentum is restricted to discrete values.

L^2 operator

The $\hat{L}^2$ operator commutes with any one of $\hat{L}_x$, $\hat{L}_y$, and $\hat{L}_z$; conventional notation chooses $\hat{L}_z$. $\hat{L}^2$ has eigenvalues of $\hbar^2 l(l+1)$, where $l = 0, 1, 2...$ and has spherical harmonics as its eigenfunctions. The range of m is $-l \le m \le l$.

Spherical harmonics

Spherical harmonics can be written as

$$Y_{lm}(\theta,\phi) = (-1)^m\sqrt{\frac{2l+1}{4\pi}\frac{(l-m)!}{(l+m)!}}P_l^m(\cos\theta)\exp(im\phi) \tag{9.34}$$

where $P_l^m(u)$ are the associated Legendre functions.

Spherical harmonics can be visualized in terms of the spatial modes of vibration of a spherical shell, with the spherical harmonics being proportional to the amplitude of the oscillations of the sphere at any given angles corresponding to a point on the sphere. The number of nodal circles is l, with m of these passing through both poles.

Chapter 10

The hydrogen atom

Prerequisites: Chapters 2–5 and Chapter 9.

The hydrogen atom is a very important problem in quantum mechanics. It is also a problem described by Schrödinger's equation that is mathematically exactly solvable (if we neglect a few small corrections that lie beyond the simple Schrödinger equation description).[1] These solutions give the basis for our conceptual understanding of atoms and molecules, as well as much of the notation. In engineering, the hydrogen atom solutions also can be used as a model to explain a phenomenon (i.e., Wannier excitonic effects) that is very important in the optical absorption in most semiconductor optoelectronic devices.

The hydrogen atom problem is additionally an excellent tutorial one to take us beyond the simple one-dimensional spatial problems we have used as examples so far. In particular, it is a concrete example of the angular momentum behavior we discussed in the previous chapter. Also, unlike all the problems dealt with so far, the hydrogen atom involves two particles (the electron and the proton), not just one. The hydrogen atom, therefore, gives us an introduction to how we can handle more than one particle (though, unfortunately, problems containing more than two particles are almost always not exactly solvable). Additionally, we use this problem to illustrate another technique that is often used in deriving the solutions for those Schrödinger problems that can be solved exactly: namely, series solution of differential equations.[2]

How should we tackle this problem of two particles? The basic approach we take to this and many other problems in quantum mechanics is to generalize the Schrödinger equation. We write quite generally for time-independent problems

$$\hat{H}\psi = E\psi \tag{10.1}$$

---

[1] The simple Schrödinger equation solution given here neglects (i) the interaction of the electron spin and the proton orbit (the spin-orbit interaction), a relativistic correction that can be viewed as an interaction between the electron spin and the magnetic field that results from the proton moving in an orbit relative to the electron and which gives rise to some splitting of the otherwise degenerate levels, known as fine structure splitting; (ii) the hyperfine interaction associated with the interaction of the electron and proton spins; and (iii) the Lamb shift, a splitting of degeneracy of specific levels resulting from quantum electrodynamics. Even these corrections are very important to physics because they test more advanced aspects of quantum mechanics extraordinarily well.

[2] This method would have been the way we would have arrived at the Hermite polynomial solutions of the harmonic oscillator problem, e.g., in Chapter 2, though we merely stated that result there rather than deduced it from first principles.

where now we mean that the Hamiltonian $\hat{H}$ is the operator representing the energy of the entire system and, similarly, ψ is the wavefunction representing the state of the entire system.

10.1 Multiple-particle wavefunctions

For the hydrogen atom, there are two particles: the electron and the proton. Each of these has a set of coordinates associated with it (x_e, y_e, and z_e for the electron and x_p, y_p, and z_p for the proton). The wavefunction, therefore, in general is a function of all six of these coordinates.

In the simple single-particle problems we considered before, we could imagine that the wavefunction at any given time was a function in ordinary three-dimensional space. Now, however, we have to move beyond that simple picture. This hydrogen atom wavefunction is a function of six spatial coordinates and is really a wavefunction in what is called *configuration space*. The dimensionality of this space would grow even further if we had more particles; three particles would require nine spatial coordinates. A small lump of a solid material would need on the order of 10^{23} spatial coordinates to describe its configuration space!

To make matters worse, it is not simply the case that the proton wavefunction is a function of three position variables and the electron wavefunction is similarly a function of three position variables. The underlying issue that we have to confront here is that the behavior of the electron and the proton cannot be completely described if we insist on describing them separately, each with wavefunctions determined only by their own position coordinates.

Why do we need this additional complexity to describe a multiple-particle system? Imagine, for example, that we had a hydrogen atom in a box. The hydrogen atom might be found anywhere in the box and, hence, the electron might be found anywhere in the box. We might think that a simple electron wavefunction describing the electron as being found anywhere in the box would be sufficient. Similarly, the proton might be found anywhere in the box and we might also imagine that we could have a simple proton wavefunction describing it as being anywhere in the box. But that would not describe the relationship between the electron and the proton. In particular, for example, it would not tell us the answer to a question such as, "If we find the electron at a particular position, what is the probability that we will find the proton within 1 Å of that position?" Obviously, if the electron and proton were in a bound state of the hydrogen atom, the answer to that question would be that there is a large chance of finding the proton nearby; if the hydrogen atom was in a very high excited state or an ionized state, that probability would be quite low. Hence, we see that simple, separate wavefunctions for electron and proton, each a function only of that particle's own coordinates, are not sufficient to describe the complete state of the pair of particles.

We have an analogous situation in ballroom dancing. We could have a probability distribution describing the chance of finding a man at any particular place on the dance floor. Because ballroom dancers typically move around the entire floor during the dance, that distribution is relatively uniform over most of the dance floor. We would have a similar probability distribution for finding a woman at any particular place. But those distributions would not tell us the probability of finding a woman dancer close to a man dancer, a probability that is high because they dance in pairs. Hence, those individual probability distributions are just not sufficient to describe the dancers.

By careful choice of coordinate systems, we can sometimes find ways of separating the problem so that we do get back to a simple description, with our six-coordinate wavefunction being a product of two functions each of only three coordinates (and that is, indeed, the case for the hydrogen atom in empty space). In general, we cannot hope to factor the wavefunction that simply, however.[3]

Note too that this configuration space is not the same as the Hilbert space we use to describe functions. Our configuration space has three dimensions for each particle (possibly four if we include time). The Hilbert space has a possibly infinite number of dimensions for each degree of freedom (e.g., for each coordinate direction for each particle). It is, of course, certainly true that the Hilbert space acquires more dimensions as we describe sets of more than one particle.[4]

10.2 Hamiltonian for the hydrogen atom problem

The electron and proton each have a mass (m_e and m_p, respectively) and we correspondingly expect kinetic energy operators associated with each of these masses. In this particular problem, there is also a potential energy corresponding to the Coulomb attraction of the electron and the proton. With all these terms and definitions, the Hamiltonian becomes

$$\hat{H} = -\frac{\hbar^2}{2m_e}\nabla_e^2 - \frac{\hbar^2}{2m_p}\nabla_p^2 + V\left(\left|\mathbf{r}_e - \mathbf{r}_p\right|\right) \tag{10.2}$$

where the Laplacian operator for the electron coordinates is

$$\nabla_e^2 \equiv \frac{\partial^2}{\partial x_e^2} + \frac{\partial^2}{\partial y_e^2} + \frac{\partial^2}{\partial z_e^2} \tag{10.3}$$

(and, similarly, for ∇_p^2), the position vector for the electron coordinates is

$$\mathbf{r}_e = x_e\mathbf{i} + y_e\mathbf{j} + z_e\mathbf{k} \tag{10.4}$$

(and, similarly, for $\mathbf{r}_p$), and the Coulomb potential energy arising from the electrostatic attraction between the electron and the proton is

$$V\left(\left|\mathbf{r}_e - \mathbf{r}_p\right|\right) = -\frac{e^2}{4\pi\varepsilon_o\left|\mathbf{r}_e - \mathbf{r}_p\right|} \tag{10.5}$$

Note that this potential energy only depends on the distance $|\mathbf{r}_e - \mathbf{r}_p|$ between the electron and proton coordinates.[5]

The Schrödinger equation with this Hamiltonian can now be written explicitly as

$$\begin{aligned}\left[-\frac{\hbar^2}{2m_e}\nabla_e^2 - \frac{\hbar^2}{2m_p}\nabla_p^2 + V\left(\left|\mathbf{r}_e - \mathbf{r}_p\right|\right)\right]\psi\left(x_e, y_e, z_e, x_p, y_p, z_p\right) \\ = E\psi\left(x_e, y_e, z_e, x_p, y_p, z_p\right)\end{aligned} \tag{10.6}$$

[3] The problem of a hydrogen atom in a box does not strictly separate in the same way – indeed it is not clear that it does separate. The problem then is that the box imposes some energy terms that depend on absolute coordinates, whereas the usual hydrogen atom problem separates because it depends only on relative positions.

[4] We return to this issue in Chapter 12 when we discuss direct product spaces.

[5] See Appendix B, Section B.2, for Coulomb's law.

10.3 Coordinates for the hydrogen atom problem

Eq. (10.6) contains all of the physical assumptions we want for this problem. Our task is now the mathematical one of solving this equation. Our first goal is to find a way of breaking up this problem into more manageable parts. We do this by an intelligent choice of coordinate systems and by separating the resulting differential equation. There are two questions in the choice of coordinate systems.

(i) The coordinates of what?

With a single particle, it was obvious to choose the coordinates of the single particle. With two particles, we might at first think we simply choose the separate coordinates of the individual particles as the coordinates with which to work. In fact, the problem does not separate conveniently if we do that. We choose center-of-mass coordinates of the pair of particles instead.

(ii) What axes?

For a problem with spherical symmetry, like the hydrogen atom, we could make an intelligent guess that the problem would be simpler using spherical polar coordinates rather than simple Cartesian ones.

Center-of-mass coordinates

The main hint as to what coordinate approach we might take is to note that the potential in this problem is only a function of the relative separation of the electron and proton, $|\mathbf{r}_e - \mathbf{r}_p|$. We might, therefore, try to choose a new set of six coordinates in which three are the relative positions

$$x = x_e - x_p \, , \; y = y_e - y_p \, , \; z = z_e - z_p \tag{10.7}$$

that is, with a relative position vector

$$\mathbf{r} = x\mathbf{i} + y\mathbf{j} + z\mathbf{k} \, , \tag{10.8}$$

from which we obtain

$$r = \sqrt{x^2 + y^2 + z^2} = |\mathbf{r}_e - \mathbf{r}_p| \tag{10.9}$$

What should we choose for the other three coordinates? Experience with related classical problems, such as a moon orbiting a planet under gravitational attraction, tells us that the position of the center of mass of the moon and the planet is not affected by the orbit of the moon around the planet – each of them executes an orbit about the center of mass. We therefore look for a similar approach here. The position $\mathbf{R}$ of the center of mass is the same as the balance point of a light-weight beam with the two masses attached at opposite ends and so is the weighted average of the positions of the two individual masses; that is,

$$\mathbf{R} = \frac{m_e \mathbf{r}_e + m_p \mathbf{r}_p}{M} \tag{10.10}$$

where M is the total mass

$$M = m_e + m_p \tag{10.11}$$

Now let us construct the differential operators we need in terms of these coordinates. With

$$\mathbf{R} = X\mathbf{i} + Y\mathbf{j} + Z\mathbf{k} \tag{10.12}$$

we have, for example, for the new coordinates in the x-direction

$$X = \frac{m_e x_e + m_p x_p}{M}, \quad x = x_e - x_p \tag{10.13}$$

and so the first derivatives in the x-direction can be written as

$$\begin{aligned}\left.\frac{\partial}{\partial x_e}\right|_{x_p} &= \left.\frac{\partial X}{\partial x_e}\right|_{x_p}\left.\frac{\partial}{\partial X}\right|_x + \left.\frac{\partial x}{\partial x_e}\right|_{x_p}\left.\frac{\partial}{\partial x}\right|_X \\ &= \frac{m_e}{M}\left.\frac{\partial}{\partial X}\right|_x + \left.\frac{\partial}{\partial x}\right|_X\end{aligned} \tag{10.14}$$

and, similarly

$$\begin{aligned}\left.\frac{\partial}{\partial x_p}\right|_{x_e} &= \left.\frac{\partial X}{\partial x_p}\right|_{x_e}\left.\frac{\partial}{\partial X}\right|_x + \left.\frac{\partial x}{\partial x_p}\right|_{x_e}\left.\frac{\partial}{\partial x}\right|_X \\ &= \frac{m_p}{M}\left.\frac{\partial}{\partial X}\right|_x - \left.\frac{\partial}{\partial x}\right|_X\end{aligned} \tag{10.15}$$

(Here, we have explicitly indicated the variables in the x-direction held constant in each partial differentiation to try to reduce confusion.)

The second derivatives become

$$\begin{aligned}\left.\frac{\partial^2}{\partial x_e^2}\right|_{x_p} &= \left.\frac{\partial}{\partial x_e}\right|_{x_p}\left(\left.\frac{\partial}{\partial x_e}\right|_{x_p}\right) = \frac{m_e}{M}\left.\frac{\partial}{\partial x_e}\right|_{x_p}\left.\frac{\partial}{\partial X}\right|_x + \left.\frac{\partial}{\partial x_e}\right|_{x_p}\left.\frac{\partial}{\partial x}\right|_X \\ &= \left(\frac{m_e}{M}\right)^2\left.\frac{\partial^2}{\partial X^2}\right|_x + \left.\frac{\partial^2}{\partial x^2}\right|_X + \frac{m_e}{M}\left(\left.\frac{\partial}{\partial x}\right|_X\left.\frac{\partial}{\partial X}\right|_x + \left.\frac{\partial}{\partial X}\right|_x\left.\frac{\partial}{\partial x}\right|_X\right)\end{aligned} \tag{10.16}$$

and, similarly

$$\left.\frac{\partial^2}{\partial x_p^2}\right|_{x_e} = \left(\frac{m_p}{M}\right)^2\left.\frac{\partial^2}{\partial X^2}\right|_x + \left.\frac{\partial^2}{\partial x^2}\right|_X - \frac{m_p}{M}\left(\left.\frac{\partial}{\partial x}\right|_X\left.\frac{\partial}{\partial X}\right|_x + \left.\frac{\partial}{\partial X}\right|_x\left.\frac{\partial}{\partial x}\right|_X\right) \tag{10.17}$$

and so (dropping the explicit statement of variables held constant)

$$\begin{aligned}\frac{1}{m_e}\frac{\partial^2}{\partial x_e^2} + \frac{1}{m_p}\frac{\partial^2}{\partial x_p^2} &= \frac{m_e + m_h}{M^2}\frac{\partial^2}{\partial X^2} + \left(\frac{1}{m_e} + \frac{1}{m_p}\right)\frac{\partial^2}{\partial x^2} \\ &= \frac{1}{M}\frac{\partial^2}{\partial X^2} + \frac{1}{\mu}\frac{\partial^2}{\partial x^2}\end{aligned} \tag{10.18}$$

where μ is the so-called reduced mass

$$\mu = \frac{m_e m_p}{m_e + m_p} \tag{10.19}$$

The same kinds of relations can be written for each of the other Cartesian directions and so we have

$$\nabla_{\mathbf{R}}^2 \equiv \frac{\partial^2}{\partial X^2}+\frac{\partial^2}{\partial Y^2}+\frac{\partial^2}{\partial Z^2} \text{ and } \nabla_{\mathbf{r}}^2 \equiv \frac{\partial^2}{\partial x^2}+\frac{\partial^2}{\partial y^2}+\frac{\partial^2}{\partial z^2} \tag{10.20}$$

Instead of Eq. (10.2), we can write the Hamiltonian in a new form with center of mass coordinates

$$\hat{H} = -\frac{\hbar^2}{2M}\nabla_{\mathbf{R}}^2 - \frac{\hbar^2}{2\mu}\nabla_{\mathbf{r}}^2 + V(\mathbf{r}) \tag{10.21}$$

which will now allow us to separate the problem.

To attempt the separation in center-of-mass coordinates, we first presume that the total wavefunction can be written as a product

$$\psi(\mathbf{R},\mathbf{r}) = S(\mathbf{R})U(\mathbf{r}) \tag{10.22}$$

(As always for such proposed separations, the justification is that the separation leads to solutions of the equation that fit all the constraints of the problem and that are sufficiently general.) Substituting this form in the Schrödinger equation (10.1) with the Hamiltonian in the form of Eq. (10.21), we obtain

$$-U(\mathbf{r})\frac{\hbar^2}{2M}\nabla_{\mathbf{R}}^2 S(\mathbf{R}) + S(\mathbf{R})\left[-\frac{\hbar^2}{2\mu}\nabla_{\mathbf{r}}^2 + V(\mathbf{r})\right]U(\mathbf{r}) = ES(\mathbf{R})U(\mathbf{r}) \tag{10.23}$$

Dividing by $S(\mathbf{R})U(\mathbf{r})$ and moving some terms, we have

$$-\frac{1}{S(\mathbf{R})}\frac{\hbar^2}{2M}\nabla_{\mathbf{R}}^2 S(\mathbf{R}) = E - \frac{1}{U(\mathbf{r})}\left[-\frac{\hbar^2}{2\mu}\nabla_{\mathbf{r}}^2 + V(\mathbf{r})\right]U(\mathbf{r}) \tag{10.24}$$

The left-hand side depends only on **R** and the right-hand side depends only on **r**, so both of these must equal some constant, which we call E_{CoM}. Hence, we now have two separated equations

$$-\frac{\hbar^2}{2M}\nabla_{\mathbf{R}}^2 S(\mathbf{R}) = E_{CoM} S(\mathbf{R}) \tag{10.25}$$

and

$$\left[-\frac{\hbar^2}{2\mu}\nabla_{\mathbf{r}}^2 + V(\mathbf{r})\right]U(\mathbf{r}) = E_H U(\mathbf{r}) \tag{10.26}$$

where

$$E_H = E - E_{CoM} \tag{10.27}$$

We can see immediately that the first of these equations, Eq. (10.25), is the Schrödinger equation for a free particle of mass M, with wavefunction solutions

$$S(\mathbf{R}) = \exp(i\mathbf{K}\cdot\mathbf{R}) \tag{10.28}$$

and eigenenergies

$$E_{CoM} = \frac{\hbar^2 K^2}{2M} \tag{10.29}$$

This solution corresponds to the center of mass of the pair of particles moving as a composite particle with mass equal to the total mass of the two particles – that is, this is the motion of the entire hydrogen atom.

The second of these equations, Eq. (10.26), corresponds to the "internal" relative motion of the electron and proton and will give us the internal states of the hydrogen atom.

To solve Eq. (10.26), we can make use of the spherical symmetry of this equation and change to spherical polar coordinates. (Henceforth, we drop the subscript **r** in the ∇^2 operator for simplicity of notation.) In spherical polars, we have

$$\nabla^2 \equiv \frac{1}{r^2}\frac{\partial}{\partial r}r^2\frac{\partial}{\partial r} + \frac{1}{r^2}\left[\frac{1}{\sin\theta}\frac{\partial}{\partial\theta}\left(\sin\theta\frac{\partial}{\partial\theta}\right) + \frac{1}{\sin^2\theta}\frac{\partial^2}{\partial\phi^2}\right] \tag{10.30}$$

where the term in square brackets is the operator $\nabla^2_{\theta,\phi} \equiv -\hat{L}^2/\hbar^2$ that we introduced in Chapter 9 in discussing angular momentum. Knowing the solutions to the angular momentum problem, we propose the separation

$$U(\mathbf{r}) = R(r)Y(\theta,\phi) \tag{10.31}$$

where $Y(\theta,\phi)$ will be the appropriate spherical harmonic function.

Bohr radius and Rydberg energy

Before proceeding to solve the complete problem mathematically, at this point it is useful to introduce appropriate units for the problem;[6] we need dimensionless forms of the equations for some of the subsequent mathematics and these units also have simple physical significance.

We presume that the hydrogen atom has some characteristic size, which is called the *Bohr radius*, a_o. We expect that the "average" Coulomb potential energy (strictly its expectation value), therefore, is

$$\left\langle E_{potential}\right\rangle \approx -\frac{e^2}{4\pi\varepsilon_o a_o} \tag{10.32}$$

For a reasonably smooth single-peaked wavefunction of characteristic size a_o, the second spatial derivative is roughly $-1/a_o^2$ times the value of the function[7] (the minus sign comes about because the function is presumably peaked in the middle and falls off toward the sides – a reasonable guess at least for a lowest state – hence, giving a negative second derivative). Remembering that the kinetic energy operator is $-(\hbar^2/2\mu)\nabla^2$, the average kinetic energy (again, strictly its expectation value), therefore, is

---

[6] Of course, it makes no real difference if we get this choice of units "right," so we can be fairly informal in our arguments to choose them. As usual, however, we know the answer and, hence, make the right choices anyway.

[7] A function rising from 0 to 1 in a distance a_o would have to have a slope of ~ $1/a_o$, and if the same function now falls from 1 to 0 in a distance a_o, it will now need a slope of ~ $-1/a_o$. Hence, over a distance ~ $2a_o$ the derivative has changed by $-2/a_o$, corresponding to a second derivative of ~ $-1/a_o^2$.

$$\langle E_{kinetic} \rangle \approx \frac{\hbar^2}{2\mu a_o^2} \tag{10.33}$$

If we are looking for the lowest state of the system to start with, we would like to minimize the total energy, in the spirit of a variational calculation. With our very simple model here, the total energy is

$$\begin{aligned} \langle E_{total} \rangle &= \langle E_{kinetic} \rangle + \langle E_{potential} \rangle \\ &\approx \frac{\hbar^2}{2\mu a_o^2} - \frac{e^2}{4\pi\varepsilon_o a_o} \end{aligned} \tag{10.34}$$

Of course, the total energy is a balance between the potential energy, which is made lower (i.e., more negative) by choosing a_o smaller, and the kinetic energy, which is made lower (i.e., less positive) by making a_o larger. For the simple model represented in Eq. (10.34), differentiation shows that the choice of a_o that minimizes the energy overall is

$$a_o = \frac{4\pi\varepsilon_o\hbar^2}{e^2\mu} \cong 0.529 \text{ Å} = 5.29 \times 10^{-11} \text{ m} \tag{10.35}$$

This equation, Eq. (10.35), is the standard definition of the Bohr radius. We see, therefore, that the hydrogen atom is approximately 1 Å in diameter.

With this choice of a_o, the corresponding total energy of the state is, from Eq. (10.34)

$$\langle E_{total} \rangle = -\frac{\hbar^2}{2\mu a_o^2} = -\frac{\mu}{2}\left(\frac{e^2}{4\pi\varepsilon_o\hbar}\right)^2 \tag{10.36}$$

We can usefully define an energy unit that we call the "Rydberg," Ry,

$$Ry = \frac{\hbar^2}{2\mu a_o^2} = \frac{\mu}{2}\left(\frac{e^2}{4\pi\varepsilon_o\hbar}\right)^2 \cong 13.6 \text{ eV} \tag{10.37}$$

in which case $\langle E_{total} \rangle = -Ry$.

Though we have produced the Bohr radius, a_o, and the Rydberg, Ry, here by informal arguments, they turn out to be rigorously meaningful quantities once we have solved the complete hydrogen atom problem. Specifically, the energy of the lowest hydrogen atom state indeed turns out to be $-Ry$.

10.4 Solving for the internal states of the hydrogen atom

This section can be omitted at a first reading, though it contains the detailed solution of the radial equation, including the technique of power-series solutions of differential equations. At least parts of this section are also required for most of the problems at the end of this chapter.

Now let us explicitly solve for the internal states of the hydrogen atom. It makes our mathematics simpler if we write the separation of Eq. (10.31) in the form

$$U(\mathbf{r}) = \frac{1}{r}\chi(r)Y(\theta,\phi) \tag{10.38}$$

where, obviously

$$\chi(r) = rR(r) \tag{10.39}$$

With this choice, we obtain the convenient simplification of the radial derivatives

$$\frac{1}{r^2}\frac{\partial}{\partial r}r^2\frac{\partial}{\partial r}\frac{\chi(r)}{r} = \frac{1}{r}\frac{\partial^2\chi(r)}{\partial r^2} \tag{10.40}$$

Hence, the Schrödinger equation (10.26) becomes

$$-\frac{\hbar^2}{2\mu}Y(\theta,\phi)\frac{1}{r}\frac{\partial^2\chi(r)}{\partial r^2} + \frac{\chi(r)}{r^3}\frac{1}{2\mu}\hat{L}^2 Y(\theta,\phi) + Y(\theta,\phi)V(r)\frac{\chi(r)}{r} = E_H\frac{1}{r}\chi(r)Y(\theta,\phi) \tag{10.41}$$

Dividing by $-\hbar^2\chi(r)Y(\theta,\phi)/2\mu r^3$ and rearranging, we have

$$\frac{r^2}{\chi(r)}\frac{\partial^2\chi(r)}{\partial r^2} + r^2\frac{2\mu}{\hbar^2}\left(E_H - V(r)\right) = \frac{1}{\hbar^2}\frac{1}{Y(\theta,\phi)}\hat{L}^2 Y(\theta,\phi) \tag{10.42}$$

In the usual manner for a separation argument, we have arranged that the left-hand side depends only on r and the right-hand side depends only on θ and ϕ, so both sides must be equal to a constant. As it happens, because we have already solved the angular momentum problem, we already know what that constant is (i.e., we already know that $\hat{L}^2 Y_{lm}(\theta,\phi) = \hbar^2 l(l+1)Y_{lm}(\theta,\phi)$; explicitly, therefore, the constant is $l(l+1)$). Hence, in addition to the $\hat{L}^2$ eigenequation that we have already solved, this separation gives us a radial equation for the hydrogen atom wavefunction

$$-\frac{\hbar^2}{2\mu}\frac{d^2\chi(r)}{dr^2} + \left(V(r) + \frac{\hbar^2}{2\mu}\frac{l(l+1)}{r^2}\right)\chi(r) = E_H\chi(r) \tag{10.43}$$

which we can write as an ordinary differential equation because all the functions and derivatives are only in one variable, r.

Hence, we have a Schrödinger-like wave equation for this radial part of the wavefunction, with an additional effective potential energy term $l(l+1)/r^2$. We remember that $l = 0, 1, 2, \ldots$, and label the solutions with the l subscript.

Note that, incidentally, though here we have a specific form for $V(r)$ in our assumed Coulomb potential in Eq. (10.5), the algebra for the separation of the wave equation works for any potential that is only a function of r (sometimes known as a *central potential*). The precise form of Eq. (10.43) will be different for different central potentials but the separation remains.

The separation allowed us to separate out the $\hat{L}^2$ angular momentum eigenequation. The solutions of this equation have specific $\hat{L}^2$ values ($\hbar^2 l(l+1)$) and L_z values ($m\hbar$). Hence, we see that we have proved that angular momentum (in the sense of $\hat{L}^2$ and L_z values) is conserved (i.e., has well-defined and constant values) in any of the eigenstates of this hydrogen atom or, indeed, in any problem with a central potential. Note that we did not start by presuming conservation of angular momentum for such problems; this conservation was a consequence of the solution to the problems.

Solution of the hydrogen radial wavefunction

We solve for the radial hydrogen wavefunction completely from first principles. Aside from giving the answer, this gives one illustration of a method used for solving such differential equations. Such methods are also used, for example, to solve the harmonic oscillator problem and other ordinary differential equation problems in quantum mechanics, typically resulting in the known named polynomial functions (e.g., Hermite polynomials).

This method has typically two basic steps. First, we look for what must be the behavior of the solution in asymptotic situations (here, we look at both $r \to 0$ and $r \to \infty$). This will suggest underlying functional forms that should be present in the full solution. We especially want to identify underlying functions that have infinite power series, such as exponentials. At this stage, we can make all kinds of assumptions and approximations – we are only trying to identify underlying functional forms, not yet solve the actual equation, so there are no rules. We construct a solution incorporating these forms, substitute it in the differential equation, and obtain a new differential equation for the remaining unknown part of the function. A goal of this part of the solution method is to end up with an equation that has a solution that can be defined as a polynomial with a finite number of terms. In the second step, we solve this new differential equation by postulating such a finite power series (polynomial) as a solution and deduce the coefficients and, hence, deduce the function. Typically, it is because the power series must have finite numbers of terms that forces the quantization to give discrete eigen solutions. The finite length of the power series also leads to finite numbers of zeros in the eigenfunctions, a very common phenomenon in quantum mechanics.

Asymptotic behavior

(i) $r \to 0$

Consider first the case as $r \to 0$. Examining Eq. (10.43), we see that we should presume that $\chi(r) \to 0$ as $r \to 0$. Both the actual Coulomb potential and the additional effective potential in this equation are tending toward infinite magnitude as $r \to 0$. The only way we could keep the equation satisfied for finite $\chi(r)$ and finite E_H as $r \to 0$ is if $\chi(r) \to 0$. (From the point of view of solving Eq. (10.43) for $\chi(r)$, it is as if there were an infinite potential barrier for $\chi(r)$ at $r = 0$.)

As we approach $r = 0$, the dominant term in the effective potential is the term in $1/r^2$; for sufficiently small *r*, it will become arbitrarily larger than the Coulomb potential term. Hence, very near to $r = 0$, we can write Eq. (10.43) as

$$-\frac{d^2\chi}{dr^2} + \frac{l(l+1)}{r^2}\chi = 0 \tag{10.44}$$

We now presume that $\chi(r)$ must be an analytic function and we can, consequently, postulate a power-series form for it (e.g., a form such as $\chi(r) = a_0 r^0 + a_1 r^1 + a_1 r^2 + \ldots$).

At small *r*, we find in the end that we are only interested in the lowest power in the power series (for sufficiently small *r*, the lowest power in the series will always dominate). We do not, however, presume that the lowest power is r^0 and instead presume that we start with some power r^p where *p* is still to be determined. Hence, we presume a form at small *r*

$$\chi(r) \sim r^p \tag{10.45}$$

Substituting this form in Eq. (10.44), we obtain

$$-p(p-1)r^{p-2} + l(l+1)r^{p-2} = 0 \tag{10.46}$$

This must be true for all (very small) *r* and so we presume that $p = l + 1$ (that presumption obviously satisfies Eq. (10.46)). Hence, we expect that for very small *r*

$$\chi(r) \sim r^{l+1} \tag{10.47}$$

(ii) $r \to \infty$

Now we explicitly use the Coulomb attraction of the electron and proton as the potential from Eq. (10.5), so Eq. (10.43) becomes

$$-\frac{\hbar^2}{2\mu}\frac{d^2\chi(r)}{dr^2}-\left(\frac{e^2}{4\pi\varepsilon_o r}-\frac{\hbar^2}{2\mu}\frac{l(l+1)}{r^2}\right)\chi(r)=E_H\chi(r) \tag{10.48}$$

For very large *r*, both the $1/r$ and the $1/r^2$ terms become arbitrarily small, so we have approximately

$$-\frac{\hbar^2}{2\mu}\frac{d^2\chi(r)}{dr^2}\cong E_H\chi(r) \tag{10.49}$$

For any bound state of the system, the eigenenergy E_H is negative.[8] This asymptotic equation, therefore, has a simple solution for large *r*, which is[9]

$$\chi \propto \exp\left(-\sqrt{-\frac{2\mu}{\hbar^2}E_H}\; r\right) \tag{10.50}$$

This solution should not surprise us; it physically corresponds to the electron tunneling away from the proton into the potential barrier, which at large *r* simply is a barrier of height $|E_H|$ for a state of (negative) energy E_H.

Recasting the mathematical problem

Now we can rewrite the mathematical problem to be solved, Eq. (10.43), using the new Rydberg and Bohr radius units and the asymptotic forms. We choose to write the eigenenergy E_H of some specific state of interest in terms of the Rydberg *Ry*

$$E_H = -\frac{Ry}{n^2} \tag{10.51}$$

Here, the parameter *n* is the parameter we try to evaluate mathematically to deduce the eigenenergy. For the moment, it is simply an arbitrary real number, though we prove later that it is an integer.

We choose to define a new, dimensionless radial distance, *s*

$$s = \alpha r \tag{10.52}$$

where the parameter α is

$$\alpha = \frac{2}{na_o} = 2\sqrt{-\frac{2\mu}{\hbar^2}E_H} \tag{10.53}$$

Using this distance unit, we can rewrite Eq. (10.48), after some rearrangements, as

---

[8] The bound states have negative energy just because of the way we defined the Coulomb potential energy, which goes from negative values toward zero at infinite distance.

[9] We have here as usual discarded the exponentially growing solution as being unphysical; it certainly does not correspond to a hydrogen atom as we know it and mathematically cannot be normalized in any reasonable way.

$$\frac{d^2\chi}{ds^2} - \left[\frac{l(l+1)}{s^2} - \frac{n}{s} + \frac{1}{4}\right]\chi = 0 \tag{10.54}$$

Now, we propose a functional form for $\chi(s)$ that incorporates what we deduced previously from the asymptotic behavior. Specifically, in terms of s, from the small r behavior of Eq. (10.47), we incorporate a factor s^{l+1} and, from the large r behavior of Eq. (10.50), we obtain a factor $\exp(-s/2)$. Hence, the proposed form of the function is

$$\chi(s) = s^{l+1}L(s)\exp(-s/2) \tag{10.55}$$

where $L(s)$ is now the function to be determined.[10]

Having taken out the two asymptotic behaviors, we hope that the function $L(s)$ can be some relatively simple function, such as a low-order polynomial. The only important restriction we have to put on $L(s)$ is that it does not grow too fast with large s, so that the function $\chi(s)$ can be normalized; this restriction is actually the source of the quantization we will obtain; hence, its importance.

Now, we substitute this form into the Eq. (10.54) and obtain an equation for $L(s)$

$$s\frac{d^2L}{ds^2} - \left[s - 2(l+1)\right]\frac{dL}{ds} + \left[n - (l+1)\right]L = 0 \tag{10.56}$$

which is the kind of form we need for the next part of this solution.

Power series solution

Next, we propose that because $L(s)$ is presumably an analytic function, it can be described as a power-series, which we write formally as

$$L(s) = \sum_{q=0}^{\infty} c_q s^q \tag{10.57}$$

where the c_q are coefficients to be determined by solving the differential equation (10.56). Substituting this power series form into Eq. (10.56), we obtain

$$\sum_{q=0}^{\infty}\left[q(q-1)c_q s^{q-1} - qc_q s^q + 2q(l+1)c_q s^{q-1} + \left[n-(l+1)\right]c_q s^q\right] = 0 \tag{10.58}$$

We can open up this sum to see explicitly how we gather terms in specific powers of s. In the middle of the summation, we have two successive terms from the sum

$$\begin{aligned}&\cdots + q(q-1)c_q s^{q-1} - qc_q s^q + 2q(l+1)c_q s^{q-1} + \left[n-(l+1)\right]c_q s^q\\ &+(q-1)(q-2)c_{q-1}s^{q-2} - (q-1)c_{q-1}s^{q-1} + 2(q-1)(l+1)c_{q-1}s^{q-2} + \left[n-(l+1)\right]c_{q-1}s^{q-1}\\ &+\cdots\end{aligned}$$

We can now gather terms of the same power of s (specifically, here for minor reasons of convenience in s^{q-1}) to rewrite Eq. (10.58)

[10] The L here has nothing to do with angular momentum. We introduce this notation because it is a standard one for the associated Laguerre polynomials that will emerge in this problem.

$$\sum_{q=1}^{\infty}\left\{\left[q(q-1)+2q(l+1)\right]c_q-\left[(q-1)+(l+1)-n\right]c_{q-1}\right\}s^{q-1}=0 \tag{10.59}$$

(We have changed the sum to start at $q=1$ rather than $q=0$, but the reader can verify by writing out the first few terms of the sum that Eqs. (10.58) and Eq. (10.59) are equivalent also for the lowest powers of s.) Of course, this relation must hold for all s. The only way this can be is if the coefficients of each power of s are zero. So what we have shown is that there is a relation between successive coefficients in the power series; that is, explicitly

$$\left[q(q-1)+2q(l+1)\right]c_q=\left[(q-1)+(l+1)-n\right]c_{q-1} \tag{10.60}$$

that is,

$$c_q=\frac{(q+l-n)}{q(q+2l+1)}c_{q-1} \tag{10.61}$$

This kind of relation between successive terms in a series is called a *recurrence relation.*

We are now very close to a solution of the mathematical problem, but one important step remains. Note that for very large q, the factors in successive terms in the series are smaller by a factor of about $1/q$, so the factors themselves are approximately proportional to $1/q!$. That is the behavior we would find in the power series for the exponential function $\exp(s)$. Such behavior would cause the function $\chi(s)$ to grow with increasing s and, hence, not to be normalizable. How do we get out of this difficulty? The answer is that we try to find the condition that will cause the series to terminate at a finite number of terms. That happens if and only if the factor $(q+l-n)=0$ for some $q\geq 1$. (We make the restriction that q must be at least 1 so that we have at least one term [the term c_0s^0] in the series – otherwise, the function is zero and not of interest to us.) Hence, we require two things for termination of the series. First, because q is an integer by definition and l is already known to be an integer from the solution of the angular momentum eigenfunction problem, n must also be an integer. Second, we require that

$$n\geq l+1 \tag{10.62}$$

otherwise there is no q for which the series terminates. This termination, therefore, is responsible mathematically for the quantization of the radial behavior of the wavefunction.

Now, given our recurrence relation and the known stopping point of the power series, we can formally construct the polynomial functions $L(s)$ that are the solutions of Eq. (10.56). Hence we now have completed the last step required to solve the hydrogen atom problem mathematically. Next, we discuss the actual polynomial functions that the recurrence relation Eq. (10.61) has defined for us.

Associated Laguerre polynomials

Not surprisingly, the polynomials that solve Eq. (10.56) and that are defined by (i) the recurrence relation Eq. (10.61) and (ii) the "stopping condition" of an integer n that satisfies Eq. (10.62), are a standard set of known polynomials, the associated Laguerre polynomials. (It is not necessary that we know this to solve the problem, but this does connect with other relations that can be helpful in making, for example, normalization integrals easier.) The only choice we need to make to define our polynomials completely is the first coefficient in the power series (c_0); because the coefficients in the polynomial are constructed progressively from the preceding coefficient, this coefficient merely multiplies the entire polynomial and is arbitrary as far as solving Eq. (10.56) is concerned. To make a connection with the standard definition of the associated Laguerre polynomials, we choose

$$c_0 = \frac{(n+l)!}{(n-l-1)!(2l+1)!} \tag{10.63}$$

With this choice, our polynomials become, using the recursive relation Eq. (10.61)

$$L_{n-l-1}^{2l+1}(s) = \sum_{q=0}^{n-l-1} (-1)^q \frac{(n+l)!}{(n-l-q-1)!(q+2l+1)!} s^q \tag{10.64}$$

where we have introduced the notation for the associated Laguerre polynomials

$$L_p^j(s) = \sum_{q=0}^{p} (-1)^q \frac{(p+j)!}{(p-q)!(j+q)!q!} s^q \tag{10.65}$$

10.5 Solutions of the hydrogen atom problem

Summarizing our approach so far for the internal states of the hydrogen atom, we started by presuming a separation $U(\mathbf{r}) = R(r)Y(\theta,\phi)$ (Eq. (10.31)). The solutions for $Y(\theta,\phi)$ became the spherical harmonics $Y_{lm}(\theta,\phi)$ introduced in Chapter 9. We chose to define a function $\chi(r) = rR(r)$ (Eq. (10.39)). In terms of the variable $s = (2/na_o)r$, we have, from Eqs. (10.55) and (10.64) with the definition Eq. (10.65) for the associated Laguerre polynomials $L_p^j(s)$, the solutions

$$\chi(s) = s^{l+1} L_{n-l-1}^{2l+1}(s) \exp(-s/2) \tag{10.66}$$

where n is an integer and $n \geq l+1$ (Eq. (10.62)).

Now we can put the whole solution together. Changing back to the variable r for the separation of the electron and proton

$$\begin{aligned} R(r = na_o s/2) &\propto \frac{1}{r} s^{l+1} L_{n-l-1}^{2l+1}(s) \exp(-s/2) \\ &\propto s^l L_{n-l-1}^{2l+1}(s) \exp(-s/2) \end{aligned} \tag{10.67}$$

Explicitly introducing a normalization constant into the radial wavefunction, we have

$$R(r = na_o s/2) = \frac{1}{A} s^l L_{n-l-1}^{2l+1}(s) \exp(-s/2) \tag{10.68}$$

The full normalization integral of the wavefunction $U(\mathbf{r}) = R(r)Y(\theta,\phi)$ would be

$$1 = \int_{r=0}^{\infty} \int_{\theta=0}^{\pi} \int_{\phi=0}^{2\pi} |R(r)Y(\theta,\phi)|^2 r^2 \sin\theta d\theta d\phi dr \tag{10.69}$$

but we have already normalized the spherical harmonics with the θ and ϕ integrals, so we are left with the radial normalization

$$1 = \int_0^{\infty} R^2(r) r^2 dr \tag{10.70}$$

It is possible to show that

$$\int_0^\infty s^{2l}\left[L_{n-l-1}^{2l+1}(s)\right]^2 \exp(-s)s^2 ds = \frac{2n(n+l)!}{(n-l-1)!} \tag{10.71}$$

from which we can therefore conclude that the normalized radial wavefunction is

$$R(r) = \left[\frac{(n-l-1)!}{2n(n+l)!}\left(\frac{2}{na_o}\right)^3\right]^{1/2} \left(\frac{2r}{na_o}\right)^l L_{n-l-1}^{2l+1}\left(\frac{2r}{na_o}\right)\exp\left(-\frac{r}{na_o}\right) \tag{10.72}$$

The following are several of the low-order radial wavefunctions. Here, we choose to write the wavefunctions using the Bohr radius a_o as the unit of radial distance for simplicity of notation, so we have a radial distance $\rho = r/a_o$, and we introduce the subscripts n and l for the quantum numbers to index the various functions $R_{n,l}$. (To get the expression in absolute distance units, use the explicit formula Eq. (10.72).) These various functions are graphed in Fig. 10.1.

$$\begin{aligned}
R_{1,0}(\rho) &= 2\exp(-\rho) \\
R_{2,0}(\rho) &= \frac{\sqrt{2}}{4}(2-\rho)\exp(-\rho/2) \\
R_{2,1}(\rho) &= \frac{\sqrt{6}}{12}\rho\exp(-\rho/2) \\
R_{3,0}(\rho) &= \frac{2\sqrt{3}}{27}\left(3-2\rho+\frac{2}{9}\rho^2\right)\exp(-\rho/3) \\
R_{3,1}(\rho) &= \frac{\sqrt{6}}{81}\rho\left(4-\frac{2}{3}\rho\right)\exp(-\rho/3) \\
R_{3,2}(\rho) &= \frac{2\sqrt{30}}{1215}\rho^2\exp(-\rho/3)
\end{aligned} \tag{10.73}$$

Note the following behaviors of the wavefunction.

(i) The overall "size" of the wavefunctions becomes larger with larger n.

(ii) The number of zeros in the wavefunction is $n-1$. The associated Laguerre polynomials in the radial wavefunction have $n-l-1$ zeros and the spherical harmonics have l nodal "circles."[11] (The radial wavefunctions appear to have an additional zero at $r=0$ for all $l \geq 1$, but note that all the spherical harmonics have at least one nodal circle for all $l \geq 1$. As $r \to 0$, that nodal circle forces the wavefunction to be zero anyway, so this zero is already counted.)

In summary of the quantum numbers, the possible solutions of our hydrogen atom equation require for the so-called principal quantum number n

$$n = 1, 2, 3, \ldots \tag{10.74}$$

and

$$l \leq n-1 \tag{10.75}$$

[11] Strictly, the spherical harmonics only have l nodal circles if we use the sin $m\phi$ and cos $m\phi$ forms rather than the exp $im\phi$ form (which is not zero for any value of ϕ). We do have the option of using the sin $m\phi$ and cos $m\phi$ forms in the hydrogen problem because all these different forms are degenerate.

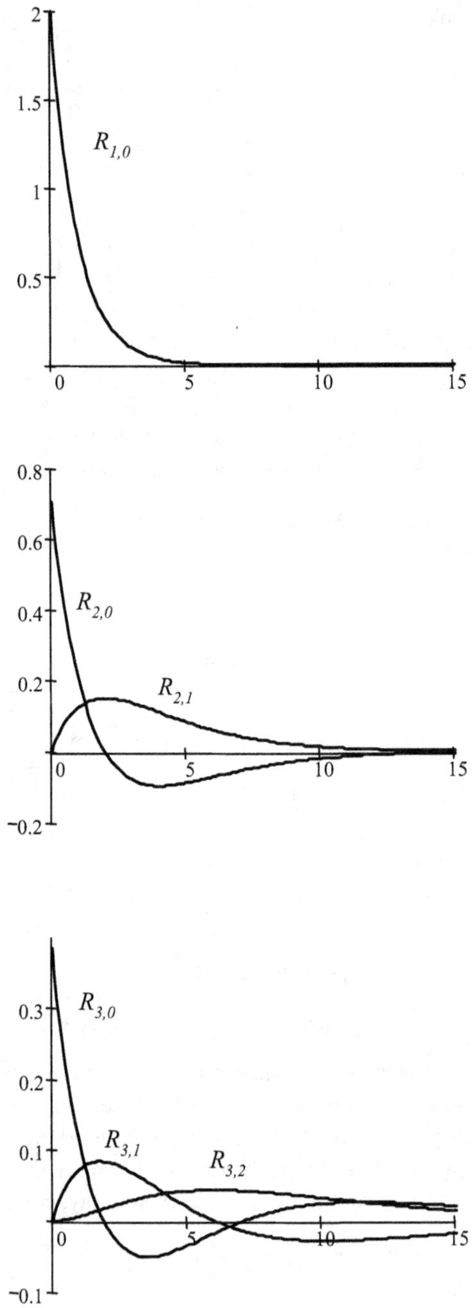

Fig. 10.1. Hydrogen atom radial wavefunctions, with radial distance units of the Bohr radius, a_o.

We previously deduced in Chapter 9 that l is a positive or zero integer. Now that we know the possible values for n, we can write immediately for the eigenenergies from Eq. (10.51)

$$E_H = -\frac{Ry}{n^2}, \quad n = 1, 2, 3, \ldots \tag{10.76}$$

This completes the solution of the hydrogen atom problem.

Problems

10.5.1* Consider an electron in a thin cylindrical shell potential. This potential is zero within the cylindrical shell and may be presumed infinite everywhere else. The shell has inner radius r_o and thickness L_r, where $r_o >> L_r$. The cylindrical shell may be presumed to be infinite along its cylindrical (z) axis.

(i) Show that the energy eigenfunctions (i.e., solutions of the time-independent Schrödinger equation) are, approximately

$$\psi(r,\phi,z) \propto \sin\frac{n\pi(r-r_o)}{L_r}\exp(im\phi)\exp(ik_z z)$$

(ii) State what the restrictions are on the values of n, m, and k (e.g., are they real, integer, limited in their range?)

(iii) Give an approximate expression for the corresponding energy eigenvalues.

Note: In cylindrical polar coordinates

$$\nabla^2 = \frac{1}{r}\frac{\partial}{\partial r}\left(r\frac{\partial}{\partial r}\right) + \frac{1}{r^2}\frac{\partial^2}{\partial\phi^2} + \frac{\partial^2}{\partial z^2}$$

10.5.2* Consider the case of a spherical potential well – that is, a structure in which the potential energy is lower and constant for all radii $r < r_o$ and higher and constant for all $r > r_o$ – and a particle of mass m_o.

For the case of an infinite potential well (i.e., one in which the potential is infinite for all $r > r_o$), find the energy of the lowest state of a particle in the well relative to the bottom of the well. (You may presume the lowest state has the lowest possible angular momentum.) (Note: Remember that it can be shown that with a radial wavefunction $R(r) = \chi(r)/r$, $\chi(0) = 0$.)

(This problem is part of the analysis of spherical semiconductor quantum dots. Such structures can be made and are commonly used in color glass filters and as fluorescent markers for biological experiments. The color of the dots or of their fluorescence is partly determined and controlled by the size of the dot through these quantum size effects. Quantum dots are also interesting for optoelectronic devices.)

10.5.3 Consider the defining differential equation for the Hermite polynomials

$$\frac{d^2H_n(x)}{dx^2} - 2x\frac{dH_n(x)}{dx} + 2nH_n(x) = 0$$

and solve it by the series solution method for functions $H_n(x)$ such that $H_n(x)\exp(-x^2/2)$ can be normalized

In your solution

(i) find a recurrence relation between the coefficients of the power-series solutions (Note: This relation will be between c_q and c_{q+2}.)

(ii) show that $H_n(x)\exp(-x^2/2)$ will not be normalizable unless the power series terminates (Note: You will have to consider two power series, one starting with c_0 and one starting with c_1, and show that neither will terminate unless n is an integer.)

(iii) choosing $c_0 =$ 0 or 1 and $c_1 = 0$ or 1, find the first five power-series solutions of the equation.

10.5.4 Consider a spherical quantum box or "dot" with an electron inside it. We presume the potential is infinite at the boundary of the dot and zero within, and that the "dot" has radius r_0.

(i) Find an expression for the eigenenergies $E_{n\ell}$, where ℓ is the usual angular momentum quantum number and n is another integer quantum number (starting at $n = 1$), expressing your result in terms of the zeros $s_{n\ell}$ of the spherical Bessel function $j_\ell(x)$, where $s_{n\ell}$ is the nth zero for a given ℓ.

(ii) Find the electron confinement energies for the nine conditions $n =$ 1, 2, 3 with $\ell = 0, 1, 2$ for each n, for the case of a 10 nm diameter semiconductor dot with electron effective mass of

$0.2m_0$. (Note: You will have to find appropriate zeros of special functions from mathematical tables or otherwise.)

Notes:

(a) The equation

$$\frac{d^2y}{dx^2}+\left[a^2+\left(\frac{1}{4}-p^2\right)\frac{1}{x^2}\right]y=0$$

has solutions

$$y=\sqrt{x}\left[AJ_p(ax)+BY_p(ax)\right]$$

where A and B are arbitrary constants, J_p is the Bessel function of order p, and Y_p is the Weber function of order p. Note that the Weber functions tend to infinity as $x \to 0$, though the Bessel functions remain finite as $x \to 0$.

(b) The spherical Bessel functions are given by

$$j_\ell(x)=\sqrt{\frac{\pi}{2x}}J_{\ell+1/2}(x)$$

and these functions can also be expressed as

$$j_\ell(x)=x^\ell\left(-\frac{1}{x}\frac{d}{dx}\right)^\ell\frac{\sin x}{x}$$

10.5.5 Find an expression for the energy eigenstates of a cylindrical quantum wire of radius r_o for which we assume there is an infinitely high barrier at radius r_o. Specify all quantum numbers and state their allowed values.

(Note: In this case, it is left as an exercise for the reader to find the necessary special functions and their properties to solve this problem.)

10.5.6 Evaluate the matrix element $\langle U_{210}|z|U_{100}\rangle$, where by $|U_{nlm}\rangle$ we mean the hydrogen atom orbital where the quantum numbers n, l, and m take their usual meanings.

10.5.7 Suppose we are considering optical transitions between different states in the hydrogen atom. We presume that the hydrogen atom is initially in a given starting state and we want to know if, for a linearly polarized oscillating electromagnetic field (i.e., one for which the optical electric field can be taken to be along the z direction) of the appropriate frequency, it can make transitions to the given final state, at least for transition rates calculated using first-order time-dependent perturbation theory in the electric dipole approximation. State for each of the following combinations whether such optical transitions are possible. Explain your method and your results. (Note: This problem requires some understanding of transition matrix elements for optical transitions as discussed in Chapter 7.)

(a) Starting state $|1,0,0\rangle$, final state $|2,1,0\rangle$

(b) Starting state $|1,0,0\rangle$, final state $|2,1,1\rangle$

(c) Starting state $|1,0,0\rangle$, final state $|2,0,0\rangle$

(d) Starting state $|2,1,0\rangle$, final state $|1,0,0\rangle$

10.5.8 (This problem can be used as a substantial assignment.) Consider the problem of a "two-dimensional" hydrogen atom. Such a situation could arise for a hydrogen atom squeezed between two parallel plates, for example. (This problem is a good limiting model for excitons in semiconductor quantum wells.) We are not concerned with the motion in the z direction perpendicular to the plates and are hence left with a Schrödinger equation for the electron and proton

$$\left[-\frac{\hbar^2}{2m_e}\nabla^2_{xye}-\frac{\hbar^2}{2m_p}\nabla^2_{xyp}-\frac{e^2}{4\pi\varepsilon_o\left|\mathbf{r}_{xye}-\mathbf{r}_{xyp}\right|}\right]\psi(x_e,y_e,x_p,y_p)=E\psi(x_e,y_e,x_p,y_p)$$

where

$$\nabla^2_{xye}\equiv\frac{\partial^2}{\partial x_e^2}+\frac{\partial^2}{\partial y_e^2}$$

where x_e and y_e are the position coordinates of the electron and, similarly, for the proton.

Solve for the complete wavefunctions and eigenenergies for all states where the electron and proton are bound to one another (you need not normalize the wavefunctions).

Give an explicit expression for the coefficients of any polynomials you derive for solutions, in terms of the lowest order coefficient in the polynomial.

Explicitly state the allowed values of any quantum numbers and give the numerical answer for the lowest energy of the system in electron volts.

Hints:

(i) This problem can be solved in a very similar fashion to the three-dimensional hydrogen atom. Use the same units as those used in that problem.

(ii) The Laplacian in two-dimensional polar coordinates is

$$\nabla^2 = \frac{1}{r}\frac{\partial}{\partial r}\left(r\frac{\partial}{\partial r}\right) + \frac{1}{r^2}\frac{\partial^2}{\partial \phi^2}$$

(iii) You should be able to get to an equation that looks something like

$$s\frac{d^2L}{ds^2} + (A - s)\frac{dL}{ds} + BL = 0$$

in solving for the radial motion, where A and B do not depend on s.

10.5.9 (This problem can be used as a substantial assignment.) Consider the effect of a small electric field on the $n = 2$ levels of the hydrogen atom. (Note: There are several such levels because of the different values of l and m possible for $n = 2$. This problem should, therefore, be approached using first-order degenerate perturbation theory. The algebra of the problem may be somewhat easier if the electric field is chosen in the z direction.)

Find how these $n = 2$ degenerate states are affected by the field. Give explicit numerical expressions for the shifts of those levels that are affected by the field F, and show how their wavefunctions are constructed as linear combinations of hydrogen wavefunctions of specific n, l, and m quantum numbers. Specify also the basis functions that can be used to describe any $n = 2$ state or states not affected by the field.

Give explicit numbers for the shifts of states for a field of 10^5 V/m. Specify energies in electron-Volts. (Note: Chapter 6 is a prerequisite for this problem.)

10.6 Summary of concepts

Generalization of Schrödinger's equation

With more than one particle or with more complicated Hamiltonians than we have considered so far, we can generalize Schrödinger's time-independent equation as

$$\hat{H}\psi = E\psi \tag{10.1}$$

where $\hat{H}$ is the energy operator for the system, E is the total energy, and ψ is the wavefunction or, more generally, the quantum mechanical state of the system.

Multiple-particle wavefunctions

In general, the wavefunction of a multiple-particle system cannot be separated into products of single-particle wavefunctions and must be written as a function of all the coordinates. For example, for the system of an electron (e) and a proton (p), in general, we have to write $\psi(x_e, y_e, z_e, x_p, y_p, z_p)$.

Center-of-mass coordinates

For a two-particle problem where the potential only depends on the relative separation of the two particles, we can separate the problem using center of mass coordinates into the relative position coordinate **r** and the center-of-mass coordinate **R**, where

$$\mathbf{R} = \frac{m_e \mathbf{r}_e + m_p \mathbf{r}_p}{M} \tag{10.10}$$

$$\mathbf{r} = x\mathbf{i} + y\mathbf{j} + z\mathbf{k} \tag{10.8}$$

with

$$x = x_e - x_p \,,\; y = y_e - y_p \,,\; z = z_e - z_p \tag{10.7}$$

and where $M = m_e + m_p$ is the total mass.

Solutions for the hydrogen atom internal states

The wavefunction solutions for the hydrogen atom internal states are

$$U_{nlm}(\mathbf{r}) = R_n(r) Y_{lm}(\theta,\phi) \qquad \text{after Eq. (10.31)}$$

where $Y_{lm}(\theta,\phi)$ are the spherical harmonics and

$$R(r) = \left[\frac{(n-l-1)!}{2n(n+l)!} \left(\frac{2}{na_o} \right)^3 \right]^{1/2} \left(\frac{2r}{na_o} \right)^l L_{n-l-1}^{2l+1} \left(\frac{2r}{na_o} \right) \exp\left(-\frac{r}{na_o} \right) \tag{10.72}$$

where the principal quantum number $n = 1, 2, 3, \ldots$, l is zero or a positive integer with $l \le n-1$ and $L_p^j(s)$ are the associated Laguerre polynomials, with associated energies $E_H = -Ry/n^2$. These wavefunctions always have $n-1$ zeros.

Chapter 11

Methods for one-dimensional problems

Prerequisites: Chapters 2 and 3, and for Section 11.4, Chapter 8, Sections 8.1–8.7.

Many interesting quantum mechanical problems can be reduced to one-dimensional mathematical problems. This reduction is often possible because the problem, though truly three-dimensional, can be mathematically separated. For example, the three-dimensional hydrogen atom also mathematically separates to leave a radial equation that looks like a one-dimensional effective Schrödinger equation. Most problems associated with electrons and planar surfaces or layered structures can be handled with one-dimensional models. Examples include field emission of electrons from planar metallic surfaces and most problems associated with semiconductor quantum well structures.

One-dimensional problems can be solved by a number of techniques. Here, we discuss one of these, the *transfer matrix technique*, and we also derive one key result of the so-called *WKB* method. We concentrate on the use of such techniques for solving tunneling problems, so we start with a brief discussion of tunneling rates.

11.1 Tunneling probabilities

Suppose we have a barrier, shown in Fig. 11.1 as a simple rectangular barrier. Electrons are incident on the barrier from the left. Some are reflected and some are transmitted. We presume that the electron energy E is less than the barrier height V_o so that we are discussing a tunneling problem. We already know how to solve this problem quantum mechanically for the simple case of a rectangular barrier, with notation as shown in the figure. We discuss now how to solve such problems for more complex forms of barrier. For any form of barrier, supposing we have found appropriate relations between the amplitudes of the incident wave (A), the reflected wave (B), and the transmitted wave (F) for some given energy of particle, how do we relate this quantum mechanical problem to actual currents of electrons?

In the full problem of tunneling emission of electrons, for example, we typically presume that we have some thermal distribution of electrons on the left of the barrier and we use a thermal argument to deduce how many electrons there are with a particular velocity v in the z direction (i.e., perpendicular to the barrier). We want to add up the results of all such tunneling currents for all of the electrons in the distribution to deduce the total emitted current. Hence, if we can find some way of deducing what fraction of the electrons traveling at some velocity v in the z direction are transmitted by the barrier, we will know the tunneling emission current.

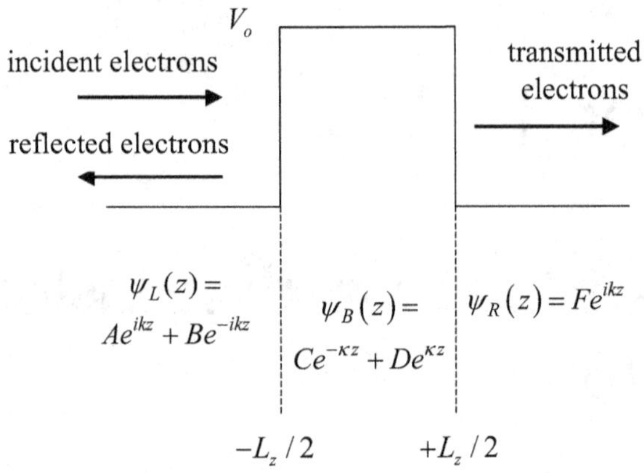

Fig. 11.1. Electrons tunneling through a simple rectangular barrier.

We discussed in Section 3.14 how to calculate the particle current quantum mechanically, concluding that the particle current density is

$$\mathbf{j}_p = \frac{i\hbar}{2m}\left(\Psi\nabla\Psi^* - \Psi^*\nabla\Psi\right) \tag{3.97}$$

where $\Psi = \Psi(\mathbf{r},t)$ is the full time-dependent wavefunction. If we presume here that we are dealing with particles of well-defined energy E, which is $E = \hbar^2k^2/2m$ in the propagating regions, the time-dependent factor $\exp(-iEt/\hbar)$ disappears in this current density equation because of the product of complex conjugates; therefore, in this case of well-defined energy, we can write

$$j_p = \frac{i\hbar}{2m}\left(\psi\nabla\psi^* - \psi^*\nabla\psi\right) \tag{11.1}$$

where ψ is now the one-dimensional spatial wavefunction $\psi(z)$. We have dropped the vector character of the particle current density $\mathbf{j}_p$ because we are considering only currents in the z direction.

If we now consider, for example, the wavefunction on the right, $F\exp(ikz)$, then we have from Eq. (11.1)

$$j_p = |F|^2\frac{\hbar k}{m} \tag{11.2}$$

The quantity $\hbar k/m$ behaves like an effective classical velocity v, with $E = \hbar^2k^2/2m = (1/2)mv^2$.[1]

For the particle current on the left, we should proceed carefully, remembering to deal with the whole wavefunction on the left in evaluating the particle current. With the wavefunction $\psi(z) = A\exp(ikz) + B\exp(-ikz)$, we have from Eq. (11.1)

[1] Note that we have taken a particularly simple situation here for this illustration in which the potential on the right of the barrier is the same as the potential on the left. If this were not the case, we would end up with a different effective classical velocity on the right, though we would still find that the overall particle currents added up correctly.

$$\begin{aligned} j_p &= \frac{i\hbar}{2m}\left\{ \begin{matrix} \left[A\exp(ikz)+B\exp(-ikz)\right]\left[-ikA^*\exp(-ikz)+ikB^*\exp(ikz)\right] \\ -\left[A^*\exp(-ikz)+B^*\exp(ikz)\right]\left[ikA\exp(ikz)-ikB\exp(-ikz)\right] \end{matrix} \right\} \\ &= \frac{\hbar k}{m}\left(|A|^2-|B|^2\right) \end{aligned} \quad (11.3)$$

because all of the spatially oscillating terms cancel. We can, therefore, consider $\hbar k|A|^2/m$ as the forward current on the left of the barrier and $\hbar k|B|^2/m$ as the reflected or backward current, adding the two to get the net current. Hence, we find that we can identify the (particle) current densities as follows, for particles of effective classical velocity $v = \hbar k/m$.

Incident current density $|A|^2 v$

Reflected current density $|B|^2 v$

Transmitted current density $|F|^2 v$

The fraction of incident particles that are transmitted by the barrier is

$$\eta = \left(|A|^2 - |B|^2\right)/|A|^2 \quad (11.4)$$

For this specific problem in which the medium on the left and the medium on the right have the same potential, we can also write Eq. (11.4) in the form

$$\eta = |F|^2/|A|^2 \quad (11.5)$$

We can use either Eq. (11.4) or Eq. (11.5), depending on which is easier to calculate. The form in Eq. (11.4) involving only A and B remains valid regardless of the form of the potential to the right and is more useful in some problems. For example, in a field-emission tunneling problem, where an electric field is applied perpendicular to a metal surface, the potential in the barrier falls off linearly with distance and there is no region on the right of uniform potential, making it harder to calculate the transmitted current directly.

It might seem that we are merely proving the obvious. Note, however, that we have not used classical notions here to deduce the results, though we have shown a connection to those notions. We have instead rigorously deduced the current densities by a first-principles quantum mechanical argument. We have been able to avoid trying to decide whether to use group velocities or phase velocities in considering the currents here, for example. This argument clears the way for practical calculations of tunneling currents, including those with more complicated barriers.

Problem

11.1.1 Consider a single barrier of thickness L, with height V much larger than the energy E of an electron wave incident on the left of the barrier. Presume that the barrier is thick enough that the amount of wave reflected back from the right-hand side of the barrier to the left is essentially negligible. Show that the fraction η of incident electrons transmitted by this barrier can be written approximately as

$$\eta \simeq \frac{16k^2\kappa^2}{\left(k^2+\kappa^2\right)^2}\exp(-2\kappa L)$$

where $k = (2m_o E/\hbar^2)^{1/2}$ and $\kappa = [2m_o(V-E)/\hbar^2]^{1/2}$. (Note: Use the expression of the form $\eta = |F|^2/|A|^2$, where F and A are the amplitudes of the transmitted and incident waves, respectively, to derive this; otherwise, the algebra becomes much more involved.)

11.2 Transfer matrix

We previously deduced how to analyze simple problems such as the potential barrier in Fig. 11.1. In principle, the simple methods we used before are extensible to barriers with multiple layers of different thicknesses and potentials in one dimension, though in practice, we need better mathematical techniques to keep track of the various coefficients. The solution of a problem with two or more layers becomes progressively more intractable unless we take another approach to handling all the various coefficients for the various waves in the different layers. Fortunately, there is a simple way of doing this, which is to introduce the transfer matrix. This enables us to turn on the power of matrix algebra to handle all of the coefficients.

To exploit this method, we presume that the potential is a series of steps. This could correspond directly to an actual step-like potential or we could be approximating some actual continuously varying potential as a series of steps, as shown in Fig. 11.2.

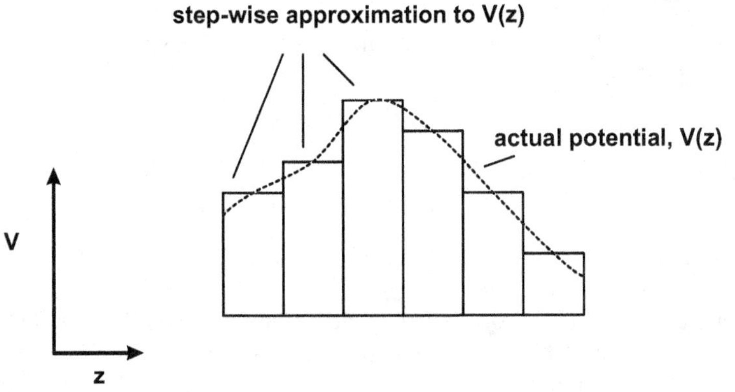

Fig. 11.2. Illustration of the approximation of a continuously varying potential by a step-wise potential.

This enables us to reduce the problem of waves in an arbitrary potential to that of waves within a simple constant potential (i.e., waves that are then either sinusoidal or exponential), together with appropriate boundary conditions to link the solutions in adjacent layers. Many actual problems, such as those involved in analyzing layered semiconductors, have such step-like potentials anyway. The concepts of the wave mechanics to analyze such a structure are closely related to those that we might use to analyze multilayer optical filters or one-dimensional waveguide structures. (The latter waveguide case is a particularly close analogy because then the optical waves also have exponentially decaying solutions past the critical reflection angle, just like the exponentially decaying solutions in the tunneling quantum mechanical problem.)

Imagine that we have a potential structure conceptually like that in Fig. 11.2. We can also imagine that we have an electron wave incident on the structure from one side, with a particular energy, E. In general, when the wave hits an interface, there is some reflected wave and some transmitted wave and we must allow for this in the formalism we set up. To set up the formalism, we consider a multilayer structure with the notation shown in Fig. 11.3 and, specifically, for one layer in the structure, as shown in Fig. 11.4. The approach we take, for each layer in the structure, is to derive a matrix that relates the forward and backward amplitudes, A_m and B_m, just to the right of the $(m - 1)$th interface, to the forward and backward amplitudes A_{m+1} and B_{m+1}, just to the right of the mth interface. By multiplying those matrices together for all of the layers, we construct a single transfer matrix for the whole structure, which enables us to analyze the entire multilayer structure.

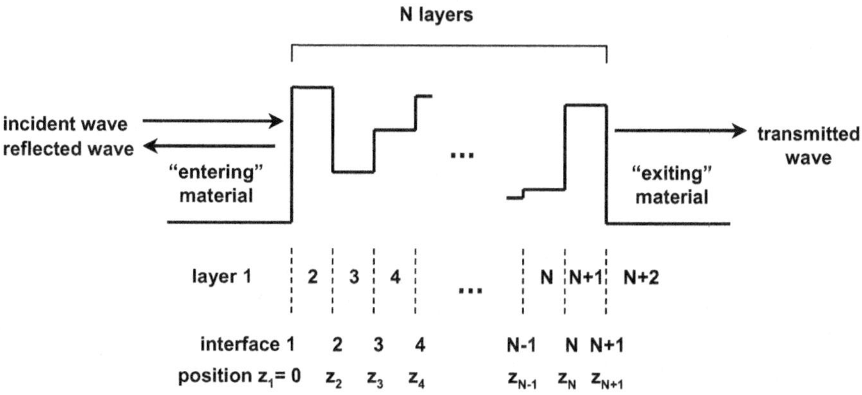

Fig. 11.3. Notation for a multiple-layer potential structure, illustrating also the case in which a wave impinges from the left (incident wave) and we have both transmitted and reflected waves.

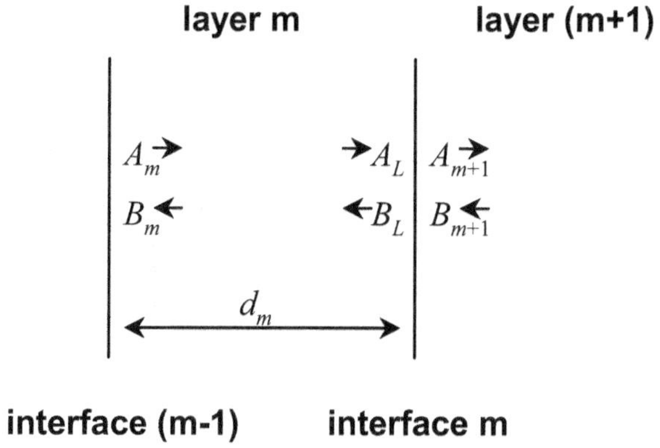

Fig. 11.4. A layer of the structure showing the forward and backward wave amplitudes just to the right of the interfaces and also the forward and backward wave amplitudes, A_L and B_L, just to the left of the mth interface.

In this formalism, each layer m has a potential energy V_m, a thickness d_m, and possibly a mass or effective mass m_{fm}. The position of the mth interface is formally, relative to the position of interface 1

$$z_m = \sum_{q=2}^{m} d_q \tag{11.6}$$

for interfaces 2 and higher (e.g., $z_2 = d_2$, $z_3 = d_2 + d_3$).

In any given layer, if $E > V_m$, we know that, in general, we have a "forward" propagating wave (i.e., propagating to the right in the figures) of the form $A = A_o \exp[ik_m(z - z_{m-1})]$ and a "backward" propagating wave of the form $B = B_o \exp[-ik_m(z - z_{m-1})]$, where A and B are complex numbers representing the amplitude of the forward and backward waves, respectively. In this case

$$k_m = \sqrt{\frac{2m_{fm}}{\hbar^2}(E - V_m)} \tag{11.7}$$

where m_{fm} is the mass of the particle in a given layer of the structure (possibly an effective mass if we are working in the effective mass approximation with a layered semiconductor structure). Similarly, if $V_m > E$, we have a forward decaying "wave" of the form $A = A_o \exp[-\kappa_m(z - z_{m-1})]$ and a backward decaying wave of the form $B = B_o \exp[\kappa_m(z - z_{m-1})]$, where

$$\kappa_m = \sqrt{\frac{2m_{fm}}{\hbar^2}(V_m - E)} \tag{11.8}$$

Now, we note that if we use only the form Eq. (11.7), not only for the situation with $E > V_m$ but also for the case $V_m > E$, we obtain imaginary k ($\equiv i\kappa$) for the $V_m > E$ case. Mathematically, as long as we choose the positive square root (either real or imaginary) in both cases, we can work only with this k. A forward "wave" can then be written in the form $\exp[ik_m(z - z_{m-1})]$ for both the $E > V_m$ and $V_m > E$ cases.

For the $V_m > E$ case, we have a "wave" $\exp[ik_m(z - z_{m-1})] \equiv \exp[-\kappa_m(z - z_{m-1})]$ with imaginary k. This simplifies our handling of the mathematics, allowing us to use the same formalism in all layers. Now, in *any* layer, we have a wave that we can write as

$$\psi(z) = A_m \exp\left[ik_m\left(z - z_{m-1}\right)\right] + B_m \exp\left[-ik_m\left(z - z_{m-1}\right)\right] \tag{11.9}$$

where k can be either real or imaginary and is given by Eq. (11.7).

Now, let us look at the boundary conditions in going from just inside one layer to the right of the boundary to just inside the adjacent layer on the left of the boundary. Using the notation of Fig. 11.4, we have, first for the continuity of the wavefunction, ψ, at the interface

$$\psi = A_L + B_L = A_{m+1} + B_{m+1} \tag{11.10}$$

In the second boundary condition, the continuity of $d\psi / dz$ gives

$$\frac{d\psi}{dz} = ik\left(A - B\right) \tag{11.11}$$

for the wave on either side of the boundary, so we have for the right boundary in Fig. 11.4

$$A_L - B_L = \Delta_m\left(A_{m+1} - B_{m+1}\right) \tag{11.12}$$

where

$$\Delta_m = \frac{k_{m+1}}{k_m} \tag{11.13}$$

with the obvious notation that subscripts m and $m + 1$ refer to the values in the corresponding layers.

In a layered semiconductor structure in the effective mass approximation, we might use continuity of $(1/m_f)d\psi / dz$ for the second boundary condition, in which case instead of Eq. (11.13), we would obtain

$$\Delta_m = \frac{k_{m+1}}{k_m}\frac{m_{fm}}{m_{fm+1}} \tag{11.14}$$

where m_{fm} is the (effective) mass in layer m, and we would use this Δ_m in Eq. (11.12) and all subsequent equations.

Using Eqs. (11.10) and (11.12) gives

$$A_L = A_{m+1}\left(\frac{1+\Delta_m}{2}\right) + B_{m+1}\left(\frac{1-\Delta_m}{2}\right) \quad (11.15)$$

and

$$B_L = A_{m+1}\left(\frac{1-\Delta_m}{2}\right) + B_{m+1}\left(\frac{1+\Delta_m}{2}\right) \quad (11.16)$$

which can be written in matrix form as

$$\begin{bmatrix} A_L \\ B_L \end{bmatrix} = \mathbf{D}_m \begin{bmatrix} A_{m+1} \\ B_{m+1} \end{bmatrix} \quad (11.17)$$

where

$$\mathbf{D}_m = \begin{bmatrix} \dfrac{1+\Delta_m}{2} & \dfrac{1-\Delta_m}{2} \\ \dfrac{1-\Delta_m}{2} & \dfrac{1+\Delta_m}{2} \end{bmatrix} \quad (11.18)$$

Having dealt with the boundary conditions, we now need to deal with the propagation that relates A_m and B_m to A_L and B_L. (We have chosen, for a minor formal reason that will become apparent, to calculate the matrices for going backward through the structure; this is not essential and makes no ultimate difference, though it does influence the signs in the following exponents). For the propagation in a given layer, m, whose layer thickness is d_m, we have

$$A_m = A_L \exp(-ik_m d_m) \quad (11.19)$$

$$B_m = B_L \exp(ik_m d_m) \quad (11.20)$$

corresponding to a matrix-vector representation

$$\begin{bmatrix} A_m \\ B_m \end{bmatrix} = \mathbf{P}_m \begin{bmatrix} A_L \\ B_L \end{bmatrix} \quad (11.21)$$

with

$$\mathbf{P}_m = \begin{bmatrix} \exp(-ik_m d_m) & 0 \\ 0 & \exp(ik_m d_m) \end{bmatrix} \quad (11.22)$$

Hence, we can now write the full transfer matrix, $\mathbf{T}$, for this structure, which relates the forward and backward wave amplitudes at the "entrance" (i.e., just to the left of the first interface) to the forward and backward wave amplitudes at the "exit" (i.e., just to the right of the last interface)

$$\begin{bmatrix} A_1 \\ B_1 \end{bmatrix} = \mathbf{T} \begin{bmatrix} A_{N+2} \\ B_{N+2} \end{bmatrix} \quad (11.23)$$

where

$$\mathbf{T} = \mathbf{D}_1\mathbf{P}_2\mathbf{D}_2\mathbf{P}_3\mathbf{D}_3 \cdots \mathbf{P}_{N+1}\mathbf{D}_{N+1} \quad (11.24)$$

Note that this transfer matrix depends on the energy E that we chose for the calculation of the k's in each layer.

Calculation of tunneling probabilities

Given that we have calculated the transfer matrix for some structure and for some energy E

$$\mathbf{T} \equiv \begin{bmatrix} T_{11} & T_{12} \\ T_{21} & T_{22} \end{bmatrix} \tag{11.25}$$

we can now deduce the fraction of incident particles at that energy that are transmitted. In such a tunneling or transmission problem, we presume that there is no wave incident from the right, so there is no backward wave amplitude on the right of the potential. Hence, for incident forward and backward amplitudes A and B, respectively, and a transmitted amplitude F

$$\begin{bmatrix} A \\ B \end{bmatrix} = \begin{bmatrix} T_{11} & T_{12} \\ T_{21} & T_{22} \end{bmatrix} \begin{bmatrix} F \\ 0 \end{bmatrix} \tag{11.26}$$

From this, we can see that

$$A = T_{11}F \tag{11.27}$$

and

$$B = T_{21}F \tag{11.28}$$

and, hence, from Eq. (11.4), the fraction of particles transmitted by this barrier is[2]

$$\eta = 1 - \frac{|T_{21}|^2}{|T_{11}|^2} \tag{11.29}$$

This technique can be used to give exact analytic results for layered potentials, though such exact results become algebraically impractical for structures with even only quite small numbers of layers. It is, however, particularly useful for numerical calculations. It is straightforward to program on a computer and, even for structures with very large numbers of layers (e.g., 100s), the calculations of transmission at a given energy can be essentially instantaneous even on a small machine. It is, therefore, a very useful practical technique for investigating one-dimensional potentials and their behavior.

Calculation of wavefunctions

Note that this method also enables us to calculate the wavefunction at any point in the structure. We can readily calculate the forward and backward amplitudes, A_m and B_m, respectively, at the left of each layer in the structure. Obviously, we have

$$\begin{bmatrix} A_{N+1} \\ B_{N+1} \end{bmatrix} = \mathbf{P}_{N+1}\mathbf{D}_{N+1} \begin{bmatrix} A_{N+2} \\ B_{N+2} \end{bmatrix} \tag{11.30}$$

and, similarly, we have, in general, for any layer within the structure

$$\begin{bmatrix} A_m \\ B_m \end{bmatrix} = \mathbf{P}_m\mathbf{D}_m \ldots \mathbf{P}_N\mathbf{D}_N\mathbf{P}_{N+1}\mathbf{D}_{N+1} \begin{bmatrix} A_{N+2} \\ B_{N+2} \end{bmatrix} \tag{11.31}$$

---

[2] We could have used the relation $A = T_{11}F$ directly to deduce $\eta = 1/|T_{11}|^2$ if the layer on the left and the layer on the right have the same uniform potential, but we use this more complete form here to include the case where the potentials are not equal on the two sides.

Given that we know the forward and backward amplitudes at the left of layer m, then the wavefunction at some point z in that layer is the sum of the forward and backward wavefunctions, as in Eq. (11.9).

Note that we could set up a calculation so that these forward and backward amplitudes at the interfaces are calculated as intermediate results if we progressively evaluate the forward and backward amplitudes for each successive layer, as in

$$\begin{bmatrix} A_m \\ B_m \end{bmatrix} = \mathbf{P}_m \mathbf{D}_m \begin{bmatrix} A_{m+1} \\ B_{m+1} \end{bmatrix} \tag{11.32}$$

rather than evaluating the transfer matrix $\mathbf{T}$ itself. We can still calculate the transmission probability η using Eq. (11.4) rather than Eq. (11.29).

Example: tunneling through a double barrier structure

Fig. 11.5 shows results calculated using this method for the tunneling probability through a double-barrier structure. This structure shows a resonance in the tunneling probability (or transmission) where the incident energy coincides with the energy of a resonance in the structure. If the barriers were infinitely thick, there would be an eigenstate of the structure approximately at the energy where this resonance occurs.

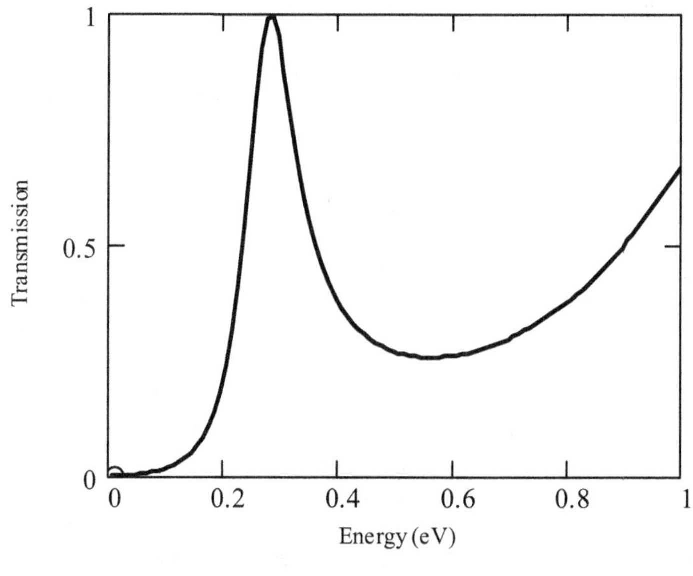

Fig. 11.5. Transmission probability as a function of incident particle energy for an electron incident on a double-barrier structure consisting of two barriers of height 1 eV and thickness 0.2 nm on either side of a 0.7 nm thick region of zero potential energy. The regions on the left and the right of the entire structure are also assumed to have zero potential energy.

Calculation of eigenenergies of bound states

It is possible to use the transfer matrix itself to find eigenstates in cases of truly bound states. For example, if the first layer (layer 1) and last layer (layer $N+2$) are infinitely thick and their potentials are $V_1 > E$ and $V_{N+2} > E$, there may be some values of E for which there are bound eigenstates. Such states would only have exponentially decaying wavefunctions into the first and last layers from the multilayer structure. Hence, $A_1 = 0$ (i.e., no exponentially growing wave going out from the left of the structure) and $B_{N+2} = 0$ (i.e., no exponentially growing wave going out from the right side of the structure). Therefore, if we have a bound eigenstate, we must have

$$\begin{bmatrix} 0 \\ B_1 \end{bmatrix} = \mathbf{T} \begin{bmatrix} A_{N+2} \\ 0 \end{bmatrix} = \begin{bmatrix} T_{11} & T_{12} \\ T_{21} & T_{22} \end{bmatrix} \begin{bmatrix} A_{N+2} \\ 0 \end{bmatrix} \tag{11.33}$$

This can be the case only if the element in the first row and first column of **T** is zero; that is,

$$T_{11} = 0 \tag{11.34}$$

This condition can be used to solve analytically for eigenenergies in simple structures or can be used in a numerical search for eigenenergies by varying E.

Tunneling resonance calculations for quasi-bound states

There are many situations of practical interest in which we do not have any truly bound states of the system, though we have strong resonances, like the transmission resonance in Fig. 11.5. If the resonances are relatively sharp, they can, in practice, have most of the characteristics of a true bound state but just with a finite lifetime. If we were to take the double-barrier structure considered for Fig. 11.5 and progressively make the barriers on either side thicker and thicker, then the transmission resonance would get sharper and sharper. The wavefunction inside the structure at the resonance energy would become closer and closer to the wavefunction of the truly bound state we would have for infinitely thick barriers. The energy of the resonance peak would also become closer and closer to the exact eigenenergy for the infinitely thick barrier case. In fact, even for the resonance in Fig. 11.5, the energy of that peak and the wavefunction within the structure at resonance are almost identical to those for the eigenstate for infinitely thick barriers.

One common situation in which we do not have truly bound states is when we apply electric fields. For any state that was originally truly bound, such as the lowest state in the hydrogen atom, once electric field is applied, there are no longer any truly bound states. Applying electric field corresponds to adding a "tilt" to the potential seen by an electron. There is always some lower energy state for the electron outside the atom on the downhill side toward the positive electrode and the electron can tunnel through the now-finite potential barrier (i.e., be field-emitted from the atom or "field-ionized"). Still, we like to think of this state as being effectively an eigenstate of the hydrogen atom but just with finite lifetime. As a practical matter, we can calculate the effective eigenstates as a function of electric field by looking for the resonances in the transmission,[3] just as in Fig. 11.5. We use the position of the center of the resonance as the effective eigenenergy and, for one-dimensional problems, we can use the wavefunction we deduce at the resonance peak energy from our transfer matrix calculation as the effective wavefunction of this "quasi-bound" state. Such an approach for calculating quasi-bound energy levels and wavefunctions is called a *tunneling resonance calculation*. This technique is used extensively for calculating the effective eigenstates of layered semiconductor structures such as semiconductor quantum wells, including shifts of the energy levels with applied fields.

---

[3] An alternative and essentially equivalent technique for finding the effective eigenenergies is to look for minima in the matrix element T_{11} of the full transfer matrix.

Problems

11.2.1* Consider a double-barrier structure consisting of two barriers of height 1 eV and thickness 0.3 nm on either side of a 1 nm thick region of zero potential energy, with the regions on the left and the right of the entire structure assumed to have zero potential energy. Construct a transfer matrix model for that structure and calculate the tunneling probability of an electron through the structure as a function of electron energy from 0 to 1 eV, graphing the resulting transmission probability.

11.2.2 Perform a similar calculation to that of Problem 11.2.1 for the case of a semiconductor structure. The material on the left, on the right, and in the middle "well" is taken to be GaAs, and we are considering an electron in the conduction band of this and the other materials. The two barriers are now taken to be the specific AlGaAs material with aluminum fraction $x = 0.3$. The conduction band barrier height relative to the GaAs conduction band is given by the relation

$$V_{AlGaAs} \simeq 0.77x \text{ eV}$$

and the electron effective masses in the GaAs and AlGaAs layers are

$$m_f \simeq (0.067 + 0.083x) m_o$$

with $x = 0$ corresponding to the GaAs case. The middle GaAs "well" layer is 9 nm thick.

Specifically, find within 0.1 meV the energy of maximum transmission of the electron through the structure in the range 0 to 0.1 eV for the following AlGaAs barrier thicknesses on either side of the well

(i) 4 nm
(ii) 6 nm
(iii) 8 nm

11.2.3 Consider a GaAs and AlGaAs structure, with a 9 nm thick GaAs layer, but with infinitely thick AlGaAs barriers on either side. Using the method of finding the energy that minimizes T_{11}, the top left element of the transfer matrix, find the bound state of this structure in the range 0 to 0.2 eV within 0.1 meV. (Note: Here, you will only have three layers in your structure altogether, with the "entering" and "exiting" materials being AlGaAs and only one GaAs intermediate layer, which makes the transfer matrix, in this case, only the product of three matrices altogether.)

11.2.4 Consider sets of one-dimensional potential wells that are 0.7 nm thick, with potential barriers separating the wells (i.e., on either side of the wells) that are 0.3 nm thick and 0.9 eV high, and consider the transmission resonances of an electron (in one dimension) in sets of these wells for energies up to 0.5 eV, presuming that the entering and exiting "materials" are the same as the potential well.

(i) Consider for reference one such well (i.e., a 0.7 nm well with two 0.3 nm barriers on either side) and find the resonance energy and sketch or graph the corresponding probability density inside the structure.

(ii) Consider two such wells (i.e., two wells and three barriers):

  (a) graph the transmission probability as a function of energy with sufficient resolution to show all the resonances
  (b) state the number of resonances
  (c) find the lowest and highest resonance energies in the range
  (d) sketch or graph the probability densities inside the structure corresponding to the lowest and highest energy resonances

(iii) Repeat part (ii) for four such wells.

(iv) Repeat part (ii) for eight such wells.

(Note: This problem illustrates the emergence of band structure in periodic systems.)

11.3 Penetration factor for slowly varying barriers

It is possible to make some analytic approximations to penetration and tunneling through barriers when the barrier potential is slowly varying over the scale of the exponential attenuation length. The rigorous analytic approach to such problems is known in quantum mechanics as the *WKB method*, though the mathematical technique predates quantum mechanics. We do not go into this analytic method here; it is largely a mathematical exercise, though it can be quite useful for obtaining analytic approximations to some problems. It can be found in many texts on quantum mechanics.[4] There is, however, one important result from the WKB approach that contains some physical insight. We derive that somewhat informally from the transfer matrix approach discussed previously, though the WKB approach is much more rigorous for deriving this result.

Suppose we have a potential as shown in Fig. 11.6 that we approximate as a series of steps, in which the potential varies slowly. For simplicity of our algebra, we choose the entering and exiting materials as having the same energy, though that is not a necessary restriction.

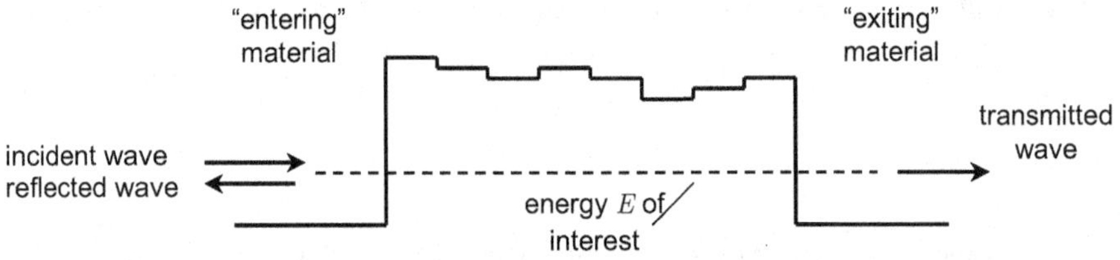

Fig. 11.6. Example of a slowly varying potential approximated as a series of steps.

For our purposes here, we presume that the energy $E << V_m$ for each layer inside the structure. We presume also that we have chosen the layers sufficiently thin in our calculation so that, at least for interfaces within the structure, $k_m \cong k_{m+1}$. Then, for interfaces within the structure, we can approximate the boundary condition matrix as the identity matrix

$$D_m \cong \begin{bmatrix} \sim 1 & \sim 0 \\ \sim 0 & \sim 1 \end{bmatrix} \tag{11.35}$$

Because all of the internal boundary condition matrices have, therefore, been approximated by identity matrices, we can omit them; so, the transfer matrix becomes

$$\mathbf{T} = \mathbf{D}_1 \mathbf{P}_2 \mathbf{P}_3 \cdots \mathbf{P}_N \mathbf{P}_{N+1} \mathbf{D}_{N+1} \tag{11.36}$$

(We have left in the boundary condition matrices for the beginning and end of the structure, where the potential may be quite discontinuous.) Because the propagation matrices are all diagonal, their product is simple, being

$$\mathbf{P}_2 \mathbf{P}_3 \cdots \mathbf{P}_N \mathbf{P}_{N+1} = \begin{bmatrix} 1/G & \sim 0 \\ \sim 0 & G \end{bmatrix} \tag{11.37}$$

where

[4] See, for example, L. I. Schiff, *Quantum Mechanics (Third Edition)* (McGraw-Hill, New York, 1968), pp. 268–279; H. Kroemer, *Quantum Mechanics for Engineering, Materials Science, and Applied Physics* (Prentice Hall, Englewood Cliffs, New Jersey, 1994), pp. 165–181; and R. L. Liboff, *Introductory Quantum Mechanics (Fourth Edition)* (Addison-Wesley, San Francisco, 2003), pp. 243–259

$$G = \prod_{q=2}^{N+1} \exp(ik_q d_q) = \prod_{q=2}^{N+1} \exp(-\kappa_q d_q) = \exp\left(-\sum_{q=2}^{N+1} \kappa_q d_q\right) \tag{11.38}$$

Now, if we have chosen the layers to be sufficiently thin, we may take the summation to be approximately equal to an integral; that is,

$$\sum_{q=2}^{N+1} \kappa_q d_q \cong \int_0^{z_{tot}} \kappa(z)\,dz \tag{11.39}$$

where z_{tot} $(= z_{N+1})$ is the total thickness of the structure (which is taken to start on the left at $z = 0$). Hence, we have

$$G \cong \exp\left(-\int_0^{z_{tot}} \kappa(z)\,dz\right) = \exp\left(-\int_0^{z_{tot}} \sqrt{\frac{2m_f}{\hbar^2}\left(V(z) - E\right)}\,dz\right) \tag{11.40}$$

where now $V(z)$ is the potential as a function of position z. If we now presume that we have an amplitude F of forward wave leaving the right of the structure and no wave arriving from the right, as in our previous example, then, using the transfer matrix of Eq. (11.36) together with the previous substitutions for the propagation matrix product, Eq. (11.37), and the form of the boundary condition matrices for the first and last interfaces, we have for the forward wave amplitude

$$A \cong \frac{(1+\Delta_1)(1+\Delta_{N+1})}{4G} F \tag{11.41}$$

from which we can evaluate the transmission probability from Eq. (11.5)[5] as

$$\eta \cong \frac{16}{\left|(1+\Delta_1)(1+\Delta_{N+1})\right|^2} \exp\left(-2\int_0^{z_{tot}} \sqrt{\frac{2m_f}{\hbar^2}\left(V(z) - E\right)}\,dz\right) \tag{11.42}$$

The prefactor before the exponential contains the boundary conditions at the beginning and end of the whole structure. The exponential factor, sometimes referred to as the *Gamow penetrability factor*, is an approximate expression for the effect of the tunneling within the barrier itself. This approximation is frequently used in tunneling calculations.

11.4 Electron emission with a potential barrier

(This section can be omitted at a first reading. Chapter 8, Sections 8.1–8.7, are also prerequisites.)

Now that we have understood what current is transmitted by a barrier for a given energy of electron and how to calculate quantum mechanically the transmission of arbitrary barriers, we can complete the basic model for electron emission in the presence of a potential barrier. The main additional element we have to introduce is the thermal distribution of electrons. The additional work here is primarily the appropriate integration over that thermal distribution. This kind of approach enables us to calculate *thermionic emission* (i.e., the simple classical

[5] The perceptive reader may notice that the expression $\eta = (|A|^2 - |B|^2)/|A|^2$ does not give the correct answer here. This is because we have not been careful enough in the matrix approximations to retain the backward wave to sufficient order. Doing so gives a much more complicated expression and prevents the simple expressions that allowed us to calculate the forward wave sufficiently accurately.

emission over a potential barrier because the particle has sufficient kinetic energy) and the corrections to that model resulting from the tunneling of particles through the barrier. Examples of such corrections are field-assisted thermionic emission, where some of the particles without sufficient energy can tunnel through the barrier, or even field emission, where the tunneling through the barrier dominates over the conventional thermionic emission over the top. To illustrate the solution to this class of problems, we set up a simple form of such problems, for the case of particles with a constant mass or effective mass, and, in the end, for the case in which the thermal distribution may be approximated by its high-energy limit. The solution of this kind of problem essentially involves no additional quantum mechanics.

The solution of problems of electron emission, such as the following

(i) from a metal into a vacuum (as is encountered, e.g., with vacuum tubes of all kinds, including cathode ray tubes and field emission displays),

(ii) from a metal into a semiconductor (as in a Schottky barrier), or

(iii) in electron emission across or "through" barriers in semiconductor devices, such as leakage of current through barriers in semiconductor devices (e.g., through the gate oxide in silicon field effect transistors)

can all be handled by essentially the same approach. A simplified form of the potential energy diagram is shown in Fig. 11.7, together with the typical terminologies for such problems.

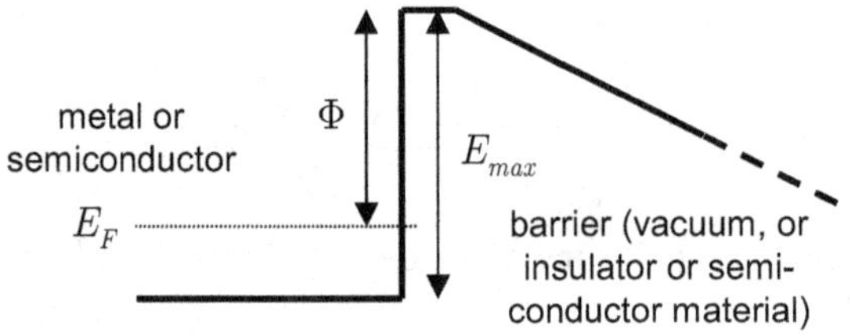

Fig. 11.7. Illustration of a simple potential barrier for emission of electrons from a metal or a semiconductor. E_{max} is the electron affinity if the barrier is the vacuum or the band discontinuity if the barrier is an insulator or a semiconductor. E_F is the Fermi energy (it is not necessarily above the bottom of the band as shown here). Φ is the work function in the case of a metal-vacuum interface (i.e., the separation between the Fermi energy and the vacuum level).

The form of the potential in the barrier is arbitrary (and not necessarily of the form shown in Fig. 11.7), though we do mean E_{max} to be the highest barrier potential seen by any electron from the left, even if it is not exactly at the interface with the metal or semiconductor on the left.

In modeling thermionic or field emission, the usual approach is to presume that any electron that makes it through the barrier never comes back and that there is negligible current of any other electrons coming back into the metal or semiconductor from the barrier side. Usually, there is some electric field present, even in the thermionic case, that sweeps any emitted electrons away from the metal or semiconductor emitter, to the right in the diagram. Such emission is not like a diffusion process, in which the electrons experience a net drift down a concentration gradient. The situation in thermionic or field emission is more severe than that – there are essentially no carriers coming back from right to left at all. The concentration gradient in going from the metal or semiconductor into the barrier is essentially infinite. The current is simply limited, then, by how fast the emitted carriers can leave the emitting material. Essentially, we figure out how many carriers are going to the right at a given energy and how fast they are going, multiply that by the transmission probability associated with that energy, and add up the resulting currents for all energies; the resulting total is the emission current.

We can construct a simple model that will be a useful starting point and, indeed, may be a sufficient model for many circumstances. We presume that the emitting metal or semiconductor on the left has an isotropic parabolic band with mass (or effective mass) m_f.[6] For simplicity, we assume that we are dealing with electrons in a conduction band. An electron in a given **k**-state, with components k_x, k_y, and k_z, has an energy (essentially a kinetic energy) relative to the bottom of the band of

$$E_{KE} = \frac{\hbar^2}{2m_f}\left(k_x^2 + k_y^2 + k_z^2\right) \tag{11.43}$$

We presume that the particles are in a thermal distribution, which, for electrons, in general is a Fermi–Dirac distribution. In a Fermi–Dirac distribution, the probability that a state of any particular energy E_{KE} is occupied is

$$P_{FD}\left(E_{KE}\right) = \frac{1}{1+\exp\left[\left(E_{KE} - E_F\right)/k_B T\right]} \tag{11.44}$$

where E_F is the Fermi energy, k_B is Boltzmann's constant (~ 1.38 x 10^{-23} J/K), and T is the temperature in Kelvin. As usual, the Fermi energy implicitly contains the information about how many electrons there are altogether (i.e., larger densities of electrons correspond to higher Fermi energies). We also presume, incidentally, that the distribution remains a thermal one even though we are continually extracting carriers in particular energy ranges. In practice, that is usually a good assumption – gases of electrons tend to thermalize very quickly within themselves (e.g., << 1 ps).

We are only interested in electrons with positive k_z and, hence, positive effective classical velocities $v_z = \hbar k_z / m_f$ (we take z to be the horizontal direction in Fig. 11.7) because those will be the only ones emitted to the right. Any electron with positive k_z just on the left of the barrier, therefore, gives a contribution to the emitted (electrical) current of $e\eta\left(\hbar^2 k_z^2 / 2m_f\right)v_z$, where, as before, $\eta\left(E_z\right)$ is the transmission probability of an electron with kinetic energy E_z associated with its motion in the z direction and e is the electronic charge (we do not worry here about the sign of the charge and the current, both of which are obvious in this problem). Electrons with negative k_z and v_z do not enter the barrier and make no contribution to the emitted current and so must not be counted in the current. Note explicitly that it is only the kinetic energy associated with the z motion that enters the calculation of the transmission probability η. The kinetic energies associated with the other directions, and the k_x and k_y values, make no difference to this transmission probability.[7] Note also, though, that the thermal occupation probability of a state does depend on the total kinetic energy, E_{KE}.

Now, we start to deduce the total emitted current density. The density of states in k space is $1/(2\pi)^3$ per unit real volume.[8] Including an additional factor of 2 for the two spin states of the electron, the total emitted current density is, therefore

---

[6] For simplicity, we also presume that whatever the mass is here, it is constant throughout the structure of interest. This is not necessarily a good approximation for the semiconductor case.

[7] This is strictly true only for parabolic bands. In general, higher momenta in other directions (e.g., x and y) lead to some change in the behavior in a given direction (e.g., z).

[8] The use of unit volume of emitting material for the calculation means that the quantum mechanical wavefunctions in the emitting material have unit amplitude when normalized to that material, which quantum mechanically keeps track of the currents correctly. If we chose another volume, we would formally have to normalize the quantum mechanical wavefunctions to that volume, which would lead to a

$$J = 2\frac{e}{(2\pi)^3} \int_{k_x=-\infty}^{+\infty} \int_{k_y=-\infty}^{+\infty} \int_{k_z=0}^{+\infty} \eta\left(\hbar^2 k_z^2 / 2m_f\right) P_{FD}\left(E_{KE}\right) v_z dk_x dk_y dk_z \tag{11.45}$$

Note explicitly that the integral is only over positive values of k_z (and, hence, over only positive values of $v_z = \hbar k_z / m_f$).

We can evaluate Eq. (11.45) now to calculate the total emission current, including thermionic and field-assisted emission effects. Applied electric fields E (technically of negative sign) simply give a potential barrier that falls off linearly to the right at a rate, $e\mathsf{E}$, resulting in a triangular barrier for tunneling.[9]

It is often (though not always[10]) the case in emission problems like this that only electrons with energies E_z near the top of the barrier or above make any substantial contribution to the emission current. The tunneling for electrons with lower energies may be totally negligible. It is also usually the case that the barrier height, E_{max}, is much larger than k_BT. In such cases, we can approximate the Fermi–Dirac distribution by neglecting the "1" on the bottom line (making it technically, then, Maxwell–Boltzmann approximation), to obtain

$$P_{FD}\left(E_{KE}\right) \simeq \exp\left(E_F / k_B T\right) \exp\left[-\frac{\hbar^2}{2m_f} \frac{\left(k_x^2 + k_y^2 + k_z^2\right)}{k_B T}\right] \tag{11.46}$$

We can now deal with the k_x and k_y integrals separately and these integrals can conveniently be evaluated mathematically by changing to polar coordinates, imagining that the integral over the $k_x - k_y$ plane is performed by adding up rings of area $2\pi k_r dk_r$ (where $k_r^2 = k_x^2 + k_y^2$), giving

$$\int_{k_x=-\infty}^{+\infty} \int_{k_y=-\infty}^{+\infty} \exp\left[-\frac{\hbar^2}{2m_f} \frac{\left(k_x^2 + k_y^2\right)}{k_B T}\right] dk_x dk_y = 2\pi \int_{k_r=0}^{\infty} \exp\left[-\frac{\hbar^2}{2m_f} \frac{k_r^2}{k_B T}\right] k_r dk_r \tag{11.47}$$

Now

$$\int_0^{\infty} \exp\left(-a^2\right) a da = \frac{1}{2} \int_0^{\infty} \exp\left(-a^2\right) d\left(a^2\right) = \frac{1}{2} \int_0^{\infty} \exp(-b) db = \frac{1}{2} \tag{11.48}$$

Therefore

$$\int_{k_x=-\infty}^{+\infty} \int_{k_y=-\infty}^{+\infty} \exp\left[-\frac{\hbar^2}{2m_f} \frac{\left(k_x^2 + k_y^2\right)}{k_B T}\right] dk_x dk_y = 2\pi \frac{2m_f k_B T}{\hbar^2} \frac{1}{2} \int_0^{\infty} \exp(-b) db = \frac{2\pi m_f k_B T}{\hbar^2} \tag{11.49}$$

---

correction factor that would exactly cancel the effect in the density of states of changing the volume. Therefore, in the end, the emitted current density does not depend on the volume of the emitting material (as long as it is large enough to avoid significant quantum-well confinement phenomena).

[9] The tunneling of particles through such triangular barriers is known as *Fowler-Nordheim tunneling* (R. H. Fowler and L. Nordheim, *Proc. Roy. Soc. London* **119A**, 173–181 (1928)).

[10] For the case in which there is negligible thermionic emission and field emission (i.e., tunneling) dominates, the Maxwell–Boltzmann approximation is likely not sufficient, especially at low temperatures, and the full Fermi–Dirac distribution should be used. This is straightforward, at least numerically.

and so

$$J = 2\frac{e}{(2\pi)^3}\frac{2\pi m_f k_B T}{\hbar^2}\exp\left(\frac{E_F}{k_B T}\right)\int_{k_z=0}^{+\infty}\eta\left(\frac{\hbar^2 k_z^2}{2m_f}\right)\exp\left[-\frac{\hbar^2}{2m_f}\frac{k_z^2}{k_B T}\right]v_z dk_z \tag{11.50}$$

We can change the variable in the integration. Noting that

$$v_z dk_z = \frac{\hbar}{m_f}k_z dk_z = \frac{\hbar}{2m_f}d\left(k_z^2\right) = \frac{1}{\hbar}dE_z \tag{11.51}$$

where

$$E_z \equiv \frac{\hbar^2 k_z^2}{2m_f} \tag{11.52}$$

is the kinetic energy associated with the z direction, we therefore have

$$J = \frac{em_f k_B T}{2\pi^2\hbar^3}\exp\left(\frac{E_F}{k_B T}\right)\int_{E_z=0}^{+\infty}\eta(E_z)\exp\left[-\frac{E_z}{k_B T}\right]dE_z \tag{11.53}$$

Because our quantum mechanical analysis enables us to work out $\eta(E_z)$ for quite arbitrary barrier forms, this expression Eq. (11.53) lets us evaluate electron emission through a barrier. This expression includes both tunneling current and the current over the top of the barrier (conventionally thought of as thermionic emission current). In principle, we have even accounted for the quantum mechanical partial reflection that can occur for particles with energy above the top of the barrier.

Finally, for reasons of notation, let us reconnect to the conventional description of thermionic emission. In the semiclassical view of thermionic emission, the transmission probability $\eta(E_z)$ is simply assumed to be unity for $E_z > E_{max}$ and zero for $E_z < E_{max}$. This view neglects not only tunneling but also the partial reflection that can occur for energies above the barrier, though it is a good starting approximation for thermionic emission, in practice. In this simple case, the thermionic emission is therefore

$$J = \frac{em_f k_B T}{2\pi^2\hbar^3}\exp\left(\frac{E_F}{k_B T}\right)\int_{E_z=E_{max}}^{+\infty}\exp\left[-\frac{E_z}{k_B T}\right]dE_z \tag{11.54}$$

that is,

$$J = \frac{m_f}{m_o}A_0 T^2\exp\left(-\frac{\Phi}{k_b T}\right) \tag{11.55}$$

where A_0 is the Richardson constant

$$A_0 = \frac{em_o k_B^2}{2\pi^2\hbar^3} \simeq 120.4\ \mathrm{A\ cm^{-2}\ K^{-2}} \tag{11.56}$$

and $\Phi = E_{max} - E_F$ is the so-called work function in the case of a metal-vacuum interface. Eq. (11.55) is known as the *Richardson–Dushman equation*. (When the quantum mechanical effects of tunneling are added, it is still common to use the Richardson constant or an effective Richardson constant in the final expressions.)

Problems

11.4.1 Consider a potential barrier of height 1 eV and thickness 1 nm. The same otherwise uniform material (taken here just to be vacuum) is on both sides of the barrier and no electric field is applied. There are electrons on the left of the barrier but none on the right. The electrons on the left are in a thermal distribution at room temperature (300K), with the Fermi energy of that distribution at -100 meV, below the lowest energy state of the electrons on the left, so we can approximately say the electrons are in a Maxwell–Boltzmann distribution. Using the approximate formula

$$\eta \simeq \frac{16k_z^2\kappa_z^2}{\left(k_z^2+\kappa_z^2\right)^2}\exp(-2\kappa_z L)$$

where $k_z = (2m_o E_z/\hbar^2)^{1/2}$ and $\kappa_z = [2m_o(V - E_z)/\hbar^2]^{1/2}$, for the transmission probability of an electron of wavevector $+k_z$ (and kinetic energy E_z) in the z direction for tunneling from the left to the right of the barrier, calculate the total tunneling emission current density (in A/cm^2) from left to right of the barrier. (Note: You may well perform an appropriate numerical integral to get this result.)

11.4.2 (Note: This problem can be used as a substantial assignment.) Consider two GaAs layers separated by an $Al_{0.3}Ga_{0.7}As$ barrier, 5 nm thick, which has a potential height for electrons of ~ 231 meV and an effective mass of ~0.0919 m_o. Presume that the electron effective mass in GaAs is 0.067 m_o. The temperature is 300K and the carrier density on the left of the barrier is such that the Fermi energy is at the bottom of the conduction band. (You may use the Maxwell–Boltzmann approximation to the thermal distribution of electrons.) There are presumed to be no carriers on the right-hand side of the barrier, though the GaAs is assumed to be at the same uniform potential as the GaAs on the left-hand side of the barrier (i.e., assume the bottom of the GaAs conduction band is at the same constant energy on both sides).

(i) Plot the transmission probability for an electron incident on the barrier from the left as a function of the energy of the electron, from 0 to 0.5 eV.
(ii) Calculate the current density of electrons (in A/cm^2) moving from the left of the barrier to the right that we would expect on a simple classical calculation (i.e., the Richardson–Dushman equation).
(iii) Calculate the current density now using the quantum mechanical calculation with the transmission probability included.
(iv) Now replace the single barrier with a pair of $Al_{0.3}Ga_{0.7}As$ barriers each 3 nm thick around an empty GaAs "well" of thickness 4 nm. Plot the transmission probability for an electron incident on the barrier from the left as a function of the energy of the electron, from 0 to 0.5 eV.
(v) With this new structure, calculate the current density now using the quantum mechanical calculation with the transmission probability included.

11.4.3 (Note: This problem can be used as a substantial assignment. It is similar to the previous problem, but here the electric field over the barrier is changed.) Consider the problem of emission of electrons from GaAs, through an AlGaAs barrier of thickness 10 nm and aluminum fraction of 30 percent, into another GaAs layer. The Fermi energy of the electrons in the conduction band of the emitting GaAs is presumed to be at the bottom of conduction band (which, incidentally, corresponds to an electron concentration of ~ 3.2×10^{17} cm^{-3}). Presume there is negligibly small carrier density in the GaAs layer on the other side of the barrier. The temperature is 300 K. You may assume the carrier distribution is approximately given by the Maxwell–Boltzmann distribution for all carriers that are important in the carrier emission. Answers for current densities need only be accurate to ~ 10 percent.

(i) Calculate the thermionic emission current density across this barrier, in A/cm^{-2}, based on the classical Richardson–Dushman equation. (The presumption is that all carriers going to the "right" across the barrier are emitted.)
(ii) Calculate the total emission current density at zero applied field across this barrier including the quantum mechanical transmission coefficient. Explain any difference in the value you get compared to the Richardson–Dushman result.
(iii) Calculate the total emission current density at a field of 20 V/micron. This field is presumed to be applied over the AlGaAs barrier only, in such a direction as to increase emission current

substantially. The GaAs layers on either side of the barrier may be presumed to have zero field in them.

(iv) Graph the total emission current density over a field range from 0 to 40 V/micron over the AlGaAs barrier.

Note: Perform this calculation by constructing a transfer matrix model for the structure and dividing the barrier into a sufficient number of layers. The effect of electric field in the barrier is incorporated by progressively changing the potential in each successive layer of the barrier structure through the addition (or subtraction) of the potential due to the electric field in the barrier. You should not need a very large number of layers to get sufficiently accurate answers. You may need to take some care in numerical integrations over the Maxwell–Boltzmann exponential function because of its strongly decaying nature, though sophisticated numerical integration techniques should not be necessary.

11.5 Summary of concepts

Transmission by a barrier

The fraction of incident particles that are transmitted by the barrier is

$$\eta = \left(|A|^2 - |B|^2\right)/|A|^2 \tag{11.4}$$

where A and B are, respectively, the amplitudes of the wave incident on the barrier and the wave reflected by it.

Transfer matrix

The transfer matrix is constructed by

(i) approximating the structure of interest by a number of layers, each with a given potential and, if necessary, effective mass

(ii) choosing an energy of interest for the incident wave

(iii) calculating boundary condition matrices for the forward and backward waves at each interface for the chosen energy

(iv) calculating propagation matrices for forward and backward waves in each layer

(v) multiplying all of these matrices together in sequence to construct a complete transfer matrix for the structure

Conventionally, the matrix is constructed backward through the structure so that by proposing only an exiting wave at the far "right" of the structure, we can deduce the relative amplitudes of the "incident" and "reflected" waves at the "left" of the structure, thus solving for the reflectivity or transmission of the wave at the chosen energy. Formally, for a matrix

$$\mathbf{T} \equiv \begin{bmatrix} T_{11} & T_{12} \\ T_{21} & T_{22} \end{bmatrix} \tag{11.25}$$

calculated in this way, the fraction of particles transmitted by this barrier is

$$\eta = 1 - \frac{|T_{21}|^2}{|T_{11}|^2} \tag{11.29}$$

For any given energy, the wavefunction amplitudes of the forward and backward waves inside the structure are available as intermediate results in the calculation. Adding the forward and backward waves together at any given point gives the wavefunction at that point.

Calculation of eigenenergies of bound states

For a structure with truly bound states, the eigenenergies are those energies for which

$$T_{11} = 0 \tag{11.34}$$

Tunneling resonance calculations for quasi-bound states

When a structure possesses sharp resonances in its transmission, the energies of those resonances can be considered as effective eigenenergies of quasi-bound states (i.e., states that have finite lifetimes), with the wavefunctions calculated at those resonances being effective eigenfunctions for those quasi-bound states.

Penetration factor for slowly varying barriers

For a barrier of potential energy $V(z) > E$ for z from 0 to some point z_{tot}, for some energy of interest E, the transmission of the barrier is

$$\eta \propto \exp\left(-2\int_0^{z_{tot}} \sqrt{\frac{2m_f}{\hbar^2}\left(V(z)-E\right)}dz\right) \qquad \text{after Eq. (11.42)}$$

Chapter 12

Spin

Prerequisites: Chapters 2–5 and Chapters 9 and 10.

Up to this point, we have essentially presumed that the state of a system, such as an electron or even a hydrogen atom, can be specified by considering a function in space and time, a function we have called the wavefunction. That description in terms of one quantity, the scalar amplitude of the wavefunction, turns out not to be sufficient to describe quantum mechanical particles. We find that we also need to specify amplitudes for the spin of the particle.

The idea that we would need additional degrees of freedom to describe a system completely is not itself unusual. For example, in classical mechanics, we might use position as a function of time to describe an object such as a football, but that would not be a complete description; we might need to add a description of the rotation of the football so that we could calculate the expected future dynamics of the ball. The rotation of the football is still something we can describe purely in terms of the position of points on the football and we might view it as being loosely analogous to the angular momentum of a quantum mechanical particle such as a hydrogen atom. We might also, however, need to add the color of the ball to its description if we were showing the ball on television and the manufacturer of the ball would care very much that the correct name was clearly displayed on the ball. Those additional attributes of the ball cannot be described simply as the positions of the points on the surface of the ball – we need to add the color of the points as an additional attribute that varies over the surface of the ball. It is an attribute that varies in space but is not itself the position of the points on the surface. Additionally, and more important for our discussion here, there might be other properties that are intrinsic to the football that would be important for its dynamics.

Elementary particles, such as electrons, neutrons, and protons, have an additional intrinsic property, called *spin*, that just like the rotation of a football can have quite measurable and important consequences for the behavior of the particle.[1] The magnitude of this spin turns out to be an intrinsic and unalterable property of each specific type of particle. In that sense, spin is not like the external rotation of the football. Spin is more analogous to having a constantly spinning gyroscope buried inside the football. Spin is, however, a very quantum mechanical property and it would be wise to leave classical analogies behind from this point on.

Spin and its consequences are extremely important in quantum mechanics. Spin turns out to be the determining quantity for deciding whether more than one particle can occupy a given state. Both chemistry and solid-state physics rely heavily on the *Pauli exclusion principle* that only

---

[1] Elementary particles have other attributes, such as "charm" and "strangeness," that do not concern us, just as the specific manufacturer of the football might not be important to the football player. These additional quantum mechanical properties do matter very much in high-energy physics and the specific manufacturer of the football does matter to the company shareholders.

one electron can occupy a given state. (We discuss the consequences of exclusion or otherwise in Chapter 13.) Magnetic effects in materials are almost entirely due to spin properties. Electron spin effects are very important in determining polarization selection rules in optical absorption and emission, both in atoms and in optoelectronic devices, for example.

12.1 Angular momentum and magnetic moments

To see just what the "spin" angular momentum of an electron is, we can look at how the energies of electron states or the positions of electrons are influenced by magnetic fields. First, we need to understand some basic concepts about angular momentum and magnetic moments. One of the most direct consequences of angular momentum is that charged particles with angular momentum have so-called magnetic moments, which means that they behave as if they were little magnets. This is easy to understand, for example, for an orbiting electron.

Classically, an electron orbiting with a velocity v in a circular orbit of radius r, as in the simple Bohr model of the hydrogen atom, has an angular momentum of magnitude

$$L = m_o vr \tag{12.1}$$

We can also write angular momentum as a vector, in which case we have, classically

$$\mathbf{L} = \mathbf{r} \times \mathbf{p} = \mathbf{r} \times m_o \mathbf{v} \tag{12.2}$$

The electron takes a time $2\pi r / v$ to complete an orbit, so the current corresponding to this orbiting electron is of magnitude $I = ev / 2\pi r$ (this is the amount of charge that will pass any point in the loop per second). The current loop corresponding to the orbit has an area πr^2.

In magnetism, we can define a quantity called the *magnetic dipole*, *magnetic moment* or *magnetic dipole moment*, a quantity that is essentially the strength of a magnet. For any closed current loop, the magnitude of the classical dipole moment is μ_d = current × area. Such an orbiting electron, therefore, classically has a magnetic dipole of magnitude

$$\mu_e = -evr / 2 = -eL / 2m_o \tag{12.3}$$

where the minus sign is present because the electron charge is negative.

Magnetic moment is a vector quantity, with the vector axis being along the polar axis of the magnet. In the full vector statement, the vector magnetic dipole moment for a current loop is

$$\boldsymbol{\mu}_d = I\mathbf{a} \tag{12.4}$$

where I is the current in the loop and **a** is a vector whose magnitude is the area of the loop and whose direction is given by the right hand rule when considering the direction of current flow around the loop. For a classical electron in a circular orbit, the magnetic dipole moment as a vector is, therefore

$$\boldsymbol{\mu}_e = -\frac{e}{2}\mathbf{r} \times \mathbf{v} = -\frac{e\mathbf{L}}{2m_o} \tag{12.5}$$

If we apply a magnetic field **B**, classically the energy of an object with a magnetic moment $\boldsymbol{\mu}_d$ changes by an amount

$$E_\mu = -\boldsymbol{\mu}_d \cdot \mathbf{B} \tag{12.6}$$

Applying a magnetic field **B** along the *z*-direction to a hydrogen atom defines the *z*-direction as the quantization axis, making the angular momentum quantized around the *z*-direction, with quantum number *m*, where the allowed values of *m* go in integer steps from -*l* to +*l*. The corresponding angular momenta are of magnitude $m\hbar$ or, in vector form, $m\hbar\hat{\mathbf{z}}$. Taking these values of angular momentum, we would expect corresponding magnetic moments for these electron orbits of

$$\boldsymbol{\mu}_e = -\frac{em\hbar\hat{\mathbf{z}}}{2m_o} \equiv -m\mu_B\hat{\mathbf{z}} \qquad (12.7)$$

hence, the name *magnetic quantum number* sometimes given to *m*. Here, we have used the natural unit, μ_B, called the *Bohr magneton*, for magnetic moment in quantized problems, where

$$\mu_B = \frac{e\hbar}{2m_o} \qquad (12.8)$$

Hence, we would expect energy changes for these states of

$$E_m = m\mu_B B \qquad (12.9)$$

as a result of applying the magnetic field.

So, in a hydrogen atom, we might expect an applied magnetic field to split the $2l+1$ degenerate levels of some state with nonzero *l* (e.g., a *P* state, with $l=1$) into $2l+1$ different energy levels. Of course, we should do that calculation properly from a quantum mechanical point of view,[2] which, in this case, would be a degenerate perturbation theory calculation of the hydrogen atom with a perturbing Hamiltonian[3] operator $\hat{H}_P = (e/2m_o)B\hat{L}_z$, but the result (neglecting spin effects) would be essentially what we expect, with $2l+1$ different energy levels appearing, with energy splittings $m\mu_B B$. The splitting of atomic levels with magnetic fields is called the *Zeeman effect* and can be seen in optical absorption spectra, for example.

Another way of seeing the magnetic moments of particles is to see how they are deflected by nonuniform magnetic fields, as in the Stern–Gerlach experiment (see Section 3.8). In such an experiment, if we prepared the particles with arbitrary initial values of *m* and repeated the experiment many different times, we should expect to see $2l+1$ different deflection angles emerging.

Hence, we expect a Zeeman splitting experiment or a Stern–Gerlach experiment to show us how many different values of the *z* component of the angular momentum are allowed in a magnetic field in the *z* direction in a given state. We can now ask, "What happens if we try to use an approach like this to look at an electron itself in a magnetic field to understand its internal angular momentum or spin?"[4] How many different values of the *z* component of the angular momentum do we see? The answer is, rather surprisingly, 2.

---

[2] See, e.g., the discussion of the Zeeman effect for the hydrogen atom in W. Greiner, *Quantum Mechanics (Third Edition)* (Springer-Verlag, Berlin, 1994), pp. 316–320.

[3] Because the electron has negative charge, a positive angular momentum in the *z* direction gives a negative dipole moment in the *z* direction; hence, even when considering the full vector form of the dipole energy, the sign in this perturbing Hamiltonian is positive as stated for a magnetic field in the *z* direction.

[4] Actual experiments may use an atom with a single electron in its outer shell in an *S* orbital state (with therefore no orbital angular momentum). The Stern–Gerlach experiment used silver atoms.

To distinguish this spin angular momentum from orbital angular momentum, we use the quantum numbers *s* rather than *l* and σ rather than *m*. If we want to reconcile this result for the electron with the quantum mechanical formalism for angular momentum we previously constructed where we had $2l+1$ different deflection angles in the Stern–Gerlach experiment, to get $2s+1=2$, we need $s=1/2$. Hence, we seem to have to assign a spin angular momentum to the electron of $\hbar s=\hbar/2$. As before, we say that σ can take values in integer steps from -*s* to +*s*, so $\sigma=-1/2$ or $+1/2$ and the corresponding angular momentum component in the *z* direction is $\pm\sigma\hbar$.

Based on our previous understanding of angular momentum from the Schrödinger equation or our creation of quantum mechanical angular momentum operators from classical angular momentum, this apparent value of ½ for the internal spin angular momentum of the electron is bizarre. When we considered angular momentum associated with wavefunctions before, we concluded that the *m* quantum number had to be an integer; otherwise, the spatial wavefunction that corresponded to it would not be single-valued after a single complete rotation about the *z* axis. How then can we have the σ quantum number be a half integer?

The answer to which we are forced is that the eigenfunctions associated with this internal angular momentum of the electron are not functions in space. We cannot describe the behavior of an electron, including its spin, only in terms of one function amplitude in space. For a complete description, we need to specify and describe another degree of freedom, the electron spin, just as for the football we needed to specify and describe another degree of freedom, the chemical composition of the surface, to handle the issue of the color of the football. It was not sufficient merely to describe the positions of the points on the surface of the football if we wanted to describe the football's color.

Incidentally, and somewhat confusingly, the magnitude of magnetic moment of the electron due to its spin is not simply $\sigma\mu_B$ as we might expect, but is instead

$$\mu_e = g\sigma\mu_B \tag{12.10}$$

where we have the so-called *gyromagnetic factor* $g \simeq 2.0023$, often approximated as a factor 2. There is also no radius of classical orbit of an electron that will give it both an angular momentum of $\hbar/2$ and a magnetic moment of $\pm g\mu_B/2$, further confirming that spin cannot be considered as corresponding to a classical orbit of any kind.

12.2 State vectors for spin angular momentum

Suppose for the moment that we are only interested in the spin properties of the electron. How do we describe the state of the electron in that case? To understand that, we can ask first how we would have described an orbital angular momentum state without describing it explicitly as a function of angle in space.

Suppose, for example, that we considered only states with a specific value of *l*. We can write such a state as $|l\rangle$ if we wish. In general, $|l\rangle$ would be some linear combination of the basis states $|l,m\rangle$ corresponding to any of the specific allowed values of *m*; that is,

$$|l\rangle = \sum_{m=-l}^{l} a_m |l,m\rangle \tag{12.11}$$

In the case of these states, each of the states $|l,m\rangle$ can also be written as one of the spherical harmonic functions in space and the resulting linear combination $|l\rangle$ can also, therefore, be written as a function of angle in space. We could also, if we wish, write $|l\rangle$ explicitly as a vector of the coefficients a_m; that is,

$$|l\rangle \equiv \begin{bmatrix} a_l \\ a_{l-1} \\ \vdots \\ a_{-l+1} \\ a_{-l} \end{bmatrix} \tag{12.12}$$

Note that the set of functions corresponding to all of the possible values of m for a given l is a complete set for describing any possible function with that value of l, including even the eigenfunctions of L_x and L_y that are oriented around the other axes.

In the case of the electron spin, we cannot write the basis functions as functions of angle in space, but we do expect that we can write them using the same kind of state and vector formalism as we use for other angular momenta. In the electron spin case, that formalism becomes very simple. Instead of l, we have s, which we know is ½, and instead of m we have σ. Just as we did for orbital angular momentum, where we could write any specific basis state as $|l,m\rangle$, we can use a notation $|s,\sigma\rangle$ for the spin case. There are, however, now only two basis states, $|1/2,1/2\rangle$ and $|1/2,-1/2\rangle$, corresponding to $\sigma = 1/2$ and $\sigma = -1/2$, respectively. Hence, if we choose to write our general spin state as $|s\rangle$, we have

$$|s\rangle = a_{1/2}|1/2,1/2\rangle + a_{-1/2}|1/2,-1/2\rangle \equiv a_{1/2}|\uparrow\rangle + a_{-1/2}|\downarrow\rangle \equiv \begin{bmatrix} a_{1/2} \\ a_{-1/2} \end{bmatrix} \tag{12.13}$$

where we have also indicated another common notation, with $|\uparrow\rangle$ being the spin-up state $|1/2,1/2\rangle$ and $|\downarrow\rangle$ being the spin-down state $|1/2,-1/2\rangle$. (The "up" and "down" refer to the z direction, conventionally.)

Any possible spin state of the electron can presumably be described this way. Rather obviously, a state with its magnetic moment in the $+z$ direction (the "spin-up" state) will be the state $\begin{bmatrix} 1 \\ 0 \end{bmatrix}$ and a state with its magnetic moment in the $-z$ direction (the "spin-down" state) will be $\begin{bmatrix} 0 \\ 1 \end{bmatrix}$. We could also multiply these states by any unit complex number and they would still be spin-up and spin-down states, respectively. The choice of unit amplitudes for these states also assures that they are normalized. Normalization, in this case, means assuring that the sum of the modulus squared of the two vector elements is equal to one; that is, $|a_{1/2}|^2 + |a_{-1/2}|^2 = 1$.

The reader might think that these vectors, $\begin{bmatrix} 1 \\ 0 \end{bmatrix}$ and $\begin{bmatrix} 0 \\ 1 \end{bmatrix}$, can represent only spin-up and spin-down states, oriented along the z axis. In fact, that is not correct; these two basis vectors can represent any possible spin state of the electron, including spin states with the magnetic moment oriented along the x or the y direction. This is readily shown once we have defined the operators for the various spin components.

12.3 Operators for spin angular momentum

Now that we have found an appropriate way of writing spin states, we need to define operators for spin angular momentum. In the case of orbital angular momentum, we started by postulating the operators associated with the components of the orbital angular momentum along the three coordinate axes, $\hat{L}_x$, $\hat{L}_y$, and $\hat{L}_z$, in terms of spatial position and spatial derivative operators, an option that we do not have for the spin operators because the spin functions are not functions of space. We also, however, were able to write commutation relations for the orbital angular momentum operators. An alternative approach to finding the characteristics of the spin operators might, therefore, be to start with the commutation relations and find a representation of spin operators that satisfy such commutation relations. In fact, some authors regard the commutation relations as being the more fundamental statement of the operator properties. If one starts with the commutation relations for angular momentum operators, one can prove that both integer and half-integer values for angular momentum are possible and these are all that are possible.[5]

By analogy with the angular momentum operators $\hat{L}_x$, $\hat{L}_y$, and $\hat{L}_z$, we can write spin angular momentum operators $\hat{S}_x$, $\hat{S}_y$, and $\hat{S}_z$ and we expect them to obey a set of commutation relations

$$\left[\hat{S}_x, \hat{S}_y\right] = i\hbar\hat{S}_z \tag{12.14}$$

$$\left[\hat{S}_y, \hat{S}_z\right] = i\hbar\hat{S}_x \tag{12.15}$$

$$\left[\hat{S}_z, \hat{S}_x\right] = i\hbar\hat{S}_y \tag{12.16}$$

It is common in discussing spin to work with the "dimensionless" operators $\hat{\sigma}_x$, $\hat{\sigma}_y$, and $\hat{\sigma}_z$ from which the spin angular momentum magnitude $\hbar/2$ has been removed; that is,

$$\hat{\sigma}_x = 2\hat{S}_x/\hbar\,,\ \hat{\sigma}_y = 2\hat{S}_y/\hbar\,,\ \hat{\sigma}_z = 2\hat{S}_z/\hbar \tag{12.17}$$

giving the set of commutation relations

$$\left[\hat{\sigma}_x, \hat{\sigma}_y\right] = 2i\hat{\sigma}_z \tag{12.18}$$

$$\left[\hat{\sigma}_y, \hat{\sigma}_z\right] = 2i\hat{\sigma}_x \tag{12.19}$$

$$\left[\hat{\sigma}_z, \hat{\sigma}_x\right] = 2i\hat{\sigma}_y \tag{12.20}$$

If we choose to represent the spin function in the vector format, then the operators become represented by matrices. One set of matrix representations of these operators is

$$\hat{\sigma}_x = \begin{bmatrix} 0 & 1 \\ 1 & 0 \end{bmatrix},\ \hat{\sigma}_y = \begin{bmatrix} 0 & -i \\ i & 0 \end{bmatrix},\ \hat{\sigma}_z = \begin{bmatrix} 1 & 0 \\ 0 & -1 \end{bmatrix} \tag{12.21}$$

Such matrix representations are known as *Pauli spin matrices*. There is more than one way we could have chosen these – in fact, there is an infinite number of ways – depending on what axis we choose for the spin. This set, which we can call the z representation, is such that the spin-up

[5] See, for example, P. A. M. Dirac, *The Principles of Quantum Mechanics (4th Edition, revised)* (Oxford, 1967), pp. 144–146.

and spin-down vectors defined previously are eigenvectors of the $\hat{\sigma}_z$ operator. The reader can check that these operators do indeed obey the commutation relations Eqs. (12.18)–(12.20).

We can, if we wish, choose to write the three different Pauli spin matrices as one entity, $\hat{\boldsymbol{\sigma}}$, which has components associated with each of the coordinate directions x, y, and z

$$\hat{\boldsymbol{\sigma}} = \mathbf{i}\hat{\sigma}_x + \mathbf{j}\hat{\sigma}_y + \mathbf{k}\hat{\sigma}_z \equiv \mathbf{i}\begin{bmatrix} 0 & 1 \\ 1 & 0 \end{bmatrix} + \mathbf{j}\begin{bmatrix} 0 & -i \\ i & 0 \end{bmatrix} + \mathbf{k}\begin{bmatrix} 1 & 0 \\ 0 & -1 \end{bmatrix} \tag{12.22}$$

For completeness in discussing the spin operators, we note that by analogy with the $\hat{L}^2$ operator, we can also define an $\hat{S}^2$ operator

$$\hat{S}^2 = \hat{S}_x^2 + \hat{S}_y^2 + \hat{S}_z^2 \tag{12.23}$$

or a $\hat{\sigma}^2$ operator

$$\hat{\sigma}^2 = \hat{\sigma}_x^2 + \hat{\sigma}_y^2 + \hat{\sigma}_z^2 \tag{12.24}$$

From the definitions for the Pauli matrices, we see that

$$\hat{\sigma}^2 \equiv 3\begin{bmatrix} 1 & 0 \\ 0 & 1 \end{bmatrix} \tag{12.25}$$

and, hence, that

$$\hat{S}^2 \equiv \frac{3}{4}\hbar^2\begin{bmatrix} 1 & 0 \\ 0 & 1 \end{bmatrix} = s(s+1)\hbar^2\begin{bmatrix} 1 & 0 \\ 0 & 1 \end{bmatrix} \tag{12.26}$$

from which we see that any spin ½ vector is an eigenvector of the $\hat{S}^2$ operator, with eigenvalue $s(s+1)\hbar^2 = (3/4)\hbar^2$. Incidentally, it is also similarly true for orbital angular momentum that any linear combination of spherical harmonics corresponding to a given l value is an eigenfunction of the $\hat{L}^2$ operator, with eigenvalue $l(l+1)\hbar^2$, so the behaviors here are still analogous to the behavior of orbital angular momentum.

Problems

See Problems 4.10.4 and 5.1.1.

12.4 The Bloch sphere

Although we have chosen to discuss spins using the z representation here, which has eigenfunctions that correspond to pure spin-up and spin-down, this representation can also describe spins oriented exactly along the x axis or exactly along the y axis or, indeed, spins oriented at any angle. Strange as it may seem, a spin pointing in the x direction can be expressed as a linear combination of spin-up and spin-down states described in the z direction. How can this be? Note that the spin vector itself is *not* a vector in ordinary geometrical space. It is a vector in a two-dimensional Hilbert space. One way to find what are the two spin vectors that correspond to a spin oriented, respectively, in the positive or negative x direction is to find the eigenvectors of the $\hat{\sigma}_x$ Pauli spin matrix, which is a simple exercise in matrix algebra, and similarly for the y direction.

The general spin state, as in Eq. (12.13), can be visualized in a particularly elegant way. There are four real numbers required to specify the electron spin vector – a real and an imaginary part (or, equivalently, a magnitude and phase) for each of the two elements of the vector. This is enough to specify the two angles and the complex amplitude (e.g., magnitude and phase) for a spin pointing in any specific direction. Because the magnitude of the vector is fixed for spin, and because we can choose the quantum mechanical phase of any single state arbitrarily without making any difference to measurable quantities, in fact, we only really need two numbers to describe a spin state. A particularly illuminating way to specify those two numbers is as a pair of angles, θ and ϕ, in terms of which we can choose to write the general spin state as

$$|s\rangle = \cos(\theta/2)|\uparrow\rangle + \exp(i\phi)\sin(\theta/2)|\downarrow\rangle \tag{12.27}$$

Because $\cos^2(\theta/2)+\sin^2(\theta/2)=1$, the magnitude of this vector is correctly guaranteed to be unity and the $\exp(i\phi)$ factor allows for any relative quantum mechanical phase between the two components.

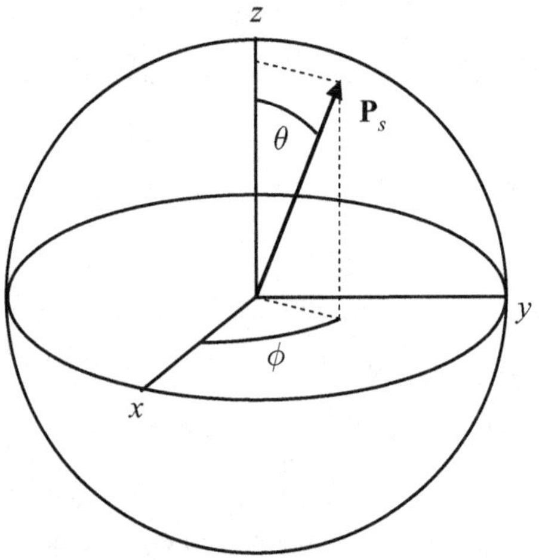

Fig. 12.1. Bloch sphere, showing the general state of a single s=1/2 spin represented as a vector $\mathbf{P}_s$.

We can now formally ask for the expectation value of the Pauli spin operator $\hat{\boldsymbol{\sigma}}$ (as defined in Eq. (12.22)) with such a state, obtaining as the result a "spin polarization" vector $\mathbf{P}_s$

$$\begin{aligned}\mathbf{P}_s &= \langle s|\hat{\boldsymbol{\sigma}}|s\rangle = \mathbf{i}\langle s|\hat{\sigma}_x|s\rangle + \mathbf{j}\langle s|\hat{\sigma}_y|s\rangle + \mathbf{k}\langle s|\hat{\sigma}_z|s\rangle \\ &= \mathbf{i}\sin\theta\cos\phi + \mathbf{j}\sin\theta\sin\phi + \mathbf{k}\cos\theta\end{aligned} \tag{12.28}$$

the algebra of which is left as an exercise for the reader. But, this vector $\mathbf{P}_s$ is simply a vector of unit length pointing from the origin out to a point on a sphere of unit radius, with angle relative to the north pole of θ and azimuthal angle ϕ. In other words, the general spin state $|s\rangle$ can be visualized (Fig. 12.1) as a vector on a unit sphere. The north pole corresponds to the state $|\uparrow\rangle$ and the south pole to state $|\downarrow\rangle$. This sphere is called the *Bloch sphere*, with the angles θ and ϕ on this sphere characterizing the spin state and the geometrical *x*, *y*, and *z* directions corresponding to the directions of the eigenvectors of the corresponding spin operators.

Problems

12.4.1* Show that for $|s\rangle = \cos(\theta/2)|\uparrow\rangle + \exp(i\phi)\sin(\theta/2)|\downarrow\rangle$

$$\langle s|\hat{\boldsymbol{\sigma}}|s\rangle = \mathbf{i}\sin\theta\cos\phi + \mathbf{j}\sin\theta\sin\phi + \mathbf{k}\cos\theta$$

12.4.2 Suppose an electron is in the spin state $|s\rangle = (1/\sqrt{2})[|\uparrow\rangle + i|\downarrow\rangle]$ and that the electron spin magnetic moment operator can be written as $\hat{\boldsymbol{\mu}}_e = g\mu_B\hat{\boldsymbol{\sigma}}$. What is the expectation value of the electron spin magnetic moment? (Note: The result is a vector).

12.4.3 Consider an arbitrary spin state $|s\rangle = \cos(\theta/2)|\uparrow\rangle + \exp(i\phi)\sin(\theta/2)|\downarrow\rangle$ and operate on it with the $\hat{\sigma}_z$ operator. Describe the result of this operation in terms of a rotation of the spin polarization vector on the Bloch sphere.

12.5 Direct product spaces and wavefunctions with spin

Thus far, we have only discussed spin in isolation – we have not also considered the spatial variation of the electron's behavior. How can we put the two of these together to obtain a description of the electron incorporating both spin and spatial behavior? The answer is that we allow the electron to have two spatial wavefunctions, one associated with spin up and the other with spin down. We can write such a wavefunction as a vector in which the components vary in space. Thus, if $|\Psi\rangle$ is to be the most complete representation of the state of the electron, including spin effects, we might write

$$|\Psi\rangle \equiv \begin{bmatrix} \psi_\uparrow(\mathbf{r},t) \\ \psi_\downarrow(\mathbf{r},t) \end{bmatrix} \equiv \psi_\uparrow(\mathbf{r},t)\begin{bmatrix} 1 \\ 0 \end{bmatrix} + \psi_\downarrow(\mathbf{r},t)\begin{bmatrix} 0 \\ 1 \end{bmatrix} \tag{12.29}$$

A function of the form $\begin{bmatrix} \psi_\uparrow(\mathbf{r},t) \\ \psi_\downarrow(\mathbf{r},t) \end{bmatrix}$ is called a *spinor*. We can also use other equivalent notations. Before doing so, it is important to understand exactly what we are doing with the Hilbert spaces.

We previously discussed a Hilbert space to represent one function, the wavefunction, in space and also, if we wished, time. For general spatial wavefunctions, for example, we needed an infinite dimensional Hilbert space, with one dimension for each basis function. The state vector in that Hilbert space was simply the vector of the amplitudes of all the basis functions required to build up the desired function in space (e.g., the vector of the Fourier amplitudes). For other more restricted problems, we can construct other Hilbert spaces. For the case in which we are only interested in electron spin, we had only a two-dimensional Hilbert space, with the dimensions labeled spin-up and spin-down. We could have constructed a Hilbert space to represent any angular function associated with a specific total orbital angular momentum l. That space would have had $2l+1$ dimensions, corresponding to the different possible values of m.

What happens, as in the present case, when we want to combine two Hilbert spaces (e.g., the spatial and temporal Hilbert space describing ordinary spatial and temporal functions and the Hilbert space for spin) to create a space that can handle any state in this more complicated problem? In the present electron spin case, we want to have a space that is sufficient to represent two spatial (or spatial and temporal) functions at once. Hence, where previously we only needed one dimension and one coefficient, associated with a particular spatial and temporal basis function, we now need two. We have doubled the number of our dimensions.

The basis functions in our new Hilbert space are all the products of the basis functions in the original separate spaces. For example, if the basis functions for the spatial and temporal function were

$$\psi_1(\mathbf{r},t),\ \psi_2(\mathbf{r},t),\ \ldots,\ \psi_j(\mathbf{r},t),\ \ldots$$

then the basis functions when we add spin into the problem are

$$\psi_1(\mathbf{r},t)\begin{bmatrix}1\\0\end{bmatrix},\ \psi_2(\mathbf{r},t)\begin{bmatrix}1\\0\end{bmatrix},\ \ldots,\ \psi_j(\mathbf{r},t)\begin{bmatrix}1\\0\end{bmatrix},\ \ldots,$$

$$\psi_1(\mathbf{r},t)\begin{bmatrix}0\\1\end{bmatrix},\ \psi_2(\mathbf{r},t)\begin{bmatrix}0\\1\end{bmatrix},\ \ldots,\ \psi_j(\mathbf{r},t)\begin{bmatrix}0\\1\end{bmatrix},\ \ldots$$

This concept of the new basis functions being the products of the elements of two basis function sets is not exclusively a quantum mechanical one. For example, if a spatial function in one dimensional box of size L_x can be represented as a Fourier series of the form

$$f(x) = \sum_n a_n \exp(i2n\pi x / L_x) \tag{12.30}$$

then a function in a two-dimensional rectangular box of sizes L_x and L_y in the respective coordinate directions can be represented as a Fourier series

$$g(x,y) = \sum_{n,p} a_{n,p} \exp(i2\pi nx / L_x)\exp(i2\pi py / L_y) \tag{12.31}$$

Here, the new basis functions are the products of the basis functions of the two Hilbert spaces associated with the two separate problems of functions in x and functions in y.

A Hilbert space formed in this way is called a *direct product space*. The spinors exist in such a direct product space formed by the multiplication (i.e., direct product)[6] of the spatial and temporal basis functions and the spin basis functions. We can also write the new basis functions using Dirac notation. In the electron spin case, we could write the basis functions as

$$|\psi_1\rangle|\uparrow\rangle,\ |\psi_2\rangle|\uparrow\rangle,\ \ldots|\psi_j\rangle|\uparrow\rangle,\ \ldots,\ |\psi_1\rangle|\downarrow\rangle,\ |\psi_2\rangle|\downarrow\rangle,\ \ldots,\ |\psi_j\rangle|\downarrow\rangle,\ \ldots$$

Here, we understand that the $|\psi_j\rangle$ kets are vectors in one Hilbert space, the space that represents an arbitrary spatial and temporal function, and the $|\uparrow\rangle$ and $|\downarrow\rangle$ kets are vectors in the other Hilbert space representing only spin functions. The products $|\psi_j\rangle|\uparrow\rangle$ and $|\psi_j\rangle|\downarrow\rangle$ are vectors in the direct product Hilbert space. We could, if we wished, also write these products using the notations $|\psi_j\,\uparrow\rangle \equiv |\psi_j\rangle|\uparrow\rangle$ and so on as the basis functions of our direct product Hilbert space

$$|\psi_1\,\uparrow\rangle,\ |\psi_2\,\uparrow\rangle,\ \ldots,\ |\psi_j\,\uparrow\rangle,\ \ldots,\ |\psi_1\,\downarrow\rangle,\ |\psi_2\,\downarrow\rangle,\ \ldots,\ |\psi_j\,\downarrow\rangle,\ \ldots$$

Direct product spaces occur any time in quantum mechanics that we add more degrees of freedom or "dynamical variables" into the problem including, for example, adding more spatial dimensions or more particles or more attributes such as spin for individual particles. With our different notations, we could also write Eq. (12.29) as

$$|\Psi\rangle = |\psi_\uparrow\rangle|\uparrow\rangle + |\psi_\downarrow\rangle|\downarrow\rangle = |\psi_\uparrow\,\uparrow\rangle + |\psi_\downarrow\,\downarrow\rangle \tag{12.32}$$

[6] This direct product can also be considered as a tensor product. Sometimes this product is marked explicitly using a sign such as ⊗, that is, writing $|\psi_1\rangle \otimes |\uparrow\rangle$ instead of $|\psi_1\rangle|\uparrow\rangle$.

12.6 Pauli equation

It is clear that with the addition of spin, the Schrödinger equation we had been using is not complete. At the very least, we should add in the energy that an electron has from the interaction with a magnetic field **B**. Classically, if we viewed the electron spin as a vector, $\boldsymbol{\sigma}$, because it has direction, just as normal angular momentum does, then we would expect an associated magnetic moment, in a simple vector generalization of Eq. (12.10)

$$\boldsymbol{\mu}_e = g\mu_B\boldsymbol{\sigma} \tag{12.33}$$

and the energy associated with that magnetic moment in the field $\mathbf{B} = \mathbf{i}B_x + \mathbf{j}B_y + \mathbf{k}B_z$, where **i**, **j**, and **k** are the unit vectors associated with the usual Cartesian coordinate directions, would be

$$E_S = \boldsymbol{\mu}_e \cdot \mathbf{B} = g\mu_B\boldsymbol{\sigma}\cdot\mathbf{B} \tag{12.34}$$

In the quantum mechanical case, as usual, we postulate the use of the operator instead of the classical quantity.

The quantum mechanical Hamiltonian corresponding to the energy E_S of Eq. (12.34) is, therefore

$$\hat{H}_S = \frac{g\mu_B}{2}\hat{\boldsymbol{\sigma}}\cdot\mathbf{B} \equiv \frac{g\mu_B}{2}B_x\begin{bmatrix}0 & 1\\ 1 & 0\end{bmatrix} + \frac{g\mu_B}{2}B_y\begin{bmatrix}0 & -i\\ i & 0\end{bmatrix} + \frac{g\mu_B}{2}B_z\begin{bmatrix}1 & 0\\ 0 & -1\end{bmatrix} \tag{12.35}$$

where we have used $\hat{\boldsymbol{\sigma}}$ as in Eq. (12.22). (The factor ½ in this expression compared to the classical one is only because we like to work with Pauli matrices with eigenvalues of unit magnitude rather than the half-integer magnitude associated with the spin itself – it does not express any other difference in the physics.) The Pauli equation includes this term. It also attempts to treat all electromagnetic effects on the electron, so it uses $\hat{\mathbf{p}} - e\mathbf{A}$ instead of just the momentum operator $\hat{\mathbf{p}} = -i\hbar\nabla$ in constructing the rest of the energy terms in the equation (this point is discussed in Appendix E). Hence, instead of the Schrödinger equation, we have the Pauli equation

$$\left[\frac{1}{2m_o}\left(\hat{\mathbf{p}} - e\mathbf{A}\right)^2 + V + \frac{g\mu_B}{2}\hat{\boldsymbol{\sigma}}\cdot\mathbf{B}\right]\Psi = i\hbar\frac{\partial\Psi}{\partial t} \tag{12.36}$$

Note here that $\Psi \equiv \begin{bmatrix}\psi_\uparrow(\mathbf{r},t)\\ \psi_\downarrow(\mathbf{r},t)\end{bmatrix}$ is a spinor. The Pauli equation is, therefore, not one differential equation but is, in general, two coupled ones. This equation is the starting point for investigating the effects of magnetic effects on electrons and can be used, for example, to derive the Zeeman effect rigorously, including the effects of spin.

12.7 Where does spin come from?

Spin appears as a bizarre concept in quantum mechanics and, thus far, we have simply said it exists. Arguably, we can do no more than that. The entirety of quantum mechanics is merely postulated as a model to explain what we see. Initially, spin and the mathematical framework of the Pauli spin matrices were postulated simply to explain experimental behavior. Later,

Dirac showed that[7] if one postulated a version of the quantum mechanics of an electron that was correct according to special relativity, in his famous Dirac equation for the electron, the spin behavior of the electron emerged naturally.[8] In special relativity, it is essential that one treats space and time on a much more equal footing. Essentially, it was not possible to construct a relativistically invariant wave equation that is a first-order differential equation in time without introducing another degree of freedom in the formulation and that additional dynamical variable is spin.

It is usually stated that spin, therefore, is a relativistic effect, though, in fact, it is only necessary[9] to require that the electron obeys a wave equation that treats time and space both with only first derivatives (rather than having time treated using a first derivative and space using a second derivative, as in the Schrödinger equation) to have the spin behavior emerge. One can, therefore, construct both relativistic and nonrelativistic wave equations that treat time and space both through first derivatives and which also have all of the solutions of the Schrödinger equation as solutions but which additionally naturally incorporate spin. If one takes this approach nonrelativistically, one obtains an equation that can also be rewritten as the Pauli equation.

Whether we were trying to construct a relativistic or nonrelativistic equation for the electron, simply put, when we postulated Schrödinger's equation, we did not get it quite right. If we postulate the correct equation, spin emerges naturally as a requirement and nature tells us we need to incorporate spin for a complete description of the electron.

12.8 Summary of concepts

Magnetic moment

Magnetic moment is a measure of the strength of a magnet and is a vector quantity pointing along the polar axis of the magnet. For a current I in a loop of area a, the magnitude of the magnetic moment is

$$\mu_d = Ia \quad \text{after Eq. (12.4)}$$

The magnetic moment corresponding to an electron in an orbit of magnetic quantum number m is of magnitude $m\mu_B$, where $\mu_B = e\hbar / 2m_o$ is the Bohr magneton.

The magnetic moment corresponding to an electron of spin σ is

$$\mu_e = g\sigma\mu_B \quad (12.10)$$

where $g \simeq 2.0023$ is the gyromagnetic factor, often approximated as a factor 2.

Spin

Spin is a property intrinsic to each elementary particle that behaves like an angular momentum but that cannot be written as a function of space. Whereas orbital angular momentum has integer values $l = 0, 1, 2, \ldots$, for the electron, the magnitude s of the spin is ½.

---

[7] See P. A. M. Dirac, *The Principles of Quantum Mechanics (4th Edition, revised)* (Oxford, 1967), Chapter XI.

[8] The same equation he derived also proposed the positron, the positive antiparticle to the electron.

[9] W. Greiner, *Quantum Mechanics (Third Edition)* (Springer-Verlag, Berlin, 1994), Chapter 13.

State vectors for electron spin

A general state of electron spin can be represented by a linear combination of two basis states, one corresponding to the spin-up state, written as $|\uparrow\rangle$, $|1/2,1/2\rangle$, or $\begin{bmatrix}1\\0\end{bmatrix}$, and the other corresponding to a spin-down state, written as $|\downarrow\rangle$, $|1/2,-1/2\rangle$, or $\begin{bmatrix}0\\1\end{bmatrix}$. The "up" and "down" refer to the z direction, conventionally, though any axis in space can be chosen. A general electron spin state can, therefore, be written as

$$|s\rangle = a_{1/2}|1/2,1/2\rangle + a_{-1/2}|1/2,-1/2\rangle \equiv a_{1/2}|\uparrow\rangle + a_{-1/2}|\downarrow\rangle \equiv \begin{bmatrix}a_{1/2}\\ a_{-1/2}\end{bmatrix} \quad (12.13)$$

Spin operators

By analogy with orbital angular momentum operators, spin operators $\hat{S}_x$, $\hat{S}_y$, and $\hat{S}_z$ can be defined with analogous commutation relations. More commonly, the operators

$$\hat{\sigma}_x = 2\hat{S}_x/\hbar\,,\ \hat{\sigma}_y = 2\hat{S}_y/\hbar\,,\ \hat{\sigma}_z = 2\hat{S}_z/\hbar \quad (12.17)$$

are used, with commutation relations

$$\left[\hat{\sigma}_x, \hat{\sigma}_y\right] = 2i\hat{\sigma}_z \quad (12.18)$$

$$\left[\hat{\sigma}_y, \hat{\sigma}_z\right] = 2i\hat{\sigma}_x \quad (12.19)$$

$$\left[\hat{\sigma}_z, \hat{\sigma}_x\right] = 2i\hat{\sigma}_y \quad (12.20)$$

These operators can be written as the Pauli spin matrices

$$\hat{\sigma}_x = \begin{bmatrix}0 & 1\\ 1 & 0\end{bmatrix},\ \hat{\sigma}_y = \begin{bmatrix}0 & -i\\ i & 0\end{bmatrix},\ \hat{\sigma}_z = \begin{bmatrix}1 & 0\\ 0 & -1\end{bmatrix} \quad (12.21)$$

Direct product spaces

When we want to combine two Hilbert spaces each corresponding to different attributes (e.g., the spatial and temporal Hilbert space describing ordinary spatial and temporal functions and the Hilbert space for spin) to create a space that can handle any state with the combined attributes, we create a direct product space in which the basis functions in our new Hilbert space are all the products of the basis functions in the original separate spaces and there is one basis function for each product. Thus, when adding spin in considering the electron, we have twice as many functions as before, with one spin-up version and one spin-down version of each spatial basis function originally used for the (spatial) wavefunctions.

Spinor

A spinor is a vector in the direct product space of the spatial (or space and time) and spin basis functions and corresponds to a vector with a possibly different spatial (or space and time) function for each spin direction; that is,

$$|\Psi\rangle \equiv \begin{bmatrix}\psi_\uparrow(\mathbf{r},t)\\ \psi_\downarrow(\mathbf{r},t)\end{bmatrix} \equiv \psi_\uparrow(\mathbf{r},t)\begin{bmatrix}1\\0\end{bmatrix} + \psi_\downarrow(\mathbf{r},t)\begin{bmatrix}0\\1\end{bmatrix} \quad (12.29)$$

A spinor can represent any possible state of a single electron, including spin.

Pauli equation

An improved version of the Schrödinger equation that includes additional terms to treat the effects of magnetic fields on electrons, including the effects of spin and orbital angular momenta, is the Pauli equation

$$\left[\frac{1}{2m_o}\left(\hat{\mathbf{p}}-e\mathbf{A}\right)^2+V+\frac{g\mu_B}{2}\hat{\boldsymbol{\sigma}}\cdot\mathbf{B}\right]\Psi=i\hbar\frac{\partial\Psi}{\partial t} \tag{12.36}$$

where $\hat{\boldsymbol{\sigma}}$ is the vector spin operator

$$\hat{\boldsymbol{\sigma}}=\mathbf{i}\hat{\sigma}_x+\mathbf{j}\hat{\sigma}_y+\mathbf{k}\hat{\sigma}_z\equiv\mathbf{i}\begin{bmatrix}0&1\\1&0\end{bmatrix}+\mathbf{j}\begin{bmatrix}0&-i\\i&0\end{bmatrix}+\mathbf{k}\begin{bmatrix}1&0\\0&-1\end{bmatrix} \tag{12.22}$$

Chapter 13

Identical particles

Prerequisites: Chapters 2–5, and Chapters 9, 10, and 12.

One aspect of quantum mechanics that is very different from the classical world is that particles can be absolutely identical – so identical that it is meaningless to say which is which. This "identicality" has substantial consequences for what states are allowed, quantum mechanically, and in the counting of possible states. Here, we examine this identicality, introducing the concepts of fermions and bosons and the Pauli exclusion principle that lies behind so much of the physics of materials.

13.1 Scattering of identical particles

Suppose we have two electrons in the same spin state,[1] electrons that, for the moment, we imagine we can label as electron 1 and electron 2. We write the spatial coordinates of electron 1 as $\mathbf{r}_1$ and those of electron 2 as $\mathbf{r}_2$. As far as we know, there is absolutely no difference between one electron and another. They are absolutely interchangeable. We might think, because of something we know about the history of these electrons, that it is more likely that we are looking at electron 1 rather than electron 2, but there is no way by making a measurement so that we can actually know for sure at which one we are looking.

We could imagine that the two electrons were traveling through space, each in some kind of wavepacket. The wavepackets might each be quite localized in space at any given time. These wavepackets, however, each extend out arbitrarily far, even though the amplitude becomes small and, hence, the wavefunctions always overlap to some degree. We may find the following argument more convincing if we imagine that the wavepackets are initially directed toward one another and that these wavepackets substantially overlap for some time as they "bounce" off one another as shown in Fig. 13.1, repelled by the electron Coulomb repulsion or even some other force as yet undetermined.[2]

On the right of the scattering region, when we measure the electrons, it is possible that we will find one near path a and another near path b. But, because two electrons of the same spin are absolutely identical, we have absolutely no way of knowing whether it is electron 1 or electron

[1] It is not even necessary that they have the same spin if we believe there is any possibility that some interaction could swap the spins.

[2] Considering the scattering may help us believe that the electrons could be exchanged through some physical process. In fact, it is not even necessary that we consider any such scattering if we accept that the wavepackets overlap to any finite extent in space. If the wavepackets do overlap, even only very weakly, then, on finding an electron at some point in space, we are never exactly sure from which wavepacket the electron came. That slight doubt is sufficient for the argument that follows.

2 that we find near any particular path. We might think we have good reason to believe, because of our understanding of the scattering process, that if electron 1 started out on path a on the left, it is relatively unlikely that electron 1 emerged into path b on the right, but we have to accept that it is possible. Let us write the wavefunction, $\psi_a(\mathbf{r})$, associated with path a, at least on the right of the scattering region and at some particular time, and similarly write $\psi_b(\mathbf{r})$ for the corresponding wavefunction on path b. Hence, we might expect that the two-particle wavefunction $\psi_{tp}(\mathbf{r}_1,\mathbf{r}_2)$ on the right can be written as some linear combination of the two possible outcomes

$$\psi_{tp}(\mathbf{r}_1,\mathbf{r}_2) = c_{12}\psi_a(\mathbf{r}_1)\psi_b(\mathbf{r}_2) + c_{21}\psi_a(\mathbf{r}_2)\psi_b(\mathbf{r}_1) \tag{13.1}$$

where c_{12} is the amplitude for the outcome that it is electron 1 on path a and electron 2 on path b, and oppositely for the amplitude c_{21}.

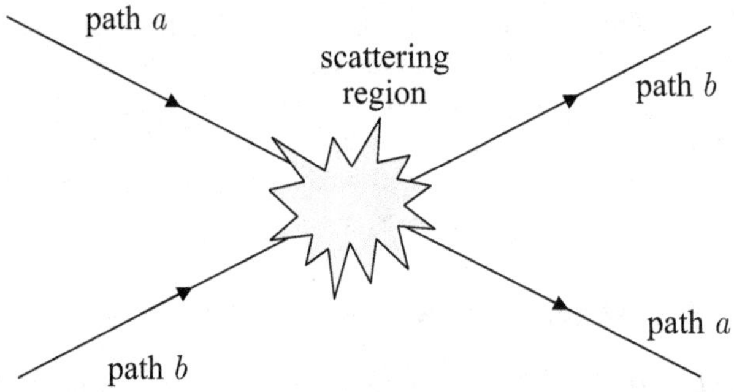

Fig. 13.1. Hypothetical scattering of two electrons. Initially, on the left of the scattering region, there is an electron wavepacket that is concentrated around path a and another concentrated around path b. After the scattering region, which is some volume where we are convinced we cannot neglect the overlap of the electron wavepackets, we expect that there is some probability of finding an electron around path a on the right of the scattering region and some probability of finding another electron around path b on the right of the scattering region.

But, we believe electrons to be absolutely identical, to the extent that it can make no difference to any measurable outcome if we swap the electrons. We can never measure the wavefunction itself (indeed, it may not have any real meaning), but we do expect to be able to measure $|\psi_{tp}|^2$. Swapping the electrons changes $\psi_{tp}(\mathbf{r}_1,\mathbf{r}_2)$ into $\psi_{tp}(\mathbf{r}_2,\mathbf{r}_1)$, so we conclude that

$$|\psi_{tp}(\mathbf{r}_1,\mathbf{r}_2)|^2 = |\psi_{tp}(\mathbf{r}_2,\mathbf{r}_1)|^2 \tag{13.2}$$

which means that

$$\psi_{tp}(\mathbf{r}_2,\mathbf{r}_1) = \gamma\psi_{tp}(\mathbf{r}_1,\mathbf{r}_2) \tag{13.3}$$

where γ is some complex number of unit magnitude. We could, of course, swap the particles again. Because the particles are absolutely identical, this swapping process produces exactly the same result if we started out with the particles the other way around, so

$$\psi_{tp}(\mathbf{r}_1,\mathbf{r}_2) = \gamma\psi_{tp}(\mathbf{r}_2,\mathbf{r}_1) \tag{13.4}$$

But, we already know $\psi_{tp}(\mathbf{r}_2,\mathbf{r}_1) = \gamma\psi_{tp}(\mathbf{r}_1,\mathbf{r}_2)$ from Eq. (13.3), so we have

$$\psi_{tp}(\mathbf{r}_1,\mathbf{r}_2) = \gamma^2\psi_{tp}(\mathbf{r}_1,\mathbf{r}_2) \tag{13.5}$$

Presuming such a pair of swaps brings the wavefunction back to its original form,[3] then

$$\gamma^2 = 1 \tag{13.6}$$

and so we have only two possibilities for γ

$$\gamma = 1 \tag{13.7}$$

or

$$\gamma = -1 \tag{13.8}$$

that is,

$$\psi_{tp}(\mathbf{r}_1, \mathbf{r}_2) = \pm\psi_{tp}(\mathbf{r}_2, \mathbf{r}_1) \tag{13.9}$$

Now we can substitute our general linear combination from Eq. (13.1) in Eq. (13.9) to get

$$c_{12}\psi_a(\mathbf{r}_1)\psi_b(\mathbf{r}_2) + c_{21}\psi_a(\mathbf{r}_2)\psi_b(\mathbf{r}_1) = \pm\left(c_{21}\psi_a(\mathbf{r}_1)\psi_b(\mathbf{r}_2) + c_{12}\psi_a(\mathbf{r}_2)\psi_b(\mathbf{r}_1)\right) \tag{13.10}$$

Rearranging, we have

$$\psi_a(\mathbf{r}_1)\psi_b(\mathbf{r}_2)[c_{12} \mp c_{21}] = \psi_a(\mathbf{r}_2)\psi_b(\mathbf{r}_1)[c_{12} \mp c_{21}] \tag{13.11}$$

But, this must hold for all $\mathbf{r}_1$ (or, for that matter, all $\mathbf{r}_2$) and, in general, $\psi_a(\mathbf{r}_1) \neq \psi_b(\mathbf{r}_1)$ (or, for that matter, $\psi_a(\mathbf{r}_2) \neq \psi_b(\mathbf{r}_2)$) because they represent different and largely separate wavepackets, so we must have

$$c_{12} \mp c_{21} = 0 \tag{13.12}$$

that is,

$$c_{12} = \pm c_{21} \tag{13.13}$$

So, given that the electrons emerge on paths a and b, we have shown that there are only two possibilities for the nature of the wavefunction on the right of the scattering volume, either

$$\psi_{tp}(\mathbf{r}_1, \mathbf{r}_2) = c\left[\psi_a(\mathbf{r}_1)\psi_b(\mathbf{r}_2) + \psi_a(\mathbf{r}_2)\psi_b(\mathbf{r}_1)\right] \tag{13.14}$$

or

$$\psi_{tp}(\mathbf{r}_1, \mathbf{r}_2) = c\left[\psi_a(\mathbf{r}_1)\psi_b(\mathbf{r}_2) - \psi_a(\mathbf{r}_2)\psi_b(\mathbf{r}_1)\right] \tag{13.15}$$

where c is, in general, some complex constant. We have therefore proved that, on the right, the amplitudes of the function $\psi_a(\mathbf{r}_1)\psi_b(\mathbf{r}_2)$ and the function $\psi_a(\mathbf{r}_2)\psi_b(\mathbf{r}_1)$ are equal in magnitude (though possibly opposite in sign).

But, the reader might say, for the electron on path a on the left, the scattering probability into path a on the right is, in general, different from the scattering probability into path b on the right (and the reader would be correct in saying that!). Therefore, how can we have the amplitudes of the two possibilities $\psi_a(\mathbf{r}_1)\psi_b(\mathbf{r}_2)$ and $\psi_a(\mathbf{r}_2)\psi_b(\mathbf{r}_1)$ being equal in magnitude? The resolution of this apparent problem is that even on the left of the scattering volume, at some time before the scattering, the wavefunction $\psi_{tpbefore}(\mathbf{r}_1, \mathbf{r}_2)$ must also have had the two possibilities $\psi_{abefore}(\mathbf{r}_1)\psi_{bbefore}(\mathbf{r}_2)$ and $\psi_{abefore}(\mathbf{r}_2)\psi_{bbefore}(\mathbf{r}_1)$ being equal in magnitude – that is specifically, corresponding to the final situation of Eq. (13.14), of the form

[3] This presumption can be regarded as a postulate of quantum mechanics. It is not required merely for measurable quantities to be unchanged by such swapping.

$$\psi_{tpbefore}(\mathbf{r}_1,\mathbf{r}_2)=c_{before}\left[\psi_{abefore}(\mathbf{r}_1)\psi_{bbefore}(\mathbf{r}_2)+\psi_{abefore}(\mathbf{r}_2)\psi_{bbefore}(\mathbf{r}_1)\right] \tag{13.16}$$

or, corresponding to the final situation of Eq. (13.15), of the form

$$\psi_{tpbefore}(\mathbf{r}_1,\mathbf{r}_2)=c_{before}\left[\psi_{abefore}(\mathbf{r}_1)\psi_{bbefore}(\mathbf{r}_2)-\psi_{abefore}(\mathbf{r}_2)\psi_{bbefore}(\mathbf{r}_1)\right] \tag{13.17}$$

Actually, we know from the basic linearity of quantum mechanical operators that this must be the case, as we can now show. If we know the wavefunction on the left before the scattering event, $\psi_{tpbefore}(\mathbf{r}_1,\mathbf{r}_2)$, we know, in general, that we could integrate the Schrödinger equation in time to deduce the wavefunction after scattering, which we can, in general, call $\psi_{tpafter}(\mathbf{r}_1,\mathbf{r}_2)$. The result of that integration is the same as some linear operator $\hat{S}$ (i.e., a time-evolution operator as discussed in Section 3.11) acting on the initial state $\psi_{tpbefore}(\mathbf{r}_1,\mathbf{r}_2)$ because the integration is just a sum of linear operations on the initial state. Hence, we can write

$$\psi_{tpafter}(\mathbf{r}_1,\mathbf{r}_2)=\hat{S}\psi_{tpbefore}(\mathbf{r}_1,\mathbf{r}_2) \tag{13.18}$$

Now, there is absolutely no difference in the effect of the Hamiltonian on the state $\psi_{abefore}(\mathbf{r}_1)\psi_{bbefore}(\mathbf{r}_2)$ and on the state $\psi_{abefore}(\mathbf{r}_2)\psi_{bbefore}(\mathbf{r}_1)$ because the particles are absolutely identical (if there were a difference, there would be a different energy associated with these two states and, hence, we could distinguish between them). Hence, because $\hat{S}$ is derived from the Hamiltonian, the same holds true for it. So, if

$$\hat{S}\psi_{abefore}(\mathbf{r}_1)\psi_{bbefore}(\mathbf{r}_2)=\psi_{aafter}(\mathbf{r}_1)\psi_{bafter}(\mathbf{r}_2) \tag{13.19}$$

then

$$\hat{S}\psi_{abefore}(\mathbf{r}_2)\psi_{bbefore}(\mathbf{r}_1)=\psi_{aafter}(\mathbf{r}_2)\psi_{bafter}(\mathbf{r}_1) \tag{13.20}$$

Now, $\hat{S}$ is a linear operator, so

$$\begin{aligned}\hat{S}\psi_{tpbefore}(\mathbf{r}_1,\mathbf{r}_2)&=\hat{S}c_{before}\left[\psi_{abefore}(\mathbf{r}_1)\psi_{bbefore}(\mathbf{r}_2)\pm\psi_{abefore}(\mathbf{r}_2)\psi_{bbefore}(\mathbf{r}_1)\right]\\&=c_{before}\hat{S}\left[\psi_{abefore}(\mathbf{r}_1)\psi_{bbefore}(\mathbf{r}_2)\pm\psi_{abefore}(\mathbf{r}_2)\psi_{bbefore}(\mathbf{r}_1)\right]\\&=c_{before}\left\{\left[\hat{S}\psi_{abefore}(\mathbf{r}_1)\psi_{bbefore}(\mathbf{r}_2)\right]\pm\left[\hat{S}\psi_{abefore}(\mathbf{r}_2)\psi_{bbefore}(\mathbf{r}_1)\right]\right\}\\&=c_{before}\left[\psi_{aafter}(\mathbf{r}_1)\psi_{bafter}(\mathbf{r}_2)\pm\psi_{aafter}(\mathbf{r}_2)\psi_{bafter}(\mathbf{r}_1)\right]\end{aligned} \tag{13.21}$$

Hence, we have shown that if we start out with a linear combination of the form $\psi_{abefore}(\mathbf{r}_1)\psi_{bbefore}(\mathbf{r}_2)\pm\psi_{abefore}(\mathbf{r}_2)\psi_{bbefore}(\mathbf{r}_1)$ on the left, we end up with a linear combination of the form $\psi_{aafter}(\mathbf{r}_1)\psi_{bafter}(\mathbf{r}_2)\pm\psi_{aafter}(\mathbf{r}_2)\psi_{bafter}(\mathbf{r}_1)$ on the right. The Schrödinger equation can just as well be integrated backward in time, starting mathematically with a wavefunction on the right of the form $\psi_{aafter}(\mathbf{r}_1)\psi_{bafter}(\mathbf{r}_2)\pm\psi_{aafter}(\mathbf{r}_2)\psi_{bafter}(\mathbf{r}_1)$, in which case we would get to an initial wavefunction of the form $\psi_{abefore}(\mathbf{r}_1)\psi_{bbefore}(\mathbf{r}_2)\pm\psi_{abefore}(\mathbf{r}_2)\psi_{bbefore}(\mathbf{r}_1)$. The action of the scattering does not change this underlying property of the wavefunction.

In the preceding argument, we have supposed the two electrons were scattering off one another. Our scattering was very generally described and the same conclusion can be drawn for any state of the pair of particles where the two particles overlap or interact, including, for example, electrons in an atom or molecule.

In our argument so far, we have discussed the pair of identical particles as if they were electrons with the same spin; but, in fact, we have not presumed any specific property of these particles other than that they are absolutely identical. Thus, we could apply the same quantum mechanical argument to protons with the same spin or neutrons with the same spin. We can also apply this argument to photons. We find that a given kind of particle always corresponds to only one of the possible choices of γ.

All particles corresponding to $\gamma = +1$ (i.e., a wavefunction for a pair of particles of the form $\psi_{tp}(\mathbf{r}_1, \mathbf{r}_2) = c[\psi_a(\mathbf{r}_1)\psi_b(\mathbf{r}_2) + \psi_a(\mathbf{r}_2)\psi_b(\mathbf{r}_1)]$) are called *bosons*. Photons and all particles with integer spin (including also, e.g., ^{4}He nuclei) are bosons. We say that such particles have a wavefunction that is symmetric in the exchange of two particles. Sometimes, loosely, we say the wavefunction is symmetric, though the symmetry we are referring to here is a symmetry in the exchange of the particles, not in the spatial distribution of the wavefunction.

All particles corresponding to $\gamma = -1$ (i.e., a wavefunction for a pair of particles of the form $\psi_{tp}(\mathbf{r}_1, \mathbf{r}_2) = c[\psi_a(\mathbf{r}_1)\psi_b(\mathbf{r}_2) - \psi_a(\mathbf{r}_2)\psi_b(\mathbf{r}_1)]$) are called *fermions*. Electrons, protons, neutrons, and all particles with half-integer spin (e.g., 1/2, 3/2, 5/2, ...) are fermions.[4] Such particles have a wavefunction that is antisymmetric in the exchange of two particles. Again, loosely, we sometimes say this wavefunction is antisymmetric, though again we are not referring to its spatial distribution and mean that it is antisymmetric with respect to exchange of the particles.

Problem

13.1.1 Suppose that the initial state of the pair of identical particles (not restricted to being electrons) on the left in Fig. 13.1 is one of the states required for identical particles; that is,

$$\psi_{tpbefore}(\mathbf{r}_1, \mathbf{r}_2) = \psi_{abefore}(\mathbf{r}_1)\psi_{bbefore}(\mathbf{r}_2) \pm \psi_{abefore}(\mathbf{r}_2)\psi_{bbefore}(\mathbf{r}_1)$$

Suppose also that, mathematically, the effect of the scattering in Fig. 13.1 on a state $\psi_{abefore}(\mathbf{r}_1)\psi_{bbefore}(\mathbf{r}_2)$ is

$$\psi_{abefore}(\mathbf{r}_1)\psi_{bbefore}(\mathbf{r}_2) \to s_{straight}\psi_{aafter}(\mathbf{r}_1)\psi_{bafter}(\mathbf{r}_2) + s_{swap}\psi_{bafter}(\mathbf{r}_1)\psi_{aafter}(\mathbf{r}_2)$$

where $s_{straight}$ and s_{swap} are constant complex numbers. Show that the resulting state after the scattering is still of the right form required for identical particles.

13.2 Pauli exclusion principle

Fermions have one particularly unusual property compared to classical particles, the Pauli exclusion principle, which follows from the simple antisymmetric definition given previously. For two fermions, we know the wavefunction is of the form Eq. (13.15). Suppose now that we postulate that the two fermions are in the same single-particle state – say, state a. Then, the wavefunction becomes

$$\psi_{tp}(\mathbf{r}_1, \mathbf{r}_2) = c\left[\psi_a(\mathbf{r}_1)\psi_a(\mathbf{r}_2) - \psi_a(\mathbf{r}_2)\psi_a(\mathbf{r}_1)\right] = 0 \tag{13.22}$$

Note that this wavefunction is zero everywhere. Hence, it is not possible for two fermions of the same type (e.g., electrons) in the same spin state to be in the same single-particle state. This is the famous Pauli exclusion principle, originally proposed to explain the occupation of atomic orbitals by electrons. Only fermions show this exclusion principle, not bosons. There is no corresponding restriction on the number of bosons that may occupy a given mode.[5]

---

[4] Why bosons are associated with integer spin and fermions with half-integer spin is not a simple story. The simplest statement we can make is that it is a consequence of relativistic quantum field theory. The arguments are reviewed by I. Duck and E. C. G. Sudarsham, "Toward an understanding of the spin-statistic theorem," *Am. J. Phys.* **66**, 284–303 (1998).

[5] See Appendix D, Section D.9, for a brief general discussion of the concept of modes.

13.3 States, single-particle states, and modes

The use of the word *state* can be confusing in discussions of Pauli exclusion and identical particles in general. A quantum mechanical system at any given time is only in one state. In one given state of the system, individual particles can be in different *single-particle states* or *modes*. To clarify the distinction between the overall state of the system and states of individual particles, in what follows in this chapter for the possible states of individual particles, we use *single-particle state* in the fermion case and *mode* in the boson case whenever there would be confusion. Though we use these two different terms, they refer to the same concept – we could equally well use *mode* for both fermions and bosons, for example, but it would just be unusual to do so.

This distinction between the *state* of the system on the one hand and *single-particle states* and *modes* on the other can be illustrated by analogy. Consider the fermion case first. The United States of America has fifty States (with a capital "S"). Each State can have a Governor. A State cannot have more than one Governor. (We can say, therefore, that there is something we could call a democratic exclusion principle at work that means we cannot have more than one Governor in a State.) It is possible that due to some tragic event, a State might have no Governor at some time. We can, therefore, write down a column vector with fifty elements that we can call the state (with a small "s") of the Governorships. The elements in that vector correspond to the States listed in alphabetical order. Depending on whether there is a Governor in a given State, we put a "1" (Governor) or a "0" (no Governor) in the corresponding element of the vector. The analogy here, then, is that the quantum mechanical state of the system of particles is like the state vector of the Governorships and that state corresponds to occupation or not of various different States by individual Governors. Here, we have a clear difference between the ideas of the "state" and the "State," and this example is closely analogous to the fermion case with Pauli exclusion. We can still use *state* to refer to an overall quantum mechanical state of a system, but we should find some other term as the analog of *State* in our Governor example. We use *single-particle state* here as the analog of *State* for fermions.

For the case of bosons, a corresponding analogy might be to consider the number of citizens in a given State (a number that is not limited) and also to have a state vector that lists the number of citizens in each State. For the case of bosons like photons or phonons, we do sometimes already correctly use the word *mode* as the analog of *State* because we already use the same mode concept to describe classical electromagnetic or vibrational fields. Because more than one boson can be in a given "mode," it can be confusing to talk about a single-particle state for bosons, so here we use *mode* as the analog of *State* for the boson case while using *state* to refer to the vector of all of the occupation numbers of the different modes.

13.4 Exchange energy

Exchange energy is another important property with no classical analog that is associated with identical particles. Suppose we have two electrons in identical spin states. They will certainly have a Coulomb repulsion,[6] so we could write the Hamiltonian in a similar fashion to the one we wrote for the hydrogen atom, except here the two particles are identical and the Coulomb potential is repulsive rather than attractive. The Hamiltonian is, therefore

---

[6] They might also have some magnetic interaction between their spins, though we neglect that here. It does not alter the essence of this argument, so we work with the simple Schrödinger equation without spin.

$$\hat{H} = -\frac{\hbar^2}{2m_o}\left(\nabla_{\mathbf{r}_1}^2 + \nabla_{\mathbf{r}_2}^2\right) + \frac{e^2}{4\pi\varepsilon_o\left|\mathbf{r}_1 - \mathbf{r}_2\right|} \tag{13.23}$$

Suppose for simplicity that we can approximate their wavefunction by a simple product form (i.e., terms like $\psi_a(\mathbf{r}_1)\psi_b(\mathbf{r}_2)$). Then, because they are fermions

$$\psi_{tp}(\mathbf{r}_1,\mathbf{r}_2) = \frac{1}{\sqrt{2}}\left[\psi_a(\mathbf{r}_1)\psi_b(\mathbf{r}_2) - \psi_a(\mathbf{r}_2)\psi_b(\mathbf{r}_1)\right] \tag{13.24}$$

where now we presume that the individual wavefunctions $\psi_a(\mathbf{r})$ and $\psi_b(\mathbf{r})$ are normalized, and the factor $1/\sqrt{2}$ now ensures that the total wavefunction normalizes to unity also. To save ourselves some writing, we can also write this in bra-ket notation as

$$\left|\psi_{tp}\right\rangle = \frac{1}{\sqrt{2}}\left(\left|1,a\right\rangle\left|2,b\right\rangle - \left|2,a\right\rangle\left|1,b\right\rangle\right) \tag{13.25}$$

where $\left|1,a\right\rangle \equiv \psi_a(\mathbf{r}_1)$ and so on.

(Note, incidentally, that the order of the products of the wavefunctions or kets does not matter in expressions such as Eqs. (13.24) and (13.25). Obviously

$$\psi_a(\mathbf{r}_1)\psi_b(\mathbf{r}_2) = \psi_b(\mathbf{r}_2)\psi_a(\mathbf{r}_1) \tag{13.26}$$

because $\psi_a(\mathbf{r}_1)$ and $\psi_b(\mathbf{r}_2)$ are each simply a number for any given value of $\mathbf{r}_1$ or $\mathbf{r}_2$. For the case of the bra-ket notation, changing the order of the kets $\left|1,a\right\rangle$ and $\left|2,b\right\rangle$ could also result in a change in the order of integration in a bra-ket expression but, in general, that makes no difference for quantum mechanical wavefunctions, so we also can state

$$\left|1,a\right\rangle\left|2,b\right\rangle = \left|2,b\right\rangle\left|1,a\right\rangle \tag{13.27}$$

Quite generally, the order of the statement of the vectors corresponding to different degrees of freedom or dynamical variables does not matter in direct product spaces.)

Now, we evaluate the expectation value of the energy of this two-particle state. We have

$$\langle E\rangle = \left\langle\psi_{tp}\right|\hat{H}\left|\psi_{tp}\right\rangle \tag{13.28}$$

that is,

$$\langle E\rangle = \frac{1}{2}\begin{bmatrix}\left\langle 1,a\right|\left\langle 2,b\right|\hat{H}\left|1,a\right\rangle\left|2,b\right\rangle + \left\langle 2,a\right|\left\langle 1,b\right|\hat{H}\left|2,a\right\rangle\left|1,b\right\rangle \\ -\left\langle 1,a\right|\left\langle 2,b\right|\hat{H}\left|2,a\right\rangle\left|1,b\right\rangle - \left\langle 2,a\right|\left\langle 1,b\right|\hat{H}\left|1,a\right\rangle\left|2,b\right\rangle\end{bmatrix} \tag{13.29}$$

The first two terms in Eq. (13.29) (which are actually equal) have a straightforward meaning. Formally evaluating these, we have, for the first one

$$\begin{aligned}\left\langle 1,a\right|\left\langle 2,b\right|\hat{H}\left|1,a\right\rangle\left|2,b\right\rangle &= \left\langle 1,a\right|\left\langle 2,b\right|\left(-\frac{\hbar^2}{2m_o}\left(\nabla_{\mathbf{r}_1}^2 + \nabla_{\mathbf{r}_2}^2\right) + \frac{e^2}{4\pi\varepsilon_o\left|\mathbf{r}_1 - \mathbf{r}_2\right|}\right)\left|1,a\right\rangle\left|2,b\right\rangle \\ &= \left\langle 1,a\right|\left\langle 2,b\right| - \frac{\hbar^2}{2m_o}\nabla_{\mathbf{r}_1}^2\left|1,a\right\rangle\left|2,b\right\rangle + \left\langle 1,a\right|\left\langle 2,b\right| - \frac{\hbar^2}{2m_o}\nabla_{\mathbf{r}_2}^2\left|1,a\right\rangle\left|2,b\right\rangle \\ &\quad + \left\langle 1,a\right|\left\langle 2,b\right|\frac{e^2}{4\pi\varepsilon_o\left|\mathbf{r}_1 - \mathbf{r}_2\right|}\left|1,a\right\rangle\left|2,b\right\rangle \\ &= E_{KEa} + E_{KEb} + E_{PEab}\end{aligned} \tag{13.30}$$

Here, E_{KEa} is the kinetic energy of an electron in single-particle state a, given by

$$\begin{aligned} E_{KEa} &= \langle 1,a|\langle 2,b| -\frac{\hbar^2}{2m_o}\nabla_{\mathbf{r}_1}^2 |1,a\rangle|2,b\rangle = \langle 1,a| -\frac{\hbar^2}{2m_o}\nabla_{\mathbf{r}_1}^2 |1,a\rangle\langle 2,b|2,b\rangle \\ &= -\frac{\hbar^2}{2m_o}\int \psi_a^*(\mathbf{r})\nabla^2\psi_a(\mathbf{r})d^3\mathbf{r} \end{aligned} \tag{13.31}$$

(Note, that $\langle 2,b|2,b\rangle = 1$ because the single-particle wavefunctions are presumed normalized.) Similarly

$$E_{KEb} = -\frac{\hbar^2}{2m_o}\int \psi_b^*(\mathbf{r})\nabla^2\psi_b(\mathbf{r})d^3\mathbf{r} \tag{13.32}$$

is the kinetic energy of an electron in single-particle state b. The final contribution in Eq. (13.30) is simply the potential energy resulting from the Coulomb interaction of the charge density from one electron in single-particle state a and the other in single-particle state b; that is,

$$E_{PEab} = \langle 1,a|\langle 2,b| \frac{e^2}{4\pi\varepsilon_o|\mathbf{r}_1 - \mathbf{r}_2|} |1,a\rangle|2,b\rangle = e^2\int \frac{|\psi_a(\mathbf{r})|^2|\psi_b(\mathbf{r}')|^2}{4\pi\varepsilon_o|\mathbf{r}-\mathbf{r}'|}d^3\mathbf{r}d^3\mathbf{r}' \tag{13.33}$$

Evaluating the second term, $\langle 2,a|\langle 1,b|\hat{H}|2,a\rangle|1,b\rangle$, in Eq. (13.29) gives exactly the same answer (i.e., the naming of the variables $\mathbf{r}_1$ and $\mathbf{r}_2$ is interchanged, but the net result is identical mathematically). Hence, we have

$$\frac{1}{2}\left[\langle 1,a|\langle 2,b|\hat{H}|1,a\rangle|2,b\rangle + \langle 2,a|\langle 1,b|\hat{H}|2,a\rangle|1,b\rangle\right] = E_{KEa} + E_{KEb} + E_{PEab} \tag{13.34}$$

This is the energy we might have expected in a semiclassical view, consisting of the kinetic energies of the two particles and the potential energy from their interaction.

But, there are more terms in Eq. (13.29). These additional terms constitute what is called the exchange energy, an energy contribution with no classical analog. We note that by the Hermiticity of the Hamiltonian

$$\langle 2,a|\langle 1,b|\hat{H}|1,a\rangle|2,b\rangle = \left[\langle 1,a|\langle 2,b|\hat{H}|2,a\rangle|1,b\rangle\right]^* \tag{13.35}$$

and so the exchange energy can be written

$$\begin{aligned} E_{EXab} &= -\frac{1}{2}\left(\langle 1,a|\langle 2,b|\hat{H}|2,a\rangle|1,b\rangle + \left[\langle 1,a|\langle 2,b|\hat{H}|2,a\rangle|1,b\rangle\right]^*\right) \\ &= -\mathrm{Re}\left[\int \psi_a^*(\mathbf{r}_1)\psi_b^*(\mathbf{r}_2)\hat{H}\psi_a(\mathbf{r}_2)\psi_b(\mathbf{r}_1)d^3\mathbf{r}_1 d^3\mathbf{r}_2\right] \end{aligned} \tag{13.36}$$

and, finally

$$\langle E\rangle = E_{KEa} + E_{KEb} + E_{PEab} + E_{EXab} \tag{13.37}$$

This exchange energy is a correction to our calculation of the energy from the Coulomb interaction, a correction that comes from the requirement of antisymmetry with respect to particle exchange. That exchange antisymmetry, therefore, changes the energy of states involving two (or more) identical fermions (and does so for any form of interaction, not just the Coulomb interaction).

This phenomenon of exchange energy is very important in, for example, the states of the helium atom, where different energy spectra result for the situations of the two electron spins being aligned (i.e., orthohelium) or antiparallel (i.e., parahelium). It is important to understand that this change in energy is not caused by the magnetic interaction between spins (though there might be a small correction from that); it results from the exchange energy, not some additional term in the Hamiltonian itself. It is also true that exchange-energy phenomena are very important in magnetism.

If exchange energy is such a real phenomenon and electrons, in general, are in these kinds of states that involve other electrons and that are antisymmetric with respect to exchange, why were the calculations we did on single electrons valid at all? The answer is that if the two or more electrons are far apart from one another, there is negligible correction from the exchange energy. We can see this by looking at the exchange-energy expression Eq. (13.36). If the function $\psi_a(\mathbf{r})$ is only substantial in a region near some point $\mathbf{r}_a$, then so also is the function $\nabla^2\psi_a(\mathbf{r})$. Similarly, if the function $\psi_b(\mathbf{r})$ is only substantial near to some point $\mathbf{r}_b$, then so also is the function $\nabla^2\psi_b(\mathbf{r})$. Hence, if the points $\mathbf{r}_a$ and $\mathbf{r}_b$ are far enough apart that there is negligible overlap of the functions $\psi_a(\mathbf{r})$ and $\psi_b(\mathbf{r})$, then

$$\int \psi_a^*(\mathbf{r}_1)\nabla_{\mathbf{r}_1}^2\psi_b(\mathbf{r}_1)d^3\mathbf{r}_1 \simeq 0 \text{ and } \int \psi_b^*(\mathbf{r}_2)\nabla_{\mathbf{r}_2}^2\psi_a(\mathbf{r}_2)d^3\mathbf{r}_2 \simeq 0 \tag{13.38}$$

Similarly, for such negligible overlap, regardless of the form of the potential energy $V(\mathbf{r}_1,\mathbf{r}_2)$ $(= e^2/(4\pi\varepsilon_o|\mathbf{r}_1-\mathbf{r}_2|)$ in the previous example)

$$\int \psi_a^*(\mathbf{r}_1)\psi_b^*(\mathbf{r}_2)V(\mathbf{r}_1,\mathbf{r}_2)\psi_a(\mathbf{r}_2)\psi_b(\mathbf{r}_1)d^3\mathbf{r}_1 d^3\mathbf{r}_2 \simeq 0 \tag{13.39}$$

simply because the functions $\psi_a(\mathbf{r})$ and $\psi_b(\mathbf{r})$ do not overlap. Hence, there is only a contribution to the exchange energy if the individual particle wavefunctions overlap.

This argument is unchanged if we add other potentials into the Hamiltonians for the individual particles, such as a confining box or a proton to give an electrostatic potential to form a hydrogen atom. In the practical absence of any significant exchange energy, the problem essentially separates again into problems that are apparently for single electrons and our previous results, such as for the hydrogen atom, are completely valid for calculating energies and wavefunctions.

Bosons also have exchange energies, though we may come across them less often, in practice. Photons, though bosons, interact so weakly that such exchange corrections are usually negligible; so, except for possible nonlinear optical interactions, we do not, in practice, need to consider such exchange energies for photons. We do, however, need to consider exchange energies for bosons such as ^{4}He (helium-4 nuclei or atoms). (See Problem 13.4.1.)

Problems

13.4.1 Suppose that we have two ^{4}He nuclei, which are bosons, and also have electric charge because they have been stripped of their electrons. Suppose that they interact with one another through a potential energy that can be written in the form $V(\mathbf{r}_1,\mathbf{r}_2)$ (this potential could just be the Coulomb repulsion between these nuclei, for example, though the precise form does not matter for this problem). Suppose also that these particles are sufficiently weakly interacting that we can approximate their wavefunctions using products of the form $\psi_a(\mathbf{r}_1)\psi_b(\mathbf{r}_2)$, though we have to remember to symmetrize the overall wavefunction with respect to exchange. Write down an expression for the exchange energy of this pair of particles in terms of the Hamiltonian $\hat{H}$ for this pair of particles. (In other words, essentially go through the preceding argument for electrons, but presume bosons instead of fermions.)

13.4.2 *The classical beamsplitter.* An optical beamsplitter is a partially reflecting and partially transmitting mirror. One common form of a beamsplitter is in the form of a cube, with the reflecting surface being a diagonal one, as in the sketch.

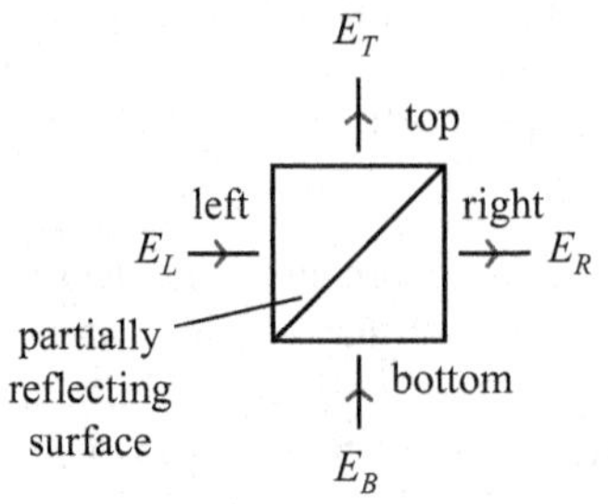

The beamsplitter is presumed loss-less, so an input beam on, say, the left face or the bottom face has its power split between the top and right output beams. E_L, E_B, E_T, and E_R are the electric field amplitudes of the various light beams as shown. For convenience, we take a complex representation for the fields, all of which we presume to be of the same frequency (i.e., $E_L = E_{Lo}\exp(-i\omega t)$ and similarly for the other beams) and of the same polarization (in the direction out of the page). The power in the left beam can, therefore, be taken to be $|E_L|^2$ (within a constant that will not matter for this problem) and similarly for the other beams. Because the beamsplitter is a linear optical component, we can write the relation between the input (left and bottom) beams and the output (right and top) beams using a matrix; that is, in general, we can write

$$\begin{bmatrix} E_T \\ E_R \end{bmatrix} = \hat{S}\begin{bmatrix} E_L \\ E_B \end{bmatrix} \quad \text{where} \quad \hat{S} = \begin{bmatrix} R_{LT} & T_{BT} \\ T_{LR} & R_{BR} \end{bmatrix}$$

where R_{LT}, R_{BR}, T_{BT}, and T_{LR} are constants for a given beamsplitter (with a self-evident notation in terms of reflection and transmission coefficients).

(i) Show for this lossless beamsplitter that

$$R_{LT}T_{BT}^* + R_{BR}^*T_{LR} = 0 \text{ and } |R_{LT}|^2 + |T_{LR}|^2 = |R_{BR}|^2 + |T_{BT}|^2 = 1$$

(Hints: Conservation of power between the total input power ($|E_L|^2 + |E_B|^2$) and the total output power must hold for (a) arbitrary phase difference between E_L and E_B; and (b) arbitrary field magnitudes $|E_L|$ and $|E_B|$.)

(ii) Given the conditions proved in part (i), show that the matrix $\hat{S}$ is unitary.

13.4.3 *The boson beamsplitter.* Consider a 50:50 optical beamsplitter; that is, one that takes an input beam in the left or bottom input port and splits it equally in power between the top and right output ports. Following Problem 13.4.2, a suitable matrix that could describe such a beamsplitter would be

$$\hat{S} = \frac{1}{\sqrt{2}}\begin{bmatrix} i & 1 \\ 1 & i \end{bmatrix}$$

where we are considering two possible input beams, "left" and "bottom," and two possible output beams, "top" and "right," all of which are presumed to have exactly the same frequency and polarization (out of the page). When viewed quantum mechanically, these different beams each represent different modes (or single-particle states). In this way of viewing a beamsplitter, it couples the quantum mechanical amplitudes as follows:

$$\begin{bmatrix} \text{amplitude in top mode} \\ \text{amplitude in right mode} \end{bmatrix} = \hat{S}\begin{bmatrix} \text{amplitude in left mode} \\ \text{amplitude in bottom mode} \end{bmatrix}$$

Hence, the state $|1,L\rangle$ corresponding to a photon 1 in the left input mode is transformed according to the rule

$$|1,L\rangle \to \frac{1}{\sqrt{2}}\left(i|1,T\rangle + |1,R\rangle\right)$$

("→" means "is transformed into") and, similarly, for photon 1 initially in the bottom input mode

$$|1,B\rangle \rightarrow \frac{1}{\sqrt{2}}\left(|1,T\rangle + i|1,R\rangle\right)$$

where L, R, T, and B refer to the left, right, top, and bottom modes, respectively.

Presume that the initial state of the system is one photon in the left input mode and one in the bottom input mode.

(i) Construct a state for these two photons in these two (input) modes that is correctly symmetrized with respect to exchange.

(ii) Using the transformation rules for the effect of the beamsplitter on any single photon state, deduce the output state after the beamsplitter, simplifying as much as possible.

(iii) Now suppose we perform a measurement on the output state and find one photon in the top mode. In what mode will we find the second photon?

(iv) In general, what can we say about the output modes in which we find the two photons?

(Note: This problem illustrates a particularly intriguing quantum mechanical behavior of a simple beamsplitter that is not predicted classically.)

13.4.4* *The fermion beamsplitter.* Imagine that a beamsplitter represented by the same matrix as in Problem 13.4.3 is operating as an electron (rather than photon) beamsplitter (for electrons of the same spin); deduce by a similar analysis to that of Problem 13.4.3:

(i) What will be the output quantum mechanical state of the two electrons if the input state is one electron in the left mode (or single-particle state) and one in the bottom mode (or single-particle state), simplifying as much as possible?

(ii) In general, what can we say about the output modes (single-particle states) in which we find the two electrons?

13.5 Extension to more than two identical particles

Suppose now that we consider more than two particles. If we had N different (i.e., not identical) particles that were approximately not interacting, at least in some region of space and time (e.g., substantially before or substantially after the scattering), then, as before, we expect that we could construct the state $|\Psi_{different}\rangle$ for those by multiplying the N single-particle states or modes; that is,

$$|\psi_{different}\rangle = |1,a\rangle|2,b\rangle|3,c\rangle\ldots|N,n\rangle \quad (13.40)$$

where the numbers and the letter N refer to the particles and the small letters refer to the single-particle states or modes that the individual particles occupy.

Now suppose the particles are identical. Even if we have many particles, it should still be true that swapping any two identical particles should make no difference to any observable. We can follow through the argument as before and find that swapping the same particles a second time should get us back to where we started; again we would, therefore, find that swapping any two particles once either multiplies the state by +1 (bosons) or −1 (fermions).

More than two bosons

If all the particles are identical bosons and we are interested in the state of the set of bosons where one particle is in mode a, another is in mode b, another is in mode c, and so on, then we can construct a state (the general symmetric state) that consists of a sum of all conceivable permutations of the particles among the modes.

$$\left|\psi_{identical\ bosons}\right\rangle \propto \sum_{\hat{P}} \hat{P}\left|\left|1,a\right\rangle\left|2,b\right\rangle\left|3,c\right\rangle\ldots\left|N,n\right\rangle\right\rangle \tag{13.41}$$

Here, $\hat{P}$ is one of the permutation operators. It is an operator, like all other operators we have been considering, that changes one function in the Hilbert space into another – in this case, by permuting the particles (numbers 1, 2, 3, ...) among the modes (letters *a*, *b*, *c*, ...). (It is a linear operator.) The meaning of the sum is that it is taken over all of the possible distinct permutations.[7] The notation Eq. (13.41) is just a mathematical way of saying we are summing over all distinct permutations of the *N* particles among the chosen set of modes. Incidentally, for this boson case, it is quite allowable for two or more of the states to be the same mode (e.g., for mode *b* to be the same mode as mode *a*) an important and general property of bosons.

The state defined by Eq. (13.41) satisfies the symmetry requirement that swapping any two particles does not change the sign of the state and leaves the state amplitude unchanged. Swapping particles merely corresponds to changing the order of the terms in the sum but leaves the sum itself unchanged.

More than two fermions

For the case of fermions, we can write the state for *N* identical fermions as

$$\left|\psi_{identical\ fermions}\right\rangle = \frac{1}{\sqrt{N!}}\sum_{\hat{P}=1}^{N!} \pm\hat{P}\left|\left|1,a\right\rangle\left|2,b\right\rangle\left|3,c\right\rangle\ldots\left|N,n\right\rangle\right\rangle \tag{13.42}$$

where now by $\pm\hat{P}$ we mean that we use the "+" sign when the permutation corresponds to an even number of pair-wise permutations of the individual particles and the "–" sign when the permutation corresponds to an odd number of pair-wise permutations of the individual particles. Note, in this case, that if two of the states are identical (e.g., if $b = a$), then the fermion state is exactly zero because for each permutation there is an identical one with the opposite sign that exactly cancels it (caused by permuting the two particles that are in the same single-particle state). This is the extension of the Pauli exclusion principle to *N* particles.

There is a particularly convenient way to write the *N* particle fermion state, which is called the *Slater determinant*. The determinant is simply another way of writing a sum of the form of Eq. (13.42)[8]; that is, we can write

$$\left|\psi_{identical\ fermions}\right\rangle = \frac{1}{\sqrt{N!}}\begin{vmatrix} \left|1,a\right\rangle & \left|2,a\right\rangle & \cdots & \left|N,a\right\rangle \\ \left|1,b\right\rangle & \left|2,b\right\rangle & \cdots & \left|N,b\right\rangle \\ \vdots & \vdots & \ddots & \vdots \\ \left|1,n\right\rangle & \left|2,n\right\rangle & \cdots & \left|N,n\right\rangle \end{vmatrix} \tag{13.43}$$

---

[7] The number of distinct permutations is a slightly complicated formula if more than one boson occupies a particular single-particle state or mode because permutations that correspond merely to swapping one or more bosons within a given mode should be omitted because they are not distinct permutations. The number of distinct permutations can be evaluated in any given case without particular difficulty, though a general formula for it is somewhat complicated, so we omit it here and we do not therefore explicitly normalize Eq. (13.41) by dividing by the number of distinct permutations.

[8] The formula for a determinant that is equivalent to the form of Eq. (13.42) is known as the *Leibniz formula*.

The reader may remember from the theory of determinants that

(i) the determinant is zero if two of the rows are identical, which here corresponds to the Pauli exclusion principle

(ii) the determinant changes sign if two of the columns are interchanged, which here corresponds to exchanging two particles

13.6 Multiple-particle basis functions

So far, we have been considering states of the multiple particles in which we presume the particles are interacting sufficiently weakly that, to a reasonable approximation, we can write the wavefunctions in terms of products of the wavefunctions of the individual particles considered on their own. In general, it is not possible or, at least, obvious that we can factor multiple-particle states this way. When we considered the hydrogen atom, for example, the resulting state of the two particles, electron and proton, was not simply the product of the eigen wavefunction of an electron considered on its own (which would be a plane wave) and the eigen wavefunction of a proton considered on its own (which would also be a plane wave), though if the particles had both been uncharged and there were no other potential energies of importance, we could have factored the state this way. How can we deal with such situations of strong interaction between the particles, yet still handle the symmetries of the wavefunction with respect to exchange?

The answer is that we can construct basis functions for the direct product space corresponding to the multiple-particle system and require the basis functions to have the necessary symmetry properties with respect to exchange of particles. If each basis function obeys the required symmetry properties with respect to exchange of particles, then the linear combination of them required to represent the state of the (possibly interacting) multiple-particle system also has the same symmetry properties with respect to exchange.

Hence, we can, for example, find some complete basis set to represent one of the particles, $\psi_i(\mathbf{r}_j) \equiv |j,i\rangle$, and we can formally construct a basis function $\Psi_{ab\cdots n}(\mathbf{r}_1,\mathbf{r}_2,\ldots\mathbf{r}_N) \equiv |\Psi_{ab\cdots n}\rangle$ for the N particle system in the direct product space. Depending on the symmetry of the particles with respect to exchange, there are different forms for this basis function:

(i) for nonidentical particles

$$\psi_{ab\cdots n}(\mathbf{r}_1,\mathbf{r}_2,\ldots\mathbf{r}_N) = \psi_a(\mathbf{r}_1)\psi_b(\mathbf{r}_2)\cdots\psi_n(\mathbf{r}_N) \tag{13.44}$$

or, equivalently

$$|\Psi_{ab\cdots n}\rangle = |1,a\rangle|2,b\rangle\cdots|N,n\rangle \tag{13.45}$$

where each of the $\psi_a(\mathbf{r})$ may be chosen to be any of the single-particle basis functions or basis modes $\psi_i(\mathbf{r})$.

(ii) for identical bosons[9]

[9] In constructing these boson basis functions, we are only interested here in how many distinguishable basis functions there are. For simplicity we do not, in general, consider the normalization of the basis functions, though that is straightforward to do in any specific case.

$$\left|\Psi_{ab\cdots n}\right\rangle \propto \sum_{\hat{P}} \hat{P}\left|1,a\right\rangle\left|2,b\right\rangle\cdots\left|N,n\right\rangle \tag{13.46}$$

(iii) for identical fermions

$$\left|\Psi_{ab\cdots n}\right\rangle = \frac{1}{\sqrt{N!}}\sum_{\hat{P}=1}^{N!} \pm\hat{P}\left|1,a\right\rangle\left|2,b\right\rangle\cdots\left|N,n\right\rangle \tag{13.47}$$

Number of basis states for nonidentical particles

In the case of nonidentical particles, there is one basis state of the form Eq. (13.45) for every choice of combination of single-particle basis functions or basis modes. If we imagine there are M possible single-particle basis functions or basis modes and there are N particles, then there are, in general, M^N such basis functions for the N particle system. Specifying a state of that N particle system involves specifying M^N expansion coefficients, and there are M^N distinct states of these N nonidentical particles (even if we now allow them to interact); that is,

$$\begin{gathered}\text{Number of states of } N \text{ nonidentical particles}\\ \text{with } M \text{ available single-particle states or modes}\\ = M^N\end{gathered} \tag{13.48}$$

Number of basis states for identical bosons

In the case of identical bosons, the N-particle basis states corresponding to different permutations of the same set of choices of basis modes are not distinct, so there are fewer basis states. For example, the state $\left|\Psi_{ab\cdots n}\right\rangle$ is not distinct from the state $\left|\Psi_{ba\cdots n}\right\rangle$ in Eq. (13.46); because all permutations of the products of the basis modes already exist in the sum, these two states are merely the same sum of products performed in a different order. The counting of these states is somewhat complicated if one tries to do it from first principles, though, fortunately, it corresponds to a standard problem in the theory of permutations and combinations in statistics, which is the problem of counting the number of combinations of M things (here, the basis modes) taken N at a time (because we always have N particles) with repetitions,[10] for which the result is $(M+N-1)!/[N!(M-1)!]$. (For example, the set of combinations of two identical particles among three states, *a*, *b*, and *c*, with repetitions is *ab*, *ac*, *bc*, *aa*, *bb*, *cc*, giving six in all, which corresponds to $(3+2-1)!/[2!(3-1)!]=6$.) Just as for the nonidentical particle case, this number of basis states is also the number of different possible states we can have for the system of particles even if we allow interactions.

$$\text{Number of states of } N \text{ identical bosons with } M \text{ available modes} = \frac{(M+N-1)!}{N!(M-1)!} \tag{13.49}$$

Number of basis states for identical fermions

In the case of identical fermions, just as in the identical boson case, we also have to avoid double counting basis states that are merely permutations of the same choice of single-particle basis states.[11] Additionally, many of these basis functions would not exist because they would

---

[10] See, e.g., E. Kreysig, *Advanced Engineering Mathematics (Third Edition)* (Wiley, New York, 1972), p. 719.

[11] Possibly permuting the order of the indices *a*, *b*, … *n* will change the sign of the basis function in the identical fermion case, but two basis functions that differ only in sign are not distinct from one another

involve more than one particle in the same state, so there are even fewer possible basis states for multiple identical fermions. Specifically, if there are M choices for the first single-particle basis state a in $|\Psi_{ab\cdots n}\rangle$, then there are $M-1$ choices for the second single-particle basis state b, and so on, down to $M-N+1$ choices for the last single-particle basis state n. Hence, instead of M^N initial choices, we have only $M(M-1)\cdots(M-N+1) = M!/(M-N)!$. We then also have to divide by $N!$ because there are $N!$ different orderings of N different entities (the different entities, in this case, are the different single-particle basis states). Hence, in the identical fermion case, there are $M!/[(M-N)!N!]$ possible basis states; hence, there is the same number of possible states altogether, even if we allow interactions between the particles. That is,

$$\begin{gathered}\text{Number of states of } N \text{ identical fermions}\\ \text{with } M \text{ available single-particle states}\\ = \frac{M!}{(M-N)!N!}\end{gathered} \tag{13.50}$$

Example of numbers of states

For example, suppose we have two particles, each of which can be in one of two different single-particle states or modes, a and b. We might imagine that these particles are in some potential such that there are two single-particle states or modes quite close in energy and all other possible single-particle states or modes are sufficiently far away in energy that for other reasons, we can approximately neglect them in our counting. We might be considering, for example, two particles in a weakly coupled pair of similar quantum boxes[12] (or, if we consider only a one-dimensional problem, coupled potential wells). Because we know for some other reason that the particles cannot have much energy (e.g., the temperature may be low), we may approximately presume that the particles can only be in one or the other of the two lowest coupled single-particle states or modes of these two wells or boxes. For each of the different situations we consider, these two single-particle states or modes might be somewhat different, for example, because of exchange energy, but that does not affect our argument here, which is only one of counting of states. For each situation (nonidentical particles, identical bosons, and identical fermions), there are only two single-particle basis functions or basis modes and, consequently, only two single-particle states or modes, a and b, from which to make up the states of the pair of particles.

Let us now write out the possible states in each case. For all of these cases, the number of possible single-particle states or modes of a particle is $M=2$ and the number of particles is $N=2$.

(i) For nonidentical particles, such as two electrons with different spin,[13] the possible distinct states of this pair of particles are

$$|1,a\rangle|2,a\rangle,\ |1,b\rangle|2,b\rangle,\ |1,a\rangle|2,b\rangle,\ |1,b\rangle|2,a\rangle \tag{13.51}$$

(e.g., they are certainly not orthogonal). Basis functions are always arbitrary anyway within a multiplying complex constant.

[12] A quantum box is simply a structure with confining potentials in all three dimensions. Presuming that it has some bound states, these bound states are discrete in energy, as is readily proved for a cubic box, e.g., by a simple extension of the rectangular potential well problem to three dimensions.

[13] Where there is no possibility of the spins exchanging between the particles.

that is, there are from Eq. (13.48) $2^2 = 4$ states of the pair of particles.

(ii) For identical bosons, such as two ^{4}He (helium-four) atoms, which turn out to be bosons because they are made from six particles each with spin ½ (i.e., two protons, two neutrons, and two electrons, which must, therefore, have a total spin that is an integer), the possible distinct states of this pair of particles are

$$|1,a\rangle|2,a\rangle,\ |1,b\rangle|2,b\rangle,\ \frac{1}{\sqrt{2}}\left(|1,a\rangle|2,b\rangle+|2,a\rangle|1,b\rangle\right) \tag{13.52}$$

(Note that, rather trivially, $|1,a\rangle|2,a\rangle+|2,a\rangle|1,a\rangle$ represents the same state as $|1,a\rangle|2,a\rangle$ because the ordering of the kets does not matter and similarly for the state with both particles in the b mode.) Note that in this case, we did introduce a factor $1/\sqrt{2}$ to normalize the symmetric combination state because we may use such states later. Here we have from Eq. (13.49) $(2+2-1)!/2!(2-1)!=3$ possible states, in contrast to the four in the case of independent particles.

(iii) For identical fermions, there is only one possible state of the pair of particles because the two particles have to be in different single-particle states and there are only two single-particle states to choose from for each particle; that is, the state is

$$\frac{1}{\sqrt{2}}\left(|1,a\rangle|2,b\rangle-|2,a\rangle|1,b\rangle\right) \tag{13.53}$$

where, again, we have normalized this particular wavefunction for possible future use. This count of only one state agrees with the formula Eq. (13.50), which gives $2!/(2!0!)=1$ state (where we remember that $0!=1$).

The differences in the number of available states in the three cases of nonidentical particles, identical bosons, and identical fermions lead to very different behavior once we consider the thermal occupation of states. For example, if we presume that we are at some relatively high temperature, such that the thermal energy, k_BT, is much larger than the energy separation of the two states a and b, then the thermal occupation probabilities of all the different two-particle states all tend to be similar. For the case of the nonidentical particles, which behave like classical particles as far as the counting of states is concerned, with the four states $|1,a\rangle|2,a\rangle$, $|1,b\rangle|2,b\rangle$, $|1,a\rangle|2,b\rangle$, $|1,b\rangle|2,a\rangle$ of Eq. (13.51), we therefore expect a probability of ~ ¼ of occupation of each state. Therefore, the probability that the two particles are in the same state is ~ ½.

For the case of the identical bosons, there are only three possible states, so the probability of occupation of any one state is now ~ 1/3. Two of the two-particle states have the particles in identical modes $|1,a\rangle|2,a\rangle$, $|1,b\rangle|2,b\rangle$; only one two-particle state, specifically, $(1/\sqrt{2})(|1,a\rangle|2,b\rangle+|2,a\rangle|1,b\rangle)$, corresponds to the particles in different modes. Hence, the probability of finding the two identical bosons in the same mode is now 2/3, larger than the 1/2 for the nonidentical particle case.

For the case of identical fermions, there is only one possible state, which, therefore, has probability ~1, and it necessarily corresponds to the two particles being in different states.

Therefore, identical bosons are more likely to be in the same states than are nonidentical (or classical) particles and identical fermions are less likely to be in the same states than are nonidentical (or classical) particles (in fact, they are never in the same states).

The most common description of the differences between bosons and fermions is that we can have as many identical bosons in the same mode as we wish, whereas for identical fermions we can only have one in each single-particle state. For nonidentical (or classical) particles, we can also have as many particles as we wish in a given mode or single-particle state. The difference between bosons and nonidentical particles is that compared to the states for nonidentical particles, there are fewer states in which identical bosons are in different modes, as we saw in the previous example.

Bank account analogy for counting states

The previous arguments on counting states may be more tangible if we consider an analogy with dollars in bank accounts.

You might be a person who does not trust banks, so you might have an antique jar (labeled *a*) in the kitchen with your spending money and a box (labeled *b*) under the bed with your savings money. You put your dollar bills, each of which is clearly labeled with a unique serial number, into one or the other of the antique jar (*a*) or the box (*b*). This situation is analogous to the quantum mechanical situation of nonidentical particles (the dollar bills) and different single-particle states or modes (*a* or *b*) into which they can be put (the antique jar or the box). If I have two dollar bills, then there are four possible situations (states of the entire system of two dollar bills in the antique jar and/or the box):

bill 1 in the box and bill 2 in the box

bill 1 in the box and bill 2 in the antique jar

bill 1 in the antique jar and bill 2 in the box

bill 1 in the antique jar and bill 2 in the antique jar

making four states altogether. This reproduces the counting we found above for the states of nonidentical particles.

Consider next that instead you have two bank accounts: a checking account (*a*) and a savings account (*b*). Because these are bank accounts, you can know how much money you have in each account, but the dollars are themselves absolutely identical in the bank accounts (dollars in bank accounts do not have serial numbers), so now there are only three possible states, which are

two dollars in the savings account

one dollar in the savings account and one in the checking account

two dollars in the checking account

Note that there are two states in which both dollars are in the same account but only one in which they are in different accounts. This bank account argument above reproduces the counting we found above for boson states.

Consider now that you have two bank accounts, a checking account (*a*) and a savings account (*b*), but you are living in the Protectorate of Pauliana, which has particularly restrictive banking laws because of past bad experiences with smart but impoverished students. In Pauliana, you may only have one dollar in each bank account. Then for your two dollars, there is only one possible state.

one dollar in the savings account and one dollar in the checking account.

This reproduces the counting we found for fermion states.

In this analogy, then, dollar bills are like classical particles or nonidentical quantum mechanical particles. Each of them is quite different – they even have different serial numbers. But dollars in bank accounts are like quantum mechanical identical particles. It is quite meaningless to ask which dollar is in which bank account. This is the sense in which identical quantum mechanical particles are identical – it is quite meaningless to ask which one is in which single-particle state or mode.

Problem

13.6.1 Suppose we have three particles, labeled 1, 2, and 3, and three single-particle states or modes, *a*, *b*, and *c*. We presume the particles are essentially not interacting, so the state of the three particles can be written in terms of products of the form $|1,a\rangle|2,b\rangle|3,c\rangle$, though we do presume that the states have to obey appropriate symmetries with respect to interchange of particles if the particles are identical. For the purposes of this problem, we are only interested in situations where each particle is in a different single-particle state or mode. For example, if the different states correspond to substantially different positions in space, we presume that we are considering states in which we would always find one and only one particle near each of these three positions if we performed a measurement.

Write out all the possible states of the three particles:

(i) if the particles are identical bosons

(ii) if the particles are identical fermions

(iii) if the particles are each different (e.g., one is an electron, one is a proton, and one is a neutron)

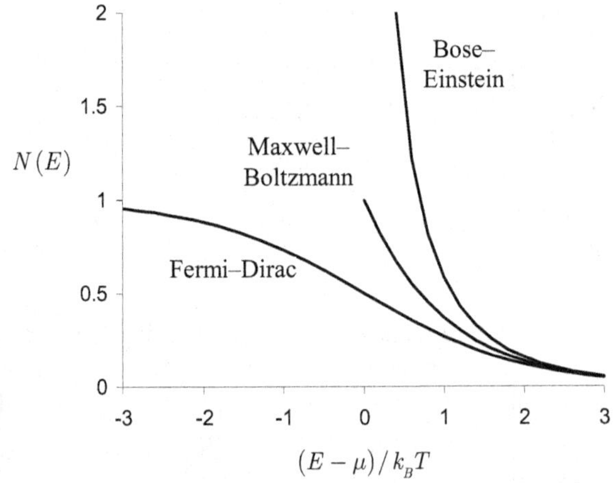

Fig. 13.2. Comparison of the Fermi–Dirac, Maxwell–Boltzmann, and Bose–Einstein distributions, showing the number of particles per state, *N*(*E*), as a function of the energy separation from the chemical potential in units of $k_B T$.

13.7 Thermal distribution functions

We do not go into the detailed behavior of thermal occupation probabilities at finite temperatures here, though the reader should be aware that identical bosons obey a thermal distribution known as the Bose–Einstein distribution, whereas identical fermions obey the Fermi–Dirac distribution, both of which are different from the classical Maxwell–Boltzmann distribution. The three distributions are compared in Fig. 13.2. Note that consistent with the previous conclusions, identical bosons are more likely to be in the same single-particle state or mode than are classical or nonidentical particles (i.e., the Bose–Einstein distribution lies above the Maxwell–Boltzmann distribution), whereas identical fermions are less likely to be in the same single-particle state or mode than are classical or nonidentical particles (i.e., the Fermi–Dirac distribution lies below the Maxwell–Boltzmann distribution). For reference, appropriate formulae for the three distributions are for the average number of particles in a single-particle state or mode of energy E at a temperature T with a chemical potential μ:

(i) Maxwell–Boltzmann

$$N(E) = \exp\left(\frac{\mu}{k_B T}\right)\exp\left(-\frac{E}{k_B T}\right) \tag{13.54}$$

(ii) Fermi–Dirac

$$N(E) = \frac{1}{1 + \exp\left[\dfrac{E - \mu}{k_B T}\right]} \tag{13.55}$$

For the Fermi–Dirac case, the chemical potential μ is often called the Fermi energy and is then written E_F.

(iii) Bose–Einstein

$$N(E) = \frac{1}{\exp\left[\dfrac{E - \mu}{k_B T}\right] - 1} \tag{13.56}$$

For the particular case of photons in a mode (or other similar bosons with only one possible mode), the chemical potential is zero.[14] The energy E of a particle is then $\hbar\omega$, so we have a special case of the Bose–Einstein distribution, known as the Planck distribution, which is

$$N(E) = \frac{1}{\exp\left[\dfrac{\hbar\omega}{k_B T}\right] - 1} \tag{13.57}$$

13.8 Important extreme examples of states of multiple identical particles

In general, the states of multiple identical particles can be quite complicated. There are, however, some important states that turn out to be quite simple. Two examples are filled bands in semiconductors (or any crystalline solid) and multiple photons in the same mode of the electromagnetic field.

[14] The chemical potential is formally the change in Helmholtz free energy per particle added. In the case of photons in a given mode, exchange of particles with a reservoir is exactly the same process as exchange of energy with the reservoir because there is only one state (and, hence, one energy) possible for the photon in that mode. If the photons are in thermal equilibrium with the reservoir, they are by definition in equilibrium with the reservoir with respect to exchange of energy because that is what thermal equilibrium means. But that means they are also in equilibrium with the reservoir with respect to exchange of particles with the reservoir. If we are in equilibrium with respect to exchange of particles, the change in Helmholtz free energy per particle added is zero; hence, the chemical potential is zero.

A filled semiconductor band

One important extreme example of a state for multiple identical fermions is a filled valence band in a semiconductor. In semiconductor physics, one usually takes a single-particle approximation in which one electron is assumed to move in an average periodic potential from the nuclei and other electrons; hence, one arrives at the Bloch state of a single electron in a particular **k** state (see Chapter 8). The various possible single-particle Bloch states of a single electron (for a given spin) correspond to all of the different possible **k** values in the band, of which there are N_c if there are N_c unit cells in the crystal. A full band, therefore, corresponds to N_c electrons of a given spin in the N_c different single-particle states. There is only one such state that obeys the antisymmetry with respect to exchange, which is the Slater determinant of all of the single-particle states in the band.

N photons in a mode

Photons are bosons with a particularly simple behavior. Photons in a mode do not have excited states of any kind and there is, therefore, no meaning to the identical photons in a given mode having more than one state from which they can choose. They are either there or they are not. Therefore, $M = 1$ and the number of possible states of the N photons in the mode is simply $(1+N-1)!/[N!(1-1)!] = 1$. That multiple-particle state is simply all the photons in the same mode.[15]

13.9 Quantum mechanical particles reconsidered

The arguments about identical particles in quantum mechanics are necessarily quite subtle and can certainly be confusing, just as the earlier discussion of wave-particle duality was bizarre. Much of this confusion comes from the fact that we insist on calling these entities particles. The confusion stems largely from what a philosopher would call an ontological problem. When we think of something called a particle, we intrinsically attach attributes to it such as size, shape, charge, mass, position, velocity, and notions of discreteness, countability, and also of uniqueness or "identity": this particle here and that particle there are different particles and always will be. These attributes and notions comprise the "ontology" of the idea of a particle (or, in dictionary definition, the "nature of its being"). When we start thinking about a quantum mechanical particle, we have to progressively delete or modify most of this ontology that ordinarily comes along with the idea of a particle. We could save ourselves a lot of time if we just did not use the word particle for these quantum mechanical entities in the first place; we could then avoid having to selectively "unlearn" the previous ontological baggage. In fact, from our list of ontological attributes and notions, about all that remains for a quantum mechanical particle like an electron is charge, mass, some intertwined version of position and velocity (or momentum) from the uncertainty principle, some kind of discreteness, and some heavily modified notions of counting. We have also had to add other attributes of wave-like interference and spin that are not possessed by classical particles.

There are probably fewer ontological problems if we consider instead levels of excitation of modes. Instead of saying there are three photons in mode a and two in mode b, we could simply say that mode a is in its third level of excitation and mode b is in its second level of

---

[15] If one thinks of photons in a particular mode as being typical bosons, one misses quite a lot of the richness of the behavior possible with bosons. Photons and other particles that are associated with the quantization of simple harmonic oscillators of various kinds, e.g., phonons, can be rather atypical bosons because of their particularly simple properties.

excitation. Certainly, the counting becomes much more obvious, as we saw in the bank-account analogy. It does not really matter if we never introduce the idea of particles – as long as we have the rules constructed by quantum mechanics for manipulating states, it is of no ultimate importance what words we use (or possibly abuse) to describe them. In the next three chapters, we set up the next level of rules for working with fermions and bosons (or whatever we would like to call them).

Of course, we are not commonly going to express our world in terms of excitation levels of modes – particles are here to stay. In fact, we might well find it disquieting to think of electrons as merely being excitation levels of modes rather than being particles. That is a psychological problem rather than a physical one, but the price we pay is a confusion about quantum mechanics that is essentially self-inflicted. The good news is that if we merely accept the rules of quantum mechanics and apply them faithfully, all of these problems go away.[16]

13.10 Distinguishable and indistinguishable particles

In this chapter, we have been rather careful so far to use the word *identical* and avoid the word *indistinguishable* or, more particularly, we have not yet used the term *distinguishable*. That is because we wish to make a distinction between these two concepts, as we now describe. This distinction between the meanings of these words is not universally applied, but the reader should, at least, be aware that there are two different ideas that should not be confused.

We believe all electrons are identical. There is no meaning to ascribing a particular name to one electron and a different name to another. They do not have separate identities, just as two dollars in a bank account do not have distinct identities or names. It is possible, however, that we might regard two specific electrons as being distinguishable from a practical point of view. If they are so far apart that their interaction is negligible, then we can regard them as distinct or distinguishable because there is no physical process by which they could be swapped over. (This is like saying we have one dollar in a bank account in, say, California, and another in a bank account in Hawaii, but for some reason there are presently no communications between the two banks.) In that case, because there is no possibility of exchange, it makes no difference to any calculation on our two electrons whether we bother to "symmetrize" the two-particle wavefunction into its correct two-fermion form. We saw this explicitly when we were discussing exchange energy for two electrons whose wavefunctions do not overlap. In this case, we can get away with treating these two distinguishable electrons as if they were nonidentical particles. It is also true that because it makes no difference to this calculation, we can still symmetrize the wavefunction properly if we want to, even though it may take more paper to write down the algebra. So, even if two particles are identical, if there is no reasonable physical process by which they could be swapped over, as a reasonable approximation, such distinguishable particles can be treated as if they were nonidentical.

It is also true that all photons are identical. This does, however, lead to the apparently absurd conclusion that a microwave photon and a gamma-ray photon are identical. Photons in different modes (e.g., different frequencies or different directions) mostly do not interact with one another – we can pass two light beams right through one another, for example. We can, therefore, in practice often regard photons in different modes as being distinguishable from one another, treating them as if they were nonidentical particles and, hence, dropping the

[16] The self-aware reader may note that we are essentially back to the brainwashing techniques promised earlier.

symmetrization of the state in such cases. We cannot, however, always do that and, if in doubt, we should symmetrize the state because that is always correct.

How could it be that the microwave photon and the gamma ray should fundamentally be viewed as identical (at least in the sense we use the term here)? Suppose that they were in a medium in which two-photon absorption was possible. Even if the photon energies did not add up correctly to correspond to that absorbing transition, there is still a nonlinear refractive effect (i.e., intensity-dependent refractive index) or, more generally, what is known as a *four-wave mixing effect* that results. We can, loosely, view that effect as corresponding to "virtual" two-photon absorption followed almost immediately by two-photon emission; in that process, we have lost track of which photon is which (in fact, it is meaningless to ask which is which), at least in terms of any labels that we might have put on the photons. Whether we like this form of words to describe a two-photon process (more generally, a $\chi^{(3)}$ process) does not matter; at least in principle, we would not get quite the right answer for calculations on such a process if we did not symmetrize the initial state of the two photons correctly. In that case, the two photons are certainly not distinguishable. One might argue that such a process is a very unusual thing and would require very special nonlinear optical materials for it to be of any important strength, and that is generally correct. It is, however, at least in principle, possible to have such a process even in the vacuum because there is a two-photon transition that can create an electron-positron pair. The correction from that process for low-energy photons is, of course, quite negligible from a practical point of view. But its existence does tell us that the only absolutely correct state for these two photons is a symmetrized one; in that sense, these two photons necessarily must be viewed as identical (in the meaning of that word as used in quantum mechanics), though they can nearly always be presumed to be distinguishable. (Note also that photons of the same frequency but in different spatial modes must quite definitely be treated as identical if those modes interact through simple linear optical components, as we see in considering the interaction of photons in a simple beamsplitter in Problem 13.4.3.)

It is, therefore, possible to say as an approximation (albeit sometimes an extremely good one) that two identical particles are distinguishable if the exchange interaction between them is negligibly small; in that case, this "distinguishability" allows us to treat them as if they were nonidentical particles for practical purposes. Conversely, if we say that two particles are indistinguishable (because of the possibility of exchange of them), then we are saying that we have to symmetrize the state properly with respect to exchange.

13.11 Summary of concepts

Fermions and bosons

For the two-particle wavefunction $\psi_{tp}(\mathbf{r}_1,\mathbf{r}_2)$ of two identical particles of positions $\mathbf{r}_1$ and $\mathbf{r}_2$, respectively, exchanging the particles either leaves the wavefunction unchanged, in which case the particles are called bosons, or it multiplies the wavefunction by -1, in which case the particles are called fermions; that is,

$$\psi_{tp}(\mathbf{r}_1,\mathbf{r}_2) = \pm\psi_{tp}(\mathbf{r}_2,\mathbf{r}_1) \quad (13.9)$$

with the "+" sign corresponding to bosons and the "-" sign to fermions. States with the "+" sign are said to be symmetric with respect to exchange and those with the "-" sign are said to be antisymmetric with respect to exchange.

All particles with integer spin are bosons (e.g., photons, phonons, ^{4}He nuclei) and all particles with half-integer spin are fermions (e.g., electrons, protons, neutrons).

Identical particles

All particles of the same species are identical (e.g., all photons are identical, all electrons are identical, all protons are identical) regardless of what single-particle state or mode they occupy.

Pauli exclusion principle

It is not possible for two fermions of the same type (e.g., electrons) in the same spin state to be in the same single-particle state.

Exchange energy

Because of the requirement that multiple-particle wavefunctions must have specific symmetries with respect to exchange of particles, systems of identical particles have additional terms that appear when we evaluate the energy of the system. These terms, which do not have an analog in the simple single-particle or classical system, give rise to corrections to the energy called exchange energy.

Multiple-particle states

When we consider states of multiple particles with identical spin states, because the particles are indistinguishable, we need to consider all possible permutations of the identical particles among the different single-particle states or modes.

For N identical bosons in identical spin states occupying modes a, b, c, ..., the wavefunction becomes

$$\left|\psi_{identical\ bosons}\right\rangle \propto \sum_{\hat{P}} \hat{P}\left|\left|1,a\right\rangle\left|2,b\right\rangle\left|3,c\right\rangle\ldots\left|N,n\right\rangle\right\rangle \tag{13.41}$$

where by the summation, we mean the sum over all the distinct permutations (given by the different permutation operators $\hat{P}$) of the particles among the modes. For N identical fermions in identical spin states occupying single-particle states a, b, c, ..., the wavefunction becomes

$$\left|\psi_{identical\ fermions}\right\rangle = \frac{1}{\sqrt{N!}}\sum_{\hat{P}=1}^{N!} \pm\hat{P}\left|\left|1,a\right\rangle\left|2,b\right\rangle\left|3,c\right\rangle\ldots\left|N,n\right\rangle\right\rangle \tag{13.42}$$

where now by the summation, we also mean that we use the + sign when the permutation corresponds to an even number of pair-wise permutations of the individual particles and the – sign when the permutation corresponds to an odd number of pair-wise permutations of the individual particles. The fermion case can also conveniently be written as a Slater determinant

$$\left|\psi_{identical\ fermions}\right\rangle = \frac{1}{\sqrt{N!}}\begin{vmatrix} \left|1,a\right\rangle & \left|2,a\right\rangle & \cdots & \left|N,a\right\rangle \\ \left|1,b\right\rangle & \left|2,b\right\rangle & \cdots & \left|N,b\right\rangle \\ \vdots & \vdots & \ddots & \vdots \\ \left|1,n\right\rangle & \left|2,n\right\rangle & \cdots & \left|N,n\right\rangle \end{vmatrix} \tag{13.43}$$

Similar expressions result also when we are considering multiple-particle basis states.

Numbers of states

For M available single-particle basis states or basis modes and N particles, the numbers of possible states are

(i) for nonidentical particles, M^N

(ii) for identical bosons, $\dfrac{(M+N-1)!}{N!(M-1)!}$

(iii) for identical fermions, $\dfrac{M!}{(M-N)!N!}$

Distinguishability and indistinguishability

Particles in specific single-particle states or modes, even if the particles are identical, can be viewed as distinguishable if the possibility of particle exchange between the two single-particle states or modes is negligible (or, equivalently, if the exchange correction to the Hamiltonian is negligible). Distinguishable particles may then be treated (approximately) as if they are nonidentical and the symmetrization of the state with respect to particle exchange can be omitted. If the exchange interaction is not negligible, then the particles must be viewed as indistinguishable and the states must be symmetrized appropriately with respect to exchange. If in doubt, it is always correct to symmetrize the states of identical particles; such symmetrization of the states makes no difference if the exchange interaction is negligible.

Chapter 14

The density matrix

Prerequisites: Chapters 2–5. Preferably also Chapters 6 and 7.

The density operator or, as it is more commonly called, the *density matrix*, is a key tool in quantum mechanics that allows us to connect the world of quantum mechanics with the world of statistical mechanics and we introduce it briefly here. This is an important connection. Just as we needed statistical ideas in complicated classical systems (e.g., large collections of atoms or molecules), we need the same ideas in complicated quantum mechanical systems.

We also examine one elementary but important application in optics: we use the density matrix to turn the extreme and unrealistic idea of infinitely sharp "δ-function" optical transitions that emerged from the simple perturbation theory model of Chapter 7 into more physical absorption lines with finite width, just as we actually see in atomic and molecular spectra, for example.

14.1 Pure and mixed states

So far in thinking about the state of a system, the only randomness we have considered is that involved in quantum mechanical measurement. We have presumed that otherwise everything else was definite. Suppose, for example, we were thinking about the state of polarization of a photon. In the way that we have so far considered states, we could write a general state of polarization as

$$|\psi\rangle = a_H |H\rangle + a_V |V\rangle \tag{14.1}$$

where $|H\rangle$ means a horizontally polarized photon state and $|V\rangle$ means a vertically polarized one. There is apparently randomness on quantum mechanical measurement; if we performed a measurement – using, for example, a polarizing beamsplitter oriented to separate horizontal and vertical polarizations to different outputs with different detectors – we expect a probability of $|a_H|^2$ of measuring the photon in the horizontal polarization and $|a_V|^2$ of measuring it in the vertical polarization.

Because we must have $|a_H|^2 + |a_V|^2 = 1$ by normalization, we could also choose to write

$$a_H = \cos\theta \quad a_V = \exp(i\delta)\sin\theta \tag{14.2}$$

The fact that $\sin^2\theta + \cos^2\theta = 1$ ensures that this way of writing the state is properly normalized. For the case where $\delta = 0$, we are describing ordinary linear polarization of a light field and θ has the simple physical meaning of the angle of the optical electric vector relative to the horizontal axis. It is also possible with light that there can be some phase delay between the horizontal and vertical components of the optical electric field and that phase difference can be expressed through the exp($i\delta$) term. When $\delta \neq 0$, the field is, in general, "elliptically polarized," which is the most general possible state of polarization of a propagating photon, and the special cases of $\delta = \pm\pi/2$ with $\theta = 45°$ give the two different kinds of circularly polarized photons (i.e., right and left circularly polarized).[1]

An important point about such a state is that we can always build a polarizing filter that will pass a photon of any specific polarization 100 percent of the time. Even if the photon was in the most general possible elliptically polarized state, we could arrange to delay only the horizontal polarization by a compensating amount $-\delta$ to make the photon linearly polarized,[2] and then orient a linear polarizer at an angle θ so that the photon was always passed through. Such polarization compensators and filters are routine optical components. When we can make such a polarization filter so that we get 100 percent transmission of the photons, we say that the photons are in a "pure" state (here, Eq. (14.1)). The states of photons or anything else we have considered so far have all been pure states in this sense; we can, at least, imagine that an appropriate filter could be made to pass any particles that are in any one such specific quantum mechanical state with 100 percent efficiency.[3]

But, such "pure" states are by no means the only ones we will encounter. Suppose that we have a beam of light that is a mixture from two different and quite independent, lasers, lasers "1" and "2", giving laser beams of possibly even different colors. Presume that laser 1 contributes a fraction P_1 of the total number of photons and laser 2 contributes a fraction P_2. Then, the probability that a given photon is from laser 1 is P_1 and, similarly, there is probability P_2 it is from laser 2. We presume also that these two lasers give photons of two possibly different polarization states, $|\psi_1\rangle$ and $|\psi_2\rangle$, respectively, in the beam.

Something is quite measurably different about this "mixed" state; no setting of our polarizing filter will, in general, pass 100 percent of the photons. If we set the polarization filter to pass all the photons in state $|\psi_1\rangle$, it will, in general, not pass all the photons in state $|\psi_2\rangle$ and vice versa. If we set the polarizing filter in any other state, it will not pass 100 percent of either set of photons. This measurable difference tells us that we cannot simply write this mixed state as some linear combination of the two different polarization states in the fashion we have used up to now for linear combinations of quantum mechanical states. If we were able to do that (e.g., in some linear combination of the form $b_1|\psi_1\rangle + b_2|\psi_2\rangle$), we would be able to construct a polarizing filter that would pass 100 percent of the photons in this state (which is a pure state) and we know for our mixed state that no such polarization filter is possible.[4]

---

[1] *Right circular polarization* means that if we were to look back toward the source of a plane wave (i.e., looking against the direction of propagation), the electric field vector would be rotating, with constant length, in a clockwise direction, and this corresponds to $\delta = \pi/2$. *Left circular polarization* similarly corresponds to $\delta = -\pi/2$ and anticlockwise rotation.

[2] A birefringent material, which has different refractive indices on two axes at right angles to one another, can create such a relative phase delay if it is oriented in the correct direction and is of the right thickness.

[3] If the reader prefers to think in electron spin states rather than photon polarization states, we can construct filters out of Stern–Gerlach–like magnet setups, with beam blocks to block the undesired spin components. Spin states can also be described in the same form as in Eqs. (14.1) and (14.2), using spin-up and spin-down components along a given axis instead of the horizontal and vertical components of the photon case. The rest of the mathematics proceeds essentially identically.

[4] It is true that if the photons from the two different lasers were actually both in the same mode (i.e., exactly the same frequency and spatial form and of defined relative phase), we could construct such a pure state by combining them, but that would violate our presumption that the lasers are independent.

The difference between pure and mixed states can have important consequences for other measurable quantities. Suppose, for example, that for some particle with mass we have a potential well, such as the simple infinite one-dimensional potential well we considered earlier, with infinitely high potential barriers around a layer of some thickness, L_z, in the z direction. Let us put the particle in a pure state that is an equal linear superposition of the lowest two states of this well, $|\psi\rangle = (1/\sqrt{2})(|\psi_1\rangle + |\psi_2\rangle)$. Then, the position of this particle, as given formally by the expectation value $\langle z \rangle$ of the $\hat{z}$ position operator, oscillates backward and forward in time in the well because of the different time-evolution factors $\exp(-iE_1 t/\hbar)$ and $\exp(-iE_2 t/\hbar)$ for the two energy eigenstates (with energies E_1 and E_2, respectively). Suppose instead we take an ensemble of identical potential wells and randomly prepare half of them with the particle in the lowest state and half of them with the particle in the second state. Statistically, because we do not know which wells are which, at least before performing any measurements, each of these wells is in a mixed state, with 50 percent probability of being in either the first or second state. Now, we evaluate the expectation value $\langle z \rangle$ of the $\hat{z}$ position operator for each potential well. In each well in this ensemble, $\langle z \rangle$ evaluates to the position of the center of the well because both these wavefunctions are equally balanced about the center (the lowest being symmetric and the second being antisymmetric). The average, $\overline{\langle z \rangle}$, of all of the expectation values from each of the different wells (what we can call an *ensemble average*) is also zero and there is no oscillation in time. Hence again the mixed state and the pure state have quite different properties. Again, it would not be correct simply to write the mixed state as a linear combination of the form $b_1|\psi_1\rangle + b_2|\psi_2\rangle$.

We could also consider a slightly more complicated version of the ensemble of potential wells, one in which each well is skewed, as would be the case if we applied electric field perpendicular to the wells for the case of a charged particle like an electron in the well. Then, the expectation values $\langle z \rangle$ of the position would be different for the first and second states of the well (and neither one would, in general, be in the center of the well), with some specific values $\langle z \rangle = z_1$ for the first state and $\langle z \rangle = z_2$ for the second state. In the case of the pure state, with a combination of these two states in each well, we would still expect oscillation. We could now instead presume a mixed state with possibly different probabilities P_1 and P_2, respectively, that we had prepared a given well in the first or second state. In this mixed state, we would still have no oscillation and our ensemble average value of the measured position would now be

$$\overline{\langle z \rangle} = P_1 z_1 + P_2 z_2 \equiv \sum_{j=1}^{2} P_j \langle \psi_j | \hat{z} | \psi_j \rangle \tag{14.3}$$

More generally for a mixed state, we expect that the ensemble average expectation value for some operator $\hat{A}$ corresponding to an observable quantity can be written

$$\overline{\langle A \rangle} = \sum_j P_j \langle \psi_j | \hat{A} | \psi_j \rangle \tag{14.4}$$

for some set of different quantum mechanical state preparations $|\psi_j\rangle$ made with respective probabilities P_j.

Note, incidentally, that there is no requirement in mixed states that the different $|\psi_j\rangle$ are orthogonal. We could be considering several different polarization states that are quite close to one another in angle. For example, there might be some fluctuation in time in the precise output polarization of some laser, perhaps because some mirror in the laser cavity is subject to vibrations. Then, we would have to consider a mixed state of many different possible, similar, but not identical polarizations.

The question now is whether we can find a convenient and powerful algebra for representing such mixed states and their properties, so that we can get to results such as $\overline{\langle A \rangle}$ by a more elegant method than simply adding up probabilities and measured values as in Eq. (14.3). Ideally, also, that method would give the correct results even if the mixed state became a pure one (i.e., if we simply had one of the $P_j = 1$ and all the others zero), giving us one unified approach for pure and mixed states. We have already concluded that the linear superposition form $b_1 |\psi_1\rangle + b_2 |\psi_2\rangle$ does not work for representing mixed states, so we may need to go beyond such a simple addition. The answer to this question is to introduce the density operator.

Problem

14.1.1 Suppose that we are measuring the value of the spin magnetic moment of electrons. We take the spin magnetic dipole moment operator to be $\hat{\boldsymbol{\mu}}_e = g\mu_B \hat{\boldsymbol{\sigma}}$. We compare the average value we measure in two different states, both of which are equal mixtures of x and y spin character for the electrons.

(i) Consider the pure spin state $|s_p\rangle = (1/\sqrt{3})(|s_x\rangle + |s_y\rangle)$. Here, $|s_x\rangle$ and $|s_y\rangle$ are, respectively, spin states oriented along the $+x$ and $+y$ directions. (Hint: See Eq. (12.27) and the associated discussion of the Bloch sphere to see how to write out these two states.)

(a) Show that this pure state is normalized.

(b) Find the expected value (which is a vector) of the spin magnetic dipole moment (i.e., the average result on measuring the magnetic dipole on multiple successive electrons all prepared in this state).

(c) What is the magnitude of this average dipole moment (i.e., the length of the vector)?

(ii) Consider now the mixed spin state, with equal probabilities of the electrons being in the pure state $|s_x\rangle$ and the pure state $|s_y\rangle$.

(a) What is the resulting average expected value (again a vector) of the spin magnetic dipole moment when measuring an ensemble of successive electrons prepared this way?

(b) What is the magnitude of this ensemble average value?

(iii) What differences are there between the measured magnetic dipole moments in the two different cases of pure and mixed states?

14.2 Density operator

However we are going to represent the mixed state, it must obviously contain the probabilities P_j and the pure states $|\psi_j\rangle$, but it must not simply be a linear combination of the states. The structure we propose instead is the density operator

$$\rho = \sum_j P_j |\psi_j\rangle\langle\psi_j| \tag{14.5}$$

We see that this is an operator because it contains the outer products of state vectors (i.e., $|\psi_j\rangle\langle\psi_j|$). Though the mathematics of operating with this operator takes on the same algebraic rules as any other operators we have been using, we have deliberately left the "hat" off the top of this operator to emphasize that its physical meaning and use are quite different from other operators we have considered: ρ is *not* an operator representing some physical observable. Rather, ρ is representing the *state* (in general, a mixed state) of the system.

If ρ is a useful way of representing the mixed state, it must allow us to calculate quantities like the ensemble average measured value $\overline{\langle A \rangle}$ for any physical observable with corresponding operator $\hat{A}$. In fact, if we can evaluate $\overline{\langle A \rangle}$ for any physically observable quantity, then ρ will be the most complete way we can have of describing this mixed quantum mechanical state because it tells the value we will get of any measurable quantity, to within our underlying statistical uncertainties.

14.3 Density matrix and ensemble average values

To see how to use the density operator, we start by writing it out in terms of some complete orthonormal basis set, $|\phi_m\rangle$. First, we expand each of the pure states $|\psi_j\rangle$ in this set, obtaining

$$|\psi_j\rangle = \sum_u c_u^{(j)} |\phi_u\rangle \tag{14.6}$$

where the superscript (j) means these are the coefficients for the expansion of the specific state $|\psi_j\rangle$. Then, we use Eq. (14.6) and its adjoint in Eq. (14.5) to obtain

$$\begin{aligned}\rho &= \sum_j P_j \left(\sum_u c_u^{(j)} |\phi_u\rangle \right) \left(\sum_v \left(c_v^{(j)}\right)^* \langle\phi_v| \right) \\ &= \sum_{u,v} \left(\sum_j P_j c_u^{(j)} \left(c_v^{(j)}\right)^* \right) |\phi_u\rangle\langle\phi_v|\end{aligned} \tag{14.7}$$

Written this way, the matrix representation of ρ is now clear. We have for a matrix element in this basis

$$\rho_{uv} \equiv \langle\phi_u|\rho|\phi_v\rangle = \sum_j P_j c_u^{(j)} \left(c_v^{(j)}\right)^* \equiv \overline{c_u c_v^*} \tag{14.8}$$

where we have also introduced and defined the idea of the ensemble average of the coefficient product $\overline{c_u c_v^*}$. Given the form Eq. (14.8), we now more typically talk of ρ as the density matrix (rather than the density operator), with matrix elements given as in Eq. (14.8). The density matrix is just a way of representing the density operator, but because we essentially always represent the density operator this way, the two terms are, in practice, used interchangeably, with density matrix being the more common usage.

There are several important properties of the density matrix we can deduce from Eq. (14.8).

(i) The density matrix is Hermitian; that is, explicitly

$$\rho_{vu} \equiv \sum_j P_j c_v^{(j)} \left(c_u^{(j)}\right)^* = \left(\sum_j P_j c_u^{(j)} \left(c_v^{(j)}\right)^* \right)^* = \rho_{uv}^* \tag{14.9}$$

Because the density matrix is Hermitian, so also is the density operator because the density matrix is just a representation of the density operator.

(ii) The diagonal elements ρ_{mm} give us the probabilities of finding the system in a specific one of the basis states $|\phi_m\rangle$. $c_m^{(j)}(c_m^{(j)})^* \equiv |c_m^{(j)}|^2$ is the probability for any specific pure state j that we will find the system in (basis) state m. Hence, adding these up with probabilities P_j gives us the overall probability of finding the system in state m in the mixed state. (The meaning of the off-diagonal elements is more subtle but quite important. They are a measure of what we could loosely think of as the "coherence" between different states in the system and we return to discuss this later.)

(iii) The sum of the diagonal elements of the density matrix is unity; that is, remembering that we can formally write the sum of the diagonal elements of some matrix or operator as the trace (*Tr*) of the matrix or operator

$$Tr(\rho) = \sum_m \rho_{mm} = \sum_m \sum_j P_j \left| c_m^{(j)} \right|^2 = \sum_j P_j \sum_m \left| c_m^{(j)} \right|^2 = \sum_j P_j = 1 \quad (14.10)$$

because (a) the state $|\psi_j\rangle$ is normalized (so $\sum_m | c_m^{(j)} |^2 = 1$) and (b) the sum of all the probabilities P_j of the various pure states $|\psi_j\rangle$ in the mixed state must be 1.

Now, we come to a key trick with the density matrix. Let us consider an operator $\hat{A}$ corresponding to some physical observable and specifically consider the product $\rho\hat{A}$; that is,

$$\rho\hat{A} = \sum_{u,v} \left(\sum_j P_j c_u^{(j)} \left(c_v^{(j)} \right)^* \right) |\phi_u\rangle\langle\phi_v| \hat{A} \quad (14.11)$$

We can, therefore, write some diagonal element of the resulting matrix as

$$\begin{aligned} \langle\phi_q|\rho\hat{A}|\phi_q\rangle &= \sum_{u,v} \left(\sum_j P_j c_u^{(j)} \left(c_v^{(j)} \right)^* \right) \langle\phi_q|\phi_u\rangle\langle\phi_v|\hat{A}|\phi_q\rangle \\ &= \sum_{u,v} \left(\sum_j P_j c_u^{(j)} \left(c_v^{(j)} \right)^* \right) \delta_{qu} \langle\phi_v|\hat{A}|\phi_q\rangle \\ &= \sum_v \sum_j P_j c_q^{(j)} \left(c_v^{(j)} \right)^* \langle\phi_v|\hat{A}|\phi_q\rangle \end{aligned} \quad (14.12)$$

Then, the sum of all of these diagonal elements is the trace of $\rho\hat{A}$, which we can write as

$$\begin{aligned} Tr\left(\rho\hat{A}\right) &\equiv \sum_q \langle\phi_q|\rho\hat{A}|\phi_q\rangle = \sum_j P_j \left(\sum_v \left(c_v^{(j)} \right)^* \langle\phi_v| \right) \hat{A} \left(\sum_q c_q^{(j)} |\phi_q\rangle \right) \\ &= \sum_j P_j \langle\psi_j|\hat{A}|\psi_j\rangle \end{aligned} \quad (14.13)$$

Note that the result here is exactly the same as the result for the ensemble average value $\overline{\langle A \rangle}$ of the expectation value of the operator $\hat{A}$ for this specific mixed state as given in Eq. (14.4). Hence, we have a key result of density matrix theory

$$\overline{\langle A \rangle} = Tr\left(\rho\hat{A}\right) \quad (14.14)$$

We have, therefore, found that the density matrix, through the use of the relation (14.14), describes any measurable ensemble average property of a mixed state. Hence, the density matrix does indeed give as full a description as possible of a mixed state.

Note that this result, Eq. (14.14), is completely independent of the basis used to calculate the trace – the basis $|\phi_m\rangle$ could be any set that is complete for the problem of interest. (This invariance of the trace with respect to any complete orthonormal basis is a general property of traces of operators.) Note also that if we have the system in a pure state $|\psi\rangle$, in which case $P = 1$ for that state and there is zero probability for any other state, then we recover the usual result for the expectation value (i.e., $Tr\left(\rho\hat{A}\right) = \langle\psi|\hat{A}|\psi\rangle = \langle A \rangle$), so the density matrix description gives the correct answers for pure or mixed states.

Problems

14.3.1* Suppose we have a set of photons in a mixed state, with probabilities $P_1 = 0.2$ and $P_2 = 0.8$, respectively, of being in the two different pure states

$$|\psi_1\rangle = |\psi_H\rangle \text{ and } |\psi_2\rangle = \frac{3}{5}|\psi_H\rangle + \frac{4i}{5}|\psi_V\rangle$$

where $|\psi_H\rangle$ and $|\psi_V\rangle$ are the normalized and orthogonal basis states representing horizontal and vertical polarization, respectively. ($|\psi_1\rangle$, therefore, is a horizontally polarized state and $|\psi_2\rangle$ is an elliptically polarized state.) Write the density matrix for this state, in the $|\psi_H\rangle$ and $|\psi_V\rangle$ basis, with $\langle\psi_H|\rho|\psi_H\rangle$ as the top left element.

14.3.2 Consider the mixed spin state, with equal probabilities of the electrons being in the pure state $|s_x\rangle$ and the pure state $|s_y\rangle$. Here, $|s_x\rangle$ and $|s_y\rangle$ are, respectively, spin states oriented along the $+x$ and $+y$ directions. (See Problem 14.1.1)

(i) Evaluate the density operator ρ on the z spin basis (i.e., $|\uparrow\rangle$ and $|\downarrow\rangle$).

(ii) Now write this density operator as a density matrix, with the term in $|\uparrow\rangle\langle\uparrow|$ in the top left element.

(iii) Taking the spin magnetic dipole moment operator to be $\hat{\boldsymbol{\mu}}_e = g\mu_B\hat{\boldsymbol{\sigma}}$, evaluate $\hat{\boldsymbol{\mu}}_e$ as a matrix on the same z spin basis (i.e., $|\uparrow\rangle$ and $|\downarrow\rangle$), with the element $\langle\uparrow|\hat{\boldsymbol{\mu}}_e|\uparrow\rangle$ in the top left corner.

(iv) Using the expression of the form $\overline{\langle A\rangle} = Tr\left(\rho\hat{A}\right)$, evaluate the ensemble average expectation value for the spin magnetic dipole moment in this mixed state. (Hint: The answer should be the same as that for Problem 14.1.1 (ii)(a).)

14.4 Time evolution of the density matrix

When we want to understand how a quantum mechanical system in some pure state $|\psi_j\rangle$ evolves, we can use the Schrödinger equation

$$\hat{H}|\psi_j\rangle = i\hbar\frac{\partial}{\partial t}|\psi_j\rangle \tag{14.15}$$

How can we describe the evolution of a mixed state? In principle, we can do this by considering each pure state in the mixture and appropriately averaging the result, but there is a more elegant approach, which is directly to calculate the time evolution of the density matrix. The density matrix, after all, contains all the available information about the mixed state. To see how to construct the appropriate equation for the density matrix, we start with the Schrödinger equation, Eq. (14.15), and substitute using the expansion, Eq. (14.6), in some basis set $|\phi_n\rangle$ to obtain

$$i\hbar\sum_n \frac{\partial c_n^{(j)}(t)}{\partial t}|\phi_n\rangle = \sum_n c_n^{(j)}(t)\hat{H}|\phi_n\rangle \tag{14.16}$$

where we have put all of the time dependence of the state into the coefficients $c_n^{(j)}(t)$. Now, operating from the left of Eq. (14.16) with $\langle\phi_m|$, we have

$$i\hbar\frac{\partial c_m^{(j)}(t)}{\partial t} = \sum_n c_n^{(j)}(t)H_{mn} \tag{14.17}$$

where $H_{mn} = \langle \phi_m | \hat{H} | \phi_n \rangle$ is a matrix element of the Hamiltonian. Also, we can take the complex conjugate of both sides of Eq. (14.17). Noting that $\hat{H}$ is Hermitian (i.e., $H_{mn}^* = H_{nm}$), we have

$$-i\hbar \frac{\partial \left(c_m^{(j)}(t)\right)^*}{\partial t} = \sum_n \left(c_n^{(j)}(t)\right)^* H_{nm} \tag{14.18}$$

or, trivially, changing indices

$$-i\hbar \frac{\partial \left(c_n^{(j)}(t)\right)^*}{\partial t} = \sum_s \left(c_s^{(j)}(t)\right)^* H_{sn} \tag{14.19}$$

But, from Eq. (14.8)

$$\frac{\partial \rho_{mn}}{\partial t} = \sum_j P_j \left(c_m^{(j)} \frac{\partial \left(c_n^{(j)}\right)^*}{\partial t} + \left(c_n^{(j)}\right)^* \frac{\partial c_m^{(j)}}{\partial t} \right) \tag{14.20}$$

So, substituting from Eqs. (14.17) and (14.19) (and changing the summation index in Eq. (14.17) from n to q), we have

$$\begin{aligned} \frac{\partial \rho_{mn}}{\partial t} &= \sum_j P_j \left(\frac{i}{\hbar} c_m^{(j)} \sum_q \left(c_q^{(j)}\right)^* H_{qn} - \frac{i}{\hbar} \left(c_n^{(j)}\right)^* \sum_s c_s^{(j)} H_{ms} \right) \\ &= \frac{i}{\hbar} \left(\sum_q \left(\sum_j P_j c_m^{(j)} \left(c_q^{(j)}\right)^* \right) H_{qn} - \sum_s H_{ms} \left(\sum_j P_j c_s^{(j)} \left(c_n^{(j)}\right)^* \right) \right) \end{aligned} \tag{14.21}$$

So, using the definition in Eq. (14.8) of the density matrix elements, we have

$$\begin{aligned} \frac{\partial \rho_{mn}}{\partial t} &= \frac{i}{\hbar} \left(\sum_q \rho_{mq} H_{qn} - \sum_s H_{ms} \rho_{sn} \right) \\ &= \frac{i}{\hbar} \left(\left(\rho \hat{H}\right)_{mn} - \left(\hat{H} \rho\right)_{mn} \right) \\ &= \frac{i}{\hbar} \left[\rho, \hat{H} \right]_{mn} \end{aligned} \tag{14.22}$$

or, equivalently

$$\frac{\partial \rho}{\partial t} = \frac{i}{\hbar} \left[\rho, \hat{H} \right] \tag{14.23}$$

Eq. (14.23) is, therefore, the equation that governs the time evolution of the density matrix and, hence, of the mixed state. This is a very useful equation in applications of the density matrix. It is sometimes known as the *Liouville equation* because it has the same form as the Liouville equation of classical mechanics.[5]

[5] The Liouville equation of classical mechanics is the equation of motion for the phase space probability distribution.

14.5 Interaction of light with a two-level "atomic" system

One simple and important example of the use of the density matrix is the interaction of light with an electron system with two energy levels. We presume these levels have energies E_1 and E_2, with corresponding energy eigenfunctions $|\psi_1\rangle$ and $|\psi_2\rangle$. We also presume for simplicity that the system is much smaller than an optical wavelength, so an incident optical field E is simply uniform across the system, and we take E to be polarized in the z direction. We take a simple "electric dipole" interaction between the light and the electron, so that the change of energy of the electron as it is displaced by an amount z is $e\mathsf{E}z$. Hence, we can take the perturbing Hamiltonian, in a semiclassical approximation for simplicity, to be

$$\hat{H}_p = e\mathsf{E}z = -\mathsf{E}\hat{\mu} \tag{14.24}$$

where $\hat{\mu}$ is the operator corresponding to the electric dipole, with matrix elements

$$\mu_{mn} = -e\langle\psi_m|z|\psi_n\rangle \tag{14.25}$$

so that

$$\left(\hat{H}_p\right)_{mn} \equiv H_{pmn} = -\mathsf{E}\mu_{mn} \tag{14.26}$$

For simplicity also, we choose the states $|\psi_1\rangle$ and $|\psi_2\rangle$ both to have definite parity, so that

$$\mu_{11} = \mu_{22} = 0 \text{ and, hence, } H_{p11} = H_{p22} = 0 \tag{14.27}$$

and we are free to choose the relative phase of the two wavefunctions such that μ_{12} is real so that we have

$$\mu_{12} = \mu_{21} \equiv \mu_d \tag{14.28}$$

Hence, the dipole operator of this system can be written as

$$\hat{\mu} = \begin{bmatrix} 0 & \mu_d \\ \mu_d & 0 \end{bmatrix} \tag{14.29}$$

and the perturbing Hamiltonian is

$$\hat{H}_p = \begin{bmatrix} 0 & -\mathsf{E}\mu_d \\ -\mathsf{E}\mu_d & 0 \end{bmatrix} \tag{14.30}$$

The unperturbed Hamiltonian $\hat{H}_o$ is just a 2 x 2 diagonal matrix on this basis, with E_1 and E_2 as the diagonal elements, so the total Hamiltonian is

$$\hat{H} = \hat{H}_o + \hat{H}_p = \begin{bmatrix} E_1 & -\mathsf{E}\mu_d \\ -\mathsf{E}\mu_d & E_2 \end{bmatrix} \tag{14.31}$$

The density matrix is also a 2 x 2 matrix because there are only two basis states under consideration here and, in general, we can write it as

$$\rho = \begin{bmatrix} \rho_{11} & \rho_{12} \\ \rho_{21} & \rho_{22} \end{bmatrix} \tag{14.32}$$

The dipole induced in this system is important for two different reasons. First, as before, we see that it is closely related to the perturbing Hamiltonian. Second, it represents the response of the system to the electric field. Formally, the polarization of a system in response to electric field is the dipole moment per unit volume and the relation between polarization and electric field allows us to define the electric susceptibilities or dielectric constants that we typically use to describe the optical properties of materials. So, we particularly want to know the expectation value or ensemble average value of the dipole. We have not yet defined what the state of this system is, but we can still use Eq. (14.14) to write

$$\overline{\langle \mu \rangle} = Tr(\rho\hat{\mu}) \tag{14.33}$$

Using Eqs. (14.29) and (14.32), we have

$$\rho\hat{\mu} = \begin{bmatrix} \rho_{11} & \rho_{12} \\ \rho_{21} & \rho_{22} \end{bmatrix} \begin{bmatrix} 0 & \mu_d \\ \mu_d & 0 \end{bmatrix} = \begin{bmatrix} \rho_{12}\mu_d & \rho_{11}\mu_d \\ \rho_{22}\mu_d & \rho_{21}\mu_d \end{bmatrix} \tag{14.34}$$

Hence

$$\overline{\langle \mu \rangle} = \mu_d(\rho_{12} + \rho_{21}) \tag{14.35}$$

Now let us try to evaluate the behavior of the density matrix in time so we can deduce the behavior of $\overline{\langle \mu \rangle}$. We have from Eq. (14.23) with the definitions of ρ from Eq. (14.32) and $\hat{H}$ from Eq. (14.31)[6]

$$\begin{aligned} \frac{d\rho}{dt} &= \frac{i}{\hbar}\left(\rho\hat{H} - \hat{H}\rho\right) \\ &= \frac{i}{\hbar}\left(\begin{bmatrix} \rho_{11} & \rho_{12} \\ \rho_{21} & \rho_{22} \end{bmatrix}\begin{bmatrix} E_1 & -\mathrm{E}\mu_d \\ -\mathrm{E}\mu_d & E_2 \end{bmatrix} - \begin{bmatrix} E_1 & -\mathrm{E}\mu_d \\ -\mathrm{E}\mu_d & E_2 \end{bmatrix}\begin{bmatrix} \rho_{11} & \rho_{12} \\ \rho_{21} & \rho_{22} \end{bmatrix}\right) \\ &= \frac{i}{\hbar}\begin{bmatrix} -\mathrm{E}\mu_d(\rho_{12} - \rho_{21}) & -\mathrm{E}\mu_d(\rho_{11} - \rho_{22}) + (E_2 - E_1)\rho_{12} \\ -\mathrm{E}\mu_d(\rho_{22} - \rho_{11}) + (E_1 - E_2)\rho_{21} & -\mathrm{E}\mu_d(\rho_{21} - \rho_{12}) \end{bmatrix} \end{aligned} \tag{14.36}$$

There are two specific useful equations we can deduce from this. The first concerns $d\rho_{12}/dt$ or $d\rho_{21}/dt$. We only need consider one of these because ρ_{12} and ρ_{21} are complex conjugates of one another because of the Hermiticity of the density matrix. We have, taking the "2 – 1" element of both sides

$$\begin{aligned} \frac{d\rho_{21}}{dt} &= \frac{i}{\hbar}\left((\rho_{11} - \rho_{22})\mathrm{E}\mu_d - (E_2 - E_1)\rho_{21}\right) \\ &= -i\omega_{21}\rho_{21} + i\frac{\mu_d}{\hbar}\mathrm{E}(\rho_{11} - \rho_{22}) \end{aligned} \tag{14.37}$$

where $\hbar\omega_{21} = E_2 - E_1$. The second equation we write relates to the change in $\rho_{11} - \rho_{22}$, which is essentially the fractional population difference between the lower and upper states.

$$\frac{d}{dt}(\rho_{11} - \rho_{22}) = 2i\frac{\mu_d}{\hbar}\mathrm{E}(\rho_{21} - \rho_{21}^*) \tag{14.38}$$

where we have used the Hermiticity of ρ (which tells us that $\rho_{12} = \rho_{21}^*$).

---

[6] We have changed from a partial derivative in time to a total derivative because there are no other variables in this problem.

This analysis and Eqs. (14.37) and (14.38) hint at various behaviors we might expect as we shine light on this "atom." For example, Eq. (14.38) shows that the relative fractional population of the upper and lower states of this atom is likely to be changing in time because of the presence of field E.

The analysis so far has made no approximations. In particular, this is not a perturbation theory analysis. Solving the pair of coupled equations, Eqs. (14.37) and (14.38), would cover any possible behavior of this idealized system. Of course, there is nothing in these equations so far that was not in the original time-dependent Schrödinger equation; solving that separately for each of the possible pure starting states $|\psi_j\rangle$ of interest and then averaging the resulting expectation values for some quantity of interest, such as the dipole moment, gives us the same results as we would get from our density matrix analysis so far.

A key benefit of the density matrix is, however, that we can model additional random processes that lie outside the idealized problem and about which we may know little. Suppose we consider the fractional population difference $\rho_{11}-\rho_{22}$ between the "lower" and "upper" states. We might know that, in thermal equilibrium, and with no optical electric field present, this difference settles down to some particular value, $(\rho_{11}-\rho_{22})_o$. Suppose, then, that perhaps due to optical absorption from the lower to the upper state, we have some specific different fractional population difference $\rho_{11}-\rho_{22}$. Experience tells us that such systems often settle back to $(\rho_{11}-\rho_{22})_o$ with an exponential decay with a time constant T_1. This settling might result from random collisions of an atom (e.g., with the walls of the box containing the atom or possibly with other atoms) or possibly just due to spontaneous emission. In this model, these processes can change the probabilities P_j of the various pure states; hence, we expect this density matrix picture could be appropriate. If we believe all that,[7] then we could hypothesize that we could add an appropriate term to Eq. (14.38); for example,

$$\frac{d}{dt}\left(\rho_{11}-\rho_{22}\right)=2i\frac{\mu_d}{\hbar}\mathsf{E}\left(\rho_{21}-\rho_{21}^*\right)-\frac{\left(\rho_{11}-\rho_{22}\right)-\left(\rho_{11}-\rho_{22}\right)_o}{T_1} \tag{14.39}$$

We can see that if there is no driving optical field E, Eq. (14.39) indeed leads to an exponential settling of the fractional population difference to its equilibrium value $(\rho_{11}-\rho_{22})_o$ with a time constant T_1.[8]

In practice, though, just changing Eq. (14.38) into Eq. (14.39) is not sufficient for modeling such systems. We have to consider a similar process also for the off-diagonal elements of the density matrix, as in Eq. (14.37). To understand this, we need to understand the meaning of the off-diagonal elements.

Within any given pure state j, the product $c_u^{(j)}\left(c_v^{(j)}\right)^*$ is something that is, in general, oscillating. There is a time dependence $\exp\left(-iE_u t/\hbar\right)$ built into $c_u^{(j)}$ and associated with basis state u. Similarly, $\left(c_v^{(j)}\right)^*$ has a time dependence $\exp\left(iE_v t/\hbar\right)$, so the product has an oscillation of the form $\exp(-i(E_u-E_v)t/\hbar)$. In our two-level system, even if we manage to prepare the system in one pure state with such an oscillation in this product, as time evolves we can imagine that our simple system can get scattered into another pure state k with some probability, possibly even one in which the fractional populations ρ_{11} and ρ_{22} are unchanged,

---

[7] Or if we are just too lazy to construct a better model, e.g., correctly including spontaneous emission from first principles.

[8] This "T_1" relaxation of populations is sometimes called a *longitudinal relaxation*.

but in which the phases of the coefficients $c_1^{(k)}$ and $c_2^{(k)}$ are different. At any given time, therefore, we may have an ensemble of different possibilities for the quantum mechanical state and, in general, these different possibilities will have different phases of oscillation. If we have sufficiently many such random phases that are sufficiently different in our mixed state, then the ensemble average of a product $c_u c_v^*$ for different u and v (i.e., $\overline{c_u c_v^*}$) averages out to zero. But this ensemble average is simply the off-diagonal density matrix element ρ_{uv}, as defined in Eq. (14.8).

Hence, we can see that these off-diagonal elements contain information about the coherence of the populations in different states. The collisions or other processes that could scatter into states with different phases for the expansion coefficients can be called *dephasing processes*. The simplest model we can construct is to postulate that such dephasing processes acting alone on the off-diagonal density matrix elements would give rise to an exponential settling of any off-diagonal element to zero, with some time constant T_2.[9] Hence, we can postulate adding a term $-\rho_{21}/T_2$ to Eq. (14.37) to obtain

$$\frac{d\rho_{21}}{dt} = -i\omega_{21}\rho_{21} + i\frac{\mu_d}{\hbar}\mathsf{E}(\rho_{11} - \rho_{22}) - \frac{\rho_{21}}{T_2} \tag{14.40}$$

The dephasing rate or loss of coherence (i.e., the T_2 process) is always comparable to or faster than the population relaxation (i.e., the T_1 process)[10] because any process that relaxes the population by such a decay also destroys the coherence.[11] In this equation, in the absence of an optical field E, ρ_{21} would execute an oscillation at approximately frequency ω_{21}, decaying to zero amplitude approximately exponentially with a time constant T_2.

Eqs. (14.39) and (14.40) now constitute our model for our two-level system, including collision, relaxation, and dephasing processes that lie outside the Hamiltonian we know about for the system. We are particularly interested in solving for the behavior of the system for the case of an oscillating electric field, for example, of the form

$$\mathsf{E}(t) = \mathsf{E}_o \cos\omega t = \frac{\mathsf{E}_o}{2}\left(\exp(i\omega t) + \exp(-i\omega t)\right) \tag{14.41}$$

To solve the equations in this case, we make a substitution and one final simplifying approximation. As is often the case, it is useful to take the underlying oscillation out of the variables of interest. Therefore, we choose to define a new "slowly varying" quantity

$$\beta_{21}(t) = \rho_{21}(t)\exp(i\omega t) \tag{14.42}$$

and substitute using this to obtain, instead of Eqs. (14.39) and (14.40)

$$\frac{d}{dt}(\rho_{11} - \rho_{22}) = i\frac{\mu_d}{\hbar}\mathsf{E}_o(\beta_{21} - \beta_{21}^*) - \frac{(\rho_{11} - \rho_{22}) - (\rho_{11} - \rho_{22})_o}{T_1} \tag{14.43}$$

$$\frac{d\beta_{21}}{dt} = i(\omega - \omega_{21})\beta_{21} + i\frac{\mu_d}{2\hbar}\mathsf{E}_o(\rho_{11} - \rho_{22}) - \frac{\beta_{21}}{T_2} \tag{14.44}$$

---

[9] This T_2 relaxation of populations is sometimes called a *transverse relaxation*.

[10] Old professors often become incoherent long before they have stopped talking.

[11] More precisely, we can write $(1/T_2) = (1/2T_1) + (1/T_2')$, where T_2' describes dephasing processes that are in addition to the population relaxation. See, e.g., M. Fox, *Quantum Optics* (Oxford, 2006), p. 181.

where we have also made the approximation of dropping all terms proportional to $\exp(\pm 2i\omega t)$. Such terms average out to zero over timescales of cycles and, hence, make relatively little contribution to the resulting values of $\rho_{11}-\rho_{22}$ and β_{21} that are obtained formally by integrating these equations, in practice.[12] Eqs. (14.43) and (14.44) are often known as the *optical Bloch equations*.[13] In terms of β_{21}, the ensemble average of the dipole moment is now, from Eq. (14.35)

$$\begin{aligned}\overline{\langle\mu\rangle} &= \mu_d\left(\beta_{12}\exp(i\omega t)+\beta_{21}\exp(-i\omega t)\right)\\ &= 2\mu_d\left[\operatorname{Re}(\beta_{21})\cos\omega t+\operatorname{Im}(\beta_{21})\sin\omega t\right]\end{aligned} \tag{14.45}$$

where we have used the fact that $\beta_{21}=\beta_{12}^*$, which follows from the definition Eq. (14.42) and the fact that the density matrix itself is Hermitian.

Now let us solve Eqs. (14.43) and (14.44) for one important and useful case. We want to know what happens in the steady state – that is, for a monochromatic field and when the system has settled down. In that case, we expect first that the fractional population difference, $\rho_{11}-\rho_{22}$, will no longer be changing, so $d(\rho_{11}-\rho_{22})/dt=0$. We also expect that any coherent responses we have from the system – as seen for example in the off-diagonal density matrix element ρ_{21} – will have settled down to following the appropriate driving field terms. Because we already took out the variation of the form $\exp(-i\omega t)$ in setting up the slowly varying element β_{21}, we therefore expect that $d\beta_{21}/dt=0$ in the steady state.

Therefore, setting the left-hand sides of both Eqs. (14.43) and (14.44) to zero, we can solve these equations. Adding Eq. (14.43) and its complex conjugate gives us an expression for the real part of β_{21} and, similarly, subtracting them gives the imaginary part. These can then be substituted into Eq. (14.44), leading to a solution for $\rho_{11}-\rho_{22}$ and, hence, for all the desired variables. Hence, we have

$$\rho_{11}-\rho_{22}=(\rho_{11}-\rho_{22})_o\frac{1+(\omega-\omega_{21})^2T_2^2}{1+(\omega-\omega_{21})^2T_2^2+4\Omega^2T_2T_1} \tag{14.46}$$

$$\operatorname{Im}(\beta_{21})=\frac{\Omega T_2(\rho_{11}-\rho_{22})_o}{1+(\omega-\omega_{21})^2T_2^2+4\Omega^2T_2T_1} \tag{14.47}$$

$$\operatorname{Re}(\beta_{21})=\frac{(\omega_{21}-\omega)\Omega T_2^2(\rho_{11}-\rho_{22})_o}{1+(\omega-\omega_{21})^2T_2^2+4\Omega^2T_2T_1} \tag{14.48}$$

where $\Omega=\mu_d\mathsf{E}_o/2\hbar$.

To understand what these equations are telling us, we now presume that we have some large number N of such systems ("atoms") per unit volume. The population difference between the number in the lower state and the number in the higher state (per unit volume) is, therefore, $\Delta N=N(\rho_{11}-\rho_{22})$ and the population difference in the absence of the optical field is $\Delta N_o=N(\rho_{11}-\rho_{22})_o$. Then, instead of Eq. (14.46), we can write

---

[12] This approximation is known as the *rotating wave approximation*.

[13] The actual Bloch equations in their simplest form were derived to describe another "two-level" system – namely, spin-1/2 nuclei in a magnetic field. The optical Bloch equations here are mathematically closely analogous. See, e.g., M. Fox, *Quantum Optics* (Oxford, 2006), for a recent introductory discussion.

$$\Delta N = \Delta N_o \frac{1+(\omega-\omega_{21})^2 T_2^2}{1+(\omega-\omega_{21})^2 T_2^2 + 4\Omega^2 T_2 T_1} \tag{14.49}$$

In general in electromagnetism, the (static) polarization P is defined as

$$\mathsf{P} = \varepsilon_o \chi \mathsf{E} \tag{14.50}$$

where χ is the susceptibility. When we have an oscillating field, the response of the medium and, hence, the polarization can be out of phase with the electric field and then it is convenient to generalize the idea of susceptibility. We can formally think of it as a complex quantity with real and imaginary parts χ' and χ'', respectively, or, equivalently, we can explicitly write the response to a real field $\mathsf{E}_o \cos \omega t$ as

$$\mathsf{P} = \varepsilon_o \mathsf{E}_o \left(\chi' \cos \omega t + \chi'' \sin \omega t\right) \tag{14.51}$$

It is also generally true in electromagnetism that the polarization is the dipole moment per unit volume. Hence, here we can also write

$$\mathsf{P} = N\overline{\langle \mu \rangle} \tag{14.52}$$

Hence, putting Eqs. (14.45), (14.47), (14.48), (14.51), and (14.52) together, we can write

$$\chi'(\omega) = \frac{\mu_d^2 T_2 \Delta N_o}{\varepsilon_o \hbar} \frac{(\omega_{21}-\omega) T_2}{1+(\omega-\omega_{21})^2 T_2^2 + 4\Omega^2 T_2 T_1} \tag{14.53}$$

$$\chi''(\omega) = \frac{\mu_d^2 T_2 \Delta N_o}{\varepsilon_o \hbar} \frac{1}{1+(\omega-\omega_{21})^2 T_2^2 + 4\Omega^2 T_2 T_1} \tag{14.54}$$

In electromagnetism, the in-phase component of the polarization (and, hence, χ' – the real part of χ) is responsible for refractive index and the quadrature (i.e., 90 degrees shifted) component (and, hence, χ'' – the imaginary part of χ) is responsible for optical absorption. These behaviors can be checked by formally solving the wave equation, as derived from Maxwell's equations, with these susceptibilities present. In this particular case, therefore, what we have found is an expression for the refractive index and absorption of an atomic absorption line.

If we consider the case where the electric field amplitude is small, then $\Omega \simeq 0$ and we have the normal "linear" refraction variation

$$\chi'(\omega) = \frac{\mu_d^2 T_2 \Delta N_o}{\varepsilon_o \hbar} \frac{(\omega_{21}-\omega) T_2}{1+(\omega-\omega_{21})^2 T_2^2} \tag{14.55}$$

and a Lorentzian absorption line

$$\chi''(\omega) = \frac{\mu_d^2 T_2 \Delta N_o}{\varepsilon_o \hbar} \frac{1}{1+(\omega-\omega_{21})^2 T_2^2} \tag{14.56}$$

associated with an atomic or two-level transition. These functions are sketched in Fig. 14.1.

Note that, finally, we have succeeded in getting rid of the δ-function behavior of absorbing transitions that we got from simple time-dependent perturbation theory. Now, we have an absorbing line with a much more reasonable shape, with a width. The parameter of the width is $1/T_2$ – that is, the more dephasing "collisions" there are with the "atom," the wider the absorption line. Incidentally, any process that leads to the recovery of the atom from its excited state back to its lower state also causes a decay of the off-diagonal elements; so even if there are no additional dephasing processes, the absorbing line still has a width. When the only recovery process is spontaneous emission, the resulting line-width is called the *natural line-width*. We can view the line-width of the absorbing transition as being consistent with the energy-time uncertainty principle: if the state (or the coherence of the state) only persists for a finite time, then the state cannot have a well-defined energy – hence, the line-width.

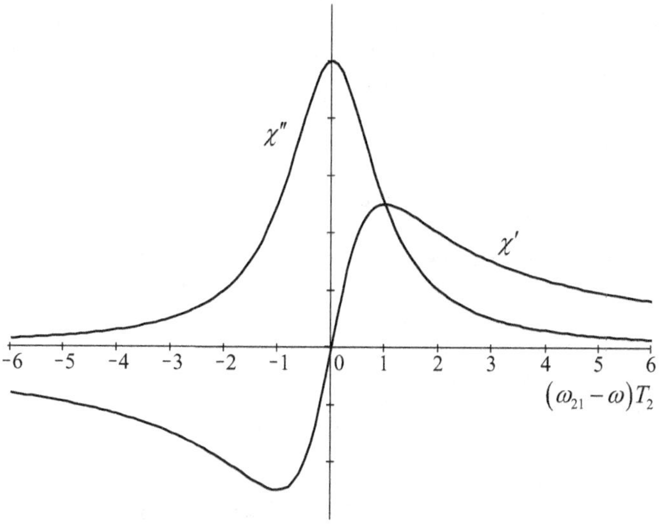

Fig. 14.1. Sketch of the Lorentzian line shape for χ'' and the corresponding line shape for χ', as in Eqs. (14.56) and (14.55), respectively.

The approach we have taken so far is not perturbation theory. Within its simplifications and the approximations of simple times T_1 and T_2 to describe the relaxation of the system, it is exact for all values of the fields. In particular, the results we have here model absorption saturation for this system. If we keep trying to absorb more and more photons into an ensemble of these atoms, there will be fewer and fewer atoms in their lower states (and also more and more in their upper states), which removes these atoms from absorption. (It also allows them to show stimulated emission from the upper to the lower state, which is a process that is exactly the opposite of absorption.) Hence, the absorption should progressively disappear as we go to higher intensities. This process is built into the model we have here. The quantity Ω^2 is proportional to the electric field squared, which, in turn, is proportional to the intensity I of the light field. Hence, we can write $4\Omega^2 T_2 T_1 \equiv I / I_S$, where I_S is called the *saturation intensity*. Hence, for example, on resonance ($\omega_{21} = \omega$), we have

$$\chi''(\omega) \propto \frac{1}{1 + I / I_S} \tag{14.57}$$

This equation is extensively used to describe the process of *absorption saturation* that is often encountered in absorbing systems illuminated with the high intensities available from lasers.

Problem

14.5.1 Suppose we have been driving an ensemble of two-level atoms with an energy separation between the two levels of $\hbar\omega_{21}$ and a (real) dipole matrix element μ_d between these two levels. The atoms have been illuminated with an electric field of the form $E_o \cos \omega t$ for a sufficiently long time that the system has reached a steady state. We work in the approximation where we have characteristic times T_1 and T_2 to describe the population recovery time and the dephasing time, respectively. For simplicity also, we presume we are at a low temperature so that the equilibrium condition (in the absence of optical electric field) is that all the atoms are in their lower state.

(i) Write down expressions for the values of the fractional difference $\rho_{11} - \rho_{22}$ between the lower and upper-state populations of the atoms and for the ensemble average $\langle \mu \rangle$ of the polarization of these atoms at time $t = 0$.

(ii) At time $t = 0$, we suddenly cut off this driving field (i.e., we suddenly set $E_o = 0$). Find expressions for the subsequent behavior of $\rho_{11} - \rho_{22}$ and $\langle \mu \rangle$. (Note: You may neglect the additional electric field that results from the polarization, for simplicity assuming that the total electric field is always zero for $t \geq 0$. You may use expressions based on the rotating wave approximation for this part.)

(iii) Presuming that $T_1 \gg T_2 \gg 1/\omega_{21}$ (e.g., $T_1 = 4T_2 = 20 \times 2\pi / \omega_{21}$) and, for simplicity, presuming we are operating far from resonance (i.e., $|\omega - \omega_{21}| \gg 1/T_2$), sketch the resulting behavior as a function of time of both the fraction of the population of atoms that are in their upper state and the ensemble average of the polarization (in arbitrary units for both these quantities), indicating all the characteristic times on your sketch. (Note: It is easier here *not* to use the slowly-varying or rotating-wave expressions for the off-diagonal elements.)

(iv) It might be that radiative recombination is the only process by which the atoms recover to their ground state (and, hence, that is the only process that contributes to T_1). You may find in part (ii) that the ensemble average of the overall magnitude of the polarization decays faster than the population recovers to its ground state. Explain why that rapid polarization decay can still be consistent with the longer overall (radiative) recovery of the population.

14.6 Density matrix and perturbation theory

The preceding discussion of a two-level system showed an exact solution of a simple problem, including both the linear response and the nonlinear response of absorption saturation. Just as in solutions of Schrödinger's equation, for more complicated systems, exact solutions are usually not possible. Again, just as for Schrödinger's time-dependent equation, we can use perturbation theory, but now with Eq. (14.22) or Eq. (14.23) for the time evolution of the density matrix instead of Schrödinger's equation. One common approach is to generalize the kinds of relaxation time approximations we introduced previously, writing instead of Eq. (14.22)

$$\frac{\partial \rho_{mn}}{\partial t} = \frac{i}{\hbar}\left[\rho, \hat{H}\right]_{mn} - \gamma_{mn}\left(\rho_{mn} - \rho_{mno}\right) \tag{14.58}$$

Here, ρ_{mno} is the value to which density matrix element ρ_{mn} settles in the absence of excitation and γ_{mn} is the *relaxation rate* of element ρ_{mn}. For our two-level system problem, if we had chosen $\gamma_{11} = \gamma_{22} = 1/T_1$ and $\gamma_{21} = \gamma_{12} = 1/T_2$, then Eq. (14.58) would lead to Eqs. (14.39) and (14.40) that we had solved essentially exactly for that system.

In the common perturbation-theory approach, one starts with the equations of the form Eq. (14.58) for each of the density matrix elements and constructs a perturbation theory just as we had done in Chapter 7 when starting with the time-dependent Schrödinger equation. In fact, this density matrix version is the one commonly used for calculating nonlinear optical coefficients, rather than the simple Schrödinger equation version of Chapter 7, because the inclusion of relaxation avoids all of the singularities that otherwise occur when the transition energy and the photon energy coincide. For reasons of space, we do not repeat this version of perturbation theory here, referring the reader to the literature on nonlinear optics.[14]

Further reading

K. Blum, *Density Matrix Theory and Applications* (Plenum, New York, 1996), gives a comprehensive and accessible discussion of the density matrix and all its many applications.

M. Fox, *Quantum Optics* (Oxford, 2006), discusses the density matrix and its application to two-level systems in a clear introductory account.

A. Yariv, *Quantum Electronics (3rd Edition)* (Wiley, New York, 1989), gives an introductory account of the density matrix applied to a two-level system, in an approach similar to that we have taken here.

14.7 Summary of concepts

Pure state

A pure state, in practice, is one that can be written as a simple linear combination of quantum mechanical states (e.g., as in the form $b_1|\psi_1\rangle + b_2|\psi_2\rangle$). (It is not necessary that $|\psi_1\rangle$ and $|\psi_2\rangle$ are orthogonal.)

Ensemble average

If we have a system that can occupy any of a set of different states with probabilities P_n, then we can imagine constructing a set or "ensemble" of N replicas of the system, with P_1N of them in state 1, P_2N in state 2 and so on. An ensemble average is the average of some quantity when measured in each of the members of this ensemble.

Mixed state

A mixed state is one in which we only know probabilities P_j that the system is in one of the different pure states $|\psi_j\rangle$. (It is not necessary that the $|\psi_j\rangle$ are orthogonal.)

Ensemble average measured value of a quantity

For a system in a quantum mechanical mixed state, the ensemble average value of a quantity that is represented by an operator $\hat{A}$ is

$$\overline{\langle A \rangle} = \sum_j P_j \langle \psi_j | \hat{A} | \psi_j \rangle \tag{14.4}$$

Density operator

The density operator is a way of representing a system that is in a mixed state and is given by

$$\rho = \sum_j P_j |\psi_j\rangle\langle\psi_j| \tag{14.5}$$

As a special case, with one particular $P_j = 1$, it can also represent a pure state, so the density operator and its associated formalism can be used for both pure and mixed states.

---

[14] See, in particular, the excellent treatment in R. W. Boyd, *Nonlinear Optics (Second Edition)* (Academic, New York, 1992), pp. 116–148.

Note that though it is an operator, in quantum mechanics it is not like other operators that represent variables, such as position, momentum, or energy; instead, it represents the state of a system. To emphasize this difference, we represent it without the "hat" used on other operators.

Density matrix

The density operator is almost always represented by a matrix, in which case the matrix elements are

$$\rho_{uv} \equiv \langle \phi_u | \rho | \phi_v \rangle = \sum_j P_j c_u^{(j)} \left(c_v^{(j)} \right)^* \equiv \overline{c_u c_v^*} \tag{14.8}$$

where the c's are the expansion coefficients of the pure states on some complete orthonormal basis $|\phi_u\rangle$

$$|\psi_j\rangle = \sum_u c_u^{(j)} |\phi_u\rangle \tag{14.6}$$

and the "bar" notation on $\overline{c_u c_v^*}$ denotes an ensemble average.

The density matrix is Hermitian (and, hence, so also is the density operator) and

$$Tr(\rho) = 1 \qquad \text{after Eq. (14.10)}$$

Ensemble average from the density matrix

For a quantity represented by the quantum mechanical operator $\hat{A}$, the ensemble average measured value of that quantity when the system is in a state represented by density matrix ρ is given by

$$\overline{\langle A \rangle} = Tr\left(\rho \hat{A}\right) \tag{14.14}$$

which applies for both pure and mixed states.

Equation of motion for the density matrix

Just as the time-dependent Schrödinger equation can give the time evolution of a system in a pure state represented by a wavefunction ψ, so also the equation

$$\frac{\partial \rho}{\partial t} = \frac{i}{\hbar}\left[\rho, \hat{H}\right] \tag{14.23}$$

can give the time evolution of a system in a mixed (or pure) state represented by the density operator ρ. This same equation can also be written explicitly for each element of the corresponding density matrix ρ as

$$\frac{\partial \rho_{mn}}{\partial t} = \frac{i}{\hbar}\left[\rho, \hat{H}\right]_{mn} \qquad \text{after Eq. (14.22)}$$

Relaxation time approximation

When the system interacts in a random fashion with other elements that are not included in the Hamiltonian $\hat{H}$, the overall behavior can often be modeled by phenomenologically adding relaxation rates to obtain, instead of Eq. (14.22)

$$\frac{\partial \rho_{mn}}{\partial t} = \frac{i}{\hbar}\left[\rho, \hat{H}\right]_{mn} - \gamma_{mn}\left(\rho_{mn} - \rho_{mno}\right) \tag{14.58}$$

where ρ_{mno} is the equilibrium value of the density matrix element ρ_{mn} in the absence of excitation.

Optical Bloch equations

A specific pair of equations that follow from the relaxation time approximation, with an additional approximation of the neglect of terms oscillating at 2ω (the rotating-wave approximation), are the optical Bloch equations

$$\frac{d}{dt}\left(\rho_{11} - \rho_{22}\right) = i\frac{\mu_d}{\hbar}\mathsf{E}_o\left(\beta_{21} - \beta_{21}^*\right) - \frac{\left(\rho_{11} - \rho_{22}\right) - \left(\rho_{11} - \rho_{22}\right)_o}{T_1} \tag{14.43}$$

$$\frac{d\beta_{21}}{dt} = i\left(\omega - \omega_{21}\right)\beta_{21} + i\frac{\mu_d}{2\hbar}\mathsf{E}_o\left(\rho_{11} - \rho_{22}\right) - \frac{\beta_{21}}{T_2} \tag{14.44}$$

which describe an optically excited two-level system. Here, $\beta_{21}(t) = \rho_{21}(t)\exp(i\omega t)$ and the system, with dipole matrix element μ_d, is being excited with an electric field of the form $\mathsf{E}_o \cos\omega t$. These equations lead to the classic Lorentzian line shape of atomic transitions and also explain absorption saturation in a simple model.

Chapter 15

Harmonic oscillators and photons

Prerequisites: Chapters 2–5, 9, 10, 12, and 13. For additional background on vector calculus, electromagnetism and modes see Appendices C and D.

In this section, we return to the harmonic oscillator and consider it in a mathematically more elegant way. This approach leads to the introduction of "raising" and "lowering" operators that take us from one harmonic oscillator state to another. The introduction of these operators allows us to rewrite the harmonic oscillator mathematics quite economically. We then show that the electromagnetic field for a given mode can also be described in a manner exactly analogous to a harmonic oscillator. In this case, we describe the states of this generalized harmonic oscillator in terms of the number of photons per mode, with that number corresponding exactly to the quantum number for the corresponding harmonic oscillator state. The raising and lowering operators are now physically interpreted as "creation" and "annihilation" operators for photons and are key operators for describing electromagnetic fields.

By this process, we can describe the electromagnetic field quantum mechanically, rather than our previous semiclassical use of quantum mechanical electron behavior but with classical electric and magnetic fields. We say we have "quantized" the electromagnetic field. This quantization is then the basis for all of quantum optics. This approach also prepares us for discussions in subsequent chapters of fermion operators and of the full description of stimulated and spontaneous emission.

15.1 Harmonic oscillator and raising and lowering operators

We remember the (time-independent) Schrödinger equation that we had constructed before for the harmonic oscillator (Eq. (2.75)). That approach was based on the idea that there was a simple, mechanical potential energy that was proportional to the square of the displacement, z; that is,

$$\hat{H}\psi = \left[-\frac{\hbar^2}{2m}\frac{d^2}{dz^2} + \frac{1}{2}m\omega^2 z^2 \right]\psi = E\psi \tag{15.1}$$

Here, we had identified ω as the angular frequency of oscillation of the classical oscillator. For mathematical convenience, we introduce a dimensionless distance, as in Eq. (2.76)

$$\xi = \sqrt{\frac{m\omega}{\hbar}}z \tag{15.2}$$

which enables us to rewrite the Schrödinger equation, as in Eq. (2.77), as

$$\frac{1}{2}\left[-\frac{d^2}{d\xi^2}+\xi^2\right]\psi=\frac{E}{\hbar\omega}\psi \tag{15.3}$$

Now, we show how to write this equation in another elegant form by introducing the raising and lowering operators.

The term $-d^2/d\xi^2+\xi^2$ reminds us of the difference of two squares of numbers

$$-a^2+b^2=b^2-a^2=(-a+b)(a+b) \tag{15.4}$$

though here we note that $d^2/d\xi^2$ is an operator, so we cannot quite use this same algebraic identity. Anyway, if we examine a product of this form for our present case, we have

$$\frac{1}{\sqrt{2}}\left(-\frac{d}{d\xi}+\xi\right)\times\frac{1}{\sqrt{2}}\left(\frac{d}{d\xi}+\xi\right)=\frac{1}{2}\left(-\frac{d^2}{d\xi^2}+\xi^2\right)-\frac{1}{2}\left(\frac{d}{d\xi}\xi-\xi\frac{d}{d\xi}\right) \tag{15.5}$$

As before, when we were considering the commutator of momentum and position, we note that for any function $f(\xi)$

$$\begin{aligned}\left(\frac{d}{d\xi}\xi-\xi\frac{d}{d\xi}\right)f(\xi)&=\frac{d}{d\xi}(\xi f(\xi))-\xi\frac{d}{d\xi}f(\xi)\\&=f(\xi)\frac{d\xi}{d\xi}+\xi\frac{d}{d\xi}f(\xi)-\xi\frac{d}{d\xi}f(\xi)=f(\xi)\end{aligned} \tag{15.6}$$

so we can write the commutation relation

$$\left(\frac{d}{d\xi}\xi-\xi\frac{d}{d\xi}\right)=1 \tag{15.7}$$

Hence, from Eq. (15.5), we have

$$\frac{1}{2}\left(-\frac{d^2}{d\xi^2}+\xi^2\right)=\frac{1}{\sqrt{2}}\left(-\frac{d}{d\xi}+\xi\right)\times\frac{1}{\sqrt{2}}\left(\frac{d}{d\xi}+\xi\right)+\frac{1}{2} \tag{15.8}$$

We can, if we wish, choose to write what we will come to call the *raising* or *creation operator*

$$\hat{a}^\dagger\equiv\frac{1}{\sqrt{2}}\left(-\frac{d}{d\xi}+\xi\right) \tag{15.9}$$

(pronounced "*a* dagger") and the *lowering* or *annihilation operator*

$$\hat{a}\equiv\frac{1}{\sqrt{2}}\left(\frac{d}{d\xi}+\xi\right) \tag{15.10}$$

Note that we have written these two operators using the notation that one is the Hermitian adjoint of the other. This is, in fact, the case. The operator $d/d\xi$ is itself anti-Hermitian, as we demonstrated earlier for the operator d/dz; that is, $\langle\phi|d/d\xi|\psi\rangle=-[\langle\psi|d/d\xi|\phi\rangle]^*$ for two arbitrary states $|\phi\rangle$ and $|\psi\rangle$.[1] Also, ξ is Hermitian (it represents the position operator, at least in the position representation and, with position being an observable, the operator must be Hermitian), so $\langle\phi|\xi|\psi\rangle = [\langle\psi|\xi|\phi\rangle]^*$. Therefore

$$\langle\phi|\hat{a}|\psi\rangle = \langle\phi|\frac{1}{\sqrt{2}}\left(\frac{d}{d\xi}+\xi\right)|\psi\rangle = \left[\langle\psi|\frac{1}{\sqrt{2}}\left(-\frac{d}{d\xi}+\xi\right)|\phi\rangle\right]^* = \left[\langle\psi|\hat{a}^\dagger|\phi\rangle\right]^* \quad (15.11)$$

and so, by definition, the operators $\hat{a}$ and $\hat{a}^\dagger$ are indeed Hermitian adjoints.[2] Incidentally, these operators are *not* Hermitian (i.e., it is *not* true that $\hat{a} = \hat{a}^\dagger$) and thus differ mathematically from many other operators we have been considering so far.

Now using the definitions for these operators, Eqs. (15.9) and (15.10), we can write from Eq. (15.3)

$$\left(\hat{a}^\dagger\hat{a}+\frac{1}{2}\right)\psi = \frac{E}{\hbar\omega}\psi \quad (15.12)$$

In other words, we can rewrite the Hamiltonian as

$$\hat{H} \equiv \hbar\omega\left(\hat{a}^\dagger\hat{a}+\frac{1}{2}\right) \quad (15.13)$$

We know from the previous solution of the harmonic oscillator that the eigenenergy associated with eigenstate $|\psi_n\rangle$ is $E_n = \hbar\omega(n+1/2)$ and so we know that

$$\hat{a}^\dagger\hat{a}|\psi_n\rangle = n|\psi_n\rangle \quad (15.14)$$

This operator $\hat{a}^\dagger\hat{a}$ obviously has the harmonic oscillator states as its eigenstates and the number of the state as its eigenvalue and is sometimes called the number operator, $\hat{N}$; that is,

$$\hat{N} \equiv \hat{a}^\dagger\hat{a} \quad (15.15)$$

with the eigenequation

$$\hat{N}|\psi_n\rangle = n|\psi_n\rangle \quad (15.16)$$

Properties of raising and lowering operators

Thus far, we have merely made a mathematical rearrangement of the solutions we already knew from the previous discussion of the harmonic oscillator, introducing two new and rather unusual operators, $\hat{a}$ and $\hat{a}^\dagger$. These operators are, however, very useful, as we will see later, and give an elegant way of looking at harmonic oscillators, a way that is particularly useful also as we generalize to considering the electromagnetic field.

---

[1] Remember, of course, that the momentum operator in the z direction, $-i\hbar d/dz$, is Hermitian and so d/dz must be anti-Hermitian.

[2] The proof that $\hat{a}$ and $\hat{a}^\dagger$ are, indeed, Hermitian conjugates of one another is obvious if one writes them out in the matrix form in the position representation, as in, e.g., Eq. (4.132). If one merely adds the position variable itself (x in that example) along the diagonal to construct the matrix for the operator $i(d/dx)+x$, it is obvious that reflecting the matrix about the diagonal and taking the complex conjugate leads to the matrix representation for the operator $-i(d/dx)+x$.

The operators $\hat{a}$ and $\hat{a}^{\dagger}$ have a very important property, easily proved from their definitions, which is their commutator. Specifically, we find

$$\left[\hat{a},\hat{a}^{\dagger}\right]=\hat{a}\hat{a}^{\dagger}-\hat{a}^{\dagger}\hat{a}=1 \tag{15.17}$$

We can use this property, together with the "number operator" property (Eq. (15.14)), to show the reason why these operators are called raising and lowering operators or creation and annihilation operators.

Suppose first we operate on both sides of Eq. (15.14) with $\hat{a}$. Then, we have

$$\hat{a}\left(\hat{a}^{\dagger}\hat{a}\right)|\psi_n\rangle = n\hat{a}|\psi_n\rangle \tag{15.18}$$

that is, regrouping the operators on the left and substituting from Eq. (15.17), we have

$$\left(\hat{a}\hat{a}^{\dagger}\right)\left(\hat{a}|\psi_n\rangle\right)=\left(1+\hat{a}^{\dagger}\hat{a}\right)\left(\hat{a}|\psi_n\rangle\right)=n\left(\hat{a}|\psi_n\rangle\right) \tag{15.19}$$

that is,

$$\hat{a}^{\dagger}\hat{a}\left(\hat{a}|\psi_n\rangle\right)=(n-1)\left(\hat{a}|\psi_n\rangle\right) \tag{15.20}$$

But this means from Eq. (15.14) that $\hat{a}|\psi_n\rangle$ is simply $|\psi_{n-1}\rangle$, at least within some normalizing constant A_n; that is,

$$\hat{a}|\psi_n\rangle = A_n|\psi_{n-1}\rangle \tag{15.21}$$

Hence, we see why the operator $\hat{a}$ is called the lowering operator because it changes the state $|\psi_n\rangle$ into the state $|\psi_{n-1}\rangle$.

We can perform a similar analysis, operating on both sides of Eq. (15.14) with $\hat{a}^{\dagger}$. The details of this are left as an exercise for the reader. The result is similar to Eq. (15.20)

$$\hat{a}^{\dagger}\hat{a}\left(\hat{a}^{\dagger}|\psi_n\rangle\right)=(n+1)\left(\hat{a}^{\dagger}|\psi_n\rangle\right) \tag{15.22}$$

Again, we conclude from Eq. (15.14) that $\hat{a}^{\dagger}|\psi_n\rangle$ must simply be $|\psi_{n+1}\rangle$, at least within some normalizing constant B_{n+1}

$$\hat{a}^{\dagger}|\psi_n\rangle = B_{n+1}|\psi_{n+1}\rangle \tag{15.23}$$

and we similarly see why the operator $\hat{a}^{\dagger}$ is called the raising operator because it changes the state $|\psi_n\rangle$ into the state $|\psi_{n+1}\rangle$.

Incidentally, one way to remember which operator is which is to think of the superscript dagger "†" as a "+" sign corresponding to raising the state. Indeed, it is quite a common alternative notation to use a superscript "+" sign.

We can now proceed one step further, which is to deduce what A_n and B_n are. Premultiplying Eq. (15.21) by $\langle\psi_{n-1}|$ gives

$$\langle\psi_{n-1}|\hat{a}|\psi_n\rangle = A_n \tag{15.24}$$

But, using the normal properties of operators and state vectors and the conjugate of Eq. (15.23) rewritten for initial state $|\psi_{n-1}\rangle$ rather than $|\psi_n\rangle$, we have

$$\langle\psi_{n-1}|\hat{a}=\left[\hat{a}^{\dagger}|\psi_{n-1}\rangle\right]^{\dagger}=\left[B_n|\psi_n\rangle\right]^{\dagger}=B_n^{*}\langle\psi_n| \tag{15.25}$$

so

$$\langle \psi_{n-1} | \hat{a} | \psi_n \rangle = A_n = B_n^* \langle \psi_n | \psi_n \rangle = B_n^* \tag{15.26}$$

Hence

$$\hat{a}^\dagger \hat{a} | \psi_n \rangle = A_n \hat{a}^\dagger | \psi_{n-1} \rangle = A_n B_n | \psi_n \rangle = |A_n|^2 | \psi_n \rangle = n | \psi_n \rangle \tag{15.27}$$

and so

$$A_n = \sqrt{n} \tag{15.28}$$

at least within a unit complex constant, which we choose to be +1.[3] Hence, we have instead of Eq. (15.21)

$$\hat{a} | \psi_n \rangle = \sqrt{n} | \psi_{n-1} \rangle \tag{15.29}$$

and instead of Eq. (15.23)

$$\hat{a}^\dagger | \psi_n \rangle = \sqrt{n+1} | \psi_{n+1} \rangle \tag{15.30}$$

Eigenfunctions using raising and lowering operators

We know that the harmonic oscillator has a lowest state, which corresponds to $n = 0$. Hence, from Eq. (15.29), we must have

$$\hat{a} | \psi_0 \rangle = 0 \tag{15.31}$$

(or, more strictly, $0 | \psi_o \rangle$ because the result is strictly a state vector of zero length, not a number).

We can use this as an alternative method of deducing $|\psi_0\rangle \equiv \psi_0(\xi)$. Using the differential operator definition of $\hat{a}$ from Eq. (15.10), we have

$$\frac{1}{\sqrt{2}} \left(\frac{d}{d\xi} + \xi \right) \psi_0(\xi) = 0 \tag{15.32}$$

from which we can immediately verify that the solution is

$$\psi_0(\xi) = \frac{1}{(\pi)^{1/4}} \exp\left(-\frac{\xi^2}{2} \right) \tag{15.33}$$

(where we have also normalized the solution), in agreement with our previous method of direct solution of the second-order differential Schrödinger equation. Now, however, we can proceed using the raising operator to construct all the other solutions for different n. Successive application of $\hat{a}^\dagger$ to $|\psi_0\rangle$ gives

[3] This choice is not totally arbitrary. If we were to consider multiple different harmonic oscillators, as we could do in considering multiple different modes of the electromagnetic field, we need to have specific commutation relations between the creation and annihilation operators for different modes. This choice does give the correct commutation relations in that case. We return to this point when considering more general boson commutation relations.

$$\left(\hat{a}^{\dagger}\right)^{n}\left|\psi_{0}\right\rangle = \sqrt{n!}\left|\psi_{n}\right\rangle \tag{15.34}$$

and so the normalized eigenstates can be written as

$$\left|\psi_{n}\right\rangle = \frac{1}{\sqrt{n!}}\left(\hat{a}^{\dagger}\right)^{n}\left|\psi_{0}\right\rangle \tag{15.35}$$

We see by this approach that each eigenfunction can be progressively deduced from the preceding ones. We would reconstruct the previously derived solutions with Hermite polynomials and Gaussians if we proceeded this way. We may use this result either actually to construct the eigenfunctions explicitly or as a substitution to allow convenient manipulations of the states by operators.

We can see that the use of these raising and lowering operators has certainly given us a compact and elegant way of stating the properties of the harmonic oscillator, one in which the properties can largely be stated without recourse to the actual wavefunction in space. This latter aspect is particularly useful as we generalize the harmonic oscillator to other results where the concept of a quantum mechanical wavefunction in space is possibly not meaningful, such as with the electromagnetic field.

Note, incidentally, that saying that the harmonic oscillator is in state n is mathematically identical to considering that the harmonic oscillator "mode" contains a number n of identical particles. Each particle has energy $\hbar\omega$. It is, of course, meaningless to say what order the particles are in – we can only say that we have n identical particles in the oscillator. This is exactly the kind of behavior we expect for identical bosons that are all in one mode. As we deduced in Chapter 13, there is only one way in which we have n identical bosons in one mode and there is only one way in which this oscillator "mode" can be in state n. For the harmonic oscillator discussed here, it may not appear particularly useful to introduce this concept of identical particles, though it is viewed as a useful concept as we discuss the electromagnetic field.

Problems

15.1.1. Prove the relation

$$\left[\hat{a},\hat{a}^{\dagger}\right] = \hat{a}\hat{a}^{\dagger} - \hat{a}^{\dagger}\hat{a} = 1$$

for the harmonic oscillator raising and lowering operators, starting from their definitions

$$\hat{a}^{\dagger} \equiv \frac{1}{\sqrt{2}}\left(-\frac{d}{d\xi}+\xi\right) \text{ and } \hat{a} \equiv \frac{1}{\sqrt{2}}\left(\frac{d}{d\xi}+\xi\right)$$

15.1.2. Given that

$$\hat{a}^{\dagger}\hat{a}\left|\psi_{n}\right\rangle = n\left|\psi_{n}\right\rangle$$

and

$$\left[\hat{a},\hat{a}^{\dagger}\right] = \hat{a}\hat{a}^{\dagger} - \hat{a}^{\dagger}\hat{a} = 1$$

show that

$$\hat{a}^{\dagger}\hat{a}\left(\hat{a}^{\dagger}\left|\psi_{n}\right\rangle\right) = (n+1)\left(\hat{a}^{\dagger}\left|\psi_{n}\right\rangle\right)$$

15.2 Hamilton's equations and generalized position and momentum

In our previous attempts at turning classical problems into quantum mechanical ones, such as the postulation of the Schrödinger equation or of the angular momentum operator, we have written a classical equation in terms of momentum and position and then substituted the operator $-i\hbar\partial/\partial x$ for the momentum p_x, and similarly for other components if needed. Can we take a similar approach for other kinds of problems, such as the electromagnetic field, and arrive at the analog of the Schrödinger equation?

The answer is that we can, but we do not do this by considering momentum and position in the sense that we do with a particle. What we do is find quantities (e.g., for the electromagnetic field) that are mathematically analogous to momentum and position for a particle, though these quantities are not physically momentum and position in the case of an electromagnetic wave. To find these analogous quantities, we go back and examine the mathematical roles of momentum and position in the Hamiltonian of a particle and look for quantities with analogous roles in the Hamiltonian of, for example, one mode of the electromagnetic field. Once we find those, we can proceed as before to quantize by substituting an appropriate operator for the quantity analogous to momentum.

This general approach to quantization of finding quantities mathematically analogous to momentum and position in the Hamiltonian and substituting the operator for the momentum works in a broad range of examples, beyond the electromagnetic case. In the end, its only justification is that it appears to work, giving us a quantum model of the world around us that agrees with what we see.

To see how we find quantities that are mathematically analogous to position and momentum in a Hamiltonian, we return momentarily to classical mechanics. In classical mechanics, the Hamiltonian, H, represents the total energy and, in the simple case of one particle in one dimension, is a function of the momentum, p, and the position, conventionally written as q. p and q are considered to be independent variables. Hence, in classical mechanics, for a particle of mass m, considering only one dimension for simplicity

$$H=\frac{p^2}{2m}+V(q) \tag{15.36}$$

where $V(q)$ is the potential energy. The force on the particle is the negative of the gradient of the potential (a particle accelerates when going downhill); that is, the force is

$$F=-\frac{dV}{dq}\equiv-\frac{\partial H}{\partial q} \tag{15.37}$$

As usual, from Newton's second law (i.e., force = rate of change of momentum), we know, therefore, that

$$\frac{dp}{dt}=-\frac{\partial H}{\partial q} \tag{15.38}$$

We note that

$$\frac{\partial H}{\partial p}=\frac{p}{m} \tag{15.39}$$

Because $p = mv$ where v is the particle velocity and, by definition, $v \equiv dq/dt$, we therefore have

$$\frac{dq}{dt} = \frac{\partial H}{\partial p} \tag{15.40}$$

The two Eqs. (15.38) and (15.40) play a central role in the Hamiltonian picture of classical mechanics and are sometimes known as *Hamilton's equations*. For our purposes, we note only that if the Hamiltonian depends on two quantities p and q and these quantities and the Hamiltonian obey Hamilton's Eqs. (15.38) and (15.40), then we have found the quantities analogous to momentum and position. It has been very successful in quantum mechanics to use these quantities as the basis for quantization by substituting a differential operator $-i\hbar\, d/dq$ for p in the corresponding Hamiltonian.

Note that in this general case, both p and q may bear little relation to the momentum or position of anything; all that matters is that they and the Hamiltonian obey Hamilton's Eqs. (15.38) and (15.40). The art in using this approach to suggest a quantization is in finding appropriate quantities in the problem that behave in the same mathematical manner as the p and q in the Hamiltonian equations.

15.3 Quantization of electromagnetic fields

For a summary of the relevant additional vector calculus and electromagnetism necessary from this point on, see Appendices C and D.

Description of a mode of the electromagnetic field

Here, we consider a very simple example of a mode of the electromagnetic field and we start by describing it in classical terms, checking the proposed classical behavior against Maxwell's equations. We imagine a box of length L in the x direction. We presume it is arbitrarily large in the other dimensions and, consequently, the mode can be described as a standing plane wave in the x direction of some wavevector k. We certainly expect that the electric field **E** is perpendicular to the x direction because both the **E** field and the magnetic field **B** are transverse to the direction of propagation for a simple plane electromagnetic wave. We choose a polarization for the mode, here selecting the electric field to be polarized in the z direction, with an appropriate amplitude, E_z. The **E** field in the other two directions is taken to be zero. We also expect that the magnetic field **B** is perpendicular to the **E** field, so we choose it polarized in the y direction, with amplitude B_y, with zero **B** field in the other two directions. Hence, we postulate that

$$E_z = p(t) D \sin kx \tag{15.41}$$

and

$$B_y = q(t) \frac{D}{c} \cos kx \tag{15.42}$$

where c is the velocity of light, introduced here for subsequent convenience; D is a constant still to be determined; and $p(t)$ and $q(t)$ are, at the moment, simply functions of time yet to be determined. We now check that these fields satisfy the appropriate Maxwell's equations, which will justify all our postulations about these classical fields and tell us some other required relations between our postulated quantities. For this, we now presume that we are in a vacuum, so there is no charge density and no magnetic materials and the permittivity and permeability are their vacuum values of ε_o and μ_o, respectively.

Using the Maxwell equation

$$\nabla \times \mathbf{E} = -\frac{\partial \mathbf{B}}{\partial t} \tag{15.43}$$

we find with our postulated electric and magnetic fields and noting that $\partial E_z / \partial y = 0$ because we have an infinite plane wave with no variation in the y direction

$$\frac{\partial E_z}{\partial x} = \frac{\partial B_y}{\partial t} \tag{15.44}$$

that is,

$$kpD \cos kx = \frac{D}{c}\frac{\partial q}{\partial t} \cos kx \tag{15.45}$$

so

$$\frac{dq}{dt} = \omega p \tag{15.46}$$

where $\omega = kc$.

Similarly, using the Maxwell equation

$$\nabla \times \mathbf{B} = \varepsilon_o \mu_o \frac{\partial \mathbf{E}}{\partial t} \tag{15.47}$$

and noting that $\partial B_y / \partial z = 0$ because the plane wave has no variation in the z direction

$$\frac{\partial B_y}{\partial x} = \varepsilon_o \mu_o \frac{\partial E_z}{\partial t} \tag{15.48}$$

that is,

$$-kq\frac{D}{c}\sin kx = \varepsilon_o \mu_o \frac{dp}{dt} D \sin kx \tag{15.49}$$

that is, using the relation from electromagnetism

$$\varepsilon_o \mu_o = \frac{1}{c^2} \tag{15.50}$$

we have

$$\frac{dp}{dt} = -\omega q \tag{15.51}$$

So, we have found that our postulated form for the mode of the radiation field does indeed satisfy the two Maxwell equations, Eqs. (15.43) and (15.47), provided we can simultaneously have the relations in Eqs. (15.46) and (15.51) between our time-varying amplitudes p and q.

We can draw one further important conclusion here. Differentiating Eq. (15.46) with respect to time t and substituting from Eq. (15.51), we find

$$\frac{d^2q}{dt^2} = -\omega^2 q \qquad (15.52)$$

which means that the electromagnetic mode does indeed behave exactly like a harmonic oscillator, with oscillation (angular) frequency ω.

Hamiltonian for the electromagnetic mode

Now, we want to construct the Hamiltonian for the mode so that we can manage to find the appropriate quantities that will behave analogously to momentum and position by obeying Hamilton's Eqs. (15.38) and (15.40). The reader will have guessed that we are striving to make p and q be those quantities, though we still have just a little work to do to make that so.

Formally, in an electromagnetic field in a vacuum, the energy density is

$$W = \frac{1}{2}\left(\varepsilon_o \mathbf{E}^2 + \frac{1}{\mu_o}\mathbf{B}^2\right) \qquad (15.53)$$

In a box of length L, then, per unit cross-sectional area, the total energy is the Hamiltonian

$$\begin{aligned} H &= \int_0^L W dx = \frac{D^2}{2}\int_0^L \left(\varepsilon_o p^2 \sin^2 kx + \frac{1}{\mu_o c^2} q^2 \cos^2 kx\right) dx \\ &= \frac{D^2 \varepsilon_o}{2}\int_0^L \left(p^2 \sin^2 kx + q^2 \cos^2 kx\right) dx = \frac{D^2 L \varepsilon_o}{4}\left[p^2 + q^2\right] \end{aligned} \qquad (15.54)$$

We now try to choose D so as to get p and q to correspond to the analogs of momentum and position with the Hamiltonian H of Eq. (15.54) by having H, p, and q obey Hamilton's Eqs. (15.38) and (15.40). We note that we already have the relation in Eq. (15.46) between dq/dt and p and the relation in Eq. (15.51) between dp/dt and q. Inspection shows that if we choose

$$H = \frac{\omega}{2}\left(p^2 + q^2\right) \qquad (15.55)$$

that is,

$$D = \sqrt{\frac{2\omega}{L\varepsilon_o}} \qquad (15.56)$$

then H, p, and q do indeed satisfy the Hamiltonian Eqs. (15.38) and (15.40), making p and q the desired analogs of momentum and position. Specifically, using Eq. (15.46)

$$\frac{\partial H}{\partial p} = \omega p = \frac{dq}{dt} \qquad (15.57)$$

and using Eq. (15.51)

$$\frac{\partial H}{\partial q} = \omega q = -\frac{dp}{dt} \qquad (15.58)$$

Quantization

Having derived an appropriate classical Hamiltonian that describes the electromagnetic mode, we now proceed to quantize it in a manner that, in the end, is the same approach that we took in postulating the Schrödinger equation for the electron. We postulate that we can substitute the operator

$$\hat{p} = -i\hbar \frac{d}{dq} \tag{15.59}$$

for the scalar quantity p of the classical Hamiltonian, obtaining a Hamiltonian operator

$$\hat{H} = \frac{\omega}{2}\left[-\hbar^2 \frac{d^2}{dq^2} + q^2\right] \tag{15.60}$$

It is more convenient to use the dimensionless unit

$$\xi = q / \sqrt{\hbar} \tag{15.61}$$

(For future use, we also can define an operator

$$\hat{\pi} = \hat{p} / \sqrt{\hbar} \equiv -i\frac{d}{d\xi} \tag{15.62}$$

which is a dimensionless form of the momentum operator.)

Use of the dimensionless unit ξ gives the Hamiltonian in the form

$$\hat{H} = \frac{\hbar\omega}{2}\left[-\frac{d^2}{d\xi^2} + \xi^2\right] \tag{15.63}$$

which is mathematically identical to the Hamiltonian we used for the harmonic oscillator (e.g., in Eq. (15.3)).

(Incidentally, we can also write

$$\hat{H} = \frac{\hbar\omega}{2}\left(\hat{\pi}^2 + \xi^2\right) \tag{15.64}$$

if we prefer this dimensionless form.)

This completes the postulation of the mathematical analogy between a mode of the electromagnetic field and a quantum harmonic oscillator. Obviously, if our basic postulation of the quantization by substitution of the operator of Eq. (15.59) is correct for describing an electromagnetic mode quantum mechanically (and it appears, in practice, that it is), we can now re-use all of the elegant formalism we developed for the harmonic oscillator to describe the electromagnetic mode, including the raising and lowering operators (i.e., creation and annihilation operators).

Because there are many possible modes of the electromagnetic field, as we think about a future generalization to considering all of those modes, we need to distinguish to which mode the operators refer. It is common to use the index λ to index the different modes (though it is not sufficient to imagine that λ is the wavelength of the mode – other attributes, such as polarization and the spatial form of the mode, are also important). With that new notation, which helps us emphasize that we have made the switch from a simple mechanical oscillator to an electromagnetic mode, for a given mode we have a frequency ω_λ, a Hamiltonian $\hat{H}_\lambda$, creation and annihilation operators $\hat{a}_\lambda^\dagger$ and $\hat{a}_\lambda$, and a number operator $\hat{N}_\lambda$. We can also label the eigenstates similarly as $|\psi_{\lambda n}\rangle$ as being the nth eigenstate associated with the mode λ. We should also change to using the coordinate ξ_λ because each different mode has its own corresponding coordinate. With this notation, we can rewrite some of our results to summarize the key relations for the electromagnetic mode case.

Analogously to Eq. (15.13)

$$\hat{H}_\lambda \equiv \hbar\omega_\lambda \left(\hat{a}_\lambda^\dagger \hat{a}_\lambda + \frac{1}{2} \right) \tag{15.65}$$

The number operator for this mode is defined analogously to Eq. (15.15) as

$$\hat{N}_\lambda \equiv \hat{a}_\lambda^\dagger \hat{a}_\lambda \tag{15.66}$$

so, analogously to Eqs. (15.14) and (15.16)

$$\hat{N}_\lambda |\psi_{\lambda n}\rangle = \hat{a}_\lambda^\dagger \hat{a}_\lambda |\psi_{\lambda n}\rangle = n_\lambda |\psi_{\lambda n}\rangle \tag{15.67}$$

Now, it is fairly obvious to make the association of n_λ with the number of photons in the mode, with an energy that grows in proportion to the number of photons by an amount $n_\lambda \hbar\omega$.

We also have the commutation relation, analogous to Eq. (15.17)

$$\left[\hat{a}_\lambda, \hat{a}_\lambda^\dagger\right] = \hat{a}_\lambda \hat{a}_\lambda^\dagger - \hat{a}_\lambda^\dagger \hat{a}_\lambda = 1 \tag{15.68}$$

We have the lowering relation, analogous to Eq. (15.29)

$$\hat{a}_\lambda |\psi_{\lambda n}\rangle = \sqrt{n_\lambda} |\psi_{\lambda n-1}\rangle \tag{15.69}$$

which we now think is taking the state with n_λ photons in mode λ and changing it into the state with $n_\lambda - 1$ photons; hence, in the electromagnetic field context, we call $\hat{a}_\lambda$ the annihilation operator for that mode. Similarly, we have the raising relation, analogous to Eq. (15.30)

$$\hat{a}_\lambda^\dagger |\psi_{\lambda n}\rangle = \sqrt{n_\lambda + 1} |\psi_{\lambda n+1}\rangle \tag{15.70}$$

which is taking the state with n_λ photons in mode λ and changing it into the state with $n_\lambda + 1$ photons, so we call $\hat{a}_\lambda^\dagger$ the creation operator for that mode. We also expect we have by analogy with Eq. (15.31)

$$\hat{a}_\lambda |\psi_{\lambda 0}\rangle = 0 \tag{15.71}$$

and, as in Eq. (15.35)

$$|\psi_{\lambda n}\rangle = \frac{1}{\sqrt{n_\lambda !}} \left(\hat{a}_\lambda^\dagger\right)^{n_\lambda} |\psi_{\lambda 0}\rangle \tag{15.72}$$

We have now essentially completed our process of quantizing one mode of the electromagnetic field, working by analogy with the quantization of the harmonic oscillator. We could argue that there is little justification for what we have done other than that we have followed a similar procedure to one that worked before and, from a fundamental point of view, that is quite a valid complaint. As always, the only justification for such a postulation is that it works, as indeed this quantization of the electromagnetic field does.

15.4 Nature of the quantum mechanical states of an electromagnetic mode

This quantization of the electromagnetic field mode has all been done by analogy, leading to a somewhat abstract set of results. Now, we investigate what are some of the consequences of these results in more tangible properties of the electromagnetic field. Based on our previous experiences with quantum mechanical wavefunctions, which were mostly functions of a spatial coordinate (though the spin functions were not), our first instinct might be to look at the wavefunction (or its squared modulus). It is instructive to do this and the wavefunction does have some meaning, though it is quite different from that of the electron spatial wavefunction, for example.

Just as before in Eq. (15.33), we have, for example, for the state with no photons in the mode

$$\psi_{\lambda 0}\left(\xi_\lambda\right) = \frac{1}{\left(\pi\right)^{1/4}} \exp\left(-\frac{\xi_\lambda^2}{2}\right) \tag{15.73}$$

but if we now work backward to find the physical interpretation of the coordinate ξ_λ, we find from Eqs. (15.50), (15.56), and (15.42)

$$B_y = \xi_\lambda \sqrt{\frac{2\mu_o \hbar \omega_\lambda}{L}} \cos kx \tag{15.74}$$

In other words, ξ_λ is, in a dimensionless form, the amplitude of the mode of the magnetic field. It is not a spatial coordinate. For example, we can interpret $|\psi_{\lambda 0}(\xi_\lambda)|^2$ as being the probability that for the lowest state of this electromagnetic field mode, the field mode will be found to have (dimensionless) amplitude ξ_λ.[4] That probability (per unit range of dimensionless field amplitude) is, therefore, the Gaussian $(1/\sqrt{\pi})\exp\left(-\xi_\lambda^2\right)$. We would find related results for the states of the mode with more photons. This kind of result tells us two behaviors that are quite different from what we might expect from classical electromagnetic fields; namely:

(i) in a state with a given number of photons, the magnetic field amplitude is not a fixed quantity but is rather described by a distribution

(ii) even with no photons in the mode, the magnetic field amplitude is not zero

Similar results can be stated for the electric-field mode amplitude. The presence of such amplitudes of the electric and magnetic fields even with no photons is called *vacuum fluctuations*. In general, the states with specific numbers of photons, which correspond to the eigenstates of the Hamiltonian or the number operator, do *not* have precisely defined electric or magnetic field amplitudes.

Though we may sometimes be interested in these distributions of magnetic or electric field amplitude, we are generally much less interested in these than we were in the probabilities of finding particles at points in space. As a result, in the quantized electromagnetic field, we make relatively little use of the wavefunctions themselves. Most of the results in which we are interested, such as processes where we are adding or subtracting photons (as in optical

---

[4] The mode amplitude is also oscillating in time, though it is really the amplitude of the oscillations in time and space that we are describing here by the coordinate ξ. Though a field oscillating in time would obviously have a distribution of its field amplitudes because the field is oscillating even with a fixed mode amplitude, we are talking here about a distribution in the amplitude of the oscillations.

emission and absorption), can more conveniently be described through the use of operators and state vectors. Typically, the basis set and the resulting state we use for electromagnetic fields are not written as functions, $\psi(\xi_\lambda)$, of the amplitudes, ξ_λ, of the fields in the modes, but are instead basis vectors corresponding to specific numbers of photons in a mode; that is, basis vectors corresponding to the energy or number eigenstates of the mode

$$|\psi_{\lambda n}\rangle \equiv |n_\lambda\rangle \tag{15.75}$$

where we now typically choose to drop the use of the ψ in our notation, simply indexing the basis states by the number of photons n_λ in the mode λ, as in the notation on the right of Eq. (15.75).

Once we start describing states as having a specific number of particles in them, we sometimes use the terms *number states* or *Fock states*.[5] The states as described in Eq. (15.75) are, therefore, number or Fock states. Representations of quantum mechanical states based on such number-state representations are also sometimes called *Fock representations*.

In the end, this choice of basis and representation is a matter of taste or convenience. We can always evaluate the average value of any measurable quantity by taking the expectation value of the associated operator in the state of interest, regardless of what basis or representation we use for the state and operator. Note that we could also have described the electron perfectly completely without explicit recourse to functions of space coordinates, using state vectors instead and using a basis other than the simple position basis (e.g., a Fourier basis).

15.5 Field operators

Given the usefulness of the creation and annihilation operators, we would like to represent other operators in this problem in terms of them. Remembering the original definitions of these operators in terms of ξ and $d/d\xi$, we have analogously to Eq. (15.9)

$$\hat{a}_\lambda^\dagger \equiv \frac{1}{\sqrt{2}}\left(-\frac{d}{d\xi_\lambda} + \xi_\lambda\right) \tag{15.76}$$

and, analogously to Eq. (15.10)

$$\hat{a}_\lambda \equiv \frac{1}{\sqrt{2}}\left(\frac{d}{d\xi_\lambda} + \xi_\lambda\right) \tag{15.77}$$

Now, we note that we can write

$$\hat{\xi}_\lambda \equiv \frac{1}{\sqrt{2}}\left(\hat{a}_\lambda + \hat{a}_\lambda^\dagger\right) \tag{15.78}$$

This reminds us, incidentally, that $\hat{\xi}_\lambda$ is really an operator, not just a coordinate. Now that we have no explicit mention of position in the equation, we can explicitly put the "hat" over this operator. We ran into the same issue in considering the position $\mathbf{r}$ in the spatial representation of electron wavefunctions, for example (see the discussion in Section 5.4). In the position representation in that case, it was quite acceptable simply to use the quantity $\mathbf{r}$ when we

[5] The Russian physicist Vladimir Fock is also renowned for the Hartree–Fock approximation used extensively in solid-state physics.

wanted to evaluate the expectation value of position in some state $\phi(\mathbf{r})$, writing $\langle \mathbf{r} \rangle \equiv \langle \phi | \mathbf{r} | \phi \rangle = \int \phi^*(\mathbf{r}) \mathbf{r} \phi(\mathbf{r}) d^3\mathbf{r}$. The ability to use the quantity as its own operator is a consequence of being in the representation based on that quantity. In this **r** representation, where we can write out a function $f(\mathbf{r})$ as a vector of its values at each of the points **r**, the operator corresponding to **r** is simply a diagonal matrix, with elements of value **r** along the diagonal.

In general, however, if we change representations, we must explicitly recognize that **r** is actually an operator, which should be written as $\hat{\mathbf{r}}$, and the corresponding matrix in any other basis is not diagonal. We should, in general, be writing $\langle \mathbf{r} \rangle = \langle \phi | \hat{\mathbf{r}} | \phi \rangle$, which is correct in any representation. The same line of argument should be followed here in understanding that ξ_λ is actually an operator. Because we are now likely to be using this operator with the photon number basis states, we have to recognize its operator nature explicitly.

We can also write the dimensionless form of the generalized momentum operator, defined as in Eq. (15.62) as

$$\hat{\pi}_\lambda = -i d / d\xi_\lambda \equiv \hat{p}_\lambda / \sqrt{\hbar} \tag{15.79}$$

in the form

$$\hat{\pi}_\lambda = \frac{i}{\sqrt{2}} \left(\hat{a}_\lambda^\dagger - \hat{a}_\lambda \right) \tag{15.80}$$

When we started looking at the electromagnetic mode classically, we wrote it in terms of standing waves and two parameters, p and q. Now that we have completed our quantization of that mode, we realize that both p and q have been replaced by operators. With our expressions Eqs. (15.78) and (15.80), and the definitions of the dimensionless operators ξ_λ and $\hat{\pi}_\lambda$, we now substitute back into the relations Eqs. (15.41) and (15.42) that defined the electric and magnetic fields in our mode. Instead of scalar quantities for the electric and magnetic fields for this mode, we now have operators

$$\hat{E}_{\lambda z} = i \left(\hat{a}_\lambda^\dagger - \hat{a}_\lambda \right) \sqrt{\frac{\hbar \omega_\lambda}{\varepsilon_o L}} \sin kx \tag{15.81}$$

and

$$\hat{B}_{\lambda y} = \left(\hat{a}_\lambda^\dagger + \hat{a}_\lambda \right) \sqrt{\frac{\mu_o \hbar \omega_\lambda}{L}} \cos kx \tag{15.82}$$

What do we mean by such field operators? Just as before, if we want to know the average value of a measurable quantity, we take the expected value of its operator, and the same is true here. For a state $|\phi\rangle$ of this mode, we would have

$$\langle E_{\lambda z} \rangle = \langle \phi | \hat{E}_{\lambda z} | \phi \rangle \tag{15.83}$$

$$\langle B_{\lambda y} \rangle = \langle \phi | \hat{B}_{\lambda y} | \phi \rangle \tag{15.84}$$

Incidentally, this postulation of operators for the electric and magnetic fields is an example of a quantum field theory. We started our discussion of quantum mechanics by quantizing the states of a particle. We have now come to a description in which we are quantizing a field and we can view the particles – in this case, photons – as emerging from the quantization of the field. It is also possible and useful to construct quantum field theories for electrons in a solid, for example, which gives a particularly elegant way of writing solid-state physics. Much of modern quantum mechanics that is concerned with elementary particles is also in the form of quantum field theories.

Uncertainty principle of electric and magnetic fields in a mode

Note that the electric and magnetic field operators do not commute – we cannot, in general,[6] simultaneously know both the electric and magnetic field exactly.

Explicitly, from Eqs. (15.81) and (15.82), we have

$$\left[\hat{E}_{\lambda z},\hat{B}_{\lambda y}\right]=i\frac{\hbar\omega_\lambda}{L}\sqrt{\frac{\mu_o}{\varepsilon_o}}\sin kx\cos kx\left[\hat{a}_\lambda^\dagger-\hat{a}_\lambda,\hat{a}_\lambda+\hat{a}_\lambda^\dagger\right] \tag{15.85}$$

that is,

$$\left[\hat{E}_{\lambda z},\hat{B}_{\lambda y}\right]=$$

$$i\frac{\hbar\omega_\lambda}{L}\sqrt{\frac{\mu_o}{\varepsilon_o}}\sin kx\cos kx\left(\hat{a}_\lambda^\dagger\hat{a}_\lambda+\hat{a}_\lambda^\dagger\hat{a}_\lambda^\dagger-\hat{a}_\lambda\hat{a}_\lambda-\hat{a}_\lambda\hat{a}_\lambda^\dagger-\hat{a}_\lambda\hat{a}_\lambda^\dagger+\hat{a}_\lambda\hat{a}_\lambda-\hat{a}_\lambda^\dagger\hat{a}_\lambda^\dagger+\hat{a}_\lambda^\dagger\hat{a}_\lambda\right)$$

$$=2i\frac{\hbar\omega_\lambda}{L}\sqrt{\frac{\mu_o}{\varepsilon_o}}\sin kx\cos kx\left[\hat{a}_\lambda^\dagger\hat{a}_\lambda-\hat{a}_\lambda\hat{a}_\lambda^\dagger\right] \tag{15.86}$$

So, using the known commutator of the creation and annihilation operators, Eq. (15.68), we have

$$\left[\hat{E}_{\lambda z},\hat{B}_{\lambda y}\right]=-2i\frac{\hbar\omega_\lambda}{L}\sqrt{\frac{\mu_o}{\varepsilon_o}}\sin kx\cos kx \tag{15.87}$$

We remember that the general form of the commutation relation, $[\hat{A},\hat{B}]=i\hat{C}$ (Eq. (5.4)) leads to the uncertainty principle $\Delta A\Delta B\geq\left|C\right|/2$ (from Eq. (5.23)), so we have for the standard deviations of the expected values of the electric and magnetic fields in this mode

$$\Delta E_{\lambda z}\Delta B_{\lambda y}\geq\frac{\hbar\omega_\lambda}{L}\sqrt{\frac{\mu_o}{\varepsilon_o}}\sin kx\cos kx \tag{15.88}$$

Problem

15.5.1* Find the commutator $[\hat{\xi}_\lambda,\hat{\pi}_\lambda]$ starting from the operator definitions

$$\hat{\xi}_\lambda\equiv\frac{1}{\sqrt{2}}\left(\hat{a}_\lambda+\hat{a}_\lambda^\dagger\right)\text{ and }\hat{\pi}_\lambda=\frac{i}{\sqrt{2}}\left(\hat{a}_\lambda^\dagger-\hat{a}_\lambda\right)$$

[6] We will see later that it is possible to know them both exactly if one of them is zero, as would happen at a node of the standing wave for one or other of the electric or magnetic field, though that is a special case.

15.6 Quantum mechanical states of an electromagnetic field mode

Many states are possible quantum mechanically for the electromagnetic mode. Nearly all of these are quite different from the fields we are used to classically. We briefly discuss two of the types of states that could be considered for electromagnetic field modes: the number states and the coherent state. The reader should be aware that there are many other states possible in principle, including ones with such intriguing titles as *squeezed states* and *photon antibunched states*.

Of these various states, only the coherent state has much relation to the kinds of fields we expect in a mode from a classical analysis. It is essentially the kind of field generated by a laser and corresponds quite closely to our classical notion of an electromagnetic field in a mode. The other kinds of states generally have little relation to any classical concept of an electromagnetic field. They are, in practice, quite difficult to generate controllably, generally requiring sophisticated nonlinear optical techniques, and have all only been demonstrated to a limited degree in the laboratory.

Number state

The eigenstates $|n_\lambda\rangle$ of the Hamiltonian would seem to be the most obvious quantum mechanically. These states correspond to a specific number n_λ photons in the mode and are known as the number states or Fock states. They are also, obviously, eigenstates of the number operator. Their properties are, however, very unlike any classical field we might expect.

As mentioned previously, in these states, the probability of measuring any particular amplitude of the $B_{\lambda y}$ field in the mode is distributed according to the square of the Hermite–Gaussian functions that are the solutions of the harmonic oscillator problem with quantum number n_λ. The $E_{\lambda z}$ amplitudes are similarly distributed.[7] The expectation values of the electric and magnetic field amplitudes are also both zero for any number state. To prove this, for example, for the electric field mode amplitude, we have

$$\begin{aligned}\langle n_\lambda|\hat{E}_{\lambda z}|n_\lambda\rangle &= i\sqrt{\frac{\hbar\omega_\lambda}{L\varepsilon_o}}\sin kx\langle n_\lambda|\hat{a}_\lambda^\dagger-\hat{a}_\lambda|n_\lambda\rangle \\ &= i\sqrt{\frac{\hbar\omega_\lambda}{L\varepsilon_o}}\sin kx\left(\sqrt{n+1}\langle n_\lambda||n_\lambda+1\rangle-\sqrt{n}\langle n_\lambda||n_\lambda-1\rangle\right) \\ &= 0\end{aligned} \qquad (15.89)$$

because the states $|n_\lambda\rangle$, $|n_\lambda-1\rangle$, and $|n_\lambda+1\rangle$ are all orthogonal, being different eigenstates of the same Hermitian operator. A similar proof can be performed for the magnetic field mode amplitude.

The reader might think it very odd that there may be energy in this mode, but yet there appears to be no field. It is not correct that there is no field in the mode, it is just that the average value

[7] There are several ways of showing this. Note, e.g., that our choice of assigning the electric field amplitude a generalized momentum character and the magnetic field amplitude a generalized position character was essentially arbitrary – we could equally well have made this assignment the other way around, in which case it would be the electric field amplitude that was the argument of the Hermite–Gaussian functions, whose square would then give the probability of measuring the mode of the electric field to have any particular amplitude.

of the mode amplitude is zero. This can be explained if we presume that the phase of the field is quite undetermined in such a number state. Any given measurement is quite likely to result in a finite amplitude for the electric or magnetic field in the mode; but, because of the possibility of the mode amplitude being positive or negative, the average is zero.

We do see, though, that the number states, while simple mathematically, bear little relation to classical fields.

Representation of time dependence - Schrödinger and Heisenberg representations

So far, in discussing the states of the electromagnetic field mode from a quantum mechanical point of view we have been dealing with the solutions to the time-independent Schrödinger equation for the mode. Note here that we use the term *Schrödinger equation* in the generalized sense where we mean that

$$\hat{H}|\phi\rangle = E|\phi\rangle \qquad (15.90)$$

is a (time-independent) Schrödinger equation for a system in an eigenstate $|\phi\rangle$ with eigenenergy E. Explicitly, for the eigenstates of our electromagnetic mode, we have

$$\hat{H}|n_\lambda\rangle = (n_\lambda + 1/2)\hbar\omega_\lambda |n_\lambda\rangle \qquad (15.91)$$

In our generalization of our earlier postulations about quantum mechanics, we also postulate here that the time-dependent generalized Schrödinger equation is valid; that is,

$$\hat{H}|\phi\rangle = i\hbar\frac{\partial}{\partial t}|\phi\rangle \qquad (15.92)$$

This postulation does appear to work.

One note about taking this approach is that we are implicitly assuming that the time-dependence of the system is described by time dependence of the state, not of the operators. It is actually completely a matter of taste or convenience whether we put the time dependence into the states or the operators. When we come to evaluate the expectation value of the operator, we obtain identical results. The time-dependent state picture is described as the Schrödinger representation and the time-dependent operator picture is described as the Heisenberg representation. Either one is valid, though in the Heisenberg representation we cannot use the time-dependent Schrödinger equation and must use a somewhat different but equivalent formalism. Here, we explicitly operate in the Schrödinger picture, adding the time dependence to the states and choosing the operators (specifically, the creation and annihilation operators) to be time-independent.

With the Schrödinger approach to describing time dependence, as before, to get the time variation of a given state, we multiply the time-independent energy eigenstate descriptions we have used so far by $\exp[-i(n_\lambda + 1/2)\hbar\omega_\lambda t/\hbar]$ (i.e., $\exp[-i(n_\lambda + 1/2)\omega_\lambda t]$ to make Eq. (15.91) consistent with Eq. (15.92)). Including the time dependence in this way, our number states become

$$\exp\left[-i(n_\lambda + 1/2)\omega t\right]|n_\lambda\rangle \qquad (15.93)$$

Now, we are ready to consider the specific case of the coherent state, which is a specific superposition of number states and has a particularly important overall time dependence.

Coherent state

The state that corresponds most closely to the classical field in an electromagnetic mode is the coherent state. We introduced the coherent state previously as an example in discussing the time dependence of the harmonic oscillator. Now, using our current notation, we can rewrite the coherent state originally proposed in Eq. (3.23) as

$$|\Psi_{\lambda\bar{n}}\rangle = \sum_{n_\lambda=0}^{\infty} c_{\lambda\bar{n}n} \exp\left[-i\left(n_\lambda + \frac{1}{2}\right)\omega_\lambda t\right]|n_\lambda\rangle \qquad (15.94)$$

where

$$c_{\lambda\bar{n}n} = \sqrt{\frac{\bar{n}^{n_\lambda}\exp(-\bar{n})}{n_\lambda!}} \qquad (15.95)$$

Here, the quantity $\bar{n}$ turns out to be the expected value of the number of photons in the mode. As before, we note that

$$|c_{\lambda\bar{n}n}|^2 = \frac{\bar{n}^{n_\lambda}\exp(-\bar{n})}{n_\lambda!} \qquad (15.96)$$

is the Poisson statistical distribution with mean $\bar{n}$ and standard deviation $\sqrt{\bar{n}}$.

We have shown by explicit illustration before (i.e., Chapter 3) that this state oscillates at frequency ω and that remains true here. In this state, the electric and magnetic fields do not have precise values, just as the position did not have precise values before in the mechanical harmonic oscillator. As the average number of photons $\bar{n}$ increases, the relative variation in the values of the electric and magnetic fields decreases (though the absolute value of the variation actually increases) and the behavior resembles a classical pair of oscillating electric and magnetic fields ever more closely.

Note that in the coherent state, the number of photons in the mode is not determined. The coefficients $|c_{\lambda\bar{n}n}|^2$ tell us the probability that we will find n_λ photons in the mode if we make a measurement. This number is now found to be distributed according to a Poisson distribution. It is, in fact, the case that the statistics of the number of photons in an oscillating "classical" electromagnetic field are Poissonian. For example, if one puts a photodetector in a laser beam, one will measure a Poissonian distribution of the arrival rates of the photons, an effect known as *shot noise.*

Note that the coherent state is not an eigenstate of any operator representing a physically observable quantity. In fact, the coherent states are the eigenstates of the annihilation operator, $\hat{a}$, the proof of which is left as an exercise for the reader. The annihilation operator is not a Hermitian operator and does not itself represent an observable physical quantity.

Problems

15.6.1 Show that the coherent state in Eq. (15.94) is an eigenstate of the annihilation operator $\hat{a}_\lambda$, with eigenvalue $\sqrt{\bar{n}}\exp(-i\omega_\lambda t)$.

15.6.2* Using the results of Problem 15.6.1, find an expression for the expectation value of the "position" ξ_λ for the coherent state in Eq. (15.94) in terms of $\bar{n}$, ω_λ, and time t.

15.6.3 Consider the uncertainty relation for position and momentum in the coherent state of Eq. (15.94).

(i) Deduce the uncertainty relation for the operators $\hat{\xi}_\lambda$ and $\hat{\pi}_\lambda$.

(ii) By evaluating $\left\langle \hat{\xi}_\lambda^2 \right\rangle - \bar{\xi}_\lambda^2$ for the coherent state, show that the standard deviation of the width of the resulting probability distribution for the "position" ξ is $1/\sqrt{2}$, independent of time.

(iii) Repeat the calculation in (ii) for "momentum," with operator $\hat{\pi}_\lambda$, instead of "position."

(iv) Deduce that the coherent state is a "minimum uncertainty" state; that is, it has the minimum possible product of the standard deviations of "position" and "momentum."

(Hints: Use the results of Probs. 15.6.1 and 15.6.2, and the general relations for uncertainty principles in Chapter 5.)

15.7 Generalization to sets of modes

Thus far, we have derived the theory considering only one specific plane-wave mode of the electromagnetic field. In any specific classical electromagnetic problem, there is a set of modes that actually form a complete set for describing all solutions of Maxwell's equations. In free space, these would be a set of propagating or standing waves. In resonators, there are, in general, resonator modes, such as the Laguerre–Gaussian modes of a typical laser cavity, that also form a complete set for describing such classical waves. We can generalize the formalism, with each such mode being a harmonic oscillator with annihilation and creation operators.

Previously, we had postulated a simple example mode that had sinusoidal spatial behavior in one direction, was plane in the other directions and could be separated into a product of a spatial and a temporal part (see Eqs. (15.41) and (15.42)). Now, we postulate a set of classical modes, each of which has the following form

$$\mathbf{E}_\lambda(\mathbf{r},t) = -p_\lambda(t) D_\lambda \mathbf{u}_\lambda(\mathbf{r}) \tag{15.97}$$

$$\mathbf{B}_\lambda(\mathbf{r},t) = q_\lambda(t) \frac{D_\lambda}{c} \mathbf{v}_\lambda(\mathbf{r}) \tag{15.98}$$

Here, $\mathbf{E}_\lambda$, $\mathbf{B}_\lambda$, $\mathbf{u}_\lambda$, and $\mathbf{v}_\lambda$ are all, in general, vectors and D_λ is a constant. (The forms of Eqs. (15.41) and (15.42) correspond to these with $\mathbf{u}_\lambda(\mathbf{r}) = -\hat{\mathbf{z}}\sin(kx)$ and $\mathbf{v}_\lambda(\mathbf{r}) = \hat{\mathbf{y}}\cos(kx)$.) We will find these forms will satisfy Maxwell's equations and the wave equation in free space[8] if we require that

$$\nabla \times \mathbf{u}_\lambda(\mathbf{r}) = \frac{\omega_\lambda}{c} \mathbf{v}_\lambda(\mathbf{r}) \tag{15.99}$$

$$\nabla \times \mathbf{v}_\lambda(\mathbf{r}) = \frac{\omega_\lambda}{c} \mathbf{u}_\lambda(\mathbf{r}) \tag{15.100}$$

$$\frac{dq_\lambda}{dt} = \omega_\lambda p_\lambda \tag{15.101}$$

$$\frac{dp_\lambda}{dt} = -\omega_\lambda q_\lambda \tag{15.102}$$

We presume that the classical electromagnetic problem has been solved with the boundary conditions of the system to yield these electromagnetic modes for the problem. We also

---

[8] It is straightforward to extend this approach to situations other than free space, e.g., a dielectric material.

presume that the spatial functions $\mathbf{u}_\lambda(\mathbf{r})$ and $\mathbf{v}_\lambda(\mathbf{r})$ are normalized over the entire volume appropriate for the problem and we presume that they are all orthogonal[9]; that is,

$$\int \mathbf{u}_{\lambda 1}(\mathbf{r}).\mathbf{u}_{\lambda 2}(\mathbf{r})d^3\mathbf{r} = \delta_{\lambda 1,\lambda 2} \tag{15.103}$$

and

$$\int \mathbf{v}_{\lambda 1}(\mathbf{r}).\mathbf{v}_{\lambda 2}(\mathbf{r})d^3\mathbf{r} = \delta_{\lambda 1,\lambda 2} \tag{15.104}$$

Now, suppose we consider that the electromagnetic system is in a classical superposition of such electromagnetic modes; that is, the electric field for some specific set of amplitudes of the oscillatory terms p_λ is

$$\mathbf{E}(\mathbf{r},t) = \sum_\lambda -p_\lambda(t)D_\lambda \mathbf{u}_\lambda(\mathbf{r}) \tag{15.105}$$

and the magnetic field is, similarly, for some specific set of amplitudes of the oscillatory terms q_λ

$$\mathbf{B}(\mathbf{r},t) = \sum_\lambda q_\lambda(t)\frac{D_\lambda}{c}\mathbf{v}_\lambda(\mathbf{r}) \tag{15.106}$$

We know the energy density of an electromagnetic field (Eq. (15.53)), so we can write for the total energy of such a field

$$\begin{aligned} H &= \int \frac{1}{2}\left(\varepsilon_o \mathbf{E}^2 + \frac{1}{\mu_o}\mathbf{B}^2\right)d^3\mathbf{r} \\ &= \frac{1}{2}\varepsilon_o \sum_{\lambda 1,\lambda 2} D_{\lambda 1}D_{\lambda 2}\int\left[p_{\lambda 1}p_{\lambda 2}\mathbf{u}_{\lambda 1}(\mathbf{r}).\mathbf{u}_{\lambda 2}(\mathbf{r}) + q_{\lambda 1}q_{\lambda 2}\mathbf{v}_{\lambda 1}(\mathbf{r}).\mathbf{v}_{\lambda 2}(\mathbf{r})\right]d^3\mathbf{r} \\ &= \frac{1}{2}\varepsilon_o \sum_\lambda D_\lambda^2\left(p_\lambda^2 + q_\lambda^2\right) \end{aligned} \tag{15.107}$$

where we have used the orthogonality of the spatial electromagnetic modes to eliminate all cross-terms in the integral and have used $1/c^2 = \varepsilon_o\mu_o$. Now, we see that we can write this total energy (or classical Hamiltonian function) as the sum of separate energies (or classical Hamiltonian functions) for each separate mode; that is,

$$H = \sum_\lambda H_\lambda \tag{15.108}$$

where

$$H_\lambda = \frac{1}{2}\varepsilon_o D_\lambda^2\left(p_\lambda^2 + q_\lambda^2\right) \tag{15.109}$$

Now, we can cast this classical mode Hamiltonian into the correct form so that we get Hamilton's equations as in Eqs. (15.38) and (15.40); that is, in the present notation, we want to obtain

---

[9] They will be orthogonal, at least if the system is lossless, because they will be the (classical) eigenfunctions of a (classical) Hermitian operator in such an electromagnetic mode problem.

$$\frac{dp_\lambda}{dt} = -\frac{\partial H_\lambda}{\partial q_\lambda} \tag{15.110}$$

and

$$\frac{dq_\lambda}{dt} = \frac{\partial H_\lambda}{\partial p_\lambda} \tag{15.111}$$

If we choose[10]

$$D_\lambda = \sqrt{\frac{\omega_\lambda}{\varepsilon_o}} \tag{15.112}$$

then we now have

$$H_\lambda = \frac{\omega_\lambda}{2}\left(p_\lambda^2 + q_\lambda^2\right) \tag{15.113}$$

Explicitly considering Eq. (15.111), for example, we have, using Eq. (15.101)

$$\frac{\partial H_\lambda}{\partial p_\lambda} = \omega_\lambda p_\lambda = \frac{dq_\lambda}{dt} \tag{15.114}$$

as required.

Hence, for our multimode case, we have come to the point that for each mode, we have a classical Hamiltonian in exactly the same form as we had for the single-mode case in Eq. (15.55) and have established, for each mode, the quantities that correspond to the momentum and position of the classical Hamiltonian. We can proceed for each mode exactly as we did before, quantizing each mode with its own annihilation and creation operators. Formally, we postulate a momentum operator for each mode, as in Eq. (15.59)

$$\hat{p}_\lambda = -i\hbar \frac{d}{dq_\lambda} \tag{15.115}$$

and thereby propose the quantum mechanical Hamiltonian for the mode as in Eq. (15.60), from Eq. (15.113)

$$\hat{H}_\lambda = \frac{\omega_\lambda}{2}\left[-\hbar^2 \frac{d^2}{dq_\lambda^{\,2}} + q_\lambda^2\right] \tag{15.116}$$

We next rewrite this Hamiltonian as in Eqs. (15.63) and (15.65) as

$$\hat{H}_\lambda = \frac{\hbar\omega_\lambda}{2}\left[-\frac{d^2}{d\xi_\lambda^{\,2}} + \xi_\lambda^{\,2}\right] = \hbar\omega_\lambda\left(\hat{a}_\lambda^\dagger \hat{a}_\lambda + \frac{1}{2}\right) \tag{15.117}$$

where we have defined dimensionless units $\xi_\lambda = q_\lambda / \sqrt{\hbar}$ as in Eq. (15.61) and have creation and annihilation operators defined as in Eq. (15.9)

[10] The only substantial difference between this choice of D_λ and the choice of D in the single-mode case is that the present mode functions are presumed already normalized, whereas we included the normalization of the sine functions in D previously.

$$\hat{a}_\lambda^\dagger \equiv \frac{1}{\sqrt{2}}\left(-\frac{d}{d\xi_\lambda} + \xi_\lambda\right) \tag{15.118}$$

and in Eq. (15.10)

$$\hat{a}_\lambda \equiv \frac{1}{\sqrt{2}}\left(\frac{d}{d\xi_\lambda} + \xi_\lambda\right) \tag{15.119}$$

Now, the total Hamiltonian for the set of modes becomes

$$\hat{H} = \sum_\lambda \hbar\omega_\lambda \left(\hat{a}_\lambda^\dagger \hat{a}_\lambda + \frac{1}{2}\right) \tag{15.120}$$

Multimode photon states

When we are considering multiple different photon modes, it is convenient to write the state of such a system in what we can call the occupation number representation. In such a representation, for each basis state we merely write down a list of the number of photons in each particular mode. For example, the state with one photon in mode k, three in mode m, and none in any other mode could be written as

$$|0_a,\ldots,0_j,1_k,0_l,3_m,0_n,\ldots\rangle$$

where we imagine we have labeled the states progressively with the lowercased letters.

Just as before, the creation and annihilation operators have the properties now specific to the given mode, analogous to Eq. (15.69)

$$\hat{a}_\lambda|\ldots,n_\lambda,\ldots\rangle = \sqrt{n_\lambda}\,|\ldots,(n_\lambda - 1)_\lambda,\ldots\rangle \tag{15.121}$$

with

$$\hat{a}_\lambda|\ldots,0_\lambda,\ldots\rangle = 0 \tag{15.122}$$

Similarly, analogous to Eq. (15.70), we have

$$\hat{a}_\lambda^\dagger|\ldots,n_\lambda,\ldots\rangle = \sqrt{n_\lambda + 1}\,|\ldots,(n_\lambda + 1)_\lambda,\ldots\rangle \tag{15.123}$$

As in Eq. (15.66), the number operator for a specific mode in the multimode case is

$$\hat{N}_\lambda \equiv \hat{a}_\lambda^\dagger \hat{a}_\lambda \tag{15.124}$$

as follows from the definitions of the effects of the creation and annihilation operators on the multimode state.

Just as in the single-mode case, we can create such a state by progressively operating with the appropriate creation operators starting mathematically with the "zero" state or completely "empty" state, often written simply as $|0\rangle$. For our example state, we could write

$$|0_a,\ldots,0_j,1_k,0_l,3_m,0_n,\ldots\rangle = \frac{1}{\sqrt{1!3!}}\hat{a}_k^\dagger \hat{a}_m^\dagger \hat{a}_m^\dagger \hat{a}_m^\dagger |0\rangle \tag{15.125}$$

where we had to introduce the square-root factor with factorials so that we could keep the state normalized, compensating for the square-root factors introduced by the creation operators. In general, we can write a state with n_1 particles in mode 1, n_2 particles in mode 2, and so on, as

$$\left|n_1,n_2,\ldots,n_\lambda,\ldots\right\rangle=\frac{1}{\sqrt{n_1!n_2!\ldots n_\lambda!\ldots}}\left(\hat{a}_1^\dagger\right)^{n_1}\left(\hat{a}_2^\dagger\right)^{n_2}\ldots\left(\hat{a}_\lambda^\dagger\right)^{n_\lambda}\ldots\left|0\right\rangle \tag{15.126}$$

Commutation relations for boson operators

With bosons, it makes no difference to the final state in what mathematical order we create particles in a mode. The result of a different order of creation could be viewed as permuting the particles (the photons) among the single-particle states (here, the modes), but any permutation of bosons among the single-particle states makes no difference to the resulting multiboson state.[11] Hence, the creation operators for different modes commute with one another.

Consequently, we can state for operators operating on any state, $\hat{a}_j^\dagger\hat{a}_k^\dagger=\hat{a}_k^\dagger\hat{a}_j^\dagger$ or

$$\hat{a}_j^\dagger\hat{a}_k^\dagger-\hat{a}_k^\dagger\hat{a}_j^\dagger=0 \tag{15.127}$$

where we have written this in the form of a commutation relation. A similar argument can be made for annihilation operators that it does not matter in what order we destroy particles, so we similarly have

$$\hat{a}_j\hat{a}_k-\hat{a}_k\hat{a}_j=0 \tag{15.128}$$

For the case of mixtures of annihilation and creation operators, if we are annihilating a boson in one mode and creating one in another, it does not matter what mathematical order we do that either. Only if we are creating and annihilating in the same mode does it matter what order we do this, with a commutation relation we have previously deduced (Eq. (15.68)). Hence, in general, we can write

$$\hat{a}_j\hat{a}_k^\dagger-\hat{a}_k^\dagger\hat{a}_j=\delta_{jk} \tag{15.129}$$

This completes the commutation relations we need for the boson operators[12] and the relations we need for working with photons themselves in a quantum mechanical manner.

Multimode field operators

It is now straightforward to construct the full multimode electric and magnetic field operators. Working from the classical definition of the multimode electric field, Eq. (15.105), as an expansion in classical field modes, we substitute the operator $\hat{p}_\lambda$ for the quantity $p_\lambda(t)$ in each mode, using also the value of D_λ from Eq. (15.112), obtaining

$$\hat{\mathbf{E}}=\sum_\lambda -\hat{p}_\lambda(t)\sqrt{\frac{\omega_\lambda}{\varepsilon_o}}\mathbf{u}_\lambda(\mathbf{r}) \tag{15.130}$$

[11] The same is not true for fermions, where interchanging two particles changes the sign of the state, which leads to quite different commutation behavior for fermions.

[12] We can now see a reason for the choice of $A_n=\sqrt{n}$ in Eq. (15.28). We had the freedom to choose it to have any complex phase at that point, but to satisfy the boson requirement of not changing the sign of the resulting wavefunction if we change the order of creating two bosons (i.e., the operator pairs $\hat{a}_j^\dagger\hat{a}_k^\dagger$ and $\hat{a}_k^\dagger\hat{a}_j^\dagger$ should give the same sign of result) we do at least need the complex phase factor to be the same for all boson creation operators and the simplest choice is +1. As we will see in the next chapter, for fermions the sign is not necessarily the same for each possible creation operator acting on given states.

Noting that, from Eqs. (15.79) and (15.80)

$$\hat{p}_\lambda = i\sqrt{\frac{\hbar}{2}}\left(\hat{a}_\lambda^\dagger - \hat{a}_\lambda\right) \quad (15.131)$$

we therefore have

$$\hat{\mathbf{E}}(\mathbf{r},t) = i\sum_\lambda \left(\hat{a}_\lambda - \hat{a}_\lambda^\dagger\right)\sqrt{\frac{\hbar\omega_\lambda}{2\varepsilon_o}}\mathbf{u}_\lambda(\mathbf{r}) \quad (15.132)$$

By a similar argument, we can start with the classical expression for a multimode magnetic field, Eq. (15.106), substituting the operator $\hat{q}_\lambda$ for the quantity $q_\lambda(t)$ in each mode, to obtain

$$\mathbf{B}(\mathbf{r},t) = \sum_\lambda \hat{q}_\lambda \frac{D_\lambda}{c}\mathbf{v}_\lambda(\mathbf{r}) \quad (15.133)$$

Using Eqs. (15.61) and (15.78), we can write

$$\hat{q}_\lambda \equiv \sqrt{\frac{\hbar}{2}}\left(\hat{a}_\lambda + \hat{a}_\lambda^\dagger\right) \quad (15.134)$$

so we obtain

$$\hat{\mathbf{B}}(\mathbf{r},t) = \sum_\lambda \left(\hat{a}_\lambda + \hat{a}_\lambda^\dagger\right)\sqrt{\frac{\hbar\omega_\lambda\mu_o}{2}}\mathbf{v}_\lambda(\mathbf{r}) \quad (15.135)$$

Problem

15.7.1 Consider a set of modes of the electromagnetic field in which the electric field is polarized along the x direction and the magnetic field is polarized in the y direction. Restricting consideration only to those modes, find the simplest expression you can for the commutation relation $\left[\hat{E}_x, \hat{B}_y\right]$ for this multimode field.

15.8 Vibrational modes

At the start of this chapter, we dealt with an abstract harmonic oscillator. There are many systems that have such behavior and we analyzed in detail the case of electromagnetic modes. Any mechanical vibrating mode can also be analyzed in this way. Such modes occur, in particular, in molecules and in crystalline solids. In a classical view, we can think of systems of atoms as being masses connected by springs. In principle, there is a spring connecting each mass to each other mass. Such a system is complicated but for any finite number N of such masses we can write down $3N$ coupled differential equations (i.e., one for each mass and each of the three spatial coordinate directions) in which the forces from the springs connecting to each other mass act on a given mass to accelerate it according to Newton's second law. The standard method of approaching such problems is to look for solutions in which every mass in the crystal is oscillating at the same frequency. This leads to a matrix equation that can be solved for the eigenvectors, with the frequency (or its square) essentially as the eigenvalues. The resulting eigenvectors correspond to the modes of the system and are known as the normal modes. The elements in the eigenvectors tell the relative amplitude and direction of the oscillation of each individual mass in that mode.

If we change to these eigenvectors as the mathematical basis, we again obtain a set of uncoupled harmonic oscillator equations. The overall amplitude of the mode's displacement from its equilibrium position behaves like the position coordinate of a harmonic oscillator, and a corresponding coordinate based on the time rate of change of the position coordinate and a mass parameter (which is a properly weighted combination of the individual masses) serves as the momentum coordinate. We then rigorously obtain equations for each mode that can be quantized using the harmonic oscillator model. Each mode then has its own creation and annihilation operators.

We do not formally go through this analysis here for such vibrations, but it is straightforward and leads to a formalism that, when expressed in terms of boson creation and annihilation operators, is identical to that from Eq. (15.108) through Eq. (15.129) of the multimode photon case. Instead of photon modes, we have the normal modes of vibration of the system. We can also think of quasi-particles occupying these modes just as we think of photons occupying the photon modes. This approach is taken extensively in solid-state physics for crystalline materials, with the resulting particles being called *phonons*.

15.9 Summary of concepts

Raising and lowering operators for a harmonic oscillator

For the nth state $|\psi_n\rangle$ of a harmonic oscillator or an equivalent physical system, the lowering (or annihilation) operator $\hat{a}$ and the raising (or creation) operator $\hat{a}^\dagger$ have the following properties:

$$\hat{a}|\psi_n\rangle = \sqrt{n}|\psi_{n-1}\rangle \tag{15.29}$$

$$\hat{a}^\dagger|\psi_n\rangle = \sqrt{n+1}|\psi_{n+1}\rangle \tag{15.30}$$

Note specifically from Eq. (15.29) that

$$\hat{a}|\psi_0\rangle = 0 \tag{15.31}$$

where here "0" is the zero vector in the Hilbert space. Also

$$\left(\hat{a}^\dagger\right)^n|\psi_0\rangle = \sqrt{n!}|\psi_n\rangle \tag{15.34}$$

or, equivalently

$$|\psi_n\rangle = \frac{1}{\sqrt{n!}}\left(\hat{a}^\dagger\right)^n|\psi_0\rangle \tag{15.35}$$

These operators have a commutator

$$\left[\hat{a},\hat{a}^\dagger\right] = \hat{a}\hat{a}^\dagger - \hat{a}^\dagger\hat{a} = 1 \tag{15.17}$$

which is the main algebraic relation used to simplify expressions in such operators.

Number operator

For the nth state $|\psi_n\rangle$ of a harmonic oscillator or an equivalent physical system, the number operator $\hat{N}$ is defined as

$$\hat{N} \equiv \hat{a}^\dagger\hat{a} \tag{15.15}$$

with the eigenequation

$$\hat{N}|\psi_n\rangle = n|\psi_n\rangle \tag{15.16}$$

Harmonic oscillator Hamiltonian

In terms of the raising (creation) and lowering (annihilation) operators and the number operator, the harmonic oscillator Hamiltonian can be written

$$\hat{H} \equiv \hbar\omega\left(\hat{a}^\dagger\hat{a} + \frac{1}{2}\right) \equiv \hbar\omega\left(\hat{N} + \frac{1}{2}\right) \quad \text{after Eq. (15.13) and using Eq. (15.15)}$$

Classical Hamiltonian and Hamilton's equations

For some problem of interest, if we can write down a function H of two variables – an effective momentum p and an effective position q – where that function H represents the energy of the system and where H, p, and q obey Hamilton's equations

$$\frac{dp}{dt} = -\frac{\partial H}{\partial q} \tag{15.38}$$

and

$$\frac{dq}{dt} = \frac{\partial H}{\partial p} \tag{15.40}$$

then we can call H the classical Hamiltonian for the problem.

Quantization using effective momentum and position

For some problem, if we can write down the classical Hamiltonian in terms of p and q, then we can postulate a quantum mechanical Hamiltonian for the problem by substituting an operator $-i\hbar\, d/dq$ for the variable p in the expression for the classical Hamiltonian.

Quantization of the electromagnetic field

For each mode of the electromagnetic field, we can show that it has a classical Hamiltonian that can be written in the form

$$H = \frac{\omega}{2}\left(p^2 + q^2\right) \tag{15.55}$$

where ω is the angular frequency of oscillation of the electromagnetic field in this mode and H, p, and q obey Hamilton's equations. For each mode we can, therefore, postulate the above procedure to quantize it and, hence, obtain quantum mechanical creation (raising) and annihilation operators, a number operator, and a Hamiltonian exactly analogous to those of the simple harmonic oscillator. Now we think of raising the mode from one state to the next one as creating a photon in the mode and, oppositely, for lowering as annihilating a photon in the mode.

Because we may be discussing many different modes, we can introduce a subscript λ on all of the operators to label which specific mode we are considering at a given time. Otherwise, we have operator relations identical to those or a simple harmonic oscillator.

Nature of quantum mechanical states of a mode

The "eigenfunctions" of the harmonic oscillator equations for an electromagnetic mode are functions not of position but of the amplitude of oscillation of the magnetic field. Just as the modulus squared of the wavefunction for a simple mechanical harmonic oscillator gave the probability of finding the mass at some particular position, now the modulus squared of the wavefunction gives the relative probability of finding the electromagnetic field oscillating with a specific amplitude of magnetic field.

Because we are not usually interested in this underlying "coordinate" of magnetic field amplitude, we typically drop explicit mention of the wavefunction and use a notation for the eigenstates of a given mode λ of $|n_\lambda\rangle \equiv |\psi_{\lambda n}\rangle$. Such a state with a specific number of photons in a mode (or specific numbers of photons in each mode) is called a number state or Fock state.

Note that just as the position of the mass in a quantum mechanical simple harmonic oscillator is not definite even in the lowest state of the oscillator (i.e., it has some spatial wavefunction), there is a finite possibility of measuring a finite amplitude for the magnetic field even if there are no photons in the mode. We will measure a complete (Gaussian) range of resulting amplitudes for the magnetic field amplitude even with no photons in the mode and such fluctuations are called vacuum fluctuations.

In quantum mechanics, the electric and magnetic fields in nearly all quantum mechanical states do not have definite values, just as the position and momentum do not have definite values in most quantum mechanical states of a particle with mass.

Electric and magnetic field operators

We can define field operators for a simple z-polarized standing wave mode in the x direction, in a box of length L as

$$\hat{E}_{\lambda z} = i\left(\hat{a}_\lambda^\dagger - \hat{a}_\lambda\right)\sqrt{\frac{\hbar\omega_\lambda}{\varepsilon_o L}}\sin kx \tag{15.81}$$

$$\hat{B}_{\lambda y} = \left(\hat{a}_\lambda^\dagger + \hat{a}_\lambda\right)\sqrt{\frac{\mu_o\hbar\omega_\lambda}{L}}\cos kx \tag{15.82}$$

The expected values of the electric or magnetic field can be evaluated by taking the expected values of these operators in any given quantum mechanical state. The electric and magnetic fields in any given mode obey an uncertainty principle, which becomes

$$\Delta E_{\lambda z}\Delta B_{\lambda y} \geq \frac{\hbar\omega_\lambda}{L}\sqrt{\frac{\mu_o}{\varepsilon_o}}\sin kx\cos kx \tag{15.88}$$

for this mode.

These operators are examples of field operators in quantum mechanics. This quantization of the electromagnetic field is an example of a quantum-field theory in quantum mechanics, in which particles (here, photons) emerge from quantization of a field.

The expectation value of the magnetic (or electric) field in a number state is zero, just as the expectation value of the position of a mass in a simple mechanical harmonic oscillator state is zero (or the center of the potential).

Coherent state

The coherent state is the state that corresponds most closely with the classical idea of an oscillating electromagnetic field. For a given mode λ, a coherent state is

$$\left|\Psi_{\lambda\bar{n}}\right\rangle = \sum_{n_\lambda=0}^{\infty} c_{\lambda\bar{n}n} \exp\left[-i\left(n_\lambda + \frac{1}{2}\right)\omega_\lambda t\right]\left|n_\lambda\right\rangle \text{ (15.94), with } c_{\lambda\bar{n}n} = \sqrt{\frac{\bar{n}^{n_\lambda}\exp(-\bar{n})}{n_\lambda!}} \quad (15.95)$$

where $\bar{n}$ expresses the average number of photons in the mode. In a coherent state, the number of photons is not determined and will be measured to have a Poisson distribution.

Multimode fields

For a multimode electromagnetic field, the Hamiltonian is the sum of the individual Hamiltonians for each mode; that is, with λ indexing the different modes

$$\hat{H} = \sum_\lambda \hbar\omega_\lambda \left(\hat{a}_\lambda^\dagger \hat{a}_\lambda + \frac{1}{2}\right) \quad (15.120)$$

and the multimode field operators are similarly the sums of the field operators for the individual modes; that is,

$$\hat{\mathbf{E}}(\mathbf{r},t) = i\sum_\lambda \left(\hat{a}_\lambda - \hat{a}_\lambda^\dagger\right)\sqrt{\frac{\hbar\omega_\lambda}{2\varepsilon_o}}\mathbf{u}_\lambda(\mathbf{r}) \quad (15.132)$$

$$\hat{\mathbf{B}}(\mathbf{r},t) = \sum_\lambda \left(\hat{a}_\lambda + \hat{a}_\lambda^\dagger\right)\sqrt{\frac{\hbar\omega_\lambda\mu_o}{2}}\mathbf{v}_\lambda(\mathbf{r}) \quad (15.135)$$

Commutation relation for boson creation and annihilation operators

For creation and annihilation operators associated with specific modes j and k

$$\hat{a}_j\hat{a}_k^\dagger - \hat{a}_k^\dagger\hat{a}_j = \delta_{jk} \quad (15.129)$$

Chapter 16

Fermion operators

Prerequisites: Chapters 2–5, 9, 10, 12, 13, and 15.

Thus far, we have worked with a quantum mechanical wave for electrons and similar particles, we have worked with classical waves for electric and magnetic fields, and we have introduced a quantum mechanical way of looking at electric and magnetic fields through the use of boson annihilation and creation operators. The use of these operators for boson modes led to the quantization of the electromagnetic field into photons. These operators naturally behaved in such a way as to give the photons the properties we expect of bosons, allowing any zero or positive integer number of photons in a mode. We will find that we can also introduce annihilation and creation operators for fermions and it is the purpose of this chapter to make this introduction. These operators lead to the natural quantization of the number of fermions possible in a fermion "mode" or single-particle state, limiting us to zero or one as required. Analogously to the boson operator description of the electromagnetic wave with field operators, we can also describe the quantum mechanical wave associated with electrons and similar particles in terms of the fermion operators.[1]

Just as the quantization of the electromagnetic wave led to the appearance in our quantum mechanics of boson particles, we can view this introduction of the fermion annihilation and creation operators as leading to the appearance in our quantum mechanics of fermion particles. We started the quantum mechanics of electrons and similar particles knowing them as particles and later discovered their wave properties. With electromagnetism, we came out of the nineteenth century believing electromagnetism to be a wave phenomenon[2] and later introduced the boson quantum theory that led to photon particles. The introduction of the fermion annihilation and creation operators can be viewed as completing a picture here because it now introduces a formalism that explicitly gives us the particles that have the fermion's characteristics (especially, Pauli exclusion).

It should be pointed out that unlike the use of boson creation and annihilation operators to quantize the classical electromagnetic field, the postulation of fermion annihilation and

---

[1] This new description of the quantum mechanical wave in terms of field operators is sometimes called *second quantization* because it is rather like taking a quantum mechanical wave and quantizing it, just as we did for the electromagnetic wave.

[2] Newton had a corpuscular theory of light, but this was largely pushed into the background following the success of wave theory in explaining diffraction phenomena of light in the early nineteenth century, phenomena that had no ready explanation with a classical particle theory of light. The failure of the classical theory of radiation to agree with thermodynamics was what led Planck to postulate quantization of the light field and Einstein to introduce the photon. The quantized wave field model introduced in the previous Chapter, of course, has both wave and particle characters, as required.

creation operators does not fundamentally add anything to the quantum mechanics we already have for fermions (at least, at the level to which we consider the quantum mechanics of electrons and similar fermion particles here). It does, however, give us a very convenient way of writing the quantum mechanics and also gives us a formalism that is of the same type as the boson creation and annihilation operators we created previously. This approach naturally includes the Pauli exclusion principle so we do not have to add it in some ad hoc fashion. Once we work in systems with many fermions, the use of fermion creation and annihilation operators is almost essential from a practical point of view so as to keep track of the fermion character of systems with many particles. Even when we are only considering a single fermion, the creation and annihilation operators give a particularly simple notation that we can use to describe other operators, such as the Hamiltonian. This kind of description is particularly useful for processes that involve collisions of fermions with one another (e.g., as in electron-electron scattering) or the interaction of fermions and bosons (e.g., as in optical absorption and emission or electron-phonon scattering).

16.1 Postulation of fermion annihilation and creation operators

The approach we take here is simply to postulate annihilation and creation operators for fermions, giving them the required properties. Later, we use them to rewrite operators involving interactions with fermions. The key property these operators require, in comparison to the boson operators, is that they correctly change the sign of the wavefunction upon exchange of particles. This leads us to a formalism similar in character to the boson operators, though we will find so-called anticommutation relations instead of the commutation relations of the boson operators. It should be emphasized that a principal reason for introducing such fermion operators is so that we never again have to worry about the details of the antisymmetry with respect to exchange of fermions; the anticommutation relations we develop take care of these details quite conveniently.

Description and ordering of multiple-fermion states

The reader will remember from our previous discussion that we can write a basis state for multiple identical fermions as a sum over all possible permutations of particles among the occupied states or, equivalently, a Slater determinant, in the forms

$$\begin{aligned}\left|\psi_{N;\,a,b,\ldots n}\right\rangle &= \frac{1}{\sqrt{N!}}\sum_{\hat{P}=1}^{N!}\pm\hat{P}\left|\left|1,a\right\rangle\left|2,b\right\rangle\left|3,c\right\rangle\ldots\left|N,n\right\rangle\right\rangle \\ &\equiv \frac{1}{\sqrt{N!}}\begin{vmatrix}\left|1,a\right\rangle & \left|2,a\right\rangle & \cdots & \left|N,a\right\rangle \\ \left|1,b\right\rangle & \left|2,b\right\rangle & \cdots & \left|N,b\right\rangle \\ \vdots & \vdots & \ddots & \vdots \\ \left|1,n\right\rangle & \left|2,n\right\rangle & \cdots & \left|N,n\right\rangle\end{vmatrix}\end{aligned} \tag{16.1}$$

Here, there are N identical fermions and they occupy single-particle basis states $a, b, \ldots, n$. Note that the single-particle basis states are specific individual possible states that a fermion can occupy and, in the notation shown here, each has a lowercased letter associated with it. For example, each possible electron state in a potential well or atom would correspond to a different basis state here. If we had a system with multiple potential wells or atoms, each possible single electron state associated with each potential well or atom would have its own unique label (here, a lowercased letter, though obviously we could run out of those quite quickly!).

Though this notation might seem to imply that each of the possible states is occupied, that is not, in general, the case. In fact, very few of the possible single-particle states are typically occupied in any given multiple-fermion basis state. We might, for example, have three electrons in a system with five potential wells and be considering a (multiple-particle) basis state in which there is one electron in the ground state of well 1, one in the second state of well 3, and one in the seventeenth state of well 4. All of the other single-particle basis states would be unoccupied in this multiple-particle basis state. The formalism also allows as a perfectly viable mathematical basis state that we might have two electrons in one well – one on the lowest state and one on the sixth state, for example. Such a state is not forbidden by Pauli exclusion and, hence, is a viable two-particle basis state, though it is not necessarily an eigenstate of the Hamiltonian (the first electron would give rise to a repulsive potential for the second electron and so the second electron would not see a simple square potential anymore).

To draw the connection with the boson case, each of the single-particle basis states can be considered as a mode (single-particle state) of the fermion field, just as the boson basis states were modes of the electromagnetic field. The boson modes could have any integer number of particles in them, though the fermion modes can only have zero or one. Just as the boson annihilation and creation operators for a given mode allowed any positive or zero integer number of bosons in the mode, so also the fermion annihilation and creation operators allow only zero or one fermion in the mode if we set them up correctly. Now that we have drawn the analogy with boson modes, rather than using the term mode for fermions we now mostly revert to use the term *single-particle state*, though the reader should understand this is merely a matter of taste and usage.

In the construction of the determinants for the multiple-fermion basis functions, we need to define, at least for the temporary purposes of our argument, one standard order of labeling of the single-particle basis states. For example, if we had a system with five potential wells, we might label sequentially all of the states in well 1, then next all of the states in well 2, and so on. We could choose some other labeling sequence, labeling all of the first states in wells 1 through 5, then all of the second states in wells 1 through 5, and so on, or we could even choose some more complicated labeling sequence. It does not matter which sequence we choose, but we have to have one standard labeling sequence to which we can refer in all of our mathematical operations. For simplicity here, we presume that we can label the single-particle basis states (or fermion modes) using the lowercased letters, and our standard order is the one in which those lowercased letters are in alphabetical order.

We might, for example, have a basis state corresponding to three identical fermions, one in state b, one in state k, and one in state m. In standard order, we would write that state as

$$\left|\psi_{3;\,b,k,m}\right\rangle = \frac{1}{\sqrt{3!}}\begin{vmatrix} |1,b\rangle & |2,b\rangle & |3,b\rangle \\ |1,k\rangle & |2,k\rangle & |3,k\rangle \\ |1,m\rangle & |2,m\rangle & |3,m\rangle \end{vmatrix} \equiv \left|0_a, 1_b, 0_c, \ldots, 1_k, 0_l, \ldots, 1_m, 0_n \ldots\right\rangle \tag{16.2}$$

Here, we have also introduced another notation to describe this basis state, which we can describe as the *occupation number notation*. This notation is similar to the boson occupation number notation introduced previously. In this notation, 0_a in the ket means that the single-particle fermion state (or fermion mode) a is empty, 1_b means state b is occupied, and so on. Because this is a fermion state, the determinant combination of the different fermions to the occupied states is, of course, understood.

We could also write a state that was not in standard order for the rows; for example,

$$\left|\psi_{3;\,k,b,m}\right\rangle = \frac{1}{\sqrt{3!}}\begin{vmatrix} |1,k\rangle & |2,k\rangle & |3,k\rangle \\ |1,b\rangle & |2,b\rangle & |3,b\rangle \\ |1,m\rangle & |2,m\rangle & |3,m\rangle \end{vmatrix} \tag{16.3}$$

To get that state into standard order for the rows, we would have to swap the first and second rows in the determinant. We know that if we swap two adjacent rows in a determinant we have to multiply the determinant by -1; so, swapping the top two rows, we have

$$\begin{aligned} \left|\psi_{3;\,k,b,m}\right\rangle &= -\frac{1}{\sqrt{3!}}\begin{vmatrix} |1,b\rangle & |2,b\rangle & |3,b\rangle \\ |1,k\rangle & |2,k\rangle & |3,k\rangle \\ |1,m\rangle & |2,m\rangle & |3,m\rangle \end{vmatrix} \\ &= -\left|0_a, 1_b, 0_c, \ldots, 1_k, 0_l, \ldots, 1_m, 0_n \ldots\right\rangle = -\left|\psi_{3;\,b,k,m}\right\rangle \end{aligned} \tag{16.4}$$

Fermion creation operators

Suppose now we postulate a fermion creation operator for fermion mode or single-particle basis state k and write it as $\hat{b}_k^\dagger$. Then, we need it to have the property that it takes any state in which single-particle basis state k is empty and turns it into one in which this state k is occupied. We also need it to have a very particular behavior with regard to the sign of the wavefunction it creates, so that operations that are equivalent to swapping two particles change the sign of the wavefunction. This sign behavior means we have to construct the operator with some care over signs, though in the end this is quite straightforward. We find, incidentally, that these sign requirements lead to a very particular kind of commutation relation for the fermion annihilation and creation operators (i.e., an anticommutation relation). In practice, these anticommutation relations are very useful algebraically. Here, we progressively build up the properties of these operators, starting with the creation operator.

For the sake of definiteness of illustration, let us suppose that we start with the state in which single-particle states b and m are occupied but state k and all other states are not. In the permutation notation, we can, therefore, propose that $\hat{b}_k^\dagger$ has the following effect on that state

$$\hat{b}_k^\dagger \frac{1}{\sqrt{2!}}\sum_{\hat{P}=1}^{2!} \pm\hat{P}\left|\,|1,b\rangle|2,m\rangle\right\rangle = \frac{1}{\sqrt{3!}}\sum_{\hat{P}=1}^{3!} \pm\hat{P}\left|\,|1,b\rangle|2,m\rangle|3,k\rangle\right\rangle \tag{16.5}$$

In other words, the action of $\hat{b}_k^\dagger$ is to add a third particle into the system and we propose for definiteness that it adds it to the end of the list in the permutation notation.[3]

[3] It actually does not really matter where we propose that we add the particle in the list, though we have to choose something consistent. We could, e.g., have chosen to add it to the front of the list as particle 1, moving all the other states to one higher particle number. In general, adding the particle at a different point in the list is equivalent to redefining our arbitrary standard order and adding the particle at the end again. The reason why it does not matter where we add the particle in the list is ultimately because we always end up using fermion operators in annihilation–creation pairs in working out any observable and the differences in sign that would result from adding the particle to the beginning of the list as opposed to the end cancel out when we use such a pair of operators as long as we are consistent. The reader might be worried about this apparent arbitrariness, but note that the creation operator turns out not to be Hermitian and, hence, expectation values of it do not correspond to physically observable quantities. In quantum mechanics, we have already encountered an unobservable quantity, the wavefunction, that has a degree of

Now, let us examine these kinds of processes in determinant notation. From this point on, for simplicity we ignore the factorial normalization factors that should precede the determinants; we concentrate on the signs and orderings only.

Adding to the end of the list in the permutation notation is equivalent to adding a row to the bottom of the determinant (and a column to the right); that is, in determinant notation, Eq. (16.5) is

$$\hat{b}_k^\dagger \begin{vmatrix} |1,b\rangle & |2,b\rangle \\ |1,m\rangle & |2,m\rangle \end{vmatrix} = \begin{vmatrix} |1,b\rangle & |2,b\rangle & |3,b\rangle \\ |1,m\rangle & |2,m\rangle & |3,m\rangle \\ |1,k\rangle & |2,k\rangle & |3,k\rangle \end{vmatrix} \tag{16.6}$$

(To go from permutation notation to determinant notation, note that the sequence in the permutation is the same as that down the leading diagonal of the determinant.)

For the particular example case we have chosen, we find now that the determinant is not written in standard order. (In fact, the only case in which this determinant would be in standard order would be if the state in which we added a particle was further along the list in standard order than all of the other occupied states.) To get this particular determinant into standard order, we need to swap the bottom two rows and in performing this one swap, we must, therefore, multiply the determinant by -1. Hence, in this case

$$\hat{b}_k^\dagger \begin{vmatrix} |1,b\rangle & |2,b\rangle \\ |1,m\rangle & |2,m\rangle \end{vmatrix} = -\begin{vmatrix} |1,b\rangle & |2,b\rangle & |3,b\rangle \\ |1,k\rangle & |2,k\rangle & |3,k\rangle \\ |1,m\rangle & |2,m\rangle & |3,m\rangle \end{vmatrix} \tag{16.7}$$

Suppose now that we add another particle, this time in state j, using the operator $\hat{b}_j^\dagger$. Then, we have

$$\begin{aligned} \hat{b}_j^\dagger \hat{b}_k^\dagger \begin{vmatrix} |1,b\rangle & |2,b\rangle \\ |1,m\rangle & |2,m\rangle \end{vmatrix} &= -\hat{b}_j^\dagger \begin{vmatrix} |1,b\rangle & |2,b\rangle & |3,b\rangle \\ |1,k\rangle & |2,k\rangle & |3,k\rangle \\ |1,m\rangle & |2,m\rangle & |3,m\rangle \end{vmatrix} \\ &= -\begin{vmatrix} |1,b\rangle & |2,b\rangle & |3,b\rangle & |4,b\rangle \\ |1,k\rangle & |2,k\rangle & |3,k\rangle & |4,k\rangle \\ |1,m\rangle & |2,m\rangle & |3,m\rangle & |4,m\rangle \\ |1,j\rangle & |2,j\rangle & |3,j\rangle & |4,j\rangle \end{vmatrix} \end{aligned} \tag{16.8}$$

To get to standard order, we have to swap the bottom j row with the adjacent m row, multiplying by -1, and then swap the j row, now second from the bottom, with the adjacent k row, multiplying again by -1; that is,

arbitrariness to it (e.g., its absolute phase). Again, in working out any observable quantity with the wavefunction, the arbitrariness cancels and disappears. When we do work out observable quantities using creation and annihilation operators, this arbitrariness similarly becomes unimportant.

$$\begin{vmatrix} |1,b\rangle & |2,b\rangle & |3,b\rangle & |4,b\rangle \\ |1,k\rangle & |2,k\rangle & |3,k\rangle & |4,k\rangle \\ |1,m\rangle & |2,m\rangle & |3,m\rangle & |4,m\rangle \\ |1,j\rangle & |2,j\rangle & |3,j\rangle & |4,j\rangle \end{vmatrix} \to -1 \begin{vmatrix} |1,b\rangle & |2,b\rangle & |3,b\rangle & |4,b\rangle \\ |1,k\rangle & |2,k\rangle & |3,k\rangle & |4,k\rangle \\ |1,j\rangle & |2,j\rangle & |3,j\rangle & |4,j\rangle \\ |1,m\rangle & |2,m\rangle & |3,m\rangle & |4,m\rangle \end{vmatrix}$$
$$\to (-1)^2 \begin{vmatrix} |1,b\rangle & |2,b\rangle & |3,b\rangle & |4,b\rangle \\ |1,j\rangle & |2,j\rangle & |3,j\rangle & |4,j\rangle \\ |1,k\rangle & |2,k\rangle & |3,k\rangle & |4,k\rangle \\ |1,m\rangle & |2,m\rangle & |3,m\rangle & |4,m\rangle \end{vmatrix} \tag{16.9}$$

and so

$$\hat{b}_j^\dagger \hat{b}_k^\dagger \begin{vmatrix} |1,b\rangle & |2,b\rangle \\ |1,m\rangle & |2,m\rangle \end{vmatrix} = - \begin{vmatrix} |1,b\rangle & |2,b\rangle & |3,b\rangle & |4,b\rangle \\ |1,j\rangle & |2,j\rangle & |3,j\rangle & |4,j\rangle \\ |1,k\rangle & |2,k\rangle & |3,k\rangle & |4,k\rangle \\ |1,m\rangle & |2,m\rangle & |3,m\rangle & |4,m\rangle \end{vmatrix} \tag{16.10}$$

Now, suppose that we do this two-particle creation operation in the opposite order. First, just as for Eq. (16.7)

$$\hat{b}_j^\dagger \begin{vmatrix} |1,b\rangle & |2,b\rangle \\ |1,m\rangle & |2,m\rangle \end{vmatrix} = - \begin{vmatrix} |1,b\rangle & |2,b\rangle & |3,b\rangle \\ |1,j\rangle & |2,j\rangle & |3,j\rangle \\ |1,m\rangle & |2,m\rangle & |3,m\rangle \end{vmatrix} \tag{16.11}$$

Next, if we operate with $\hat{b}_k^\dagger$, we obtain

$$\hat{b}_k^\dagger \hat{b}_j^\dagger \begin{vmatrix} |1,b\rangle & |2,b\rangle \\ |1,m\rangle & |2,m\rangle \end{vmatrix} = - \begin{vmatrix} |1,b\rangle & |2,b\rangle & |3,b\rangle & |4,b\rangle \\ |1,j\rangle & |2,j\rangle & |3,j\rangle & |4,j\rangle \\ |1,m\rangle & |2,m\rangle & |3,m\rangle & |4,m\rangle \\ |1,k\rangle & |2,k\rangle & |3,k\rangle & |4,k\rangle \end{vmatrix} \tag{16.12}$$

Now, however, we only have to swap adjacent rows once, not twice as in Eq. (16.9), to get the determinant into standard order; that is, swapping the bottom two rows and multiplying by -1, we obtain

$$\hat{b}_k^\dagger \hat{b}_j^\dagger \begin{vmatrix} |1,b\rangle & |2,b\rangle \\ |1,m\rangle & |2,m\rangle \end{vmatrix} = \begin{vmatrix} |1,b\rangle & |2,b\rangle & |3,b\rangle & |4,b\rangle \\ |1,j\rangle & |2,j\rangle & |3,j\rangle & |4,j\rangle \\ |1,k\rangle & |2,k\rangle & |3,k\rangle & |4,k\rangle \\ |1,m\rangle & |2,m\rangle & |3,m\rangle & |4,m\rangle \end{vmatrix} \tag{16.13}$$

This result is -1 times the result from that of the operators in the order $\hat{b}_j^\dagger \hat{b}_k^\dagger$ in Eq. (16.10). Note that this behavior corresponds exactly to what we want for fermion creation operators. Swapping two particles must give a change in sign for the overall fermion wavefunction. Creating two particles in one order rather than the other must give a result that is equivalent to swapping the two particles in the resulting state.

This behavior of obtaining opposite signs for the result if the particles are created in opposite order is a general one and it does not matter what the initial state is or what specific states the particles are being created in (as long as the particles are being created in states that are different and that are initially empty). For example, we would get exactly the same difference in sign in the result if we had considered the pairs of operators $\hat{b}_a^\dagger \hat{b}_k^\dagger$ and $\hat{b}_k^\dagger \hat{b}_a^\dagger$ or the pairs $\hat{b}_j^\dagger \hat{b}_n^\dagger$ and $\hat{b}_n^\dagger \hat{b}_j^\dagger$. The key point is that one of the pairs of operators always results in one more swap of adjacent rows than the other because it encounters one more row to be swapped. In the pair $\hat{b}_j^\dagger \hat{b}_k^\dagger$, the row added by the $\hat{b}_j^\dagger$ operator has to be swapped past the row corresponding to a particle in state k, whereas the row added by the $\hat{b}_k^\dagger$ operator in the pair $\hat{b}_k^\dagger \hat{b}_j^\dagger$ does not have to be swapped past the row added by the $\hat{b}_j^\dagger$. This asymmetry is because one of the two states in the pair has to be ahead of the other in the standard order.

Hence, we have the result, valid for the operators operating on any state in which single-particle states j and k are initially empty

$$\hat{b}_j^\dagger \hat{b}_k^\dagger + \hat{b}_k^\dagger \hat{b}_j^\dagger = 0 \tag{16.14}$$

In fact, this relation in Eq. (16.14) is universally true for any state. To see this, we note first that for any state in which state k is initially occupied, the fermion creation operator for that state must have the property that

$$\hat{b}_k^\dagger \left| \ldots, 1_k, \ldots \right\rangle = 0 \tag{16.15}$$

because we cannot create two fermions in one single-particle state. Hence, for any state for which the single-particle state k is occupied, trivially we have

$$\hat{b}_j^\dagger \hat{b}_k^\dagger \left| \ldots, 1_k, \ldots \right\rangle = 0 \text{ and } \hat{b}_k^\dagger \hat{b}_j^\dagger \left| \ldots, 1_k, \ldots \right\rangle = 0 \tag{16.16}$$

Hence, our relation Eq. (16.14) still works here because each individual term is zero. We get an exactly similar result if the initial state is such that the single-particle state j is occupied. We also trivially get the same result for any initial state if $j = k$ because we are trying to create at least two fermions in the single-particle state (three if it is already occupied), so we also get zero for both terms. Hence, we conclude that (16.14) is valid for any starting state.

A relation of the form of Eq. (16.14) is called an *anticommutation relation*. It is like a commutation relation between operators but with a plus sign in the middle rather than the minus sign of a commutation relation. A notation sometimes used for an anticommutator of two operators, taking the expression of Eq. (16.14) as an example, is

$$\hat{b}_j^\dagger \hat{b}_k^\dagger + \hat{b}_k^\dagger \hat{b}_j^\dagger \equiv \left\{ \hat{b}_j^\dagger, \hat{b}_k^\dagger \right\} \tag{16.17}$$

Here, we progressively develop a family of anticommutation relations for the fermion operators. They turn out to be the principal relations we use for simplifying expressions using fermion operators and they are often quite convenient and useful.

To proceed further, let us generalize and formalize the definition of the creation operator and the resulting signs. We see, with our choice that we add the particle in state k initially to the end of the list or, equivalently, to the bottom of the determinant and then swap it into place, that the number of swaps we have to perform is the number S_k of occupied states that are above the state k of interest in the standard order. With this definition, we can write formally[4]

[4] Note that S does depend on the occupation of the other single-particle states in the multiple-particle state in question. This may seem an unusual concept, though it is quite consistent with the annihilation

$$\hat{b}_k^\dagger |\ldots,0_k,\ldots\rangle = (-1)^{S_k} |\ldots,1_k,\ldots\rangle \tag{16.18}$$

Fermion annihilation operators

Now, we can proceed to define annihilation operators. From Eq. (16.18), we can see that

$$\langle\ldots,1_k,\ldots|\hat{b}_k^\dagger|\ldots,0_k,\ldots\rangle = (-1)^{S_k}\langle\ldots,1_k,\ldots|\ldots,1_k,\ldots\rangle = (-1)^{S_k} \tag{16.19}$$

Let us now take the complex conjugate or, actually, the Hermitian adjoint, of both sides of Eq. (16.19) and use the known algebra for Hermitian adjoints of products; that is,

$$\begin{aligned}\langle\ldots,1_k,\ldots|\hat{b}_k^\dagger|\ldots,0_k,\ldots\rangle^\dagger &= \left(\hat{b}_k^\dagger|\ldots,0_k,\ldots\rangle\right)^\dagger\left(\langle\ldots,1_k,\ldots|\right)^\dagger\\ &= \langle\ldots,0_k,\ldots|\hat{b}_k|\ldots,1_k,\ldots\rangle\\ &= (-1)^{S_k}\end{aligned} \tag{16.20}$$

(where we use the fact that the Hermitian adjoint of a Hermitian adjoint takes us back to where we started; i.e., $(\hat{b}_k^\dagger)^\dagger = \hat{b}_k$). From Eq. (16.20), we deduce, therefore, that

$$\hat{b}_k|\ldots,1_k,\ldots\rangle = (-1)^{S_k}|\ldots,0_k,\ldots\rangle \tag{16.21}$$

Hence, whereas $\hat{b}_k^\dagger$ creates a fermion in single-particle state k (provided that state was empty), $\hat{b}_k$ annihilates a fermion in single-particle state k provided that state was full and is called the *fermion annihilation operator* for state k.

We can think of the action of the annihilation operator on the Slater determinant as progressively swapping the row corresponding to state k in the determinant with the one below it until that row gets to the bottom of the determinant, in which case we remove it (and the last column) of the determinant, in an inverse fashion to the process with the creation operator discussed previously. By a similar set of arguments, we arrive at the anticommutation relation for the annihilation operator, valid for all states and for $j = k$

$$\hat{b}_j\hat{b}_k + \hat{b}_k\hat{b}_j = 0 \tag{16.22}$$

where we also used the relation, analogous to Eq. (16.15)

$$\hat{b}_k|\ldots,0_k,\ldots\rangle = 0 \tag{16.23}$$

which merely states that if the single-particle state k is empty to start with, we cannot annihilate another particle from that state.

Mixtures of creation and annihilation operators

The final set of properties that we require are those for mixtures of annihilation and creation operators. We can proceed in a similar fashion as before. Suppose that we are initially in the state with single-particle states b, j, and m occupied and we operate on this state first with the annihilation operator $\hat{b}_j$. Then, we have

operators being linear quantum mechanical operators. This interrelation with the occupation of other single-particle states is an intrinsic and classically bizarre property of fermion states.

$$\hat{b}_j \begin{vmatrix} |1,b\rangle & |2,b\rangle & |3,b\rangle \\ |1,j\rangle & |2,j\rangle & |3,j\rangle \\ |1,m\rangle & |2,m\rangle & |3,m\rangle \end{vmatrix} = -\begin{vmatrix} |1,b\rangle & |2,b\rangle \\ |1,m\rangle & |2,m\rangle \end{vmatrix} \tag{16.24}$$

where the minus sign arises because we had to swap the j and m rows to get the j row to the bottom. Now, we operate with $\hat{b}_k^\dagger$, obtaining

$$\begin{aligned} \hat{b}_k^\dagger \hat{b}_j \begin{vmatrix} |1,b\rangle & |2,b\rangle & |3,b\rangle \\ |1,j\rangle & |2,j\rangle & |3,j\rangle \\ |1,m\rangle & |2,m\rangle & |3,m\rangle \end{vmatrix} &= -\hat{b}_k^\dagger \begin{vmatrix} |1,b\rangle & |2,b\rangle \\ |1,m\rangle & |2,m\rangle \end{vmatrix} \\ &= \begin{vmatrix} |1,b\rangle & |2,b\rangle & |3,b\rangle \\ |1,k\rangle & |2,k\rangle & |3,k\rangle \\ |1,m\rangle & |2,m\rangle & |3,m\rangle \end{vmatrix} \end{aligned} \tag{16.25}$$

where the minus sign is cancelled because we had to swap the k row from the bottom with the m row. Next, let us consider applying these operators in the opposite order

$$\hat{b}_k^\dagger \begin{vmatrix} |1,b\rangle & |2,b\rangle & |3,b\rangle \\ |1,j\rangle & |2,j\rangle & |3,j\rangle \\ |1,m\rangle & |2,m\rangle & |3,m\rangle \end{vmatrix} = -\begin{vmatrix} |1,b\rangle & |2,b\rangle & |3,b\rangle & |4,b\rangle \\ |1,j\rangle & |2,j\rangle & |3,j\rangle & |4,j\rangle \\ |1,k\rangle & |2,k\rangle & |3,k\rangle & |4,k\rangle \\ |1,m\rangle & |2,m\rangle & |3,m\rangle & |4,m\rangle \end{vmatrix} \tag{16.26}$$

where the minus sign has arisen because we had to swap the k row from the bottom with the m row. Applying the $\hat{b}_j$ operator now gives

$$\begin{aligned} \hat{b}_j \hat{b}_k^\dagger \begin{vmatrix} |1,b\rangle & |2,b\rangle & |3,b\rangle \\ |1,j\rangle & |2,j\rangle & |3,j\rangle \\ |1,m\rangle & |2,m\rangle & |3,m\rangle \end{vmatrix} &= -\hat{b}_j \begin{vmatrix} |1,b\rangle & |2,b\rangle & |3,b\rangle & |4,b\rangle \\ |1,j\rangle & |2,j\rangle & |3,j\rangle & |4,j\rangle \\ |1,k\rangle & |2,k\rangle & |3,k\rangle & |4,k\rangle \\ |1,m\rangle & |2,m\rangle & |3,m\rangle & |4,m\rangle \end{vmatrix} \\ &= -\begin{vmatrix} |1,b\rangle & |2,b\rangle & |3,b\rangle \\ |1,k\rangle & |2,k\rangle & |3,k\rangle \\ |1,m\rangle & |2,m\rangle & |3,m\rangle \end{vmatrix} \end{aligned} \tag{16.27}$$

In operating with $\hat{b}_j$, two swaps are required because we have to swap past both the m and k rows. As before, we find an additional row swap required with one order of operators rather than the other (similar behavior would be found if we considered the pairs $\hat{b}_j^\dagger \hat{b}_k$ and $\hat{b}_k \hat{b}_j^\dagger$). The result Eq. (16.27) is minus the result Eq. (16.25). Hence, we see that at least when operating on states when single-particle state j is initially full and single-particle state k is initially empty

$$\hat{b}_j \hat{b}_k^\dagger + \hat{b}_k^\dagger \hat{b}_j = 0 \tag{16.28}$$

Again, if state j is initially empty, then both pairs of operators lead to a zero result and similarly if state k is initially full. Hence, as long as states j and k are different states, Eq. (16.28) is universally true.

The only special case we have to consider more carefully here is for $j = k$. Suppose we consider the case in which single-particle state k is initially full. Then, we have

$$\hat{b}_k\hat{b}_k^\dagger \begin{vmatrix} |1,b\rangle & |2,b\rangle & |3,b\rangle \\ |1,k\rangle & |2,k\rangle & |3,k\rangle \\ |1,m\rangle & |2,m\rangle & |3,m\rangle \end{vmatrix} = 0 \tag{16.29}$$

because $\hat{b}_k^\dagger$ operating on this state gives zero. For the other order of operators, we have

$$\hat{b}_k^\dagger\hat{b}_k \begin{vmatrix} |1,b\rangle & |2,b\rangle & |3,b\rangle \\ |1,k\rangle & |2,k\rangle & |3,k\rangle \\ |1,m\rangle & |2,m\rangle & |3,m\rangle \end{vmatrix} = -\hat{b}_k^\dagger \begin{vmatrix} |1,b\rangle & |2,b\rangle \\ |1,m\rangle & |1,m\rangle \end{vmatrix}$$
$$= \begin{vmatrix} |1,b\rangle & |2,b\rangle & |3,b\rangle \\ |1,k\rangle & |2,k\rangle & |3,k\rangle \\ |1,m\rangle & |2,m\rangle & |3,m\rangle \end{vmatrix} \tag{16.30}$$

It is left as an exercise for the reader to repeat this derivation for the situation in which state k is initially empty. In both cases, the result is the same; one or the other of the pairs returns the original state and the other pair returns zero. Hence, we can say that

$$\hat{b}_k\hat{b}_k^\dagger + \hat{b}_k^\dagger\hat{b}_k = 1 \tag{16.31}$$

Putting Eq. (16.31) together with Eq. (16.28), we can write the anticommutation relation

$$\hat{b}_j\hat{b}_k^\dagger + \hat{b}_k^\dagger\hat{b}_j = \delta_{jk} \tag{16.32}$$

Finally, it only remains for us to note that $\hat{b}_k^\dagger\hat{b}_k$ is the fermion number operator for the state k; that is, it tells us the number of fermions occupying state k. If the state is initially empty, it returns the value zero and if the state is initially full, it returns the value 1. We can write this as

$$\hat{N}_k = \hat{b}_k^\dagger\hat{b}_k \tag{16.33}$$

Problems

16.1.1* Consider a system that has two possible single-fermion states, 1 and 2, and can have anywhere from zero to two particles in it. There are, therefore, four possible states of this system: $|0_1,0_2\rangle$ (the state with no particles in either single-fermion state, a state we could also write as the empty state $|0\rangle$), $|1_1,0_2\rangle$, $|0_1,1_2\rangle$, and $|1_1,1_2\rangle$. (We also choose the standard ordering of the states to be in the order 1, 2.) Any state of the system could be described as a linear combination of these four basis states; that is,

$$|\Psi\rangle = c_1|0_1,0_2\rangle + c_2|1_1,0_2\rangle + c_3|0_1,1_2\rangle + c_4|1_1,1_2\rangle$$

which we could also choose to write as a vector

$$|\Psi\rangle = \begin{bmatrix} c_1 \\ c_2 \\ c_3 \\ c_4 \end{bmatrix}$$

(i) Construct 4 x 4 matrices for each of the operators $\hat{b}_1^\dagger$, $\hat{b}_1$, $\hat{b}_2^\dagger$, and $\hat{b}_2$.

(ii) Explicitly verify by matrix multiplication the anticommutation relations

$$\hat{b}_1^\dagger\hat{b}_1 + \hat{b}_1\hat{b}_1^\dagger = 1$$
$$\hat{b}_2^\dagger\hat{b}_2 + \hat{b}_2\hat{b}_2^\dagger = 1$$
$$\hat{b}_1^\dagger\hat{b}_2^\dagger + \hat{b}_2^\dagger\hat{b}_1^\dagger = 0$$

$$\hat{b}_1^\dagger \hat{b}_1^\dagger + \hat{b}_1^\dagger \hat{b}_1^\dagger = 0$$

16.1.2 Prove the relation

$$\hat{b}_j \hat{b}_k + \hat{b}_k \hat{b}_j = 0$$

by considering the swapping of rows in determinants (i.e., follow similar arguments to the analogous relation for creation operators in the previous section), considering all relevant initial states and choices of *j* and *k*.

16.2 Wavefunction operator

In this occupation number representation, it is very convenient to have an operator that represents the quantum mechanical wavefunction itself. One major use of it is to provide a simple way to postulate operators, such as Hamiltonians, that, in addition to having the spatial character they had before, also now have the creation and annihilation properties we desire for fermions. We use these later to rewrite Hamiltonians and other operators so that they have the fermion character built into them.

The simplest situation we could consider would be a wavefunction operator in which we have a single particle. Then, we can propose an operator for the wavefunction at position $\mathbf{r}$

$$\hat{\psi}(\mathbf{r}) = \sum_j \hat{b}_j \phi_j(\mathbf{r}) \tag{16.34}$$

where the $\phi_k(\mathbf{r})$ are some complete basis set for describing functions of space. Suppose for example that we had a situation in which the single particle of interest here was in state *m*; that is, the state with wavefunction $\phi_m(\mathbf{r})$. We can also write that state as

$$|\ldots 0_l, 1_m, 0_n, \ldots\rangle \equiv \hat{b}_m^\dagger |0\rangle \tag{16.35}$$

where here by $|0\rangle$, we mean the state with no fermions present in any single-particle state.[5] Then, we find that

$$\begin{aligned} \hat{\psi}(\mathbf{r})|\ldots 0_l, 1_m, 0_n, \ldots\rangle &= \hat{\psi}(\mathbf{r})\hat{b}_m^\dagger |0\rangle \\ &= \sum_j \phi_j(\mathbf{r}) \hat{b}_j \hat{b}_m^\dagger |0\rangle \end{aligned} \tag{16.36}$$

Now, we use the anticommutation relation in Eq. (16.32), obtaining

$$\hat{\psi}(\mathbf{r})|\ldots 0_l, 1_m, 0_n, \ldots\rangle = \sum_j \phi_j(\mathbf{r})\left(\delta_{jm} - \hat{b}_m^\dagger \hat{b}_j\right)|0\rangle \tag{16.37}$$

But

$$\hat{b}_j |0\rangle = 0 \tag{16.38}$$

because an attempt to annihilate a particle that is not there results in a null result. Hence, we have

---

[5] Note, incidentally, that just as for the empty state we encountered with bosons, the empty state $|0\rangle$ for fermions is a perfectly well-defined state of the system. It is one of the possible basis states for a multiple-fermion system. In Hilbert space, it is a vector of unit length, just like any other basis state. It does *not* have zero length.

$$\hat{\psi}(\mathbf{r})|\ldots 0_l, 1_m, 0_n, \ldots\rangle = \phi_m(\mathbf{r})|0\rangle \quad (16.39)$$

We can see, then, that this operator has successfully extracted the amplitude $\phi_m(\mathbf{r})$ as we would hope for a system in single-particle state m. We have also acquired the ket $|0\rangle$ in the final result, which might seem odd, but remember that we should have a state vector here because the result of operating on a state vector should be a state vector (i.e., a matrix times a vector is a vector), so this is required for formal purposes.

This illustration with the wavefunction operator also shows a very typical algebraic operation with creation and annihilation operators. To simplify an expression, one progressively rearranges it using the commutation or anticommutation relations to push an annihilation operator to the extreme right, operating on the $|0\rangle$ state or a state with no particle in the single-particle state associated with the annihilation operator. Such a term then vanishes, leaving a simpler expression.

We can also see by a simple extension of this algebra that if the particle is initially not in a specific single-particle state but rather in a linear superposition – that is,

$$|\psi_S\rangle = \sum_k c_k |\ldots, 1_k, \ldots\rangle \quad (16.40)$$

where by $|\ldots, 1_k, \ldots\rangle$, we mean the state with one particle in state k and no other single-particle states occupied – then

$$\hat{\psi}(\mathbf{r})|\psi_S\rangle = \left(\sum_k c_k \phi_k(\mathbf{r})\right)|0\rangle \quad (16.41)$$

which has now extracted the linear superposition of wavefunctions we would have desired.

The next more complex case is to propose a wavefunction operator for a two-fermion state. We propose

$$\hat{\psi}(\mathbf{r}_1, \mathbf{r}_2) = \frac{1}{\sqrt{2}} \sum_{j,n} \hat{b}_n \hat{b}_j \phi_j(\mathbf{r}_1) \phi_n(\mathbf{r}_2) \quad (16.42)$$

(The $1/\sqrt{2}$ term is to ensure normalization of the final result.)

It is left as an exercise for the reader to demonstrate that such an operator, operating on a state with two different single-particle states occupied, leads to a linear combination of products of wavefunctions that is correctly antisymmetric with respect to exchange of these two particles. In other words, if this operator acts on a state that has one fermion in single-particle state k and an identical fermion in single-particle state m (i.e., the state $|\ldots, 1_k, \ldots, 1_m, \ldots\rangle \equiv \hat{b}_k^\dagger \hat{b}_m^\dagger |0\rangle$) then

$$\hat{\psi}(\mathbf{r}_1, \mathbf{r}_2)|\ldots, 1_k, \ldots, 1_m, \ldots\rangle = \frac{1}{\sqrt{2}}\left[\phi_k(\mathbf{r}_1)\phi_m(\mathbf{r}_2) - \phi_k(\mathbf{r}_2)\phi_m(\mathbf{r}_1)\right]|0\rangle \quad (16.43)$$

We can propose to extend such wavefunction operators to larger numbers of particles, postulating

$$\hat{\psi}(\mathbf{r}_1, \mathbf{r}_2, \ldots \mathbf{r}_N) = \frac{1}{\sqrt{N}} \sum_{a,b,\ldots,n} \hat{b}_n \ldots \hat{b}_b \hat{b}_a \phi_a(\mathbf{r}_1) \phi_b(\mathbf{r}_2) \ldots \phi_n(\mathbf{r}_N) \quad (16.44)$$

with the expectation that these operators also extract the correct sum of permutations.

Having now postulated the fermion annihilation and creation operators and the wavefunction operator, we are ready to make use of them. In addition to their obvious use as a way of writing fermion states, their other use is to represent operators (especially Hamiltonians) and we deal with this next.

Problem

16.2.1 Consider the two-particle wavefunction operator

$$\hat{\psi}(\mathbf{r}_1,\mathbf{r}_2)=\frac{1}{\sqrt{2}}\sum_{j,n}\hat{b}_n\hat{b}_j\phi_j(\mathbf{r}_1)\phi_n(\mathbf{r}_2)$$

and a state $|\ldots,1_k,\ldots,1_m,\ldots\rangle \equiv \hat{b}_k^\dagger \hat{b}_m^\dagger |0\rangle$ that has one fermion in single-particle state k and an identical fermion in single-particle state m. Show that

$$\hat{\psi}(\mathbf{r}_1,\mathbf{r}_2)|\ldots,1_k,\ldots,1_m,\ldots\rangle=\frac{1}{\sqrt{2}}\left[\phi_k(\mathbf{r}_1)\phi_m(\mathbf{r}_2)-\phi_k(\mathbf{r}_2)\phi_m(\mathbf{r}_1)\right]|0\rangle$$

(i.e., this operator correctly constructs the combination of wavefunction products that is antisymmetric with respect to exchange of identical particles).

16.3 Fermion Hamiltonians

So far, we have treated fermions using a Schrödinger wave equation and have explicitly forced the wavefunctions to have the necessary fermion character (i.e., we have forced them to be antisymmetric with respect to exchange of particles). The Schrödinger equation itself did not force the fermion character on the wavefunctions and the solution of the Schrödinger equation did not itself give rise to the particles in any way. By contrast, with bosons, we have found an equation (which can also be viewed as a Schrödinger equation in the general sense) whose solution did itself give rise to the particles (e.g., photons) and this equation also was made up using operators that enforced the correct behavior of the states with respect to boson exchange.

In the previous sections, we constructed the fermion annihilation and creation operators. Our purpose in this section is now to use the fermion operators to represent other operators, especially Hamiltonians, in a fashion analogous to that developed for boson Hamiltonians, obtaining the necessary particle-like behavior. This development is necessarily somewhat abstract, but the reader should trust that this will be worth the effort. It leads to a particularly elegant and simple way of representing fermion properties that is valid for both single- and multiple-particle systems. This formalism works together with the boson formalism developed previously to result in a particularly simple way of describing fermion–boson interactions (e.g., optical absorption of photons by electron systems). The interactions of fermions and bosons will be the subject of the next chapter.

The key to creating the Hamiltonians and other operators in the form we want is the use of the wavefunction operator to substitute where the wavefunction itself appeared before. This is a postulation, but we will see that it does correspond to our previous results, as well as giving us a convenient way of handling other problems such as systems of multiple identical fermions or fermion–boson interactions. We consider progressively more sophisticated cases of fermion operators, showing how these can be written using creation and annihilation operators.

Single-particle fermion Hamiltonian with a single fermion

First, we consider the simplest case of a Hamiltonian for a single fermion. Previously, we had a simple Hamiltonian such as the simplest Schrödinger equation for a single particle

$$\hat{H}_{\mathbf{r}} = -\frac{\hbar^2}{2m}\nabla_{\mathbf{r}}^2 + V(\mathbf{r}) \tag{16.45}$$

The expected value for energy was, then, for any given state $|\psi\rangle$ (presuming for simplicity here that the state can also be described by a spatial wavefunction $\psi(\mathbf{r})$)

$$\langle E\rangle = \langle\psi|\hat{H}_{\mathbf{r}}|\psi\rangle = \int \psi^*(\mathbf{r})\left[-\frac{\hbar^2}{2m}\nabla_{\mathbf{r}}^2 + V(\mathbf{r})\right]\psi(\mathbf{r})d^3\mathbf{r} \tag{16.46}$$

Now, we postulate a new Hamiltonian operator, one that has the fermion particle character we hope to add into the Hamiltonian operator. Our postulated method for constructing such an operator is to substitute for the wavefunction in an equation such as Eq. (16.46) with the wavefunction operator, generating our desired new fermion operator instead of the expectation value. Hence, we obtain for our single-particle Hamiltonian operator

$$\hat{H} \equiv \int \hat{\psi}^\dagger(\mathbf{r})\left[-\frac{\hbar^2}{2m}\nabla_{\mathbf{r}}^2 + V(\mathbf{r})\right]\hat{\psi}(\mathbf{r})d^3\mathbf{r} \tag{16.47}$$

For the moment, we presume for simplicity that the single-particle basis states with spatial wavefunctions $\phi_m(\mathbf{r})$ are the eigenstates of this single-particle Hamiltonian, with corresponding eigenenergies E_m; we will discuss the more general case later. Now we can formally rewrite Eq. (16.47) using the definition of the wavefunction operator, Eq. (16.34), obtaining

$$\begin{aligned}\hat{H} &= \int \sum_{j,k} \hat{b}_j^\dagger \hat{b}_k \phi_j^*(\mathbf{r})\left[-\frac{\hbar^2}{2m}\nabla_{\mathbf{r}}^2 + V(\mathbf{r})\right]\phi_k(\mathbf{r})d^3\mathbf{r} \\ &= \sum_{j,k} \hat{b}_j^\dagger \hat{b}_k E_k \int \phi_j^*(\mathbf{r})\phi_k(\mathbf{r})d^3\mathbf{r} \\ &= \sum_{j,k} \hat{b}_j^\dagger \hat{b}_k E_k \delta_{jk}\end{aligned} \tag{16.48}$$

that is,

$$\hat{H} = \sum_j E_j \hat{b}_j^\dagger \hat{b}_j \equiv \sum_j E_j \hat{N}_j \tag{16.49}$$

To show that this corresponds to results from our previous approach, let us evaluate the expectation value of the energy. Suppose that the system was in some state $|\psi\rangle$ that was a linear superposition of the basis states; then, we could write, as usual, in the **r** representation

$$|\psi\rangle = \sum_m c_m \phi_m(\mathbf{r}) \tag{16.50}$$

or, equivalently, in the number state notation

$$|\psi\rangle = \sum_m c_m \hat{b}_m^\dagger |0\rangle \tag{16.51}$$

where we have used the notation $\hat{b}_m^\dagger|0\rangle$ as a convenient way of writing the basis state in which only the single-particle state m is occupied. Note, incidentally, that we can take the Hermitian conjugate of both sides to obtain

$$\langle\psi| = \sum_m c_m^* \langle 0|\hat{b}_m \tag{16.52}$$

Now, let us formally evaluate the expectation value of the energy in this state using the Hamiltonian Eq. (16.49). We obtain

$$\langle E \rangle = \langle \psi | \hat{H} | \psi \rangle = \sum_{m,n,j} c_m^* c_n E_j \langle 0 | \hat{b}_m \hat{b}_j^\dagger \hat{b}_j \hat{b}_n^\dagger | 0 \rangle \tag{16.53}$$

Now, we simplify the expression $\hat{b}_m \hat{b}_j^\dagger \hat{b}_j \hat{b}_n^\dagger |0\rangle$ using the anticommutation relations for fermion operators. Our standard algebraic approach is to use the anticommutation relations to push annihilation operators to the right; that will lead to disappearance of terms because an annihilation operator acting on the empty state $|0\rangle$ gives a zero result. Therefore, we keep making substitutions of the form

$$\hat{b}_m \hat{b}_j^\dagger = \delta_{mj} - \hat{b}_j^\dagger \hat{b}_m \tag{16.54}$$

which is simply the anticommutation relation for operators $\hat{b}_m$ and $\hat{b}_j^\dagger$. Hence, we have

$$\begin{aligned} \hat{b}_m \hat{b}_j^\dagger \hat{b}_j \hat{b}_n^\dagger |0\rangle &= \left(\delta_{mj} - \hat{b}_j^\dagger \hat{b}_m\right)\left(\delta_{nj} - \hat{b}_n^\dagger \hat{b}_j\right)|0\rangle \\ &= \left(\delta_{mj} - \hat{b}_j^\dagger \hat{b}_m\right)\delta_{nj}|0\rangle \\ &= \delta_{mj}\delta_{nj}|0\rangle \end{aligned} \tag{16.55}$$

Hence, substituting back into Eq. (16.53), we have

$$\begin{aligned} \langle E \rangle &= \sum_{m,n,j} c_m^* c_n E_j \delta_{mj} \delta_{nj} \langle 0 | 0 \rangle \\ &= \sum_j |c_j|^2 E_j \end{aligned} \tag{16.56}$$

which is exactly the result we would have expected based on our previous approach as in Eq. (16.46), for example. Hence, this new approach to Hamiltonians does appear to work, at least for this simple example. Incidentally, except for the other simple anticommutation relations (Eqs. (16.14) and (16.22)) that merely change the sign on swapping fermion creation operators or fermion annihilation operators, the algebraic approach used here is essentially the main algebraic manipulation one has to perform with fermion operators and is very powerful.

Single-particle fermion Hamiltonians with multiple-particle states

Now, let us take a somewhat more complicated example, which is that of single-particle fermion Hamiltonians with multiple fermions. What we now mean by a single-particle Hamiltonian is that the Hamiltonian of any one fermion can be written entirely in terms of that fermion's properties and coordinates; that is, that fermion's energy does not depend on the other fermions in the system, or, equivalently, the fermions are noninteracting.

We might imagine this would be a good starting point for multiple uncharged fermions, such as a set of neutrons, possibly so dilute that they do not interact with one another in any other important way (i.e., negligible collisions of any kind). The reader might object that this is a rather artificial example and it certainly is. A much more important and real example is the case of electrons in a crystalline solid. The reader may remember (from Chapter 8) that the starting approximation for modeling such electrons is that each one is presumed to move in the average potential, $V(\mathbf{r})$, created by all the other electrons and nuclei, with this $V(\mathbf{r})$ being presumed the same for all electrons. By making that approximation, we have now decoupled all the Hamiltonians for all the different electrons and have a total Hamiltonian that is simply the sum of the single-electron Hamiltonians. Hence, the case we are about to analyze is actually a very important and useful one.

Suppose, then, that we have N identical fermions. Fermion i is presumed to have a single-particle Hamiltonian in the original $\mathbf{r}$ form such as

$$\hat{H}_{\mathbf{r}i} = -\frac{\hbar^2}{2m}\nabla_{\mathbf{r}_i}^2 + V(\mathbf{r}_i) \tag{16.57}$$

with the total Hamiltonian for the set of N fermions, consequently, being, in the original $\mathbf{r}$ form

$$\hat{H}_{\mathbf{r}} = \sum_{i=1}^{N} \hat{H}_{\mathbf{r}i} \tag{16.58}$$

What we now show is that even for the multiple-fermion case, we can still write the total Hamiltonian operator exactly as in Eq. (16.49). Now that we write the Hamiltonian in the new form with creation and annihilation operators, we do not have to change the Hamiltonian for noninteracting fermions regardless of how many particles there are in the system (i.e., we do not have to write a sum like Eq. (16.58) over all the particles). We begin here to see some of the power of the annihilation and creation operator form for multiple-fermion systems. Here, we illustrate that this Hamiltonian Eq. (16.49) works for the case of two fermions ($N = 2$).

Suppose, then, that we have a specific two-fermion state with one fermion in single-particle state k and one in single-particle state m. We can write that state as

$$|\psi_{TP}\rangle = |\ldots,1_k,\ldots,1_m,\ldots\rangle = \hat{b}_k^\dagger \hat{b}_m^\dagger |0\rangle \tag{16.59}$$

We want to evaluate the expectation value for the energy in this particular two-particle state presuming that we can use the Hamiltonian in its occupation number representation as in Eq. (16.49). Then, we have

$$\begin{aligned}\langle E\rangle = \langle\psi_{TP}|\hat{H}|\psi_{TP}\rangle &= \sum_j \left(\hat{b}_k^\dagger \hat{b}_m^\dagger|0\rangle\right)^\dagger E_j \hat{b}_j^\dagger \hat{b}_j \hat{b}_k^\dagger \hat{b}_m^\dagger |0\rangle \\ &= \sum_j E_j \langle 0|\hat{b}_m \hat{b}_k \hat{b}_j^\dagger \hat{b}_j \hat{b}_k^\dagger \hat{b}_m^\dagger|0\rangle\end{aligned} \tag{16.60}$$

Now, we have a small exercise in creation and annihilation operator algebra to simplify this expression. We take the standard approach of using the anticommutation relation between annihilation and creation operator pairs to push the annihilation operators to the right. We then can eliminate each term involving an annihilation operator on the right because $\hat{b}_i|0\rangle = 0$ for any i. We have

$$\begin{aligned}\hat{b}_m \hat{b}_k \hat{b}_j^\dagger \hat{b}_j \hat{b}_k^\dagger \hat{b}_m^\dagger|0\rangle &= \hat{b}_m\left(\delta_{jk} - \hat{b}_j^\dagger \hat{b}_k\right)\left(\delta_{jk} - \hat{b}_k^\dagger \hat{b}_j\right)\hat{b}_m^\dagger|0\rangle \\ &= \left(\delta_{jk}\hat{b}_m\hat{b}_m^\dagger - \delta_{jk}\hat{b}_m\hat{b}_k^\dagger\hat{b}_j\hat{b}_m^\dagger - \delta_{jk}\hat{b}_m\hat{b}_j^\dagger\hat{b}_k\hat{b}_m^\dagger + \hat{b}_m\hat{b}_j^\dagger\hat{b}_k\hat{b}_k^\dagger\hat{b}_j\hat{b}_m^\dagger\right)|0\rangle \\ &= \Big[\delta_{jk}\left(1-\hat{b}_m^\dagger\hat{b}_m\right) - \delta_{jk}\left(\delta_{mk} - \hat{b}_k^\dagger\hat{b}_m\right)\left(\delta_{mj} - \hat{b}_m^\dagger\hat{b}_j\right) \\ &\quad - \delta_{jk}\left(\delta_{mj} - \hat{b}_j^\dagger\hat{b}_m\right)\left(\delta_{mk} - \hat{b}_m^\dagger\hat{b}_k\right) + \left(\delta_{mj} - \hat{b}_j^\dagger\hat{b}_m\right)\left(1-\hat{b}_k^\dagger\hat{b}_k\right)\left(\delta_{mj} - \hat{b}_j^\dagger\hat{b}_m\right)\Big]|0\rangle\end{aligned} \tag{16.61}$$

Now, we have annihilation operators on the far right on every expression involving creation and annihilation operators, so all of those terms disappear. Hence, we have

$$\hat{b}_m\hat{b}_k\hat{b}_j^\dagger\hat{b}_j\hat{b}_k^\dagger\hat{b}_m^\dagger|0\rangle=\left(\delta_{jk}-\delta_{jk}\delta_{mk}\delta_{mj}-\delta_{jk}\delta_{mj}\delta_{mk}+\delta_{mj}\right)|0\rangle \tag{16.62}$$

But, by choice, m and k are different states so δ_{mk} never has any value other than zero.[6] Hence, we have

$$\hat{b}_m\hat{b}_k\hat{b}_j^\dagger\hat{b}_j\hat{b}_k^\dagger\hat{b}_m^\dagger|0\rangle=\left(\delta_{jk}+\delta_{mj}\right)|0\rangle \tag{16.63}$$

Substituting back into Eq. (16.60), we have

$$\langle E\rangle=\sum_j E_j\left(\delta_{jk}+\delta_{mj}\right)\langle 0|0\rangle=E_k+E_m \tag{16.64}$$

which is exactly what we would expect for noninteracting fermions, one in state k and one in state m. Hence, this illustration shows how the Hamiltonian Eq. (16.49) also works for multiple-particle states. Unlike the **r** representation of the Hamiltonian, we do not have to add separate Hamiltonians for each identical fermion; hence, we have an elegant form of Hamiltonian for multiple-fermion systems.

Representation of general single-particle fermion operators

The Hamiltonian is a rather special case because for a single-particle operator, the occupation number states were chosen as eigenstates of the Hamiltonian. How would we represent other single-particle fermion operators (e.g., the momentum operator or the position operator) in this annihilation and creation operator formalism? Having a suitable approach for this is practically quite useful; we may need it, for example, if we are to handle the position **r** in the electric dipole interaction $\mathbf{E}\cdot\mathbf{r}$ or, similarly, to handle the momentum in the $\mathbf{A}\cdot\mathbf{p}$ interaction in other formulations of the interaction of an electron with an electromagnetic field.

Here, we consider the general case of a system with N fermions. In the **r** representation of some operator, $\hat{G}_\mathbf{r}$ (e.g., the momentum operator) for a multiple-fermion system, we would have to add all of the operators corresponding to the coordinates of each particle; that is,

$$\hat{G}_\mathbf{r}=\sum_{i=1}^{N}\hat{G}_{\mathbf{r}i} \tag{16.65}$$

where $\hat{G}_{\mathbf{r}i}$ is the operator for a specific particle (e.g., it might be the momentum operator $\hat{\mathbf{p}}_{\mathbf{r}i}=-i\hbar\nabla_{\mathbf{r}i}$). In the annihilation and creation operator formalism, we postulate instead that

$$\hat{G}=\int\hat{\psi}^\dagger\hat{G}_\mathbf{r}\hat{\psi}\;d^3\mathbf{r}_1d^3\mathbf{r}_2\ldots d^3\mathbf{r}_N \tag{16.66}$$

where $\hat{\psi}$ is the N-particle fermion wavefunction operator. Substituting the N-particle fermion wavefunction operator into Eq. (16.66), we obtain

$$\begin{aligned}\hat{G}=&\frac{1}{N}\sum_{i=1}^{N}\sum_{\substack{a,b,\ldots n\\a',b',\ldots n'}}\hat{b}_{a'}^\dagger\hat{b}_{b'}^\dagger\ldots\hat{b}_{n'}^\dagger\hat{b}_n\ldots\hat{b}_b\hat{b}_a\\&\times\int\phi_{a'}^*(\mathbf{r}_1)\phi_{b'}^*(\mathbf{r}_2)\ldots\phi_{n'}^*(\mathbf{r}_N)\hat{G}_{\mathbf{r}i}\phi_a(\mathbf{r}_1)\phi_b(\mathbf{r}_2)\ldots\phi_n(\mathbf{r}_N)\,d^3\mathbf{r}_1d^3\mathbf{r}_2\ldots d^3\mathbf{r}_N\end{aligned} \tag{16.67}$$

---

[6] We also know that $\hat{b}_k^\dagger\hat{b}_k^\dagger=0$ because we cannot create two fermions in the same state. Hence, we could have used this property to replace $\hat{b}_k^\dagger\hat{b}_m^\dagger$ formally with $(1-\delta_{mk})\hat{b}_k^\dagger\hat{b}_m^\dagger$ as a first step in this algebra. The product $(1-\delta_{mk})\delta_{mk}=0$ under all conditions, so all terms containing δ_{mk} vanish.

where each of the $a,b,\ldots n$ and each of the $a',b',\ldots n'$ ranges over all possible single-particle fermion states.

Now, all the spatial integrals, except the one over $\mathbf{r}_i$, lead to Kronecker deltas of the form $\delta_{k'k}$, forcing $a' = a$, $b' = b$, and so on, except for particle i. Hence, we have

$$\hat{G} = \frac{1}{N}\sum_{i=1}^{N} \sum_{a,b,\ldots,i1,i2,\ldots n} G_{i1i2}\hat{b}_a^\dagger \hat{b}_b^\dagger \ldots \hat{b}_{i1}^\dagger \ldots \hat{b}_n^\dagger \hat{b}_n \ldots \hat{b}_{i2} \ldots \hat{b}_b \hat{b}_a \tag{16.68}$$

where

$$G_{i1i2} = \int \phi_{i1}^*(\mathbf{r}_i)\hat{G}_{\mathbf{r}i}\phi_{i2}(\mathbf{r}_i)d^3\mathbf{r}_i \tag{16.69}$$

We can use the anticommutation relation $\hat{b}_j\hat{b}_k + \hat{b}_k\hat{b}_j = 0$ progressively to swap the operator $\hat{b}_{i2}$ from the right to the center; similarly, we use the anticommutation relation $\hat{b}_j^\dagger\hat{b}_k^\dagger + \hat{b}_k^\dagger\hat{b}_j^\dagger = 0$ progressively to swap the operator $\hat{b}_{i1}^\dagger$ from the left to the center. Each such application of an anticommutation relation results in a sign change, but there are equal number of swaps from the left and from the right, so there is no net sign change in this operation. Hence, we have

$$\hat{G} = \frac{1}{N}\sum_{i=1}^{N} \sum_{a,b,\ldots,i1,i2,\ldots n} G_{i1i2} \underbrace{\hat{b}_a^\dagger \hat{b}_b^\dagger \ldots \hat{b}_n^\dagger}_{omitting\ \hat{b}_{i1}^\dagger} \hat{b}_{i1}^\dagger \hat{b}_{i2} \underbrace{\hat{b}_n \ldots \hat{b}_b \hat{b}_a}_{omitting\ \hat{b}_{i2}^\dagger} \tag{16.70}$$

In practice, with any operator, we are, in the end, always working out matrix elements for the operator. Any two operators with identical matrix elements are equivalent operators. We can consider two, possibly different, N-fermion basis states, $|\psi_{1N}\rangle$ and $|\psi_{2N}\rangle$ and consider matrix elements of the operator $\hat{G}$ between such states. Because of Pauli exclusion, the only strings of operators that can survive in matrix elements for legal fermion states are those in which all the operators $\hat{b}_a, \hat{b}_b, \ldots \hat{b}_n$ are different from each other (i.e., correspond to annihilation operators for different single-particle states) and are each different from both $\hat{b}_{i1}$ and $\hat{b}_{i2}$; otherwise, we would be trying either to annihilate two fermions from the same state or create two fermions in the same state, both of which are impossible. Hence, for these states, because no two single-particle states in our string of creation operators or our string of annihilation operators can be identical, not only do the pairs of annihilation operators anticommute and the pairs of creation operators anticommute as usual, so also do all the pairs of creation and annihilation operators with different subscripts (other than possibly the pair $\hat{b}_{i1}^\dagger\hat{b}_{i2}$). Hence, we can take the creation operator $\hat{b}_a^\dagger$ and swap it all the way from the left until we get to the left of the corresponding annihilation operator $\hat{b}_a$, only acquiring minus signs as we do so. Actually, however, we acquire an even number of minus signs because the number of swaps taken to get to the middle is equal to the number to get from the middle to the final position, so there is no change in sign in all these swaps. We can repeat this procedure for each creation operator (other than $\hat{b}_{i1}^\dagger$, which we do not need to move anyway), so we have

$$\hat{G} = \frac{1}{N}\sum_{i=1}^{N} \sum_{a,b,\ldots,i1,i2,\ldots n} G_{i1i2}\hat{b}_{i1}^\dagger \hat{b}_{i2} \underbrace{\hat{N}_n \ldots \hat{N}_b \hat{N}_a}_{omitting\ \hat{b}_{i1}^\dagger \hat{b}_{i2}} \tag{16.71}$$

When this operator operates on a specific N-fermion basis state $|\psi_{1N}\rangle$, the only terms in the summation that can survive are those for which the list of states $a,b,\ldots n$ corresponds to occupied states in $|\psi_{1N}\rangle$, so the sum over $a,b,\ldots n$ (omitting $i1$ and $i2$) and the number operators can be dropped without changing any matrix element. Hence, we can write

$$\hat{G} = \frac{1}{N}\sum_{i=1}^{N}\sum_{i1,i2} G_{i1i2}\hat{b}_{i1}^{\dagger}\hat{b}_{i2} \tag{16.72}$$

It makes no difference which fermion we are considering – G_{i1i2} is the same for every fermion – and so we can write, finally, simplifying notation by substituting j for $i1$ and k for $i2$

$$\hat{G} = \sum_{j,k} G_{jk}\hat{b}_j^{\dagger}\hat{b}_k \tag{16.73}$$

which is the general form for a single-particle fermion operator. The Hamiltonian, Eq. (16.49), is just a special case for a diagonal operator. Hence, we have found a very simple form for the single-particle fermion operator valid for any number of fermions.

Two-particle fermion Hamiltonians

The analysis for single-particle fermion Hamiltonians is useful for situations in which the fermions do not interact with one another. Fermions such as electrons do, however, interact quite strongly (e.g., through their Coulomb repulsion) and so it is quite common to have to deal with such interactions. For such cases, we need to have two-particle operators (again, usually Hamiltonians, though not necessarily). Here, we find out how to approach such processes using the annihilation and creation operators.

Consider what happens, for example, when we have two interacting fermions. In the $\mathbf{r}$ form, we might have an operator $\hat{D}_{\mathbf{r}}(\mathbf{r}_1,\mathbf{r}_2)$ that depends on the coordinates of both particles. Then we would postulate that we could create the corresponding new operator in the annihilation and creation operator form in an analogous way to our previous approach for single-particle operators; that is, we postulate

$$\hat{D} = \int \hat{\psi}^{\dagger}(\mathbf{r}_1,\mathbf{r}_2)\hat{D}_{\mathbf{r}}(\mathbf{r}_1,\mathbf{r}_2)\hat{\psi}(\mathbf{r}_1,\mathbf{r}_2)d^3\mathbf{r}_1 d^3\mathbf{r}_2 \tag{16.74}$$

with the two-fermion wavefunction operator $\hat{\psi}(\mathbf{r}_1,\mathbf{r}_2) = (1/\sqrt{2})\Sigma_{j,k}\hat{b}_k\hat{b}_j\phi_j(\mathbf{r}_1)\phi_k(\mathbf{r}_2)$. Substituting this form into Eq. (16.74), we have

$$\hat{D} = \frac{1}{2}\sum_{a,b,c,d}\hat{b}_a^{\dagger}\hat{b}_b^{\dagger}\hat{b}_d\hat{b}_c \int \phi_a^*(\mathbf{r}_1)\phi_b^*(\mathbf{r}_2)\hat{D}_{\mathbf{r}}(\mathbf{r}_1,\mathbf{r}_2)\phi_c(\mathbf{r}_1)\phi_d(\mathbf{r}_2)d^3\mathbf{r}_1 d^3\mathbf{r}_2 \tag{16.75}$$

or, equivalently

$$\hat{D} = \frac{1}{2}\sum_{a,b,c,d} D_{abcd}\hat{b}_a^{\dagger}\hat{b}_b^{\dagger}\hat{b}_d\hat{b}_c \tag{16.76}$$

where

$$D_{abcd} = \int \phi_a^*(\mathbf{r}_1)\phi_b^*(\mathbf{r}_2)\hat{D}_{\mathbf{r}}(\mathbf{r}_1,\mathbf{r}_2)\phi_c(\mathbf{r}_1)\phi_d(\mathbf{r}_2)d^3\mathbf{r}_1 d^3\mathbf{r}_2 \tag{16.77}$$

Note, incidentally, that the order of the suffices on the chain of operators $\hat{b}_a^{\dagger}\hat{b}_b^{\dagger}\hat{b}_d\hat{b}_c$ is *not* a,b,c,d. The ordering is in the opposite sense for the annihilation operators. This different ordering here has emerged from the wavefunction operators and the properties of Hermitian conjugation.

We presume that we would be able to show that the two-particle fermion operator of Eq. (16.76) would remain unchanged as we changed the system to have more than two fermions in it. The arguments would be very similar to those given before that showed that the single-particle fermion operator Eq. (16.73) was unchanged as we considered multiple-fermion states, so we presume that (16.76) is a general statement for a two-particle fermion operator in this annihilation and creation operator approach.

Example – electrons interacting through the Coulomb potential

The meaning of the two-particle fermion operator becomes clearer if we consider a simple example. Consider two electrons (of the same spin) interacting through the Coulomb repulsion. The Hamiltonian in the **r** form would be, as in Eq. (13.23) when we considered the exchange interaction between identical electrons

$$\hat{H}_{\mathbf{r}}(\mathbf{r}_1,\mathbf{r}_2) = -\frac{\hbar^2}{2m_o}\left(\nabla^2_{\mathbf{r}_1} + \nabla^2_{\mathbf{r}_2}\right) + \frac{e^2}{4\pi\varepsilon_o|\mathbf{r}_1 - \mathbf{r}_2|} \tag{16.78}$$

Hence, our two-particle operator formalism gives us the new operator

$$\hat{H} = \frac{1}{2}\sum_{a,b,c,d} H_{abcd}\hat{b}_a^\dagger\hat{b}_b^\dagger\hat{b}_d\hat{b}_c \tag{16.79}$$

where H_{abcd} is defined analogously to Eq. (16.77).

Suppose, specifically, that we have the two-fermion state where one electron is in the basis state $\phi_k(\mathbf{r})$ and the other is in the state $\phi_m(\mathbf{r})$; that is, the two-particle state can be written

$$|\psi_{TP}\rangle = \hat{b}_k^\dagger\hat{b}_m^\dagger|0\rangle \tag{16.80}$$

Now, let us evaluate the expectation value of the energy in this two-particle state using the Hamiltonian Eq. (16.79). We have

$$\langle\psi_{TP}|\hat{H}|\psi_{TP}\rangle = \frac{1}{2}\langle 0|\sum_{a,b,c,d} H_{abcd}\hat{b}_m\hat{b}_k\hat{b}_a^\dagger\hat{b}_b^\dagger\hat{b}_d\hat{b}_c\hat{b}_k^\dagger\hat{b}_m^\dagger|0\rangle \tag{16.81}$$

Now

$$\begin{aligned}&\langle 0|\hat{b}_m\hat{b}_k\hat{b}_a^\dagger\hat{b}_b^\dagger\hat{b}_d\hat{b}_c\hat{b}_k^\dagger\hat{b}_m^\dagger|0\rangle \\ &= \delta_{ak}\delta_{bm}\delta_{ck}\delta_{dm} + \delta_{am}\delta_{bk}\delta_{cm}\delta_{dk} - \delta_{am}\delta_{bk}\delta_{ck}\delta_{dm} - \delta_{ak}\delta_{bm}\delta_{cm}\delta_{dk}\end{aligned} \tag{16.82}$$

the proof of which is left as an exercise for the reader (as usual, this merely requires repeated application of the anticommutation relation of the form $\hat{b}_p\hat{b}_q^\dagger = \delta_{pq} - \hat{b}_q^\dagger\hat{b}_p$ to push the annihilation operators to the right). Hence, we have for the energy expectation value

$$\langle\psi_{TP}|\hat{H}|\psi_{TP}\rangle = \frac{1}{2}\left(H_{kmkm} + H_{mkmk} - H_{mkkm} - H_{kmmk}\right) \tag{16.83}$$

Explicitly, we have

$$H_{kmkm} = H_{mkmk} = \int\phi_k^*(\mathbf{r}_1)\phi_m^*(\mathbf{r}_2)\hat{H}_{\mathbf{r}}\phi_k(\mathbf{r}_1)\phi_m(\mathbf{r}_2)d^3\mathbf{r}_1d^3\mathbf{r}_2 \tag{16.84}$$

and

$$H_{kmmk} = H^*_{mkkm} = \int\phi_k^*(\mathbf{r}_1)\phi_m^*(\mathbf{r}_2)\hat{H}_{\mathbf{r}}\phi_m(\mathbf{r}_1)\phi_k(\mathbf{r}_2)d^3\mathbf{r}_1d^3\mathbf{r}_2 \tag{16.85}$$

These terms are exactly the same as we previously calculated using the **r** formalism in Chapter 13. H_{kmkm} (or, equivalently, $(1/2)(H_{kmkm} + H_{mkmk})$) is the sum of the kinetic energies for the two particles and the Coulomb potential energy for two electrons, as in Eqs. (13.30) through (13.34) (and is, therefore, the energy we would calculate if the particles were not identical);

$-(1/2)(H_{mkkm} + H_{kmmk})$ is the exchange energy of Eq. (13.36). Hence, this approach does reproduce the results of our previous **r** formalism. Importantly, this formalism, in which we never have to explicitly introduce the antisymmetry of the wavefunction for two identical fermions, has correctly introduced the exchange-energy terms. This exchange term has emerged naturally through the use of the anticommutation relations for the fermion operators.

Problems

16.3.1 For identical fermions, prove

$$\langle 0|\hat{b}_m\hat{b}_k\hat{b}_a^\dagger\hat{b}_b^\dagger\hat{b}_d\hat{b}_c\hat{b}_k^\dagger\hat{b}_m^\dagger|0\rangle = \delta_{ak}\delta_{bm}\delta_{ck}\delta_{dm} + \delta_{am}\delta_{bk}\delta_{cm}\delta_{dk} - \delta_{am}\delta_{bk}\delta_{ck}\delta_{dm} - \delta_{ak}\delta_{bm}\delta_{cm}\delta_{dk}$$

16.3.2 Consider electrons and protons and construct the Hamiltonian in the **r** form for one electron and one proton assuming Coulomb interaction only.

(i) Transform this Hamiltonian into an occupation number form with creation and annihilation operators for both electrons and protons. (Hint: Use separate wavefunction operators for the electron wavefunction and for the proton wavefunction, noting any commutation relation between the operators corresponding to different kinds of particles.)

(ii) By presuming that the system is in a state $|\psi\rangle$ in which the electron is in some electron basis state k and the proton is in some proton basis state m, formally evaluate the expectation value of the energy of this system of one electron and one proton using this occupation number form of the Hamiltonian. Make use of the commutation and/or anticommutation relations appropriate for the operators (you need not actually evaluate any integrals), showing that, in this case, there is no exchange-energy term in the resulting energy expectation value.

16.3.3* (i) Using a complete orthonormal spatial basis set $\phi_n(\mathbf{r})$, write a general expression in terms of creation and annihilation operators for the position operator $\hat{\mathbf{r}}$ of a fermion.

(ii) Suppose now the fermion is explicitly in a one-dimensional potential well with infinitely high potential barrier on each side. Write an explicit expression for the position operator using the energy eigenfunctions $\phi_n(z)$ of this one-dimensional potential well problem as a basis. You may need the result

$$\int_0^\pi (x - \pi/2)\sin(nx)\sin(mx)\,dx = \frac{-4nm}{(n-m)^2(n+m)^2}, \text{ for } n+m \text{ odd}$$
$$= 0, \text{ for } n+m \text{ even}$$

16.4 Summary of concepts

Fermion annihilation and creation operators

We define fermion creation $\hat{b}_k^\dagger$ and annihilation $\hat{b}_k$ operators for a given single-particle state k, respectively, through the relations

$$\hat{b}_k^\dagger|\ldots,0_k,\ldots\rangle = (-1)^{S_k}|\ldots,1_k,\ldots\rangle \tag{16.18}$$

$$\hat{b}_k^\dagger|\ldots,1_k,\ldots\rangle = 0 \tag{16.15}$$

$$\hat{b}_k|\ldots,1_k,\ldots\rangle = (-1)^{S_k}|\ldots,0_k,\ldots\rangle \tag{16.21}$$

$$\hat{b}_k|\ldots,0_k,\ldots\rangle = 0 \tag{16.23}$$

where S_k is the number of occupied single-particle states ahead of state k in the arbitrary standard ordering we have chosen for labeling the single-particle states.

Anticommutator

The anticommutator is denoted with curly brackets and for two operators, $\hat{c}$ and $\hat{d}$, is defined as the sum

$$\hat{c}\hat{d} + \hat{d}\hat{c} \equiv \{\hat{c}, \hat{d}\} \qquad \text{after Eq. (16.17)}$$

Anticommutation relations

Fermion creation and annihilation operators have the following anticommutation relations:

$$\hat{b}_j^\dagger \hat{b}_k^\dagger + \hat{b}_k^\dagger \hat{b}_j^\dagger = 0 \tag{16.14}$$

$$\hat{b}_j \hat{b}_k + \hat{b}_k \hat{b}_j = 0 \tag{16.22}$$

$$\hat{b}_k^\dagger \hat{b}_j + \hat{b}_j \hat{b}_k^\dagger = \delta_{jk} \tag{16.32}$$

Fermion number operator

The fermion number operator for a single-particle state k is defined as

$$\hat{N}_k = \hat{b}_k^\dagger \hat{b}_k \tag{16.33}$$

Wavefunction operators

We can define a single-particle wavefunction operator

$$\hat{\psi}(\mathbf{r}) = \sum_j \hat{b}_j \phi_j(\mathbf{r}) \tag{16.34}$$

For a system in a specific single-particle state m, we therefore have, for example,

$$\hat{\psi}(\mathbf{r})|\ldots 0_l, 1_m, 0_n, \ldots\rangle = \phi_m(\mathbf{r})|0\rangle \tag{16.39}$$

so the wavefunction operator has successfully "extracted" the wavefunction of that state and it similarly extracts linear superpositions of wavefunctions for a particle in a linear superposition of single-particle states.

We can similarly define two-particle and N-particle wavefunction operators, respectively

$$\hat{\psi}(\mathbf{r}_1, \mathbf{r}_2) = \frac{1}{\sqrt{2}} \sum_{j,n} \hat{b}_n \hat{b}_j \phi_j(\mathbf{r}_1) \phi_n(\mathbf{r}_2) \tag{16.42}$$

$$\hat{\psi}(\mathbf{r}_1, \mathbf{r}_2, \ldots \mathbf{r}_N) = \frac{1}{\sqrt{N}} \sum_{a,b,\ldots,n} \hat{b}_n \ldots \hat{b}_b \hat{b}_a \phi_a(\mathbf{r}_1) \phi_b(\mathbf{r}_2) \ldots \phi_n(\mathbf{r}_N) \tag{16.44}$$

Single-particle fermion Hamiltonian

For noninteracting fermions (or ones approximated as noninteracting, like electrons in a solid in a single-electron approximation), we can write the Hamiltonian as

$$\hat{H} = \sum_j E_j \hat{b}_j^\dagger \hat{b}_j \equiv \sum_j E_j \hat{N}_j \tag{16.49}$$

where the states j have been chosen to be the eigenstates of this single-particle Hamiltonian. We call this a "single-particle" Hamiltonian, but we understand that this same Hamiltonian remains valid no matter how many such noninteracting particles there are in the system; we no longer have to add Hamiltonians for all the different particles. It is a single-particle operator in the sense that there are no interactions between particles involved.

General single-particle fermion operator

When the single-particle basis states are not necessarily the eigenstates of the operator, a single-particle fermion operator can be written as

$$\hat{G} = \sum_{j,k} G_{jk} \hat{b}_j^\dagger \hat{b}_k \tag{16.73}$$

where the matrix elements are taken over the single-particle basis wavefunctions; that is,

$$G_{i1i2} = \int \phi_{i1}^* (\mathbf{r}_i) \hat{G}_{\mathbf{r}i} \phi_{i2} (\mathbf{r}_i) d^3\mathbf{r}_i \tag{16.69}$$

Again, this Hamiltonian remains the same regardless of how many such noninteracting particles are in the system.

General two-particle Fermion operator

With similar definitions as before, an operator that expresses the interaction between two fermions, such as the Coulomb interaction Hamiltonian between two electrons, can be written as

$$\hat{D} = \frac{1}{2} \sum_{a,b,c,d} D_{abcd} \hat{b}_a^\dagger \hat{b}_b^\dagger \hat{b}_d \hat{b}_c \tag{16.76}$$

The use of such an operator automatically gives rise to the exchange terms, for example, based solely on the properties of the fermion creation and annihilation operators. Even for large numbers of particles, this operator remains unchanged.

Chapter 17

Interaction of different kinds of particles

Prerequisites: Chapters 2–7, 9, 10, 12, 13, 15, and 16.

So far in the treatment of creation and annihilation operators, we have considered operators, including especially Hamiltonians, in which we are concerned with only one kind of particle (i.e., either identical bosons or identical fermions). Many important phenomena involve interactions of different kinds of particles (e.g., interactions of photons or phonons with electrons). Here, we discuss how to handle operators for such situations. As a specific, and particularly useful example, we discuss the electron–photon interaction. This leads us through perturbation theory in this operator formalism to a proper quantum mechanical treatment of absorption and stimulated and spontaneous emission.

17.1 States and commutation relations for different kinds of particles

The approach is an extension of what we have done before. We need two additions. First, though we continue to work in the occupation number representation, we must include the description of the occupied single-particle states for each different particle in the overall description of the states. Second, we need commutation relations between operators corresponding to different kinds of particles.

In considering the occupation number basis states – for example, for a system with two different kinds of particles – we simply have to list which states are occupied for each different kind of particle. Suppose that we have a set of identical electrons and a set of identical bosons (e.g., photons). Then, for a state with one fermion in fermion state k and one in state q, and one photon in photon mode λ_d and three in photon mode λ_s, we could write the state either in the form of a list or using creation operators acting on the empty state; that is,

$$\begin{aligned}&\left|\ldots,0_j,1_k,0_l,\ldots 0_p,1_q,0_r,\ldots;\ \ldots,0_{\lambda c},1_{\lambda d},0_{\lambda e},\ldots,0_{\lambda r},3_{\lambda s},0_{\lambda t},\ldots\right\rangle \\ &\equiv\left|N_{fm};N_{bn}\right\rangle\equiv\frac{1}{\sqrt{3!}}\hat{b}_k^\dagger\hat{b}_q^\dagger\hat{a}_{\lambda d}^\dagger\left(\hat{a}_{\lambda s}^\dagger\right)^3\left|0\right\rangle\end{aligned} \tag{17.1}$$

By N_{fm}, we simply mean the mth possible list of occupied fermion states (here, the list $\ldots,0_j,1_k,0_l,\ldots 0_p,1_q,0_r,\ldots$) and, similarly, by N_{bn}, we mean the nth possible list of occupied boson states (here, the list $\ldots,0_{\lambda c},1_{\lambda d},0_{\lambda e},\ldots,0_{\lambda r},3_{\lambda s},0_{\lambda t},\ldots$). Note now that the empty state $\left|0\right\rangle$ is one that is empty both of this kind of fermion and this kind of boson.

As for the commutation relations, we simply postulate that creation and annihilation operators corresponding to nonidentical particles commute under all conditions. Specifically, then, for the boson and fermion operators, we would have

$$\begin{aligned} \hat{b}_j^\dagger \hat{a}_\lambda^\dagger - \hat{a}_\lambda^\dagger \hat{b}_j^\dagger &= 0 \quad \hat{b}_j \hat{a}_\lambda - \hat{a}_\lambda \hat{b}_j = 0 \\ \hat{b}_j^\dagger \hat{a}_\lambda - \hat{a}_\lambda \hat{b}_j^\dagger &= 0 \quad \hat{b}_j \hat{a}_\lambda^\dagger - \hat{a}_\lambda^\dagger \hat{b}_j = 0 \end{aligned} \tag{17.2}$$

Note that such relations also would hold for annihilation and creation operators corresponding to two different kinds of fermions, such as electrons and protons or, for that matter, two different kinds of bosons, such as photons and phonons.

17.2 Operators for systems with different kinds of particles

The basic approach for constructing operators corresponding to interactions between different kinds of particles is progressively to apply the methods appropriate for each particle. Because of the commutation relations for annihilation and creation operators for different particles, there is no particular necessary order for applying these methods. The construction of such operators is probably best understood by illustration.

The case of most interest to us here is that of the interaction of electrons and photons. The basic approach is to use the field operators instead of the classical fields and use the fermion wavefunction operators to transform the fermion position (or momentum) operator to the occupation number form.

Suppose, for example, that we consider the Hamiltonian of the electric dipole interaction between electrons and electromagnetic modes. We presume, first, that we had "turned off" (mathematically, at least) any interaction between the electron and the photons. Then, because there is no interaction, the resulting Hamiltonian would simply be the sum of the separate fermion (electron) and boson (photon) Hamiltonians; that is,

$$\hat{H}_o = \sum_j E_j \hat{b}_j^\dagger \hat{b}_j + \sum_\lambda \hbar\omega_\lambda \hat{a}_\lambda^\dagger \hat{a}_\lambda \tag{17.3}$$

As before, the sum over j is over all possible single-particle fermion states and the sum over λ is over all possible photon modes.

Now, let us consider the interaction between the electrons and the photons. Previously, for the electric dipole interaction, we had from a semiclassical view of the energy of an electron at position $\mathbf{r}_i$ in an electric field $\mathbf{E}$

$$\hat{H}_{scedr} = e\,\mathbf{E}\cdot\mathbf{r} \tag{17.4}$$

(where $\mathbf{r}$ is actually the position operator). Substituting the multimode electric field operator of Eq. (15.132) for the classical field $\mathbf{E}$ gives, for any specific electron i

$$\hat{H}_{edri} = \sqrt{-1}\, e \sum_\lambda \left(\hat{a}_\lambda - \hat{a}_\lambda^\dagger\right) \sqrt{\frac{\hbar\omega_\lambda}{2\varepsilon_o}} \mathbf{u}_\lambda(\mathbf{r}_i).\mathbf{r}_i \tag{17.5}$$

For N electrons, we have to add all of these Hamiltonians; that is,

$$\hat{H}_{edr} = \sum_{i=1}^{N} \sqrt{-1}\, e \sum_\lambda \left(\hat{a}_\lambda - \hat{a}_\lambda^\dagger\right) \sqrt{\frac{\hbar\omega_\lambda}{2\varepsilon_o}} \mathbf{u}_\lambda(\mathbf{r}_i).\mathbf{r}_i \tag{17.6}$$

Now, we want to transform this Hamiltonian in **r** form into the fermion occupation number form also. To do so, we formally use the N-fermion field operators. Because the fermion and boson operators commute with one another, the boson operators also commute with the (fermion) wavefunction operators, so we can write

$$\begin{aligned}\hat{H}_{ed} &= \int \hat{\psi}^\dagger \hat{H}_{ed\mathbf{r}} \hat{\psi} d^3\mathbf{r}_1 d^3\mathbf{r}_2 \ldots d^3\mathbf{r}_N \\ &= \int \hat{\psi}^\dagger \left[\sum_{i=1}^{N} \sqrt{-1}\, e \sum_\lambda \left(\hat{a}_\lambda - \hat{a}_\lambda^\dagger\right) \sqrt{\frac{\hbar\omega_\lambda}{2\varepsilon_o}} \mathbf{u}_\lambda(\mathbf{r}_i).\mathbf{r}_i \right] \hat{\psi}\, d^3\mathbf{r}_1 d^3\mathbf{r}_2 \ldots d^3\mathbf{r}_N \\ &= \sum_\lambda \left(\hat{a}_\lambda - \hat{a}_\lambda^\dagger\right) \int \hat{\psi}^\dagger \left[\sum_{i=1}^{N} \sqrt{-1}\, e \sqrt{\frac{\hbar\omega_\lambda}{2\varepsilon_o}} \mathbf{u}_\lambda(\mathbf{r}_i).\mathbf{r}_i \right] \hat{\psi}\, d^3\mathbf{r}_1 d^3\mathbf{r}_2 \ldots d^3\mathbf{r}_N\end{aligned} \tag{17.7}$$

But, this is a situation we anticipated before. Inside this expression, we have a single-particle fermion operator with a multiple-fermion state, as in Eq. (16.65), which we can write in the **r** form as

$$\hat{H}_{ed\lambda\mathbf{r}} = \sum_{i=1}^{N} \hat{H}_{ed\lambda\mathbf{r}i} \tag{17.8}$$

where here

$$\hat{H}_{ed\lambda\mathbf{r}i} = \sqrt{-1}\, e \sqrt{\frac{\hbar\omega_\lambda}{2\varepsilon_o}} \mathbf{u}_\lambda(\mathbf{r}_i).\mathbf{r}_i \tag{17.9}$$

This is a single-particle fermion operator because each $\hat{H}_{ed\lambda\mathbf{r}i}$ only depends on the coordinates of one fermion.

With this notation, Eq. (17.7) becomes

$$\hat{H}_{ed} = \sum_\lambda \left(\hat{a}_\lambda - \hat{a}_\lambda^\dagger\right) \int \hat{\psi}^\dagger \hat{H}_{ed\lambda\mathbf{r}} \hat{\psi}\, d^3\mathbf{r}_1 d^3\mathbf{r}_2 \ldots d^3\mathbf{r}_N \tag{17.10}$$

Now, we can use our specific previous results for single-particle fermion operators with multiple-fermion states, which allow us to write (as in Eq. (16.73))

$$\int \hat{\psi}^\dagger \hat{H}_{ed\lambda\mathbf{r}} \hat{\psi}\, d^3\mathbf{r}_1 d^3\mathbf{r}_2 \ldots d^3\mathbf{r}_N = \sum_{j,k} H_{ed\lambda jk} \hat{b}_j^\dagger \hat{b}_k \tag{17.11}$$

where (as in Eq. (16.69))

$$\begin{aligned}H_{ed\lambda jk} &= \int \phi_j^*(\mathbf{r}_i) \hat{H}_{ed\lambda\mathbf{r}i} \phi_k(\mathbf{r}_i) d^3\mathbf{r}_i \\ &= \sqrt{-1}\, e \sqrt{\frac{\hbar\omega_\lambda}{2\varepsilon_o}} \int \phi_j^*(\mathbf{r}_i)\left[\mathbf{u}_\lambda(\mathbf{r}_i).\mathbf{r}_i\right] \phi_k(\mathbf{r}_i) d^3\mathbf{r}_i\end{aligned} \tag{17.12}$$

Hence, substituting back into expression Eq. (17.7) for this perturbing electric-dipole Hamiltonian, we have

$$\hat{H}_{ed} = \sum_{j,k,\lambda} H_{ed\lambda jk} \hat{b}_j^\dagger \hat{b}_k \left(\hat{a}_\lambda - \hat{a}_\lambda^\dagger\right) \tag{17.13}$$

In such an expression, all of the details of the specific form of the single-particle fermion states and of the electromagnetic modes are contained within the constants $H_{ed\lambda jk}$.

The annihilation and creation operators identify specific processes that could occur given appropriate starting states. We can open up the creation and annihilation operator expression

$$\hat{b}_j^\dagger \hat{b}_k \left(\hat{a}_\lambda - \hat{a}_\lambda^\dagger\right) = \hat{b}_j^\dagger \hat{b}_k \hat{a}_\lambda - \hat{b}_j^\dagger \hat{b}_k \hat{a}_\lambda^\dagger \tag{17.14}$$

Hence, for example, if fermion state k was occupied and fermion state j was empty, and we had at least one photon in mode λ, then we could have a process corresponding to the operators $\hat{b}_j^\dagger \hat{b}_k \hat{a}_\lambda$, which involves annihilating a photon in mode λ and changing an electron from state k to state j. That is, we are describing an absorption process in which absorption of a photon takes an electron from one state to another. Similarly, the process corresponding to the operators $\hat{b}_j^\dagger \hat{b}_k \hat{a}_\lambda^\dagger$ is one of emission of a photon as an electron goes from state k to state j. We return to evaluating transition rates for such processes once we have discussed time-dependent perturbation theory for this formalism.

Problem

17.2.1 Suppose that we are going to describe a number of particles in each case using the basis set of plane waves of different directions and/or different wavevector magnitudes, indexed by the wavevector $\mathbf{k}$. Suppose we have one or more electrons (all of the same spin), with the annihilation operator $\hat{b}_{\mathbf{k}}$ corresponding to the annihilation of an electron in basis plane-wave state of wavevector $\mathbf{k}$. Similarly, consider one or more protons (all with the same spin) with annihilation operators $\hat{d}_{\mathbf{k}}$, and one or more photons (all with the same polarization in any given plane-wave state) with annihilation operators $\hat{a}_{\mathbf{k}}$. For each of the following sets of operators, describe the process being represented by the operators and if that process necessarily cannot ever actually happen because of basic rules such as Pauli exclusion, say so. (E.g., the operators $\hat{b}_{\mathbf{k}1}^\dagger \hat{b}_{\mathbf{k}2} \hat{a}_{\mathbf{k}3}$ correspond to the process of absorption of a photon from the plane-wave state with wavevector $\mathbf{k}_3$ and changing an electron from the plane wave state with wavevector $\mathbf{k}_2$ to that with wavevector $\mathbf{k}_1$.)

(i) $\hat{d}_{\mathbf{k}1}^\dagger \hat{d}_{\mathbf{k}2} \hat{a}_{\mathbf{k}3}$

(ii) $\hat{b}_{\mathbf{k}1}^\dagger \hat{b}_{\mathbf{k}2} \left(\hat{a}_{\mathbf{k}3}\right)^2$

(iii) $\hat{d}_{\mathbf{k}1}^\dagger \hat{d}_{\mathbf{k}2}^\dagger \hat{d}_{\mathbf{k}3} \hat{d}_{\mathbf{k}4}$

(iv) $\left(\hat{b}_{\mathbf{k}1}^\dagger\right)^2 \hat{b}_{\mathbf{k}3} \hat{b}_{\mathbf{k}4}$

(v) $\hat{b}_{\mathbf{k}1}^\dagger \hat{d}_{\mathbf{k}1}^\dagger \hat{b}_{\mathbf{k}3} \hat{d}_{\mathbf{k}4}$

(vi) $\hat{b}_{\mathbf{k}1}^\dagger \hat{a}_{\mathbf{k}2}^\dagger \hat{b}_{\mathbf{k}3} \hat{a}_{\mathbf{k}4}$

17.3 Perturbation theory with annihilation and creation operators

The time-dependent perturbation theory we derived remains valid as we change the way we write the Hamiltonian and the states in the occupation number and annihilation and creation operator form. We have not changed the underlying nature of the quantum mechanics and are still interested in the time evolution of the amplitudes of the various possible states of the system.

When we use perturbation theory for states and operators in this occupation number form, we are usually considering transitions caused by interactions between different particles. We have an unperturbed Hamiltonian, $\hat{H}_o$, such as the one for noninteracting fermions and bosons in Eq. (17.3). Then, we consider the interactions between particles, such as the electric dipole interaction discussed for electrons and photons, Eq. (17.13), as a perturbation. Note, incidentally, that this approach works for any kinds of particles; we could, for example, apply it to electron–electron scattering. It is, of course, not necessary that the interaction is between different kinds of particles for us to apply perturbation theory. For definiteness here, we discuss a system in which there is one kind of fermion (which we can think of as electrons) and one kind of boson (which we can think of as photons).

For completeness, we briefly write out the key steps in first-order time-dependent perturbation theory in the notation of our current approach. First, we note that the quantum mechanical states we are using are those of the entire system. Previously, we might have considered only the electron state, treating the perturbation, such as an electric dipole perturbation, as being from something external to the quantum system, such as a classical field. Now, our basis states must describe both the occupation of each single-particle electron state and the occupation of each boson mode. Hence, we write our basis states as in Eq. (17.1). Specifically, the mth state of this entire (noninteracting) fermion–boson system can be written as $|N_{fm};N_{bm}\rangle$, where N_{fm} is the list of all the occupation numbers of each possible single-particle fermion state and, similarly, N_{bm} is the list of all the occupation numbers of each possible boson mode. These states are the eigenstates of the unperturbed Hamiltonian, which we take as the $\hat{H}_o$ of Eq. (17.3) that is simply the sum of the separate fermion and boson Hamiltonians. Analogous to Eq. (7.3), we therefore have

$$\hat{H}_o|N_{fm};N_{bm}\rangle = E_m|N_{fm};N_{bm}\rangle \tag{17.15}$$

where E_m would be the energy of this fermion–boson system in state m in the absence of any interaction between the fermions and bosons.

The actual state of the system is some linear superposition $|\psi\rangle$, which we expand in the previous basis; that is, analogous to Eq. (7.4), we have

$$|\psi\rangle = \sum_m c_m \exp(-iE_m t/\hbar)|N_{fm};N_{bm}\rangle \tag{17.16}$$

where we have explicitly added the time-varying factors $\exp(-iE_m t/\hbar)$ so that we can leave them out of the states $|N_{fm};N_{bm}\rangle$.[1] Note again that in contrast to previous approaches that treated perturbations as external phenomena, E_m is the energy of the complete (unperturbed) fermion–boson system in this state m, not merely the energy of the fermion.

The state $|\psi\rangle$ is presumed to obey the time-dependent Schrödinger equation with the complete Hamiltonian, including the perturbation $\hat{H}_p$; that is, analogous to Eq. (7.2)

$$i\hbar\frac{\partial}{\partial t}|\psi\rangle = \left(\hat{H}_o + \hat{H}_p\right)|\psi\rangle \tag{17.17}$$

Substituting Eq. (17.16) in Eq. (17.17), eliminating terms on both sides using Eq. (17.15), and premultiplying by the bra for state q of the fermion–boson system, $\langle N_{fq};N_{bq}|$, gives analogously to Eq. (7.7)

$$i\hbar\dot{c}_q \exp(-iE_q t/\hbar) = \sum_m c_m \exp(-iE_m t/\hbar)\langle N_{fq};N_{bq}|\hat{H}_p|N_{fm};N_{bm}\rangle \tag{17.18}$$

[1] We have also used c rather than a for the expansion coefficients to avoid confusion with boson operators.

Taking the usual perturbation approach of basing the first-order change in wavefunctions on the zeroth-order state (i.e., on the unperturbed wavefunctions), we have, analogously to Eq. (7.10)

$$\dot{c}_q^{(1)} \simeq \frac{1}{i\hbar}\sum_m c_m^{(0)} \exp\left[-i\left(E_m - E_q\right)t/\hbar\right]\left\langle N_{fq};N_{bq}\left|\hat{H}_p\right|N_{fm};N_{bm}\right\rangle \tag{17.19}$$

Now, in our new notation, we are ready to use time-dependent perturbation theory. We use as an example the important process of optical emission and absorption, including spontaneous emission.

Problem

17.3.1* Suppose we have electrons (which are fermions with annihilation operator $\hat{b}$) and ^{4}He nuclei (which are bosons [here with an annihilation operator $\hat{c}$] because they consist of an even number of fermions and which are also charged because they are ionized). The Coulomb interaction between these two particles can be described by a (perturbing) Hamiltonian of the form

$$\hat{H}_{Cen} = \sum_{j,k,\lambda,\mu} H_{Cjk\lambda\mu}\hat{b}_j^\dagger\hat{c}_\lambda^\dagger\hat{b}_k\hat{c}_\mu$$

Evaluate using the creation and annihilation operator formalism the matrix element $\left\langle N_{fq};N_{bq}\left|\hat{H}_{Cen}\right|N_{fm};N_{bm}\right\rangle$ for an initial state (i.e., $\left|N_{fm};N_{bm}\right\rangle$) corresponding to one electron in single-particle state u with no other electron states occupied and one ^{4}He nucleus in mode α with no other ^{4}He states occupied, and a final state $\left|N_{fq};N_{bq}\right\rangle$ corresponding to one electron in state v with no other electron states occupied and one ^{4}He nucleus in state β with no other ^{4}He states occupied. (Note: The answer to this is rather simple and even obvious, but this problem gives an elementary exercise in using the commutation properties of the various operators.)

17.4 Stimulated emission, spontaneous emission, and optical absorption

Optical emission and absorption are particularly common and practically important processes that can be understood using time-dependent perturbation theory in the occupation number and annihilation and creation operator form. Spontaneous emission is the process responsible for the vast majority of all photons we see yet can only be understood using the quantized electromagnetic field. The annihilation and creation operator approach gives a particularly straightforward way of describing and comparing spontaneous and stimulated emission of photons and also optical absorption.

We consider a simple situation to expose the key mechanisms and their behavior. Suppose that we start with the electron–photon system in some specific basis state s, so that $c_s^{(0)} = 1$ and all other such coefficients are zero. Then, Eq. (17.19) becomes

$$\dot{c}_q^{(1)} \simeq \frac{1}{i\hbar}\exp\left[i\left(E_q - E_s\right)t/\hbar\right]\left\langle N_{fq};N_{bq}\left|\hat{H}_p\right|N_{fs};N_{bs}\right\rangle \tag{17.20}$$

Now, let us take as our perturbing Hamiltonian the electric dipole interaction of Eq. (17.13)

$$\hat{H}_p = \hat{H}_{ed} = \sum_{j,k,\lambda} H_{ed\lambda jk}\hat{b}_j^\dagger\hat{b}_k\left(\hat{a}_\lambda - \hat{a}_\lambda^\dagger\right) \tag{17.21}$$

For simplicity here, we presume we have only one electron and that there are only two single-particle states of interest for this electron:

State 1 – the lowest energy state of the electron, with energy E_1

State 2 – the upper state of the electron, with energy E_2

We consider the three possible processes of photon absorption, spontaneous emission and stimulated emission.

Absorption

Consider first that the electron is initially in state 1 (the lower state), there is one photon in mode λ_1, and there are no photons in any other modes. Then, we can write the initial state as

$$\left|N_{fs};N_{bs}\right\rangle = \hat{b}_1^\dagger \hat{a}_{\lambda_1}^\dagger \left|0\right\rangle \tag{17.22}$$

This state has an energy

$$E_s = E_1 + \hbar\omega_{\lambda_1} \tag{17.23}$$

(Here and in the following arguments, we omit all of the additional $\hbar\omega_\lambda/2$ contributions to the energy that we usually acquire from the zero point energy of the harmonic oscillator. This merely corresponds to a choice of energy origin and here, anyway, we would have to know how many modes altogether we were going to consider if we were to leave this term in the energy expressions.) Now, we have

$$\hat{H}_p\left|N_{fs};N_{bs}\right\rangle = \sum_{j,k,\lambda} H_{ed\lambda jk}\hat{b}_j^\dagger \hat{b}_k\left(\hat{a}_\lambda - \hat{a}_\lambda^\dagger\right)\hat{b}_1^\dagger \hat{a}_{\lambda_1}^\dagger\left|0\right\rangle \tag{17.24}$$

Examining the sequence of operators, we have

$$\begin{aligned}
\hat{b}_j^\dagger \hat{b}_k\left(\hat{a}_\lambda - \hat{a}_\lambda^\dagger\right)\hat{b}_1^\dagger \hat{a}_{\lambda_1}^\dagger\left|0\right\rangle &= \hat{b}_j^\dagger \hat{b}_k \hat{b}_1^\dagger\left(\hat{a}_\lambda \hat{a}_{\lambda_1}^\dagger - \hat{a}_\lambda^\dagger \hat{a}_{\lambda_1}^\dagger\right)\left|0\right\rangle \\
&= \hat{b}_j^\dagger\left(\delta_{k1} - \hat{b}_1^\dagger \hat{b}_k\right)\left(\delta_{\lambda\lambda_1} + \hat{a}_{\lambda_1}^\dagger \hat{a}_\lambda - \hat{a}_\lambda^\dagger \hat{a}_{\lambda_1}^\dagger\right)\left|0\right\rangle \\
&= \hat{b}_j^\dagger \delta_{k1}\left(\delta_{\lambda\lambda_1} - \hat{a}_\lambda^\dagger \hat{a}_{\lambda_1}^\dagger\right)\left|0\right\rangle \\
&= \delta_{k1}\delta_{\lambda\lambda_1}\hat{b}_j^\dagger\left|0\right\rangle - \delta_{k1}\hat{b}_j^\dagger \hat{a}_\lambda^\dagger \hat{a}_{\lambda_1}^\dagger\left|0\right\rangle
\end{aligned} \tag{17.25}$$

When we come to form $\left\langle N_{fq};N_{bq}\middle|\hat{H}_p\middle|N_{fs};N_{bs}\right\rangle$ as needed in Eq. (17.20), we only get nonzero answers if our final state also has either $\hat{b}_j^\dagger\left|0\right\rangle$ or $\hat{b}_j^\dagger \hat{a}_\lambda^\dagger \hat{a}_{\lambda_1}^\dagger\left|0\right\rangle$ in it. Hence, there are only two possible basis state choices for the final state $\left|N_{fq};N_{bq}\right\rangle$ that give nonzero results.

(i) One possibility is

$$\left|N_{fq};N_{bq}\right\rangle = \hat{b}_j^\dagger\left|0\right\rangle \tag{17.26}$$

which is the state with one electron in state j and no photons in any modes. This state has energy

$$E_q = E_j \tag{17.27}$$

which leads to

$$\begin{aligned}
\dot{c}_q^{(1)} &\simeq \frac{1}{i\hbar}\exp\left[i\left(E_j - E_1 - \hbar\omega_{\lambda_1}\right)t/\hbar\right]\sum_{k,\lambda} H_{ed\lambda jk}\delta_{k1}\delta_{\lambda\lambda_1}\left\langle 0\middle|\hat{b}_j\hat{b}_j^\dagger\middle|0\right\rangle \\
&= \frac{1}{i\hbar}\exp\left[i\left(E_j - E_1 - \hbar\omega_{\lambda_1}\right)t/\hbar\right]H_{ed\lambda_1 j1}
\end{aligned} \tag{17.28}$$

Now, we integrate over time. This leads to a similar result to what we obtained before in time-dependent perturbation theory for an oscillating perturbation (i.e., Fermi's Golden Rule), including eventually a Dirac δ-function that enforces conservation of energy for steady transition rates. By definition, we choose $c_q^{(1)}(t=0)=0$ because we regard the system as starting in the specified initial state at $t = 0$. Hence, integrating from $t = 0$ to t_o, we have

$$\begin{aligned} c_q^{(1)}(t_o) &= -\frac{H_{ed\lambda_1 j1}}{E_j - E_1 - \hbar\omega_{\lambda_1}}\left\{\exp\left[i\left(E_j - E_1 - \hbar\omega_{\lambda_1}\right)t_o/\hbar\right]-1\right\} \\ &= -2iH_{ed\lambda_1 j1}\exp\left[i\left(E_j - E_1 - \hbar\omega_{\lambda_1}\right)t_o/2\hbar\right]\frac{\sin\left[\left(E_j - E_1 - \hbar\omega_{\lambda_1}\right)t_o/2\hbar\right]}{E_j - E_1 - \hbar\omega_{\lambda_1}} \end{aligned} \quad (17.29)$$

So

$$\begin{aligned} \left|c_q^{(1)}(t_o)\right|^2 &= 4\left|H_{ed\lambda_1 j1}\right|^2\frac{\sin^2\left[\left(E_j - E_1 - \hbar\omega_{\lambda_1}\right)t_o/2\hbar\right]}{\left(E_j - E_1 - \hbar\omega_{\lambda_1}\right)^2} \\ &= \frac{2\pi}{\hbar}t_o\left|H_{ed\lambda_1 j1}\right|^2\left\{\frac{1}{t_o}\frac{2\hbar}{\pi}\frac{\sin^2\left[\left(E_j - E_1 - \hbar\omega_{\lambda_1}\right)t_o/2\hbar\right]}{\left(E_j - E_1 - \hbar\omega_{\lambda_1}\right)^2}\right\} \end{aligned} \quad (17.30)$$

Now, the function in curly brackets ($\{\ldots\}$) is a sharply peaked function near $E_j - E_1 - \hbar\omega_{\lambda_1} = 0$ and it has unit area when integrated over this energy argument (note that $\int_{-\infty}^{\infty}[(\sin^2 x)/x^2]\,dx = \pi$). Hence, in the limit of large t_o, it can be replaced by the Dirac δ-function; that is,

$$\left|c_q^{(1)}(t_o)\right|^2 = \frac{2\pi}{\hbar}t_o\left|H_{ed\lambda_1 j1}\right|^2\delta\left(E_j - E_1 - \hbar\omega_{\lambda_1}\right) \quad (17.31)$$

which we see gives an occupation probability steadily rising with time t_o for this state q. Hence the transition rate is

$$w_q \simeq \frac{2\pi}{\hbar}\left|H_{ed\lambda_1 j1}\right|^2\delta\left(E_j - E_1 - \hbar\omega_{\lambda_1}\right) \quad (17.32)$$

Now, for $j = 1$, the δ-function vanishes for any finite $\hbar\omega_{\lambda_1}$ – we simply cannot get the energies to match if the final electron state is the same as the initial electron state – so the only final state q that gives a nonzero transition rate is the state $j = 2$, with the corresponding restriction that

$$E_2 - E_1 = \hbar\omega_{\lambda_1} \quad (17.33)$$

Hence, our process is:

We start with one photon in mode λ_1 and the electron in state 1.

We finish with no photons and the electron in state 2.

which describes a normal absorption process, correctly now requiring the destruction of the photon in the process, with transition rate given by Eq. (17.32).

(ii) The other possibility for the final state that we have to consider, given Eq. (17.25), is

$$\left|N_{fq};N_{bq}\right\rangle = \hat{b}_j^\dagger\hat{a}_\lambda^\dagger\hat{a}_{\lambda_1}^\dagger\left|0\right\rangle \quad (17.34)$$

with a corresponding energy

$$E_q = E_j + \hbar\omega_\lambda + \hbar\omega_{\lambda_1} \tag{17.35}$$

But, $E_q - E_s = E_j - E_1 + \hbar\omega_\lambda$ cannot be close to zero because $E_j - E_1 \geq 0$ and $\hbar\omega_\lambda$ is also positive. Hence, on integrating over time as before, this term does not give rise to any steady transition rate. (As we will see, this term would actually correspond to photon emission, but we cannot emit a photon because we are starting in the lowest energy electron state; hence, there is no lower state to which we can emit.) Hence, this possibility can be discarded here.

Hence, having started with the electron in the lower state and one photon in a mode, the only process we are left with is the optical absorption process described in (i).

Spontaneous emission

Suppose now that the electron is initially in state 2 (the upper state) and there are no photons in any mode. This situation is not like any we have considered before in our semiclassical analysis. Indeed, semiclassically it would be trivial; with no electromagnetic field, there would be no transitions. The result now, though, is different.

Our starting state now is

$$|N_{fs}; N_{bs}\rangle = \hat{b}_2^\dagger |0\rangle \tag{17.36}$$

with a corresponding energy

$$E_s = E_2 \tag{17.37}$$

Now, in forming $\hat{H}_p |N_{fs}; N_{bs}\rangle$, we encounter the string of operators

$$\begin{aligned}\hat{b}_j^\dagger \hat{b}_k \left(\hat{a}_\lambda - \hat{a}_\lambda^\dagger\right) \hat{b}_2^\dagger |0\rangle &= \hat{b}_j^\dagger \hat{b}_k \hat{b}_2^\dagger \left(\hat{a}_\lambda - \hat{a}_\lambda^\dagger\right) |0\rangle \\ &= \hat{b}_j^\dagger \left(\delta_{2k} - \hat{b}_2^\dagger \hat{b}_k\right)\left(\hat{a}_\lambda - \hat{a}_\lambda^\dagger\right)|0\rangle \\ &= -\delta_{2k} \hat{b}_j^\dagger \hat{a}_\lambda^\dagger |0\rangle\end{aligned} \tag{17.38}$$

So that we get a nonzero result for $\langle N_{fq}; N_{bq} | \hat{H}_p | N_{fs}; N_{bs}\rangle$, we must, therefore, choose for state q

$$|N_{fq}; N_{bq}\rangle = \hat{b}_j^\dagger \hat{a}_\lambda^\dagger |0\rangle \tag{17.39}$$

which is the state with the electron now in state j and a photon in mode λ. This state q has energy

$$E_q = E_j + \hbar\omega_\lambda \tag{17.40}$$

Hence, we now have for this state q

$$\begin{aligned}\dot{c}_q^{(1)} &\simeq \frac{1}{i\hbar} \exp\left[i\left(E_j - E_2 + \hbar\omega_\lambda\right)t/\hbar\right] \sum_k H_{ed\lambda jk} \delta_{k2} \langle 0 | \hat{a}_\lambda \hat{b}_j \hat{b}_j^\dagger \hat{a}_\lambda^\dagger |0\rangle \\ &= \frac{1}{i\hbar} \exp\left[i\left(E_j - E_2 + \hbar\omega_\lambda\right)t/\hbar\right] H_{ed\lambda j2}\end{aligned} \tag{17.41}$$

Integrating and taking $\left|c_q^{(1)}\right|^2$ to get the transition rate gives

$$w_q = \frac{2\pi}{\hbar}\left|H_{ed\lambda j2}\right|^2 \delta\left(E_j - E_2 + \hbar\omega_\lambda\right) \tag{17.42}$$

As before, for any finite $\hbar\omega_\lambda$, the only possible choice is $j = 1$ for the final state if there is to be any transition rate, with the requirement

$$E_2 - E_1 = \hbar\omega_\lambda \tag{17.43}$$

that is, we have

$$w_q = \frac{2\pi}{\hbar}\left|H_{ed\lambda 12}\right|^2 \delta\left(E_1 - E_2 + \hbar\omega_\lambda\right) \tag{17.44}$$

This transition process is spontaneous emission. The electron starts in its higher state 2 with no photons present and ends in its lower state 1 with one photon present. This photon can be in any mode λ that has the correct photon energy to match the energy separation $E_2 - E_1$ (and for which the coefficient $H_{ed\lambda 12}$ is not formally zero for some other reason). This process has emerged naturally as a consequence of quantizing the electromagnetic field, requiring essentially no additional physics other than that quantization.

Stimulated emission

We have one final and important process to consider, which is stimulated emission. This process is strong in laser light, though it is also present in small amounts all the time and is necessary in order to make the statistical mechanics of light agree with observation.[2]

Suppose now that we have a photon in mode λ_1 and also have an electron in its upper state 2. The initial state is, therefore

$$\left|N_{fs}; N_{bs}\right\rangle = \hat{b}_2^\dagger \hat{a}_{\lambda_1}^\dagger \left|0\right\rangle \tag{17.45}$$

with an energy

$$E_s = E_2 + \hbar\omega_{\lambda_1} \tag{17.46}$$

Then, with algebra similar to Eq. (17.25)

$$\hat{b}_j^\dagger \hat{b}_k \left(\hat{a}_\lambda - \hat{a}_\lambda^\dagger\right)\hat{b}_2^\dagger \hat{a}_{\lambda_1}^\dagger \left|0\right\rangle = \delta_{k2}\delta_{\lambda\lambda_1}\hat{b}_j^\dagger\left|0\right\rangle - \delta_{k2}\hat{b}_j^\dagger \hat{a}_\lambda^\dagger \hat{a}_{\lambda_1}^\dagger\left|0\right\rangle \tag{17.47}$$

The first term here is simply the absorption term, but this vanishes because there is no electron state into which we can absorb, given that we are starting in the upper state. The second term has two possibilities in the summation, which we now consider.

(i) Suppose $\lambda \neq \lambda_1$. Then, for some specific λ, to get a nonzero result for $\left\langle N_{fq}; N_{bq}\right|\hat{H}_p\left|N_{fs}; N_{bs}\right\rangle$, the final state has to be

$$\left|N_{fq}; N_{bq}\right\rangle = \hat{b}_j^\dagger \hat{a}_\lambda^\dagger \hat{a}_{\lambda_1}^\dagger \left|0\right\rangle \tag{17.48}$$

with energy

$$E_q = E_j + \hbar\omega_\lambda + \hbar\omega_{\lambda_1} \tag{17.49}$$

---

[2] This is Einstein's famous "A and B" coefficient argument. See, e.g., H. Haken, *Light (Vol. 1)*, (North-Holland, Amsterdam, 1981), pp. 58–62.

corresponding to a state with the electron in level j and a photon in each of the different modes λ and λ_1. We have, for some specific λ and j

$$\begin{aligned}\dot{c}_q^{(1)} &\simeq \frac{-1}{i\hbar}\exp\left[i\left(E_j - E_2 + \hbar\omega_\lambda\right)t/\hbar\right]H_{ed\lambda j2}\langle 0|\hat{a}_{\lambda_1}\hat{a}_\lambda\hat{b}_j\hat{b}_j^\dagger\hat{a}_\lambda^\dagger\hat{a}_{\lambda_1}^\dagger|0\rangle \\ &= \frac{-1}{i\hbar}\exp\left[i\left(E_j - E_2 + \hbar\omega_\lambda\right)t/\hbar\right]H_{ed\lambda j2}\end{aligned} \tag{17.50}$$

leading to a transition rate

$$w_q = \frac{2\pi}{\hbar}\left|H_{ed\lambda j2}\right|^2 \delta\left(E_j - E_2 + \hbar\omega_\lambda\right) \tag{17.51}$$

for which the only possibility here for a nonzero transition rate is $j = 1$, and

$$E_2 - E_1 = \hbar\omega_\lambda \tag{17.52}$$

with a transition rate

$$w_q = \frac{2\pi}{\hbar}\left|H_{ed\lambda 12}\right|^2 \delta\left(E_1 - E_2 + \hbar\omega_\lambda\right) \tag{17.53}$$

This process is just spontaneous emission into mode λ and the transition rate Eq. (17.53) is identical to that of Eq. (17.44). The only point of this current derivation is to show explicitly that the presence of a photon in another mode has no influence on spontaneous emission. (In fact, spontaneous emission does not care how many photons are in any other modes.)

(ii) Suppose that now we consider the case $\lambda = \lambda_1$. Now to get a nonzero result for $\langle N_{fq}; N_{bq} | \hat{H}_p | N_{fs}; N_{bs}\rangle$, the final state has to be

$$\left|N_{fq}; N_{bq}\right\rangle = \frac{1}{\sqrt{2!}}\hat{b}_j^\dagger\left(\hat{a}_{\lambda_1}^\dagger\right)^2|0\rangle \tag{17.54}$$

with energy

$$E_q = E_j + 2\hbar\omega_{\lambda_1} \tag{17.55}$$

Note that to have a normalized state here, we have had to introduce the factor $1/\sqrt{2!}$ (see, for example, Eq. (15.72) or Eq. (15.126)). Hence, we obtain a term in

$$\left\langle N_{fq}; N_{bq}\left|\hat{H}_p\right|N_{fs}; N_{bs}\right\rangle = \left\langle N_{fq}; N_{bq}\left|\sum_{j,k,\lambda} H_{ed\lambda jk}\hat{b}_j^\dagger\hat{b}_k\left(\hat{a}_\lambda - \hat{a}_\lambda^\dagger\right)\right|N_{fs}; N_{bs}\right\rangle \tag{17.56}$$

that is

$$\begin{aligned}&H_{ed\lambda_1 j2}\langle 0|\frac{1}{\sqrt{2!}}\left(\hat{a}_{\lambda_1}\right)^2\hat{b}_j\hat{b}_j^\dagger\left(\hat{a}_{\lambda_1}^\dagger\right)^2|0\rangle \\ &= \sqrt{2!}\,H_{ed\lambda_1 j2}\langle 0|\frac{1}{\sqrt{2!}}\left(\hat{a}_{\lambda_1}\right)^2\hat{b}_j\frac{1}{\sqrt{2!}}\hat{b}_j^\dagger\left(\hat{a}_{\lambda_1}^\dagger\right)^2|0\rangle \\ &= \sqrt{2}H_{ed\lambda_1 j2}\end{aligned} \tag{17.57}$$

The $\sqrt{2}$ is very important – it shows we are getting a larger amplitude for this process than we did for the spontaneous emission term just calculated. This $\sqrt{2}$ can be traced back to the fact that we started with one photon in this mode λ_1 and created another one.

Hence, for this process, we have

$$\dot{c}_q^{(1)} \simeq \frac{-1}{i\hbar}\exp\left[i\left(E_j - E_2 + \hbar\omega_{\lambda_1}\right)t/\hbar\right]\sqrt{2}\, H_{ed\lambda_1 j2} \tag{17.58}$$

leading to a transition rate into this final state of

$$w_q = \frac{2\pi}{\hbar} 2\left|H_{ed\lambda_1 j2}\right|^2 \delta\left(E_j - E_2 + \hbar\omega_{\lambda_1}\right) \tag{17.59}$$

for which the only possibility for finite transition rate is with $j = 1$ and

$$E_2 - E_1 = \hbar\omega_{\lambda_1} \tag{17.60}$$

with a corresponding transition rate, finally, of

$$w_q = \frac{2\pi}{\hbar} 2\left|H_{ed\lambda_1 12}\right|^2 \delta\left(E_1 - E_2 + \hbar\omega_{\lambda_1}\right) \tag{17.61}$$

Note, in particular, the additional factor of 2 that has appeared in (17.61). The additional process going on here is stimulated emission into mode λ_1. Note that other things being equal (e.g., matrix elements and energies), the transition rate into the mode already occupied with a photon is twice as high as the spontaneous emission into an unoccupied mode. Bosons want to go into modes that are already occupied.

A conventional way to look at this is to say that the spontaneous emission process into mode λ_1 is still going on and it accounts, on the average, for half of the factor of 2 in Eq. (17.61), with the additional stimulated emission process accounting for the other half of the factor of 2. Thus, with one input photon per mode, the spontaneous and stimulated emission rates into the mode are equal on the average. Of course, if they are transitions into the same mode, the resulting photons are completely indistinguishable in the end and the algebra here does not care how we choose to re-express it in words.

Multiple-photon case

It is left as an exercise for the reader to analyze the case of n_{λ_1} photons initially in mode λ_1.

Stimulated emission

The result for stimulated emission in that case is

$$w_q = \frac{2\pi}{\hbar}\left(n_{\lambda_1} + 1\right)\left|H_{ed\lambda_1 12}\right|^2 \delta\left(E_1 - E_2 + \hbar\omega_{\lambda_1}\right) \tag{17.62}$$

with the transition rate into the mode λ_1 being $n_{\lambda_1} + 1$ times larger than the spontaneous rate into an otherwise similar mode. Again, we can view the "1" here in the factor $n_{\lambda_1} + 1$ as being from spontaneous emission and the n_{λ_1} as being the additional transition rate from stimulated emission.

Spontaneous emission

The spontaneous emission in any other mode is unaffected by the presence of n_{λ_1} photons in mode λ_1, as can be shown by directly considering the multiple-photon case.

Absorption

The result for absorption with n_{λ_1} photons initially in mode λ_1 can similarly be shown to be

$$w_q = \frac{2\pi}{\hbar} n_{\lambda_1} \left| H_{ed\lambda_1 12} \right|^2 \delta\left(E_2 - E_1 - \hbar\omega_{\lambda_1}\right) \tag{17.63}$$

Note specifically that we wrote the matrix element $H_{ed\lambda_1 12}$, not the matrix element $H_{ed\lambda_1 21}$ in Eq. (17.63). Given the definition of $H_{ed\lambda_1 jk}$ (Eq. (17.12)), we see that

$$H_{ed\lambda_1 12} = H^*_{ed\lambda_1 21} \tag{17.64}$$

and so the squared moduli are the same. (This is a general property for any kind of perturbing Hamiltonian because Hamiltonians must be Hermitian.)

The relation between the absorption and stimulated emission strengths is fundamental, as is their relation to the spontaneous emission strengths. There is only one set of matrix elements involved in all of these processes.[3]

Total spontaneous emission rate

To complete this discussion of spontaneous emission and to enable us to calculate actual rates of the decay of systems from higher to lower (electron) states, we need to add up the spontaneous emission rates for all possible modes. In the course of this, we will formally manage to get rid of the delta function, getting a finite answer for the complete problem of spontaneous emission. We presume that we start off with the electron in an excited state (here, it must be state 2 by assumption) and no photons in any modes (though such photons do not make any difference to the spontaneous emission calculation – we would just also have to deal with stimulated emission). The total spontaneous transition rate for interaction with all possible modes is the sum of the transition rates into all possible final states q through spontaneous emission

$$W_{spon} = \sum_q w_q \tag{17.65}$$

where w_q is the spontaneous emission rate into a specific mode λ, as given by Eq. (17.44). Because we know that the electron must start in state 2 and end in state 1, the sum over final states reduces to summing over all possible photon modes λ. Hence, we have

$$W_{spon} = \frac{2\pi}{\hbar} \sum_\lambda \left| H_{ed\lambda 12} \right|^2 \delta\left(E_1 - E_2 + \hbar\omega_\lambda\right) \tag{17.66}$$

To get our final result, we need to be able to evaluate $H_{ed\lambda 12}$ and to perform the sum over the modes. First, we look at evaluating the matrix element.

As is common with electric dipole interactions, we presume that the field is approximately uniform over the scale of the quantum system of interest, so we can replace $\mathbf{u}_\lambda(\mathbf{r})$ by $\mathbf{u}_\lambda(\mathbf{r}_o)$ where $\mathbf{r}_o$ is the approximate position of the quantum system of interest. Hence, for the matrix element, we need here (Eq. (17.12))

$$H_{ed\lambda jk} = i\, e\sqrt{\frac{\hbar\omega_\lambda}{2\varepsilon_o}} \int \phi_j^*(\mathbf{r})\left[\mathbf{u}_\lambda(\mathbf{r})\cdot\mathbf{r}\right]\phi_k(\mathbf{r})\, d^3\mathbf{r} = i\, e\sqrt{\frac{\hbar\omega_\lambda}{2\varepsilon_o}}\mathbf{u}_\lambda(\mathbf{r}_o)\cdot\mathbf{r}_{jk} \tag{17.67}$$

---

[3] Einstein had derived these relations by consideration of thermodynamic equilibrium and statistical mechanics with only relatively minor assumptions about atomic transitions, predicting in the process the necessity of the concept of stimulated emission, in his famous "A and B coefficients" argument. See, e.g., H. Haken, *Light (Vol. 1)* (North-Holland, Amsterdam, 1981), pp. 58–62.

where

$$\mathbf{r}_{jk} = \int \phi_j^* (\mathbf{r}) \mathbf{r} \phi_k (\mathbf{r}) d^3\mathbf{r} \tag{17.68}$$

Next, we need to choose the form of the modes. For our calculation here, we presume that the modes of the electromagnetic field of interest to us are all plane waves in what is essentially unbounded, free space. This is a standard assumption in most calculations of spontaneous emission, though the reader should note that it is not always the correct assumption. For example, if the electron system is within some resonator, especially a small resonator, the modes of interest are those of the resonator and the following result does not necessarily apply. It is quite possible to inhibit spontaneous emission in such a resonator by making sure that the modes of the resonator do not coincide either in energy or in field amplitude distribution with electronic states and it is also possible to enhance emission using resonators by lining up the resonator modes with the frequencies of transitions between levels.

Given that we are going to use plane waves, we need a normalizable form of the modes. We imagine we have a cubic box of volume V_b. It is common for mathematical convenience to use running waves (which, technically, come from the somewhat unphysical periodic boundary conditions encountered also in solid-state physics for electron waves), though one could use standing waves and get the same result for a large box. The resulting modes have the form

$$\mathbf{u}_\lambda (\mathbf{r}) = \mathfrak{e} \frac{1}{\sqrt{V_b}} \exp(i\mathbf{k}_\lambda.\mathbf{r}) \tag{17.69}$$

where $\mathfrak{e}$ is a unit vector in the polarization direction of the electric field. These modes are readily seen to be normalized over the box of volume V_b. As usual for such modes, the allowed values of – for example – k_x are spaced by $2\pi / L_x$, where L_x is the length of the box in the x direction and similarly for the y and z directions, leading to a density of modes in $\mathbf{k}$-space of $V_b / (2\pi)^3$. For such propagating waves, we also have two distinct polarization directions, though we handle polarization properties directly here rather than adding them into the density of states. As is usual in the case of a large box, we approximate the sum over the modes λ by an integral over $\mathbf{k}$ with this density of states and also formally a sum over the two possible polarizations; that is,

$$\sum_\lambda \ldots \to \sum_{polarizations} \int \frac{V_b}{(2\pi)^3} d^3\mathbf{k}_\lambda \ldots \tag{17.70}$$

In considering the polarizations, we can choose two polarization directions at right angles to one another and at right angles to $\mathbf{k}_\lambda$. Specifically, we need to choose them relative to the vector matrix element $\mathbf{r}_{12}$. Fig. 17.1 shows one possible choice, which is to choose one of the polarization directions $\mathfrak{p}$ to be in the plane of the vectors $\mathbf{k}_\lambda$ and $\mathbf{r}_{12}$.

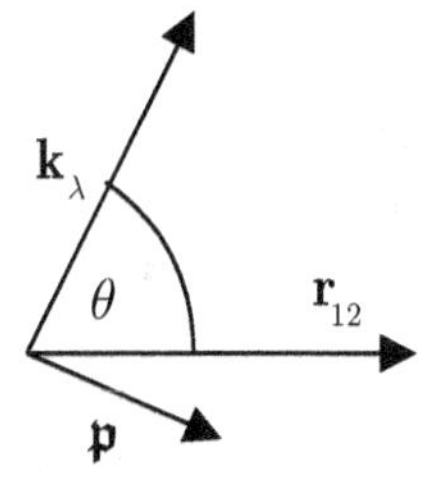

Fig. 17.1. Vectors for polarization effects in dipole interactions.

With this choice, the other polarization direction is always perpendicular to $\mathbf{r}_{12}$, so the vector dot product $\mathbf{u}_\lambda(\mathbf{r}_o)\cdot\mathbf{r}_{12}$ vanishes for all modes with this polarization. Hence, we can drop the sum over polarizations. For this choice, we therefore find that

$$\mathbf{u}_\lambda(\mathbf{r}_o)\cdot\mathbf{r}_{12} = u_\lambda(\mathbf{r}_o) r_{12}\sin\theta \tag{17.71}$$

(where the nonbold quantities refer to the magnitude of the corresponding vectors).

Now, we can use all these results to rewrite the total spontaneous transition rate of Eq. (17.66)

$$W_{spon} = \frac{2\pi}{\hbar}\int \frac{V_b}{(2\pi)^3}\left| i\, e\sqrt{\frac{\hbar\omega_\lambda}{2\varepsilon_o}}\frac{1}{\sqrt{V_b}}\exp(i\mathbf{k}_\lambda.\mathbf{r}_o) r_{12}\sin\theta\right|^2 \delta(E_1 - E_2 + \hbar\omega_\lambda) d^3\mathbf{k}_\lambda \tag{17.72}$$

That is,

$$\begin{aligned} W_{spon} &= \frac{e^2 |r_{12}|^2}{8\pi^2\varepsilon_o}\int \omega_\lambda \sin^2\theta\, \delta(E_1 - E_2 + \hbar\omega_\lambda) d^3\mathbf{k}_\lambda \\ &= \frac{e^2 |r_{12}|^2}{8\pi^2\varepsilon_o}\int_{k_\lambda=0}^{\infty}\int_{\theta=0}^{\pi} \omega_\lambda\, \delta(E_1 - E_2 + \hbar\omega_\lambda)\sin^2\theta\, 2\pi\sin\theta k_\lambda^2 d\theta dk_\lambda \end{aligned} \tag{17.73}$$

where we have chosen to integrate in spherical polar coordinates, noting there is variation only in the θ angle. Now, noting that $ck_\lambda = \omega_\lambda$ and changing variables to $\hbar ck_\lambda \equiv \hbar\omega_\lambda$, we have

$$W_{spon} = \frac{e^2 |r_{12}|^2}{4\pi\varepsilon_o\hbar^4 c^3}\int_{\hbar\omega_\lambda=0}^{\infty} \hbar\omega_\lambda\, \delta(E_1 - E_2 + \hbar\omega_\lambda)(\hbar\omega_\lambda)^2 d\hbar\omega_\lambda \int_0^{\pi}\sin^3\theta d\theta \tag{17.74}$$

Given that

$$\int_0^{\pi}\sin^3\theta d\theta = \frac{4}{3} \tag{17.75}$$

we finally have that the total spontaneous emission rate is

$$W_{spon} = \frac{e^2 |r_{12}|^2 \omega_{12}^3}{3\pi\varepsilon_o\hbar c^3} \tag{17.76}$$

where

$$\omega_{12} = (E_2 - E_1)/\hbar \tag{17.77}$$

Such a rate gives rise to a natural lifetime, τ_{nat}, for a state

$$\tau_{nat} = 1/W_{spon} \tag{17.78}$$

A quantum mechanical system sitting in empty space in an excited state decays on average over this timescale to its lower state, emitting a photon. The direction of the mode into which the photon is emitted is random (though weighted somewhat by the polarization effects).

Problems

17.4.1 Consider an electron that may be in one of two states, state 1 with electron energy E_1 or state 2 with electron energy E_2, where $E_2 > E_1$. This electron is assumed to interact with the electromagnetic field through the electric-dipole interaction

$$\hat{H}_{ed} = \sum_{j,k,\lambda} H_{ed\lambda jk}\hat{b}_j^\dagger \hat{b}_k \left(\hat{a}_\lambda - \hat{a}_\lambda^\dagger\right)$$

where λ indexes the photon modes and j and k refer to electron states. The initial state of this system is presumed to be that the electron is in state 2, there are $n_{\lambda 1}$ photons in mode λ_1, and no photons in other modes.

(i) Show that the stimulated emission transition rate into the state q_{stim} with $n_{\lambda 1}+1$ photons in mode λ_1 is

$$w_{qstim} = \frac{2\pi}{\hbar}\left(n_{\lambda 1}+1\right)\left|H_{ed\lambda_1 21}\right|^2 \delta\left(E_2 - E_1 - \hbar\omega_{\lambda 1}\right)$$

(ii) Show that the spontaneous emission transition rate into a state q_{spon} with $n_{\lambda 1}$ photons still in mode λ_1 and one photon in another mode λ is

$$w_{qspon} = \frac{2\pi}{\hbar}\left|H_{ed\lambda 21}\right|^2 \delta\left(E_2 - E_1 - \hbar\omega_\lambda\right)$$

(iii) Suppose now that the initial state is still with $n_{\lambda 1}+1$ photons in the mode λ_1, but that the electron is in state 1. Show that the absorption transition rate into the state with electron in state 2 and $n_{\lambda 1}$ photons in mode λ_1 is identical to the rate w_{qstim} calculated above for the stimulated emission.

17.4.2 (This problem can be used as a substantial assignment.) Consider a mass m on a spring. The mass is constrained in such a way that it can only move along one axis (e.g., the z axis) and, because of the linear restoring force of the spring, the mass has a natural oscillation (angular) frequency of Ω, behaving like a simple harmonic oscillator. The mass has a charge q and, consequently, it can interact with electromagnetic radiation.

(i) Write an expression for the quantum mechanical Hamiltonian for this simple harmonic oscillator (neglecting any interaction with the electromagnetic field) in terms of raising and lowering (annihilation and creation) operators. (Note: To avoid confusion with annihilation and creation operators for photons, you may want to use a different letter for these, for example, $\hat{d}^\dagger$ and $\hat{d}$.)

(ii) Give an expression for the position operator for the mass in terms of raising and lowering (annihilation and creation) operators of the harmonic oscillator formed by the mass and its spring. (Note: Do not write the position operator in terms of fermion annihilation and creation operators for the mass. That makes this problem much harder and, anyway, we have not even specified here whether this mass is a fermion or a boson.)

(iii) The electromagnetic field and the charged mass are presumed to interact through the electric dipole interaction. Give an expression for the Hamiltonian for this interaction with the electromagnetic field in terms of raising and lowering and/or annihilation and creation operators, considering all possible modes of the electromagnetic field. (You should make the approximation that the field amplitude of the mode is taken to be approximately constant over the size of the oscillator and can be replaced by its value at some specific point in space in the region of the oscillator.)

(iv) Presume that the harmonic oscillator is initially in its lowest state. Considering for the moment only electromagnetic modes with electric field polarized in the z direction, describe the form of the optical absorption spectrum as a function of frequency. You may do this by considering the interaction with an electromagnetic mode initially containing one photon.

(v) Now consider a situation where the harmonic oscillator is in its first excited state and consider the interactions with all possible electromagnetic modes, all presumed initially empty of photons. Derive an expression for the spontaneous emission lifetime of this first excited state of the harmonic oscillator (i.e., how long will it take on average to emit a photon and decay to its lowest state).

(vi) Suppose that $m = 10^{-26}$ kg (roughly, the mass of a small atom), $q = +e$ (corresponding to an ion with a positive charge), and $\Omega = 2\pi \times 10^{14}\ \text{s}^{-1}$ (a typical frequency for some kinds of vibration modes in solids). In this case, what is the spontaneous emission lifetime?

(vii) Describe qualitatively in words the angular dependence of the emission rate of photons from such a system if the harmonic oscillator is initially in its first excited state.

17.5 Summary of concepts

Commutators for different kinds of particles

We postulate that the creation and annihilation operators for different kinds of particles commute; that is,

$$\begin{aligned} \hat{b}_j^\dagger \hat{a}_\lambda^\dagger - \hat{a}_\lambda^\dagger \hat{b}_j^\dagger = 0 \quad \hat{b}_j \hat{a}_\lambda - \hat{a}_\lambda \hat{b}_j = 0 \\ \hat{b}_j^\dagger \hat{a}_\lambda - \hat{a}_\lambda \hat{b}_j^\dagger = 0 \quad \hat{b}_j \hat{a}_\lambda^\dagger - \hat{a}_\lambda^\dagger \hat{b}_j = 0 \end{aligned} \tag{17.2}$$

Such commutation applies to any different kinds of particles (e.g., fermions and bosons), two different kinds of fermions, or two different kinds of bosons.

Interaction Hamiltonian

To write a Hamiltonian involving the interaction of different kinds of particles, we progressively apply the methods for each particle to transform into the creation and annihilation operator form. For example, the electric dipole Hamiltonian that represents the energy of an electron in an electric field (i.e., $\hat{H}_{scedr} = e\,\mathbf{E}\cdot\mathbf{r}$ [Eq. (17.4)]) becomes, after substituting the multimode electric field operator for **E** and using the wavefunction operator to transform **r**

$$\hat{H}_{ed} = \sum_{j,k,\lambda} H_{ed\lambda jk} \hat{b}_j^\dagger \hat{b}_k \left(\hat{a}_\lambda - \hat{a}_\lambda^\dagger\right) \tag{17.13}$$

where the constants $H_{ed\lambda jk}$ contain the appropriate integrals over the specific forms of the electron basis states and the electromagnetic modes (Eq. (17.12)).

Processes identified by creation and annihilation operators

With the form of Hamiltonian for the interaction of two particles as in Eq. (17.13), we can identify specific processes by considering the action of the creation and annihilation operators. For example, the sequence of operators $\hat{b}_j^\dagger \hat{b}_k \hat{a}_\lambda$ describes a process in which absorption of a photon (annihilation of a photon in mode λ) takes an electron from one state (annihilation of an electron in state k) to another (creation of an electron in state j).

Perturbation theory with creation and annihilation operators

Perturbation theory is essentially unchanged when we move to the representation of operators with creation and annihilation operators. We merely substitute the perturbing Hamiltonian in the new form and remember that we now have to write the complete quantum mechanical state (e.g., of electrons and photons) in all matrix elements (instead of just, e.g., the electron wavefunction). See, for example, Eq. (17.19).

Absorption, stimulated emission, and spontaneous emission

When the electromagnetic field is quantized using annihilation and creation operators, the processes of absorption, stimulated emission, and spontaneous emission all emerge automatically when considering the interaction between the electromagnetic field and some other system such as an electron. Absorption is the normal process of absorption of light. Spontaneous emission is the most common process of light emission and is the dominant process found in nearly all sources of illumination, such as light bulbs. Stimulated emission is the process found in lasers, where additional photons are created specifically in a mode in which there are already photons.

For a two-level electron system and for photons n_{λ_1} initially in a specific mode λ_1, the transition rates for the various processes are

absorption
$$w_q = \frac{2\pi}{\hbar} n_{\lambda_1} \left| H_{ed\lambda_1 12} \right|^2 \delta\left(E_2 - E_1 - \hbar\omega_{\lambda_1}\right) \tag{17.63}$$

stimulated emission
$$w_q = \frac{2\pi}{\hbar} \left(n_{\lambda_1} + 1\right) \left| H_{ed\lambda_1 12} \right|^2 \delta\left(E_1 - E_2 + \hbar\omega_{\lambda_1}\right) \tag{17.62}$$

spontaneous emission
$$w_q = \frac{2\pi}{\hbar} \left| H_{ed\lambda_1 12} \right|^2 \delta\left(E_1 - E_2 + \hbar\omega_{\lambda_1}\right) \qquad \text{after (17.44)}$$

Note that (i) all three processes have the same matrix element $H_{ed\lambda_1 12}$ and so the same underlying strength; (ii) absorption has an additional factor of n_{λ_1}, the number of photons in mode λ_1; and (iii) stimulated emission has an additional factor $n_{\lambda_1} + 1$. (Note: It is also possible to view the "1" in this factor as being the spontaneous emission into the same mode.) Spontaneous emission is independent of the number of photons in any mode.

Total spontaneous emission rate

To calculate the total spontaneous emission rate, we need to calculate the interaction matrix element for each possible mode of light with the system of interest and add up all of the resulting spontaneous emission rates into each mode. In a vacuum for a two-level system with an electric dipole interaction between the system and the electromagnetic field, we would obtain the total transition rate

$$W_{spon} = \frac{e^2 \left| r_{12} \right|^2 \omega_{12}^3}{3\pi\varepsilon_o \hbar c^3} \tag{17.76}$$

where

$$\mathbf{r}_{12} = \int \phi_1^* (\mathbf{r}) \mathbf{r} \phi_2 (\mathbf{r}) d^3\mathbf{r} \tag{17.68}$$

for wavefunctions $\phi_1(\mathbf{r})$ and $\phi_2(\mathbf{r})$ for the two states in the system. Associated with such a transition rate is the natural lifetime, τ_{nat}, for the upper state 2

$$\tau_{nat} = 1/W_{spon} \tag{17.78}$$

Chapter 18

Quantum information

Prerequisites: Chapters 2–5, 9, 10, 12, and 13.

When we think of processing information, we are typically used to a classical world in which we represent information in terms of the classical state of an object. For example, we could represent a number as the length of some particular rod in meters or the value of some electrical potential in volts; these would be analog representations. More typically in information processing, we represent numbers and other information digitally, in binary form as a sequence of "bits" that are either "1" or "0". We can use all sorts of physical representations for the 1 and 0, such as an object being "up" or "down," a device being "on" (e.g., passing current) or "off" (e.g., not passing current), or a voltage being "high" or "low."

In the quantum mechanical world, we have additional options in representing information; in particular we can use quantum mechanical superpositions, such as a superposition of "up" and "down." Though we could easily also have the equivalent of a superposition in a classical world for one physical system (a system that was half up and half down could simply be represented by it being horizontal), in quantum mechanics, we can have kinds of superpositions of multiple systems (so-called *entangled states*) that have no classical analog and the act of measurement on a quantum mechanical system in a superposition can have quite a different result from that in a classical system (i.e., the process of "collapse" into an eigenstate). These additional processes of entanglement and collapse under measurement open up quite different opportunities in handling and processing information and have led to the field known as quantum information. Here, we briefly introduce a few of the most basic ideas in applications including quantum cryptography, quantum computing, and quantum teleportation.

18.1 Quantum mechanical measurements and wavefunction collapse

Let us review first how we believe we can use the results of a quantum mechanical calculation. If that calculation says that the state of the system should be $|\psi\rangle$, then the average value we will measure for some quantity A is given by evaluating

$$\langle A\rangle = \langle\psi|\hat{A}|\psi\rangle \tag{18.1}$$

where $\hat{A}$ is the operator associated with the quantity A. Sometimes it may take us some effort to deduce what the operator $\hat{A}$ is, but once we have done that, we do get the predicted answers for the average values of A. The measurement is a statistical process – we must repeat the experiment many times from the start (i.e., including the process that puts the quantum mechanical system into the state $|\psi\rangle$) and take the average answer. We also find that every measurement we make returns a value corresponding to one of the possible eigenvalues, A_n, of $\hat{A}$. Not every measurement returns the same value. If we decompose the state into a linear combination of the normalized eigenstates $|\psi_n\rangle$ of the operator $\hat{A}$, that is,

$$|\psi\rangle = \sum_n a_n |\psi_n\rangle \tag{18.2}$$

then we find that the probability of measuring a particular value A_n is given by $|a_n|^2$. Furthermore, if we make any subsequent measurements[1] on this system, presuming that no external influence is applied to the system in the meantime, we will always subsequently get the same answer A_n on measuring the quantity A. This behavior is called the *collapse of the wavefunction*. The act of measurement of a quantity A appears to force it into one of its eigenstates. As far as we know empirically, this collapse is totally random.

18.2 Quantum cryptography

We can use this randomness for a specific practical application, which is the secure distribution of information. In conventional cryptography, if we do not have the key, we cannot decode a message because it would practically take us too long to do the calculations. We worry, though, that other people just might find a way to do those calculations (possibly with a quantum computer). Hence, they could decode our messages. Quantum cryptography, however, relies on fundamental properties of quantum mechanics that allow the exchange of information with apparently absolute security. No amount of calculation could crack the code. To see how this works, we must first show that it is impossible to clone a quantum mechanical state reliably.

No-cloning theorem

Quantum cryptography is secure because no one can guarantee to make an exact replica of an arbitrary quantum mechanical state of a system. For example, we might want to take an electron that is in a particular spin state and make another electron have exactly the same spin state, leaving the first electron in its original state. Equivalently, we might want to take a photon in a particular polarization state and make another photon with exactly the same polarization state, leaving the first photon in its original polarization state.

In the case of photons, the two polarization basis states can be, respectively, horizontally polarized and vertically polarized. A general linear combination of those two states is an elliptically polarized photon – that is, some specific ratio of amplitudes of horizontal and vertical polarization with some phase difference between the amplitudes. In both the spin and the photon polarization cases, we can, if we wish, write the state as a simple two-element vector with complex elements.

We can show, in a very simple proof,[2] that starting from the first system in an arbitrary state $|\psi_a\rangle_1$ and the second system in some prescribed starting state, $|\psi_s\rangle_2$, we cannot, in general, create the second system in the state $|\psi_a\rangle_2$, leaving the first system in state $|\psi_a\rangle_1$.[3]

---

[1] I.e., measurements after we have collapsed the system onto an eigenstate.

[2] W. K. Wootters and W. H. Zurek, "A Single Quantum Cannot Be Cloned," *Nature* **299** 802–803 (1982); D. Dieks, "Communication by EPR Devices," *Phys. Lett. A* **92**, 271–272 (1982); see also, P. W. Milonni and M. L. Hardies, "Photons Cannot Always be Replicated," *Phys. Lett. A* **92**, 321–322 (1982).

In this proof, our initial state of the two systems is, therefore, the (direct product) state $|\psi_a\rangle_1|\psi_s\rangle_2$. We then imagine that we have some operation that, over time, turns this state into the state $|\psi_a\rangle_1|\psi_a\rangle_2$. This operation is just some time evolution that we can describe by a (unitary) linear operator $\hat{T}$, such as the time-evolution operator we devised in Chapter 3

$$\hat{T} = \exp\left[-i\hat{H}\left(t-t_o\right)/\hbar\right] \tag{18.3}$$

where t is the time at the end of the operation and t_o is the time when we started. To get the desired cloning behavior, we presume that we have, of course, engineered our cloning system and, hence, its Hamiltonian $\hat{H}$, to give $\hat{T}$ the required properties. Specifically, we need at least two properties for $\hat{T}$. First, we know we want $\hat{T}$ to perform the operation

$$|\psi_a\rangle_1|\psi_a\rangle_2 = \hat{T}|\psi_a\rangle_1|\psi_s\rangle_2 \tag{18.4}$$

cloning the state a of system 1 also into system 2. Of course, we want to have this work for any initial state of system 1. Suppose we choose some other state $|\psi_b\rangle_1$ as the initial state of system 1 (we choose this orthogonal to $|\psi_a\rangle_1$). Then, we also want $\hat{T}$ now to perform the operation

$$|\psi_b\rangle_1|\psi_b\rangle_2 = \hat{T}|\psi_b\rangle_1|\psi_s\rangle_2 \tag{18.5}$$

cloning state b into system 2. There is, in general, no problem, in principle, with constructing such an operator with these two properties.

The problem comes when we want to clone a linear superposition state. Suppose, for example, that the initial state of system 1 is the linear superposition

$$|\psi_{Sup}\rangle_1 = \frac{1}{\sqrt{2}}\left(|\psi_a\rangle_1 + |\psi_b\rangle_1\right) \tag{18.6}$$

Hence, the initial state of the pair of systems is

$$\frac{1}{\sqrt{2}}\left(|\psi_a\rangle_1 + |\psi_b\rangle_1\right)|\psi_s\rangle_2 = \frac{1}{\sqrt{2}}\left(|\psi_a\rangle_1|\psi_s\rangle_2 + |\psi_b\rangle_1|\psi_s\rangle_2\right) \tag{18.7}$$

By postulation in quantum mechanics, the operators are linear – operating on a linear superposition must give the linear superposition of the operations; that is,

$$\begin{aligned}\hat{T}\frac{1}{\sqrt{2}}\left(|\psi_a\rangle_1|\psi_s\rangle_2 + |\psi_b\rangle_1|\psi_s\rangle_2\right) &= \frac{1}{\sqrt{2}}\left(\hat{T}|\psi_a\rangle_1|\psi_s\rangle_2 + \hat{T}|\psi_b\rangle_1|\psi_s\rangle_2\right) \\ &= \frac{1}{\sqrt{2}}\left(|\psi_a\rangle_1|\psi_a\rangle_2 + |\psi_b\rangle_1|\psi_b\rangle_2\right)\end{aligned} \tag{18.8}$$

This is not the result we wanted for our cloning operation, which was

$$\frac{1}{2}\left(|\psi_a\rangle_1 + |\psi_b\rangle_1\right)\left(|\psi_a\rangle_2 + |\psi_b\rangle_2\right) \tag{18.9}$$

Such a result would have cloned the superposition state of system 1 into system 2, leaving system 1 in its original superposition state. The linearity properties of the quantum mechanical

---

[3] Throughout this chapter, we use subscripts outside the bras and kets to indicate which particle the bra or ket is describing.

operator determined that if we got the cloning properties we wanted for the individual states a and b, we did not get the cloning result we wanted for the superposition. This result is not special to the particular superposition we chose – any other superposition would give us a similar "wrong" answer. Hence, though we could, in principle, make a device that could clone specific basis states, that device could not clone superpositions of those basis states; hence, we cannot make a device that will clone an arbitrary quantum state.[4] Backup of quantum states is impossible!

With some cunning, we can use this particular no-cloning property to ensure the security of communications through what is known as *quantum cryptography*.

A simple quantum encryption scheme

A well-known and simple scheme for quantum encryption using single photons was devised by Bennett and Brassard in 1984[5] and demonstrated later by them and their colleagues.[6] Schemes like this have subsequently been demonstrated in increasingly practical conditions including, for example, demonstrations over a 48 km optical fiber network[7] and through the atmosphere over 1.6 km.[8] Such a horizontal atmospheric distance is thought to be a good test of whether one could send quantum encrypted signals upward through the atmosphere to a satellite and back. The communication rate in such schemes is often not high (e.g., 100s of bits per second) though it is fast enough that such schemes could practically be used to send simple messages or secret cryptographic keys for use with classical cryptography.

---

[4] The reader might ask why it is that the process of stimulated emission does not clone arbitrary quantum mechanical states and it was this question that in part triggered the original discussions by Wootters, Zurek, Milonni, and Hardies. Consider a photon that has two accessible basis states corresponding to the two different polarizations, x and y, for example. For simplicity of illustration, imagine that these two polarizations interact equally strongly with some atomic system that is in its "upper" state; such a condition is required if we are to be able to clone photons in the specific x and y polarization states with equal strength. The mechanistic answer to why no reliable cloning is possible in that case is that with one photon in a mode (and, hence, in a specific polarization state), the process of stimulated emission and that of spontaneous emission are equally likely (see Chapter 17). Because the interaction with the mode of the orthogonal polarization is by design equally strong, the spontaneous photon that emerges on the average is just as likely to be in the same polarization as the incident photon as it is to be in the orthogonal one. That prevents reliable cloning of a photon in the same mode as the incident photon; the additional emerging photon does not have a predictable polarization state and, hence, is not predictably in the desired mode or the undesired mode (of the "wrong" polarization). When we include the unavoidable process of spontaneous emission, this attempt to use the stimulated emission process, therefore, does not even reliably clone photons in either of the two basis states, let alone cloning superpositions reliably.

[5] C. H. Bennett and G. Brassard, "Quantum Cryptography: Public Key Distribution and Coin Tossing," in *Proceedings of the IEEE International Conference on Computers, Systems, and Signal Processing*, New York, 1984 (IEEE) pp. 175–179.

[6] C. H. Bennett, F. Bessette, G. Brassard, L. Salvail, and J. Smolin, "Experimental Quantum Cryptography," *J. Cryptol.* **5**, 3–28 (1992).

[7] R. J. Hughes, G. L. Morgan, and C. G. Peterson, "Quantum Key Distribution over a 48 km Optical Fiber Network," *J. Mod. Opt.* **47**, 533–547 (2000).

[8] W. T. Buttler, R. J. Hughes, S. K. Lamoreaux, G. L. Morgan, J. E. Nordholt, and C. G. Peterson, "Daylight Quantum Key Distribution over 1.6 km," *Phys. Rev. Lett.* **84**, 5652–5655 (2000).

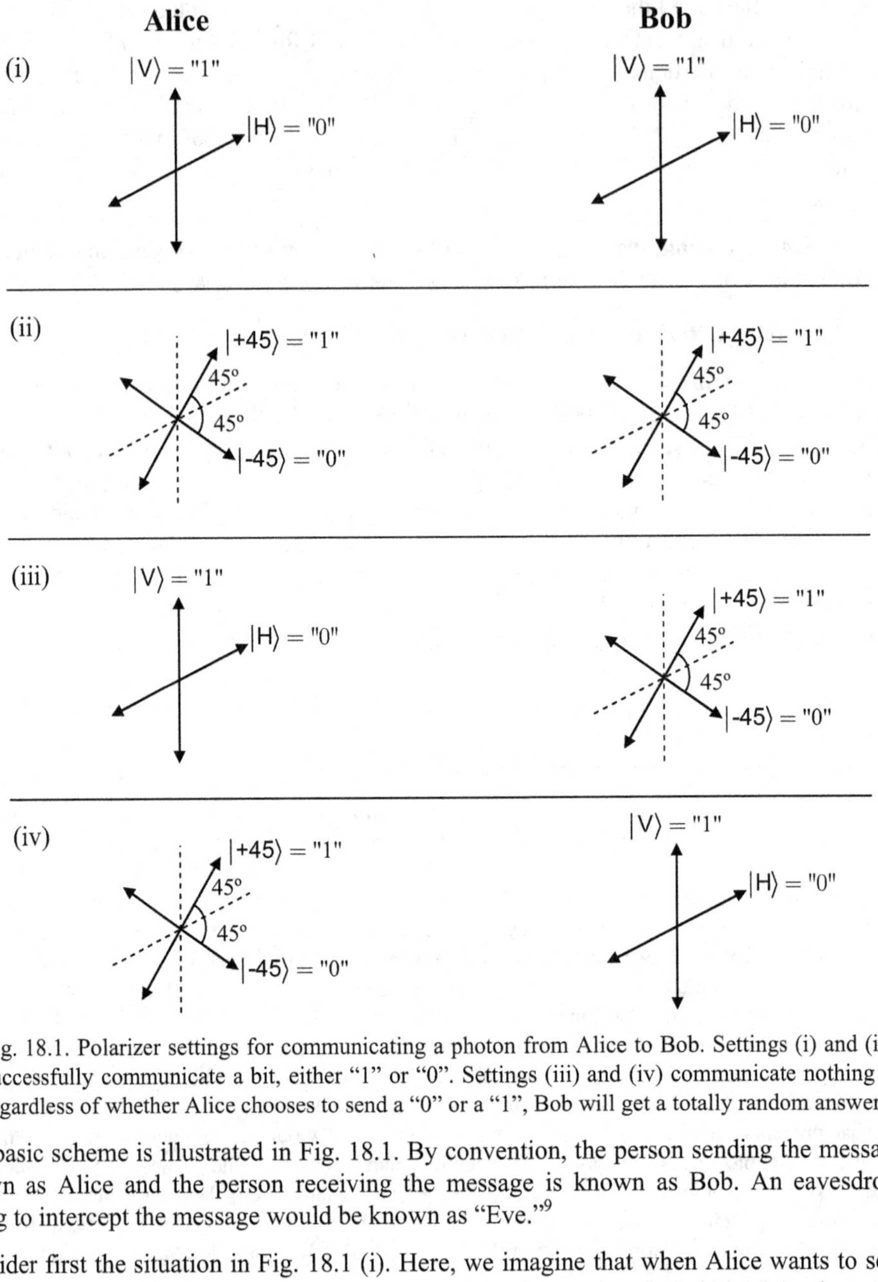

Fig. 18.1. Polarizer settings for communicating a photon from Alice to Bob. Settings (i) and (ii) successfully communicate a bit, either "1" or "0". Settings (iii) and (iv) communicate nothing – regardless of whether Alice chooses to send a "0" or a "1", Bob will get a totally random answer.

The basic scheme is illustrated in Fig. 18.1. By convention, the person sending the message is known as Alice and the person receiving the message is known as Bob. An eavesdropper trying to intercept the message would be known as "Eve."[9]

Consider first the situation in Fig. 18.1 (i). Here, we imagine that when Alice wants to send a bit of information with the value "1", she sends a photon that is polarized in the vertical direction (i.e., a photon in the state $|V\rangle$), and if she wants to send a "0", she sends a horizontally polarized photon (i.e., a photon in the state $|H\rangle$). Bob, we presume, has set up a detection apparatus that separates the two polarizations out to different single-photon detectors. He can do this with a polarizing beamsplitter oriented to separate horizontal and vertical polarizations, with appropriate photodetectors attached to it. When he receives a photon in his

[9] Physicists have a weakness for puns.

vertical polarization detector, he records a "1", and when he receives a photon in his horizontal polarization detector, he records a "0".

A scheme such as Fig. 18.1 (i) is not itself secure. Eve could insert a detection system exactly like Bob's into the path, receive the photon from Alice, write down the answer, and then, using a transmission system exactly like Alice's, retransmit the photon on to Bob, with Alice and Bob being unaware of her interception.[10]

To see how we make this system secure, imagine first that we rotate Alice's apparatus, for example by 45°. Now Alice transmits a "1" using the state $|+45\rangle$ and a "0" using the state $|-45\rangle$. If Bob leaves his apparatus unchanged, as in Fig. 18.1 (iv), he will receive no information at all. To see this, note that the state $|+45\rangle$ can be written as a linear superposition of the horizontal and vertical states; that is,

$$|+45\rangle = \frac{1}{\sqrt{2}}\left(|\mathsf{H}\rangle + |\mathsf{V}\rangle\right) \tag{18.10}$$

Measuring this state now with Bob's apparatus will give the answer $|\mathsf{H}\rangle$ half the time and the answer $|\mathsf{V}\rangle$ half the time (because the expansion coefficients in these two basis states are each $1/\sqrt{2}$, leading to probabilities of $1/2$, in what we believe to be a totally random process in quantum mechanics). Similarly

$$|-45\rangle = \frac{1}{\sqrt{2}}\left(|\mathsf{H}\rangle - |\mathsf{V}\rangle\right) \tag{18.11}$$

which also will give the answer $|\mathsf{H}\rangle$ half the time and the answer $|\mathsf{V}\rangle$ half the time. Hence it does not matter what Alice transmits.

So now let us presume that Alice and Bob each rotate their apparatus by 45°, as in Fig. 18.1 (ii). Then, they can send information just as before. If, however, Eve interposes her apparatus, still oriented horizontally and vertically, she will receive no information and, furthermore, Bob and Alice will quickly deduce that their message is being intercepted. Alice and Bob monitor for errors in their communication channel, occasionally calling one another up over the telephone and checking (quite openly and publicly) to see that they are sending and receiving the same bits on some specific test bits they choose. If Eve has interposed herself in this horizontal and vertical way, half the bits apparently received by Bob will turn out to be wrong and Alice and Bob will know to discard all of the bits and to send out a search party to find Eve and her apparatus.

Of course, Eve might soon realize that she must have her apparatus set incorrectly (perhaps alerted by the approach of the search party) and she can retreat to come back another day, possibly then setting her apparatus in the 45° fashion. Alice and Bob might, by that time, have changed back, but there is a 50 percent chance that Eve could set her apparatus appropriately on any given day and a 50 percent chance of interception is quite likely to be unacceptable to Alice and Bob.[11]

---

[10] Alice and Bob could notice some unexpected time delay in the transmission, though that would not be considered here to be a strong enough protection against eavesdropping.

[11] Incidentally, if Eve sets her apparatus at some other angle not exactly aligned to the horizontal-vertical or 45° directions, she may manage to get some information but she will also generate some errors that can also be detected by Alice and Bob. By monitoring the error rate, Alice and Bob can put some upper limit on how much information is being intercepted and, by sending somewhat more bits than actually needed

The trick to thwarting Eve is that Alice and Bob, for each time they want to try to communicate a bit, each randomly choose between the horizontal-vertical setting of their apparatus and the 45° one. This leads to four possibilities, as shown in Fig. 18.1 (i)–(iv). In two of these, their transmission is meaningful; in the other two, no information will be exchanged. All that is necessary now for successful secure information exchange by Alice and Bob is for them again to call one another on the telephone (quite openly and publicly) and figure out for which bits they had their polarizers set identically (a phone call they can make without ever revealing what information was exchanged in each such case).[12] Then, at last, Alice and Bob have a string of bits that they both know and that is known to no one else.

There is now no strategy that Eve can follow to choose her apparatus orientation that can possibly work more than half the time and, for the other half, she again generates errors half the time, a fraction easily noticed by Alice and Bob (and stimulating the search party again). Eve might have wanted to solve this problem by cloning the incoming photon, passing the clone on to Bob, who would then not notice any errors being introduced by Eve. If Eve could do the cloning, she could on the average deduce half of the information Alice was sending to Bob with random choices of her apparatus orientation. But, we know that Eve cannot perform that cloning by the no-cloning theorem proved previously. So, finally, Alice and Bob have a way of sending bits so that no one can intercept them without Alice and Bob finding out.

If Alice and Bob find that their message is being intercepted because they notice a large error rate, then again they discard the bits sent and send out the search party to find Eve. One might think that Eve then has, at least, intercepted some of their message, but that problem is easily overcome by the strategy of not sending the actual message and, instead, sending only what is known to cryptographers as the key. Once Alice and Bob have the set of shared bits that they are sure are secret, they then send their actual message. A very simply way of doing that is for Alice now to call up Bob again, on the public telephone, and tell him her actual message is, for example, bits 1, 4, 5, 9, 11, and 16 on their shared list. Only Alice and Bob then know what that message is. Provided they only use each shared bit once in their message, their actual message is totally secure.[13]

There are many other quantum cryptographic techniques[14] and there is much discussion in the research literature as to how secure such techniques are against very sophisticated kinds of attacks. The understanding at the time of this writing, however, is that when augmented by known classical information theory and cryptographic techniques to handle errors, monitor error rates, and exclude consequences from partial interception of the message, these kinds of quantum encryption protocols are very secure. As we have seen for this example, their security

---

for their messages, can use classical coding and cryptographic techniques to protect their information from this partial interception (the so-called privacy amplification).

[12] On this same phone call, they can also deal with loss of bits in transmission. In fact, likely a very large fraction of bits will be completely lost during transmission of these single photons. Alice and Bob simply discard all the situations in which Alice sent a photon and Bob received nothing.

[13] In this case, the set of shared secret bits is known as a *one-time pad.* A more sophisticated way of using a one-time pad is for Alice to send a string of random bits to Bob to make up the one-time pad and then for Alice and Bob to send, publicly, a string of bits that is the exclusive OR of the desired message with the one-time pad. Because the one-time pad is a string of random bits, the transmitted message also looks like random bits and only Bob can decode it using his copy of the one-time pad.

[14] For a review, see, e.g., N. Gisin, G. Ribordy, W. Tittel, and H. Zbinden, "Quantum Cryptography," *Rev. Mod. Phys.* **74** 145–95 (2002).

derives from the randomness of quantum mechanical measurement for superposition states and from the no-cloning theorem that is itself a very simple consequence of the linearity of quantum mechanics.

18.3 Entanglement

Up till now, we have nearly always been considering the state of one particle at a time. What if we have two particles? What then are the possible states? Suppose that we think of particle 1 as being in one of a set of possible states $|\psi_m\rangle_1$ and particle 2 as being in one of a set of possible states $|\phi_n\rangle_2$. The particles could be photons, for example. Particle 1 might be a photon going to the left in a particular spatial mode (beam shape) and with a specific frequency, and the different possible states might be different polarization states, such as vertical or horizontal polarization. Similarly, particle 2 might be a photon going to the right. Alternatively, the particles might be electrons going along different paths, with the spin-up and spin-down states being the different possible states for each particle.[15]

For the sake of definiteness, we can consider the photon case. Then, appropriate basis states for particle 1 (the left-going photon) would be $|\mathsf{H}\rangle_1$ and $|\mathsf{V}\rangle_1$, where H and V refer to horizontal and vertical polarization, respectively. Similarly, for particle 2 (the right-going photon) we would have $|\mathsf{H}\rangle_2$ and $|\mathsf{V}\rangle_2$ as basis states.

A possible state of these two photons is, for example, $|\mathsf{H}\rangle_1|\mathsf{V}\rangle_2$, which is the left-going photon horizontally polarized and the right-going photon vertically polarized. Other examples include $|\mathsf{H}\rangle_1|\mathsf{H}\rangle_2$, $|\mathsf{V}\rangle_1|\mathsf{V}\rangle_2$, and $|\mathsf{V}\rangle_1|\mathsf{H}\rangle_2$ with obvious meanings. We can express other polarizations of a given photon as linear combinations of horizontal and vertical. For example, the state $(1/\sqrt{2})(|\mathsf{H}\rangle_1+|\mathsf{V}\rangle_1)$ describes a left-going photon polarized at an angle of 45°. Hence, a state like $(1/\sqrt{2})(|\mathsf{H}\rangle_1+|\mathsf{V}\rangle_1)|\mathsf{H}\rangle_2$ describes a left-going photon polarized at 45° and a right-going photon horizontally polarized. In each of the states in this paragraph, we can assign a definite polarization to each photon. Though we have expressed these states in the quantum mechanical notation and we mean them to be quantum mechanical states, these state descriptions could also correspond to a classical description of a particle: in a classical view each photon has a well-defined polarization that is a property of that photon. But, these states are not the only ones allowed by quantum mechanics.

For example, consider the following state of the two photons

$$|\Phi^+\rangle_{12}=\frac{1}{\sqrt{2}}\left(|\mathsf{H}\rangle_1|\mathsf{H}\rangle_2+|\mathsf{V}\rangle_1|\mathsf{V}\rangle_2\right) \tag{18.12}$$

A pair of photons in a state like this is sometimes called an EPR pair (after Einstein, Podolsky, and Rosen) and the state itself is sometimes called a Bell state.[16] This state is a linear superposition of two of the states we considered already. According to quantum mechanics, such a state is a valid state of the system. Just as for all the states we have considered so far in

---

[15] We presume that the two photons (or the two electrons) are practically distinguishable particles in the sense we introduced in Section 13.10 because they are moving on very distinct paths and there is no practical possibility of exchange. Hence, we can omit the symmetrization of the wave functions with respect to exchange, which greatly simplifies our algebra in discussing entangled states. We emphasize that the entangled states we are discussing here are *not* merely the states one gets as a result of symmetrization with respect to exchange.

[16] We have a lot more to say about such states when we discuss Bell's inequalities in Chapter 19.

this section, we can view it as a vector in the four-dimensional Hilbert space that describes the polarization state of two photons, a direct product space in which $|H\rangle_1|H\rangle_2$, $|V\rangle_1|V\rangle_2$, $|H\rangle_1|V\rangle_2$, and $|V\rangle_1|H\rangle_2$ are appropriate orthonormal basis vectors.

It is relatively straightforward with modern techniques, such as spontaneous optical parametric down-conversion (a second-order nonlinear optical technique),[17] to generate pairs of photons in EPR states. There are several ways of generating EPR states; in each of them, there is something intrinsic to the generating process that guarantees that though we may not know the state of the individual particles, each member of the pair has to be found in the same (or possibly the opposite) polarization state as the other member when we make a measurement. (In the case of electron spin EPR pairs, the electrons may be guaranteed to be in opposite spin states, for example.)

There is, however, something very unusual and nonclassical about the state in Eq. (18.12). It cannot be factorized into a product of a state of particle 1 and a state of particle 2. States that like this one[18] cannot be factorized into a product of the states of individual systems on their own are said to be *entangled.* A consequence of such an entangled state is that, in such a state, particle 1 does not have a definite state of its own independent of the state of particle 2. Imagine, for example, that we make a measurement on this state – say, of the polarization of the left-going photon (photon 1) and we find a horizontal polarization, which means we have collapsed the overall state of the system into one that now only has terms in $|H\rangle_1$. After that measurement, the state of the whole system now is $|H\rangle_1|H\rangle_2$; we have also collapsed the state of the second (right-going) photon into a horizontal polarization (even though we did not "touch" it). The state of the right-going photon depends on the state we measure for the left-going photon, even though both results are possible for the measurement of the left-going photon.

There are, incidentally, three other states like Eq. (18.12) that together constitute the four Bell states; specifically

$$|\Phi^-\rangle_{12} = \frac{1}{\sqrt{2}}\left(|H\rangle_1|H\rangle_2 - |V\rangle_1|V\rangle_2\right) \tag{18.13}$$

$$|\Psi^+\rangle_{12} = \frac{1}{\sqrt{2}}\left(|H\rangle_1|V\rangle_2 + |V\rangle_1|H\rangle_2\right) \tag{18.14}$$

$$|\Psi^-\rangle_{12} = \frac{1}{\sqrt{2}}\left(|H\rangle_1|V\rangle_2 - |V\rangle_1|H\rangle_2\right) \tag{18.15}$$

These four Bell states are orthogonal and are a complete basis for describing any such two-particle system with two available basis states per particle (here, $|H\rangle$ and $|V\rangle$), the proofs of which are left as exercises for the reader.

There are many bizarre and important consequences of this kind of entangled behavior for the meaning of quantum mechanics and we return to these in the Chapter 19 when we discuss Bell's inequalities. For the moment, we simply want to emphasize that once we consider the states of more than one quantum system at a time, there is a whole additional range of states in

---

[17] P. G. Kwiat, E. Waks, A. G. White, I. Appelbaum, and P. H. Eberhard "Ultrabright Source of Polarization-Entangled Photons," *Phys. Rev. A* **60**, R773–R776 (1999).

[18] The state in Eq. (18.8) in the proof of the no-cloning theorem is also an entangled state.

quantum mechanics, these entangled states, that have no analog in the classical view of the world.

For the two particles considered here, the space is four-dimensional as we said previously. The most general quantum mechanical state of these two photons is, therefore

$$|\psi\rangle = c_{HH}|H\rangle_1|H\rangle_2 + c_{HV}|H\rangle_1|V\rangle_2 + c_{VH}|V\rangle_1|H\rangle_2 + c_{VV}|V\rangle_1|V\rangle_2 \tag{18.16}$$

where we obviously have needed four (generally complex) coefficients, the c's, to specify the state of just two photons. Classically, we would have needed, at most, two complex numbers to specify the polarization state of two photons (one complex number is enough to specify the relative amplitude and phase of the two polarization components of one photon).

As we increase the number of particles, even restricting ourselves to particles with only two basis states of interest, the dimensionality of the resulting direct-product Hilbert space and, hence, the number of required expansion coefficients – the c's – rises quickly. For three particles, we would have eight required coefficients to express the quantum mechanical state, for four particles, sixteen coefficients, and so on, leading to 2^N coefficients for N particles. 300 particles would therefore require 2^{300} coefficients, a number that may be larger than the number of atoms in the observable universe!

Problems

18.3.1 State whether each of the following states is entangled (i.e., can it be factored into a product of states of the individual particles, in which case it is not entangled):

(i) $\frac{1}{\sqrt{2}}\left(|H\rangle_1|V\rangle_2 - |H\rangle_1|H\rangle_2\right)$

(ii) $\frac{1}{\sqrt{2}}\left(|H\rangle_1|V\rangle_2 - |V\rangle_1|H\rangle_2\right)$

(iii) $\frac{3}{5}|H\rangle_1|V\rangle_2 + \frac{4i}{5}|V\rangle_1|V\rangle_2$

(iv) $\frac{1}{2}\left(|H\rangle_1|H\rangle_2 + |H\rangle_1|V\rangle_2 + |V\rangle_1|H\rangle_2 + |V\rangle_1|V\rangle_2\right)$

18.3.2 Show that the Bell states $|\Phi^-\rangle_{12}$ and $|\Psi^+\rangle_{12}$ are orthogonal.

18.3.3* Show that the set of four Bell states is complete as a basis for two-particle states where each particle has two available basis states (i.e., show that the general state of Eq. (18.16) can be written in terms of the Bell states and write that state out in terms of Bell states and the coefficients c_{HH}, c_{HV}, c_{VH}, and c_{VV}.) (Hint: Note that, e.g., $|H\rangle_1|H\rangle_2$ can be written as a sum of two different Bell states.)

18.3.4 (i) Consider the Bell state $|\Phi^-\rangle_{12} = (1/\sqrt{2})(|H\rangle_1|H\rangle_2 - |V\rangle_1|V\rangle_2)$ of two photons and suppose now that we wish to express it not on a basis of horizontal ($|H\rangle$) and vertical ($|V\rangle$) polarized states, but instead on a basis rotated by 45°; that is, a new basis

$$|{+45}\rangle = (1/\sqrt{2})(|H\rangle + |V\rangle) \qquad |{-45}\rangle = (1/\sqrt{2})(|H\rangle - |V\rangle)$$

Show that expressed on this particular basis, the resulting state is still a Bell state.

(ii) Repeat part (i) but with the Bell state $|\Phi^+\rangle_{12} = (1/\sqrt{2})(|H\rangle_1|H\rangle_2 + |V\rangle_1|V\rangle_2)$. What difference do you note between the two results?

See also Problems 13.4.2–13.4.4.

18.4 Quantum computing

The basic idea of quantum computing is to make a machine that operates on a quantum state rather than a classical one. We can see a reason why that might be interesting from the previous discussion of the quantum mechanical states of many particles. A machine that only had to deal with N two-level quantum mechanical systems, at least as the input to the machine, could actually be performing an operation on 2^N numbers at once and essentially in parallel. No classical machine can do that for even moderate N. For $N = 300$ binary inputs, there are not enough atoms even to store 2^{300} numbers at one atom per number. Of course, such a statement does not prove that a quantum computer can be made or whether it is better at some specific problem than a classical one. There is, however, a strong motivation to explore such quantum computing because it is well known in classical computing that the difficulty of solving certain rather important problems grows so fast with the size of the problem that no classical computer could possibly solve them once they get above a certain size.

The full simulation of a 300 element quantum system would be an example of such a problem because we could not even store the necessary coefficients classically. Finding the factors of a large number is another particularly famous problem that is known to be similarly hard. One major use of such factorization would be to crack codes, which are often based on the practical impossibility of factoring large numbers. After important initial proof-of-concept proposals that a quantum computer could, at least, solve some problem faster than a classical one (in the sense of how the computation scaled for large versions of the problem),[19] it was the proof[20] that a quantum computer could solve the factorization problem in a way that scaled better than a classical machine that led to a rapid growth in the field. A second important proof was that some database searching could also scale better on a quantum computer.[21]

A classical computer takes some information in a classical state – some data as a set of bits, for example – processes it in a "black box," and gives an output again as another classical state containing bits of information. A quantum computer at this level is similar. Instead of bits that can take a value of "0" or "1", a quantum computer could work with "qubits" that are represented as the linear superposition of two states of a quantum mechanical object. The object might be an electron spin in a linear superposition of spin-up and spin-down, a photon in a linear superposition of two different polarization states, an atom in a linear superposition of two possible states, or any of various other quantum mechanical systems that could exist in such a two-state linear superposition. In general, the state of that qubit can be written as

$$|\psi\rangle = c_0|0\rangle + c_1|1\rangle \equiv \begin{bmatrix} c_0 \\ c_1 \end{bmatrix} \tag{18.17}$$

where $|0\rangle$ is the quantum mechanical state chosen to represent a logical "0", for example, a horizontal polarization state $|\mathrm{H}\rangle$ of a photon, a spin-down state $|\downarrow\rangle$ of an electron, or a ground state $|g\rangle$ of an atom and, similarly, $|1\rangle$ could be represented by vertical polarization $|\mathrm{V}\rangle$, spin-up $|\uparrow\rangle$, or an excited atomic state $|e\rangle$. (Because of normalization, $|c_0|^2 + |c_1|^2 = 1$.)

---

[19] D. Deutsch, in *Proc. Roy. Soc. London A* **400**, 97 (1985).

[20] P. W. Shor, in *Proc. of the 35th Annual Symposium on Foundations of Computer Science* (ed. S. Goldwasser), IEEE Computer Society Press (1994), p. 124.

[21] L. K. Grover, in *Proc. of the 28th Annual ACM Symposium on the Theory of Computing (STOC)*, Philadelphia, PA, May 1996, p. 212; *Phys. Rev. Lett.* **79**, 325 (1997).

To run the quantum computer, an input quantum state is prepared. At least conceptually, it is fed into the quantum black box (known rather poetically as the "oracle" in quantum computing). In practice, inputting the quantum state may well take the form of carefully initializing the quantum states of the various quantum elements (e.g., electrons with spins, or atoms) in the box to specific quantum "starting conditions." Then, we might "turn on" the quantum computer and let its quantum mechanical state evolve in time because of the various designed interactions between the different quantum systems. Sometimes, such running of the computer involves shining specific pulses of light onto specific quantum elements at specific times – for example, in a quantum version of "clocking" the computer – to trigger the various required quantum operations. Then, finally, we would read out the state of the system or some subset of it. That readout is a quantum mechanical measurement process and in that process we do necessarily throw away some of the information about the final quantum state of the system; that loss is one of the issues in designing and running quantum computers.

In conceptualizing a quantum computer, there are ideas that are loosely analogous to the ideas of gates in classical computers. In a classical computer, we know that any classical logic system could be made entirely from 2-input NOR gates with "fan-out" of 2 (i.e., capable of driving the inputs of two subsequent such gates). That does not mean that we would necessarily make a computer that way, but the demonstration of such a NOR gate would be a "completeness" proof that such a classical computer could be made.

Gates in quantum computers are necessarily different. One major difference is that they are reversible, which ordinary classical logic gates are not because they deliberately throw away information – knowledge of the output state of a classical NOR gate is not sufficient to tell you what the input state was. (It is possible, though, to make a classical computer with classically reversible gates.) The reversibility goes along with the idea that the quantum computer works through the evolution of quantum mechanical states; that evolution is a unitary process that is, in principle, reversible using the inverse unitary operator.

One way of expressing the necessary basic operations for a quantum computer is in terms of four different operations. Three of these are operations on a single qubit. We can write these operations as 2 x 2 matrices representing the corresponding unitary operators. One possible set is

$$U_H = \frac{1}{\sqrt{2}}\begin{bmatrix} 1 & 1 \\ 1 & -1 \end{bmatrix},\ U_Z = \begin{bmatrix} 1 & 0 \\ 0 & -1 \end{bmatrix},\ U_{NOTX} = \begin{bmatrix} 0 & 1 \\ 1 & 0 \end{bmatrix} \tag{18.18}$$

These unitary operators can act on a given qubit and are known as Hadamard (U_H), Z (U_Z) and NOT X (U_{NOTX}) operators, respectively. We can think of the qubit using the Bloch sphere picture introduced in Chapter 12 (see Fig. 12.1). That Bloch sphere picture can be used to represent the complete quantum mechanical state of any two-state system. If, then, the qubit is represented as a vector pointing from the center of a sphere to its surface (of unit radius), these various operations correspond to rotations of various kinds of that vector. In practice, single qubit operations like these can be achieved, for example, by appropriate pulses of magnetic fields in given directions in the case of spins or corresponding pulses of electromagnetic fields in the case of two-level "atomic" systems. Manipulating a photon representation of a qubit can simply involve changing the polarization state of the photon,[22] which can be done arbitrarily with various well-known polarization components that can be arranged to correspond to the

[22] The Bloch sphere is closely analogous to the Poincaré sphere used to represent arbitrary photon polarization states in optics.

single-qubit operators. Such single-qubit manipulations have been known and understood for some time, long before qubits and quantum computing were proposed.

The fourth operation, however, involves an interaction between two qubits, an interaction called a Controlled-NOT (C-NOT). One of these qubits is called the control and the other is called the target. If the control is $|0\rangle$, the target qubit is passed through unchanged, but if the control is $|1\rangle$, the target qubit is inverted – that is, a target qubit of state $|0\rangle$ is changed to state $|1\rangle$ and a target qubit of $|1\rangle$ is changed to state $|0\rangle$ – hence the name *Controlled-NOT.*

A two-qubit state is a vector in a four-dimensional Hilbert space; that is, like Eq. (18.16)

$$\begin{aligned}|\psi\rangle &= c_{00}|0\rangle_{control}|0\rangle_{target} + c_{01}|0\rangle_{control}|1\rangle_{target} \\ &+ c_{10}|1\rangle_{control}|0\rangle_{target} + c_{11}|1\rangle_{control}|1\rangle_{target} \\ &= \begin{bmatrix} c_{00} \\ c_{01} \\ c_{10} \\ c_{11} \end{bmatrix}\end{aligned} \tag{18.19}$$

and so the corresponding operator can be written

$$U_{CNOT} = \begin{bmatrix} 1 & 0 & 0 & 0 \\ 0 & 1 & 0 & 0 \\ 0 & 0 & 0 & 1 \\ 0 & 0 & 1 & 0 \end{bmatrix} \tag{18.20}$$

For example, the input state with the control as a logic 0 and the target as a logic 1 is the state with $c_{00} = 0$, $c_{01} = 1$, $c_{10} = 0$, and $c_{11} = 0$. Writing that state as a column vector and operating with U_{CNOT} gives

$$\begin{bmatrix} 0 \\ 1 \\ 0 \\ 0 \end{bmatrix} = \begin{bmatrix} 1 & 0 & 0 & 0 \\ 0 & 1 & 0 & 0 \\ 0 & 0 & 0 & 1 \\ 0 & 0 & 1 & 0 \end{bmatrix} \begin{bmatrix} 0 \\ 1 \\ 0 \\ 0 \end{bmatrix} \tag{18.21}$$

which is just the state we started with; that is, as intended, the target qubit passes through unchanged if the control qubit is logic 0. But, if we choose to have the control qubit be a logic 1 – that is, if we choose the input state as $c_{00} = 0$, $c_{01} = 0$, $c_{10} = 0$, and $c_{11} = 1$ – then

$$\begin{bmatrix} 0 \\ 0 \\ 1 \\ 0 \end{bmatrix} = \begin{bmatrix} 1 & 0 & 0 & 0 \\ 0 & 1 & 0 & 0 \\ 0 & 0 & 0 & 1 \\ 0 & 0 & 1 & 0 \end{bmatrix} \begin{bmatrix} 0 \\ 0 \\ 0 \\ 1 \end{bmatrix} \tag{18.22}$$

The resulting output state is, therefore, $c_{00} = 0$, $c_{01} = 0$, $c_{10} = 1$, and $c_{11} = 0$, which is the state with the target qubit now a logic 0 – that is, it has been "flipped" – and the control bit remaining at logic 1.

A key point about the physical process that implements such an operator is that there must be interaction between two systems representing the two qubits for this to work. If the system representing the control qubit is in its $|1\rangle$ state, we would want it to affect the system representing the target qubit. Now, one common approach in running such a gate is that we might have a pulse (e.g., a light pulse) that shines on the gate (and especially on the target qubit system) every "cycle" of operation. This pulse itself is not carrying information and might loosely be analogous to a "clock" pulse in a conventional digital system. If the control qubit system is in its $|0\rangle$ state, then this clock pulse does nothing to the target qubit system (e.g., perhaps the clock pulse then has the wrong optical frequency to affect the target qubit system). If, however, the control qubit is in its $|1\rangle$ state, perhaps it then changes some transition frequency in the target qubit system, through some interaction between the control and target qubit systems. With this change in transition energy, the target qubit system could then be sensitive to the clock pulse in such a way that the clock pulse then causes an inversion of the target qubit state. This kind of system would implement the Controlled-NOT function.

Many different systems have been proposed for implementing quantum computing and various simple ones have been demonstrated. Example systems include ions in ion traps, superconducting flux and charge qubits, quantum dots, and spins in semiconductor impurities.[23]

A major challenge for quantum computing is that it is difficult to isolate the quantum mechanical systems enough from their environment. The consequence of that is that, at least, the phase of the quantum mechanical system keeps being disturbed, which destroys the fidelity of the quantum mechanical states being used; quantum computing does rely on the phase of the quantum mechanical states being undisturbed for sufficiently long times. Essentially, we need systems with long dephasing or decoherence times (like the T_2 dephasing times discussed in Chapter 14). One possible solution is to perform quantum error correction to restore the fidelity of the state, though that itself requires quantum computing gates and so we would still need to be able to get above some threshold number of quantum operations without dephasing. At the time of writing, the difficulties of dephasing prevent any serious quantum computer from being constructed, but the idea remains a very intriguing one that could ultimately have substantial practical implications.

18.5 Quantum teleportation

The idea of quantum teleportation is to transfer a quantum state from one place to another without actually transferring the specific carrier of that state.[24] For example, we might have a photon (photon 1) in an unknown superposition of horizontal and vertical polarization; we want to manage to have a different (distinguishable) photon (photon 3) be in the same superposition state somewhere else, but without sending photon 1 there. In fact, we may even destroy (absorb) photon 1 in the process.

---

[23] For a critical review, see, e.g., T. P. Spiller, W. J. Munro, S. D. Barrett, and P. Kok, "An Introduction to Quantum Information Processing: Applications and Realizations," *Contemporary Physics* **46**, 407–436 (2005).

[24] The common science fiction notion of teleportation is that we manage to transfer matter from one place to another; that is not what is meant by teleportation here, unfortunately. We are only transporting the quantum mechanical state that the matter might have had. In transferring or "beaming down" the starship captain to the planet surface, if we wanted him or her to be in the same quantum mechanical state as they had been on the starship, then quantum teleportation in the sense being discussed here would have to be part of the mechanism. Whether it is important to transfer the quantum mechanical state or only the classical state in such a transfer of a living entity is an interesting philosophical question, but one that is certainly beyond us here.

This would appear to be an extremely tricky task to accomplish. We know from the no-cloning theorem that we cannot simply clone photon 1 to produce another photon (photon 3) in the same arbitrary superposition. We also know from our previous discussion that simply making a measurement on photon 1 – for example, with a polarizing beamsplitter or filter of some kind together with photodetectors – will not reliably tell us the full quantum state of photon 1; we end up statistically "collapsing" the state and unavoidably throwing away information about the original quantum state of the photon.

The key to making this work is to "share entanglement," which is achieved through sharing a pair of photons (an EPR pair) that are in an entangled state in the form of a Bell state. The block diagram[25] of the quantum teleportation apparatus is shown in Fig. 18.2.

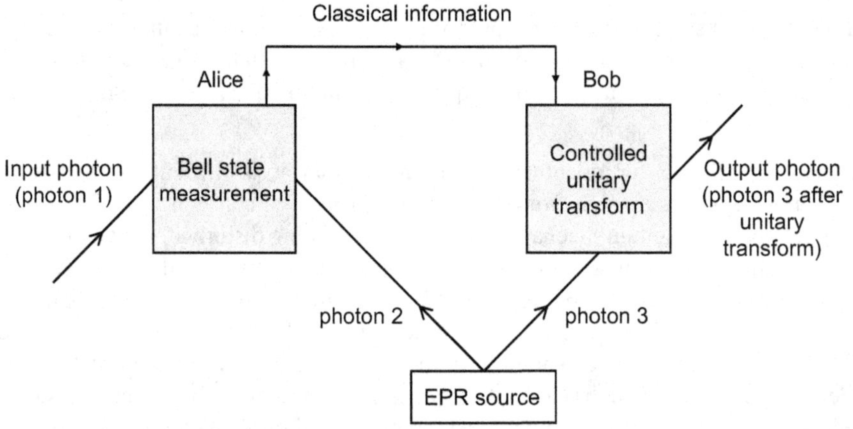

Fig. 18.2. Apparatus for quantum teleportation.

The EPR photon pair is presumed to be in the Bell state of Eq. (18.15)[26]

$$|\Psi^-\rangle_{23} = \frac{1}{\sqrt{2}}\left(|\mathrm{H}\rangle_2|\mathrm{V}\rangle_3 - |\mathrm{V}\rangle_2|\mathrm{H}\rangle_3\right) \tag{18.23}$$

We presume that the input photon is in some general state that is a linear superposition of the horizontal and vertical polarizations; that is,

$$|\psi\rangle_1 = c_{\mathrm{H}}|\mathrm{H}\rangle_1 + c_{\mathrm{V}}|\mathrm{V}\rangle_1 \tag{18.24}$$

The state of all three photons, therefore, can be written as

$$|\Psi\rangle_{123} = \frac{1}{\sqrt{2}}\left(c_{\mathrm{H}}|\mathrm{H}\rangle_1 + c_{\mathrm{V}}|\mathrm{V}\rangle_1\right)\left(|\mathrm{H}\rangle_2|\mathrm{V}\rangle_3 - |\mathrm{V}\rangle_2|\mathrm{H}\rangle_3\right) \tag{18.25}$$

A core trick in the teleportation is to note that this state can be rewritten as

---

[25] Alice and Bob appear here as the "names" associated with the boxes for historical reasons from quantum cryptography, though it is important to emphasize that there is, in fact, no human intervention required here to make this work. In fact, neither Alice nor Bob acquires any knowledge about the state of photon 1 or photon 3. The only information they receive here is about the EPR pair, but that is information "internal" to this system.

[26] This particular Bell state is readily produced by parametric down-conversion.

$$|\Psi\rangle_{123} = \frac{1}{2}\Big[|\Phi^+\rangle_{12}\left(c_H|V\rangle_3 - c_V|H\rangle_3\right) + |\Phi^-\rangle_{12}\left(c_H|V\rangle_3 + c_V|H\rangle_3\right)$$
$$+|\Psi^+\rangle_{12}\left(-c_H|H\rangle_3 + c_V|V\rangle_3\right) - |\Psi^-\rangle_{12}\left(c_H|H\rangle_3 + c_V|V\rangle_3\right)\Big] \quad (18.26)$$

Note that we have managed to write the state in terms of Bell states of photons 1 and 2 (Eqs. (18.12)–(18.15)). If we now make a measurement in Alice's Bell state measurement box of the Bell state of this pair of photons 1 and 2, we collapse the state into just one of those terms. For example, suppose we made the measurement of the Bell state and got the answer $|\Phi^-\rangle_{12}$, an answer we can know classically because it is the result of a measurement; then, the overall system of three photons would now be in the state

$$|\Psi\rangle_{123} = \frac{1}{2}|\Phi^-\rangle_{12}\left(c_H|V\rangle_3 + c_V|H\rangle_3\right) \quad (18.27)$$

Because Alice can tell Bob the result of her measurement by communication over an ordinary classical channel (e.g., a telephone line), Bob now knows that photon 3 is in the state $c_H|V\rangle_3 + c_V|H\rangle_3$. This is not the same as the original state of the photon (which was by definition $c_H|H\rangle_1 + c_V|V\rangle_1$), but that is easily fixed. In this specific case, Bob could rotate the polarization of the photon by 90° clockwise, turning vertical polarization into horizontal and horizontal into –vertical (i.e., $|V\rangle \to |H\rangle$ and $|H\rangle \to -|V\rangle$) and insert a half wave plate to delay the vertical polarization by 180° relative to the horizontal, turning c_V to $-c_V$. Then, photon 3 will be in exactly the same state as photon 1 was, *without either Alice or Bob ever knowing what that state was* (i.e., without them knowing the coefficients c_H and c_V). For other results from Alice's Bell state measurement, Bob has to implement other polarization manipulations, but those present no fundamental problem (e.g., he could use electrically controlled phase shifters). In general terms, Bob implements a specific unitary transformation on photon 3 (a combination here of phase delays and polarization rotations) that depends on the outcome of Alice's Bell state measurement; hence, for any result from Alice, Bob can put photon 3 into exactly the same state as photon 1 originally had, thus completing the teleportation of the quantum mechanical state.

The reader should feel that there is something very strange going on here. The measurement of photons 1 and 2 by Alice has apparently changed the state of photon 3. This particular strangeness is at the core of the EPR paradox and the consequences that follow from it, such as Bell's inequalities, both of which are discussed in the next chapter.

Further reading

M. Fox, *Quantum Optics* (Oxford, 2006), Chapters 12–14, gives a clear tutorial introduction to quantum information, more extensive than is possible here.

M. A. Nielsen and I. L. Chang, *Quantum Computation and Quantum Information* (Cambridge, 2000) gives an extensive discussion of the subject.

D. Bouwmeester, A. Ekert, and A. Zeilinger (eds.) *The Physics of Quantum Information* (Springer 2000) introduces the subject and gives an extended discussion of research with contributions from a broad range of participants in the field.

N. Gisin, G. Ribordy, W. Tittel, and H. Zbinden, "Quantum Cryptography," *Rev. Mod. Phys.* **74**, 145–195 (2002) gives a clear and comprehensive discussion of quantum cryptography.

T. P. Spiller, W. J. Munro, S. D. Barrett, and P. Kok, "An Introduction to Quantum Information Processing: Applications and Realizations," *Contemporary Physics* **46**, 407–436 (2005) reviews the various aspects of quantum information and the approaches to demonstrating it.

18.6 Summary of concepts

No cloning theorem

It is not possible to make a quantum mechanical apparatus that can clone a quantum mechanical state in an arbitrary superposition. The proof relies essentially on the linearity of quantum mechanical operators.

Entangled states

An entangled state of two or more particles is one that cannot be factored into a product of states for each particle.

Bell states

Bell states are classic examples of entangled states for pairs of particles, each of which has two accessible basis states (e.g., horizontal $|\mathrm{H}\rangle$ and vertical $|\mathrm{V}\rangle$ polarization of a photon or spin-up and spin-down of an electron). The four Bell states for a pair of photons are

$$|\Phi^{+}\rangle_{12} = \frac{1}{\sqrt{2}}\left(|\mathrm{H}\rangle_1 |\mathrm{H}\rangle_2 + |\mathrm{V}\rangle_1 |\mathrm{V}\rangle_2\right) \tag{18.12}$$

$$|\Phi^{-}\rangle_{12} = \frac{1}{\sqrt{2}}\left(|\mathrm{H}\rangle_1 |\mathrm{H}\rangle_2 - |\mathrm{V}\rangle_1 |\mathrm{V}\rangle_2\right) \tag{18.13}$$

$$|\Psi^{+}\rangle_{12} = \frac{1}{\sqrt{2}}\left(|\mathrm{H}\rangle_1 |\mathrm{V}\rangle_2 + |\mathrm{V}\rangle_1 |\mathrm{H}\rangle_2\right) \tag{18.14}$$

$$|\Psi^{-}\rangle_{12} = \frac{1}{\sqrt{2}}\left(|\mathrm{H}\rangle_1 |\mathrm{V}\rangle_2 - |\mathrm{V}\rangle_1 |\mathrm{H}\rangle_2\right) \tag{18.15}$$

Chapter 19

Interpretation of quantum mechanics

Prerequisites: Chapters 2–5, 9, 10, 12, and 13, and Chapter 18, Sections 18.1 and 18.3.

Quantum mechanics as we have discussed it here works remarkably well. When we can figure out how to calculate something using quantum mechanics, we know of no situation where it gives the wrong answer. At a pragmatic level, it therefore works. It does, however, have some aspects about it that are very different from the classical view of the world. Interpreting what quantum mechanics says about reality is a very tricky subject with strange consequences and truly bizarre proposals about how reality actually works. Here, we try to give a short introductory summary of some of those key ideas. It is necessarily brief and incomplete and is certainly not conclusive. Indeed, this field of the interpretation of quantum mechanics can be considered unresolved, even if protagonists of various interpretations might attempt to convince us otherwise. Fundamentally, we do not know how to resolve some of the more philosophical aspects because we have no experiment to discriminate between them, though some points previously believed to be only philosophical have been resolved by experiments, with remarkable consequences.

19.1 Hidden variables and Bell's inequalities

Is quantum mechanics truly random? Perhaps the solution to the apparent randomness of quantum mechanics, with states collapsing into eigenstates with only statistical weights, is that quantum mechanics as stated is incomplete, in the sense that classical statistical mechanics is incomplete. Statistical mechanics discusses the most likely outcomes from a statistical point of view of processes based on, for example, the collisions of atoms or molecules in a gas. It is a good way of calculating thermodynamic behavior, such as relations between pressure and temperature in gases.

In a classical view of the world, though, there is an underlying deterministic behavior of colliding mechanical particles, such as billiard-ball–like collisions of atoms or molecules, and the use of the statistical treatment just helps us avoid dealing with all those collisions for some calculations. For other calculations, we presume there is that underlying deterministic theory we could use if we wanted to. If we only use statistical mechanics, these underlying variables (i.e., the actual positions and momenta of each atom or molecule) are hidden from us, though, at least in a classical world, we presume that they exist. Perhaps quantum mechanics rests also on such hidden variables and if we could figure out what they were, the apparent randomness of quantum mechanics would disappear as a fundamental aspect of nature.

Einstein, in particular, was not happy with either the randomness of quantum mechanics as stated nor the collapse of the wavefunction. With his colleagues, Podolsky and Rosen, he came up with a famous thought-experiment that he believed demonstrated the absurdity of the randomness and the wavefunction collapse and the reasonableness of a hidden variable approach, a thought-experiment known since then as the *EPR paradox*.[1]

Essentially, the core of the EPR paradox is that it is possible to create two distinguishable particles (an EPR pair) in a quantum mechanical superposition state of the form of one of the Bell states discussed in Chapter 18; for example, for two photons 1 and 2 going in different directions, like

$$|\Phi^+\rangle_{12} = \frac{1}{\sqrt{2}}\left(|H\rangle_1|H\rangle_2 + |V\rangle_1|V\rangle_2\right) \tag{19.1}$$

that is, a linear superposition of the state where the two photons are both horizontally polarized and the state where the two photons are both vertically polarized.[2]

In such a state, if one measures one of the photons to be horizontally polarized (i.e., in a state $|H\rangle$), according to quantum mechanics, the state of both particles is forced to collapse into the one element $|H\rangle_1|H\rangle_2$ in the linear superposition; a measurement on the other photon is now bound to give the result $|H\rangle$ also if the apparatuses have aligned polarizers. Similarly, measuring the result $|V\rangle$ for one photon will lead, according to quantum mechanics, to the inescapable conclusion that the other photon will also be in the state $|V\rangle$.

Also according to quantum mechanics, neither photon has a defined polarization until it is measured and so one seems to be forced to conclude that somehow the measurement of one photon leads to a change in the other one's state. This is, indeed, a bizarre notion, especially if one arranges that the photons are very widely separated at the time either of them is measured – so far apart that there is no possibility during the time of measurement that any signal can be conveyed, even at the velocity of light,[3] between the two measurement apparatuses. Einstein referred to such a change in state of the other particle as "spooky action at a distance." He considered such behavior strong evidence that there should be hidden variables.

A potential way out of the absurdity would seem to be to assert that, in fact, each photon actually does have a specific polarization at the time it leaves the apparatus that creates it, but that we merely are unaware of it until the measurement takes place. The fact that the photons had actual polarizations at all times would be a variable hidden from quantum mechanics, in the sense that the actual positions and momenta of gas atoms are hidden from classical statistical mechanics. This supposed hidden variable would presumably be quite real in some more complete theory. This information on the polarization of the photon would presumably be carried with the photon as a local property of the photon; hence, such a hidden variable is called a local hidden variable. This particular hidden variable theory, with polarization as the hidden variable, will certainly not work, as will become clear in the following discussion.

---

[1] A. Einstein, B. Podolsky, and N. Rosen, *Phys. Rev.* **47**, 777–780 (1935).

[2] This is not actually the state proposed by EPR, but it is more convenient for our discussion and has the correct basic character.

[3] Though the influence would apparently travel faster than light in the quantum mechanical view, no useful information can be conveyed this way because we still cannot choose the outcome of the measurement that the distant observer will find; we merely know what result that observer will see.

Einstein apparently held the opinion for the rest of his life that there were, however, some local hidden variables behind quantum mechanics, but he, and everyone else, believed that there was no experimental test that could answer the question as to whether local hidden variables actually existed. This whole discussion, therefore, remained in the unprovable world of philosophical speculation.

The appeal of the idea of local hidden variables is obvious. It restores to us the belief that though we do not yet have a theory that explains exactly how such variables behave, the world actually does agree with our apparent local determinate view (i.e., things have actual well-defined states given by properties that exist in some region of space near the things).

In 1964, Bell[4] proposed a way of distinguishing between local hidden variable theories and the predictions of quantum mechanics. He showed that in a particular kind of behavior that could be seen in an EPR experiment, any local hidden variable theory would give a result that would be different from the predictions of quantum mechanics. This remarkable prediction was tested with increasing sophistication.[5] The result is the same in all tests. The experiments agree with quantum mechanics and disagree with any local hidden variable theory.

The specific results that discriminate between the local hidden variable theories and quantum mechanics relate to the correlation we would see between the results from the two different apparatuses for measuring the polarization of the two different photons. In particular, the key aspect is what happens when the two apparatuses have their axes rotated at an angle. For certain ranges of angles, the two theories disagree substantially and measurably and the results for local hidden variable theories obey a set of inequalities (i.e., Bell's inequalities). The behavior in the quantum case turns out to give correlations that are different from those that are possible in any local hidden variable theory.

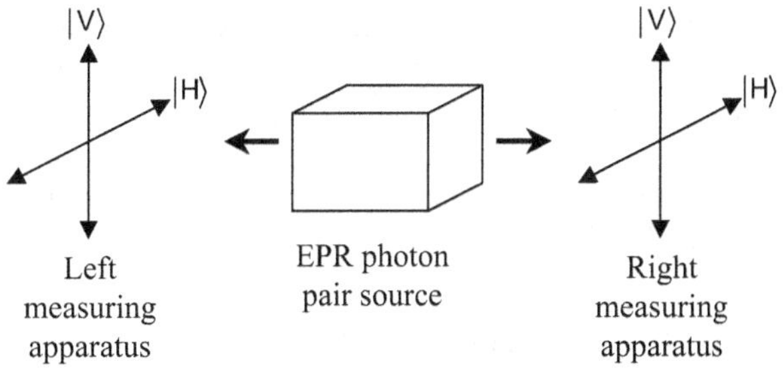

Fig. 19.1. Illustration of an EPR photon pair source and two aligned apparatuses to detect horizontal or vertical polarization.

We can easily see that we run into trouble with local hidden variable approaches. Suppose, for example, that the photons in the EPR pair are heading off to two different measuring apparatuses with their axes aligned as in Fig. 19.1. As we mentioned previously, quantum

---

[4] J. S. Bell, "On the Einstein-Podolsky-Rosen paradox," *Physics* **1**, 195–200 (1964); reprinted in J. S. Bell, *Speakable and Unspeakable in Quantum Mechanics* (Cambridge University Press, 1993), pp. 14–21.

[5] J. F. Clauser and A. Shimony, "Bell's Theorem. Experimental Tests and Implications," *Rep. Prog. Phys.* **41**, 1881–1927 (1978); A. Aspect, P. Grangier, and G. Roger, "Experimental Realization of Einstein-Podolsky-Rosen-Bohm *Gedankenexperiment*: A New Violation of Bell's Inequalities," *Phys. Rev. Lett.* **49**, 91–94 (1982); A. Aspect, "Bell's inequality test: more ideal than ever," *Nature* **398**,189–190 (1999).

mechanics predicts that for an EPR pair in, for example, the Bell state of Eq. (19.1), if we measure one photon to be horizontal, then we will find the other photon also to be horizontal; similarly, if we measure one photon to be vertical, the other photon will also be measured to be vertical. This is the behavior we find also in experiments; with their polarization axes aligned, both apparatuses always measure the same polarization for the two photons.

Suppose we asserted that, in fact, the two photons had the same polarization, with that actual polarization being the hidden variable, though we just did not know what it was until we measured it. In that case, the polarization of each photon would be what is called a *local hidden variable* – a quantity whose value is presently unknown but that is a property that is "carried along with" the photon; in this sense, it is local to the photon.

If that polarization is not aligned with either the horizontal or vertical axis, then each photon has a probability of being measured to be either horizontal or vertical, and many times, therefore, we will see the two photons being measured to have different polarizations. But this is not what we observe in an experiment. With such EPR pairs of photons, we always find the two photons to have the same measured polarization (even if we choose the angles of the polarizers *after* the photons have been created – a so-called "delayed choice" experiment).

We can fix this particular hidden variable theory to get the same answer as the experiment for aligned polarizers, though we have to introduce attributes and behavior that bear no relation to any physical model we had thought about for photons or electromagnetic radiation. (For example, we can simply say that each photon has some additional local attribute that causes it to emerge on a particular axis from the polarizer and, because both photons have the same attribute, they both emerge always on the same axis.)

It turns out, however, as Bell proved, that once we misalign the two measuring apparatuses (i.e., rotate the polarizer axes of one apparatus with respect to the other), there are inequalities relating the correlations between the measured results on the two different apparatuses that must be obeyed for *any* local hidden variable theory. Those limiting correlations can be tested against experiment. It is found that the experimental results violate these inequalities, so no local hidden variable theory agrees with experiment. (It is also true that, so far as we can tell, the experimental results do agree with the quantum mechanical prediction for such EPR states.)

The stunning conclusion, then, is that reality is not local. We cannot describe reality as we see it based only on local properties. To be forced to this conclusion, we do not even need to believe that quantum mechanics is correct. Even if quantum mechanics is wrong and its agreement with the experimental results is merely a coincidence, it is still not possible to construct a local hidden variable theory that agrees with the experimental results.

A simple version of a Bell's inequality

The general proof of Bell's inequalities can be difficult to follow. We can examine a simpler version here, a version that needs very little mathematics.[6] Consider the consequences of a deterministic local hidden variable theory, where the hidden variables are definite properties of each particle on its own and, hence, can be said to be local to the particle. The particle may also have various attributes or definite values of variables that are not hidden. The net result of

[6] This is based on an argument given in J. S. Bell "Bertlmann's Socks and the Nature of Reality," in *Speakable and Unspeakable in Quantum Mechanics* (Cambridge, 1987), pp. 139–158, amended to discuss photons and extended to include a Venn diagram approach.

the combination of the hidden and nonhidden variables, which we can call the local state of the particle, determines the outcome of any measurement on the particle. Measuring devices acting on other particles we therefore expect do not matter. In such a theory, they cannot possibly matter if those measurements are made sufficiently far away and at such times that no information can get to our particle in time to influence the outcome of our measurement.

In particular, then, in such a local deterministic view of reality, the values of the hidden and nonhidden variables for a photon (i.e., the local state of the photon) determine from which port of a polarizing beamsplitter the photon will emerge because this is, by choice, a deterministic theory. So we can, if we wish, draw a Venn diagram, as in Fig. 19.2. Each combination of local variables that corresponds to a particular measurable set of behaviors with polarizers at any angle is represented by a point on this Venn diagram. This figure shows three possibly overlapping regions. One region includes all possible local states that lead to the photon passing through a polarizer set at 0°. A second region includes all possible local states that lead to the photon passing through a polarizer at 22.5°. A third region includes all possible local states for which a photon does *not* pass through a polarizer at 45°. All of these three regions can overlap, pair-wise and overall and still be in agreement with our observations on what happens with single photons and polarizers.

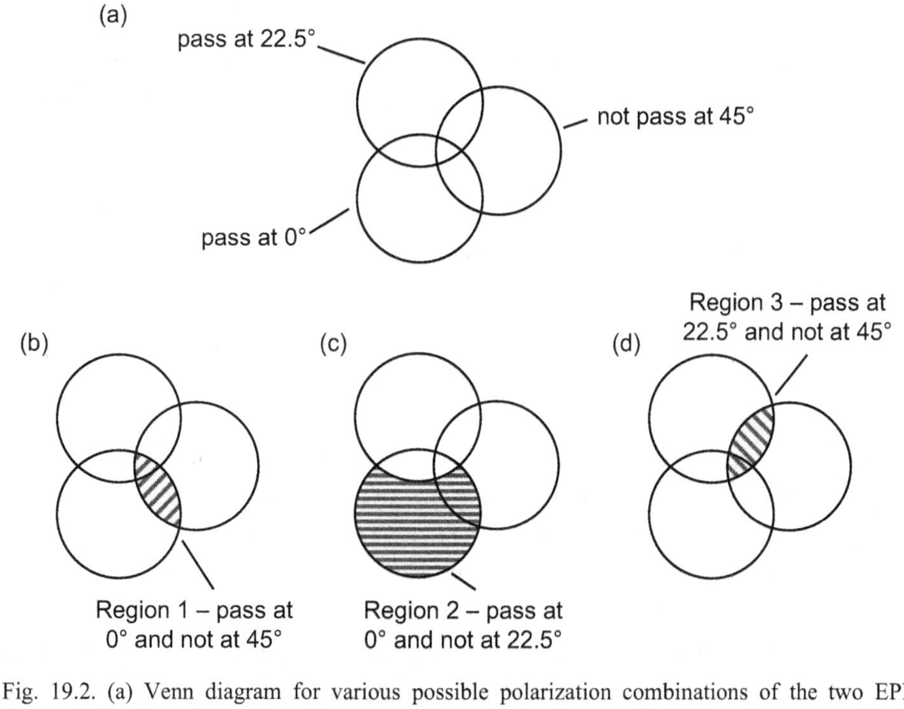

Fig. 19.2. (a) Venn diagram for various possible polarization combinations of the two EPR photons. (b), (c), and (d) Venn diagrams with specific regions marked. Note that every point in Region 1 lies within either Region 2 or Region 3.

Now, we have to restrict ourselves to performing only one test on a given photon (with a polarizer set at 0°, 22.5°, or 45°) because that test or measurement may change the state of the photon in some way. However, with our EPR photon pair source, we have two photons to use in two different experiments (one on the left and one on the right) and we already know that photons prepared this way always behave identically if the polarizers are set identically. Hence, on the presumption that their behavior when they get to the polarizer is determined by local information they carry with them, we conclude that the local states of these two photons are identical in that they lie at the same point on this Venn diagram. So now we can consider the overlap regions of the Venn diagram, which correspond to those sets of local states in which, for example, one photon passes a polarizer at 0° and the other passes a polarizer at 22.5°.

We can identify three particular regions of the Venn diagram of interest to us here. Region 1 is the set of all local states in which one photon will pass a polarizer at 0° while the other will not pass a polarizer at 45°. Region 2 is the set where one photon will pass at 0° and the other will not pass at 22.5° and Region 3 is the set where one photon will pass at 22.5° and the other will not pass at 45°.

It is obvious from the Venn diagram that Regions 2 and 3 taken together encompass all of the states in Region 1 (i.e., Region 1 is a subset of the union of Regions 2 and 3). Hence, the sum of the number of local states in Regions 2 and 3 taken together always either equals or exceeds the number of local states in Region 1. We presume that we can perform this experiment many times with random local states. (We can verify the randomness by performing experiments on just one of the photons to see that they give random behavior independent of polarizer angle.) In such a case, therefore, the probability that the local state is found to lie in Region 1 must be less than or equal to the probability that the local state is found to lie in Region 2 or Region 3; that is,

the probability that one photon will pass at 0° and the other will not pass at 22.5°

+

the probability that one photon will pass at 22.5° and the other will not pass at 45°

≥

the probability that one photon will pass at 0° and the other will not pass at 45°

(19.2)

This is a specific example for a Bell's inequality. If we, therefore, find in an experiment that this inequality is violated, then we have to throw out any deterministic local hidden variable theory.

But experiments do violate this inequality. Therefore, deterministic local hidden variable theories cannot explain reality.

Note that this argument does not even mention quantum mechanics. Bell's inequalities in this sense have nothing to do with quantum mechanics. They show that because of the results of experiments, reality cannot be explained by local variables, hidden or otherwise.

Of course, we find that the results of the experiments do also agree with the predictions of quantum mechanics, which, therefore, gives another strong argument for quantum mechanics, but at a great price – we are forced to abandon a local view of reality. Particles cannot simply be described deterministically by properties they carry along with them.

The quantum mechanical calculation of these probabilities is straightforward. Consider a photon linearly polarized at an angle θ to the horizontal. We can then describe the polarization state of that photon as

$$|\psi(\theta)\rangle = \cos\theta|H\rangle + \sin\theta|V\rangle \tag{19.3}$$

where $|H\rangle$ and $|V\rangle$ are the horizontal and vertical polarization basis states, respectively. We take the probability that such a photon will pass a horizontally oriented beamsplitter to be

$$P_H = |\langle H|\psi(\theta)\rangle|^2 = \cos^2\theta \tag{19.4}$$

Now, we consider two different photon modes, one propagating to the left (L) and the other to the right (R). We presume that we have one photon in each mode and that this pair of photons is in an "EPR" (i.e., Bell) state with the two orthogonal polarization states now generalized to be at angles θ and $\theta+\pi/2$ instead of just horizontal and vertical. Hence, we can have the form

$$|\psi_{EPR}\rangle = \frac{1}{\sqrt{2}}\left[|\psi(\theta),L\rangle|\psi(\theta),R\rangle + |\psi(\theta+\pi/2),L\rangle|\psi(\theta+\pi/2),R\rangle\right] \tag{19.5}$$

where $|\psi(\theta),L\rangle$ means that the photon in the left-going mode is polarized with angle θ to the horizontal axis, and so on. We now consider examining this state with a horizontal polarizer on the left (or, at least, the horizontal arm of a polarizing beamsplitter) and a polarizer at angle ϕ to the horizontal on the right. For such an examination, we have an amplitude

$$\begin{aligned}
A &= \left(\langle H,L|\langle\psi(\phi),R|\right)|\psi_{EPR}\rangle \\
&= \frac{1}{\sqrt{2}}\left(\langle H,L|\langle\psi(\phi),R|\right)\begin{pmatrix}|\psi(\theta),L\rangle|\psi(\theta),R\rangle \\ +|\psi(\theta+\pi/2),L\rangle|\psi(\theta+\pi/2),R\rangle\end{pmatrix} \\
&= \frac{1}{\sqrt{2}}\begin{bmatrix}\langle H,L|\psi(\theta),L\rangle\langle\psi(\phi),R|\psi(\theta),R\rangle \\ +\langle H,L|\psi(\theta+\pi/2),L\rangle\langle\psi(\phi),R|\psi(\theta+\pi/2),R\rangle\end{bmatrix}
\end{aligned} \tag{19.6}$$

Note that we can write

$$\langle\psi(\phi)|\psi(\theta)\rangle = \langle H|\psi(\theta-\phi)\rangle \tag{19.7}$$

because the inner product of these two vectors does not change if we rotate both of them by an angle $-\phi$. Hence, we have

$$\begin{aligned}
A &= \frac{1}{\sqrt{2}}\left[\cos\theta\cos(\theta-\phi) + \cos(\theta+\pi/2)\cos(\theta+\pi/2-\phi)\right] \\
&= \frac{1}{\sqrt{2}}\left[\cos\theta\cos(\theta-\phi) + \sin(\theta)\sin(\theta-\phi)\right] \\
&= \frac{1}{\sqrt{2}}\frac{1}{2}\left[\cos(\phi) + \cos(2\theta-\phi) + \cos(\phi) - \cos(2\theta-\phi)\right] \\
&= \frac{1}{\sqrt{2}}\cos\phi
\end{aligned} \tag{19.8}$$

independent of the angle θ of the polarization axis of the EPR pair. Hence, we can conclude that the probability of the "left" photon passing the left polarizer at angle 0 (horizontal) and the "right" photon passing the right polarizer at angle ϕ is

$$P = |A|^2 = \frac{1}{2}\cos^2\phi \tag{19.9}$$

If a photon on the right passes at an angle ϕ, then it fails to emerge from the other arm of a polarization beamsplitter, an arm that passes a photon of polarization angle $\phi - \pi/2$. Hence, we can finally conclude that the probability of the left photon passing the left polarizer at angle 0 (horizontal) and the right photon failing to pass the right polarizer at angle ϕ (i.e., passing a right polarizer at angle $\phi - \pi/2$) is

$$P_\phi = \frac{1}{2}\cos^2(\phi - \pi/2) = \frac{1}{2}\sin^2\phi \tag{19.10}$$

The choice of the polarizer orientation on the left as horizontal is arbitrary and so this expression applies when the angle between the two polarizers is ϕ. Hence, we have

the probability that one photon will pass at 0° and the other will not pass at 22.5°
$((1/2)\sin^2(22.5°) \simeq 0.0732$

+

the probability that one photon will pass at 22.5° and the other will not pass at 45°
$(1/2)\sin^2(22.5°) \simeq 0.0732$

$\simeq 0.1464$

But

the probability that one photon will pass at 0° and the other will not pass at 45°
$(1/2)\sin^2(45°) = 0.2500$

Hence, because $0.2500 > 0.1464$, Bell's inequality, Eq. (19.2) for this case, is violated by the quantum mechanical calculation, a calculation that also appears to agree well with experiment.

Hence, no local hidden variable theory can explain the results of the EPR experiment with misaligned polarizers. Hence, if quantum mechanics is to explain the results of experiments (and it does), quantum mechanics cannot be explained by local hidden variable theories either. This is a remarkable conclusion.

Problem

19.1.1 (i) Consider the Bell state $|\Phi^+\rangle_{12} = (1/\sqrt{2})(|H\rangle_1|H\rangle_2 + |V\rangle_1|V\rangle_2)$ of two photons. Express it now not on a basis of horizontal ($|H\rangle$) and vertical ($|V\rangle$) polarized states but instead on a basis $|\theta\rangle$ and $|\theta + \pi/2\rangle$ rotated by an angle θ in a positive (i.e., anticlockwise) direction relative to the horizontal axis. (Hint: On such a basis, $|V\rangle = \sin\theta|\theta\rangle + \cos\theta|\theta + \pi/2\rangle$ and a similar expression exists for $|H\rangle$.)

(ii) Hence, show that the two photons in such a state will always come out of the same arm of each polarizer when two aligned polarizers are used to examine the pair of photons.

19.2 The measurement problem

Thus far, in discussing quantum mechanics, we have presumed that we can make a measurement and it is this measurement that forces quantum mechanical systems into eigenstates of the quantity being measured (i.e., the collapse of the wavefunction). There is, however, a major problem. We do not know what a measurement is in quantum mechanics and we cannot construct a model of it in the quantum mechanics we have constructed with linear operators acting on states.

The proof of this difficulty is very simple. Suppose we have some measurement apparatus that is going to act on a quantum mechanical state to give the measured value. Suppose, then, that this apparatus is described by the quantum mechanical operator $\hat{M}$, which must, therefore, be a linear operator. Suppose that the system starts out in one of the eigenstates of the quantity being measured by the apparatus. It is sufficient here to deal with a quantity that only has two eigenstates (e.g., electron spin). Hence, for the initial eigenstate $|\uparrow\rangle$, we have the result of the measurement being the state

$$M|\uparrow\rangle = |\uparrow\rangle \tag{19.11}$$

that is, the system, when measured in an eigenstate, stays in that eigenstate. Similarly, for the other possible initial eigenstate

$$M|\downarrow\rangle = |\downarrow\rangle \tag{19.12}$$

So far, this agrees with our observation. But suppose instead that the system starts in a linear superposition state $a_\uparrow|\uparrow\rangle + a_\downarrow|\downarrow\rangle$. Then, on operating on that state, because of the linearity of $\hat{M}$, we have

$$\hat{M}\left(a_\uparrow|\uparrow\rangle + a_\downarrow|\downarrow\rangle\right) = a_\uparrow\hat{M}|\uparrow\rangle + a_\downarrow\hat{M}|\downarrow\rangle = a_\uparrow|\uparrow\rangle + a_\downarrow|\downarrow\rangle \tag{19.13}$$

Note that the resulting state is a linear superposition also.[7] The "measurement" operation has not been able to collapse the resulting state into an eigenstate, in disagreement with our understanding of measurement. Therefore, there is no way of describing measurement using the quantum mechanics we have constructed.

19.3 Solutions to the measurement problem

The measurement problem has existed since, at least, the late 1920s. There is no proposed solution to it that does not appear either absurd in some way or that invokes something outside quantum mechanics for which we have no explanation. Also, however, there is no experiment that we know of that can discriminate between the different proposals and, therefore, all of them remain in the realm of metaphysical philosophical speculation. We cannot even use their relative absurdity (even if we had a standard for that) as a way of deciding which one is the right one. We have already seen that the experimentally testable aspects of quantum mechanics lead us to conclusions that are quite absurd but nonetheless correct. Here, we briefly mention some of the contenders for solutions to the measurement problem. This is not a complete list (the actual list is very long, especially if one gets into variations of the contenders presented here). It does, however, give some flavor of the speculations in this field.

The Standard Interpretation

The so-called Standard Interpretation of quantum mechanics is that we merely separate out the measurement process as being something different and use our simple rules to work out the probability of various outcomes. It is essentially the view advocated by von Neumann and by Dirac. From an engineering point of view, this presents no apparent problems as long as it keeps working for every problem we ask it to solve. It is, however, very unsatisfactory from a scientific or philosophical point of view because we have no description of how the measurement process works and at what point we make the boundary between the quantum

[7] It might be argued that we ought to include the state of the measurement apparatus in the overall quantum mechanical state being considered, but that does not get us out of this problem, still leaving us in a linear superposition state when we start with a linear superposition state.

mechanical world and the (presumably) macroscopic world of measurements. We take the approach that the quantum system keeps evolving in superposition states until we measure it, but we do not know what measurement is. The Standard Interpretation does not solve the measurement problem but does acknowledge that it exists.

The Standard Interpretation is also an example of theory in which reality (or, specifically, the determinate experience we have of things being in specific states rather than superpositions) is created by the act of observation, whatever that is. The classic illustration of the absurdity of the Standard Interpretation is Schrödinger's cat, which is presumed to be trapped within a box that has a device that will go off randomly, with a quantum mechanical process such as radioactive decay triggering the death of the cat. The Standard Interpretation would say that if we do not open the box and make a measurement, the cat will continue to "exist" in a linear superposition of alive and dead which, the criticism would say, is absurd. Whether or not it is absurd, and whether absurdity should be given any weight in deciding the validity of a quantum theory provided the theory agrees with experiment, remain matters of opinion.

The Copenhagen Interpretation

The Copenhagen Interpretation is essentially due to Bohr. We know that we must accept the wave-particle duality – electrons have both wave and particle character. Such a duality requires us to accept two views that, in a classical view, at least, are contradictory (or, in the terminology of the Copenhagen Interpretation, complementary). We also note that certain variables have a complementarity to them, such as position and momentum – we can accurately only know one of these at a time, as given by the uncertainty principle. In the Copenhagen Interpretation, quantum mechanical particles or systems have no properties until we measure them and, in so doing, we destroy the possibility of measuring their complementary variable, just as a position measurement leaves us with no possibility of an answer to the momentum (consistent with the uncertainty principle). Because in this interpretation there are no actual properties that a particle has until it is measured, there are no hidden variables. Thus, Bell's inequalities are not a problem in this interpretation (though it does still leave the absurdity of "spooky action at a distance").

In the Copenhagen Interpretation view, perhaps we should extend this complementarity principle to other aspects of quantum mechanics that are also apparently contradictory, such as the "complementarity" of the linear superposition aspects of quantum mechanical states (having character seen also in waves with their linear superposition) and the apparent definite states seen as a result of measurements (having character seen also in particles with their definite discrete existence).

Modern opinion on this approach is sometimes that generations of physicists were misled into believing that Bohr had indeed somehow solved the problem of the interpretation of quantum mechanics, including the measurement problem, though no one was exactly sure how precisely he had done this (see, e.g., Gell-Mann[8]). Whatever Bohr was advocating is not clearly defined from a mathematical point of view once we go beyond aspects like the uncertainty principle and wave-particle duality. This approach also does not appear to offer a way of resolving the difficulties of the interpretation of quantum mechanics (though we may have to accept that

---

[8] M. Gell-Mann, "What Are the Building Blocks of Matter?" in D. Huff and O. Prewitt (eds.), *The Nature of the Physical Universe* (Wiley, New York, 1979), p. 29. The quote from Gell-Mann is "The fact that an adequate philosophical presentation has been so long delayed is no doubt caused by the fact that Niels Bohr brainwashed a whole generation of theorists into thinking that the job was done fifty years ago."

there is no such resolution between the classical concepts of definite reality and the quantum mechanical world). A common problem with the Copenhagen Interpretation is that one cannot find a consistent or clear definition in the literature as to what it actually is, especially as one extends the ideas of complementarity to new domains.

Bohm's pilot wave

David Bohm[9] noticed that Schrödinger's equation for a particle like an electron can be written in a way that looks exactly like a particular way of writing the classical equation of motion of such a particle, with only one difference. Specifically, he adds a new "quantum potential" Q to the classical potential.

The algebra of this is quite straightforward. We start with the time-dependent Schrödinger equation, as usual

$$\frac{-\hbar^2}{2m}\nabla^2\psi + V\psi = i\hbar\frac{\partial\psi}{\partial t} \tag{19.14}$$

and then we make a mathematical choice to write

$$\psi(\mathbf{r},t) = R(\mathbf{r},t)\exp(iS(\mathbf{r},t)) \tag{19.15}$$

where R and S are real quantities. Any complex function can be represented in this way. If we substitute the form in Eq. (19.15) into Eq. (19.14), then, after a little algebra, we can deduce the equation

$$\frac{\partial S}{\partial t} + \frac{(\nabla S)^2}{2m} + V + Q = 0 \tag{19.16}$$

where

$$Q = -\frac{\hbar^2}{2m}\frac{\nabla^2 R}{R} \tag{19.17}$$

Now, Eq. (19.16) without the quantum potential Q is known as the Hamilton–Jacobi equation of classical mechanics. That equation is an alternative formulation of classical mechanics for such a particle and it reproduces all the usual classical behavior. In that case, S is known as the "action" or Hamilton's principal function and the momentum is $\mathbf{p} = \nabla S$. As usual for classical mechanics, Eq. (19.16) (at least without Q) is, therefore, a completely deterministic equation in which position and momentum are both simultaneously well defined. In considering the correspondence between the classical and quantum worlds, for a large wavepacket and a large mass, then the quantum potential is a very small correction and, hence, even using the full form of Eq. (19.16), we obtain the classical behavior with which we are familiar.

Bohm's core point, though, is that because we can use definite position and momentum as meaningful concepts in the classical version of Eq. (19.16), so also can we use them when we add in a finite quantum potential Q. We are merely adding in the potential from another field – here, a potential derived from the solution to Schrödinger's equation – to the Hamilton–Jacobi

[9] The original papers are D. Bohm, *Phys. Rev.* **85**, 166–193 (1952) and D. Bohm, *Phys. Rev.* **89**, 458–466 (1953). A comprehensive discussion of all of Bohm's ideas is given in D. Bohm and B. J. Hiley, *The Undivided Universe* (Routledge, New York, 1993). The argument here is adapted primarily from Chapter 3 of that book.

equation, just as we might add in another potential from, say, an electromagnetic field that is the solution of Maxwell's equations. This quantum potential also acts, together with the other potentials in the system, to guide the particle. As a result, this potential, or the wavefunction that generates it, can be regarded as a *pilot wave*.

As far as Schrödinger's equation is concerned, there is nothing apparently wrong with Bohm's assertion. Eq. (19.16) is derived from Schrödinger's equation, so it must be consistent with it. The randomness of quantum mechanics in this picture comes from the ordinary randomness of the initial positions and momenta of the particles.

This picture does correctly reproduce the results of classic experiments like diffraction through two slits. In this case, the wavefunction through the quantum potential gives the necessary additional potential to guide the particles, one by one, so that an ensemble of them with suitable random starting conditions will reproduce the diffraction pattern we see, with the zeros at specific points as required. If we block one of the slits, of course, we also block the wavefunction from passing through that slit; as a result the wavefunction does not have the diffraction pattern corresponding to two slits. Hence neither does the quantum potential and the ensemble of particles, therefore, will not show that two-slit diffraction pattern either, just as we expect. Note, though, that in this picture, the particle does definitely go through one slit or the other and any individual particle does have definite position and momentum at all times. Because of this definiteness, there is no collapse of the wavefunction required to explain definite measurements.

There have been many objections to this picture. In this simple form as presented here, it is not relativistically correct and it only applies to one particle. Approaches to addressing such problems do exist, however.[10] It does also appear to give a special status to position, whereas the conventional description of quantum mechanics treats position on an equal footing with any other dynamical variable. Bohm's approach has, however, never disappeared from the discussion of interpretations of quantum mechanics and remains a contender. It may also be that Bohm's approach did not receive the attention it deserved initially because Bohm was excluded from working at any major U.S. institution in the 1950s.[11]

This kind of approach is also not a local description of reality because the wavefunction and the potential Q exist at all positions, not merely the current position of the particle; they are nonlocal hidden variables. Hence, this kind of approach can be consistent with the consequences of Bell's inequalities, which only exclude *local* hidden variables.[12]

Nonlinearity

Our proof that there is a measurement problem is based on the linearity of quantum mechanics. Perhaps quantum mechanics is not actually exactly linear and the effects of that nonlinearity

---

[10] See, e.g., D. Bohm and B. J. Hiley, *The Undivided Universe* (Routledge, New York, 1993).

[11] Bohm was summoned to the House Un-American Activities Committee in 1949, on the basis of allegations of communist sympathies that were never apparently substantiated. He refused on principle to testify and was found in contempt of Congress. He was required to leave Princeton University and could obtain no other position in the United States. After various positions in other countries, he settled in the United Kingdom. He was subsequently cleared of the contempt of Congress charges. See D. Z. Albert, "Bohm's Alternative to Quantum Mechanics," *Scientific American*, May 1994, pp. 58–67.

[12] J. S. Bell, *Speakable and Unspeakable in Quantum Mechanics* (Cambridge University Press, 1993), pp. 29–39.

become strong as we move to the macroscopic world in which we appear to exist. Then, mathematically, we could discount the kind of proof we gave previously and propose that some state collapse is possible.

This approach does not seem to get a lot of attention, presumably because no one has been able to come up with a theory that both satisfactorily explains the microscopic world and yet gives the kind of state collapse we think we see. (Bell briefly mentions this possibility,[13] for example.) In principle, if quantum mechanics is slightly nonlinear, it ought to be possible to find such behavior experimentally, so it does, at least, offer hope of resolving the problem conclusively.

Distinction between matter and mind

Perhaps the process of measurement and collapse of the wavefunction occurs when a conscious mind (whatever that is) intervenes. Consciousness is then viewed as the aspect that lies outside quantum mechanics as we have stated it so far. Such a supposition might allow us to test experimentally for the presence of consciousness in a chain of events; consciousness would collapse the wavefunction and lead to an evolution after that point that was different from the evolution we would have seen for a true superposition. If the system that is supposed to be conscious is a relatively complex one (e.g., a human brain), however, it is doubtful that we could calculate the evolution of a true superposition in interacting with that system and so we probably could not distinguish it from the evolution of a collapsed state of a complex system; hence this experiment is perhaps discouragingly difficult. Also, from a theoretical point of view, because we have no clear definition of consciousness to work with, it is difficult to make much progress with such a theory. It does raise the hypothesis that maybe consciousness has something to do with nonlinearity in quantum mechanics but does not propose any serious way of making progress with that conjecture either.

Many-worlds hypothesis

Everett[14] proposed in 1957 that there was no collapse of the wavefunction. Rather, each possible outcome of a measurement actually exists but in different "worlds." Performing a measurement causes reality to split into multiple branches or worlds, each corresponding to a different possible result. Multiple replicas of the observer then exist, one for each world, and in each world the observer believes a different outcome happened. In this approach, an observer can be a machine, and its main characteristic is that it writes down results (e.g., in a register of some kind). For each possible answer the machine might write down, there is a different world. An alternative version would have the observer have multiple different minds, one for each outcome, in which case it is known as a *many-minds hypothesis*.

This approach is truly bizarre, but does not obviously violate any laws of quantum mechanics and claims not to require anything other than linear quantum mechanics to describe everything, including the observer. It does, however, require a remarkable amount of splitting of the universe. It is apparently believed by quite large numbers of distinguished physicists.[15]

---

[13] J. S. Bell, *Speakable and Unspeakable in Quantum Mechanics* (Cambridge University Press, 1993), p 190

[14] H. Everett III, "'Relative State' Formulation of Quantum Mechanics," *Reviews of Modern Physics* **29**, 454–62 (1957).

[15] Perhaps it is only in this world that those physicists believe it and in another world where Everett's paper was rejected for publication no one even considers it! Also, remember that for every advocate of

Problem

19.3.1 Derive the relations

$$\text{(i)}\ \frac{\partial S}{\partial t}+\frac{(\nabla S)^2}{2m}+V+Q=0 \ \text{and (ii)}\ \frac{\partial R^2}{\partial t}+\nabla\cdot\left(R^2\frac{\nabla S}{m}\right)=0$$

from Schrödinger's time-dependent equation, where

$$Q=-\frac{\hbar^2}{2m}\frac{\nabla^2 R}{R}$$

and the wavefunction is written in the form

$$\psi(\mathbf{r},t)=R(\mathbf{r},t)\exp(iS(\mathbf{r},t)).$$

(Note: The second relation (ii) corresponds to conservation of particles or probability density because $(\nabla S)/m$ can be interpreted as velocity; hence, $R^2(\nabla S)/m$ represents the flow of probability density.)

19.4 Epilogue

In the end, this discussion of the meaning of quantum mechanics and the measurement problem in particular has perhaps merely confused the reader and caused the reader (unwisely?) to doubt quantum mechanics as the way to understand the world. Perhaps instead we should have taken the advice of Willis Lamb (a Nobel Laureate): "I have taught graduate courses in quantum mechanics for over 20 years ..., and for almost all of them I have dealt with measurement in the following manner. On beginning the lectures I told the students, 'You must first learn the rules of calculation in quantum mechanics, and then I will tell you about the theory of measurement and discuss the meaning of the subject.' Almost invariably, the time allotted to the course ran out before I had time to fulfill my promise."[16]

Further reading

The subject of the interpretation of quantum mechanics is a particularly intriguing one and there are many quite readable discussions at a variety of different levels. Several of these are listed here.

Popular accounts

D. Z. Albert, *Quantum Mechanics and Experience* (Harvard University Press, 1992). Albert, an active philosopher of quantum mechanics, presents an accessible account of the issues and possible solutions to the measurement problem.

D. Deutsch, *The Fabric of Reality* (Penguin Books, 1997). Deutsch, a major scientific contributor to fields such as quantum computing, discusses a broad range of ideas in quantum mechanics and other areas, with a strong emphasis on a many-worlds view of quantum mechanics, for which Deutsch is an impassioned advocate.

N. Herbert, *Quantum Reality* (Anchor Books, 1985). This is a readable account, accessible to those without deep understanding of quantum mechanics, of many of the ideas regarding quantum mechanics, measurement, and reality.

---

the many-worlds hypothesis in this world, presumably in some other world he or she is saying exactly the opposite, so how much weight should we give to their opinions?

[16] J. A. Wheeler and W. H. Zurek (eds.), *Quantum Theory and Measurement* (Princeton University Press, 1983), pp. xviii–xix.

A. Rae, *Quantum Physics – Illusion or Reality? (Second edition)* (Cambridge, 2004) This is a clear and readable account of the various issues in the interpretation of quantum mechanics, presented with minimal mathematics.

Bell's inequalities

J. S. Bell, *Speakable and Unspeakable in Quantum Mechanics* (Cambridge University Press, 1993). This collects Bell's writings on quantum measurement and reality, including the original paper (which was published in a journal that may be difficult to find in some libraries) and also some other more accessible treatments by Bell (including "Bertlmann's Socks and the Nature of Reality" [pp. 139–158]).

R. A. Bertlmann and A. Zeilinger (eds.), *Quantum [Un]Speakables – From Bell to Quantum Information* (Springer, 2002), is a collection of essays, many at a fairly technical level, in honor of John Bell, primarily covering many aspects of Bell's inequalities and the work following from them, including quantum information.

Specific popular accounts of Bell's inequalities can be found in the following two articles.

B. d'Espagnat, "The Quantum Theory and Reality," *Scientific American*, **241**, Nov. 1979, pp. 150–181.

D. Mermin, "Is the Moon There When Nobody Looks? Reality and the Quantum Theory," *Physics Today* **38**, No. 4, April 1985, pp. 38–47.

Many-worlds theories

J. A. Barrett, *The Quantum Mechanics of Minds and Worlds* (Oxford University Press, 1999). This book is an account of the many versions of the many-worlds view of quantum mechanics. It is a scholarly work but is generally readable for those with a basic understanding of quantum mechanics.

Collections of papers

J. A. Wheeler and W. H. Zurek, *Quantum Theory and Measurement* (Princeton University Press, 1983). Many of the key papers in the field of quantum measurement including, for example, the original EPR paper, Bell's paper, and Everett's are collected in this anthology.

19.5 Summary of concepts

Bell's inequalities and local hidden variable theories

If we perform an EPR experiment on a pair of photons in an appropriate EPR state, when we measure a horizontal polarization for one of the pair of photons, then we will always find a horizontal polarization in the other photon and, similarly, for vertical polarizations. The measurement of one photon appears to influence the result of the other measurement, regardless of how far away the other measurement is.

An apparent solution to this would be to postulate that there is some other unknown variable that carries the necessary information along with each photon in the first place, a so-called local hidden variable. But Bell's inequalities allow an experimental test, a test that disproves any such local hidden variable theory, independent of quantum mechanics. Hence, we are apparently forced to conclude that reality is not local (i.e., reality at some point is not describable by properties found only at that point) and can be influenced by remote events.

The measurement problem

There is no linear operator that can describe the measurement process; that is, no linear operator can, in general, give the collapse of the wavefunction onto an eigenstate of the quantity being measured. Hence, we cannot apparently describe a measurement apparatus quantum mechanically. Proposed solutions to this problem include the Copenhagen Interpretation, making a distinction between mind and matter, Bohm's pilot wave, adding nonlinearity to quantum mechanics, the many-worlds hypothesis, and the many-minds hypothesis, though there is no universal consensus on the solution.

Appendix A

Background mathematics

In this appendix, we summarize the core background mathematics that we presume in the rest of the book. A major purpose here is to clarify the mathematical notations and terminology. Readers coming from different backgrounds may be more used to one notation or another and other books that the reader may consult may use different notation and terms, so we clarify the ones we use here and their relations to others. This appendix may also serve as a refresher or to patch over some holes temporarily in the reader's knowledge until the reader has more time to study the relevant mathematics in more detail. This short discussion here is certainly not a complete one and no attempt is made to give rigorous and complete mathematics.

Quantum mechanics is sometimes presented as being a very mathematical subject. It is true that many aspects of quantum mechanics can only be well defined using a mathematical vocabulary. In fact, we can assure the reader that the mathematics of quantum mechanics is not harder than that found in classical physics or any analytic branch of engineering and the required background is essentially the same as in those fields. Because quantum mechanics is very fundamentally based on linear mathematics, quantum mechanics, in practice, is arguably simpler mathematically than many other areas of science and engineering.

A.1 Geometrical vectors

Cartesian coordinates

Geometrical vectors are entities that, in ordinary space, have length and also direction. We most often think of (geometrical) vectors in Cartesian (i.e., "x, y, z") coordinates, where all the axes are at right angles. Conventionally, we construct unit vectors (i.e., vectors of length 1 unit of distance) in these coordinate directions, **i**, **j**, and **k**, respectively, for the x, y, and z directions.[1] We will also often write a vector **F** in terms of its Cartesian components; that is,

$$\mathbf{F} = F_x\mathbf{i} + F_y\mathbf{j} + F_z\mathbf{k} \tag{A.1}$$

---

[1] In this book, we mostly use **i**, **j**, and **k** as the unit vectors for Cartesian directions because this is a standard and relatively simple mathematical notation. We will also, however, have to use **k** for another purpose as the wavevector because that is the most standard notation for that concept in physics. If there is any confusion, we will use $\hat{\mathbf{x}}$, $\hat{\mathbf{y}}$, and $\hat{\mathbf{z}}$ for the unit vectors in the Cartesian directions, which is another common notation. That notation could cause some confusion with our use otherwise of the "hat" notation, as in $\hat{A}$, to denote operators. Some overlap of notations here unfortunately seems unavoidable if we are to stay with notations that are otherwise reasonably conventional for mathematics and/or physics. Hopefully, there will be no confusion in practice and we try to be explicit where any possible confusion might arise.

where F_x, F_y, and F_z are all real numbers. The quantity $F_x\mathbf{i} \equiv \mathbf{i}F_x$, for a positive number F_x, means a vector of length F_x in the direction of the (unit) vector **i** (i.e., in the x direction). If F_x is a negative number, the vector $F_x\mathbf{i}$ is of length $|F_x|$ and the vector is pointing in the negative x direction. The quantities F_x, F_y, and F_z can also be viewed as the *projections* of the vector **F** onto the x, y, and z directions. The length of the vector **F**, which we can write as $|\mathbf{F}|$, is

$$|\mathbf{F}| = \sqrt{F_x^2 + F_y^2 + F_z^2} \tag{A.2}$$

Often, if there is no confusion, we write F instead of $|\mathbf{F}|$ for the length of vector **F**.

We conventionally always work in so-called *right-handed axes* as shown in Fig. A.1, and such axes obey the *right-hand rule*.

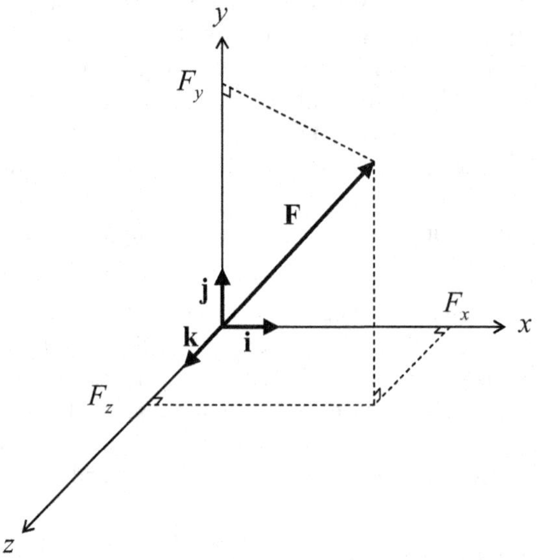

Fig. A.1. Right-handed coordinate axes, showing the unit vectors **i**, **j**, and **k** along the x, y, and z axes, respectively, and a vector **F** with its x, y, and z components F_x, F_y, and F_z, respectively.

Right-hand rule

If the thumb of your right hand points to the right, if the index finger points upward, and if the middle finger is bent inward at a right angle to the index finger to point toward you, then the thumb, index finger, and middle finger give the relative directions of the x-, y-, and z-axes in a right-handed coordinate system. The thumb is along the x-axis, the index finger is along the y-axis, and the middle finger is along the z-axis. Hence, if the x-axis is horizontal and the y-axis is vertical, the z-axis points toward the viewer.

Note also that cyclic permutations of the axes defined this way are also right handed. For example, if the index finger above is the x direction and the middle finger is the y direction, then the thumb is in the z direction.

Another way of remembering the right-hand rule is through the right-hand screw rule. If one imagines one is turning a screwdriver clockwise to drive in an ordinary (right-handed) screw and one is turning from the x to the y axis in doing so, then the z axis is the one pointing in the direction in which the screw is being driven into the material.

Vector dot product

For two geometrical vectors $\mathbf{a} = a_x\mathbf{i} + a_y\mathbf{j} + a_z\mathbf{k}$ and $\mathbf{b} = b_x\mathbf{i} + b_y\mathbf{j} + b_z\mathbf{k}$, where the a's and b's here are real numbers, the vector dot product can be written

$$\mathbf{a} \cdot \mathbf{b} = a_x b_x + a_y b_y + a_z b_z \tag{A.3}$$

or, equivalently

$$\mathbf{a} \cdot \mathbf{b} = |\mathbf{a}||\mathbf{b}|\cos\theta \equiv ab\cos\theta \tag{A.4}$$

where θ is the angle between the two vectors. Note that the vector dot product is an operation involving two vectors that generates a scalar result. In that sense, it is an example of what is sometimes called an *inner product.*

The vector dot product is *commutative*; that is,

$$\mathbf{a} \cdot \mathbf{b} = \mathbf{b} \cdot \mathbf{a} \tag{A.5}$$

Note that the length of a vector can be deduced from its dot product with itself; that is,

$$|\mathbf{a}| \equiv a = \sqrt{\mathbf{a} \cdot \mathbf{a}} \tag{A.6}$$

The components of a vector can formally be deduced by taking the dot product of the vector with the unit vectors of the coordinate axes; for example, for the vector **F**

$$F_x = \mathbf{F} \cdot \mathbf{i} \tag{A.7}$$

Vector cross product

For two geometrical vectors $\mathbf{a} = a_x\mathbf{i} + a_y\mathbf{j} + a_z\mathbf{k}$ and $\mathbf{b} = b_x\mathbf{i} + b_y\mathbf{j} + b_z\mathbf{k}$, where the a's and b's here are real numbers, the vector cross product can be written

$$\mathbf{a} \times \mathbf{b} = \left(a_y b_z - a_z b_y\right)\mathbf{i} + \left(a_z b_x - a_x b_z\right)\mathbf{j} + \left(a_x b_y - a_y b_x\right)\mathbf{k} \tag{A.8}$$

It can also be written

$$\mathbf{a} \times \mathbf{b} = \mathbf{n}|\mathbf{a}||\mathbf{b}|\sin\theta = \mathbf{n}ab\sin\theta \tag{A.9}$$

where **n** is a unit vector perpendicular to the plane in which **a** and **b** lie. In this definition, we still need to clarify which of two possible directions to choose for **n**. If we have chosen to work in right-handed coordinates, then **n** is given by a right-hand rule; here, because **a** and **b** are not necessarily at right angles to one another, the right-hand screw rule is most useful. If we arrange to look at the vectors **a** and **b** from the direction such that turning from **a** to **b** corresponds to turning a screwdriver in a clockwise direction, then the direction **n** is the direction a right-handed screw would move (i.e., away from us). Another way to say this is that the coordinate systems (**a**, **b**, **a** × **b**) or (**a**, **b**, **n**) are right handed (though **a** and **b** are not necessarily at right angles to one another).

The form in Eq. (A.8) can be written in a more readily memorable form as a determinant (see Section A.9 for a discussion of determinants)

$$\mathbf{a} \times \mathbf{b} = \begin{vmatrix} \mathbf{i} & \mathbf{j} & \mathbf{k} \\ a_x & a_y & a_z \\ b_x & b_y & b_z \end{vmatrix} \tag{A.10}$$

Note that the cross product of two vectors is a vector. Note also that it is *not* commutative. In fact, changing the order of the vectors changes the sign of the cross product; that is, $\mathbf{b} \times \mathbf{a} = -\mathbf{a} \times \mathbf{b}$. We can view this change of sign as being a result of us having now to rotate in the opposite direction in going from **b** to **a**, compared to going from **a** to **b**.

A.2 Exponential and logarithm notation

The number *e*, shown here to the first twenty decimal places

$$e \simeq 2.71828\ 18284\ 59045\ 23536 \tag{A.11}$$

is the base of the natural logarithms. In raising it to a power, we most commonly use the "exp" notation

$$\exp(x) \equiv e^x \tag{A.12}$$

to minimize confusion with the use of the letter *e* to represent the magnitude of the charge on the electron and to avoid the use of small superscript characters.

Remember that the product of two exponentials is given by the sum of the exponents; that is,

$$\exp(a)\exp(b) = \exp(a+b) \tag{A.13}$$

and that

$$\frac{1}{\exp(a)} = \exp(-a) \tag{A.14}$$

In discussing logarithms to the base *e*, the notations

$$\ln x \equiv \log x \equiv \log_e x \tag{A.15}$$

are all equivalent; we prefer the first (i.e., $\ln x$) here.

The logarithm is the inverse function of the exponential and vice versa

$$\ln\left[\exp(a)\right] = a \tag{A.16}$$

and

$$\exp\left[\ln(a)\right] = a \tag{A.17}$$

Remember also that the log of a product is the sum of the logs

$$\ln(ab) = \ln(a) + \ln(b) \tag{A.18}$$

and that

$$\ln(1/a) = -\ln(a) \tag{A.19}$$

A.3 Trigonometric notation

In terms of sines and cosines, we have the definitions

$$\tan x = \frac{\sin x}{\cos x} \tag{A.20}$$

$$\cot x = \frac{1}{\tan x} = \frac{\cos x}{\sin x} \tag{A.21}$$

Sines, cosines, and tangents all use a notation[2] of the form

$$\sin^2(x) \equiv [\sin(x)]^2 \tag{A.22}$$

The notation on the right would be less confusing but by convention the notation on the left is commonly used. This convention is *only* used for such trigonometric functions and is *only* used for positive powers. For functions in general, a notation such as $f^2(x)$, if it is used at all, means the "function of the function" of x – that is, $f^2(x) \equiv f(f(x))$ – which is *not* the same as the trigonometric power notation; that is, in general, $\sin(\sin(x)) \neq [\sin(x)]^2$.

The inverse function of the sine can be written as arcsin or $\sin^{-1}$ and has the meaning that

$$\arcsin(\sin(\theta)) \equiv \sin^{-1}(\sin(\theta)) = \theta \tag{A.23}$$

and, similarly, for the other trigonometric inverse functions. In this case, the notation does agree with the general notation for functions, where $f^{-1}(y)$ is the inverse function of $f(x)$ such that $f^{-1}(f(x)) = x$, and the notation does *not* mean "sin(x) raised to the power -1," that is, in general

$$\sin^{-1}(x) \neq \frac{1}{\sin(x)} \tag{A.24}$$

A.4 Complex numbers

Complex numbers exploit the mathematics associated with the square root of the number minus one. There are two notations used to represent this square root

$$i = \sqrt{-1} \tag{A.25}$$

$$j = \sqrt{-1} \tag{A.26}$$

The "i" notation is used in physics. Typically, electrical engineering uses the "j" notation.[3] We use the "i" notation exclusively here, as is common in quantum mechanics. As with other square roots, both $+i$ and $-i$ are square roots of -1 and they are the only square roots of -1.

A number that can be written as

---

[2] The hyperbolic functions sinh, cosh, and tanh also use this form of notation.

[3] Physics and engineering also have different conventions for complex representations of oscillations and propagating waves. In physics, one often writes $\exp(-i\omega t)$ for an oscillation at (angular) frequency ω and $\exp[-i(\omega t - kz)]$ for a forward-propagating wave in the z direction; in quantum mechanics, one *always* makes this choice because of a choice made in the definition of the time-dependent Schrödinger equation. In electrical engineering, one commonly writes $\exp(j\omega t)$ for an oscillation and $\exp[j(\omega t - kz)]$ for a forward-propagating wave. Because of these conventions, some authors suggest we should think of $j = -i$, though that is not always a reliable conversion.

$$f = ib \tag{A.27}$$

(by which we mean i times b) where b is a real number, is said to be an *imaginary* number. A number that can be written

$$g = a + ib \tag{A.28}$$

where a and b are both real numbers, is called a *complex* number. a is the *real part* of g, which can be written as

$$a = \mathrm{Re}(g) \tag{A.29}$$

and b is the *imaginary part* of g, which can be written as

$$b = \mathrm{Im}(g) \tag{A.30}$$

Quantities or entities that can be represented by a single real, imaginary, or complex number are sometimes called *scalars* or *scalar quantities*. This term distinguishes them from vectors and other entities requiring more than one number to specify them.

The *complex conjugate* of a complex number has the sign of the imaginary part reversed and is usually indicated by the notation[4] of a superscript asterisk; that is, for our complex number g, the complex conjugate is written g^*, with

$$g^* = a - ib \tag{A.31}$$

The quantity

$$|g|^2 \equiv g^* g = g g^* \tag{A.32}$$

is called the *modulus squared* of g. It is always a positive real number. The *modulus* or *magnitude* of g is the positive square root

$$|g| = +\sqrt{|g|^2} \tag{A.33}$$

and is also always a positive real number. Note that

$$|g|^2 = (a + ib)(a - ib) = a^2 + b^2 \tag{A.34}$$

The relation $a^2 + b^2 = (a + ib)(a - ib)$ in Eq. (A.34) is a commonly used algebraic identity. Another common manipulation with complex numbers is to change a complex number in the denominator so it can be written in the form of Eq. (A.28) by multiplying top and bottom lines by the complex conjugate of the denominator; that is,

$$g = \frac{1}{c + id} = \frac{1}{(c + id)}\frac{(c - id)}{(c - id)} = \frac{c - id}{c^2 + d^2} = \frac{c}{c^2 + d^2} - i\frac{d}{c^2 + d^2} \tag{A.35}$$

Note that any complex number can always be written in the form of Eq. (A.28).

---

[4] When we consider the extension of the idea of complex conjugates to matrices and operators, for an operator $\hat{A}$, we define the Hermitian conjugate or Hermitian adjoint as $\hat{A}^\dagger$. A complex number is a trivial case of an operator so, technically, for a complex number g, we could write its complex conjugate as either g^* or $g^\dagger$.

Euler's formula relates exponentials, cosines, and sines

$$\exp(i\theta) = \cos\theta + i\sin\theta \tag{A.36}$$

Occasionally, authors use the notation $\exp(i\theta) \equiv \text{cis}(\theta)$, whose form mimics the expression on the right ("cis" ≡ "*cos* i *sin*"), but we do not use that notation here. Euler's formula leads to another very convenient way of representing any complex number, which is to write

$$g = |g|\exp(i\theta) \tag{A.37}$$

This is sometimes known as the *polar form* of a complex number. In this case, θ is sometimes known as the *argument* or *phase* of the complex number. Complex numbers are conveniently visualized on a two-dimensional diagram, known as an *Argand diagram*, on a plane known as the *complex plane* (Fig. A.2).

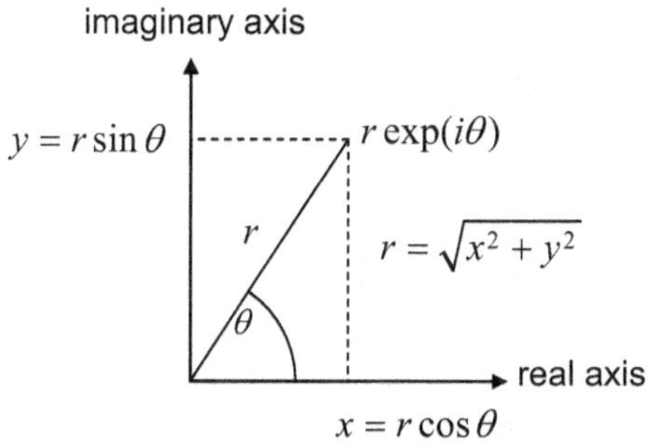

Fig. A.2. Argand diagram of a complex number $c = x + iy = r\exp(i\theta)$ on the complex plane.

One important result in the algebra of complex numbers is that for any two complex numbers g and h, the complex conjugate of the product is the product of the complex conjugates; that is,

$$(gh)^* = g^* h^* \tag{A.38}$$

Also

$$\left(\frac{1}{gh}\right)^* = \frac{1}{g^* h^*} \tag{A.39}$$

and

$$\left(\frac{g}{h}\right)^* = \frac{g^*}{h^*}$$

These three results are very easy to prove using the complex exponential form.[5]

The exponential notation for complex numbers, Eq. (A.37), makes it particularly simple to write down the *nth root of unity*, which is that number that, raised to the *n*th power, gives unity. Obviously, and trivially, the number 1 has this property for any power *n*, but less

[5] E.g., with $g = |g|\exp(i\theta_g)$ and $h = |h|\exp(i\theta_h)$

then $gh = |g|\exp(i\theta_g)|h|\exp(i\theta_h) = |g||h|\exp[i(\theta_g + \theta_h)]$

so $(gh)^* = |g||h|\exp[-i(\theta_g + \theta_h)] = |g|\exp(-i\theta_g)|h|\exp(-i\theta_h) = g^* h^*$

trivially and more usefully for many purposes, because $\exp(2\pi i) = 1$, we can write the nth root of unity as

$$\sqrt[n]{1} = \exp(2\pi i / n) \tag{A.40}$$

Complex numbers are always a purely mathematical concept in that there is no actual physically measurable quantity that is imaginary or complex.[6] The use of complex numbers can allow great simplifications of algebra, but at the end in relating to the physical world, we have to return to real numbers. In much of the use of complex numbers in engineering and in classical physics (e.g., for classical electromagnetic waves), one starts by writing, for example, $\exp(i\omega t)$ instead of $\cos(\omega t)$, because the algebra of complex exponentials is simpler than that of sines and cosines, and in the end one simply takes the real part of the final result. In the use of complex numbers in quantum mechanics, however, the underlying internal quantities in quantum mechanical calculations are chosen to be expressed in complex numbers from the start. To get to the real numbers of interest for physically measurable quantities, one more commonly uses the modulus squared of the final result.

A.5 Differential calculus

We presume that the reader is familiar with elementary differential calculus, including the derivatives of common functions and the various rules for derivatives of combinations of functions of a single variable (many of these formulae are listed for reference in Appendix G). We discuss the notation and definitions underlying differential calculus in this section, in part because we need to use the fundamental definitions of derivatives in the book and also to prepare the reader for differential equations.

First derivative

The first derivative of a function $y(x)$ with respect to x is written as

$$y'(x) = \frac{dy}{dx} \tag{A.41}$$

Here, we use exclusively the "dy/dx" or "Leibniz" notation because of its intrinsic elegance and its intuitive relation to the formal definition of the derivative. To be precise in the Leibniz notation, we may have to specify the value of x at which we are taking the derivative; for a specific point x_o, we can write

$$y'(x_o) = \left.\frac{dy}{dx}\right|_{x=x_o} \tag{A.42}$$

by which we mean the derivative taken at the point $x = x_o$. (There is a similar notation used for a different purpose in partial derivatives [discussed later] though, in practice, there is usually no confusion. One can more generally view this "vertical line and subscript" notation as giving some additional necessary specifications or restrictions on the derivative.)

---

[6] A good electrical engineer might well object to this because he or she is quite comfortable operating in the complex plane for amplitude and phase; but, in the end, the quantities of amplitude and phase (or of quadrature components) that the engineer measures are real numbers.

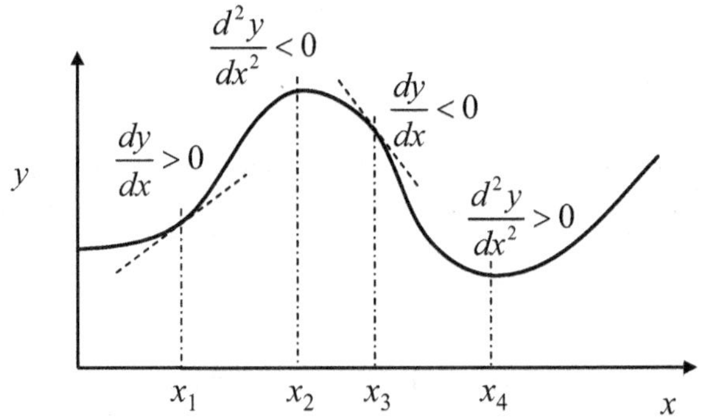

Fig. A.3. Sketch of the function $y(x)$ (solid curve), showing (i) a positive first derivative at point x_1, corresponding to the positive slope of the tangent dashed line; (ii) a negative second derivative at point x_2, corresponding to the curve bending progressively from upward toward downward with increasing x; (iii) a negative first derivative at point x_3, corresponding to the negative slope of the tangent dashed line; and (iv) a positive second derivative at point x_4, corresponding to the curve bending progressively from downward toward upward with increasing x.

We can formally define the derivative as

$$\left.\frac{dy}{dx}\right|_{x=x_o} = \lim_{\Delta x \to 0} \frac{y(x_o + \Delta x) - y(x_o)}{\Delta x} = \lim_{\Delta x \to 0} \frac{y\left(x_o + \frac{\Delta x}{2}\right) - y\left(x_o - \frac{\Delta x}{2}\right)}{\Delta x} \tag{A.43}$$

where the "lim" notation here means the limit of the expression to the immediate right as Δx is made arbitrarily small. We have shown the more standard "forward difference" limit first and also a symmetrical "central difference" limit second, either of which is valid. The derivative can be thought of as the rate of change of y with x. If we draw a graph of $y(x)$ as a function of x, then the derivative is just the slope of the graph at the point of interest, as sketched in Fig. A.3.

Second derivative

The second derivative is similarly the rate of change of the first derivative. Hence, we can write for the second derivative, taking the difference of two first derivatives at points spaced by Δx symmetrically about x_o

$$y''(x_o) \equiv \left.\frac{d^2y}{dx^2}\right|_{x=x_o} = \lim_{\Delta x \to 0} \frac{y'\left(x_o + \frac{\Delta x}{2}\right) - y'\left(x_o - \frac{\Delta x}{2}\right)}{\Delta x} \tag{A.44}$$

Rewriting using the symmetrical central difference form of the first derivative from Eq. (A.43), we have

$$y''(x_o) \equiv \left.\frac{d^2y}{dx^2}\right|_{x=x_o} = \lim_{\Delta x \to 0} \frac{y(x_o + \Delta x) - 2y(x_o) + y(x_o - \Delta x)}{(\Delta x)^2} \tag{A.45}$$

The second derivative graphically is the amount of "bend" in the curve, as illustrated in Fig. A.3. A positive second derivative corresponds to a curve that is bending upward.

Higher order derivatives can be defined in a similar fashion, with analogous notations.

Partial differentiation

Frequently, we have to deal with functions of more than one variable. For example, we might have the function $h(x,y)$ representing the height of a hill, with the x and y coordinates representing the positions in the "east" and "north" directions, respectively. The rate at which our height would increase if we walked east would be specified by the partial derivative

$$\frac{\partial h}{\partial x} \equiv \left.\frac{\partial h}{\partial x}\right|_y \tag{A.46}$$

The symbol ∂ is a curved "d" and the expression on the left of Eq. (A.46) is sometimes pronounced "partial d h by d x." The more explicit notation on the right of Eq. (A.46), with a vertical line and a subscript, is clarifying specifically that this derivative is taken with the y variable held constant. The simpler notation on the left is more common but where there is any possibility of confusion, the notation on the right is used. Some authors use the notation $\partial_x h$ for the partial derivative in Eq. (A.46), though we do not use that in this book.

Just as for ordinary differentiation, we can also take higher derivatives, such as

$$\frac{\partial^2 h}{\partial x^2} \equiv \left.\frac{\partial^2 h}{\partial x^2}\right|_y \tag{A.47}$$

which is the curvature of the hill in the x direction. We can also find cross-derivatives of the form

$$\frac{\partial^2 h}{\partial x \partial y} \equiv \left.\frac{\partial}{\partial x}\right|_y \left.\frac{\partial h}{\partial y}\right|_x \tag{A.48}$$

which are slightly more difficult to visualize; this specific one is the rate of change as we move east of the slope in the northerly direction. Note that, provided all of the various first derivatives and the two cross-derivatives in the two different orders both exist, we can interchange the order of the partial differentiations in the cross-derivative; that is,

$$\frac{\partial^2 h}{\partial x \partial y} = \frac{\partial^2 h}{\partial y \partial x} \tag{A.49}$$

In quantum mechanics, we often come across functions of all three Cartesian coordinate directions, x, y, and z, such as some function $\phi(x,y,z)$ or functions of the spatial coordinates and of time, such as $\Phi(x,y,z,t)$, so we frequently need to use partial derivatives.

Differentials and changing coordinate systems

This particular topic does not come up until Chapter 9, so it can be omitted on first reading.

If we imagined that we walked a very small distance δx east and another very small distance δy north, then we would expect that the resulting change in height, δh, would be given by

$$\delta h \simeq \left.\frac{\partial h}{\partial x}\right|_y \delta x + \left.\frac{\partial h}{\partial y}\right|_x \delta y \tag{A.50}$$

We are saying that the change in height, δh, is the sum of two parts: first, the change we get from going east by δx, which is the slope in the easterly direction ($\partial h / \partial x$) times the distance in the easterly direction (δx); and, second, the change we get from going north by δy, which is the slope in the northerly direction ($\partial h / \partial y$) times the distance in the northerly direction (δy). We can take the limit of very small δx and δy, in which case we can write this as what is called a *differential* (or sometimes an *exact differential*)

$$dh = \left.\frac{\partial h}{\partial x}\right|_y dx + \left.\frac{\partial h}{\partial y}\right|_x dy \tag{A.51}$$

Suppose we knew what path we were taking on the hill (i.e., we know both x and y as a function of time t), even if the path was not along some specific compass direction. We can then talk about the total derivatives dx/dt and dy/dt, which are the velocities we have in the x and y directions, respectively, on the path on which we are walking. Then, in a time dt, we would move an amount $dx = (dx/dt)dt$ in the x direction, and, similarly, in the y direction, so we could then write

$$dh = \left.\frac{\partial h}{\partial x}\right|_y \left(\frac{dx}{dt}\right) dt + \left.\frac{\partial h}{\partial y}\right|_x \left(\frac{dy}{dt}\right) dt \tag{A.52}$$

Alternatively, we could now write the rate our height was changing in terms of the *total derivative* dh/dt, which would be

$$\frac{dh}{dt} = \left.\frac{\partial h}{\partial x}\right|_y \left(\frac{dx}{dt}\right) + \left.\frac{\partial h}{\partial y}\right|_x \left(\frac{dy}{dt}\right) \tag{A.53}$$

Even for a function of more than one variable, total derivatives (i.e., ones written with "*d*" rather than "∂") can still have meaning, so the distinction between the two kinds of derivatives is important.

Another important use of partial derivatives and the differential is in changing from one coordinate system to another. Suppose that we wanted instead to express the slopes of the hill along some other coordinate directions – for example, southeast (a) and northeast (b) – but we only knew the slopes along the east and north coordinate directions. We do, however, know how far we move along each of the old coordinate directions as we move along each of the new coordinate directions; that is, we can easily write down that as we move south-east by one unit, the amount we move in the easterly direction is $1/\sqrt{2}$ units (because $\cos 45° = 1/\sqrt{2}$), so

$$\left.\frac{\partial x}{\partial a}\right|_b = \frac{1}{\sqrt{2}} \text{ and similarly } \left.\frac{\partial y}{\partial a}\right|_b = -\frac{1}{\sqrt{2}},\ \left.\frac{\partial x}{\partial b}\right|_a = \frac{1}{\sqrt{2}} \text{ and } \left.\frac{\partial y}{\partial b}\right|_a = \frac{1}{\sqrt{2}} \tag{A.54}$$

Suppose then we make a small movement along the a (southeast) direction, in which case, of course, we are making no movement along the b (northeast) direction that is at right angles to it. So, we can write

$$\delta x \simeq \left.\frac{\partial x}{\partial a}\right|_b \delta a \tag{A.55}$$

and, similarly, for δy. So, we have, from Eq. (A.50)

$$\delta h \simeq \left.\frac{\partial h}{\partial x}\right|_y \left.\frac{\partial x}{\partial a}\right|_b \delta a + \left.\frac{\partial h}{\partial y}\right|_x \left.\frac{\partial y}{\partial a}\right|_b \delta a \tag{A.56}$$

or, because all of this is performed at constant b, we can divide both sides by δa and take the limit of small δa to get the partial derivative

$$\left.\frac{\partial h}{\partial a}\right|_b = \left.\frac{\partial h}{\partial x}\right|_y \left.\frac{\partial x}{\partial a}\right|_b + \left.\frac{\partial h}{\partial y}\right|_x \left.\frac{\partial y}{\partial a}\right|_b \tag{A.57}$$

More generally, because this works for any function of x and y,[7] we can write

$$\left.\frac{\partial}{\partial a}\right|_b = \left.\frac{\partial x}{\partial a}\right|_b \left.\frac{\partial}{\partial x}\right|_y + \left.\frac{\partial y}{\partial a}\right|_b \left.\frac{\partial}{\partial y}\right|_x \tag{A.58}$$

which shows us how we can change partial derivatives from one coordinate system to another.

A.6 Differential equations

Differential equations in one variable

A *differential equation* is an equation that involves one or more derivatives of some function. A very simple differential equation would be

$$\frac{dy}{dx} = ay \tag{A.59}$$

where a is some real number. The solution to this equation is

$$y = A\exp(ax) \tag{A.60}$$

Here, A is an arbitrary constant. The reader can check that this exponential function is indeed the solution. No matter what the constant A is, the function in Eq. (A.60) still solves Eq. (A.59) as stated.

Unlike in simple differential calculus, there is no simple general set of rules to work out the solution to differential equations; in that sense, differential equations are more like integral calculus, where one essentially has to use intelligent guesswork to deduce the analytic form of an integral. Fortunately, at least in understanding quantum mechanics, in practice, we end up using a relatively small range of differential equations and, for these equations, we usually know the solutions already or have some standard techniques to work them out.[8] As in any analytic subject, for problems beyond idealized ones, we typically have to use numerical or approximation techniques to find the solutions.[9]

First-order differential equations

This particular equation, Eq. (A.59), is an example of a *first-order* differential equation. A first-order differential equation contains no derivatives higher than the first order (i.e., it does not contain the second derivative, d^2y/dx^2, or any higher derivatives).

A first-order equation typically has one undetermined constant (here, A) in its general solution. A solution of a differential equation in which we still have the maximum possible number

[7] At least, for suitable continuous functions of x and y for which all of these derivative exist.

[8] We present in the book any such techniques we use to work out solutions to some equations, techniques such as series solution, and the reader need not know these in advance.

[9] We discuss some such techniques in the book, e.g., perturbation theory and transfer matrix methods, but it is not necessary that the reader know these in advance.

(here, one) of undetermined constants is called a *general solution*. Often, we know more about the problem being solved and, if so, we may be able to be more specific about the value of A. One common situation is where we know some *boundary condition*; a boundary condition is some additional piece of knowledge we have about the solution at some point. That point is often one that coincides with some physical boundary in the physical problem being solved – hence, the name boundary condition – though the term *boundary condition* is often still used even if there is no physical boundary at that point.

For example, we might know that some particular ramp has a relation between its height, y, and the horizontal distance, x, that obeys the equation $dy/dx = 0.4y$, a specific example of a relation of the form of Eq. (A.59) with $a = 0.4$. One kind of boundary condition is where we know the actual value of the solution at some point – for example, we might know that the height of the ramp at the horizontal distance $x = 0$ was $y = 1.5$. (This position $x = 0$ might be the start of the ramp, in which case the term *boundary condition* makes obvious sense.) Then, because $\exp(0) = 1$, we can immediately conclude that $A = 1.5$, and our full solution, obeying this additional boundary condition, is $y = 1.5\exp(0.4x)$. Another kind of boundary condition is where we know the slope (first derivative) of the solution at some point. For example, we might know that the slope of the ramp at $x = 0$ was $dy/dx = 0.6$ (a slope corresponding to 0.6 m rise in 1 m or an angle θ for which $\tan\theta = 0.6$; i.e., an angle of ~ 31°, sloping upward). Then using our solution Eq. (A.60) with $a = 0.4$ (i.e., we know $y = A\exp(0.4x)$) and our original differential equation, Eq. (A.59), we must have $dy/dx = 0.6 = 0.4A\exp(0)$ (i.e., $A = 1.5$).

Note that we can also write an equation such as Eq. (A.59) but with an imaginary coefficient ib (where b is a real number); that is,

$$\frac{dy}{dx} = iby \tag{A.61}$$

which has the general solution

$$y = A\exp(ibx) \equiv A(\cos bx + i\sin bx) \tag{A.62}$$

Note, incidentally, that neither $\cos b$ nor $\sin b$ is, on its own, a solution of this equation.[10]

Second-order differential equations

A second-order differential equation contains no derivatives higher than the second order. It may contain the first derivative (e.g., dy/dx) and must contain the second derivative (e.g., d^2y/dx^2) to be called a second-order differential equation but must contain no higher derivatives, such as a third derivative d^3y/dx^3.

In practice, second-order differential equations cover a very broad range of problems in science and engineering, including most simple oscillations and waves. In this book, we do not need any differential equations beyond second order.

A simple second-order differential equation is

$$\frac{d^2y}{dt^2} = -\omega^2 y \tag{A.63}$$

[10] This point is quite important in the time-dependent Schrödinger equation.

where ω is a real number. This equation is sometimes called the *simple harmonic oscillator equation*. Any of the functions $\exp(i\omega t)$, $\exp(-i\omega t)$, $\cos(\omega t)$, or $\sin(\omega t)$ is a solution to this equation. All of these functions describe oscillatory behavior in time t, at an angular frequency ω (or a conventional frequency $\omega / 2\pi$). To write the general solution of a second-order equation, we need two undetermined constants; two equivalent forms of the general solution of this simple harmonic oscillator equation are

$$y = A\cos(\omega t) + B\sin(\omega t) \tag{A.64}$$

and

$$y = C\exp(i\omega t) + D\exp(-i\omega t) \tag{A.65}$$

where A, B, C, and D are (possibly complex) undetermined constants. These two forms are equivalent because we can write the complex exponential in terms of sines and cosines, as in Eq. (A.36), or we can write sines and cosines in terms of complex exponentials, as in the useful formulae

$$\cos\theta = \frac{1}{2}\left[\exp(i\theta) + \exp(-i\theta)\right] \tag{A.66}$$

and

$$\sin\theta = \frac{1}{2i}\left[\exp(i\theta) - \exp(-i\theta)\right] \tag{A.67}$$

Note that, mathematically, we can replace ω by $-\omega$ in the general solutions in Eqs. (A.64) and (A.65) and still have equally valid general solutions.

Another common second-order differential equation is the one-dimensional Helmholtz wave equation, which is written

$$\frac{d^2 y}{dx^2} + k^2 y = 0 \tag{A.68}$$

where k is a real number. This equation can describe the spatial form of, for example, simple waves on a string at any given time.

The reader may note that, in fact, the simple harmonic oscillator equation and the one-dimensional Helmholtz wave equation are actually identical mathematically. Merely exchanging k and ω and x and t in these equations turns one into the other. Hence, the solutions of the one-dimensional Helmholtz wave equation and the simple harmonic oscillator equation are mathematically identical. We can write the general solutions of the Helmholtz wave equation by repeating Eqs. (A.64) and (A.65) with k substituted for ω and x substituted for t.

These second-order equations illustrated here show another important feature of some differential equations; these equations, Eqs. (A.63) and (A.68), are both *linear*. The most important defining characteristic of a linear equation is that if f and g are both functions that are solutions of the equation, then so also is the function $f + g$. It is easily seen by direct substitution in Eq. (A.63) that any sum of the functions $\exp(i\omega t)$, $\exp(-i\omega t)$, $\cos(\omega t)$, or $\sin(\omega t)$ is also a solution to this equation. For linear equations, it is also the case that if f is a solution, so also is cf where c is some constant, which is also easily verified to be the case for any of the functions $\exp(i\omega t)$, $\exp(-i\omega t)$, $\cos(\omega t)$, or $\sin(\omega t)$ or for the general solutions in Eqs. (A.64) or (A.65). (The specific first-order equations, Eqs. (A.59) and (A.61), are also linear, though, because they only have one functional form of solution, the linearity criterion is restricted to the simpler case where, if f is a solution, so also is cf, for some constant c.) A key factor in ensuring the linearity of these various differential equations is that y and its various derivatives (e.g., dy/dt and d^2y/dt^2) only appear to the first power (e.g., there are no terms in y^2, in $(dy/dt)^2$, in $(d^2y/dt^2)^2$, or any higher powers). The differential equations in this book are all first- or second-order linear differential equations. Because the concept of linearity is so central to quantum mechanics, it is also discussed at some length in the main text of this book.

For second-order differential equations, we can see by the preceding example that there are two undetermined coefficients in the general solutions. Hence, we need two boundary conditions of some kind to get to a specific solution with no undetermined coefficients. These conditions could be, for example, values of the function y and its first derivative at one specific point or the values of the function y itself at two points.

Partial differential equations

Just as we can have equations involving the simple derivatives of functions of one variable, we can also have equations involving the partial derivatives of functions of more than one variable. Such equations are called partial differential equations. A classic example of such an equation is the one-dimensional wave equation

$$\frac{\partial^2\phi(z,t)}{\partial z^2}-\frac{1}{c^2}\frac{\partial\phi(z,t)}{\partial t^2}=0 \quad \text{(A.69)}$$

Note that in writing the equation this way, it is implicit that the second (partial) derivative here with respect to time, $\partial^2\phi/\partial t^2$, is taken at constant position, z; similarly, the second (partial) derivative with respect to position, $\partial^2\phi/\partial z^2$, is taken at constant time, t. This is considered sufficiently obvious that it is not necessary to make this explicit by using the vertical-line-and-subscript notation.

This particular wave equation is very common in classical physics. For waves propagating in the z direction at some wave propagation velocity c, with some simplifying assumptions, this equation describes waves on a string, plane acoustic waves in general, and plane electromagnetic waves.[11]

For a wave on a string, ϕ could, for example, be the upward displacement of a horizontal string. We can simply rewrite Eq. (A.69) as $\partial^2\phi/\partial t^2=c^2(\partial^2\phi/\partial z^2)$; the equation is then telling us that at any given point on the string, the acceleration of the string in the upward direction ($\partial^2\phi/\partial t^2$) is proportional to the (upward) "bend" ($\partial^2\phi/\partial z^2$) of the string at that point, with a proportionality constant of c^2.

The reader can verify by direct substitution that any function $f(z-ct)$ and any function $g(z+ct)$ (with continuous first and second derivatives in both cases) are solutions of this equation. $f(z-ct)$ corresponds to a wave propagating, with constant shape, in the $+z$ direction at velocity c; similarly, $g(z+ct)$ corresponds to one propagating in the $-z$ direction. The general solution can be written

$$\phi(z,t)=f(z-ct)+g(z+ct) \quad \text{(A.70)}$$

where f and g are arbitrary functions with appropriately continuous derivatives.

---

[11] The time-dependent Schrödinger equation for quantum mechanical waves is *not* of this form, however.

Note, incidentally, that it is not the absolute sign of z or ct in $f(z-ct)$ or $g(z+ct)$ that determines the direction of the propagating wave; only the relative sign of z and ct matters. Thus, $f(ct-z)$ would still be a wave propagating in the $+z$ direction, though its shape would be the mirror image of the wave given by $f(z-ct)$.

Partial differential equations like this may typically require *initial conditions* to specify the solution completely, such as the specific form of the shape of the string at some initial time,[12] and might also involve boundary conditions, such as the specification that the string position is fixed at both ends, as in a violin or a guitar.

In many cases, we are only interested in waves where the amplitude at any given point on the wave is oscillating sinusoidally at one specific (angular) frequency ω; that is, where the time-dependence of the wave is of the form $\exp(i\omega t)$, $\exp(-i\omega t)$, $\cos(\omega t)$, $\sin(\omega t)$ or any combination of these. For example, we might know the wave to be of the form $\phi(z,t) = h(z)\sin(\omega t)$. Then, $\partial^2\phi / \partial t^2 = -\omega^2\phi$ and the one-dimensional wave equation reduces to the one-dimensional Helmholtz equation, Eq. (A.68) with $k^2 = \omega^2 / c^2$, which explains why the Helmholtz equation is called a wave equation.

Frequently, we also deal with functions that depend on all three spatial coordinates, so we simultaneously deal with partial derivatives in all three directions. In that case, other shorthand notations are often introduced.

Laplacian operator

The first example we need of a shorthand notation for partial derivatives in multiple directions is the so-called Laplacian operator, ∇^2, also sometimes called *del squared* or, simply, the *Laplacian*. This is also one of the operators in so-called vector calculus, though we defer a more detailed discussion of vector calculus, in general, to Appendix C. In Cartesian coordinates, the Laplacian can be written

$$\nabla^2 \equiv \frac{\partial^2}{\partial x^2} + \frac{\partial^2}{\partial y^2} + \frac{\partial^2}{\partial z^2} \tag{A.71}$$

Some authors use the notation Δ instead of ∇^2 for the Laplacian, though we do not use that notation in this book.

The wave equation in three spatial dimensions can be written

$$\frac{\partial^2\phi(x,y,z,t)}{\partial x^2} + \frac{\partial^2\phi(x,y,z,t)}{\partial y^2} + \frac{\partial^2\phi(x,y,z,t)}{\partial z^2} - \frac{1}{c^2}\frac{\partial\phi(x,y,z,t)}{\partial t^2} = 0 \tag{A.72}$$

or, using the Laplacian notation

$$\nabla^2\phi(x,y,z,t) - \frac{1}{c^2}\frac{\partial\phi(x,y,z,t)}{\partial t^2} = 0 \tag{A.73}$$

[12] Actually, to get a unique solution to the wave equation, not only would we have to know the values of the displacement of the string at some time but also the velocities. E.g., for some pulse-like distortion in the middle of a long string, if we do not know the vertical velocity of the points on the string, we do not know whether the pulse is moving left or right. In fact, if one merely distorts the string in some fixed pulse-shape (i.e., with no initial velocity), two pulses will head off in opposite directions. This point is quite important in understanding why the time-dependent Schrödinger equation is constructed the way that it is.

Often, in working with such equations, we regard lists of variables such as x, y, z, and t as being obvious and omit them, giving

$$\nabla^2\phi - \frac{1}{c^2}\frac{\partial\phi}{\partial t^2} = 0 \qquad \text{(A.74)}$$

This three-dimensional wave equation is also used extensively in classical physics. With some simplifying assumptions, it describes a broad range of acoustic and electromagnetic waves.

Partial differential equations in two or three spatial dimensions are often solved by making specific assumptions about the solutions on an entire bounding surface. A boundary condition in which the value of the function is set to some specific values (often zero) on some boundary is called a *Dirichlet boundary condition.* A boundary condition where the derivative or gradient of the function is set to zero on the boundary is called a *Neumann boundary condition.* We do not, however, have to make use of this notation in this book.

Gradient

At this point, we can also introduce another vector calculus operator, the *gradient* of a scalar function $f(x, y, z)$, which we can write as

$$\nabla f \equiv \text{grad}\, f = \mathbf{i}\frac{\partial f}{\partial x} + \mathbf{j}\frac{\partial f}{\partial y} + \mathbf{k}\frac{\partial f}{\partial z} \qquad \text{(A.75)}$$

If we were thinking of the function $h(x, y)$ that was the height of a hill, with x and y as the east and north coordinate directions, for example, then the gradient in two dimensions

$$\nabla_{xy} h = \mathbf{i}\frac{\partial h}{\partial x} + \mathbf{j}\frac{\partial h}{\partial y} \qquad \text{(A.76)}$$

would be a vector whose magnitude at any point was telling us the magnitude of the slope of the hill (actually, the maximum slope in any direction) and whose direction was telling us the direction in which the hill was sloping (i.e., the direction at that point in which the uphill slope was maximum). We can see, therefore, why this operation is called the gradient. Note that this operation of taking the gradient turns a scalar function into a vector function; though there is no "bold" character in it, ∇f is a vector.

Note that with the symbol ∇ (known as *del* or *nabla*) and also with other similar symbols such as the Laplacian (∇^2) above, we do allow ourselves to put subscripts on the symbol if we need to clarify exactly which coordinates we are considering.

Notation for derivatives with respect to time

Simple or partial differential equations very often involve derivatives with respect to time. In those cases, a shorthand notation is sometimes used, with the number of dots above a variable indicating the order of differentiation with respect to time; that is,

$$\dot{a} \equiv \frac{da}{dt} \text{ and } \ddot{a} = \frac{d^2a}{dt^2} \qquad \text{(A.77)}$$

in simple differential equations and a similar notation but with partial derivatives in partial differential equations. It is presumably obvious from the context whether the notation is meant to represent simple or partial differentiation.

A.7 Summation notation

Instead of writing

$$S = a_1 + a_2 + a_3 + a_4 \tag{A.78}$$

for the sum of the four numbers a_1, a_2, a_3, and a_4, we can instead use the shorthand of the summation notation, in any of the equivalent forms

$$S = \sum_{j=1}^{4} a_j \equiv \sum_{j=1,2,3,4} a_j \equiv \sum_{j=1,\dots,4} a_j \tag{A.79}$$

which in words could be stated "the sum from j equal to 1 to four of a_j." Where it is already obvious what the range of j is or if we mean to state that the relation holds no matter what the range of j is, we might omit the explicit range in the sum, writing

$$S = \sum_j a_j \tag{A.80}$$

We can also have sums over more than one index. For example, we might have some second list of numbers b_1, b_2, and b_3 and we might want to add up all the products of all of the numbers a and b. Written out explicitly, that sum would be

$$\begin{aligned} R = &\, a_1 b_1 + a_1 b_2 + a_1 b_3 \\ &+ a_2 b_1 + a_2 b_2 + a_2 b_3 \\ &+ a_3 b_1 + a_3 b_2 + a_3 b_3 \\ &+ a_4 b_1 + a_4 b_2 + a_4 b_3 \end{aligned} \tag{A.81}$$

We could write Eq. (A.81) as a pair of nested sums where we might imagine we perform the sum on the right first; that is, adding the terms in the order shown in Eq. (A.81)

$$R = \sum_{j=1}^{4} \left(\sum_{k=1}^{3} a_j b_k \right) \tag{A.82}$$

But we know that, at least for such a finite sum of presumably finite numbers, it does not matter what order we do the addition, so there is no need to specify that order. Hence, we can write any of the following (or other variants in the spirit of Eqs. (A.79) or (A.80))

$$R = \sum_{j=1}^{4} \sum_{k=1}^{3} a_j b_k = \sum_{k=1}^{3} \sum_{j=1}^{4} a_j b_k \equiv \sum_{j,k} a_j b_k \equiv \sum\nolimits_{j,k} a_j b_k \tag{A.83}$$

where in the last two forms in Eq. (A.83), we are presuming that the range of j and of k is already known or understood. As long as the ranges of the *summation indices* (i.e., the quantities like j and k that index the terms in the sum) are understood or are implicit, any reasonably clear notation can be used here for the sums. It is also quite possible, of course, to have a sum over a single term with multiple indices, such as $\Sigma_{j,k} c_{jk}$, as might occur, for example, if we wanted to sum all the elements in a matrix (see Section A.9). Some authors use a convention that a sum is automatically understood to be performed over any repeated index

$$a_{jk} b_{km} \equiv \sum_k a_{jk} b_{km} \tag{A.84}$$

but we do not use that notation in this book.

A.8 Integral calculus

Integration in one variable

We expect that the reader is familiar with elementary integral calculus, at least in one variable. The reader will remember that the formal concept of integration can be understood as the area under a curve representing a function, at least if the function is always positive. The simple picture of integration in this way is shown in Fig. A.4, where the area is essentially the sum of the areas of a set of thin rectangles or bars. In the way we use integration, if the function $f(x)$ becomes negative, we assign negative area to the bars and it is quite possible (and very useful in quantum mechanics) for integration defined this way to result in zero net area. In this view of integration, we are formally defining the integral in Fig. A.4 as

$$\int_{x_1}^{x_2} f(x)\,dx \equiv \lim_{\Delta x \to 0}\left(\sum_j f_j \Delta x\right) \tag{A.85}$$

Here, we understand that for a specific value Δx, there are $N_j = (x_2 - x_1)/\Delta x$ values of j. The area of a specific rectangular bar is $f_j \Delta x$, where f_j is, for example, the average of the function at the two edges of the bar. For such a definition, we can at least intuitively see that the sum converges to some measure of the area of the function (including the concept of negative areas).[13] The entity dx is called an *infinitesimal* (because it is infinitesimally small).

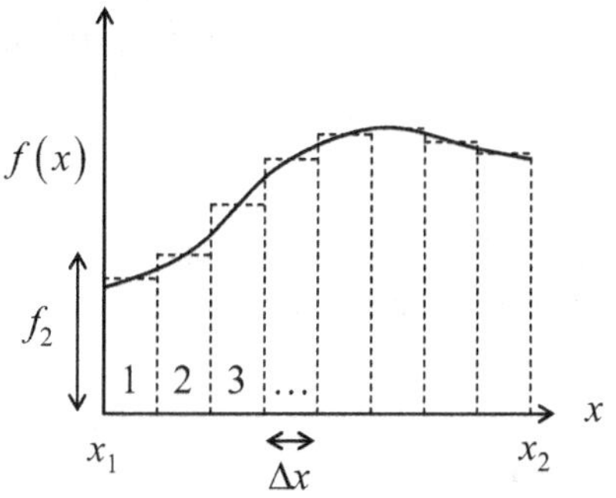

Fig. A.4. Picture of integration as a sum of the areas of rectangular bars right up against one another, each of width Δx and of heights $f_1, f_2, f_3, \ldots$ and so on chosen to approximate the height of the function near the center of the bar. In this picture, the integral from x_1 to x_2 is the limit of the sum of the areas of the bars as the width Δx is shrunk to be arbitrarily small.

It is quite common in simple calculus implicitly to consider the integral sign ($\int$ in Eq. (A.85)) and its associated infinitesimal (dx in Eq. (A.85)) as functioning as "brackets" around the

---

[13] This form of integration by summing up the areas of vertically oriented bars in such a graph is known as *Riemann integration*. It does technically have a minor problem, which is that if one of the bars happens to be right up against a *singularity* (i.e., a point where the function becomes infinite), the area of that particular bar would be infinite, no matter how small we took Δx to be. This could be the case even for functions that do actually have a finite total area despite their singularities. The formal solution to this is instead to perform the integration by taking horizontal slices rather than vertical ones and this is the essence of *Lebesgue integration*. Lebesgue integration is the formal type of integration used to decide whether some function is or is not integrable (loosely, does it have finite area). The term *Lebesgue integrable function* comes up a lot in the formal theory of quantum mechanics, though we mostly avoid it here. For a clarification of what Lebesgue and Riemann integration mean, see, e.g., M. Reed and B. Simon *Functional Analysis* (Academic Press, San Diego, 1980), pp. 12–14.

expression between them, effectively enclosing everything that has to be integrated ($f(x)$ in Eq. (A.85)). This is not, however, a consistent notation. Some authors would consider

$$\int_{x_1}^{x_2} f(x)\,dx \equiv \int_{x_1}^{x_2} dx f(x) \tag{A.86}$$

as equivalent statements. In fact, there are good reasons why sometimes we have to use the notation on the right.[14] The most consistent notation is to consider everything that is a function of the variable being integrated (e.g., of the variable x here) as being included in the integration, regardless of where it occurs in the expression.

An integral with limits (e.g., x_1 as the lower limit and x_2 as the upper limit in Eq. (A.85)) is called a *definite integral*. Integration and differentiation are inverse operations of one another, so that, specifically

$$\int_a^b \left(\frac{df}{dx}\right) dx = f(b) - f(a) \tag{A.87}$$

which is known as the *fundamental theorem of calculus*. Because of this inverse relation, the integral is sometimes known as the *antiderivative*, though we do not use that notation in this book.

Sometimes it is possible for an integral to have meaning even if we do not specify the limits, as when we can write an analytic result or formula for the integral, such as

$$\int x^2 dx = \frac{1}{3}x^3 + C \tag{A.88}$$

Because we have not specified limits, we cannot evaluate an actual number for this integral and it is called an *indefinite integral*. The result is said to be arbitrary within a *constant of integration*, which here is C.

The quantity inside the integration, such as the function $f(x)$ in Eq. (A.85) or the x^2 in Eq. (A.88), is called the *integrand*.[15]

Quite often in quantum mechanics, we may be evaluating integrals of complex numbers, though we are usually performing the integral with respect to a real variable, such as position. Such integrals pose no additional mathematical problems and could be considered as two

---

[14] This notation becomes necessary when we want to consider integral operators. Suppose we want to write an expression that covers the idea that we will integrate everything to the right that is a function of x regardless of what those functions might be or before we specify what those functions are. A way to write that is to use the notation for this integral operator of the form $\int dx$, with the understanding that anything we put to the right that is a function of x is included in the integration.

[15] *Integrand* is an example of a Latin "gerundive" in English. Gerundives all have the sense in English of "requiring to be ..," so *integrand* means "requiring to be integrated." There are many other examples that occur in mathematics and it may be easier to remember and understand them if one realizes they are all gerundives. Other mathematical examples include *operand*, which means "requiring to be operated on," *subtrahend*, which means "requiring to be subtracted," *dividend*, which means "requiring to be divided," and *addend* which means "requiring to be added." *Addend* is really the same as the word *addendum* that is used in nonmathematical contexts to describe something that needs to be added; in that case, the (singular) ending *um* that would be present in the actual Latin gerundives has survived into the English.

integrals, one of the real part and one of the imaginary part of the integrand, which would bring us back to two integrals of real integrands.

Volume integration

In quantum mechanics, we very often have to perform integrals over volumes. A simple example of a volume integral would be one that evaluates the magnitude of some volume, V. Then, we could imagine that we divided the volume up into very small "bricks" or, more mathematically, rectangular-sided boxes or *cuboids*,[16] with lengths Δx, Δy, and Δz. Such boxes would each have volume $\Delta V = \Delta x \Delta y \Delta z$. If we merely counted up how many such boxes we could fit, exactly side by side, into the volume V and multiplied by ΔV, we would have evaluated an approximation to the volume. In the limit as we made ΔV very small, we would expect our answer to become progressively more accurate. Mathematically, we could write this as

$$V = \int_V dV = \lim_{\Delta V \to 0} \left(\sum_j \Delta V \right) \tag{A.89}$$

In this case, we might be imagining that all of our bricks had sequential numbers painted on them. We then progressively put the bricks into the volume in order. Then, the index j is just the number on the bricks. We keep adding bricks until we can fit no more bricks in and that is the limit on j in the sum. For each brick we add, we add ΔV to our sum.

The notation in Eq. (A.89) is undoubtedly confusing. We have used V to be the name of the volume, so the notation $\int_V dV$ in words is *the integral over the volume* V, and we are also using V as the value (e.g., in cubic meters) of the volume. This kind of practice is very common, however.

In practice, authors use various different notations instead of (or as well as) dV for the infinitesimally small volume element. Several equivalent ways of expressing a volume integral are, therefore

$$\int_V dV \equiv \int_V d\mathbf{r} \equiv \int_V d^3\mathbf{r} \tag{A.90}$$

Here, the "vector" $\mathbf{r}$ is used as a shorthand to remind us we are working with three dimensions. The middle notation could be criticized for losing track of the physical dimensions (i.e., meters cubed) of the integral. The middle and the right notations could be criticized for appearing to imply that the result of the integral is a vector when, in fact, it is a scalar. Nonetheless, all of these are in use. We mostly use the one on the right. The reason for not always using the simple dV is that we often use different $\mathbf{r}$ variables to describe the positions of different particles in our integrals and we need to keep track of those in the integrations. Though those particles have different positions, they are probably all in the same volume.

Volume integrals are, of course, also used less trivially to perform the integrals of other quantities over the volume. A simple and useful example is the integral of density, $\rho(\mathbf{r})$ (e.g., kilograms per cubic meter) – a quantity that might well be different at different points $\mathbf{r}$ – over volume to give the total mass m_{tot} of the material in the volume; that is,

[16] Rectangular-sided boxes or cuboids are also known as *rectangular prisms* or *rectangular parallelepipeds*. We mostly use the word *cuboid* here.

$$m_{tot} = \int_V \rho(\mathbf{r}) d^3\mathbf{r} \tag{A.91}$$

We perform this kind of integral often in quantum mechanics, where we are often concerned with *probability density* rather than mass density.

In practice, with volume integrals, to evaluate them more conveniently, we would like to be able to reduce them to the one-dimensional integrals for which we have so many analytic and numerical techniques. For example, if the volume V was itself a cuboid, with sides of length x_L, y_L, and z_L, then we could convert the volume integral into a *multiple integral* ; that is three "nested" or *iterated* one-dimensional integrals that we could perform one after another to evaluate the volume, as follows

$$\int_V dV = \int_{x_c}^{x_c+x_L} \int_{y_c}^{y_c+y_L} \int_{z_c}^{z_c+z_L} dzdydx = \int_{x_c}^{x_c+x_L} \int_{y_c}^{y_c+y_L} z_L dydx = \int_{x_c}^{x_c+x_L} z_L y_L dx = z_L y_L x_L = V \tag{A.92}$$

Though it is highly desirable to break down volume integrals this way, whether it is easy to do so depends on the shape of the volume. For an arbitrary shape of volume, the length in the x direction depends on the y and z values, so the limits on the individual integrals are not constants. Though such nonconstant limits can be handled, at least, numerically, they are generally awkward. For regular shaped objects, such as spheres or cylinders, such problems can usually be solved by changing to spherical or cylindrical coordinates and we discuss such coordinate systems in Appendix C.

One mathematical question is whether it is possible to change the order of the integrations in multiple integrals. In practice, for physical problems, we normally interchange the order whenever we want, though there are some formal mathematical restrictions on this.[17]

We can, of course, also describe surface integrals in a similar way to the volume integrals and we can use a similar set of notations. In that case, we could be considering a set of tiles on a surface instead of a set of bricks in a volume. We would be using dA as a surface area element and would end up with double rather than triple integrals when converting them to iterated integral form.

A.9 Matrices

A *matrix* is a rectangular array of numbers. An $M\times N$ matrix has M rows and N columns. Rows are horizontal and columns are vertical. When written out explicitly, the array is enclosed in square brackets. Commonly, when we want to write a symbol to represent a matrix, we choose

[17] In principle, care must be taken with the order of multiple integrals and, in the most general case, the result can change if the order is changed. For the cases we consider in quantum mechanics and other physical systems, we can mostly ignore this formal difficulty. Certainly, if we are integrating continuous, finite functions of multiple variables over finite ranges, there is no problem in interchanging the order of the integrals. Not all multiple integrals in quantum mechanics are of this type, but to the extent we believe the integrals of interest to us could be approximated to an arbitrary degree of accuracy by integrals of continuous, finite functions over finite, if large, ranges, then we can exchange order whenever we want. The formal criterion for being able to exchange the order of integrals is given by Fubini's theorem in mathematics.

simple capital letters (e.g., A), though there is no requirement to do so. In this book, we nearly always put a "hat" over any symbol we use for a matrix, as in the $\hat{A}$ in Eq. (A.93).[18]

The following is a 2×3 *rectangular* matrix.

$$\hat{A} = \begin{bmatrix} 2 & 1 & -3 \\ 6 & -5 & 4 \end{bmatrix} \tag{A.93}$$

Because all matrices are, by definition, rectangular, when we say a matrix is rectangular, we almost always mean it is not a *square* matrix (i.e., one that has equal numbers of rows and columns). The following matrix is a 2×2 square matrix.

$$\hat{B} = \begin{bmatrix} 1.5 & -0.5i \\ 0.5i & 1.5 \end{bmatrix} \tag{A.94}$$

Note that the numbers or *elements* in the matrix can be real, imaginary, or complex numbers. In quantum mechanics, we work almost exclusively with matrices that contain complex numbers and we work almost exclusively with square matrices.

We index the elements in a matrix in "row-column" order and we count from the top left corner. So, we could write that A_{13} is the element in the first (top) row and third (from the left) column (i.e., the number -3) in matrix $\hat{A}$. It is common, though not universal, to use the same letter (A) for the elements as for the matrix. Sometimes the lowercased version is used instead (e.g., a_{13}), though there is no fixed convention and other symbols can be used.

The *leading diagonal* or just the *diagonal* of a matrix is the diagonal from top left to bottom right. The diagonal is only really a meaningful concept for square matrices. The elements of value 1.5 in matrix $\hat{B}$ are, therefore, on the diagonal and are called the *diagonal elements*. The elements that are not on the diagonal are often collectively called the *off-diagonal elements*. The elements $B_{12} = -0.5i$ and $B_{21} = 0.5i$ in Eq. (A.94) are, therefore, off-diagonal elements.

The particular case of a matrix that has only one row or only one column is called a *vector*. This vector concept is a generalization of the concept of a vector as used in geometry. A geometrical vector can be represented by a list of three numbers, these being its components along three coordinate axes. When we consider this generalized form of a vector, it can have any number of elements and these also may be complex numbers (in contrast to geometrical vector components). We also have to specify whether the vector is a *row vector*, in which case it is written as a matrix with one row, for example,

$$c = [4, -2, 5, 7] \tag{A.95}$$

or a *column vector*, in which case it is written as a matrix with one column, for example,

$$d = \begin{bmatrix} 2+3i \\ -5+2i \\ 4-i \\ -7-6i \end{bmatrix}$$

[18] We also use this "hat" notation to refer to operators in this book. There is no confusion there because matrices are operators in the sense we use them in this book. Operators are discussed in greater detail in Chapter 4.

Technically, a vector is a matrix, but we almost exclusively use the term *vectors* for such single-row or single-column matrices. Because of the way we use such vectors in quantum mechanics, we think of them differently and use a different set of symbols for them that we introduce in Chapter 4.[19]

An important manipulation with matrices and vectors is the *transpose*. It is most easily visualized as a reflection of the matrix or vector around a diagonal line running at 45° from top left to bottom right. It is usually notated with a superscript "*T*". For example, the transposes of the preceding vectors and matrices are

$$\hat{A}^T = \begin{bmatrix} 2 & 6 \\ 1 & -5 \\ -3 & 4 \end{bmatrix} \quad \hat{B}^T = \begin{bmatrix} 1.5 & 0.5i \\ -0.5i & 1.5 \end{bmatrix} \quad c^T = \begin{bmatrix} 4 \\ -2 \\ 5 \\ 7 \end{bmatrix} \quad d^T = [2+3i \;\; -5+2i \;\; 4-i \;\; -7-6i] \tag{A.96}$$

In this book, we if ever use the transpose, however. Instead, we use a generalization of it called the *Hermitian adjoint*, the *Hermitian transpose*, the *conjugate transpose,* or sometimes simply just the *adjoint*. This Hermitian adjoint is the most useful generalization of the transpose when matrix and vector elements may be complex. To take the Hermitian adjoint, one takes the transpose and also takes the complex conjugate of all of the elements. The Hermitian adjoint is notated with a superscript "†", where this symbol is called a "dagger." As a way of remembering it, one can think of the dagger symbol as some kind of combination of the "*T*" for the transpose and the asterisk used to denote complex conjugation. For matrices and vectors with real-numbered elements, the Hermitian adjoint and the transpose are the same operation.

If we consider the elements of the matrix, then taking the Hermitian adjoint of a square matrix simply involves swapping the indices and taking the complex conjugate; the element in the *n*th row and *m*th column of the matrix $\hat{B}^\dagger$ (i.e., $(\hat{B}^\dagger)_{nm}$) is

$$(\hat{B}^\dagger)_{nm} = \hat{B}^*_{mn} \tag{A.97}$$

For the matrix and vector with complex elements, the Hermitian adjoints are

$$\hat{B}^\dagger = \begin{bmatrix} 1.5 & -0.5i \\ 0.5i & 1.5 \end{bmatrix} \quad d^\dagger = \left[2-3i \;\; -5-2i \;\; 4+i \;\; -7+6i\right] \tag{A.98}$$

Note, incidentally, that for the particular matrix $\hat{B}$, the Hermitian adjoint of $\hat{B}$ is actually the same as $\hat{B}$ (i.e., $\hat{B}^\dagger = \hat{B}$). A matrix that has this property is said to be *Hermitian*. Note that only some complex matrices have this property, but it is a particularly important property in quantum mechanics and has some useful consequences (which we discuss in Chapter 4).

Matrix algebra

As discussed so far, matrices are simply tables of numbers. What makes matrices useful is the algebra associated with them. The algebra of matrices is related to that of numbers but has crucially important differences.

---

[19] We do not think of vectors as operators when we use them in quantum mechanics and we do not, therefore, put a hat over the symbols representing them. In fact, we use a different notation, so-called *Dirac notation*, to discuss them and we introduce that notation in Chapter 4.

Addition and subtraction

If and only if two matrices are of the same size (i.e., have the same numbers of rows and columns) then we can add them by simply adding the corresponding numbers. For example, for the two matrices, $\hat{F}$ and $\hat{G}$

$$\hat{F} = \begin{bmatrix} 1 & i \\ 2 & 1-3i \end{bmatrix} \quad \hat{G} = \begin{bmatrix} 5 & 4i \\ -6 & 7+8i \end{bmatrix} \tag{A.99}$$

we would have

$$\hat{K} = \hat{F} + \hat{G} = \begin{bmatrix} 1+5 & i+4i \\ 2-6 & 1+7-3i+8i \end{bmatrix} = \begin{bmatrix} 6 & 5i \\ -4 & 8+5i \end{bmatrix} \tag{A.100}$$

In general, for the elements of the matrices, we have

$$K_{ij} = F_{ij} + G_{ij} \tag{A.101}$$

where i and j range over the number of rows and columns, respectively (i.e., i is 1 or 2, j is 1 or 2).

Subtraction follows the same procedure but subtracting rather than adding.

Multiplication

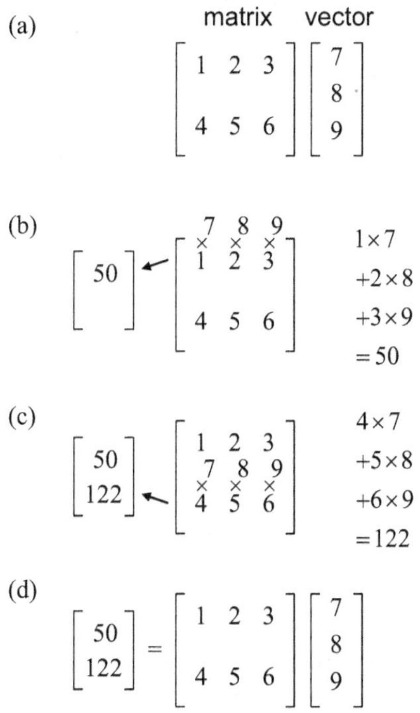

Fig. A.5. Illustration of process of matrix-vector multiplication.

Multiplication of matrices is the most important aspect of matrix algebra and it requires some definition. It is most easily understood by first considering multiplication of a vector by a matrix. Suppose we have a column vector on the right of a matrix (see Fig. A.5(a)). Suppose the column vector has N "rows" or elements in it (here, 3). For the multiplication by a matrix to be defined, this number must match the number of columns in the matrix, so the matrix must be an $M \times N$ (here, 2×3) matrix for some number of rows M (here, 2). To multiply the vector by the matrix, we can imagine that we first "pick up" the vector, then turn it anticlockwise by 90° and lay it on top of the top row of the matrix (Fig. A.5(b)). Then, we multiply each element of the vector by the corresponding element of the top row of the matrix and add up all the results. We write that sum in the top element of the vector that will be the result of this multiplication operation. We then move the vector down one row in the matrix (Fig. A.5(c)) and repeat the multiplication and addition, writing the answer in the next element of the "output" vector, and so on, until we have performed this operation for all of the rows of the matrix to get the final result (Fig. A.5(d)).

Note that the result of multiplying an N element column vector by an $M \times N$ matrix is to generate an M element column vector. Note also that we do not use any "multiplication sign" in matrix algebra. We simply put the two quantities to be multiplied beside one another.

We can also write this matrix-vector multiplication in a compact form using summation notation in which case, for a vector c and a matrix $\hat{A}$, we have, for a "result" vector d

$$d_m = \sum_n A_{mn} c_n \tag{A.102}$$

Note that with the "row-column" order of the indices, we end up summing over two adjacent, identical indices (here, n). In our example, n runs from 1 to 3 and m runs from 1 to 2.

If we are multiplying a matrix by a matrix, we simply repeat this operation for each column of the matrix on the right, working from left to right, for example, and write down the resulting columns in the resulting matrix, also working from left to right, for example. Hence, extending the above example, we would have

$$\begin{bmatrix} 50 & 14 \\ 122 & 32 \end{bmatrix} = \begin{bmatrix} 1 & 2 & 3 \\ 4 & 5 & 6 \end{bmatrix} \begin{bmatrix} 7 & 1 \\ 8 & 2 \\ 9 & 3 \end{bmatrix} \tag{A.103}$$

because $1 \times 1 + 2 \times 2 + 3 \times 3 = 14$ and $4 \times 1 + 5 \times 2 + 6 \times 3 = 32$, and so on.

For an $N \times P$ matrix $\hat{A}$ being multiplied by an $M \times N$ matrix $\hat{B}$ to give a resulting matrix $\hat{R}$

$$\hat{R} = \hat{B}\hat{A} \tag{A.104}$$

we can write this in summation notation as

$$R_{mp} = \sum_n B_{mn} A_{np} \tag{A.105}$$

Note again that we have summed over the "internal" index n.

Vector-vector multiplication is also meaningful provided the vectors have the same length and provided that one vector is a row vector and the other vector is a column vector. This rule follows directly from the discussion of matrix-matrix multiplication when we consider the multiplication of $1 \times N$ and $N \times 1$ matrices (which are, by definition, both vectors). There are two quite different possibilities for the form of the result depending on the order of the multiplication. Specifically, multiplying a $1 \times N$ matrix on the left by an $N \times 1$ matrix on the right gives a single number (the same thing as a 1×1 matrix) as the result; for example,

$$\begin{bmatrix} 1 & 2 & 3 \end{bmatrix} \begin{bmatrix} 4 \\ 5 \\ 6 \end{bmatrix} = 32 \tag{A.106}$$

In summation form, such a multiplication of two vectors c and d would reduce to

$$f = \sum_n c_n d_n \tag{A.107}$$

where f is a number. Because such a multiplication reduces two vectors to a single number, it is sometimes referred to as an *inner product*.

Multiplying an $N\times 1$ on the left by a $1\times N$ matrix on the right, however, gives an $N\times N$ matrix; for example,

$$\begin{bmatrix} 4 \\ 5 \\ 6 \end{bmatrix} \begin{bmatrix} 1 & 2 & 3 \end{bmatrix} = \begin{bmatrix} 4 & 8 & 12 \\ 5 & 10 & 15 \\ 6 & 12 & 18 \end{bmatrix} \tag{A.108}$$

In this case, in summation form, there is nothing to sum over – each of the elements in the matrix on the right is just the result of a single multiplication of two numbers and we would have, for the elements of the resulting matrix $\hat{F}$

$$F_{mp} = c_m d_p \tag{A.109}$$

Because such a multiplication generates an entire matrix from just two vectors, it is sometimes referred to as an *outer product*.

Incidentally, if we multiply a column vector d by its own Hermitian adjoint $d^\dagger$, the result $d^\dagger d$ is the sum of the modulus squared of the elements; that is,

$$d^\dagger d = \sum_n d_n^* d_n = \sum_n |d_n|^2 \tag{A.110}$$

which we can view as the square of the "length" of the vector d. This is a particularly useful property of the Hermitian adjoint and we discuss this also in Chapter 4.

Commutative, associative and distributive properties

Just like multiplication of numbers, matrix multiplication is *associative*

$$(\hat{C}\hat{B})\hat{A} = \hat{C}(\hat{B}\hat{A}) \tag{A.111}$$

and it is *distributive*

$$\hat{A}(\hat{B} + \hat{C}) = \hat{A}\hat{B} + \hat{A}\hat{C} \tag{A.112}$$

but it is *not*, in general, *commutative*; that is, in general

$$\hat{B}\hat{A} \neq \hat{A}\hat{B} \tag{A.113}$$

There can be specific matrices for which multiplication does commute (i.e., is independent of order) and such matrices have very specific and useful properties (as we see in Chapters 4 and 5), but we cannot, in general, simply interchange the order of matrix multiplication. This point is extremely important for quantum mechanics.

Multiplication of a matrix or vector by a scalar (a number) simply means we multiply all the elements in the matrix or vector by the number. We can move that number about to anywhere we want inside any chain of matrix-matrix, matrix-vector, or vector-vector products because it does not matter to the final result which of the matrices or vectors is multiplied by the number; the net result is still to multiply the entire result by the number. For example, for some matrices $\hat{A}$ and $\hat{B}$, some vector c, and some number α

$$\alpha \hat{A}\hat{B}c = \hat{A}\alpha\hat{B}c = \hat{A}\hat{B}\alpha c \tag{A.114}$$

Hermitian adjoint of a product

A particularly useful algebraic manipulation for matrices that we use extensively in the book (i.e., from Chapter 4 onward) is that the Hermitian adjoint of a product is the product of the Hermitian adjoints *with the order reversed in the multiplication*. That is,

$$\left(\hat{A}\hat{B}\right)^\dagger = \hat{B}^\dagger\hat{A}^\dagger \tag{A.115}$$

The proof of this is straightforward in the summation notation.[20]

Inverse

Another major difference with matrices compared to numbers is that there is no such operation as division in matrix algebra. Sometimes, a matrix can have an *inverse* (or, equivalently, the matrix is *invertible*.) The inverse of a matrix $\hat{A}$, if it exists, is written $\hat{A}^{-1}$ and it has the property

$$\hat{A}^{-1}\hat{A} = \hat{I} \tag{A.116}$$

where $\hat{I}$ is the *identity matrix* (of the appropriate size). The identity matrix is the matrix with the number "1" for all of its diagonal elements and "0" for all of its off-diagonal elements. It is the formal analog for matrices of the number "1" in the algebra of ordinary numbers. The 3×3 identity matrix is, for example,

$$\hat{I} = \begin{bmatrix} 1 & 0 & 0 \\ 0 & 1 & 0 \\ 0 & 0 & 1 \end{bmatrix} \tag{A.117}$$

The identity matrix (provided it is the correct size in each case to make the multiplication legal) formally has the property for any matrix $\hat{A}$

$$\hat{A}\hat{I} = \hat{I}\hat{A} = \hat{A} \tag{A.118}$$

which can be viewed as a definition of the identity matrix. That definition might seem trivial, but, just as the number "1" is important for ordinary algebra, so the identity matrix is important for matrix algebra.

---

[20] Suppose $\hat{R} = \hat{A}\hat{B}$, so that the element $R_{mp} = \sum_n A_{mn}B_{np}$. Hence,

$$(\hat{R}^\dagger)_{pm} = R^*_{mp} = \sum_n (A_{mn}B_{np})^* = \sum_n A^*_{mn}B^*_{np} = \sum_n (\hat{A}^\dagger)_{nm}(\hat{B}^\dagger)_{pn} = \sum_n (\hat{B}^\dagger)_{pn}(\hat{A}^\dagger)_{nm} .$$

But this last sum is just the matrix element in the pth row and mth column in the matrix product $\hat{B}^\dagger\hat{A}^\dagger$. Hence, $\hat{R}^\dagger = (\hat{A}\hat{B})^\dagger = \hat{B}^\dagger\hat{A}^\dagger$.

Numbers and ordinary algebraic expressions do have the property analogous to Eq. (A.116) – that is, we can write, for any nonzero number x, the equation $x^{-1}x = 1$ and in that case we can also write $x^{-1} = 1/x$ – but we can never write anything analogous to the reciprocal (i.e., $1/x$) in matrix algebra. In matrix algebra, instead of dividing by a matrix, we have to multiply by the inverse (if there is one). The conditions under which a matrix has an inverse are quite restrictive and many matrices do not have inverses. We also discuss this point when we discuss determinants.

Linear equations and matrices

Matrices give a particularly elegant way of writing systems of linear equations. For example, if we have two equations[21] for straight lines

$$\begin{aligned} A_{11}x + A_{12}y &= c_1 \\ A_{12}x + A_{22}y &= c_2 \end{aligned} \qquad (A.119)$$

we can rewrite these as the matrix equation

$$\hat{A}\begin{bmatrix} x \\ y \end{bmatrix} = \begin{bmatrix} A_{11} & A_{12} \\ A_{21} & A_{22} \end{bmatrix}\begin{bmatrix} x \\ y \end{bmatrix} = \begin{bmatrix} c_1 \\ c_2 \end{bmatrix} \qquad (A.120)$$

and, if we write b_1 instead of x and b_2 instead of y, we could also write them in summation form as

$$\sum_{n=1}^{2} A_{mn}b_n = c_m \qquad (A.121)$$

If we knew the inverse $\hat{A}^{-1}$ of the matrix $\hat{A}$, then we would be able to solve the linear equations Eq. (A.119) because premultiplying both sides by $\hat{A}^{-1}$, we have

$$\hat{A}^{-1}\hat{A}\begin{bmatrix} x \\ y \end{bmatrix} = \hat{A}^{-1}\begin{bmatrix} c_1 \\ c_2 \end{bmatrix} \qquad (A.122)$$

so, using the definition of the inverse, Eq. (A.116) and the definition of the identity operator, Eq. (A.118) (i.e., that multiplying by the identity operator makes no change to any matrix or vector), we therefore have

$$\begin{bmatrix} x \\ y \end{bmatrix} = \hat{A}^{-1}\begin{bmatrix} c_1 \\ c_2 \end{bmatrix} \qquad (A.123)$$

so that we can calculate the values of x and y that are the solutions of these two linear equations, Eq. (A.119). In this case, that solution means the point at which the two straight lines cross (if they do cross; if they do not cross, the matrix will have no inverse).

Hence, we can now understand that the operation of finding the inverse of a matrix is equivalent mathematically to solving a system of linear equations – if we know how to solve such systems of equations, we know how to calculate the inverse of a matrix. Conversely,

---

[21] If the reader is more used to the form $y = mx + c$, it is straightforward to cast either of these two straight line equations in that form. For example, the upper equation in Eq. (A.119) could be rewritten as $y = (-A_{11}/A_{12})x + (c_1/A_{12})$.

knowing whether a matrix has an inverse tells us whether the corresponding system of linear equations is solvable.

Determinant

The determinant is a number that can be calculated for any square matrix (and it can only be calculated for square matrices). Determinants have many interesting and useful mathematical properties. The single most useful and important property of determinants for us is the following:

> If the determinant of a matrix is not zero, then the matrix has an inverse and if a matrix has an inverse, the determinant of the matrix is not zero. That is, a nonzero determinant is a necessary and sufficient condition for a matrix to be invertible.

For this reason, determinants are also very useful for deciding whether systems of linear equations can have solutions. The determinant of a matrix $\hat{A}$ is written as

$$\det\left(\hat{A}\right) = \begin{vmatrix} A_{11} & A_{12} & \cdots & A_{1N} \\ A_{21} & A_{22} & \cdots & A_{2N} \\ \vdots & \vdots & \ddots & \vdots \\ A_{N1} & A_{N2} & \cdots & A_{NN} \end{vmatrix} \tag{A.124}$$

Some authors use the two vertical lines to denote the determinant even when using the symbol for the matrix (e.g., $|\hat{A}|$), but this notation can be confused with the modulus[22] so we avoid it.

There are at least two equivalent general formulae for calculating determinants (i.e., Leibniz's formula and Laplace's formula). It would take some space to explain those and we do not need them for most of the book,[23] so we omit them here. Also, in most numerical calculations, other techniques are used to calculate determinants.[24] It is, however, very useful to know the explicit formulae for 2×2 and 3×3 matrices, so we state them here.

$$\det\left(\hat{A}\right) = \begin{vmatrix} A_{11} & A_{12} \\ A_{21} & A_{22} \end{vmatrix} = A_{11}A_{22} - A_{12}A_{21} \tag{A.125}$$

$$\begin{aligned} \det\left(\hat{A}\right) &= \begin{vmatrix} A_{11} & A_{12} & A_{13} \\ A_{21} & A_{22} & A_{23} \\ A_{31} & A_{32} & A_{33} \end{vmatrix} \\ &= A_{11}\left(A_{22}A_{33} - A_{23}A_{32}\right) - A_{12}\left(A_{21}A_{33} - A_{23}A_{31}\right) + A_{13}\left(A_{21}A_{32} - A_{22}A_{31}\right) \\ &= A_{11}A_{22}A_{33} - A_{11}A_{23}A_{32} - A_{12}A_{21}A_{33} + A_{12}A_{23}A_{31} + A_{13}A_{21}A_{32} - A_{13}A_{22}A_{31} \end{aligned} \tag{A.126}$$

---

[22] It can also be confused with some notations for matrix norms, though we do not use those here.

[23] We do actually use Leibniz's formula in Chapter 13 in connection with identical particle states, however.

[24] A standard approach is to use the numerical technique of Gaussian elimination to turn the matrix into a so-called triangular matrix (i.e., a matrix with zeros for all elements below the diagonal [an upper triangular matrix] or for all elements above the diagonal [a lower triangular matrix]) or LU (lower/upper) decomposition to turn the matrix into the product of upper and lower triangular matrices. For triangular matrices, the determinant is the product of the diagonal elements; and, for the product of two matrices, the determinant is the product of the individual determinants.

This 3×3 determinant is often visualized as in Fig. A.6 below, which is based on the expression in the second row of Eq. (A.126). Multiplications proceed from the top row toward the bottom row in the black elements. The multiplications corresponding to arrows going down and to the left have a minus sign associated with them. Note that the sign associated with the top elements is negative for the middle element.

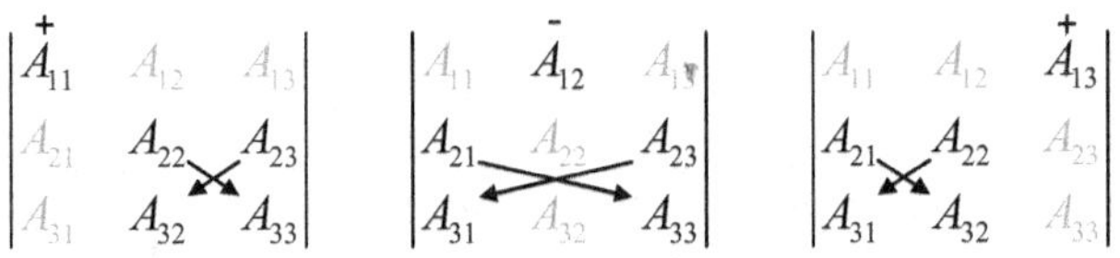

Fig. A.6. Visualization of the process for multiplying elements for the determinant of a 3×3 matrix.

Note, incidentally, that the determinant contains every product of elements from different rows in which all the elements in the each product are also in different columns,[25] as can be seen from the expression in the third row of Eq. (A.126).

Eigenvectors and eigenvalues

A particularly interesting and useful class of equations is the *eigenequation*. For matrices, such an equation is of the form

$$\hat{A}d = \lambda d \tag{A.127}$$

where d is a vector, λ is a number, and $\hat{A}$ is a square matrix. When such an equation has solutions, they occur for specific values of λ, known as *eigenvalues*, and corresponding specific vectors d known as *eigenvectors*.

Now, we can rewrite Eq. (A.127) as, first

$$\hat{A}d = \lambda \hat{I} d \tag{A.128}$$

where we have inserted the identity matrix (of the same size as the matrix $\hat{A}$). We can always do this because the identity matrix makes no change to the expression. Hence, we can further rewrite Eq. (A.127) as

$$\hat{B}d = 0 \tag{A.129}$$

where

$$\hat{B} = \hat{A} - \lambda \hat{I} \tag{A.130}$$

The matrix $\hat{B}$ is just the matrix $\hat{A}$ with the number λ subtracted off every diagonal element. Incidentally, though it is common to use the notation as in Eq. (A.129), the "0" on the right-

[25] If we always write the products in the order of the rows from top to bottom (i.e., in order of the first index in each element), then the sign of each product can be deduced from the ordering of the second index in the elements (i.e., the indices corresponding to the columns). If that order corresponds to an even number of "swaps" of the digits 1, 2, 3, …, from their original sequential order, then the sign of the product is positive (a so-called even permutation), but if that order corresponds to an odd number of swaps (an odd permutation), the sign is negative. This, in fact, is the essence of Leibniz's formula for the determinant and can be used to construct determinants of matrices of any size.

hand side is actually not the number zero but is, in fact, a zero vector with the same number of elements as the vector d (i.e., a vector with zeros for all of its elements).

For Eq. (A.129) to have a solution for some nonzero vector d and for some λ, the matrix $\hat{B}$ must *not* have an inverse. Suppose $\hat{B}$ did have an inverse, $\hat{B}^{-1}$. Then, we could multiply both sides by that inverse to obtain

$$d = \hat{B}^{-1}0 \tag{A.131}$$

which would mean that multiplying the zero vector by the matrix $\hat{B}^{-1}$ would give a nonzero vector d as the result. But, there is no matrix (with finite elements) that can do this; any matrix multiplying the zero vector gives the zero vector as the result. Hence, by reductio ad absurdum, the matrix $\hat{B}$ cannot have an inverse.

The fact that $\hat{B}$ cannot have an inverse, therefore, gives a very simple condition for whether there is a solution to the eigenequation Eq. (A.127). Using a key property of the determinant, as discussed before, we must have

$$\det\left(\hat{A} - \lambda\hat{I}\right) = 0 \tag{A.132}$$

Because we can write down an algebraic expression to evaluate the determinant, based only on the elements of the matrix, we can, therefore, tell directly from this expression, which is known as a *secular equation*, whether there is a solution. In practice, we use this secular equation to find the eigenvalues (if there are any) for which the equation has a solution and then use those eigenvalues to deduce the eigenvectors.

For example, for the matrix in Eq. (A.94), Eq. (A.132) becomes

$$\begin{gathered}\det\left(\begin{bmatrix} 1.5 & -0.5i \\ 0.5i & 1.5 \end{bmatrix} - \lambda\begin{bmatrix} 1 & 0 \\ 0 & 1 \end{bmatrix}\right) = \det\left(\begin{bmatrix} 1.5 & -0.5i \\ 0.5i & 1.5 \end{bmatrix} - \begin{bmatrix} \lambda & 0 \\ 0 & \lambda \end{bmatrix}\right) \\ = \det\left(\begin{bmatrix} 1.5-\lambda & -0.5i \\ 0.5i & 1.5-\lambda \end{bmatrix}\right) = (1.5-\lambda)^2 - (0.5i)(-0.5i) = (1.5-\lambda)^2 - 0.25 = 0\end{gathered} \tag{A.133}$$

That is, multiplying out, we have the quadratic equation

$$\lambda^2 - 3\lambda + 2 = 0 \tag{A.134}$$

which is the secular equation for this problem. The roots of this equation are, by the usual quadratic solution

$$\lambda_1 = 1 \text{ and } \lambda_2 = 2 \tag{A.135}$$

Now that we know the eigenvalues, we substitute them back into the eigenequation and deduce the corresponding eigenvectors. Our eigenequation here is, as in Eq. (A.127)

$$\begin{bmatrix} 1.5 & -0.5i \\ 0.5i & 1.5 \end{bmatrix}\begin{bmatrix} d_1 \\ d_2 \end{bmatrix} = \lambda\begin{bmatrix} d_1 \\ d_2 \end{bmatrix} \tag{A.136}$$

where d_1 and d_2 are the components of the vectors we are trying to find. We can rewrite this as

$$\begin{bmatrix} 1.5-\lambda & -0.5i \\ 0.5i & 1.5-\lambda \end{bmatrix}\begin{bmatrix} d_1 \\ d_2 \end{bmatrix} = \begin{bmatrix} 0 \\ 0 \end{bmatrix} \tag{A.137}$$

So, using the first eigenvalue, $\lambda_1 = 1$, we have, explicitly

$$\begin{bmatrix} 0.5 & -0.5i \\ 0.5i & 0.5 \end{bmatrix} \begin{bmatrix} d_1 \\ d_2 \end{bmatrix} = \begin{bmatrix} 0 \\ 0 \end{bmatrix} \tag{A.138}$$

Writing this out again in the form of linear equations, we have two equations

$$\begin{aligned} 0.5d_1 - 0.5id_2 &= 0 \\ 0.5id_1 + 0.5d_2 &= 0 \end{aligned} \tag{A.139}$$

From either one of these equations, we can directly now deduce that in this eigenvector associated with this eigenvalue

$$d_2 = id_1 \tag{A.140}$$

so, choosing the value 1 for d_1, we can write the first eigenvector

$$v_1 = \begin{bmatrix} 1 \\ i \end{bmatrix} \tag{A.141}$$

Note that we can multiply the eigenvector by any constant we want (or, equivalently, as we have done here, we can choose one of the components of the eigenvector arbitrarily) and it is still an eigenvector, so the "length" of the eigenvector is arbitrary. This property is obvious from the original eigenequation, where we see that the eigenvector is still a solution if we multiply it (on both sides) by any constant.

By similar algebra, but using the second eigenvalue $\lambda_2 = 2$, we can find the second eigenvector, which we can write as

$$v_2 = \begin{bmatrix} 1 \\ -i \end{bmatrix} \tag{A.142}$$

If we extend to larger matrices, we can see that the secular equation becomes a polynomial (actually of the same order of polynomial as the size N of the $N \times N$ of the matrix). For example, a 3×3 matrix can have 3 eigenvalues and 3 associated eigenvectors.

There are many other interesting and useful properties of matrices and determinants, but the properties discussed above are more than sufficient to start the study of quantum mechanics here. We introduce other properties as needed throughout the book

A.10 Product notation

By analogy with the previous summation notation, instead of writing

$$P = a_1 a_2 a_3 a_4 \tag{A.143}$$

we can instead write

$$P = \prod_{j=1}^{4} a_j \tag{A.144}$$

where the uppercased Greek "pi" (Π) stands for "product." Similar conventions apply to the indices here as in the case of the summation notation.

A.11 Factorial

For a positive integer[26] n greater than or equal to 1, the *factorial* function is defined as

$$n! \equiv \prod_{j=1}^{n} j \tag{A.145}$$

where $n!$ is pronounced "n factorial," or, written out

$$n! \equiv n(n-1)(n-2)\ldots\times 2\times 1 \tag{A.146}$$

and we also make the additional definition that

$$0! = 1 \tag{A.147}$$

so that the factorial function is then defined for n being an integer or zero.

[26] Factorials can also be defined for noninteger (i.e., real) positive numbers. See the discussion of the gamma function in Appendix G.

Appendix B

Background physics

In this appendix, we summarize the background elementary classical physics that we need to start the study of quantum mechanics in this book. Other physics is introduced as required throughout the book. Probably, the student has already seen most of the physics here before, but this appendix reviews the key items and notations.

B.1 Elementary classical mechanics

As we study quantum mechanics, some of the concepts and relations from elementary classical mechanics remain quite useful. In particular, the ideas of energy, momentum, and mass remain central concepts in quantum mechanics, especially in the nonrelativistic quantum mechanics that is the subject of this book.[1]

Elementary classical mechanics is sometimes called *Newtonian* classical mechanics. This name distinguishes it from *Hamiltonian* or *Lagrangian* classical mechanics (which are mathematically more sophisticated formulations of the same underlying physics), from relativistic (classical) mechanics, and from quantum mechanics. In such elementary, Newtonian mechanics, for a particle of mass m, the classical *momentum* $\mathbf{p}$, which is a vector quantity because it has direction, is

$$\mathbf{p} = m\mathbf{v} \tag{B.1}$$

where $\mathbf{v}$ is the (vector) velocity. The *kinetic energy*, which is the energy associated with motion, is given by

$$K.E. = \frac{1}{2}mv^2 \tag{B.2}$$

In quantum mechanics, momentum is a more useful quantity than velocity,[2] and we prefer to think of classical kinetic energy as

---

[1] By "nonrelativistic," we mean here that any particles with mass must effectively be moving much slower than the velocity of light. Photons, though they travel at the velocity of light, have no mass and we can still handle most of the physics of photons in quantum mechanics without having to deal with explicit relativistic physics.

[2] There are many reasons why we prefer momentum, but one simple one is that photons have momentum even though they have no mass and they travel at the velocity of light.

$$K.E. = \frac{p^2}{2m} \tag{B.3}$$

where strictly by p^2 we mean the vector dot product $\mathbf{p}\cdot\mathbf{p}$. At least in classical mechanics, Eqs. (B.2) and (B.3) are, of course, equivalent.

Another important energy in elementary classical mechanics is the *potential energy*, usually denoted by the letter V when it is used in quantum mechanics.[3] Potential energy is defined as *energy due to position*, so we usually write V as a function of position $\mathbf{r}$ (i.e., as $V(\mathbf{r})$). We can only talk about potential energy if, indeed, that energy depends only on position and not how we got there; fields that have this property, such as the ordinary gravitational field for objects near the surface of the earth, are called *conservative* fields or *irrotational* fields. Another way of stating this same conservative property is to say that that change in potential energy is zero around any closed path – that is, if we move an object around some path that brings it back to where it started, its potential energy will be the same as when it started. We are seldom concerned with gravitational fields in quantum mechanics, but we do often consider electrostatic fields and those fields are conservative.[4]

The total energy is the sum of the potential and kinetic energies. When this energy is expressed as a function of position and momentum, it is sometimes called the *Hamiltonian*. For a single classical particle, the total energy or Hamiltonian can, therefore, be written

$$H = \frac{p^2}{2m} + V(\mathbf{r}) \tag{B.4}$$

There is a more sophisticated form of classical mechanics that is built around the use of the Hamiltonian considered as a function of position and momentum, though we defer discussing that until Chapter 15. For the present, at least when we are thinking only about particles with mass, we can simply think of the Hamiltonian as the total energy, the sum of kinetic and potential energies.

Note, incidentally, that the "zero" or "origin" we use for potential energy is always arbitrary. We can choose it to be what we want as long as we are consistent. In both classical mechanics and quantum mechanics, there is no absolute origin for potential energy and we only really work with differences in potential energy between one position and another. It makes no difference to any problem in classical or quantum mechanics at what position we say the potential energy is equal to zero. We do, however, typically make some specific choice in practice, at least to make it easy to do the algebra in the problem of interest.

In Newtonian classical mechanics, we often use the concept of force. Indeed, Newton's second law tells us that

$$\mathbf{F} = m\mathbf{a} \tag{B.5}$$

where **F** is the force and **a** is the acceleration.

---

[3] Note that, although we use V for potential energy, we are not meaning an electrostatic potential in Volts, but rather an energy in Joules. This can be confusing because, quite often in the problems of interest to us, the potential energies are a result of electrostatic forces or energies. For a particle of charge q, the electrostatic potential energy in Joules is simply $q \times$ *electrostatic potential energy in Volts*, so this is not in practice a difficult conversion.

[4] Effects associated with static magnetic fields are, however, often *not* conservative.

It is also true that in Newtonian classical mechanics we can write force as being equal to the rate of change of momentum; that is,

$$\mathbf{F} = \frac{d\mathbf{p}}{dt} \tag{B.6}$$

In quantum mechanics, we do find motion of particles that corresponds to Newton's second law, but we cannot use that particular expression as a way to start out in quantum mechanics. Indeed, we seldom if ever use the idea of force, or Newton's second law, directly in quantum mechanics. This is in part because we can also express force as the gradient of a potential energy; hence, knowledge of the potential energy everywhere contains all we need to know about forces.[5]

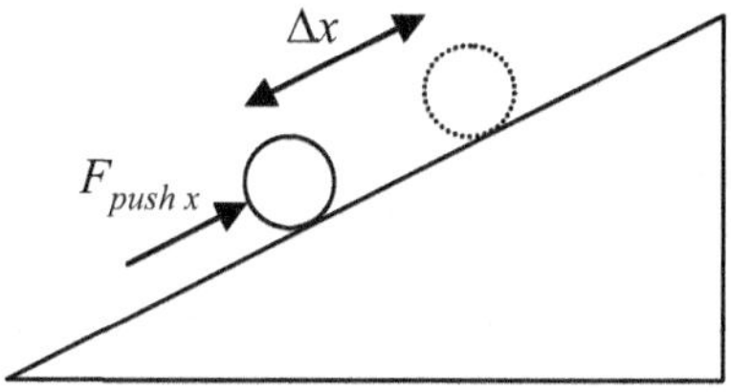

Fig. B.1. Illustration of a ball pushed slowly up a plane by a force of magnitude $F_{push\ x}$ through a distance Δx. The x direction is parallel to the inclined plane.

To understand the relation between force and potential energy, suppose we are trying to change the potential energy of a ball by pushing it slowly (and frictionlessly) up a hill, as shown in Fig. B.1. In this case, we get a change, ΔV, in potential energy V resulting from exerting a force of magnitude $F_{push\ x}$ in the x-direction (here, the direction along the surface of the hill; i.e., directly up hill) on the body through a distance Δx. As usual, the change ΔV is the product of force times distance

$$\Delta V = F_{push\ x} \Delta x \tag{B.7}$$

or, equivalently

$$F_{push\ x} = \frac{\Delta V}{\Delta x} \tag{B.8}$$

or, in the limit of very small changes in energy and correspondingly small distances

$$F_{push\ x} = \frac{dV}{dx} \tag{B.9}$$

This is the force we have to apply to the ball to push it slowly uphill. This force is the opposite of the force F_x that is trying to push the ball downhill. If we stopped pushing the ball, it would accelerate downhill, pushed by this force

$$F_x = -\frac{dV}{dx} \tag{B.10}$$

[5] At least, when we are considering only conservative fields such as electrostatic fields.

that is a result solely of the gradient, dV / dx, of the potential energy. Hence, we can see that knowing the potential energy everywhere means we do not separately need the concept of force (at least, force from potential energies).

We can readily generalize an argument like this to construct the force vector, **F**, considering inclines in three directions at right angles, to obtain

$$\mathbf{F} = -\nabla V \equiv \left[\frac{\partial V}{\partial x}\mathbf{i} + \frac{\partial V}{\partial y}\mathbf{j} + \frac{\partial V}{\partial z}\mathbf{k} \right] \tag{B.11}$$

where **i**, **j**, and **k** are unit vectors in the three usual Cartesian coordinate directions and ∇ is called the *gradient* or "grad" operator, as discussed in Appendix A.

This discussion of elementary classical mechanics is sufficient to start the discussion of quantum mechanics.

B.2 Electrostatics

Force and potential in uniform electric field

In a uniform electric field **E**, the force on a charge q is

$$\mathbf{F} = q\mathbf{E} \tag{B.12}$$

Note that a positive charge is pushed in the direction of the electric field. For an electric field E_z in the z direction, as the positively charged particle moves in the positive z direction, it is as if it is going downhill, so its potential energy ϕ decreases. Using this fact and the fact that work is force times distance, for a simple one-dimensional case, we have a potential energy

$$\phi(z) = -\int qE_z dz \tag{B.13}$$

For a constant field, the potential energy relative to that at $z = 0$ is, therefore

$$\phi(z) = -qE_z z \tag{B.14}$$

Coulomb's law

For two point charges of value Q_1 and Q_2 in free space, the force between them has a value

$$F = -\frac{Q_1 Q_2}{4\pi\varepsilon_o R^2} \tag{B.15}$$

where R is the distance between them and ε_o is the electric constant (otherwise known as the permittivity of free space). The force is repulsive (i.e., pushing them apart) if the signs of the charges are the same (which is why we have included the minus sign in Eq. (B.15)) and attractive if the charges have opposite signs. This relation, Eq. (B.15), is known as *Coulomb's law*.

Suppose initially that these two charges are a very large distance L apart. Then, we imagine bringing the two charges together so that they end up with a separation distance r. The energy we require to do that is the integral of the product of force times distance from separation L to separation r; that is,

$$\phi(r) = \int_L^r F dz = \int_L^r -\frac{Q_1 Q_2}{4\pi\varepsilon_o z^2} dz = \frac{Q_1 Q_2}{4\pi\varepsilon_o}\left(\frac{1}{r} - \frac{1}{L}\right) \tag{B.16}$$

Because L is chosen to be very large (or if we imagine the limit as the charges start infinitely far apart), then we can write that the electrostatic potential energy associated with the two charges being a distance r apart is

$$\phi(r) = \frac{Q_1 Q_2}{4\pi\varepsilon_o r} \tag{B.17}$$

We are always free to choose the zero reference for any potential energy and, implicitly, here we are choosing the zero potential energy when the charges are very far apart.

B.3 Frequency units

Frequency is the number of complete cycles per unit time (i.e., cycles per second) of an oscillation and can also be described with the unit Hz (Hertz), which is equivalent to cycles per second. It is often indicated with the symbol f or, more commonly in physics and quantum mechanics in particular, with the Greek letter ν ("nu").

Angular frequency, often indicated with the Greek letter ω ("omega"), is equal to $2\pi\nu$ for a frequency ν. It is a common notation because it saves us writing the factor 2π so many times in the algebra of oscillating systems. In fundamental angular units, by definition there are 2π radians in a circle (equivalent to 360 degrees). When thinking about oscillations, we are often thinking about something rotating in a circular motion. When viewed on an Argand diagram in the complex plane (see Appendix A), the function $\exp(i\omega t)$, for example, corresponds to a point moving around a circle of unit radius at a rate of ω radians per unit time t. Angular frequency can, therefore, be thought of as the number of radians per second and sometimes the units for it are written as rad/s. Angular frequency is also correctly and equivalently expressed in units of s^{-1} ("per second"). It is *never* expressed in Hz.

B.4 Waves and diffraction

We are familiar from daily life with waves propagating on the surface of water, for example. Classical waves like water waves can move in some direction even though there is no overall movement of the medium (e.g., water) itself in that direction. They can carry energy and, especially for acoustic or electromagnetic waves, information. Waves have *wave-fronts* or *phase-fronts* that are, loosely, the peaks that appear to be moving in some direction. The "height" of a wave at a given time and a given point in space is called the *wave amplitude*.

The most ideal and simplest of waves are *plane waves*, in which the wave-fronts or phase-fronts are straight lines in the case of waves on a two-dimensional surface like water or are plane surfaces in the case of waves in three dimensions, like acoustic waves in air or electromagnetic waves in a vacuum or air. In the simplest cases, the phase-fronts are perpendicular to the direction of propagation of the waves.

Ideal classical waves in some uniform medium can move essentially without change in their shape in time (e.g., some sort of pulse shape), in which case we say the wave propagation has no *dispersion* and would behave the same way regardless of the wave amplitude (in which case the wave is behaving in a *linear* fashion). Such ideal classical waves can usually be quite well described using the wave equations presented mathematically in Appendix A.

It is often useful to consider *monochromatic waves*, which are waves in which every point on the wave is oscillating at the same frequency. In that case, factoring out the sinusoidal oscillation in time, the remaining spatial variation of the waves can usually be described by a Helmholtz wave equation, as discussed in Appendix A.

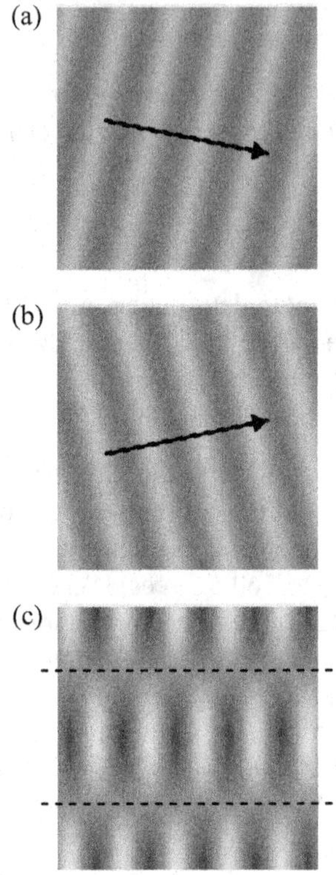

Fig. B.2. Interference of two incident waves. (a) Snapshot of one incident plane wave on its own, with its propagation direction shown by the arrow. (b) Similar snapshot of a second plane wave on its own. (c) Result when both waves are present at once. Along the dashed lines, there is never any net wave amplitude – the waves interfere destructively to cancel out. At other points between the lines, constructive interference is seen, with larger amplitudes than for either wave on its own.

Interference and standing waves

A particularly important aspect of the behavior of waves is that when two or more waves propagate into the same space, they show *interference*. This is sketched in Fig. B.2. Here, two monochromatic waves of the same frequency are propagating in the same space; *destructive interference* is seen with exact cancellation of the waves along the dashed lines, and *constructive interference* is also seen with larger total amplitudes at other points between the lines.

In general, with classical waves, the power per unit area or *intensity* in the wave is typically proportional to the time average of the square of the amplitude of the wave. Note that even though each of the two waves has a positive, nonzero intensity, the net result of adding the two waves can be to generate regions of space where the intensity is higher than the sum of the two individual intensities and other regions of space where it is lower than either intensity, including even regions where the intensity is zero. Hence, with waves one cannot simply add intensities. This point may be obvious for classical waves but the analogous process with quantum mechanical waves or, more generally, quantum mechanical amplitudes, is extremely important and is responsible for many behaviors in quantum mechanics that are very surprising.

The interference pattern of these two waves in Fig. B.2(c) is an example of a *standing wave* pattern. Even though both of the waves that have created the pattern are themselves propagating waves, the pattern in the vertical direction of Fig. B.2(c) appears to be "standing" in space – it will not move in any direction – even though the individual peaks and valleys in the constructive interference regions oscillate and move to the right.

Another very simple and common example is standing waves in the one-dimensional case, as would be found on waves on a string. In standing waves on a string, the waves are reflected off the two fixed ends of the string, therefore guaranteeing equal and oppositely propagating waves to give the resulting standing wave. For example, adding equal and oppositely propagating cosine waves in one dimension gives

$$A\cos(\omega t + kz) + A\cos(\omega t - kz) = 2A\cos(\omega t)\cos(kz) \tag{B.18}$$

(where we have used the identity $\cos a + \cos b = 2\cos[(a+b)/2]\cos[(a-b)/2]$), which has the form of a standing wave in the z direction of form $\cos(kz)$, with each point oscillating at frequency ω. Such a wave has zeros or *nodes* at the points where $\cos(kz)$ is zero (i.e., at points $z = \pi/2k, 3\pi/2k, 5\pi/2k, \ldots$ and so on) and maxima or *antinodes* in the amplitude of the oscillations at the points where $\cos(kz)$ is $+1$ or -1 (i.e., at points $z = 0, \pi, 2\pi, \ldots$ and so on). We can, incidentally, use the same terminology to refer to the positions of the dashed lines in Fig. B.2(c) as *nodal lines*, which are entire lines at which the amplitude is zero at all times.

Diffraction

One simple view of light is to consider light as being made up of rays, which are straight lines whose direction only changes when they encounter some mirror, prism, or lens. In that picture, a parallel bunch of rays would proceed in parallel forever. Such a ray picture can be a useful approximate description if every object of interest is very much larger than the wavelengths involved and if we do not try to propagate too far. With objects of scales comparable to the wavelengths involved, or where we look very closely at objects within lengths of the order of a few wavelengths or shorter, or where we try to propagate light beams over long distances, the ray picture breaks down and we have to use a wave equation to describe what is happening.

Very loosely, diffractive effects could be described as all those phenomena of waves that cannot be explained by a ray-like picture. Diffraction is certainly the unavoidable spreading of light beams or of light from apertures and it is certainly what limits the size of the smallest spot of light we can form with a given wavelength. Diffraction applies to all forms of waves obeying the wave equations of Appendix A. The spreading from diffraction is always less for smaller wavelengths (or higher frequencies) or for larger beams or apertures. Diffraction explains, for example, why in an audio system we only need one low-frequency loudspeaker (the "woofer") but we use two high-frequency loudspeakers (the "tweeters"). The wavelength of low-frequency sound is so large that it diffracts in all directions from any reasonable size of loudspeaker, but high-frequency sound is very directional, even from quite small loudspeakers. Hence, the directional information for stereo perception of sound can be adequately conveyed from the two small "tweeters".

We can write down a simple formula for the minimum total angle θ of spreading of a beam from an aperture, for example, which is approximately

$$\theta \sim \lambda / d \tag{B.19}$$

where d is the width of the aperture. The precise proportionality constant here depends on one's definition of the "width" of the beam (because the beam will not have hard edges) and on the precise geometry (e.g., slit, circular hole, square hole) of the aperture, but the proportionality constant is generally of order unity.

In general, diffraction is not simply one effect that one can calculate from a formula and, as mentioned previously, is instead really all the effects that occur when solving the wave equation and that could not be explained by rays, especially those involving apertures, edges, and periodic arrays of scattering objects. One might call interference a separate effect, though interference is involved all the time in diffractive phenomena.

All the effects of diffraction are automatically included if we solve the wave equation for the situation of interest. In general, however, diffraction effects can be quite difficult to calculate exactly because wave equations are difficult to solve exactly when one tries to include boundary conditions for objects. One simple principle that can often give a useful first approximation to diffractive effects is *Huygens's Principle*, which states that one can calculate the next wave-front by presuming that each point on the previous wave-front is a source of spherically expanding waves. Though not absolutely exact,[6] it is actually quite an accurate principle (especially if we choose to make some minor adjustments to it) and we use it in the main text to discuss elementary diffractive effects.

Incidentally, diffraction should not be confused with *refraction*, which is the bending of light rays when they enter materials with different refractive index. Refraction exists both in the simple "ray" picture and in the more complete wave picture. Diffraction should also not be confused with *dispersion*. Dispersion is most commonly the different wave velocity seen for different wavelengths, frequencies, or colors of waves as a result of wavelength or frequency dependence of material properties. A prism shows dispersion because it separates out the different colors of light to different angles. Such dispersion can be understood from a ray picture of light as coming from the different refractive indices seen by the rays of different colors.[7]

---

[6] Huygens's principle was originally published in 1690 (a publication date that was substantially delayed because Huygens had wanted to translate the work from French to Latin before publishing it). Huygens's principle is not quite correct because it predicts backward waves that do not exist. This problem was addressed in an *ad hoc* fashion by Fresnel and the resulting Huygens–Fresnel theory of diffraction works very well for many simple wave problems. A rigorous solution, involving two different kinds of sources on the wavefront, is given by Helmholtz's (and later Kirchhoff's) integration of the wave equation. More recently (D. A. B. Miller, *Optics Lett.* **16**, 1370–1372, (1991)) it has been shown that a different kind of single source works very well, restoring Huygens's original concept. The simple spherical wave is, though, a good approximation for small angles if we are only interested in the wave propagation in the forward direction.

[7] It is also true that structures made from nondispersive materials can show dispersion, especially if the material is patterned on a scale comparable to the wavelength, as in a diffraction grating, for example; in that case, diffraction can be viewed as giving rise to dispersion. A complete wave picture automatically includes all dispersion and diffraction effects, including dispersions other than angular ones, such as different phase shifts and time delays for different frequencies or pulse shapes.

Appendix C

Vector calculus

In this appendix, we summarize key definitions and relations in vector calculus. Some of these (i.e., the gradient and the Laplacian in Cartesian coordinates) have already been briefly introduced in Appendix A, Section A.6, but we give a uniform and extended set of definitions here.

C.1 Vector calculus operators

Del or nabla operator

In vector calculus, it is convenient to think of the ∇ operator, known as *del* or *nabla*, as

$$\nabla \equiv \mathbf{i}\frac{\partial}{\partial x} + \mathbf{j}\frac{\partial}{\partial y} + \mathbf{k}\frac{\partial}{\partial z} \tag{C.1}$$

This gives a useful shorthand for writing down the other important vector calculus operators in Cartesian coordinates.

Gradient

The *gradient* operator operates on a scalar function $f(x, y, z)$ to give a vector whose magnitude and direction are the slope or gradient of the scalar function at the point of interest. In Cartesian coordinates

$$\text{grad}\, f = \nabla f = \mathbf{i}\frac{\partial f}{\partial x} + \mathbf{j}\frac{\partial f}{\partial y} + \mathbf{k}\frac{\partial f}{\partial z} \tag{C.2}$$

Laplacian

The *Laplacian* operator, also known as *del squared*, operates on a scalar function, giving a scalar result. It occurs in many physical problems, including electromagnetism and quantum mechanics. It is written in Cartesian coordinates as

$$\nabla^2 f = \frac{\partial^2 f}{\partial x^2} + \frac{\partial^2 f}{\partial y^2} + \frac{\partial^2 f}{\partial z^2} \tag{C.3}$$

It is also meaningful to have the operator $\nabla \cdot \nabla$, sometimes also written as ∇^2, operate on a vector function, in which case, in Cartesian coordinates, we have

$$(\nabla \cdot \nabla)\mathbf{F} = \mathbf{i}\frac{\partial^2 F_x}{\partial x^2} + \mathbf{j}\frac{\partial^2 F_y}{\partial y^2} + \mathbf{k}\frac{\partial^2 F_z}{\partial z^2} \tag{C.4}$$

that is, it can be thought of as "del squared" operating on each of the vector components separately, leaving their vector directions.

Vector field

If we can associate a vector quantity **F** with every point (x, y, z) in space, then we can call **F** a *vector field.*

There are many examples of vector fields in physics. Many are associated with the vector representing a force. For example, the gravitational field is the force vector (i.e., the magnitude and the direction) on a hypothetical mass of 1 kg at any point in space and the electric field **E** is the force vector on a hypothetical charge of +1 Coulomb. Other examples are associated with flow. We can have a velocity vector field **v** that describes the velocity (magnitude and direction) of flow of some fluid, such as water or air. We can have flux fields, where flux is generally the "amount" of something,[1] such as mass in a fluid, crossing unit area per second, where the vector direction of the field is the direction of the flow. A common example of a "flux" vector field is electric current density **J** (A/m^2). Another class of examples is particle fluxes,[2] such as the flux of electrons (i.e., number of electrons per unit area per second)) or atoms.

Divergence

In Cartesian coordinates, the divergence of a vector **F** is defined as

$$\operatorname{div}\mathbf{F} = \nabla\cdot\mathbf{F} = \frac{\partial F_x}{\partial x} + \frac{\partial F_y}{\partial y} + \frac{\partial F_z}{\partial z} \tag{C.5}$$

We can visualize this Cartesian version in terms of the flux **F** of some quantity such as mass or charge through a small cuboidal box of sides δx, δy, and δz centered at some point (x_o, y_o, z_o) as shown in Fig. C.1. Because **F** is the vector representing the flow of the quantity per unit area, an amount $F_x(x_o + \delta x/2, y_o, z_o)\delta y\delta z$ leaves the box on the right. Here, we note that the area of the right face of the box is $\delta y\delta z$. The quantity $F_x(x_o + \delta x/2, y_o, z_o)\delta y\delta z$ is the component of the flux in the x-direction multiplied by the area perpendicular to the x-direction. We can also think of this quantity as

$$F_x\left(x_o + \frac{\delta x}{2}, y_o, z_o\right)\delta y\delta z \equiv \mathbf{F}\left(x_o + \frac{\delta x}{2}, y_o, z_o\right)\cdot\delta\mathbf{A}_{yz} \tag{C.6}$$

where $\delta\mathbf{A}_{yz}$ is a vector whose magnitude is the area of the right surface of the box and whose direction is outward from the box. The amount arriving into the box on the left face is similarly $F_x(x_o - \delta x/2, y_o, z_o)\delta y\delta z$, so the net amount leaving the box through the left or right faces is

---

[1] Historically, electric and especially magnetic fields are sometimes described in texts as *fluxes*, though in fact there is no specific physical quantity that we use for any other purpose that is flowing here. Such fields are sometimes visualized with *flux lines*, in which case the number of such lines crossing unit area perpendicular to the lines is the magnitude of the field at that point.

[2] Such particle fluxes often use **j** as the letter to indicate them, though this should not be confused with electric current (nor with the unit vector in the z direction). The flux may well be carrying electric current, but the particle flux is dealing with numbers not charge.

$$
\begin{aligned}
&F_x\left(x_o+\frac{\delta x}{2},y_o,z_o\right)\delta y\delta z-F_x\left(x_o-\frac{\delta x}{2},y_o,z_o\right)\delta y\delta z \\
&=\frac{F_x\left(x_o+\frac{\delta x}{2},y_o,z_o\right)-F_x\left(x_o-\frac{\delta x}{2},y_o,z_o\right)}{\delta x}\delta x\delta y\delta z \qquad (C.7)\\
&\simeq\frac{\partial F_x}{\delta x}\delta x\delta y\delta z
\end{aligned}
$$

where we are assuming a very small box. We can repeat this analysis for each of the other two pairs of faces; so, adding all three such equations of the form of Eq. (C.7) together, we can write for the total amount of flow leaving the small box, per unit volume of the box (i.e., dividing by $\delta V = \delta x\delta y\delta z$) the expression Eq. (C.5).

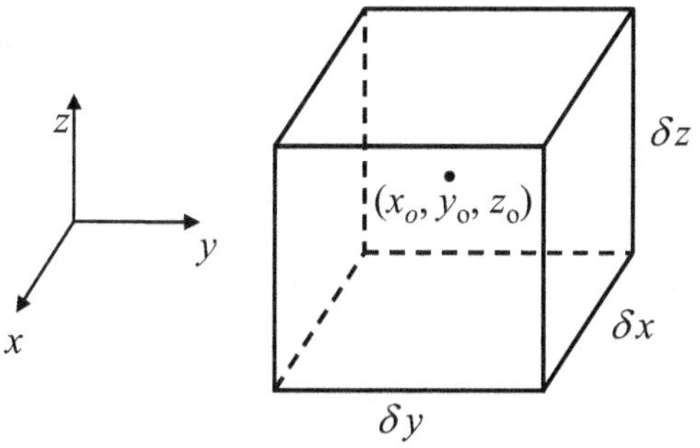

Fig. C.1. Illustration of divergence as net flow in and out of a small box.

We can also write this result in terms of the dot product with the area elements. Note that, generally, we define area vectors as having direction pointing outward to a surface, so the area vector for the left wall of the box points in the negative x-direction. Hence, the flux entering the box through the left wall can be written as $-\mathbf{F}(x_o-\delta x/2, y_o, z_o)\cdot\delta\mathbf{A}_{yz}$. This sign choice conveniently allows us to add up all such flux contributions for the six surfaces to obtain the same total as we would by adding up the three versions of Eq. (C.7) for the three different coordinate directions. We can then write the sum formally as a surface integral, and formally take the limit of a small volume to obtain the most general definition of the divergence

$$\nabla\cdot\mathbf{F}\equiv\lim_{\delta V\to 0}\frac{\oiint_S \mathbf{F}\cdot d\mathbf{A}}{\delta V} \qquad (C.8)$$

where in the integral, we mean here that we are integrating over the closed surface S that bounds the volume δV. In general, the divergence is essentially the total flow out of a very small volume around the point of interest, divided by that volume.

Note that the result of the divergence of a vector is a scalar quantity.

A vector field that has zero divergence is sometimes called *solenoidal* or *divergenceless*. Examples include the magnetic field **B**, which is divergenceless because there are thought to be no magnetic monopoles and the mass flow of an incompressible fluid (because the fluid cannot be compressed, the amount of mass in any small volume cannot be changed and so the amount of mass that leaves the volume must equal the amount that arrives into the volume).

Gauss's theorem

This theorem, also known as the *divergence theorem*, states that

$$\int_V (\nabla \cdot \mathbf{F})\, dV = \oiint_S \mathbf{F}.d\mathbf{A} \tag{C.9}$$

where S is the surface bounding the volume V, where V is now not necessarily a small volume. This theorem means that the total surface flow out of some volume V (and, therefore, through its surface S) is equal to the volume integral of the divergence. Given the definitions of the divergence, it can be visualized in terms of a set of small volumes or "bricks" that constitute the total volume, with the flows in and out of the adjacent walls of the bricks cancelling each other and, therefore, leaving the total net flow out of the volume depending only on the flow in and out of the "last" or outermost surfaces. Hence, the addition of all of the divergences for the infinitesimal volumes is the same as the surface integral of the flow.

Continuity equation

If we have a conserved quantity that can flow, such as mass in a fluid, then it must obey a continuity equation; that is, if the flux of the quantity is $\mathbf{j}$ and the density of the quantity is ρ, then

$$\nabla \cdot \mathbf{j} = -\frac{\partial \rho}{\partial t} \tag{C.10}$$

This merely states that the amount of the quantity (e.g., mass) that flows out of a small volume per unit time must equal the reduction of the amount of the quantity in the small volume.

Any particle flux field will obey a continuity equation if the particles cannot be created or destroyed (i.e., the number of particles is conserved); in this case, ρ is the particle density (number of particles per unit volume). Electrical current densities obey continuity equations because charge cannot be created or destroyed, in which case ρ is the charge density.

Curl

In Cartesian coordinates

$$\operatorname{curl}\mathbf{F} \equiv \nabla \times \mathbf{F} = \left(\frac{\partial F_z}{\partial y} - \frac{\partial F_y}{\partial z}\right)\mathbf{i} + \left(\frac{\partial F_x}{\partial z} - \frac{\partial F_z}{\partial x}\right)\mathbf{j} + \left(\frac{\partial F_y}{\partial x} - \frac{\partial F_x}{\partial y}\right)\mathbf{k} \tag{C.11}$$

or in the determinant shorthand form, which is often easier to remember for curls

$$\nabla \times \mathbf{F} = \begin{vmatrix} \mathbf{i} & \mathbf{j} & \mathbf{k} \\ \dfrac{\partial}{\partial x} & \dfrac{\partial}{\partial y} & \dfrac{\partial}{\partial z} \\ F_x & F_y & F_z \end{vmatrix} \tag{C.12}$$

The curl can be visualized by considering the work done on an object pushed by a force $\mathbf{F}$ as we move the object around some very small closed path C. This can be illustrated in two dimensions (see Fig. C.2). Here, we show a closed path made of four straight-line segments aligned with the x and y axes. On any given path segment, we can define a path vector whose length is the length of the path and whose direction is along the path in a sense given by a chosen direction (here, anti-clockwise) around this whole closed path.

For any given path segment, the work done is given by the dot product $\mathbf{F} \cdot \delta\mathbf{s}$ of the force $\mathbf{F}$ and the path segment vector $\delta\mathbf{s}$. For example, on the horizontal path, the path element vector is of length δx and is in the +ve x direction; that is, the path vector can be written $\delta x\, \mathbf{i}$. Hence, the work done on this path is $\mathbf{F}(x_o, y_o - \delta y/2)\cdot\mathbf{i}\delta x = F_x(x_o, y_o - \delta y/2)\,\delta x$. We can add the work done on all the other paths, noting that the top path is heading in the negative x-direction and so will have a negative sign as we add these up. Similarly, the left path is heading in a negative y-direction so it will have a negative sign in our sum. Hence, for the total work done in moving an object around this path, we have

$$\begin{aligned}\delta W \simeq\ & F_x\left(x_o, y_o - \frac{\delta y}{2}\right)\delta x + F_y\left(x_o + \frac{\delta x}{2}, y_o\right)\delta y \\ & - F_x\left(x_o, y_o + \frac{\delta y}{2}\right)\delta x - F_y\left(x_o - \frac{\delta x}{2}, y_o\right)\delta y\end{aligned} \tag{C.13}$$

which we can rewrite as

$$\delta W \simeq \frac{\partial F_y}{\partial x}\delta x \delta y - \frac{\partial F_x}{\partial y}\delta x \delta y \tag{C.14}$$

Hence, if we consider the curl now as being a vector perpendicular to the area (i.e., here, the xy plane) in the sense given by the right-hand rule for the path (here, therefore, pointing out of the paper) and with a length given by the "work" δW done in taking an object around this path, dividing finally by the area $\delta A = \delta x \delta y$ enclosed by the path, we would have the vector quantity, here pointing in the z direction

$$\mathbf{w} = \left(\frac{\partial F_y}{\partial x} - \frac{\partial F_x}{\partial y}\right)\mathbf{k} \tag{C.15}$$

We can see now that this is simply the z-direction component of the three-dimensional formula for the curl in Eq. (C.11), consistent with this being the curl of a field that varies only in the xy plane. Hence, the curl of $\mathbf{F}$ is the work done by the "force" $\mathbf{F}$ on an object in taking an object around a small closed path, divided by the area enclosed by the path and given a vector direction determined by the right-hand rule for the path.

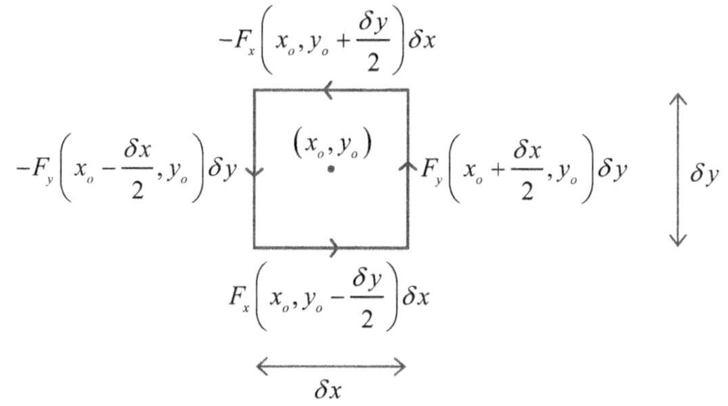

Fig. C.2. Illustration of the contributions to the work done around a small rectangular path by a force $\mathbf{F}$ on an object taken around the path shown.

The most general definition of the curl is in terms of the path integral around a closed path and can be stated

$$(\nabla \times \mathbf{F}).\hat{\mathbf{n}} \equiv \lim_{\delta A \to 0} \frac{\oint_C \mathbf{F}\cdot d\mathbf{s}}{\delta A} \tag{C.16}$$

where the integral here is around the closed path C that surrounds the area δA and $\hat{\mathbf{n}}$ is a unit vector perpendicular to the area and in the direction given by the right-hand rule and the path C.

Stokes theorem

For a closed path S that is the perimeter of some surface A

$$\oint_C \mathbf{F} \cdot d\mathbf{s} = \iint_A (\nabla \times \mathbf{F}) \cdot d\mathbf{A} \tag{C.17}$$

Here, the symbols $\oint_C d\mathbf{s}$ signify an integral taken along the closed path C that surrounds the surface A, with the chain of infinitesimal vectors $d\mathbf{s}$ forming the path. The symbol $\iint_A d\mathbf{A}$ similarly indicates a surface integral, in which the infinitesimal tiles $d\mathbf{A}$ stitched together fill the whole surface A. The surface need not be flat. The vector direction assigned to the tiles is perpendicular to the surface and is in the same sense as the right-hand rule gives as we move around the path S. That is, if we are looking at the path S in such a way that going around the path corresponds to a clockwise direction, then the sense of the elements $d\mathbf{A}$ is pointing outward from the far side of the surface.

It can be conceptually useful to visualize the Stokes theorem in terms of the curls as little loops that are stitched together, with their adjacent edges canceling, so that the final sum effect of all of them corresponds to only the effect of the final outer edge.

This theorem is also sometimes known as the *Kelvin–Stokes theorem*.

C.2 Spherical polar coordinates

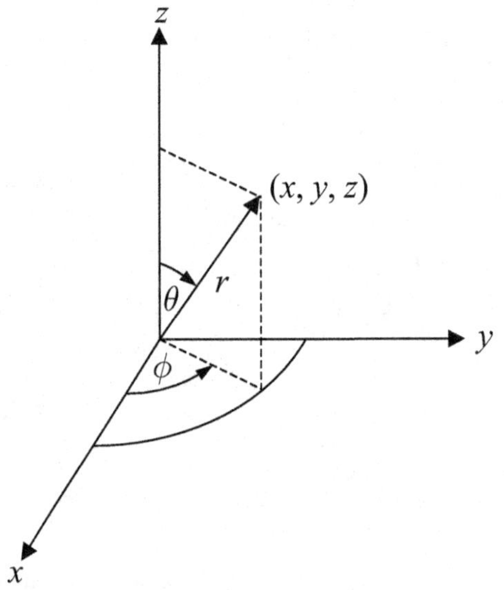

Fig. C.3. Coordinate definitions for spherical (r, θ, ϕ) coordinates.

In spherical polar coordinates, a point in space is defined in terms of its distance r from the origin of the coordinate system and in terms of two angles, θ and ϕ. We have chosen here to show the notation more common in physical science and technology for spherical polar coordinates, in which the θ is the zenith angle (i.e., the angle from the z axis) and ϕ is the azimuthal angle (i.e., the angle from the x axis, in the x-y plane).[3] It is also common to use the opposite assignment of θ and ϕ for the angles. The reader, therefore, will have to look carefully when spherical polar coordinates are being used to see which convention is implied.

In spherical polar coordinates, we can define orthogonal unit vectors $\hat{\mathbf{r}}$ (in the direction of radius outward from the origin of the coordinate system), $\hat{\theta}$ and $\hat{\phi}$, where the directions of these angular unit vectors can be chosen to be the directions shown by the arrows on the angles in Fig. C.3; in this order, they give a right-handed coordinate system. Note that these vectors are always orthogonal to one another but, unlike in Cartesian coordinates, for different points in space, these unit vectors have generally different directions.

Gradient

$$\nabla f = \frac{\partial f}{\partial r}\hat{\mathbf{r}} + \frac{1}{r}\frac{\partial f}{\partial \theta}\hat{\theta} + \frac{1}{r\sin\theta}\frac{\partial f}{\partial \phi}\hat{\phi} \tag{C.18}$$

Laplacian

$$\nabla^2 f = \frac{1}{r^2}\frac{\partial}{\partial r}\left(r^2\frac{\partial f}{\partial r}\right) + \frac{1}{r^2\sin\theta}\frac{\partial}{\partial\theta}\left(\sin\theta\frac{\partial f}{\partial\theta}\right) + \frac{1}{r^2\sin^2\theta}\frac{\partial^2 f}{\partial\phi^2} \tag{C.19}$$

Divergence

$$\nabla\cdot\mathbf{F} = \frac{1}{r^2}\frac{\partial}{\partial r}\left(r^2 F_r\right) + \frac{1}{r\sin\theta}\frac{\partial}{\partial\theta}\left(F_\theta\sin\theta\right) + \frac{1}{r\sin\theta}\frac{\partial F_\phi}{\partial\phi} \tag{C.20}$$

Curl

$$\nabla\times\mathbf{F} = \begin{vmatrix} \dfrac{\hat{\mathbf{r}}}{r^2\sin\theta} & \dfrac{\hat{\theta}}{r\sin\theta} & \dfrac{\hat{\phi}}{r} \\ \dfrac{\partial}{\partial r} & \dfrac{\partial}{\partial\theta} & \dfrac{\partial}{\partial\phi} \\ F_r & rF_\theta & r\sin\theta F_\phi \end{vmatrix} \tag{C.21}$$

Volume integral

The infinitesimal volume element for volume integrals in spherical coordinates is $r^2\sin\theta dr d\theta d\phi$ and the angular ranges for integration are 0 to π for the angle θ and 0 to 2π for the angle ϕ. Hence, a volume integral over a sphere of radius r_o is of the form

$$I = \int_{r=0}^{r_o}\int_{\theta=0}^{\pi}\int_{\phi=0}^{2\pi} r^2\sin\theta dr d\theta d\phi \tag{C.22}$$

If there is no other integrand to include in this integral, this integral then correctly evaluates to the volume of a sphere of radius r_o, which is $(4/3)\pi r_o^3$.

---

[3] To remember which angle is which between zenith and azimuthal, it may be simplest to note that zenith generally refers to the direction directly overhead or upward (here, the z axis, as in z for zenith) or, in common English, the peak or highest point or "best." Azimuthal angle is simply the other angle.

C.3 Cylindrical coordinates

In cylindrical polar coordinates, a point in space is defined in terms of its projected distance ρ in the x-y plane from the origin of the coordinate system, its projected distance from the origin in the z direction, and the angle ϕ of the radius ρ from the x axis, as shown in Fig. C.4. Note that ρ here is different from the radius r in spherical polars; ρ is the radius *in the plane*. For simplicity, we use the same angle convention in cylindrical coordinates (ρ, ϕ, z) as for spherical ones. Our cylindrical coordinate formulae are, therefore, expressed using ϕ for the azimuthal angle.[4]

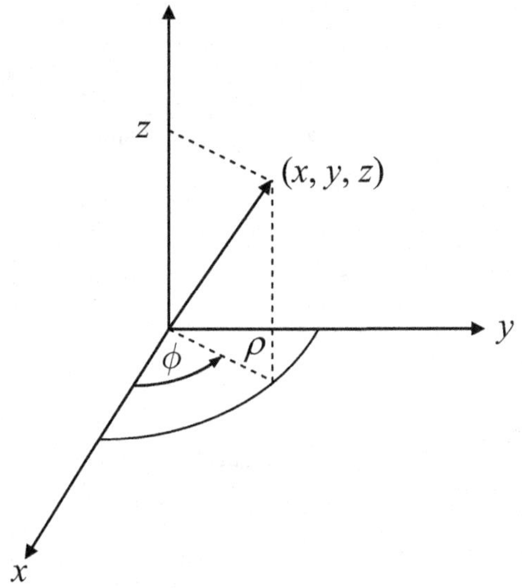

Fig. C.4. Coordinate definitions for cylindrical (ρ, ϕ, z) coordinates.

In cylindrical polar coordinates, we can define orthogonal unit vectors $\hat{\rho}$ (in the direction of radius outward in the x-y plane from the origin of the coordinate system), $\hat{\phi}$, where the direction of this angular unit vector can be chosen to be the direction of the arrow on this angle in Fig. C.4, and $\hat{\mathbf{z}}$, in the direction of the z axis[5]; in this order, they give a right-handed system. Note that these vectors are always orthogonal to one another but, unlike in Cartesian coordinates, for different points in space, the $\hat{\rho}$ and $\hat{\phi}$ unit vectors have generally different directions.

Gradient

$$\nabla f = \frac{\partial f}{\partial \rho}\hat{\rho} + \frac{1}{\rho}\frac{\partial f}{\partial \phi}\hat{\phi} + \frac{\partial f}{\partial z}\hat{\mathbf{z}} \tag{C.23}$$

---

[4] It is more common in cylindrical coordinates to use θ for the (azimuthal) angle. Because there is only one angle involved in cylindrical coordinates, there is no confusion caused by the use of ϕ instead of θ for cylindrical coordinates, however, and it may be less confusing here in remembering the spherical coordinates to use ϕ for the azimuthal angle in both cases.

[5] We use $\hat{\mathbf{z}}$ here rather than $\mathbf{k}$ for the unit vector in the z direction because we are also using "hatted" unit vectors for the other directions.

Laplacian

$$\nabla^2 f = \frac{1}{\rho}\frac{\partial}{\partial \rho}\left(\rho \frac{\partial f}{\partial \rho}\right) + \frac{1}{\rho^2}\frac{\partial^2 f}{\partial \phi^2} + \frac{\partial^2 f}{\partial z^2} \tag{C.24}$$

Divergence

$$\nabla \cdot \mathbf{F} = \frac{1}{\rho}\frac{\partial\left(\rho F_\rho\right)}{\partial \rho} + \frac{1}{\rho}\frac{\partial F_\phi}{\partial \phi} + \frac{\partial F_z}{\partial z} \tag{C.25}$$

Curl

$$\nabla \times \mathbf{F} = \begin{vmatrix} \dfrac{\hat{\rho}}{\rho} & \hat{\phi} & \dfrac{\hat{\mathbf{z}}}{\rho} \\ \dfrac{\partial}{\partial \rho} & \dfrac{\partial}{\partial \phi} & \dfrac{\partial}{\partial z} \\ F_\rho & rF_\phi & F_z \end{vmatrix} \tag{C.26}$$

Volume integral

The infinitesimal volume element for volume integrals in cylindrical coordinates is $rdrd\phi dz$ and the angular range for integration is from 0 to 2π for the angle ϕ. Hence, a volume integral over a cylinder of radius r_o and height z_o is of the form

$$I = \int_{r=0}^{r_o}\int_{\phi=0}^{2\pi}\int_{z=0}^{z_o} rdrd\phi dz \tag{C.27}$$

If there is no other integrand to include in this integral, this integral then correctly evaluates to the volume of a cylinder of radius r_o and height z_o, which is $\pi r_o^2 z_o$.

C.4 Vector calculus identities

The following are important vector identities often used in electricity and magnetism and elsewhere

$$\nabla \cdot \left(\nabla \times \mathbf{F}\right) = 0 \tag{C.28}$$

$$\nabla \times \nabla f = 0 \tag{C.29}$$

The identity in Eq. (C.29) means, incidentally, that any field that can be written as $\mathbf{F} = \nabla f$ – that is, that can be derived from the gradient of a scalar function $f(x, y, z)$ of position only (a *scalar potential*) – has a curl of zero and is, therefore, by definition *irrotational*, which, in turn, means that the integral of the field around a closed path is zero, which, in turn, means that the field is also *conservative*.

$$\nabla \times \left(\nabla \times \mathbf{F}\right) = \nabla\left(\nabla \cdot \mathbf{F}\right) - \left(\nabla \cdot \nabla\right)\mathbf{F} \tag{C.30}$$

The identity in Eq. (C.30) is used in the derivation of wave equations in electromagnetism.

$$\nabla \cdot \left(\mathbf{F} \times \mathbf{G}\right) = -\mathbf{F} \cdot \left(\nabla \times \mathbf{G}\right) + \mathbf{G} \cdot \left(\nabla \times \mathbf{F}\right) \tag{C.31}$$

The identity in Eq. (C.31) is used in the derivation of the Poynting vector in electromagnetism.

Another useful algebraic identity is

$$\begin{aligned}\Psi\nabla^2\Psi^* - \Psi^*\nabla^2\Psi &= \Psi\nabla^2\Psi^* + \nabla\Psi\nabla\Psi^* - \nabla\Psi\nabla\Psi^* - \Psi^*\nabla^2\Psi \\ &= \nabla\cdot\left(\Psi\nabla\Psi^* - \Psi^*\nabla\Psi\right)\end{aligned} \tag{C.32}$$

Appendix D

Maxwell's equations and electromagnetism

In this appendix, we summarize some of the more advanced concepts in electromagnetism that are needed at various points in the book.

D.1 Polarization of a material

When we apply an electric field to a material, we pull on the electrons and, in the opposite direction, on the positively charged nuclei of the atoms. The resulting movement[1] of the electrons and nuclei in response to this electric field is the *polarization* of the material.[2]

To be more precise in the definition of polarization, we should first discuss the idea of *dipole moment*. A typical way in which we view a dipole moment is to imagine that we have a pair of equal and opposite charges, of value $+q$ and $-q$. We can separate them by some distance d – for example in the z direction – pushing the $+q$ charge to larger (or more positive) z and the $-q$ charge to smaller (or more negative) z. In general, any such pair of equal and opposite charges separated by some distance is called a *dipole*. Such a separation is exactly what we would get if we applied an electric field that pointed in the +ve z direction. Then, we can say that we have a dipole moment, of magnitude $\mu = qd$, pointed in the +ve z direction. In general, if the positive charge is displaced from the negative charge by a vector amount $\mathbf{r}$, then the dipole moment is

$$\boldsymbol{\mu} = q\mathbf{r} \tag{D.1}$$

Note that the dipole moment $\boldsymbol{\mu}$ is a vector quantity.

---

[1] It is also possible to have a "built-in" polarization in a material that does not result from any applied external field. This happens in, for example, electrets that can hold a polarization at least semi-permanently (electrets are commonly used in inexpensive microphones), ferroelectric materials that can have a permanent polarization, and the spontaneous polarization of materials with specific kinds of crystal symmetries, e.g., GaN in its wurtzite crystal phase.

[2] We should state explicitly here that the term *polarization* in electromagnetism unfortunately also has a second quite different meaning, which is the vector direction of the electric (or sometimes magnetic) field in an electromagnetic wave. We discuss that second meaning later in this appendix. We can distinguish the two, if necessary, by talking about the polarization of the material in the first case and the polarization of the wave in the second case.

Incidentally, though we do typically view a dipole moment in terms of equal and opposite charges, it is not actually necessary to have two kinds of charge in order to have a dipole moment. We could view the movement of a single charge of value q by a vector amount $\mathbf{r}$ as corresponding to the addition of a dipole; we can put the $-q$ end of the dipole on top of the original position of the charge, thus canceling the original charge, and the $+q$ end of the dipole now sits at the new position of the charge. As far as the electrostatics are concerned, there is no difference between moving the charge q by $\mathbf{r}$ or creating the dipole $\boldsymbol{\mu} = q\mathbf{r}$.

In an actual material, we do not have just two simple point charges. Instead, we have many small elements of charge, possibly a continuous distribution in fact, each element of which is being moved as we apply the electric field. Often, we do not even know exactly how the charge is distributed to start with in the material. All we really know is that, at least when looked at on some scale well above the size of the atoms, there is some dipole moment induced when we apply an electric field. One of the very convenient aspects about dipole moments is that we can add them as vectors to get one larger effective dipole moment. In fact, at our macroscopic scale, we really have no idea what all the small dipole moments are – all we see is one effective dipole moment for some small volume. For some volume that is large compared to the atoms but small compared to macroscopic dimensions, we can simply consider this vector sum of dipole moments as being the effective dipole moment of that volume. It is then useful to define another quantity, which is the dipole moment per unit volume, and that vector quantity is called the polarization **P**.

Often, especially for small applied electric fields **E**, the polarization **P** is approximately proportional to the electric field and then we can define a proportionality constant χ (the Greek letter "chi") called the *susceptibility*.[3] Formally, we write

$$\mathbf{P} = \varepsilon_o \chi \mathbf{E} \tag{D.2}$$

This notation is also useful when we consider nonlinear materials, in which case we may expand χ in a power series. The constant ε_o is known as the *permittivity of free space* or the *electric constant*. It has no real physical meaning and is there for historical reasons associated with the definition of units.

In electromagnetism, another field **D**, the *electric displacement*, is defined as

$$\mathbf{D} = \varepsilon_o \mathbf{E} + \mathbf{P} \tag{D.3}$$

This definition leads to the definition of the *permittivity*, ε, through the relation

$$\mathbf{D} = \varepsilon \mathbf{E} \equiv \varepsilon_r \varepsilon_o \mathbf{E} \tag{D.4}$$

where ε_r is called the *relative permittivity*. We see from these definitions that

$$\varepsilon_r = 1 + \chi \tag{D.5}$$

Both ε_r and χ are dimensionless quantities.

[3] In general, the polarization **P** need not be in the same direction as the electric field **E** that creates it. Such a situation can arise in materials with particular symmetries. In that case, χ has to be considered as a tensor rather than a scalar, though we do not have to consider such situations in this book.

D.2 Maxwell's equations

Maxwell's equations unify the electric and magnetic fields and are commonly written as

$$\nabla \cdot \mathbf{D} = \rho \tag{D.6}$$

$$\nabla \cdot \mathbf{B} = 0 \tag{D.7}$$

$$\nabla \times \mathbf{E} = -\frac{\partial \mathbf{B}}{\partial t} \tag{D.8}$$

$$\nabla \times \mathbf{H} = \mathbf{J} + \frac{\partial \mathbf{D}}{\partial t} \tag{D.9}$$

where

D is the *displacement* or *electric displacement*

B is the *magnetic flux density* or *magnetic induction* or sometimes just the *magnetic field*

E is the *electric field*

H is the *magnetic field*

ρ is the *free electric charge density*

J is the *free current density*

The fact that there is a zero on the right-hand side of Eq. (D.7) expresses the fact that as far as we know, there are no *magnetic monopoles*.

The relation between **E** and **D** is shown in Eqs. (D.3) and (D.4), where we also introduced the polarization **P**. Typically, polarization is viewed as the separation of bound charges and currents are viewed as the motions of free charges. This distinction between bound and free charge often is practically useful, but it is ultimately arbitrary. We can use either current or polarization to describe the response of the material, in fact, and this choice is a matter of taste. For example, the movement of charge to create a polarization is a current and we can describe that current, if we wish, as a rate of change of polarization.

We can also, in general, write the relation between **B** and **H** as

$$\mathbf{B} = \mu_o \left(\mathbf{H} + \mathbf{M} \right) \tag{D.10}$$

where **M** is the *magnetization*. The magnetization is the magnetic response of the material to the applied magnetic field **H**,[4] and μ_o is the *magnetic constant* (or the *permeability of free space*). Just like the electric constant above, it has no real physical meaning and is present for historical reasons of the definition of units. Similarly to the relation between **D** and **E**, we can define a *permeability* μ and a *relative permeability* μ_r through the relations

$$\mathbf{B} = \mu \mathbf{H} \equiv \mu_r \mu_o \mathbf{H} \tag{D.11}$$

---

[4] Some authors introduce a magnetic current density, just as **J** is an electric current density, to make the third and fourth Maxwell equations have the same form. The magnetic current density is not an unphysical concept (magnetic monopoles are not needed to have magnetic current), but just as we could choose to use either polarization or electric current or both in the electric case, so can we choose to use magnetization or magnetic current or both in the magnetic case.

Maxwell's equations can also be written in integral forms using Gauss's theorem and Stokes's theorem.

D.3 Maxwell's equations in free space

In free space, where there is no material to give a charge, a current, a polarization, or a magnetization, we can write Maxwell's equations as

$$\nabla \cdot \mathbf{E} = 0 \tag{D.12}$$

$$\nabla \cdot \mathbf{B} = 0 \tag{D.13}$$

$$\nabla \times \mathbf{E} = -\frac{\partial \mathbf{B}}{\partial t} \tag{D.14}$$

$$\nabla \times \mathbf{B} = \varepsilon_o \mu_o \frac{\partial \mathbf{E}}{\partial t} \tag{D.15}$$

In a (nonmagnetic) dielectric with no free charge and no free currents, we can simply replace ε_o in Eq. (D.15) with $\varepsilon = \varepsilon_r \varepsilon_o$.

D.4 Electromagnetic wave equation in free space

Using the identity

$$\nabla \times (\nabla \times \mathbf{F}) = \nabla(\nabla \cdot \mathbf{F}) - (\nabla \cdot \nabla)\mathbf{F} \tag{C.30}$$

we have, from Eqs. (D.14) and (D.15)

$$\begin{aligned} \nabla \times \nabla \times \mathbf{E} &= -\frac{\partial}{\partial t} \nabla \times \mathbf{B} \\ &= -\varepsilon_o \mu_o \frac{\partial^2 \mathbf{E}}{\partial t^2} \\ &= \nabla(\nabla \cdot \mathbf{E}) - (\nabla \cdot \nabla)\mathbf{E} \end{aligned} \tag{D.16}$$

Using Eq. (D.12) and writing $\nabla \cdot \nabla \equiv \nabla^2$ (which is easy to see in Cartesian coordinates), we have the wave equation

$$\nabla^2 \mathbf{E} - \varepsilon_o \mu_o \frac{\partial^2 \mathbf{E}}{\partial t^2} = 0 \tag{D.17}$$

and we can derive a similar equation for **B**. We recognize this as being of the same form as wave equations we have studied previously (in Appendices A and B) hence, we see that the quantity

$$\varepsilon_o \mu_o \equiv 1/c^2 \tag{D.18}$$

where c is the velocity of light in free space.

If we consider a nonmagnetic isotropic dielectric with no free charge and no free currents, we can derive a similar wave equation, but with $\varepsilon_r \varepsilon_o \mu_o$ instead of $\varepsilon_o \mu_o$ in Eq. (D.17). Hence, the wave propagation velocity is now

$$v = \frac{1}{\sqrt{\varepsilon_r \varepsilon_o \mu_o}} = \frac{c}{n} \tag{D.19}$$

where

$$n = \sqrt{\varepsilon_r} \tag{D.20}$$

is the *refractive index*.

D.5 Electromagnetic plane waves

A common simple situation is where we have a plane wave; that is, one of the form

$$\mathbf{E} = \mathbf{E}_o \exp\left[i\left(\mathbf{k}\cdot\mathbf{r} - \omega t\right)\right] \tag{D.21}$$

for some constant vector $\mathbf{E}_o$. (When representing real quantities by a complex exponential, the vector $\mathbf{E}_o$ may have a complex amplitude so it can represent different phases of the wave. At the end, we take the real part of the vector.) For a wave such as this, we can deduce directly, by "following" some peak (e.g., in the real part) of the function in space as we change time, that the velocity of this wave is of magnitude

$$v = \frac{\omega}{k} \tag{D.22}$$

and is in the direction of **k**. We can also write a similar expression for **B**.

For the purposes of discussion and without loss of generality, we can choose Cartesian coordinates so the direction of **k** is the z direction and $\mathbf{k}\cdot\mathbf{r} \equiv kz$. Note that nothing is changing in the x and y directions and, in particular, any derivatives $\partial/\partial x$ and $\partial/\partial y$ give zero results. Also, note that now $\partial\mathbf{E}/\partial z \equiv ik\mathbf{E}$. Now, when we consider $\nabla\times\mathbf{E}$, we therefore have

$$\nabla\times\mathbf{E} = \begin{vmatrix} \hat{\mathbf{x}} & \hat{\mathbf{y}} & \hat{\mathbf{z}} \\ \dfrac{\partial}{\partial x} & \dfrac{\partial}{\partial y} & \dfrac{\partial}{\partial z} \\ E_x & E_y & E_z \end{vmatrix} \equiv \begin{vmatrix} \hat{\mathbf{x}} & \hat{\mathbf{y}} & \hat{\mathbf{z}} \\ 0 & 0 & ik \\ E_x & E_y & E_z \end{vmatrix} \equiv i\mathbf{k}\times\mathbf{E} \tag{D.23}$$

and similarly $\nabla\times\mathbf{B} = i\mathbf{k}\times\mathbf{B}$. We note also that $\partial\mathbf{B}/\partial t \equiv -i\omega\mathbf{B}$ and $\partial\mathbf{E}/\partial t \equiv -i\omega\mathbf{E}$. Hence, from Eq. (D.14), we have

$$i\mathbf{k}\times\mathbf{E} = i\omega\mathbf{B} \tag{D.24}$$

which means, from the basic properties of the vector cross-product, that the magnetic field **B** is perpendicular to the direction of propagation and to the electric field. Similarly, from Eq. (D.15)

$$i\mathbf{k}\times\mathbf{B} = -i\omega\varepsilon_o\mu_o\mathbf{E} \tag{D.25}$$

so the electric field is also perpendicular to the direction of propagation. Hence, for a plane electromagnetic wave in free space, the electric and magnetic fields are perpendicular to the direction of propagation and to each other. (The same results hold for waves in a uniform isotropic dielectric medium.)

D.6 Polarization of a wave

A simple situation for a plane electromagnetic wave is where the electric field always is in one specific direction; for example, the x direction for a wave propagating in the z direction. We would then say that the wave is *linearly polarized* in the x direction.[5] Any possible state of polarization of a plane wave can be described by the component of the electric field in the x direction and that in the y direction. If the two polarization components have the same phase, then the wave is linearly polarized but at some angle relative to the x direction. It is also possible for the two components x and y to have different phases, in which case the wave is, in general, *elliptically polarized.* If the x and y components have the same amplitude but differ in phase by 90°, then the wave is *circularly polarized.* There are two different senses of circular polarization (i.e., left or right circularly polarized) depending on which component is 90° ahead of the other. The two different circular polarizations can also be used to describe any possible state of polarization of a plane wave. (See Section 14.1 for further discussion of polarizations.)

D.7 Energy density

If we imagine that we pull some charges apart from one another, because of the force between the charges, we change the energy of the system overall. When we stretch a spring, we consider the energy to be stored in the spring and we similarly consider that the energy associated with separating or bringing together charges is stored in the electric field. We can take a similar view for the energies associated with bringing magnets together or of turning on and off electromagnets. We can construct a consistent picture of these various energies, at least for isotropic materials, if we assign energy densities[6]

$$U_E = \frac{1}{2}\varepsilon E^2 \tag{D.26}$$

to the electric field and

$$U_H = \frac{1}{2}\mu H^2 \tag{D.27}$$

to the magnetic field.

D.8 Energy flow

To understand energy flow in electromagnetic fields, we consider the so-called *Poynting vector*

---

[5] If we do not specify whether we are considering electric or magnetic field direction in discussing polarization, then we are discussing electric fields. We can also discuss magnetic field polarization if we wish, however, and in some more complicated situations (e.g., in some waveguides) only one or other of the electric or magnetic fields may have a definite polarization perpendicular to the propagation direction.

[6] See, e.g., J. A. Stratton, *Electromagnetic Theory* (McGraw-Hill, New York, 1941), pp. 104–125 or P. Lorrain and D. Corson, *Electromagnetic Fields and Waves, Second Edition* (Freeman, San Francisco, 1970), pp. 72–80 and pp. 351–354

$$\mathbf{S} = \mathbf{E} \times \mathbf{H} \tag{D.28}$$

Using the identity $\nabla \cdot (\mathbf{F} \times \mathbf{G}) = -\mathbf{F} \cdot (\nabla \times \mathbf{G}) + \mathbf{G} \cdot (\nabla \times \mathbf{F})$ (Eq. (C.31)), we can consider the divergence of this vector $\mathbf{S}$. Hence, assuming isotropic materials with no free currents, and using equations (D.4), (D.8), (D.9), and (D.11)

$$\begin{aligned} \nabla \cdot \mathbf{S} &= -\mathbf{E} \cdot (\nabla \times \mathbf{H}) + \mathbf{H} \cdot (\nabla \times \mathbf{E}) \\ &= -\mathbf{E} \cdot \varepsilon \frac{\partial \mathbf{E}}{\partial t} - \mathbf{H} \cdot \mu \frac{\partial \mathbf{H}}{\partial t} \\ &= -\frac{\partial}{\partial t}\left(\frac{1}{2}\varepsilon E^2 + \frac{1}{2}\mu H^2 \right) \end{aligned} \tag{D.29}$$

The quantity on the right is just minus the rate of change of energy density. Hence, with our usual interpretation of the divergence, the vector **S** represents the flow of energy per unit area in the direction of **S**.

This vector can be used to deduce the relation between electromagnetic field amplitudes and intensity (i.e., power per unit area) in electromagnetic waves. Suppose we have a monochromatic plane wave, linearly polarized in the x direction and propagating in the z direction in a nonmagnetic dielectric with no free currents. Then choosing a cosine form (the actual phase of the wave does not matter in the end for this calculation), we have

$$\mathbf{E} = E_o \hat{\mathbf{x}} \cos(kz - \omega t) \tag{D.30}$$

Then

$$\nabla \times \mathbf{E} = -\hat{\mathbf{y}} E_o k \sin(kz - \omega t) \tag{D.31}$$

We already know from the discussion of plane waves that the **B** field must be polarized in the y direction and that it has the same phase as the **E** field; that is, it must be of the form $B_o \hat{\mathbf{y}} \cos(kz - \omega t)$. Hence, using the Maxwell Eq. (D.8), we therefore have

$$\nabla \times \mathbf{E} = -\frac{\partial \mathbf{B}}{\partial t} = -\hat{\mathbf{y}} B_o \frac{\partial}{\partial t} \cos(kz - \omega t) = -\hat{\mathbf{y}} B_o \omega \sin(kz - \omega t) \tag{D.32}$$

Equating our two results in Eqs. (D.31) and (D.32) for $\nabla \times \mathbf{E}$, and using our two expressions for v, Eqs. (D.19) and (D.22), we can substitute $\omega / k = c / n$, and we have

$$B_o = \frac{n}{c} E_o \tag{D.33}$$

Hence, with $\mathbf{B} = \mu_o \mathbf{H}$ by definition in our nonmagnetic medium and using Eq. (D.18) to get rid of the μ_o in the expression, the Poynting vector becomes

$$\mathbf{S} = \hat{\mathbf{z}} \varepsilon_o n c E_o^2 \sin^2(kz - \omega t) \tag{D.34}$$

which, as we would expect, corresponds to energy flowing in the z direction. We can time-average this over a cycle (the average of the $\sin^2$ term is ½) and finally write the intensity (power per unit area) flowing across a surface perpendicular to the z direction as

$$I = \frac{1}{2} n c \varepsilon_o E_o^2 \tag{D.35}$$

This expression has a simple physical interpretation. The time-averaged energy density in the electric field is $(1/2)\times(1/2)\varepsilon E_o^2$ and there is an exactly equal amount of energy in the magnetic field. Hence, because $\varepsilon = \varepsilon_r \varepsilon_o = n^2 \varepsilon_o$, if we view this energy density as moving forward at a velocity c/n, we get the intensity of Eq. (D.35).

We have derived the expression in Eq. (D.35) presuming that we have an electric field of the form

$$\mathbf{E} = \hat{\mathbf{x}} E_o \cos(kz - \omega t) \equiv \mathrm{Re}\left(E_o \exp\left[i(kz - \omega t)\right]\right) \quad \text{(D.36)}$$

which is a common approach in electrical engineering. Sometimes, especially in physics texts, an equivalent approach is to presume a field of the form

$$\mathbf{E} = \frac{E_o}{2}\left\{\exp\left[i(kz - \omega t)\right] + \exp\left[-i(kz - \omega t)\right]\right\}\hat{\mathbf{x}} \equiv E_o \cos(kz - \omega t)\hat{\mathbf{x}} \quad \text{(D.37)}$$

or, more generally, allowing for some phase angle in the field by allowing E_o to be a complex constant

$$\begin{aligned}\mathbf{E} &= \left\{\frac{E_o}{2}\exp\left[i(kz - \omega t)\right] + \frac{E_o^*}{2}\exp\left[-i(kz - \omega t)\right]\right\}\hat{\mathbf{x}} \\ &\equiv \frac{E_o}{2}\exp\left[i(kz - \omega t)\right]\hat{\mathbf{x}} + c.c.\end{aligned} \quad \text{(D.38)}$$

where the notation "*c.c.*" stands for *complex conjugate*. The expression for intensity remains unchanged, however.

D.9 Modes

One concept that appears at various points in this book is the idea of modes. This concept is one that arises in many different parts of engineering and physics, though it can be difficult to find a clear and broad definition in textbooks.[7] We need the idea here especially when discussing electromagnetic fields and bosons, so we briefly discuss it here.

Modes occur throughout classical physics. Perhaps the most common use there is in describing oscillations. A tightly stretched string, such as a violin or guitar string, can oscillate in a way that gives a stable, standing wave pattern (at least, in the idealized case where we neglect any losses in the system). The different possible standing waves each have a sinusoidal form in space (exactly like the solutions of the particle in a box problem in Chapter 2). A characteristic of such a standing-wave solution is that all points on the string are oscillating at the same frequency. Of course, we know mathematically from Chapter 2 that such a problem is an eigenvalue problem. The standing waves are the eigenfunctions and the various corresponding frequencies are related to the eigenvalues. (It is often the square of the frequency that is the actual eigenvalue in such classical oscillator problems.)

These different distinct ways of oscillating are called modes in such classical oscillating systems or, sometimes more specifically, *normal modes*. The term *normal* here can be taken to

---

[7] When asked, many well-trained scientists and engineers will say that they understand what a mode is, but will be unable to define the idea of modes and will also be unable to remember where they learned the idea!

refer to the fact that these modes can be mathematically orthogonal functions. Mechanical systems of many kinds have such oscillating modes (at least, when we idealize them to be loss-less oscillators with no energy dissipation). Examples include musical instruments that give notes of definite pitch or frequency, mechanical objects like girders and bridges, and all sorts of combinations of masses and springs. The vibrations of atoms in solids can also be described this way. Specifically, atoms in crystalline solids show a set of (normal) modes of oscillation that are known as the *phonon modes* of the crystal. These phonon modes themselves are typically calculated as the results of a classical model based on masses connected by springs. So far, then, modes (of oscillation) are distinct (actually, mathematically orthogonal) ways in which a system can oscillate in which each point in the system is oscillating at the same frequency.

Modes occur in one other important classical context, which are modes of propagation of waves. Such a propagation mode can be a stable shape in which the wave propagates, such as in the modes of a waveguide. There, waves in the form of such a mode of a given waveguide retain exactly the same cross-sectional shape as they move down the waveguide. Specific examples of such propagating modes include the modes of glass optical fibers and of metallic microwave transmission lines and hollow waveguides (at least, if we idealize these waveguides to be loss-less). The underlying mathematics of such modes is similar to the oscillating modes, but this time the eigenvalue is a "propagation constant" rather than a frequency and the eigenfunction is the cross-sectional shape of the wave. (Typically, such a mode is also associated with a specific frequency, though there can be many different propagating modes with different propagation constants for a given frequency.)

We can see from this discussion that the common description that underlies modes is that they can be described as eigen problems. We could choose to make the following definition

A mode is an eigenfunction of a Hermitian operator describing a linear physical system.

Mathematically, we could write, using Dirac notation merely as a way of writing linear algebra and not specifically implying any quantum mechanical aspect

$$\hat{A}|\psi\rangle = \lambda|\psi\rangle \tag{D.39}$$

Here, $\hat{A}$ is the Hermitian operator describing the physical system, λ is an eigenvalue, and $|\psi\rangle$ is a corresponding eigenfunction. (The use of λ as the eigenvalue is very common in mathematical texts and should not be confused with the use in physics to represent the wavelength.) More commonly, in mathematical texts such an equation will be written

$$A\psi = \lambda\psi \tag{D.40}$$

Because the operator is Hermitian, the resulting modes have the very useful properties that they are orthogonal, they are complete, and they have real eigenvalues. Hence, they are very useful as basis sets.

When we deal with quantization of the electromagnetic field in Chapter 15, we presume that the classical electromagnetic problem of the relevant modes has already been solved for the situation of interest. We do not go into such solutions here; such problems are solved in depth elsewhere.[8]

---

[8] See, e.g., A. Yariv, *Quantum Electronics (3rd Edition)* (Wiley, New York, 1989), for an introductory discussion of optical waveguide and laser resonator modes.

The definition of modes can be generalized somewhat from the previous discussion,[9] and this generalized definition covers nearly all uses of the idea of modes.[10]

[9] Slightly more generally, modes can be the solutions of the generalized eigenvalue equation, of the form $A\psi = \lambda B\psi$, where A is Hermitian and B is an operator also. If B is a positive operator (i.e., one for which all of the eigenvalues are real and > 0), then this generalized problem can always be recast in the form of a simple eigenvalue equation again. In that case, we can see that, for example, in the representation in which B is diagonal we can easily define another diagonal operator $\sqrt{B}$ such that $\sqrt{B}\sqrt{B} = B$ (the diagonal elements of $\sqrt{B}$ are just the square roots of the eigenvalues of B) and also a diagonal inverse operator $(\sqrt{B})^{-1}$ with diagonal elements that are the reciprocals of the square roots of the eigenvalues. Simple algebra then shows that the equation $A\psi = \lambda B\psi$ can be rewritten as $C\phi = \lambda\phi$, where $\phi = \sqrt{B}\psi$ and $C = (\sqrt{B})^{-1} A(\sqrt{B})^{-1}$ and C is also Hermitian (as is easily proved, because B, $\sqrt{B}$, and $(\sqrt{B})^{-1}$ are all Hermitian). Such generalized eigenequations often occur when there is some nonuniformity in the physical problem. For example, the normal modes of a pair of different masses on springs have this kind of equation and an electromagnetic problem with a spatially varying dielectric constant, such as in waveguides or in photonic crystal or other dielectric photonic nanostructures, can also have this form. See also G. Strang, *Linear Algebra and Its Applications (3rd Edition)* (Saunders/Harcourt Brace Jovanovich, New York, 1988), pp. 343–345.

[10] In situations with loss, e.g., a lossy waveguide or a damped oscillator, the operator describing such a system, though linear, is typically not Hermitian. In that case, the eigenfunctions may not be orthogonal. Provided the loss is not too strong, however, the modes may well still be nearly orthogonal and, as such, they may still be useful for approximate results. E.g., the modes of a guitar string are lossy because sound is radiated from them if for no other reason. Nonetheless, the concept of the modes of the string is still a useful one, mathematically and conceptually.

Appendix E

Perturbing Hamiltonian for optical absorption

Here, we explain the origin of the relation

$$\hat{H}_p(\mathbf{r},t) \cong \frac{e}{m_o}\mathbf{A}\cdot\hat{\mathbf{p}} \tag{E.1}$$

for the perturbing Hamiltonian when we have an electromagnetic field, described by the *vector potential* **A**, interacting with an electron. There are four stages to this derivation. First, we need to write the classical Hamiltonian for this interaction and show why it is correct classically. This Hamiltonian is, for a particle of charge e

$$H = \frac{1}{2m_o}(\mathbf{p}-e\mathbf{A})^2 + e\phi \tag{E.2}$$

Second, we have to change this classical Hamiltonian into a quantum mechanical one. Third, we have to make a choice (i.e., the choice of the Coulomb gauge in the vector potential). Fourth, and finally, we have to make an approximation (i.e., the neglect of terms $\sim \mathbf{A}^2$).

E.1 Justification of the classical Hamiltonian

The justification of the classical Hamiltonian comes from classical mechanics and from the *Lorentz force* from elementary electromagnetism, which, for a charge of e traveling with velocity **v**, is

$$\mathbf{F} = e\mathbf{E} + e\mathbf{v}\times\mathbf{B} \tag{E.3}$$

To proceed, we need first to note or define the relation between **E** and **B** and scalar (ϕ) and vector (**A**) potentials. From the Maxwell equation $\nabla\cdot\mathbf{B}$, we know we can choose to write for some vector function **A**

$$\mathbf{B} = \nabla\times\mathbf{A} \tag{E.4}$$

because, for any vector function **A** (with continuous derivatives), $\nabla\cdot(\nabla\times\mathbf{A}) = 0$. From the Maxwell equation $\nabla\times\mathbf{E} = -(\partial\mathbf{B}/\partial t)$, we have

$$\nabla\times\mathbf{E} = -\nabla\times\left(\frac{\partial\mathbf{A}}{\partial t}\right) \tag{E.5}$$

or, equivalently

$$\nabla\times\left(\mathbf{E}+\frac{\partial\mathbf{A}}{\partial t}\right)=0 \tag{E.6}$$

If the curl of some vector function is zero, it must be possible to express it as the gradient of a scalar function, ϕ. Hence, we obtain

$$\mathbf{E}=-\frac{\partial\mathbf{A}}{\partial t}-\nabla\phi \tag{E.7}$$

Hence, we have shown that **E** and **B** can be expressed in terms of scalar (ϕ) and vector (**A**) potentials. (We show below that ϕ can be identified with the normal scalar electrostatic potential.)

Given the Lorentz force and these two definitions, Eqs. (E.4) and (E.7), there are at least two ways of deriving the Hamiltonian Eq. (E.2). One argument from Lagrangian classical mechanics (see, e.g., Leech[1]) justifies the replacement of the momentum **p** by the "generalized momentum" $\mathbf{p}-e\mathbf{A}$. The second (see, e.g., Haken,[2] who gives a relatively complete proof) formally shows the consistency through Hamiltonian classical mechanics. We do not repeat these arguments here, but the reader should understand that Eqs. (E.3), (E.4), and (E.7) are all the additional information required to justify this classical Hamiltonian.

E.2 Quantum mechanical Hamiltonian

The transition to quantum mechanics can be made by replacing the classical momentum **p** with the operator $-i\hbar\nabla$. We also now add any additional potential energy terms, so we have the potential energy V instead of the purely electrostatic potential $e\phi$. This form, leaving **A** as a normal vector quantity, is known as the *semiclassical treatment of radiation*; it is also possible to quantize the radiation field as well at this point, which would lead to a theory that would, of itself, predict spontaneous emission properly, for example, though we do not take that path here. The Hamiltonian, therefore, becomes

$$\hat{H}=\frac{1}{2m}\left(-i\hbar\nabla-e\mathbf{A}\right)^2+V \tag{E.8}$$

where we note now that this Hamiltonian is an operator rather than a scalar quantity. As always, this transition from classical mechanics to quantum mechanics cannot be proved in any formal sense; the justification is that it works.

Now, we can "multiply out" this Hamiltonian. We have to retain the order of all "multiplications" carefully because we are dealing with operators, not scalars, and operators do not necessarily commute (i.e., the order of operators matters). We obtain

$$\begin{aligned}\hat{H}&=\frac{1}{2m_o}\left(-i\hbar\nabla-e\mathbf{A}\right)\cdot\left(-i\hbar\nabla-e\mathbf{A}\right)+V\\&=\frac{-\hbar^2}{2m_o}\nabla^2+\frac{i\hbar e}{2m_o}\left(\mathbf{A}\cdot\nabla+\nabla\cdot\mathbf{A}\right)+\frac{e^2}{2m_o}\mathbf{A}^2+V\end{aligned} \tag{E.9}$$

[1] J. W. Leech, *Classical Mechanics* (Methuen, 1965).

[2] H. Haken, *Light (Vol. 1)*, (North-Holland, Amsterdam, 1981).

The term in $\nabla \cdot \mathbf{A}$ requires some care. Remember that in any use of this Hamiltonian operator, it is operating on some wavefunction ψ (i.e., it occurs in the form $\hat{H}\psi$). Therefore, this particular term occurs in the form $\nabla \cdot (\mathbf{A}\phi)$. But, we can rewrite this as

$$\nabla \cdot (\mathbf{A}\psi) = \psi[\nabla \cdot \mathbf{A}] + \mathbf{A} \cdot \nabla \psi \tag{E.10}$$

where we have used square brackets to make a distinction that $[\nabla \cdot \mathbf{A}]$ is simply a number or an ordinary function, not something in which the gradient or divergence operator operates on anything to the right. Hence, from Eqs. (E.9) and (E.10), we obtain

$$\hat{H} = \hat{H}_o + \frac{i\hbar e}{m_o}\mathbf{A} \cdot \nabla + \frac{i\hbar e}{2m_o}[\nabla \cdot \mathbf{A}] + \frac{e^2}{2m_o}\mathbf{A}^2 \tag{E.11}$$

where $\hat{H}_o$ is the "unperturbed" Hamiltonian (i.e., the Hamiltonian for $\mathbf{A} = 0$)

$$H_o = -\frac{\hbar^2}{2m_o}\nabla^2 + V \tag{E.12}$$

E.3 Choice of gauge

Though the relations Eqs. (E.4) and (E.7) uniquely determine the fields **B** and **E** from the potentials **A** and ϕ, they are not sufficient to specify the potentials uniquely; this lack of uniqueness goes beyond simply adding constant vectors or numbers to these potentials (though, obviously, we can do that because the potentials only appear in derivative forms in these field equations). In particular, we can choose new potentials

$$\mathbf{A}_{new} = \mathbf{A}_{old} - \nabla \xi \tag{E.13}$$

$$\phi_{new} = \phi_{old} + \frac{\partial \xi}{\partial t} \tag{E.14}$$

where ξ is any scalar function (with continuous derivatives) that we choose. (Choosing this function is known, for historical reasons, as choosing the *gauge*.) The new potentials $\mathbf{A}_{new}$ and ϕ_{new} give the same fields **E** and **B** as the old potentials $\mathbf{A}_{old}$ and ϕ_{old}, as is easily checked by substitution into Eqs. (E.4) and (E.7) (note that $\nabla \times (\nabla \xi) = 0$ for all ξ). Note that, in particular,

$$\nabla \cdot \mathbf{A}_{new} = \nabla \cdot \mathbf{A}_{old} - \nabla^2 \xi \tag{E.15}$$

which means we can set $\nabla \cdot \mathbf{A}$ to any value or function we want by appropriate choice of gauge.

The most common choice of gauge is the so-called Coulomb gauge, in which

$$\nabla \cdot \mathbf{A} = 0 \tag{E.16}$$

With this choice, starting from the Maxwell equation $\nabla \cdot \mathbf{E} = \rho / \varepsilon_o$, where ρ is the charge density and ε_o is the permittivity of free space, we obtain, using

$$\nabla \cdot \mathbf{E} = -\nabla \cdot \left(\frac{\partial \mathbf{A}}{\partial t} \right) - \nabla^2 \phi = -\frac{\partial}{\partial t}(\nabla \cdot \mathbf{A}) - \nabla^2 \phi = -\nabla^2 \phi = \frac{\rho}{\varepsilon_o} \tag{E.17}$$

That is, we then have that the scalar potential is the one that solves Poisson's equation and it can, therefore, be identified with the electrostatic potential. This choice of the gauge is the one we want; indeed, we have arguably already made this choice when we subsumed ϕ into the potential V in Eq. (E.8) because we intend V to be only a function of position.

Note that if we choose to define a monochromatic electromagnetic field at frequency ω entirely in terms of the vector potential $\mathbf{A}$, which we sometimes do in calculations because of this $\mathbf{A}\cdot\hat{\mathbf{p}}$ Hamiltonian, then the resulting magnitude of the electric field is, from Eq. (E.7), $|\mathbf{E}| = \omega|\mathbf{A}|$, which we can substitute in expressions for intensity in terms of electric field.

E.4 Approximation to linear system

We are only interested here in the linear optical properties of the system; we can, therefore, choose to consider only small amplitudes of $\mathbf{A}$, neglecting terms in $\mathbf{A}^2$. Hence, with this approximation and with the Coulomb gauge, we can rewrite Eq. (E.11) as

$$\hat{H} = \hat{H}_o + \frac{i\hbar e}{m_o}\mathbf{A}\cdot\nabla \tag{E.18}$$

or, formally defining

$$\hat{H} = H_o + \hat{H}_p \tag{E.19}$$

and substituting $\hat{\mathbf{p}} = -i\hbar\nabla$, we obtain the expression in Eq. (E.1).

Appendix F

Early history of quantum mechanics

The essential aspects of the quantum mechanics introduced in the first half of this book mostly come from the conceptual development of quantum mechanics from ~1900 to the early 1930s. The following is a very brief (and substantially incomplete) chronology of some of the most cited early development. Of course, the whole history is much richer than this and is worthy of a much deeper treatment.

1900 Max Planck postulates that the energy in light comes in quanta of size $h\nu$, where h is Planck's constant. This solves a famous problem of classical physics, the "ultraviolet catastrophe," in which otherwise the thermal distribution of energy in light should keep on increasing without limit at ever shorter wavelengths.

1905 Albert Einstein postulates the photon to explain the photoelectric effect. This now clearly introduces the concept of wave-particle duality for light.

1913 Neils Bohr proposes the quantization of angular momentum, which gives the Bohr model of the hydrogen atom that, especially when further developed by Sommerfeld in 1916 and by Debye, successfully explains major features of the hydrogen atom's energy levels and spectra, though it leaves other aspects unexplained, especially why it is that the orbiting electrons of this model are not continuously emitting radiation.

1922 Otto Stern and Walther Gerlach show in their experiment the quantized nature of electron spin and additionally expose the key difficulty of the measurement problem in quantum mechanics.

1924 Louis de Broglie proposes his hypothesis that the wavelength associated with a quantum mechanical particle is $\lambda = h/p$, where p is the particle momentum.

1925 Werner Heisenberg proposes matrix mechanics as a mathematical basis for quantum mechanics.

1926 Erwin Schrödinger proposes his wave equation, which explains hydrogen atom structure. (After some active debate, Schrödinger subsequently proves that the Heisenberg and Schrödinger pictures are equivalent.)

1927 Werner Heisenberg proposes the uncertainty principle.

1927 Clinton Davisson and Lester Germer perform an experiment that confirms the wave nature of the electron and de Broglie's hypothesis.

1928 Paul Dirac gives the first solution of quantum mechanics consistent with special relativity, explaining spin.

1932 John von Neumann publishes a book on the mathematics of quantum mechanics that gives firm mathematical foundations to the subject.

Appendix G

Some useful mathematical formulae

This appendix lists many of the more useful mathematical formulae for quantum mechanical problems. For convenience, we have also collected here various of the less common formulae that are introduced in the book.

G.1 Elementary mathematical expressions

Quadratic equations

$$a^2 - b^2 = (a+b)(a-b) \tag{G.1}$$

The solutions to the general quadratic equation

$$ax^2 + bx + c = 0 \tag{G.2}$$

are

$$x = \frac{-b \pm \sqrt{b^2 - 4ac}}{2a} \tag{G.3}$$

Taylor and Maclaurin series (power-series expansion)

The Taylor series

$$f(x) = f(a) + \frac{(x-a)}{1!}\left.\frac{df}{dx}\right|_a + \frac{(x-a)^2}{2!}\left.\frac{d^2 f}{dx^2}\right|_a + \ldots + \frac{(x-a)^n}{n!}\left.\frac{d^n f}{dx^n}\right|_a + \ldots \tag{G.4}$$

gives a useful way of approximating a function near to some specific point $x = a$, giving a power-series expansion in $(x-a)^n$ for the function near that point. For sufficiently small departures from the point (i.e., sufficiently small $x - a$), we can retain just the first few (often just the first two) terms in the series. The largest n for which we retain the term in this series would be the *order* of the power-series approximation, so retaining just the first two terms would give a first-order approximation, for example.

The Maclaurin[1] series

---

[1] The distinguished Scottish mathematician, Colin Maclaurin, also took a prominent part in the ultimately unsuccessful defense of the city of Edinburgh when it was under attack in the Jacobite rebellion of 1745.

$$f(x) = f(0) + \frac{x}{1!}\frac{df}{dx}\bigg|_0 + \frac{x^2}{2!}\frac{d^2 f}{dx^2}\bigg|_0 + \ldots + \frac{x^n}{n!}\frac{d^n f}{dx^n}\bigg|_0 + \ldots \tag{G.5}$$

is a special case of the Taylor series where we are expanding around the point $x = 0$.

Power-series expansions of common functions

For small a, the Maclaurin expansions of various common functions are, to first order

$$\sqrt{1+a} \simeq 1 + a/2 + \ldots \tag{G.6}$$

$$\frac{1}{1+a} \simeq 1 - a + \ldots \tag{G.7}$$

$$\sin a \simeq a + \ldots \tag{G.8}$$

$$\tan a \simeq a + \ldots \tag{G.9}$$

$$\cos a \simeq 1 - \frac{a^2}{2} + \ldots \tag{G.10}$$

$$\exp a \simeq 1 + a + \ldots \tag{G.11}$$

G.2 Formulae for sines, cosines, and exponentials

Sine and cosine addition and product formulae

$$\sin^2(\alpha) + \cos^2(\alpha) = 1 \tag{G.12}$$

$$\sin(\alpha \pm \beta) = \sin(\alpha)\cos(\beta) \pm \cos(\alpha)\sin(\beta) \tag{G.13}$$

$$\sin(2\alpha) = \sin(\alpha)\cos(\alpha) - \cos(\alpha)\sin(\alpha) \tag{G.14}$$

$$\cos(\alpha \pm \beta) = \cos(\alpha)\cos(\beta) \mp \sin(\alpha)\sin(\beta) \tag{G.15}$$

$$\cos(2\alpha) = \cos^2(\alpha) - \sin^2(\alpha) = 2\cos^2(\alpha) - 1 = 1 - 2\sin^2(\alpha) \tag{G.16}$$

$$\cos^2(\alpha) = \frac{1}{2}\left[1 + \cos(2\alpha)\right] \tag{G.17}$$

$$\sin^2(\alpha) = \frac{1}{2}\left[1 - \cos(2\alpha)\right] \tag{G.18}$$

$$\cos(\alpha)\cos(\beta) = \frac{1}{2}\left[\cos(\alpha - \beta) + \cos(\alpha + \beta)\right] \tag{G.19}$$

$$\sin(\alpha)\sin(\beta) = \frac{1}{2}\left[\cos(\alpha - \beta) - \cos(\alpha + \beta)\right] \tag{G.20}$$

$$\sin(\alpha)\cos(\beta) = \frac{1}{2}\left[\sin(\alpha - \beta) + \sin(\alpha + \beta)\right] \tag{G.21}$$

$$\cos(\alpha)+\cos(\beta)=2\cos\left(\frac{\alpha+\beta}{2}\right)\cos\left(\frac{\alpha-\beta}{2}\right) \tag{G.22}$$

$$\sin(\alpha)+\sin(\beta)=2\sin\left(\frac{\alpha+\beta}{2}\right)\cos\left(\frac{\alpha-\beta}{2}\right) \tag{G.23}$$

$$\cos(\alpha)-\cos(\beta)=-2\sin\left(\frac{\alpha+\beta}{2}\right)\sin\left(\frac{\alpha-\beta}{2}\right) \tag{G.24}$$

$$\sin(\alpha)-\sin(\beta)=2\cos\left(\frac{\alpha+\beta}{2}\right)\sin\left(\frac{\alpha-\beta}{2}\right) \tag{G.25}$$

Formulae of differential calculus

Product rule

$$\frac{d}{dx}(uv)=u\frac{dv}{dx}+v\frac{du}{dx} \tag{G.26}$$

Quotient rule

$$\frac{d}{dx}\left(\frac{u}{v}\right)=\frac{v\dfrac{du}{dx}-u\dfrac{dv}{dx}}{v^2} \tag{G.27}$$

Chain rule

$$\frac{d}{dx}f(g(x))=\left(\frac{df}{dg}\right)\times\left(\frac{dg}{fx}\right) \tag{G.28}$$

Derivatives of elementary functions

$$\frac{d}{dx}x^n=nx^{n-1} \tag{G.29}$$

$$\frac{d}{dx}\exp(ax)=a\exp(ax) \tag{G.30}$$

$$\frac{d}{dx}\ln(x)=\frac{1}{x} \tag{G.31}$$

$$\frac{d}{dx}\sin(x)=\cos(x) \tag{G.32}$$

$$\frac{d}{dx}\cos(x)=-\sin(x) \tag{G.33}$$

$$\frac{d\sin^{-1}x}{dx}=\frac{1}{\sqrt{1-x^2}} \tag{G.34}$$

$$\frac{d\tan^{-1}x}{dx}=\frac{1}{1+x^2} \tag{G.35}$$

Integration by parts

$$\int_a^b f(x)\left(\frac{dg(x)}{dx}\right)dx = \left[f(x)g(x)\right]_a^b - \int_a^b \left(\frac{df(x)}{dx}\right)g(x)\,dx \tag{G.36}$$

where we use the common notation

$$\left[h(x)\right]\Big|_a^b = h(b) - h(a) \tag{G.37}$$

and, specifically, here

$$\left[f(x)g(x)\right]_a^b = f(b)g(b) - f(a)g(a) \tag{G.38}$$

Some definite integrals

$$\int_0^\pi \sin^2(nx)\,dx = \frac{\pi}{2} \tag{G.39}$$

$$\begin{aligned}\int_0^\pi (x-\pi/2)\sin(nx)\sin(mx)\,dx &= \frac{-4nm}{(n-m)^2(n+m)^2}, \text{ for } n+m \text{ odd} \\ &= 0, \text{ for } n+m \text{ even}\end{aligned} \tag{G.40}$$

$$\int_0^\pi \sin(\theta)\cos(2\theta)\,d\theta = -2/3 \tag{G.41}$$

$$\int_0^\pi \sin(2\theta)\cos(\theta)\,d\theta = 4/3 \tag{G.42}$$

$$\int_0^\pi \sin^3\theta\,d\theta = \frac{4}{3} \tag{G.43}$$

$$\int_0^\infty t^{1/2}\exp(-t)\,dt = \frac{\sqrt{\pi}}{2} \tag{G.44}$$

$$\int_{-\infty}^\infty \frac{\sin x}{x}\,dx = \pi \tag{G.45}$$

$$\int_{-\infty}^\infty \left(\frac{\sin x}{x}\right)^2 dx = \pi \tag{G.46}$$

$$\int_{-\infty}^\infty \exp(-x^2)\,dx = \sqrt{\pi} \tag{G.47}$$

$$\int_{-\infty}^\infty \frac{1}{1+x^2}\,dx = \pi \tag{G.48}$$

G.3 Special functions

Here, we summarize some of the key mathematics related to some of the special functions that come up often in quantum mechanical problems, including the defining differential equations, some definitions, and some important expressions for the resulting solutions. For greater detail and a definitive reference on a broad range of special functions, see M. Abramowitz and I. A. Stegun, *Handbook of Mathematical Functions* (National Bureau of Standards, Washington, 1972).

Bessel functions

The standard defining equation for Bessel functions is

$$x^2\frac{d^2y}{dx^2}+x\frac{dy}{dx}+\left(x^2-p^2\right)y=0 \tag{G.49}$$

with the solutions to this equation being $J_p(x)$, a Bessel function of the first kind, and $Y_p(x)$, a Bessel function of the second kind (also sometimes called a Neumann function or a Weber function [though that name is also sometimes used for a different function]). The parameter p is called the *order* of the Bessel function; it can be any real or complex number. The Bessel function of the second kind goes to ∞ at $r=0$ and can often, therefore, be discounted in the solution of many physical problems. The Hankel function is a complex combination of Bessel functions of the first and second kinds.

We can write a power-series expansion for Bessel functions of the first kind for integer or zero values of p

$$J_p(x)=\sum_{m=0}^{\infty}\frac{(-1)^m}{m!\Gamma(m+p+1)}\left(\frac{x}{2}\right)^{2m+p} \tag{G.50}$$

Note also that

$$J_{-p}(x)=(-1)^p J_p(x) \tag{G.51}$$

If one starts out with an equation of the form $\nabla^2 f=Cf$, writes the Laplacian operator in cylindrical coordinates, and separates variables, one can end up with an equation of the form of Eq. (G.49) for the radial part of the solution with p as an integer. So integer-order Bessel functions occur often in cylindrically or circularly symmetric problems in physics and engineering, including waves on circular membranes such as drum heads, for example.

The equation

$$\frac{d^2y}{dx^2}+\left[a^2+\left(\frac{1}{4}-p^2\right)\frac{1}{x^2}\right]y=0 \tag{G.52}$$

has solutions $y=\sqrt{x}J_p(ax)$ and $y=\sqrt{x}Y_p(ax)$.

Airy functions

The differential equation

$$\frac{d^2y}{dx^2}-xy=0 \tag{G.53}$$

has solutions $Ai(x)$ and $Bi(x)$, which are called Airy functions. (See Chapter 2, Section 2.11, for a graph of the resulting two solutions.) Airy functions can formally be written in terms of so-called modified Bessel functions of $\pm 1/3$ fractional orders. They occur in problems in which a potential is varying linearly with distance, for example.

Spherical Bessel functions

The solutions of the equation

$$x^2 \frac{d^2 y}{dx^2} + 2x \frac{dy}{dx} + \left[x^2 - n(n+1) \right] y = 0 \tag{G.54}$$

are spherical Bessel functions of the first and second kinds, respectively

$$j_n(x) = \sqrt{\frac{\pi}{2x}} J_{n+1/2}(x) \quad \text{and} \quad y_n(x) = \sqrt{\frac{\pi}{2x}} Y_{n+1/2}(x) \tag{G.55}$$

and a corresponding one of the second kind. These functions are called *spherical Bessel functions*. The first two spherical Bessel functions of the first kind are

$$j_0(x) = \frac{\sin x}{x} \tag{G.56}$$

$$j_2(x) = \frac{\sin x}{x^2} - \frac{\cos x}{x} \tag{G.57}$$

Associated Legendre functions

For the differential equation

$$\frac{1}{\sin\theta} \frac{d}{d\theta} \left(\sin\theta \frac{d}{d\theta} \right) \Theta(\theta) - \frac{m^2}{\sin^2\theta} \Theta(\theta) - l(l+1)\Theta(\theta) = 0 \tag{G.58}$$

there are solutions for $l = 0, 1, 2, 3, \ldots$ with $-l \le m \le l$ (m integer), which are

$$\Theta(\theta) = P_l^m(\cos\theta) \tag{G.59}$$

where $P_l^m(x)$ are the associated Legendre functions. (See Chapter 9, Section 9.2, for more discussion.)

Spherical harmonics

The spherical harmonics $Y_{lm}(\theta, \phi)$ are the solutions to the equation

$$\nabla^2_{\theta,\phi} Y_{lm}(\theta, \phi) = -l(l+1) Y_{lm}(\theta, \phi) \tag{G.60}$$

where

$$\nabla^2_{\theta,\phi} = \left[\frac{1}{\sin\theta} \frac{\partial}{\partial\theta} \left(\sin\theta \frac{\partial}{\partial\theta} \right) + \frac{1}{\sin^2\theta} \frac{\partial^2}{\partial\phi^2} \right] \tag{G.61}$$

constitutes the θ and ϕ parts of the Laplacian in spherical coordinates. The spherical harmonics can be written in a normalized form as

$$Y_{lm}(\theta,\phi)=(-1)^m\sqrt{\frac{2l+1}{4\pi}\frac{(l-m)!}{(l+m)!}}P_l^m(\cos\theta)\exp(im\phi) \quad (G.62)$$

(See Chapter 9, Section 9.2, for more discussion.)

Hermite polynomials

The solutions to the differential equation

$$\frac{d^2y}{dx^2}-2x\frac{dy}{dx}+2ny=0 \quad (G.63)$$

for $n = 0, 1, 2, \ldots$ are the Hermite polynomials

$$y=H_n(x) \quad (G.64)$$

(See Chapter 2, Section 2.10, for more discussion.)

Associated Laguerre polynomials

The solutions to the differential equation

$$s\frac{d^2L}{ds^2}-\left[s-2(l+1)\right]\frac{dL}{ds}+\left[n-(l+1)\right]L=0 \quad (G.65)$$

where $l = 0, 1, 2, 3, \ldots$ and the integer $n \geq l+1$, are the associated Laguerre polynomials

$$L_p^j(s)=\sum_{q=0}^{p}(-1)^q\frac{(p+j)!}{(p-q)!(j+q)!q!}s^q \quad (G.66)$$

(See Chapter 10, Section 10.4, for more discussion.)

Gamma function

The Gamma function is defined for $z > 0$ as

$$\Gamma(z)=\int_0^\infty t^{z-1}\exp(-t)\,dt=2\int_0^\infty t^{2z-1}\exp(-t^2)\,dt \quad (G.67)$$

It is a function of a continuous variable that is very closely related to the factorial function. In fact, for n an integer or zero

$$\Gamma(n)=(n-1)! \quad (G.68)$$

and, in general,

$$\Gamma(z+1)=z\Gamma(z) \quad (G.69)$$

A particularly useful result is that

$$\Gamma(1/2)=\sqrt{\pi} \quad (G.70)$$

and, hence, also

$$\Gamma(3/2) = \frac{\sqrt{\pi}}{2} \equiv \left(\frac{1}{2}\right)! \tag{G.71}$$

which is a useful integral in normalizing Gaussian functions (see Eq. (G.44)) and which also therefore allows us to define factorials for half-integers. Hence, for example, we can write

$$\left(\frac{5}{2}\right)! = \sqrt{\pi} \times \frac{1}{2} \times \frac{3}{2} \times \frac{5}{2} \tag{G.72}$$

Appendix H

Greek alphabet

Upper Case	Lower Case	Name
Α	α	alpha
Β	β	beta
Γ	γ	gamma
Δ	δ	delta
Ε	ε	epsilon
Ζ	ζ	zeta
Η	η	eta
Θ	θ	theta
Ι	ι	iota
Κ	κ	kappa
Λ	λ	lambda
Μ	μ	mu
Ν	ν	nu
Ξ	ξ	xi
Ο	ο	omicron
Π	π	pi
Ρ	ρ	rho
Σ	σ	sigma
Τ	τ	tau
Υ	υ	upsilon
Φ	φ	phi
Χ	χ	chi
Ψ	ψ	psi
Ω	ω	omega

Appendix I

Fundamental constants

Constant Name	*Symbol*	*Numerical Value*	*Δ*	*Units*
Bohr magneton	μ_B	$9.274\ 008\ 99 \times 10^{-24}$	37	$J\ T^{-1}$
Boltzmann constant	k_B	$1.380\ 6503 \times 10^{-23}$	24	$J\ K^{-1}$
Electric constant	ε_o	$8.854\ 187\ 817... \times 10^{-12}$	-	$F\ m^{-1}$
Electron g factor	g_e	2.002 319 304 3737	82	-
Electron rest mass	m_e	$9.109\ 381\ 88 \times 10^{-31}$	72	kg
Elementary charge	e	$1.602\ 176\ 462 \times 10^{-19}$	63	C
Fine structure constant	α	$7.297\ 352\ 533 \times 10^{-3}$	27	-
Magnetic constant	μ_o	$4\pi \times 10^{-7}$	-	$H\ m^{-1}$
Planck's constant	h	$6.626\ 068\ 76 \times 10^{-34}$	52	J s
Planck's constant/2π	$\hbar$	$1.054\ 571\ 596 \times 10^{-34}$	82	J s
Proton rest mass	m_p	$1.672\ 621\ 58 \times 10^{-27}$	13	kg
Proton-electron mass ratio	m_p / m_e	1 836.152 6675	39	-
Rydberg constant	R_∞	10 973 731.568 549	83	m^{-1}
	$R_\infty hc / e$	13.605 691 72	53	eV
Speed of light in vacuum	c	299 792 458	-	$m\ s^{-1}$

The "Δ" quoted is the absolute value of the uncertainty in the last two digits of the quoted numerical value corresponding to one standard deviation from the numerical value given. Hence, for example, the possible values of Planck's constant within one standard deviation of the best estimate shown lie between 6.626 068 24 and 6.626 069 28 J s.

The speed of light in vacuum has been chosen to have the exact value shown because the meter is now defined as the length of the path traveled by light in vacuum during the time interval of 1/299 792 458 of a second. The magnetic constant (also known as the permeability of free space) is chosen to have the value shown because it is an arbitrary constant that arises from the choice of the system of units and the electric constant (also known as the permittivity of free space) then follows from it and the (chosen) velocity of light because, by definition, $c = 1/\sqrt{\varepsilon_o \mu_o}$, so all three of these quantities have no uncertainty by definition. The Bohr magneton is $\mu_B = e\hbar / 2m_e$. The fine structure constant is $\alpha = e^2 / 4\pi\varepsilon_o c\hbar$.

These values are the CODATA Internationally recommended values as of 1998. Reference http://physics.nist.gov/cuu/Constants/index.html .

Bibliography

This list represents particular books that may be especially useful additional texts or references on the material presented in this book. There are, of course, many other excellent books on the various subjects listed. The list is by no means comprehensive, but it reflects particular books that I have found useful.

Quantum mechanics texts

There are many good texts on the physics of quantum mechanics. Not only can these provide alternative approaches to various topics covered in this book, but they also cover a broad range of other aspects and uses of quantum mechanics.

C. Cohen-Tannoudji, B. Diu, and F. Laloë, *Quantum Mechanics (Vols. 1 and 2)* (Wiley, New York, 1977).

This is a particularly comprehensive physics text.

P. A. M. Dirac, *The Principles of Quantum Mechanics (4th Edition, revised)* (Oxford, 1967).

This text, by one of the founders of quantum mechanics, starts from an abstract mathematical view, so is not the best first text but it is clear with many insights.

R. P. Feynman, *Feynman Lectures on Physics, Vol. 3* (Addison-Wesley, 1970).

This is a readable and different text with some interesting explanations.

W. Greiner, *Quantum Mechanics (Third Edition)* (Springer-Verlag, Berlin, 1994).

This is an introductory physics text that also is strong on the historical development of quantum mechanics.

H. Haken, *Light (Vol. 1)* (North-Holland, Amsterdam, 1981).

This text introduces most of elementary quantum mechanics, with a strong emphasis on light, and has particularly good introductory descriptions of the quantum mechanics of light. (The treatment of the quantization of the electromagnetic field in the present book largely follows Haken's approach.)

W. A. Harrison, *Applied Quantum Mechanics* (World Scientific, Singapore, 2000).

This book takes a different and refreshing approach compared to many other texts and is particularly strong in discussions of solid-state physics.[1]

H. Kroemer, *Quantum Mechanics for Engineering, Materials Science, and Applied Physics* (Prentice Hall, Englewood Cliffs, New Jersey, 1994).

This is an accessible intermediate-level text, biased toward applications.

[1] Walt Harrison is also my next-door neighbor, making us possibly the only two next-door neighbors ever to have written quantum mechanics texts.

R. L. Liboff, *Introductory Quantum Mechanics (Fourth Edition)* (Addison-Wesley, San Francisco, 2003).

This comprehensive text also includes many modern example applications.

J. J. Sakurai, *Modern Quantum Mechanics (Revised Edition)* (Addison Wesley, 1994).

This is a more advanced physics text.

L. I. Schiff, *Quantum Mechanics (Third Edition)* (McGraw-Hill, New York, 1968).

This is a classic introductory physics text. (The treatment of perturbation theory in the present book largely follows Schiff's.)

Background mathematics

E. Kreysig, *Advanced Engineering Mathematics (Ninth Edition)* (Wiley, New York, 2005).

This is an excellent reference for the background mathematics.

G. B. Arfken and H. J. Weber, *Mathematical Methods for Physicists (Sixth Edition)* (Academic, New York, 2005).

This book goes more deeply into the mathematics required for physics, including specifically that for quantum mechanics, in an accessible treatment.

Background physics

D. Halliday, R. Resnick, and J. Walker, *Fundamentals of Physics (7th Edition)* (Wiley, New York, 2004).

This is a comprehensive introduction to college physics and covers all the background topics required here, as well as much other physics.

Nonlinear and quantum optics

R. W. Boyd, *Nonlinear Optics (Second Edition)* (Academic Press, New York, 1992).

This is a clear and substantial introduction to nonlinear optics.

M. Fox, *Quantum Optics* (Oxford, 2006).

This is a good introductory text on optics from a quantum mechanical viewpoint, including also excellent introductory discussions of quantum information topics.

Y. R. Shen, *The Principles of Nonlinear Optics* (Wiley, New York, 1984).

This is a classic text on nonlinear optics.

A. Yariv, *Quantum Electronics (3rd Edition)* (Wiley, New York, 1989).

This is a comprehensive introductory text for lasers and nonlinear optics.

Solid-state physics

N. W. Ashcroft and N. D. Mermin, *Solid State Physics* (Saunders College Publishing, 1976).

C. Kittel, *Introduction to Solid State Physics (8th Edition)* (Wiley, New York, 2004).

These two texts, by Kittel and by Ashcroft and Mermin, are the classic introductory texts in solid-state physics.

O. Madelung, *Introduction to Solid State Theory* (Springer-Verlag, Berlin, 1978).

Madelung gives a clear and accessible discussion of the quantum mechanics of solid-state physics.

S. L. Chuang, *Physics of Optoelectronic Devices* (Wiley, New York, 1995).

This book mostly covers optoelectronic device physics but also includes a good treatment of $\mathbf{k} \cdot \mathbf{p}$ band theory.

Density matrices

K. Blum, *Density Matrix Theory and Applications* (Plenum, New York, 1996).

This is a comprehensive and deep discussion of the specific topic of the density matrix, from basic principles through to sophisticated applications.

Electromagnetism

There are many excellent texts on electromagnetism and most any one that covers Maxwell's equations is a sufficient background reference.

J. A. Stratton, *Electromagnetic Theory* (McGraw-Hill, New York, 1941).

This is a classic early text, with many deep explanations.

P. Lorrain and D. Corson, *Electromagnetic Fields and Waves, Second Edition* (Freeman, San Francisco, 1970) .

This is a reasonably modern introductory text at the boundary between engineering and physics.

U. S. Inan and A. S. Inan, *Engineering Electromagnetics* (Addison Wesley Longman, Menlo Park, CA, 1999) and U. S. Inan and A. S. Inan, *Electromagnetic Waves* (Prentice Hall, Upper Saddle River, NJ, 2000).

The first of these two books gives a comprehensive engineering introduction to electromagnetism and the second covers electromagnetic waves in an accessible and deep treatment.

Quantum information

M. A. Nielsen and I. L. Chang, *Quantum Computation and Quantum Information* (Cambridge, 2000).

This textbook gives an extensive discussion of the subject.

D. Bouwmeester, A. Ekert, and A. Zeilinger (eds.), *The Physics of Quantum Information* (Springer, 2000).

This edited volume introduces the subject and gives an extended discussion of research with contributions from a broad range of participants in the field.

Interpretation of quantum mechanics

See the list at the end of Chapter 19 for more details and references

J. A. Barrett, *The Quantum Mechanics of Minds and Worlds* (Oxford University Press, 1999).

J. S. Bell, *Speakable and Unspeakable in Quantum Mechanics* (Cambridge University Press, 1993).

A. Rae, *Quantum Physics – Illusion or Reality? (Second edition)* (Cambridge, 2004).

Classical mechanics

H. Goldstein, *Classical Mechanics (2nd edition)* (Addison-Wesley, 1980).

This book is the classic text in classical mechanics, intended to serve as a strong introduction to the subject for physicists before or while learning quantum mechanics.

J. W. Leech, *Classical Mechanics* (Methuen, 1965).

This short text is an alternative to Goldstein.

Mathematical reference

M. Abramowitz and I. A. Stegun, *Handbook of Mathematical Functions* (National Bureau of Standards, Washington, 1972).

This is the definitive tabular reference for special functions, giving graphs and extensive lists of formulae and relations.

I. S. Gradshteyn and I. M. Ryzhik, *Table of Integrals, Series, and Products (Sixth Edition)* (Academic, New York, 2000).

This is the definitive source for the answers to integrals.

Memorization list

These are the formulae most worth learning by heart from the book, grouped by chapters and listed by their description rather than by the formulae themselves so the reader can test his or her memorization of them.

Chapter 2

De Broglie's formula relating wavelength and momentum for a particle with mass (Eq. **(2.1)**).

The time-independent Schrödinger equation (Eq. **(2.13)**).

The normalization integral for a wavefunction (Eq. **(2.20)**).

The energy eigenvalues (Eq. **(2.26)**) and the corresponding normalized eigenfunctions (Eq. **(2.29)**) for the simple "infinite" particle in a box problem.

The expression defining orthonormality of functions (Eq. **(2.35)**).

The expression for the expansion of a function on a basis set of functions (Eq. **(2.36)**) and the expression for the corresponding expansion coefficients (Eq. **(2.37)**).

The energy eigenvalues for the harmonic oscillator (Eq. **(2.81)**).

Chapter 3

The relation between energy and frequency in quantum mechanics (Eq. **(3.1)**).

The time-dependent Schrödinger equation (Eq. **(3.2)**).

The definition of group velocity (Eq. **(3.31)**).

Probability of finding a quantum mechanical system in a given eigenstate, in terms of the expansion on the set of eigenstates of the quantity being measured (Eq. **(3.46)**).

The expectation value of energy defined using the Hamiltonian operator and the wavefunction (Eq. **(3.56)**).

The definition of the momentum operator (Eq. **(3.74)**).

The position-momentum uncertainty principle (Eq. **(3.88)**).

Chapter 4

The expansion of a function on an orthonormal basis set, expressed in Dirac notation (Eq. **(4.17)**).

The matrix corresponding to an operator expressed on a particular orthonormal basis (Eq. (4.59), noting in particular which way around the matrix elements are defined).

The bilinear expansion of an operator in Dirac notation (Eq. **(4.68)**).

The identity operator expressed on a complete orthonormal basis in Dirac notation (Eq. **(4.73)**).

The expression for the complex conjugate of an inner product in Dirac notation (Eq. **(4.36)**).

The expressions for the Hermitian adjoint of the product of two matrices or operators (Eq. **(4.97)**) and for the product of an operator (or matrix) and a (state) vector (Eq. **(4.98)**).

The defining equation for an inverse operator (Eq. **(4.90)**).

The defining equation(s) for a unitary operator (Eq. **(4.92)** or Eq. **(4.108)**).

The formula for transforming a vector to a new basis using a unitary transformation operator (Eq. **(4.94)**) and the corresponding formula for transforming an operator to a new basis (Eq. **(4.112)**) noting in particular which way around this second formula is).

The defining equation for a Hermitian operator (Eq. **(4.114)**).

Chapter 5

The definition of the commutator of two operators (Eq. **(5.2)**).

The general form of the uncertainty principle (Eq. **(5.23)**) in terms of the remainder of commutation (Eq. **(5.4)**).

The energy-time uncertainty principle (relation **(5.33)**).

The rule for transitioning from a sum to an integral (Eq. **(5.41)**).

For the optional Section 5.4 on continuous eigenvalues and delta functions, the following are particularly useful relations:

The operational definition of the Dirac delta function (Eq. **(5.46)**).

The representation of the delta function

using the sin (or, equivalently, the sinc) function (Eq. **(5.48)**)

using the complex exponential function (Eq. **(5.50)**

in terms of a complete set of functions (the closure relation) (Eq. **(5.66)**).

The normalization integral for normalizing to a delta function (Eq. **(5.72)**).

Chapter 6

The first-order (time-independent) perturbation correction to the energy (Eq. **(6.32)**) and to the wavefunction (Eq. **(6.38)**).

Chapter 7

The formula for the time rate of change of the first-order correction to the expansion coefficients in time-dependent perturbation theory (Eq. **(7.10)**) and, by extension, the formula for the $(p + 1)$th-order correction in terms of the pth-order correction (Eq. **(7.13)**).

Fermi's Golden Rule (No. 2) (Eq. **(7.31)** and, in Dirac δ-function form, Eq. **(7.32)**).

Chapter 8

The Bloch theorem (Eq. **(8.20)**) for the form of a wavefunction in a periodic system.

The density of states in k-space in three dimensions (Eq. **(8.24)**).

Chapter 9

The definition of the angular momentum operator in terms of the position and momentum operators (Eq. **(9.4)**).

The commutation relations for the angular momentum operators (Eqs. **(9.8)** – **(9.10)**).

The eigenequation for the z angular momentum operator (Eq. **(9.18)**) and the eigenfunctions (Eq. **(9.19)**).

The definition of the $\hat{L}^2$ operator (Eq. **(9.20)**).

The allowed values of the l and m quantum numbers from the solution of the $\hat{L}^2$ eigenvalue problem (Eqs. **(9.29)** and **(9.30)**).

Chapter 10

The allowed values of the principal quantum number n (Eq. **(10.74)**) and the relation between n and the allowed values of the l quantum number (Eq. **(10.75)**).

The eigenenergies for the hydrogen atom (Eq. **(10.76)**).

Chapter 11

The expression for the wavevector as a function of energy inside a layer of uniform potential (valid for real or imaginary results) (Eq. **(11.7)**).

The Gamow penetrability factor (Eq. **(11.40)**) for tunneling through a barrier.

Chapter 12

The definition of the Pauli spin matrices (Eq. **(12.21)**).

The commutation relations for the spin operators and/or the Pauli spin matrices (Eqs. **(12.14)** - **(12.16)** and/or Eqs. **(12.18)** – **(12.20)**).

The general spin state expressed in terms of angles on the Bloch sphere (Eq. **(12.27)**).

Chapter 13

The form of the wavefunction of two identical bosons, symmetrized with respect to exchange (Eq. **(13.14)**), and the corresponding form for two identical fermions (Eq. **(13.15)**).

The definition of the N particle fermion state in the form of a Slater determinant (Eq. **(13.43)**).

Chapter 14

The definition of the density operator (Eq. **(14.5)**).

The definition of an element of the density matrix (Eq. **(14.8)**).

The (ensemble average) expectation value of an observable evaluated using the density matrix and the operator corresponding to the observable (Eq. **(14.14)**).

The expression for the time evolution of the density matrix in terms of the Hamiltonian (Eq. **(14.23)**).

Chapter 15

The Hamiltonian for a harmonic oscillator, expressed in terms of raising (creation) and lowering (annihilation) operators (Eq. **(15.13)**).

The definition of the number operator (Eq. **(15.15)**) and its effect on a number state (Eq. **(15.16)**).

The effects of the raising (creation) and lowering (annihilation) operators on a harmonic oscillator state $|\psi_n\rangle$, including their effects on the lowest state $|\psi_0\rangle$ (Eqs. **(15.29)**, **(15.30)**, and **(15.31)**).

The commutation relation for raising (creation) and lowering (annihilation) operators (Eq. **(15.17)**) and its extended form for a multimode electromagnetic field (Eq. **(15.129)**).

Chapter 16

The definition of the anticommutator of two operators (Eq. **(16.17)**).

The anticommutation relation for fermion annihilation and creation operators (Eq. **(16.32)**).

The definition of the fermion number operator (Eq. **(16.33)**).

The definition of the single-particle fermion wavefunction operator (Eq. **(16.34)**) and its effect on operating on a single-particle state (Eq. **(16.39)**).

The definition of the two-fermion wavefunction operator (Eq. **(16.42)**).

The definition of the single-particle fermion Hamiltonian (expressed on its eigen basis) (Eq. **(16.49)**).

The definition of a single-particle fermion operator, expressed on an arbitrary basis (Eq. **(16.73)**).

The definition of a two-particle fermion operator, expressed on an arbitrary basis (Eq. **(16.76)**) (noting particularly the order of the operators in the sum).

Chapter 17

The commutation relations for annihilation and creation operators corresponding to different kinds of particles (Eq. **(17.2)**).

Chapter 18

The expressions for the Bell states of two photons (Eqs. (18.12) – **(18.15)**).

Index

A

absorption, 188–90, 195, 226, 231–37, 253, 257, 298, 348, 412–14, 417, 519–22
  saturation, 349
  two-photon, 231
adjoint, 96, 480, 484
Airy functions, 42–43, 529–30
Ångstrom, 15
angular momentum, 242–55
  classical, 242
  operators, 250
  spin, 303
antibonding state, 176
anticommutation relation, 384, 389, 390, 392, 397
anticommutator, 389
antiderivative, 476
arcsin. *See* sine, inverse
Argand diagram, 463, 495
associated Laguerre polynomials, 269–70, 531
associated Legendre functions, 249, 252, 255, 530
associative, 102, 483

B

band gap energy, 215
band structure, 216
basis, 24, 25, 103, 140–41
  changing, 115–16, 149
  direct product, 303–6
  multiple particle, 323–28
  normalized to a delta function, 142–44
  subset, 160
beamsplitter, 320–21
Bell state, 431, 432, 438, 439, 442
Bell's inequalities, 6
Bessel functions, 43, 529
  spherical. *See* spherical Bessel functions
bilinear expansion, 108–10
Bloch equations, 347
Bloch sphere, 303–4, 435
Bloch theorem, 212
Bohm's pilot wave, 451–52
Bohr magneton, 534
Bohr model, 523
Bohr radius, 264
Boltzmann constant, 291, 534
bonding state, 176
Bose–Einstein distribution, 328
boson, 315
bound state, 33
boundary condition, 19, 38
  Born–von Karman, 210
  differential equation, 469
  Dirichlet, 473
  finite step, 26–27
  matrix, 283
  Neumann, 473
  periodic, 146, 211, 419
bra, 96, 98
bra-ket notation. *See* Dirac notation
Bravais lattice, 208
Brillouin zone, 214, 233

C

Cartesian coordinates. *See* coordinates, Cartesian
center of mass. *See* coordinates, center of mass
central potential, 265
charm, 297
chemical potential, 329
classical mechanics, 491
  Hamiltonian, 491
  Lagrangian, 491, 520
  Newtonian, 10, 491
classical turning point, 45
closure, 140–41
coherence, 346
coherent state, 63–65, 66, 74, 370, 372
collapse of the wavefunction, 5, 74, 76, 424–25, 442, 448, 449, 453, *see also* measurement, quantum mechanical
color, 1–2
commutation
  of operators, 129–31
  remainder of, 130
commutation relation, 130
  angular momentum, 244
  boson operators, 377
  electromagnetic field operators, 378
  operators for different particles, 406–7
  photon creation-annihilation, 365
  position-momentum, 355
  spin, 302

commutative, 99, 102, 106, 483
commutator, 129
completeness
  of eigenfunctions, 122
  of quantum mechanics, 5–6
  of sets, 23–24
complex conjugate, 462
complex number, 461–64
complex plane, 463
conduction band, 215
configuration space, 258
conjugate transpose, 480
conservative field, 492, 507
constant of integration, 476
continuity equation, 502
continuum of states, 37
coordinates
  Cartesian, 457–59
  center of mass, 260–62
  cylindrical, 506–7
  spherical, 245, 263, 504–5
Copenhagen interpretation of quantum mechanics, 450
Coulomb's law, 259, 494
crystal, 207–9
cubic lattice, 208
cuboid, 477
curl, 502–4
  cylindrical coordinates, 507
  spherical coordinates, 505
cylindrical coordinates. *See* coordinates, cylindrical

D

Davisson–Germer experiment, 9, 523
de Broglie's hypothesis, 8, 9, 10, 523
degeneracy, 20
del, 499
delta function, 136–49
density matrix, 335–51
  time evolution, 341–42
density of states, 136, 142, 213–14, 219–25
density operator. *See* density matrix
dephasing, 346
derivative
  first, 464–65
  matrix representation of, 125
  partial, 466–68
  second, 465
  total, 467
determinant, 486–87
  Laplace formula, 486
  Leibniz formula, 322, 486
  Slater, 322, 330, 384, 390
  vector cross product, 459
diamond lattice, 208
difference frequency generation, 196, 201
differential, 467
differential calculus, 464–68
  chain rule, 527
  derivative of common functions, 527
  product rule, 527
  quotient rule, 527
differential equation, 468–73
  first order, 468–69
  general solution, 469
  in one variable, 468–71
  linear, 470
  partial, 471–73
  second order, 469–71
  series solution, 268–70
diffraction, 9, 498
  Huygens–Fresnel, 498
  single slit, 14
  two slit, 15
  two-slit, 12
dipole moment, 192, 193, 343–48, 509
Dirac delta function. *See* delta function
Dirac equation, 308
Dirac notation, 95–98, 480
direct gap, 215
direct product space, 303–6, 317, 323
dispersion, 68, 217, 495, 498
distinguishable particles, 331–32
distributive, 483
divergence, 500–502
  cylindrical coordinates, 507
  spherical coordinates, 505
  theorem, 501
divergenceless field, 501
dot product. *See* vector, dot product
double barrier, 285

E

effective mass, 38, 72, 87, 281
effective mass theory, 216–19
eigenenergy, 20
eigenequation, 20
  matrix, 487–89
eigenfunction, 20
  energy, 20
  momentum, 137, 149
  position, 147–48, 148
eigenvalue, 20
  continuous, 146
  matrix, 487–89

eigenvector
matrix, 487–89
electric constant, 510, 534
electric dipole approximation, 185, 232, 407
electric displacement, 510, 511
electric field, 511
electromagnetic field
mode, 361–63, 366–67, 373–74
operators, 367–69, 377–78
quantization of, 361–78
electron g factor, 534
electron rest mass, 10, 534
electron-photon interaction, 407–9
electron-volt, 22
electrostatics, 494
elementary charge, 534
emission
spontaneous, 188, 415, 417, 418–20
stimulated, 188, 415–17
energy density of electric and magnetic fields, 363, 514
energy eigenfunction. *See* eigenfunction, energy
ensemble average, 339–40, 351
entanglement, 6, 7, 431–33
envelope function, 218, 222
EPR pair, 431, 438, 442, 443, 447
EPR paradox, 6, 439, 442
Euler's formula, 463
even function. *See* parity, even
exchange, 315, 384, 394, 395, 402
exchange energy, 316–19, 403
expansion
coefficient, 25, 99
in eigenfunctions, 23
in energy eigenstates, 61
expectation value, 73–74, 78
exponential, 460

F

factorial function, 490
Fermi energy, 291
Fermi–Dirac distribution, 291
fermion, 315
Fermi's Golden Rule, 186–90, 231–37
field emission, 279, 290
fine structure constant, 534
finite basis subset method, 157–60, 173, 179, 227
finite matrix method. *See* finite basis subset method
Fock state. *See* number state
force, 492
Fourier analysis, 23, 24, 84
Fourier series, 23–24, 97, 306
Fourier transform, 93, 104, 106, 145
of a Gaussian, 83
four-wave mixing, 195, 196, 199
free current density, 511
free electric charge density, 511
frequency, 495
angular, 495
function, 94
fundamental theorem of calculus, 476

G

Gamma function, 531–32
Gamow penetrability factor, 289
gauge, 521–22
Gauss's theorem, 501, 512
Gaussian, 70, 71, 72, 83, 85, 139
Gaussian elimination, 486
Goos–Hänchen shift, 29
gradient, 473, 494, 499
cylindrical coordinates, 506
spherical coordinates, 505
Gram–Schmidt orthogonalization, 26, 177
group velocity, 69
group-velocity dispersion, 71

H

Hamiltonian, 77–81, 125, 134, 186
boson, 356, 365, 376
classical, 492
classical electromagnetic, 363, 519
electric dipole, 408
electromagnetic mode, 365
fermion, 395–403
harmonic oscillator, 356
hydrogen atom, 259, 262
multimode, 376
perturbing, 162, 166, 182, 185, 198, 232, 299, 408, 410, 411
spin, 307
Hamilton's equations, 360–61
harmonic generation, 196, 199, 202
harmonic oscillator, 18, 39–41, 354–59, 470
time evolution, 62–66
Heaviside function, 140
Heisenberg's uncertainty principle. *See* uncertainty principle
Helmholtz free energy, 329
Helmholtz wave equation. *See* wave equation, Helmholtz
Hermite polynomials, 40, 531
Hermitian adjoint, 96, 114, 119, 480
Hermitian conjugate, 96

Hermitian matrix. *See* matrix, Hermitian
Hermitian operator. *See* operator, Hermitian
Hermitian transpose, 96, 114, 480
heterostructure, 218
Hilbert space, 103, 305, 306
hole, 216, 217, 235
Huygens's principle, 13, 15, 498
hydrogen atom, 18, 55, 257–71, 299
hyperfine interaction, 257

I

identical particles, 5, 311–32
identity matrix. *See* matrix, identity
imaginary number, 462
indirect gap, 215
indistinguishable particles, 331–32
infinitesimal, 475
initial condition, 472
inner product, 97, 101, 102, 459, 483
integral
  definite, 476, 528
  equation, 93, 106
  indefinite, 476
integral calculus, 475–78
  in one variable, 475–77
  integration by parts, 528
  surface integral, 478
  volume integral, 477–78, 505, 507
integrand, 476
intensity, 496, 515–16, 522
interaction picture, 183
interference, 71, 497
  constructive, 496
  destructive, 496
inverse matrix. *See* matrix, inverse
irrotational field, 492, 507

J

joint density of states, 189, 235

K

k.p method, 157, 231
Kane model, 227
Kelvin–Stokes theorem. *See* Stokes theorem
Kerr effect, 196
ket, 95, 96, 97
kinetic energy, 11, 491
Kramers degeneracy, 216
Kramers–Kronig relations, 195
Kronecker delta, 25
k-space, 213

L

L^2 function, 17, 137
Laguerre polynomials
  associated. *See* associated Laguerre polynomials
Lamb shift, 257
Laplace determinant formula. *See* determinant, Laplace formula
Laplacian, 10, 247, 472–73, 499–500
  Cartesian coordinates, 472
  cylindrical coordinates, 507
  spherical coordinates, 505
lattice vector, 207, 212, 239
Lebesgue integration, 17, 137, 475
Legendre functions
  associated. *See* associated Legendre functions
Legendre polynomials, 26
Leibniz notation, 464
Levi–Civita symbol, 244
Liebniz determinant formula. *See* determinant, Leibniz formula
linear equation, 485
linearity
  linear superposition, 59–60, 102
  multiplying by a constant, 16, 102
  of quantum mechanics, 59–60
linearly varying potential, 18, 42–50
Liouville equation, 342
logarithm, 460
logical positivism, 7
longitudinal relaxation, 345
Lorentz force, 519
Lorentzian, 139, 349
Luttinger–Kohn model, 227
Luttinger–Kohn representation, 227

M

Maclaurin series, 525–26
magnet, 298
magnetic constant, 511, 534
magnetic current density, 511
magnetic dipole moment, 298–99
magnetic field, 511
magnetic flux density, 511
magnetic induction, 511
magnetic monopole, 511
magnetic quantum number, 299
magnetic vector potential. *See* vector potential
magnetism, 3, 319
magnetization, 511
many-minds hypothesis, 453
many-worlds hypothesis, 5, 453

matrix, 478–89
  addition, 481
  determinant. *See* determinant
  diagonal elements, 479
  element, 107–8, 479
  Hermitian, 480
  Hermitian adjoint, 480
    of a product, 484
  identity, 484
  inverse, 484–86
  invertibility, 486
  leading diagonal, 479
  mechanics, 523
  multiplication, 481–84
  off-diagonal elements, 479
  rectangular, 479
  square, 479
  subtraction, 481
  transfer, 279–86
  transpose, 114, 480
Maxwell–Boltzmann distribution, 292
Maxwell's equations, 4, 509
measurement, quantum mechanical, 4, 5, 6, 7, 74, 75, 76, 132, 133, 311, 371, 424–25, 446–54, *see also* collapse of the wavefunction
mind/matter distinction, 453
modes, 64, 188, 250, 316, 330–31, 359, 361–79, 406–9, 418–20, 516–18
modulus, 462
molecular bonding, 175
momentum
  classical, 491
  operator, 81, 124, 131, 137, 149, 364, 368, 399
monochromatic waves, 496

N

nabla, 499
Newton's Laws, 4, 492
no-cloning theorem, 425–27
nodal line, 497
nonlinear optical coefficients, 195–202, 351
nonlinear refraction, 195, 199
nonlocality, 6, 7, 444
normalization, 17–18
  to a delta function, 67, 141–42
*n*th root of unity, 463
number state, 367, 371

O

odd function. *See* parity, odd
one-electron approximation, 209
ontology, 330
operator, 77, 103–25
  angular momentum. *See* angular momentum, operators
  annihilation, 354–79, 383–403
  boson, 354–79
  creation, 354–79, 383–403
  electromagnetic field. *See* electromagnetic field, operators
  fermion, 383–403
  Hermitian, 110, 119–22
  identity, 110, 113
  inverse, 110, 113
  linear, 104–6
  lowering, 354–59
  momentum. *See* momentum operator
  position. *See* position operator
  projection, 113
  raising, 354–59
  spin. *See* spin operator
  time-evolution. *See* time-evolution operator
  unitary, 110, 114–18
  wavefunction. *See* wavefunction operator
orthogonality, 24
  of eigenfunctions, 24–25
orthohelium, 319
orthonormality, 25
outer product, 483

P

parabolic band, 217, 291
parahelium, 319
paraxial approximation, 16
parity, 20, 40, 202
  even, 20
  odd, 20
Parseval's theorem, 143, 145
particle current, 85–87, 278
particle in a box, 3, 18–22, 221
  finite depth, 18, 32–37
particles, identical. *See* identical particles
Pauli equation, 307
Pauli exclusion principle, 297, 311, 315–16, 322, 383, 385, 400
Pauli spin matrices, 119, 131, 302, 303, 307
penetration factor, 288–89
permeability, 511
  of free space. *See* magnetic constant
  relative, 511
permittivity, 510
  of free space. *See* electric constant
  relative, 510
perturbation theory
  Brillouin–Wigner, 169

degenerate, 157, 299
density matrix, 350
first order, 164–65, 168, 183, 186, 192, 198
Löwdin, 227
nondegenerate, 164
oscillating perturbations, 185–90
Rayleigh–Schrödinger, 169
second order, 166, 168, 196, 198
stationary. *See* perturbation theory, time-independent
third order, 168, 196, 198
time-dependent, 182–90, 409–11
time-independent, 161–72
phase, 463
front, 495
velocity, 67, 69
phonon, 209, 379
photoelectric effect, 523
photon, 1, 18, 39, 55, 330, 354–79, 523
Pikus–Bir model, 227
Planck's constant, 9, 523, 534
plane wave, 495
electromagnetic, 513
Pockels effect, 196
Poisson distribution, 64, 372
polarizability, 180
polarization
circular, 514
elliptical, 514
linear, 232, 514
of a material, 170, 192, 196, 509–10
of a wave, 160, 232, 361, 419, 425, 514
selection rules, 231, 253, 298
position operator, 82, 149, 243, 356, 399
potential
energy, 492
scalar, 507
vector. *See* vector potential
potential well
coupled, 172
finite. *See* particle in a box, finite depth
infinite. *See* particle in a box
rectangular, 19
skewed infinite, 47, 155–60, 155–60, 179
square, 19
time evolution, 61
triangular, 46–47
with field, 47–50
Poynting vector, 514–18
principal quantum number, 271
probability amplitude, 12
probability density, 11–12, 478
product notation, 489
proton rest mass, 534
proton-electron mass ratio, 534
pure state, 335–38

Q

quadratic equation, 525
quantum box, 222, 273
quantum computing, 6, 433–37, 454
quantum cryptography, 6, 425–31
quantum encryption. *See* quantum cryptography
quantum mechanical amplitude, 12, 497
quantum mechanical measurement. *See* measurement, quantum mechanical
quantum mechanics, completeness. *See* completeness, of quantum mechanics
quantum well, 3, 18, 38, 219, 210–11, 286
quantum wire, 222, 225, 274
quasi-bound states, 286

R

Raman amplification, 196
reciprocal lattice, 213, 214
rectangular parallelepiped, 477
rectangular prism, 477
recurrence relation, 269
reduced mass, 261
reflection high-energy electron diffraction (RHEED), 15
refraction, 498
refractive index, 192–95, 199, 513
relaxation time approximation, 350
representation, 24, 98
Richardson constant, 293
Richardson–Dushman equation, 293
Riemann integration, 475
right-hand rule, 458–59
right-handed axes, 458
rotating wave approximation, 347
Rydberg, 264, 267
Rydberg constant, 534

S

scalar, 462
Schottky barrier, 290
Schrödinger's cat, 4
Schrödinger's equation, 8, 523
generalized, 257
time-dependent, 54–60
time-independent, 8–11
second derivative. *See* derivative, second
secular equation, 488
self-adjoint, 119

separability, 18, 23
separation of variables, 222, 248, 262, 263
sinc function, 14, 138, 142, 187, 188, 189
sine
  inverse, 461
slowly varying envelope approximation, 217
solenoidal field, 501
soliton, 195
span, 103
speed of light in vacuum, 534
spherical Bessel functions, 274, 530
spherical coordinates. *See* coordinates, spherical
spherical harmonics, 250–52, 263, 270, 530
spin, 3, 99, 216, 220, 228, 231, 297–308, 523
  half-integer, 315
  integer, 315
spin operator, 302–3
spinor, 305
spin-orbit interaction, 257
standard deviation, 83
standard interpretation of quantum mechanics, 449
standing wave, 21, 30, 31, 45, 71, 368, 369, 373, 497
state vector, 99–100, 117
Stern–Gerlach experiment, 76, 299, 523
Stokes theorem, 504, 512
strangeness, 297
subband, 223
sum frequency generation, 201
summation notation, 474
superposition state, 54
surface integral. *See* integral calculus, surface integral
susceptibility, 192, 197, 199, 348, 510
  nonlinear, 197, 202

T

Taylor series, 525–26
tensor product, 306
thermionic emission, 289
  field-assisted, 290
three-wave mixing, 196
tight-binding model, 176
time evolution, 54
time-evolution
  operator, 78–81, 314
total derivative. *See* derivative, total
total internal reflection, 29
trace, 112–13
transfer matrix. *See* matrix, transfer
transpose. *See* matrix, transpose
transverse relaxation, 346
trigonometric addition and multiplication formulae, 526–27
tunneling, 3, 4, 7, 29, 72
  probability, 277–79, 284
two-level system, 343–49
two-slit experiment. *See* diffraction, two-slit

U

ultraviolet catastrophe, 2, 523
uncertainty principle, 4, 7, 82–85, 135, 523
  diffraction angle, 85
  electromagnetic field, 369
  energy-time, 134
  frequency-time, 84, 135
  position-momentum, 84, 133
unit cell, 207, 214
unitary operator. *See* operator, unitary
unitary transformation, 114

V

valence band, 215
variational method, 179
vector
  calculus
    identities, 507, 508
    operators, 500
  column, 479
  dot product, 98, 101, 459
  field, 500
  generalized, 479
  length, 102
  multiplication, 483
  norm, 102
  potential, 186, 232, 519
  projection, 458
  row, 479
  space, 103, 100–103
vertical transition, 235
vibrational mode, 378–79
volume integral. *See* integral calculus, volume integral

W

wave amplitude, 495
wave equation
  electromagnetic, 512
  Helmholtz, 10, 470, 472
  one-dimensional, 472
  three-dimensional, 472–73
wave front, 495
wavefunction, 9

meaninglessness, 92
operator, 392–403
wavelength
electron, 9
wavepacket, 64, 72, 217, 311
wave-particle duality, 9, 523
WKB method, 288

Y

Young's slits. *See* diffraction, two slit

Z

Zeeman effect, 299
zero point energy, 21, 40
zinc-blende structure, 208, 228

Printed in the United States
by Baker & Taylor Publisher Services